Liquid Crystals and Ordered Fluids

Volume 2

Liquid Crystals and Ordered Fluids

Volume 1 – Proceedings of an American Chemical Society Symposium, New York City, September, 1969

Volume 2 – Selected papers from a symposium of the Division of Colloid and Surface Chemistry, Chicago, August, 1973

A Continuation Order Plan is available for this series. A continuation order will bring delivery of each new volume immediately upon publication. Volumes are billed only upon actual shipment. For further information please contact the publisher.

Liquid Crystals and Ordered Fluids

Volume 2

Edited by
JULIAN F. JOHNSON
Institute of Materials Science
University of Connecticut
Storrs, Connecticut

and

ROGER S. PORTER
Materials Research Laboratory
University of Massachusetts
Amherst, Massachusetts

PLENUM PRESS • NEW YORK-LONDON

Library of Congress Cataloging in Publication Data

Main entry under title:

Liquid crystals and ordered fluids.

Papers, from a symposium of the Division of Colloid and Surface Chemistry held in Chicago during the national meeting of the American Chemical Society, August 1973, of the 3d of a series of meetings; papers of the 1st are entered under the title: Ordered fluids and liquid crystals; papers of the 2d are entered under: Symposium on Ordered Fluids and Liquid Crystals, 2d, New York, 1969.

1. Liquid crystals—Congresses. I. Johnson, Julian Frank, 1923- ed. II. Porter, Roger Stephen, 1928- ed. III. American Chemical Society. Division of Colloid and Surface Chemistry.

QD923.L56 548′.9 74-1269

ISBN 0-306-35182-X

Selected papers from a symposium of the Division of Colloid and Surface Chemistry held in Chicago during the national meeting of the American Chemical Society, August, 1973

A Division of Plenum Publishing Corporation
227 West 17th Street, New York, N. Y. 10011

United Kingdom edition published by Plenum Press, London
A Division of Plenum Publishing Company, Ltd.
4a Lower John Street, London W1R 3PD, England

Printed in the United States of America

PREFACE

This volume represents a collection of selected papers from a symposium of the Division of Colloid and Surface Chemistry held in Chicago during the national meeting of the American Chemical Society, August, 1973. The response was remarkable to this "By Invitation" symposium on Ordered Fluids and Liquid Crystals. The size alone expresses the growth of the field. The number of contributions assembled here, for example, is approximately twice that at each of the two previous American Chemical Society symposia on this subject. Contributions from eleven countries were presented and this volume contains more than this number of papers from abroad.

The increased attention to liquid crystals has brought some interesting trends in the kinds of systems, the experimental methods, and the nature of the laboratories involved. There has, for example, been an impressive increase in the number of academic studies on liquid crystals. The works herewith published also represent an impressive variety of traditional and novel experimental techniques for the study of liquid crystals. These include rheology, infrared spectroscopy, dielectrics, ultrasonics, pulsed NMR, the Kerr effect, plus thermal and electrical conductivity.

The volume includes cohesive sets of papers in several distinct areas. Included are groupings of papers on both polymers and on aqueous systems. An additional set involves the new emphasis on studies of specific subclasses of smectic mesophases. A collection of papers on cholesteric structures is also included. The predominant set of studies continues to be in the area of nematic mesophases. This, of course, is due to the spectacular features of this mesophase type - with both realized and potential applications. The nematic compositions now under wide study thus commonly exhibit mesophase behavior near ambient temperature as the result of either special syntheses and/or compound blending. Consequently, the effects of electromagnetic fields on nematic mesophases continue to receive wide attention as a result of the optical features which make them suitable for visual displays.

This volume represents a comprehensive extension of Volume I which was published in 1970 by the same editors and publishing house. A unifying subject index is provided at the back of each volume. These books thus hopefully provide an overview of the continuingly impressive crescendo of activity in the field of Ordered Fluids and Liquid Crystals.

Roger S. Porter, Head
Polymer Science and Engineering
and Materials Research Laboratory
University of Massachusetts
Amherst, Massachusetts

Julian F. Johnson
Department of Chemistry
and Associate Director
Materials Research Institute
University of Connecticut
Storrs, Connecticut

Symposium Co-Chairmen
and Editors

September 20, 1974

CONTENTS

THE BINDING OF DIVALENT IONS TO THE PHOSPHOGLYCOPROTEIN PHOSVITIN

Kärt Grizzuti and Gertrude E. Perlmann*

The Rockefeller University

Studies of viscosity, optical rotatory dispersion and circular dichroism on the phosphoglycoprotein phosvitin revealed that this protein in some of its properties resembles a polyelectrolyte (1, 2). As pointed out previously, this is easily understood if one considers that fifty-four per cent of all amino acid residues of phosvitin are phosphorylated (3). At neutral pH, where two hydroxyls of each phosphate group are ionized, the net electronic charge corresponds to about one negative charge per amino acid residue. Furthermore, we have shown that the presence of monovalent ions has an effect on the optical rotatory dispersion and circular dichroism of phosvitin. Since in biological processes cations, especially divalent ions, play an important role in maintaining a well-defined conformation necessary for the catalytic activity of certain enzymes, the binding studies have been extended to divalent cations with the aid of equilibrium dialysis techniques. In addition, an investigation of the effect of ion binding on the conformational characteristics of this protein has been initiated using optical methods, i.e. optical rotatory dispersion and circular dichroism.

Binding Studies.- Our first esperiments were designed to establish the extent of binding of Ca^{2+}, Mg^{2+}, Mn^{2+}, Co^{2+} and Sr^{2+}. Figure 1 summarizes the results

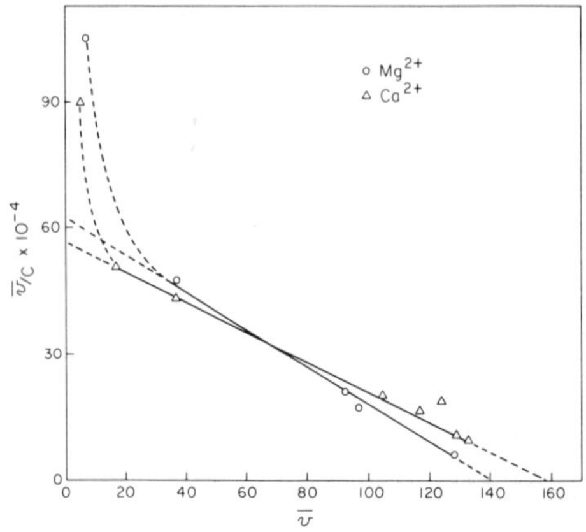

Fig. 1 Scatchard plots for the binding of Mg^{2+} (o) and Ca^{2+} (Δ) at pH 6.5 and 25°.

obtained for Ca^{2+} and Mg^{2+} at pH 6.5 and 25° using equilibrium dialysis. In this Figure the values of $\bar{\nu}/C$ are plotted against $\bar{\nu}$ (4). Here $\bar{\nu}$ represents the number of ions bound per mole of protein, C is the molar concentration of "free" ions in the solution. The plots of $\bar{\nu}/C$ vs $\bar{\nu}$ for Ca^{2+} and Mg^{2+} give essentially straight lines, indicating that in the concentration range of 0.0002 to 0.002 M at pH 6.5 the binding sites for these cations are of similar affinity. Since binding data at higher concentrations of Ca^{2+} and Mg^{2+} can not be obtained due to precipitation of phosvitin by these ions, linear extrapolation of this plot to the abscissa reveals that n = 140 and n = 157 for Mg^{2+} and Ca^{2+}, respectively.

We have previously reported that phosvitin, in addition to 136 phosphoamino acids, has also 31 dicarboxylic amino acid residues which, at pH 6.8, are fully ionized (3). Thus n = 157 for Ca^{2+} and n = 140 for Mg^{2+} is in fair agreement with the analytical data derived from amino acid analyses.

Extrapolation of the Scatchard plot to the ordinate shows an upward trend of the plot which may be caused by the presence of a few binding sites of very high affinity.

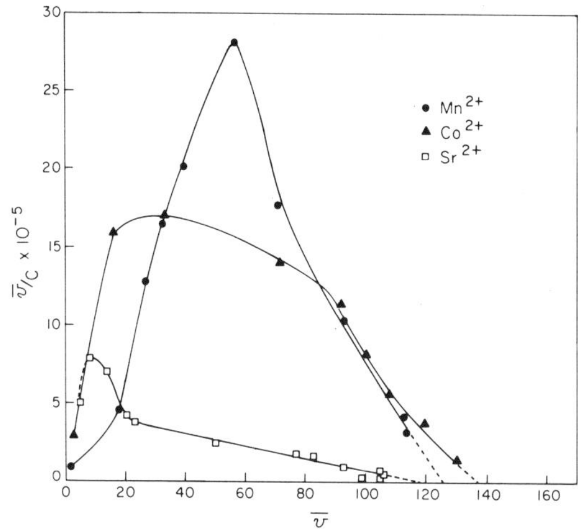

Fig. 2 Scatchard plots for the binding of Mn^{2+} (●), Co^{2+} (▲) and Sr^{2+} (□) at pH 6.5 and 25°.

In contrast to Ca^{2+} and Mg^{2+}, the Scatchard plot for the binding of Mn^{2+} to phosvitin is remarkable (Fig. 2). Not only is it different from a straight line but at low concentrations of bound Mn^{2+} (less than $\bar{\nu}$ = 50) it has a positive slope with a downward curvature. These features can not be interpreted in terms of multiple classes of independent binding sites, nor by the effect of competition of these sites, since in such cases the plots of $\bar{\nu}/C$ vs $\bar{\nu}$ are curves with an upward trend near the ordinate with a negative slope (cf. Fig. 1). The only interpretation of the observed feature is that at low concentration of Mn^{2+} an interaction between binding sites of the protein occurs. Thus, in the absence of aggregation of phosvitin, a possibility which we have not yet investigated, the binding of Mn^{2+} is cooperative and involves a structural change of the macromolecule (5, 6, 7).

At higher concentrations of bound Mn^{2+} no cooperativity is present. Between $\bar{\nu}$ = 50 and $\bar{\nu}$ = 110 a transition region exists. As pointed out above, the upward curvature with a negative slope of the Scatchard plot indicates the influence of at least one other type of binding site with a different association constant.

Finally, as illustrated in Figure 2, if more than 100 Mn^{2+} are bound by phosvitin, the plot is linear and extrapolates to n = 128, a value lower than that obtained for Mg^{2+} and Ca^{2+}.

In Figure 2 are further included the results of the binding of Co^{2+} and Sr^{2+} to phosvitin. The plot of $\bar{\nu}/C$ vs $\bar{\nu}$ for Co^{2+} manifests similar features as those found for Mn^{2+}, again indicating an interaction between binding sites of phosvitin. However, in contrast to Mn^{2+} the transition region of bound Co^{2+} per mole of phosvitin is shifted to a lower concentration of bound Co^{2+}, i.e. between $\bar{\nu}$ = 20 and $\bar{\nu}$ = 100. Thus, in this instant, the cooperative pattern is no longer present above $\bar{\nu}$ = 100.

Lastly, the binding of Sr^{2+} has been investigated. As shown with the aid of the Scatchard plot given in Figure 2, except at very low concentrations of Sr^{2+},

Table I

Number of Divalent Cation Binding Sites to Phosvitin

Cations	Number of Sites	
	weak (independent)	strong (cooperative)
Ca^{2+}	157	-
Mg^{2+}	140	-
Mn^{2+}	128	50
Co^{2+}	138	30
Sr^{2+}	120	(10)

phosvitin appears to have essentially one class of independent binding site for this cation. The experimental points above $\overline{\nu}$ = 20 fall on a straight line which intersects the horizontal axis at n = 120.

Table I summarizes the results obtained for binding sites of phosvitin for different divalent cations taken from Figures 1 and 2. Thus the nature of the divalent ions thus far studied gives the following efficiency:

for cooperativity: $Mn^{2+} > Co^{2+} > Sr^{2+}$

for weak binding sites: $Ca^{2+} > Mg^{2+} > Co^{2+} > Mn^{2+} > Sr^{2+}$

Optical Measurements. - In a previous article we have shown that the presence of high concentrations (1 M) of NaCl affects considerably the optical rotatory dispersion (ORD) patterns and circular dichroism (CD) spectra of phosvitin (2). To detect structural alterations of the protein that may occur in the presence of divalent cations, ORD and CD measurements were carried out in the presence of Ca^{2+}, Mg^{2+}, Mn^{2+}, Co^{2+} and Sr^{2+}. In an attempt to obtain a direct comparison of the optical measurements with the binding experiments, phosvitin was dissolved in Na cacodylate buffer of pH 6.8 and 0.02 ionic strength and was dialyzed against this buffer containing varying amounts of the divalent cations to be studied. This experimental setup enabled us to determine the number of Ca^{2+}, Mg^{2+}, Mn^{2+}, Co^{2+} and Sr^{2+} bound to phosvitin and to perform the ORD and CD measurements on the same solution.

As already reported, at neutral pH the ORD patterns below 300 nm are characterized by a deep trough at 205 nm and a reduced mean residue rotation $[m']_{205}$ = -10,000. The CD spectra below 300 nm show a strong negative band at 198 nm with a reduced mean residue ellipticity $[\Theta']_{198}$ = -30,000. In addition, a positive band at 220 nm and a negative one at 233 nm were recorded. In preliminary experiments we found that, if the number of Mg^{2+} bound is increased from 0 to 103 Mg^{2+} per mole of phosvitin, [m'] becomes more levorotatory. It increases from -10,500 to -15,000. In contrast, on binding of 134 Ca^{2+} per phosvitin molecule, [m'] becomes less

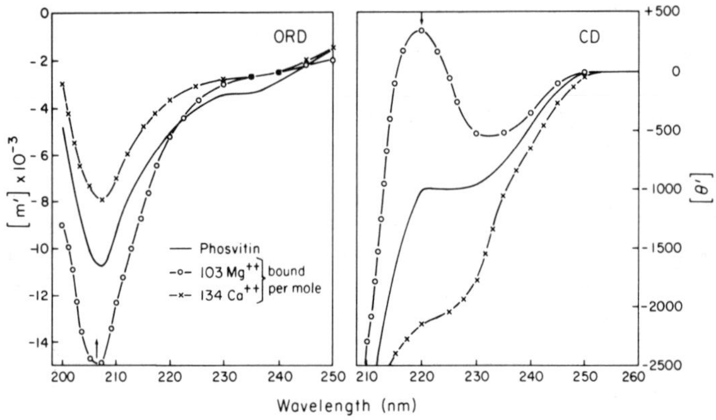

Fig. 3 Effect of the binding of Mg^{2+} and Ca^{2+} on the optical rotatory dispersion (A) and circular dichroism (B) of phosvitin at pH 6.5. Phosvitin dialyzed against Na cacodylate, Γ/2 0.02 (——); $MgCl_2$ (o——o) 103 Mg^{2+} bound and $CaCl_2$ (x——x) 134 Ca^{2+} bound.

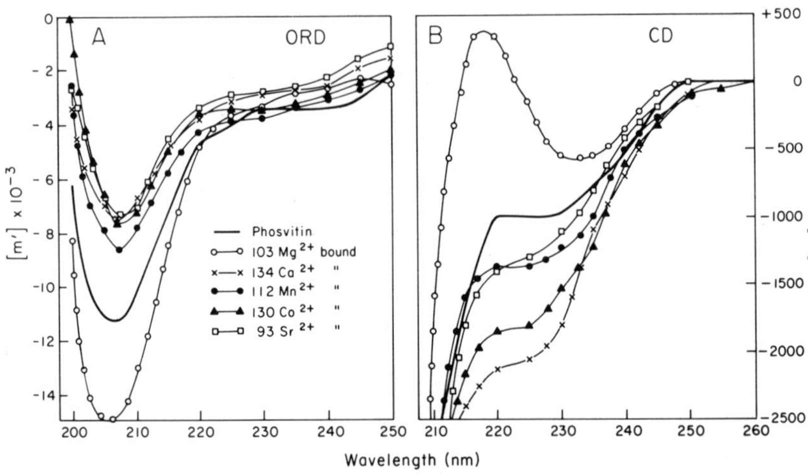

Fig. 4 Effect of the binding of Mg^{2+}, Ca^{2+}, Mn^{2+}, Co^{2+} and Sr^{2+} on the optical rotatory dispersion (A) and circular dichroism (B) of phosvitin at pH 6.5. Phosvitin dialyzed against Na cacodylate, Γ/2 0.02 (——); $MgCl_2$ (o——o) 103 Mg^{2+} bound; $CaCl_2$ (x——x) 134 Ca^{2+} bound; $MnCl_2$ (●——●) 112 Mn^{2+} bound: $CoCl_2$ (▲——▲) 130 Co^{2+} bound and $SrCl_2$ (□——□) 93 Sr^{2+} bound.

levorotatory and increases to -7500 (Fig. 3A). Due to the high absorption of cacodylate in the far ultraviolet, the CD spectra could only be recorded to 210 nm. However here too a marked difference exists between the binding of Mg^{2+} and Ca^{2+}. The results of the CD measurements are shown in Figure 3B.

Figure 4A and B illustrates the effect of bound Mn^{2+}, Co^{2+} and Sr^{2+} on the ORD patterns and CD spectra of phosvitin and also includes those obtained for Ca^{2+} and Mg^{2+} taken from Figure 3A and B. The values recorded for $[m']_{205}$ and $[\Theta']_{217}$ are given in Table II.

Although the ORD and CD measurements with Ca^{2+}, Mn^{2+}, Co^{2+} and Sr^{2+} containing phosvitin solutions yielded ORD and CD measurements which differ markedly from phosvitin in water or in the cacodylate buffer and also somewhat from each other, the effect of Mg^{2+} bound to

Table II

Effect of the binding of Mg^{2+}, Ca^{2+}, Mn^{2+}, Co^{2+} and Sr^{2+} on the reduced mean residue rotation [m'] and reduced mean residue ellipticity [Θ'] of phosvitin at pH 6.5

Nature of cation (bound)	Number of ions bound per mole of phosvitin	$[m']_{205}$	$[\Theta']_{217}$
none	none	-11,200	-1,000
Mg^{2+}	103	-15,000	480
Ca^{2+}	134	-7,500	-2,100
Mn^{2+}	112	-8,600	-1,400
Co^{2+}	130	-7,600	-1,850
Sr^{2+}	93	-7,300	-1,400

phosvitin is more drastic and the changes observed in $[m']_{205}$ and $[\Theta']_{217}$ are of opposite nature to those found for the other divalent cations. From these results we can conclude that divalent ions bound to phosvitin induce a conformational change and that these cations have different effects on the spatial structure of the side chains of the phosvitin molecule.

In conclusion, we would like to state that a difference exists in the number of Ca^{2+}, Mg^{2+}, Mn^{2+}, Co^{2+} and Sr^{2+} bound to phosvitin. There appear to exist two types of binding sites in this protein. At higher cation concentrations the "weak and independent binding" sites prevail and the interaction is of the ordinary electrostatic type. Furthermore, it has been possible to estimate the number of binding sites from the slopes of the Scatchard plots (Figures 1 and 2). In the presence of low concentrations of Mn^{2+} and Co^{2+}, however, an interaction between binding sites of the protein occurs and a cooperative pattern is present. One has to keep in mind that phosvitin has sequences in which 3 to 8 phosphoserine residues are linked covalently (8, 9), which could easily introduce differences in the binding of ions.

A connection between the results of the binding studies and of the optical measurements is not apparent. However, it is clear that the nature of the counterions is the determining factor in controlling the polypeptide backbone conformation and not the actual ion binding.

Acknowledgement

This work was supported by the National Science Foundation Grant GB-24896, and by a grant from the Research Corporation.

References

1. Perlmann, G.E. and Grizzuti, K. (1970) in Liquid Crystals and Ordered Fluids, Plenum Press, p.69.

2. Grizzuti, K. and Perlmann, G.E. (1970) J. Biol. Chem. 245, 2573.

3. Allerton, S.E. and Perlmann, G.E. (1965) J. Biol. Chem. 240, 3892.

4. Scatchard, G. (1949) Ann. N.Y. Acad. Sci. 51, 660.

5. Danchin, A. and Guéron, M. (1970) Eur. J. Biochem. 16, 532.

6. Danchin, A. (1972) Biopolymers 11, 1317.

7. Giancotti, V., Quadrifoglio, F. and Crescenzi, V. (1973) Eur. J. Biochem. 35, 78.

8. Williams, J. and Sanger, F. (1959) Biochim. Biophys. Acta 33, 294.

9. Shainkin, R. and Perlmann, G.E. (1971) J. Biol. Chem. 246, 2278.

* To whom inquiries should be directed.

LIQUID CRYSTALLINE BEHAVIOR OF BIOLOGICALLY IMPORTANT LIPIDS

POLYUNSATURATED CHOLESTEROL ESTERS AND PHOSPHOLIPIDS

D.M.Small, C.Loomis, M. Janiak, G.G.Shipley

Biophysics Section, Department of Medicine

Boston University School of Medicine

Boston, Mass. 02118

Lipids are biologically important molecules. They give structure to plasma membranes and intracellular organelles, provide a source of metabolic energy and play a role in many essential biological processes such as transmission of information across membranes, transmission of nerve impulses, antigen-antibody reactions and thrombogenesis. Lipids have also been implicated in pathological processes, for instance the lipid storage diseases (1) gallstones and (2) atherosclerosis (3).

The physical properties of some individual lipid classes (e.g. phospholipids, fatty acids, bile acids, triglycerides, sterols) have been extensively studied (4-9) and the interactions between certain lipid classes, for example phospholipid-cholesterol, have been examined both in bulk aqueous phases and at various interfaces (8,10,11). However biologically important cholesterol esters have received less attention considering the early demonstration of their behavior as liquid crystals (12). Recently the phase behavior of the biologically-important, long-chain, saturated and unsaturated esters have been studied by microscopy, calorimetric and x-ray diffraction methods (7,13). Of interest is the demonstration that the melting and liquid-crystal transitions of the cholesterol esters of the C_{18} fatty acids are dependent upon the chain unsaturation (stearate > oleate > linoleate > linolenate). (7,14) Furthermore the thermodynamic stability of the different liquid-crystal phases may be important in terms of their structural role in biological environments.

In this paper we summarize the behavior of 2 biologically important polyunsaturated esters of cholesterol, cholesteryl linoleate and cholesteryl linolenate with the phospholipid lecithin.

METHODS

a) Chemicals

Egg lecithin (Grade 1:Lipid Products, Surrey, England),cholesteryl linoleate and cholesteryl linolenate (Hormel Institute, Austin, Minn.) were judged to be greater than 99% pure by a number of chromatographic methods (15,16). The fatty acid composition of the egg lecithin was determined by gas liquid chromatography. The major fatty acids were: C16:0 41.8%, C16:1 0.4%, C18:0 2.0%, C18:1 52.1%, C18:2 2.2%, others 1.2%. The molecular weight calculated from this fatty acid composition was 765.

b) Preparation of Mixtures.

Appropriate proportions of chemicals were taken up in organic solvent (chloroform-methanol), mixed and the solvent removed under vacuum. In some mixtures an appropriate amount of water was added. All tubes were flushed with nitrogen, sealed, and equilibrated at an appropriate temperature.

Equilibration was facilitated by centrifuging the sample back and forth through a constriction in the tube during the incubation period. A sample which appeared homogeneous by microscopy and gave reproducible transitions on repeated heating and cooling was determined to be equilibrated. Each sample was then observed with respect to fluidity and homogeneity. The tube was opened and samples taken for polarizing light microscopy, differential scanning calorimetry (DSC), x-ray diffraction and in some cases proton magnetic resonance.

c) Techniques

Polarizing microscopy on a heating-cooling stage was carried out as described previously (7,15,17). DSC was carried out in a Dupont900 Differential Thermal Analyser fitted with a scanning calorimeter (7). x-ray diffraction studies were performed on sealed samples in a special heating chamber utilizing nickel-filtered Cu K_α radiation from an Elliot GX-6 rotating anode generator. Two x-ray focusing cameras were used: toroidal mirror optics (18) or double mirror optics (19). Proton magnetic resonance studies were carried out in a Perkin-Elmer R-10 NMR Spectrometer fitted with a variable temperature probe.

RESULTS AND DISCUSSION

a) Polyunsaturated Cholesterol Esters

The transitions of pure cholesteryl linolenate and cholesteryl linoleate are given below. The symbols used are as follows:

C_1 = crystal having a melting point higher than the highest liquid crystal transition
C_2 = crystals having melting point lower than the highest liquid crystal transition
Cholesteric = cholesteric mesophase
Smectic = smectic mesophase
Isotropic = isotropic liquid melt
(   ) refer to monotropic phases

The temperature above or beside the arrow indicates the temperature transition from one state to another as determined by optical or calorimetric methods. A dotted arrow indicates a transition occurring with supercooling.

Cholesteryl linolenate melts from a crystal (C_1) to an isotropic liquid (oil) at 36°C. Two monotropic liquid-crystal phases are formed on cooling. The cholesteric phase forms from the oil at 29.5°C and at 29.0°C a smectic phase forms from the cholesteric phase. The liquid-crystal phases are metastable and with time crystals (C_1) will grow from either the liquid-crystal phases or from the supercooled oil. If the smectic phase is rapidly cooled a second crystalline state (C_2) will grow which melts at 6°C to the smectic phase (7).

<u>Cholesteryl linolenate</u> (cholesteryl cis 9-10,12-13,15-16 octadecatrienoate)

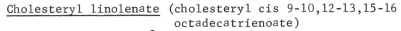

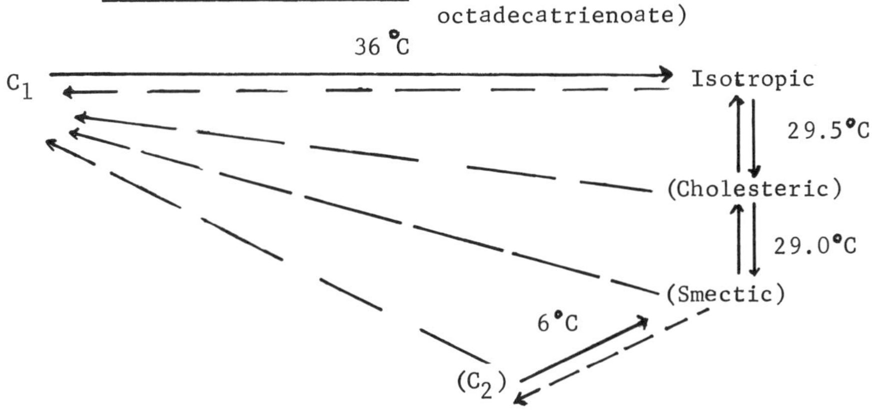

Crystal forms C_1 and C_2 give quite different x-ray diffraction patterns. In the narrow angle range the smectic phase gives a single diffraction at about 35 Å while the cholesteric and isotropic phases give only scattering.

Crystalline cholesteryl linoleate (C_1) melts to the isotropic liquid at 42°C. Two reversible monotropic phase transformations to the cholesteric phase and smectic phase occur at 36.5°C and at 34°C. On cooling to room temperature, recrystallization from the smectic phase to C_1 occurs very slowly (only after days or weeks). If the sample is cooled rapidly to -40°C, a second crystalline form, C_2, will be formed which melts at 3°C to the smectic mesophase.

Cholesteryl linoleate (cholesteryl cis 9-10,12-13 octadecadienoate)

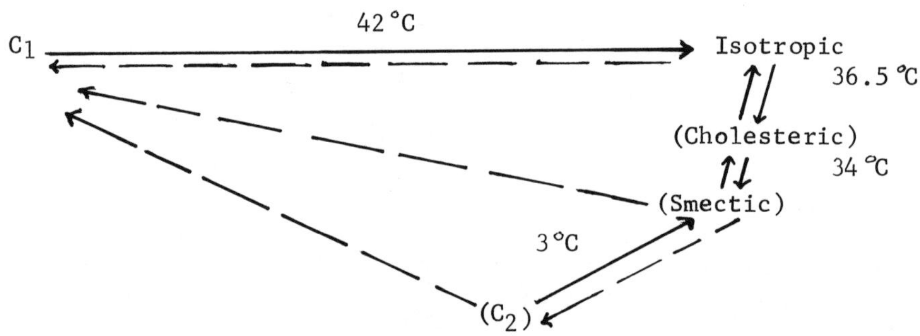

Like cholesteryl linolenate, cholesteryl linoleate has 2 crystal forms C_1 and C_2 which give different diffraction patterns. The smectic phase gives a single narrow angle diffraction at about 35.5 Å, while the cholesteric and isotropic phases give only scattering.

The proton magnetic spectra of pure cholesteryl linoleate at different temperatures are given in Figure 1. The cholesteric phase (Figure 1a) shows a fairly high resolution spectra whereas the smectic phase (Figure 1b,c,d) shows a rather broadened spectra, The spectra for the crystalline phase are very broad and would appear as a flat horizontal line. In Figure 2 the peak width at half peak height is plotted against the temperature. The dotted vertical lines represent the position of the isotropic-cholesteric transition at 36.5°C and the cholesteric-smectic transition at 34°C as determined by microscopy and DSC. The marked decrease in peak widths in these three phases indicate a sharp increase in molecular motion as one passes from smectic (S) to cholesteric (C) to isotropic liquid (I).

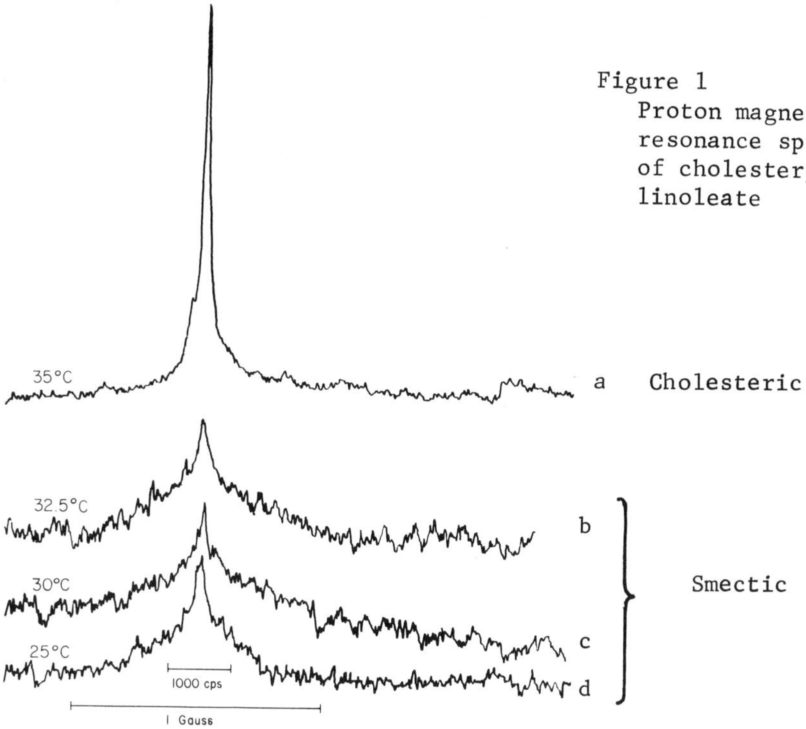

Figure 1
Proton magnetic resonance spectra of cholesteryl linoleate

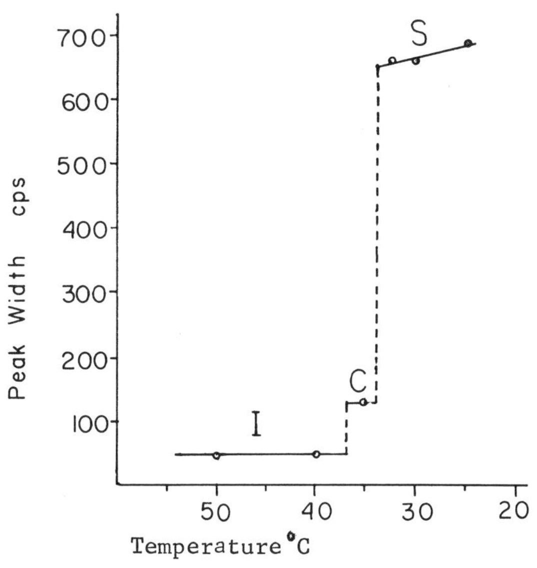

Figure 2

S = Smectic

C = Cholesteric

I = Isotropic

b. Cholesteryl Linolenate - Lecithin Interactions.

The partial condensed phase diagram for cholesteryl linolenate -lecithin systems is given in Figure 3. Pure lecithin, containing about 1% bound water by weight, undergoes several phase transitions on heating (Figure 3, right side). Egg yolk lecithin is birefringent crystalline material having a waxy appearance at low temperatures below 45°C. Many sharp x-ray diffraction lines confirm the presence of crystalline lecithin . On heating, the sample softens at about 45°C marking the transition from the rigid crystalline state to a viscous liquid-crystalline state characterized in the microscope by a "mosaic" texture (20). X-ray diffraction of this

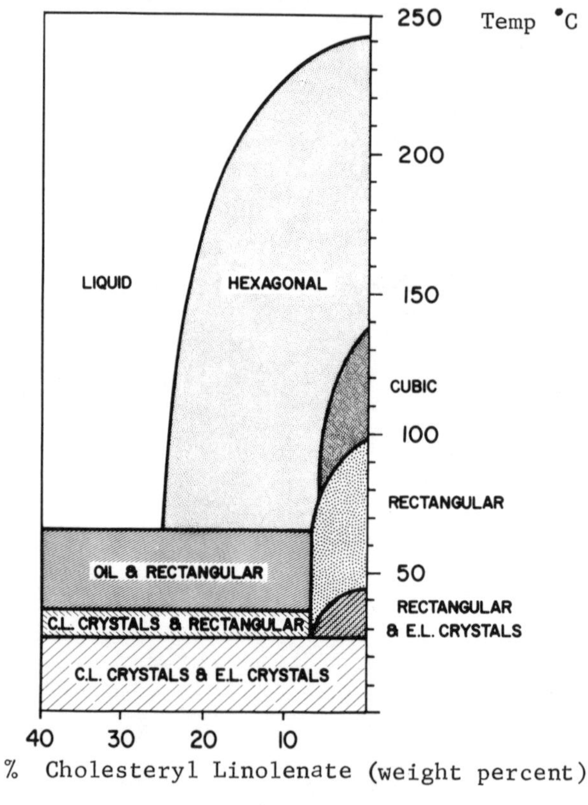

Fig. 3 - Condensed phase diagram Cholesteryl linolenate - lecithin as a function of temperature.
CL = Cholesteryl linolenate crystals
EL = Egg lecithin crystals

For further explanation see text.

phase gave low-angle reflections similar to those observed by Luzzati et al.(29), which were indexed according to a two dimensional centered rectangular lattice. The wide-angle diffraction pattern consisted of a sharp reflection at 4.7Å and two diffuse bands, at approximately 6.5Å and 4.7Å. The calculated cell parameters of the rectangular cell, a = 117.2Å; b = 44.2Å, differ only slightly from those observed by Tardieu et al. (30).

This rectangular phase disappears between 93 and 98°C and a viscous isotropic phase forms whose diffractions give a cubic symmetry. From 98°to 137°C, the cubic (viscous isotropic)phase remains. Abruptly at 130°C a birefringent phase develops which has a middle soap texture similar to that described for the middle phase of soaps (20). This phase gives diffraction consistent with the packing of lipid rods of indefinite length in a two dimensional hexagonal lattice (5). The distance between the center of the rods is about 43Å. At about 240°C this hexagonal phase melts sharply to an isotropic liquid.

Cholesteryl linolenate can be incorporated into all liquid crystalline and liquid phases formed by lecithin. However the amount incorporated in each phase varies. Thus the rectangular phase incorporates 7%, the cubic phase 5% and the hexagonal phase 25%. The two components are completely miscible at the appropriate temperature in the liquid phase.

The presence of cholesterol ester in the liquid crystalline phases of lecithin decreases the transition temperatures when compared to pure lecithin. For instance the incorporation of the maximum amount of ester lowers the melting of the rectangular phase from about 98°to about 60°C , the cubic phase from 137°C to about 80°C and the hexagonal phase from 240°C to about 70°C. Thus the effects of added cholesterol ester are similar to the effects of disordering the hydrocarbon parts of the liquid crystalline lattice. For instance increasing the unsaturation of the fatty acids of lecithin (or other lipids) decreases the transition temperatures. We therefore suggest that the cholesteryl linolenate incorporated into these liquid crystalline phases of lecithin is situated in the hydrophobic hydrocarbon parts of the liquid crystalline lattice.

The fact that cholesterol ester and lecithin are mutually soluble above the melting point of the liquid crystalline phases is interesting from a biological point of view. For instance, structural studies of both high density lipoproteins (HDL_2) and low density lipoproteins(22) suggest the presence of apolar regions composed of phospholipid and cholesterol ester from which water may be excluded. Furthermore reconstitution experiments by Scanu and Tardieu(22) indicated that the phospholipid from serum HDL_2 would not incorporate cholesterol ester in an aqueous system. This is related to the very limited solubility of cholesterol esters in lecithin in an aqueous environment (see below). However, addition of the apoprotein fraction permits incorporation of cholesterol ester. The formation of the lipoprotein complex appears to provide an apolar region from which water is excluded and in which phospholipid and cholesterol ester can interact, perhaps in a fashion analogous to the liquid or one of the liquid crystalline phases described above.

C. The Interaction of Water with Cholesterol Ester-Lecithin Systems.

Since most biological systems are aqueous systems we studied the effect of water on two systems, cholesteryl linolenate-lecithin and cholesteryl linoleate-lecithin. Three component phase diagrams of these systems at 37°C are given in Figures 4 and 5. Both diagrams are rather similar in appearance, only one zone (shaded) of a single phase is present. This phase is the lamellar liquid crystalline phase formed by lecithin alone in water (6,17), more recently called the L_α phase by Luzzati (5). This phase can incorporate very limited amounts of cholesteryl linolenate, the limit being 2% of the weight of lecithin. This corresponds to a molar ratio of about 1 cholesteryl linolenate to 40 lecithin molecules. Cholesteryl linoleate appears to have a slightly higher solubility in the lamellar liquid crystalline phase of lecithin, that is, about 3.5% of the weight of lecithin or 1 molecule of cholesteryl linoleate to 25 molecules of lecithin. Mixtures having compositions falling outside the 1 phase zone separate one or more phases. These other phases are pure water and pure cholesterol ester. In Figure 4 the cholesterol linolenate is present as an isotropic liquid. Lecithin, in the presence of water, is not soluble in this oily liquid. Cholesteryl linoleate separates as a metastable smectic liquid crystal or a crystalline phase (C_1 crystals) at 37°C and lecithin is not soluble in these phases. Thus the presence of water has 2 major effects on the cholesterol ester-lecithin systems: 1) it converts lecithin into a lyotropic lamellar liquid crystal in which the ester has very limited solubility and 2) it hydrates the polar groups of the lecithin which prevent the lecithin molecules from entering the hydrophobic oily cholesterol ester phases. While lecithin and cholesterol ester enjoy mutual solubility at appropriate temperatures in the dry state (see Figure 3), small amounts of water hydrate the polar groups of the lecithin and cause it to separate from the liquid as the lamellar liquid crystalline phase.

For the purpose of comparison the cholesterol-lecithin-water phase diagram (10,23) is shown in Figure 6. Free (unesterified) cholesterol, behaves very differently from its long chained polyunsaturated esters. It spreads to form a stable monolayer at an

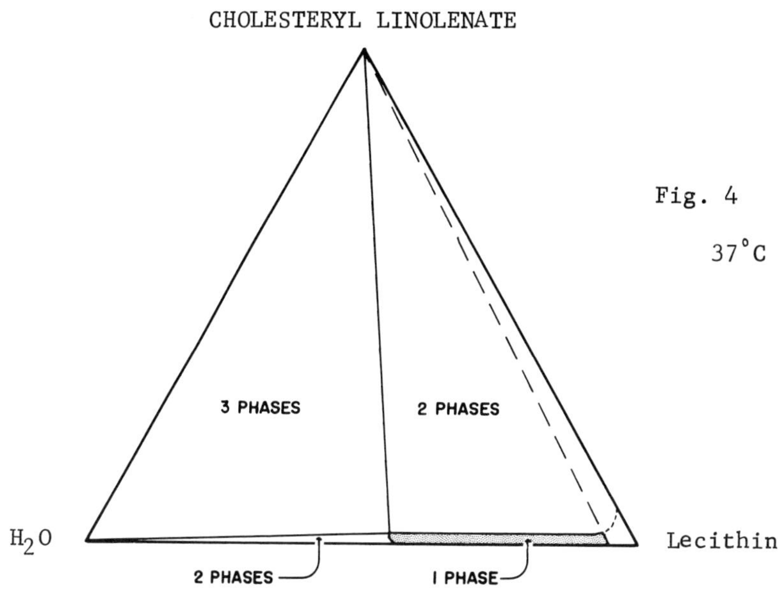

Fig. 4

37°C

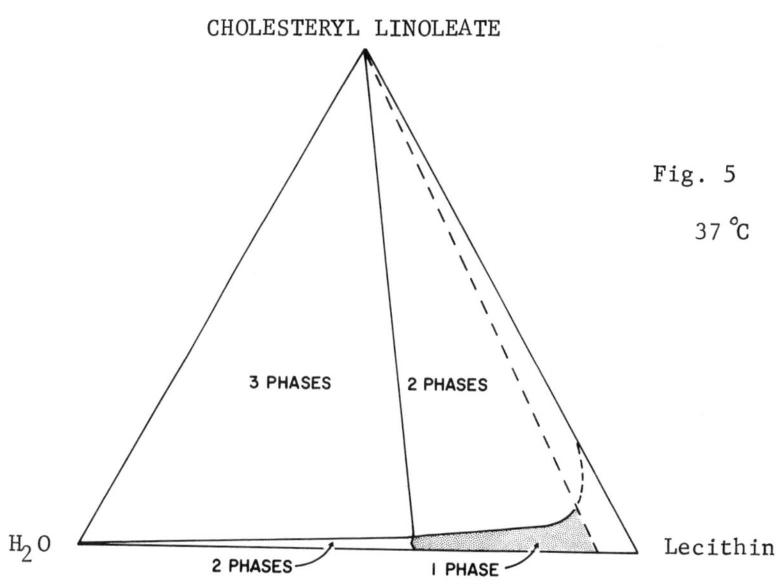

Fig. 5

37 °C

air-water interface (24) whereas cholesterol esters of long chained fatty acids do not spread (25). Cholesterol is quite soluble in the lamellar liquid crystalline phase formed by lecithin. Up to one molecule per molecule of lecithin may be solubilized by egg lecithin (10,23), by dipalmitoyl lecithin (26) or in fact by sphingomyelin (27) and even by mixtures of phospholipids (28). This maximum molar solubility ratio contrasts sharply with the 1:40 or 1:25 found for cholesteryl linolenate and cholesteryl linoleate respectively. Since biological membranes rarely contain more than traces of cholesterol esters perhaps membrane structure is in general terms physically similar to the lamellar liquid crystal system formed by phospholipids.

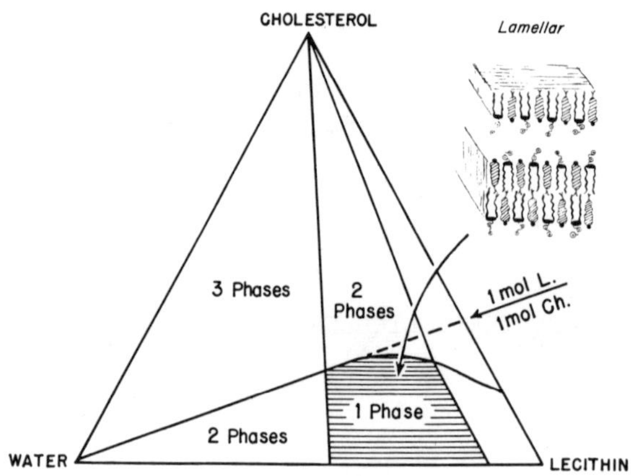

Figure 6. The three component system Lecithin-Cholesterol-Water, 25°C (10,23) The lamellar liquid crystalline phase is shown in upper right. Lecithin Cholesterol

ACKNOWLEDGEMENT

The authors thank Ms. Irene Miller for help in the preparation of the final manuscript. This research is supported by U. S. Public Health Service Grant AM 11453.

REFERENCES

1. "Lipid Storage Diseases" (J.Bernsohn and H.J.Grossman eds.) Academic Press, New York and London, (1971).
2. Small, D.M., Advances in Internal Medicine, 16,243-264(1970).
3. "Atherosclerosis" (R.J.Jones ed.) Springer-Verlag, New York, Heidelberg, Berlin, (1970).
4. Small, D.M., J.Am.Oil Chem. Soc., 45,108-119(1968).
5. Luzzati, V., In "Biological Membranes" (D.Chapman, ed.),Academic Press, New York and London, pp.71-123(1968).
6. Chapman, D. and Wallach, D.F.H., In "Biological Membranes" (D. Chapman, ed.),Academic Press, New York and London,pp.125-202(1968).
7. Small, D.M., In "Surface Chemistry of Biological Systems" (M. Blank, ed.) Plenum Press, New York, pp.55-83(1970).
8. Phillips, M.C., In "Progress in Surface and Membrane Science" (J.F.Danielli,M.D.Rosenberg and D. .Cadenhead, eds.)Academic Press, New York and London, 5,139-221(1972).
9. Shipley, G.G., In "Biological Membranes" (D.Chapman and D.F.H. Wallach, eds.) Academic Press, New York and London,2,1-89(1973)
10. Bourges, M.,Small, D.M. and Dervichian, D.G.,Biochim. Biophys. Acta, 137,157-167(1967).
11. Lecuyer, H. and Dervichian, D.G., J. Mol. Biol. 45,39(1969).
12. Friedel, G., Ann. Physique, 18.273-274(1922).
13. Shipley, G.G., Halks, M. and Small, D.M.,(unpublished results)
14. Davis, G.J., Porter, R.S., Steiner, J.W. and Small, D. M., Molec. Crystals and Liq. Crystals, 10,331-336(1970).
15. Small, D.M., Bourges, M., and Dervichian, D.G., Biochim. Biophys. Acta, 125,563-580(1966).
16. Wuthier, R.E., J. Lipid Res., 7,544-550(1966).
17. Small, D.M., J. Lipid Res., 8,551-557(1967).
18. Elliot, A., J. Sci. Instrum., 42,312-316(1965).
19. Franks, A., Br. J. Appl. Phys., 9,349-352(1958).
20. Rosevear, F.B., J. Amer.Oil Chem. Soc., 31,628-329(1954).
21. Shipley, G.G., Atkinson, D. and Scanu, A.M., J. Supramol. Struc., 1,98-103(1972).
22. Skipski, V.P., In "Blood Lipids and Lipoproteins:Quantitation, Composition, and Metabolism" (G.J.Nelson, ed.), Wiley-Inter-Science, Toronto, pp.471-583(1972).
23. Small, D.M., Bourges, M., and Dervichian, D.G., Nature, 211, 816-818(1966).
24. Adam, N.K., "The Physics and Chemistry of Surfaces", Oxford Univ. Press, London, (1941).
25. Kwong, C.N., Heikkila, R.E., and Cornwell, D.G., J. Lipid Res. 12,31(1971).
26. Ladbrooke, B.D., Williams, R.M., and Chapman, D., Biochim. Biophys. Acta, 150,333(1968).

27. Shipley, G.G., Avecilla, L. and Small, D.M., unpublished observations.
28. Rand, R. P., and Luzzati, V., Biophys. J., 8,125(1968).
29. Luzzati, V., Gulik-Krzywicki, T., Tardieu, A., Nature, 218, 1031-1034(1968).
30. Tardieu, A., Luzzati, V., and Reman, F.C., J. Mol. Biol., 75, 711-733(1973).

AN E.P.R. INVESTIGATION OF THE ALIGNMENT OF TWO SMECTIC A LIQUID CRYSTALS

Arthur Berman and Edward Gelerinter

Physics Department and Liquid Crystal Institute
Kent State University, Kent, Ohio 44242

George C. Fryburg

Lewis Research Center, National Aeronautics and
Space Administration, Cleveland, Ohio

Glenn H. Brown

Liquid Crystal Institute, Kent State University
Kent, Ohio 44242

INTRODUCTION

The purpose of this work is to extend our studies of liquid crystals that display both nematic and smectic A phases, using the technique of electron paramagnetic resonance (epr). When the liquid crystal is in the nematic phase, an application of a magnetic field of the magnitude required for our X-band spectrometer (3.3 KG) is more than sufficient to uniformly align the director throughout the sample. In other words the preferred direction is the same throughout the entire sample. When the sample temperature is lowered through the nematic→smectic transition, the picture is somewhat more complicated. The forces between the liquid crystal molecules and the surface of the sample tube are now considerably greater. In addition, the lateral forces between the smectic molecules are also fairly large, so that any surface alignment that does take place may penetrate deep into the bulk of the sample. These effects oppose the efforts of the magnetic field to align the sample. In this investigation we study the smectic A alignment as a function of both the cooling rate through the transition point and the size of the aligning magnetic field. Since the molecules are already aligned parallel to the field when the liquid crystal is in the nematic phase, surface forces, if dominant, would require a finite time to realign the sample. One can detect the effects of surface forces by varying the cooling rate. In addition, if the magnetic field is large enough, it will be the dominant aligning force. We have observed evidence of both these effects.

The compounds studied in this investigation are 4-octyloxybenzylidene-4'-ethylaniline (obea) and 4-butyloxybenzylidene-4'-acetoaniline (bbaa). These compounds have been previously [1] studied using vanadyl acetylacetonate (vaac) as the paramagnetic probe. The vaac probe did not give good agreement with the theory due to the fact that its tumbling was inhibited by viscous effects. This was verified by experiments with other liquid crystals. [2] (Previous work has established that a probe that does not tumble freely will not provide a temporal average of its surroundings. [3,4]) In order to minimize the viscous effects we chose to use nitroxides as paramagnetic probes for this study. Two probes were used, these being piperdinoxy-4-hydroxy-2,2,6,6-tetramethyl,p-toluene sulfonate and spiro(5α-cholestane-3,2'-oxazolidin)-3'-yloxy 4,4' dimethyl. They are referred to as nitroxides 3 and 4 for historical reasons. [5] Other authors [6] have studies bbaa using nitroxide 4. Their results will be compared to ours in the discussion section.

THEORETICAL

The theoretical epr line positions for paramagnetic probes dissolved in both isotropic and anisotropic media [1,3,6,7] have been calculated so that only a short discussion will be given here. For the spin probes used in this investigation the components of the "g" tensor are virtually identical, and the hyperfine tensor is very nearly axial. The effect of the quadrupole coupling is rather small. There are two reasons for this. First the coupling itself is small and second, it only affects the line positions in second order. In our work we found good agreement when the quadrupole coupling was set to zero. Other authors [6] have also found these effects to be small. Hence, the quadrupole terms will be omitted in this development and the interested reader is referred to the literature. [6,7]

If one assumes isotropic "g" and axial hyperfine tensors and follows the development of Berman in reference 7, one obtains the following for the average line separations,

$$\langle a \rangle = \left\{ \bar{A}_{\perp}^2 + \left[\bar{A}_{\parallel}^2 - \bar{A}_{\perp}^2 \right] \mathrm{Cos}^2\, \beta \right\}^{\frac{1}{2}} . \qquad (1)$$

Here β is the angle between the director of the smectic A liquid crystal and the d.c. magnetic field of the spectrometer. $\bar{A}_{\perp}$ and $\bar{A}_{||}$ are the partially averaged components of the hyperfine tensor and can be written in terms of the components of the hyperfine tensor as

$$\bar{A}_{||} = a + \frac{2}{3}(A_{||} - A_{\perp})\sigma \;, \quad \text{and}$$
$$\bar{A}_{\perp} = a - \frac{1}{3}(A_{||} - A_{\perp})\sigma \;. \tag{2}$$

Here σ is the order parameter i.e., $<3 \cos^2 \theta - 1>/2$ where θ is the angular deviation of the molecule from the director. If $\beta = 0$, then equation 1 becomes

$$<a> = a + \frac{2}{3}(A_{||} - A_{\perp})\sigma \;, \tag{3}$$

where "a", the isotropic part of the hyperfine tensor, is 14.8±0.1 gauss. This value is obtained from the liquid crystal in the isotropic phase. For methyl protected nitroxides such as those used in these [8] studies $A_{\perp} = 0.188A_{||}$. This is combined with $a = (A_{||} + 2A_{\perp})/3$ to obtain $A_{||}$ and $A_{\perp}$. Once these quantities are known, the order parameter can be measured when the liquid crystal is in the smectic A phase. In the nematic phase, the molecules align with their preferred direction parallel to the field, so that the order can be calculated in a similar manner.

The data for the smectic A rotation experiments will be compared to a theoretical curve calculated from equation 1 using the order parameter calculated from the $\beta = 0$ spectrum.

EXPERIMENTAL

Obea has a clearing temperature $T_{ni} = 72.5°C$, a nematic to smectic A transition at $T_{sn} = 71°C$ and a smectic A to crystal transition at $T_{sc} = 62.5°C$. Note that the nematic range is only 1.5° wide so that it is difficult to obtain the order versus temperature in this narrow range. The corresponding transition temperature of bbaa are $T_{ni} = 111°C$, $T_{sn} = 99°C$ and $T_{cs} = 84°C$. This compound has a considerably broader nematic range making it feasible to measure the order versus temperature.

The sample temperature was held constant to better than a few tenths of a degree using a temperature controller that has been previously described. [5] Thermal gradients of a few degrees occurred across the bbaa sample. This was indicated by the ability to observe spectra from two phases simultaneously over a range of a few degrees. The effect was not present for obea, so that we estimate thermal gradients of less than 1°C for this liquid crystal. The sample was contained in a 4mm o.d. quartz tube after standard [7] preparation. The spectra were obtained using a model V4500 Varian X-band spectrometer. The sample was aligned in the nematic phase and then the sample temperature was lowered into the smectic region using varying cooling rates and aligning magnetic fields. The hyperfine splitting was measured as a function of the orientation of the aligned smectic A relative to the d.c. magnetic field. Spectra were observed at 10° intervals between 0 and 180°.

RESULTS AND DISCUSSION

Obea was studied using nitroxide 3 as the paramagnetic probe. Typical spectra obtained from obea in its different phases are shown in figure one. The contraction of the spectrum as the obea goes from its isotropic phase to its nematic phase and the further contraction as the smectic A phase is entered are indicative of the ordering of the probe by the liquid crystal. The value of "a" is obtained from the isotropic spectrum and the value of $\langle a \rangle$ is obtained from the liquid crystal spectrum. These quantities are used to calculate the order in conjunction with equation 3. The spectral lines are fairly narrow and symmetric. This indicates that viscous effects are not a problem for this probe, i.e. the probe tumbles reasonably freely.

One of the experiments performed was to vary the cooling rate through the transition point from several seconds to several hours while using aligning magnetic fields from 3 to 12 KG. In figure two the result of aligning in a 12 KG field is illustrated. Note that the points agree very well with the theoretical curve for both fast and slow cooling. In fact, the fast and slow cooling curves are nearly identical (the upper curve has been translated up 1 gauss for clarity). Our measurements indicate that the cooling rate does not significantly affect these results when aligning magnetic fields between 6 and 12 KG are employed. We illustrate the case for an aligning field of 6 KG in figure three. One again observes good agreement with the theoretical curves and nearly identical results for fast and slow cooling. (Here again the upper curve is translated 1 gauss upward for clarity.) At 12 KG the 0° hyperfine splitting is approximately 13.4 gauss while at 6 KG the splitting is 13.7 gauss. This splitting and hence the order varies very little in this regime. In figure four this point is illustrated by a plot of the order versus aligning field for obea. The smectic A is reproducibly aligned if fields of 6 KG or above are used. We have also observed similar results when using vaac as a paramagnetic probe.

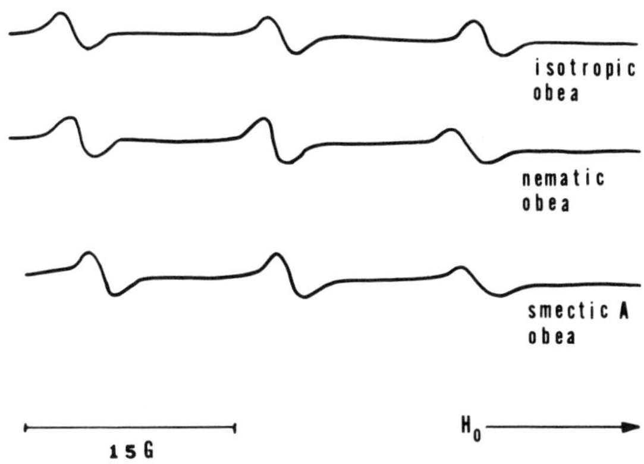

Fig. 1. Typical E.P.R. spectra of the probe nitroxide 3 in obea.

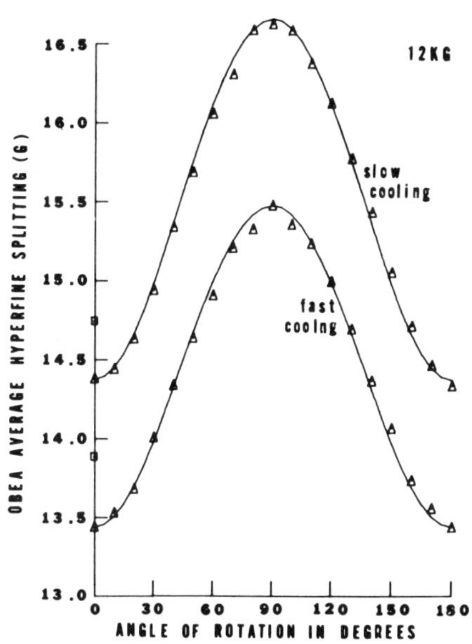

Fig. 2. Rotation curve for nitroxide 3 in obea. Vertical axis displaced (see text).

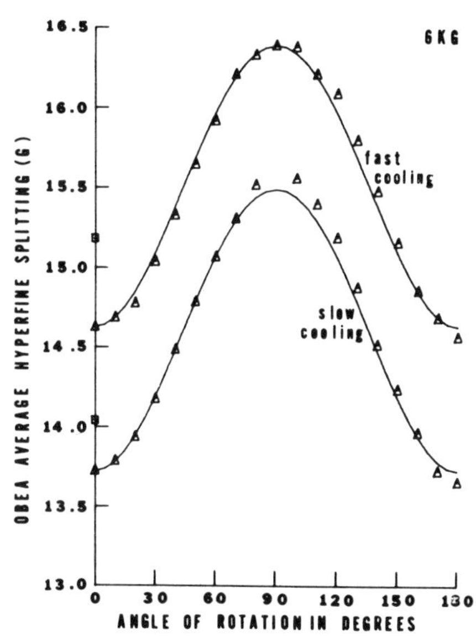

Fig. 3. Rotation curve for nitroxide 3 in obea. Vertical axis displaced (see text)

When aligning fields below 6 KG are used the situation is somewhat more complicated. The order observed by epr is dependent upon the cooling rate. This is illustrated in figure five where three different cooling rates were used with an aligning field of 4 KG. (In this figure the fast cooling curve is displaced downward ½ gauss; the slow cooling curve is displaced ½ gauss upward and the very slow cooling curve is displaced upwards by 1 gauss.) The results are surprising in that the faster the cooling, the greater the observed order. One should note that the local order of the smectic A is dependent only upon the material. Epr measures the bulk order. This is affected by both the local order and the uniformity of the sample alignment, i.e. how uniform the preferred direction is throughout the sample. When wall effects are important, there are regions in the sample that are not aligned parallel to the aligning magnetic field. This results in a lower observed order. We conclude that rapid cooling results in a better aligned sample. Presumably, the sample enters the smectic A phase before the walls have time to realign the sample. The aligning effects of the field and the walls must be comparable at 4 KG. When the aligning field is above 6 KG it is presumably strong enough to overcome wall effects, so that the cooling rate is unimportant. A similar effect has been reported by Uhrich et al. [9]. They observe that in a smectic H phase, the molecular alignment by the walls of the sample container can penetrate into the bulk of the sample. They also observe that the wall alignment can be overcome if sufficiently large aligning fields are used.

If aligning fields of 3 KG or less are used we find that it is impossible to align the smectic A at any cooling rate. In these cases the walls of the sample tube exert greater aligning forces than does the magnetic field.

It was not feasible to study bbaa using nitroxide 3 since this probe was ordered only slightly by the liquid crystal. Instead, the considerably larger nitroxide 4 probe was used. Typical spectra of nitroxide 4 in the different phases of bbaa are shown in figure six. The ordering of the probe by the liquid crystal is evidenced by the contraction of the line separation as the material goes from the isotropic to nematic to smectic A phase. The lines are closer together than those for nitroxide 3 in obea. This indicates that the nitroxide 4 probe is more ordered. The lines are slightly broader due to a slightly greater inhibition of the probes tumbling. This is not surprising since the nitroxide 4 probe is considerably larger than the 3 probe. These increased viscous effects were not large enough to cause any problems, however.

It is of interest that this is the same probe and liquid crystal that has been studied by other authors [6], so that we

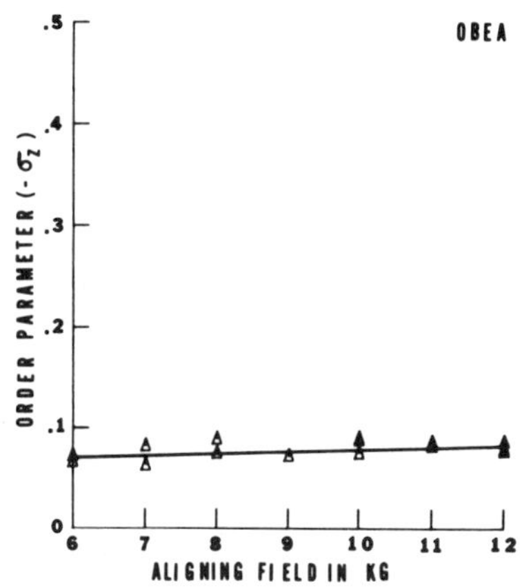

Fig 4. -σ vr. the aligning m magnetic field for nitroxide 3 in obea.

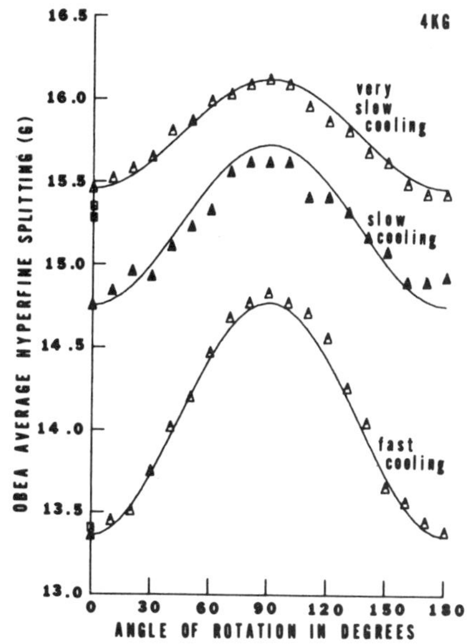

Fig 5. Rotation curve for nitroxide 3 in obea. Vertical axes displaced (see text).

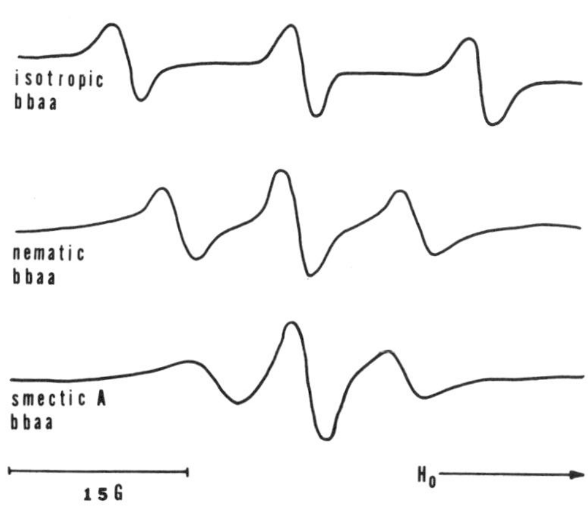

Fig. 6. Typical E.P.R. spectra of the probe nitroxide 4 in bbaa.

will compare our results with theirs. When the bbaa sample was cooled slowly in a 3.3 KG aligning field extra lines were observed in the spectra as shown in figure seven. These lines indicate that the smectic A was not well aligned. When the sample was cooled quickly in the 3.3 KG field, the extra lines were not present (see figure seven) indicating that the smectic A is now better aligned. Just as was found for obea, rapid cooling produced better alignment. Luckhurst and Setaka report observing the extra lines upon rapid cooling. This is contrary to our observations, but it is really not clear what constitutes rapid cooling. In our case, it is limited by the thermal time constant of the system to several seconds, perhaps 10 or 20. We agree with these authors that the extra lines are caused by wall alignment. However, we find that the effect is eliminated by rapid cooling.

Luckhurst and Setaka report good alignment when using an aligning field of 6.5 KG and a plastic coated sample tube to limit wall alignment. We obtain equally good alignment at 6 KG using an untreated sample tube as is shown in figure eight. This good alignment was obtained at all cooling rates. This indicates that the 6 KG aligning field is strong enough to overcome wall effects, so that no wall treatment is required. In the same figure we also present the rotation curve for an aligning field of 12 KG. (Here the 6 KG curve has been displaced upwards 2 gauss.) The 12 and 6 KG alignments are quite similar.

We have measured the order of nitroxide 4 by heating from an ordered smectic phase through the nematic to the isotropic phase. We have also measured the order while cooling from the isotropic through the nematic phase. The results are presented in figure nine. Note the discontinuity in order as the nematic smectic boundary is crossed. This is good agreement with the data reported by Luckhurst and Setaka. When vaac is used as the probe, this discontinuity is not observed. [1] The above authors attribute this to incomplete alignment of the probe. In view of the evidence presented in this paper we feel that the smectic A phase was well aligned. Other experiments indicate that the observed order was influenced by the inhibited tumbling of the vaac probe. [5,10] In a previous paper [2] we present evidence to show that the deviation of the experimental points from the theoretical vaac rotation curve is also caused by the inhibited tumbling of the probe.

SUMMARY

In summary we found that aligning fields of greater than 6 KG are required to reliably overcome the aligning effects of the sample holder walls for our geometry. We expect that this value would be different for different sample geometries. When the

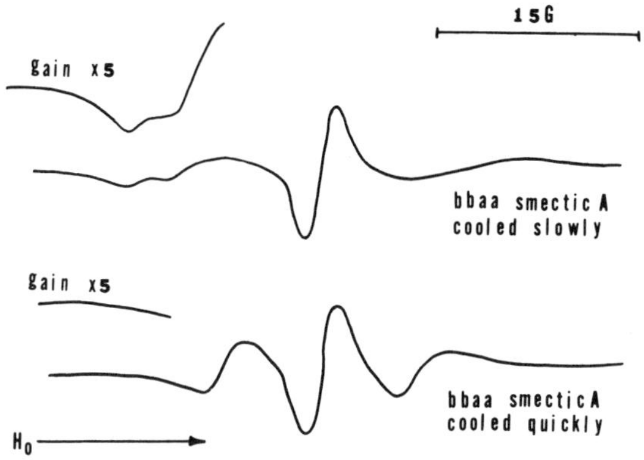

Fig. 7. Line shapes of the nitroxide 4 probe in bbaa in smectic A phases formed by fast and slow cooling in the presence of a 4 KG aligning field.

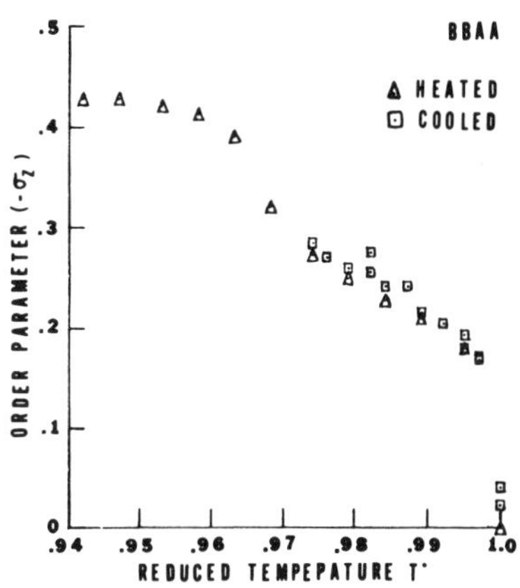

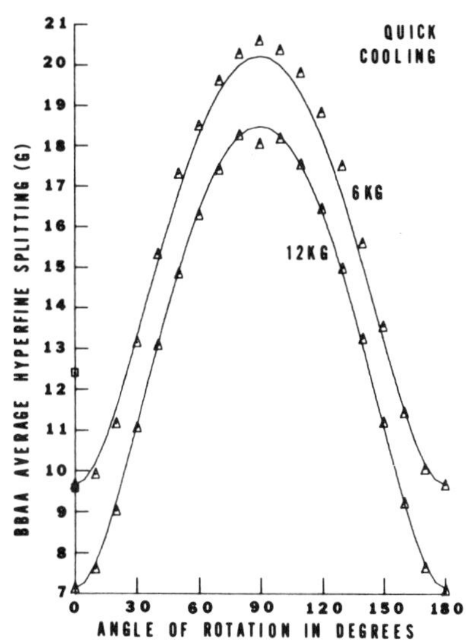

Fig. 8. Rotation curve for nitroxide 4 in bbaa. Vertical axis displaced (see text)

Fig. 9. -σ vr. T* for nitroxide 4 in bbaa. $T_{N \to S_A}$ = .97.

aligning field is not large enough to completely overcome wall effects, the uniformity of alignment depends upon the sample's cooling rate. We have observed the rather surprising result that rapid cooling is required for good alignment. We have also made the not so surprising observation, that magnetic fields below 3 KG are not adequate to align the smectic phase.

ACKNOWLEDGEMENT

We would like to acknowledge support from the Air Force Office of Scientific Research under Grant #F 44620-69-C-0021 and the National Science Foundation under Grant #GH 34164X. We would like to thank Dr. D. L. Fishel and Dr. C. F. Sheley for providing the liquid crystals and probes.

REFERENCES

1. G.C. Fryburg, E. Gelerinter and D.L. Fishel, Mol. Cryst., Liquid Cryst. 16, 39 (1972).

2. G.C. Fryburg and E. Gelerinter, Accepted for publication in Mol. Cryst., Liquid Cryst.

3. G.C. Fryburg and E. Gelerinter, J. Chem. Phys. 52, 3378 (1970).

4. P. Diehl and C.F. Schwerdtfeger, Mol. Phys. 17, 417 (1969); 421 (1969).

5. W.E. Shutt, Kent State University, M.S. Thesis (1972). W.E. Shutt, E. Gelerinter, G.C. Fryburg and C.F. Sheley, Accepted for publication in J. Chem. Phys.

6. G.R. Luckhurst and M. Setaka, Mol. Cryst., Liquid Cryst. 19, 179 (1972).

7. A.L. Berman, Kent State University, M.S. Thesis (1973).

8. J. Seelig, J. Am. Chem. Soc. 92, 3881 (1970).

9. D.L. Uhrich, Y.Y. Hsu, D.L. Fishel and J.M. Wilson, Accepted for publication in Mol. Cryst., Liquid Cryst.

10. J.I. Kaplan, E. Gelerinter and G.C. Fryburg, Accepted for publication in Mol. Cryst., Liquid Cryst.
Presented at the Fourth International Liquid Crystal Conference, Kent, Ohio, August 21-25, 1972.

PRETRANSITIONAL BEHAVIOR IN THE ISOTROPIC PHASE OF HOMOLOGOUS COMPOUNDS SHOWING NEMATIC AND SMECTIC C TYPE ORDER

T. R. Steger, Jr., J. D. Litster and W. R. Young†

M.I.T., Cambridge, Mass., and †IBM Watson Research Center, Yorktown Heights, N. Y.

We report here the preliminary results of our study of pretransitional behavior in the isotropic phase of three homologous compounds showing smectic C type order. We have measured the intensity and spectrum of scattered light in the isotropic phase of dialkoxyazoxybenzenes with terminal octyl (C_8H_{17}), decyl ($C_{10}H_{21}$), and undecyl ($C_{11}H_{23}$) groups[1]. All three compounds have a smectic C phase. The octyl and decyl compounds also have nematic phases over a range of 18°K and 2°K, respectively. The undecyl compound has a direct smectic C-isotropic transition. Previous light scattering experiments in liquid crystals just above the nematic-isotropic[2] and the cholesteric-isotropic[3] transitions have shown the existence of pretransitional phenomena in the isotropic phase. Both the static and dynamic behavior are in good agreement with de Gennes' phenomenological theory of short-range orientational order[4]. The purpose of our experiment is to investigate the effect of nearby smectic order on the fluctuations in the orientational order in the isotropic phase.

In our experimental system the sample was placed in an aluminum oven whose temperature was controlled to 2 mdeg and the temperature was measured with a platinum resistance thermometer. The light source was a Spectra-Physics 119 He-Ne laser. Intensity measurements were made using two photomultipliers: one to detect the scattered light at 90°, and one to detect the transmitted light. This allowed a ratio measurement, increasing the sensitivity and cancelling laser drifts and changes in sample turbidity with temperature. Absolute calibration was made using a known concentration of polystyrene spheres suspended in water, for which the Rayleigh ratio was calculated[5]. Spectral measurements were made using a spherical Fabry-Perot interferometer whose full-width at half-height

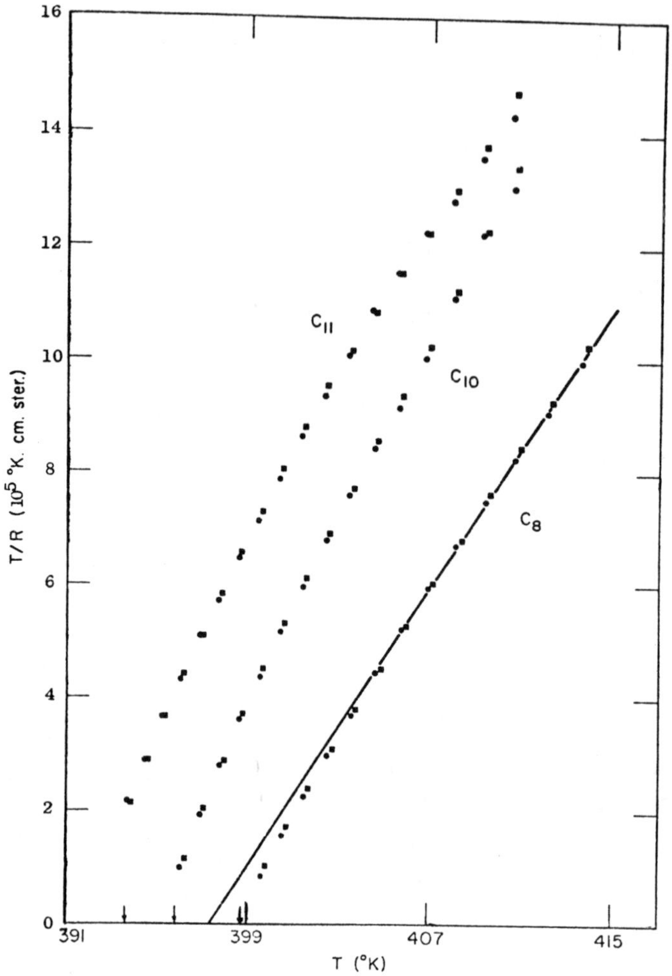

Figure 1. Results of intensity measurements for three homologous compounds, with the temperature divided by the Rayleigh ratio plotted against temperature. The arrows indicate the respective transitions temperatures T_K, and the straight line shows behavior observed above the nematic-isotropic transition and predicted by de Gennes' theory.

was 3 MHz. The spectra were analyzed by numerically convolving the instrumental profile with various Lorentzians until a fit was obtained. Our samples contained a small amount of particulate impurities. To eliminate the effects of these impurities, all of our measurements were of the depolarized component of the scattered light.

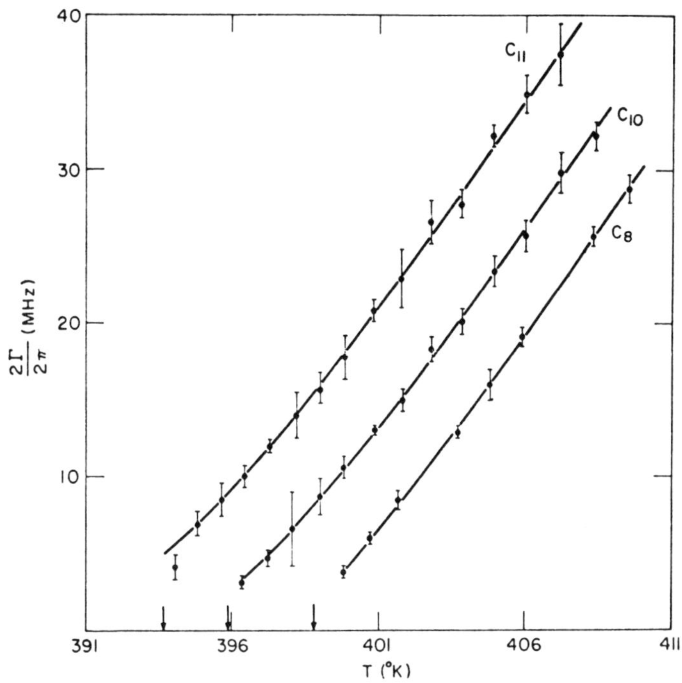

Figure 2. Results of spectral measurements for three homologous compounds showing linewidth vs. temperature. The arrows indicate the respective transition temperatures T_K, and the lines are a guide to the eye.

The intensity of the scattered light in the isotropic phase measures the amplitude of fluctuations in the orientational order. The results of our measurements are shown in Fig. 1 where we plot the temperature divided by the Rayleigh ratio and the arrows indicate the respective transition temperatures T_K. All three compounds clearly show a divergence of the fluctuations as the phase transition is approached. Previous intensity measurements above the nematic-isotropic transition have shown that de Gennes' theory correctly predicts the temperature dependence of the Rayleigh ratio to be of the form[2]

$$R \sim T/\left(T - T_c^*\right)$$

where $T_K - T_c^* \sim 1°K$. The C_8 data shows the closest agreement with this prediction, as shown by the straight line, while C_{11} shows a distinct departure from linear behavior. We believe this departure to be a result of fluctuations in short-range smectic order.

From the spectrum of the scattered light in the isotropic

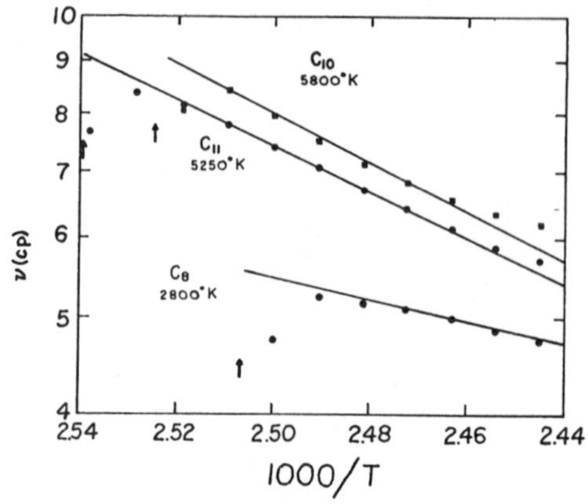

Figure 3. Arrhenius plots of phenomenological viscosity for three homologous compounds. The straight lines are a fit to exp(w/kT) and values of w/k are given. The arrows indicate the respective transition temperatures.

phase we have determined the time dependence of the fluctuations in the orientational order. The results of our spectral measurements are shown in Fig. 2. The spectra appeared to be Lorentzian within our experimental accuracy, and we plot the full-width at half-height for each temperature. Again the pretransitional behavior is clear — a critical slowing of the fluctuations as the phase transition is approached. De Gennes' theory[4] predicts the spectrum of the orientational fluctuations (when the small coupling to shear waves is ignored) to be Lorentzian with a width that varies as

$$\Gamma \sim (T - T_c^*)/\nu$$

where ν is a phenomenological viscosity. From the results of the intensity and linewidth measurements we may deduce the temperature dependence of the transport coefficient ν

$$\nu \sim (T/R)(1/\Gamma) .$$

Making the reasonable assumption that ν varies as exp(w/kT) with w an activation energy, we show Arrhenius plots of ν in Fig. 3. The straight lines are a best fit and values of w/k are given for each compound. The main cause of error is uncertainty in Γ, which is roughly constant over the temperature range. Because Γ becomes small at the lower temperatures, that is where the percentage error is the largest. We do not interpret the deviations from linear behavior near the transition temperature as being significant.

We conclude from our preliminary results that nearby smectic order does affect pretransitional behavior in the isotropic phase. McMillan[6] has used x-ray scattering to measure the short-range translational or smectic order in liquid crystals having smectic A and cholesteric phases. He observes an increase in the short-range order as the smectic A-cholesteric transition is approached. He has also proposed a model for the smectic A phase which requires coupling between the orientational and translational order. We intend to use a similar model to interpret our intensity data and explain the different temperature dependence in terms of fluctuations in the short-range smectic order. The results of these calculations will be presented in a later publication.

REFERENCES

1. For thermodynamic data see J. F. Johnson, R. S. Porter, and E. M. Barrall, II, Mol. Cryst. and Liq. Cryst. 8, 1 (1969).
2. T. W. Stinson and J. D. Litster, Phys. Rev. Letters 25, 503 (1970); T. W. Stinson, J. D. Litster, and N. A. Clark, J. Phys. (Paris) Suppl. 33, C1-69 (1972).
3. C. C. Yang, Phys. Rev. Letters 28, 955 (1972).
4. P. G. de Gennes, Mol. Cryst. and Liq. Cryst. 12, 193 (1971).
5. H. Denmon, W. Pagonis, and H. Heller, Angular Scattering Functions for Spheres (Wayne State University Press, Detroit 1966).
6. W. L. McMillan, Phys. Rev. A6, 936 (1972).

ELECTRIC FIELD EFFECTS IN THE NEMATIC AND SMECTIC PHASES OF P-N-NONYLOXYBENZOIC ACID*

L. S. Chou and E. F. Carr

Department of Physics, University of Maine

Orono, Maine 04473

ABSTRACT

Samples of p-n-nonyloxybenzoic acid have shown a positive conductivity anisotropy in the nematic phase immediately after heating from the solid, but after a number of hours the anisotropy changed to negative. A possible explanation for this change is the formation of clusters of molecules by polymerization. The clusters could be linear polymers involving hydrogen bonds. A study of the temperature dependence of the conductivity anisotropy indicated that other mechanisms are involved such as the formation of cybotactic groups involving dimers. Molecular alignment due to ionic conductivity is discussed in both the nematic and smectic phases. The dielectric anisotropy which is positive in the nematic phase decreases with decreasing temperature and reverses sign as the material enters the smectic phase.

INTRODUCTION

Previous investigations involving effects due to ionic conduction have been primarily concerned with materials exhibiting a positive conductivity anisotropy (conductivity is a maximum parallel to the nematic director). Molecular alignment due to a negative conductivity anisotropy has been reported[1] in the nematic phase of 4, 4' - di-n-heptyloxyazoxybenzene (HOAB). This work was related to that of De Vries who used X-ray techniques and suggested[2] a cybotactic structure in the nematic phase of HOAB. A temperature dependence of the conductivity anisotropy has been reported by Rondelez[3] for HOAB. Molecular alignment in a smectic A phase of a liquid crystal exhibiting a negative conductivity anisotropy has also been observed.[4]

The nematic phase of p-n-nonyloxybenzoic acid (NOBA) exhibits a positive dielectric anisotropy and normally shows a negative conductivity anisotropy. Molecular alignment due to its negative conductivity anisotropy was reported earlier.[5] Recent results[6,7] in NOBA showed a positive conductivity anisotropy in the nematic phase after melting from the solid but changed to negative after a few hours. A slow polymerization process was suggested as a possible explanation. When equilibrium is established the sample may contain small clusters where the molecules form linear polymers with hydrogen bonds. A temperature dependence of the conductivity anisotropy has been investigated for both heating from the solid and cooling from the normal liquid. A discussion of these results will be included in this article.

EXPERIMENTAL

Measurements of the dielectric loss at a microwave frequency of 24.5 GHz were used to study the molecular alignment due to electric fields, and the experimental techniques associated with these measurements were similar to those discussed earlier.[8] The externally applied electric field which orders the sample is always applied parallel to the microwave electric field when employing these techniques. The conductivity was measured with a Keithley Model 610C electrometer using dc voltages of approximately 1 volt. The samples of NOBA were purchased from Frinton Organics and purified by recrystallization and chromatographic methods. The resistivity was approximately 10^9 ohm-cm for the pure samples in the nematic phase. NOBA exhibits a smectic C phase (94-117°C) and a nematic phase (117-143°C).

RESULTS AND DISCUSSION

Conductivity Anisotropy

P-n-nonyloxybenzoic acid (NOBA) normally shows a positive conductivity anisotropy ($\sigma_{\parallel}/\sigma_{\perp} > 1$) when heated from the solid to the nematic, but after heating for a few hours the anisotropy changes to negative ($\sigma_{\parallel}/\sigma_{\perp} < 1$). $\sigma_{\parallel}$ and $\sigma_{\perp}$ represent the conductivity parallel and perpendicular to the nematic director respectively. Results for a particular sample at 134°C are shown in Fig. 1. After two hours the anisotropy changed from positive to negative and became stable after six hours. The lower curve in Fig. 1 shows the time dependence of the conductivity anisotropy at 134°C after this particular sample had been cooled to

134°C from the normal liquid. The anisotropy was negative and stabilized in a few minutes. The results can be changed some by changing the procedure as will be discussed later. The results also appear to depend on the purity of the sample.

Figure 2 shows a temperature dependence of the conductivity anisotropy in NOBA. The data for the heating curve were obtained by heating from the solid to the nematic phase at 120° and allowing 7 hours for the sample to stabilize. The heating and cooling rates were approximately 10 minutes per degree, and the sample remained in the normal liquid approximately 1 hour. The results in Figures 1 and 2 indicate that the structure in the nematic phase is not only time **dependent**, but involves more than one mechanism. The dip ($\sigma_{\parallel}/\sigma_{\perp} < 1$) in the heating (Fig. 2) curve at 141°C implies a pretransitional effect in this range. Although the absolute values of ($\sigma_{\parallel}/\sigma_{\perp}$) varied some for different samples, the dip in the neighborhood of 141°C was reproducable for the heating curve and a small dip in this range was quite often observed for the cooling curve. These results also suggest the possibility of pretransitional effects in the lower temperature range of the nematic phase.

It has been generally accepted[9] that the molecules of p-n-nonyloxybenzoic acid form linear dimers as shown in Fig. 3 (a). In view of the X-ray[2] and conductivity anisotropy results[1,3] in HOAB, the possibility of dimers forming cybotactic groups must be

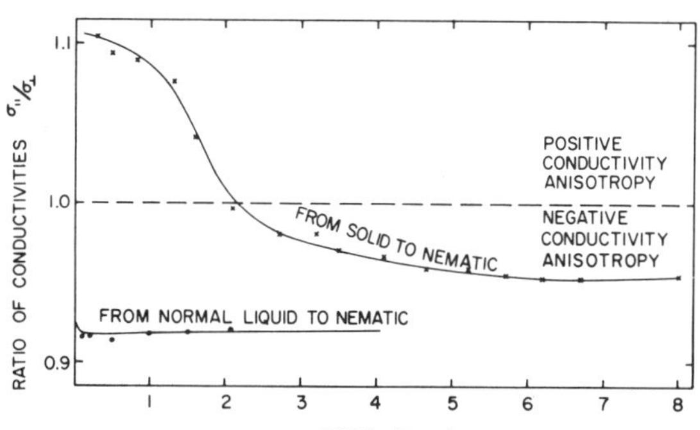

Figure 1. Time dependence of the ratio of dc conductivities for NOBA at 134°C.

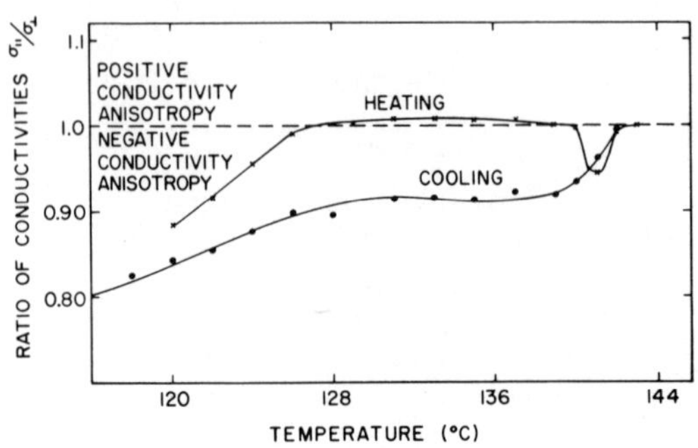

Figure 2. Temperature dependence of the ratio of dc conductivities for NOBA.

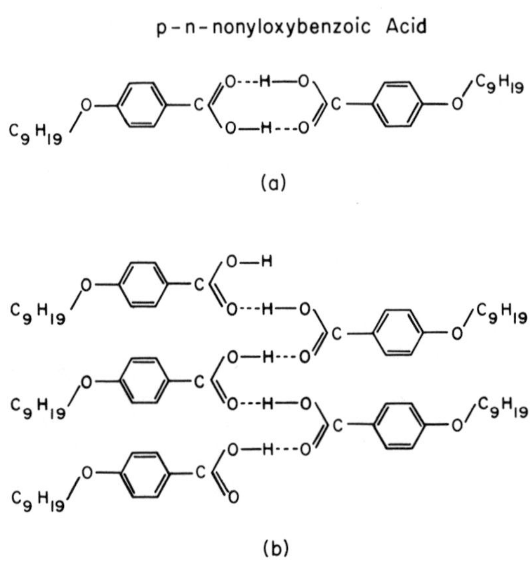

Figure 3. Polymerization of NOBA.

considered. An attempt was made to check the time dependence for the possible formation of a cybotactic structure in HOAB, but the results were negative. This implies that the time for the structure to form which gives rise to a negative conductivity anisotropy in HOAB is minutes or less. In order to explain the time dependence in Fig. 1, a structure shown in Fig. 3 (b) was suggested which reaches equilibrium very slowly.[6] This structure appears simple in two dimensions, but it can become much more complicated in three dimensions and this may be helpful in explaining the results of Figures 1 and 2. The planar layered structure is a possible explanation of the negative conductivity anisotropy. In order to explain the results in Figures 1 and 2 we should consider the possibility of both mechanisms (cybotactic groups and polymerization) acting simultaneously but one mechanism may predominate in a given temperature range. The anisotropy at 141°C does not appear to show as long a time dependence as at lower temperatures, therefore, one might expect the cybotactic structure to play an important role here.

A surprising phenomena was observed when the sample was heated quickly to 150°C (7 degrees above the clearing point) and cooled quickly to 134°C. If the conductivity anisotropy was positive it would normally return to positive and if negative it would return to negative. If the sample was held at 150°C for at least 1/2 hour it would return with a negative conductivity anisotropy regardless of the anisotropy prior to heating. Dielectric loss measurements for this rate of heating showed a temperature lag in the sample of about 3 degrees so that a temperature corresponding to that of the isotropic liquid should have been reached in all cases. This suggests a time dependent mechanism in the isotropic liquid. The existence of short-range orientational order in the isotropic liquid phase of nematic liquid crystals has been established[10] and has recently been investigated by a number of other investigators. A mechanism that is time dependent and involves many minutes seems a little unreasonable for the isotropic liquid phase, but we do not have a better explanation for these results.

Molecular Alignment Due to a 100 Hz

Electric Field in the Nematic Phase of NOBA

NOBA exhibits a positive dielectric anisotropy[5] and the molecules align with the nematic director parallel to an electric field in the dielectric regime but in the conduction regime the preferred direction is perpendicular to the field.

Figure 4 shows the dielectric loss at 24.5 GHz versus a 100 Hz external electric field at a temperature of 120°C. Magnetic fields of different values are applied parallel to the microwave

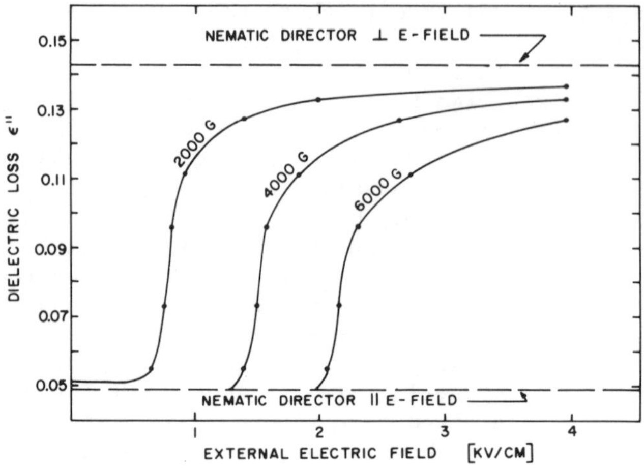

Figure 4. Dielectric loss at 24.5 GHz versus a 100 Hz external electric field for NOBA at T=120°C. The individual curves are for different magnetic fields parallel to the electric field.

electric field. Earlier work[5] showed that maximum and minimum loss was observed for the nematic director oriented perpendicular and parallel to the microwave field respectively. The nematic director was aligned by the magnetic field parallel to the microwave electric field. When the electric field increases and reaches a certain value, the molecules begin to turn away from the parallel direction. As the electric field is increased the dielectric loss rises sharply and finally reaches a saturation value, which means the preferred direction of nematic director changes from parallel to perpendicular to the magnetic field. In the region of 2 kG to 6 kG the sample was well-behaved. If we can assume the director, which is associated with various clusters, rotates in a plane, the E/H ratio corresponding to a random orientation of the molecules, $\varepsilon'' = \frac{1}{2}(\varepsilon''_{\parallel} + \varepsilon''_{\perp})$, is equal to 0.66 volt per cm-gauss. Since it is known that the torque on a cluster of molecules is proportional to H^2, and the E/H ratio does not show much variation with H, it can be assumed that the torque due to the ionic conduction is proportional to E^2 for this range of fields. This is consistent with work previously reported for materials which exhibit positive conductivity and negative dielectric anisotropies.[7]

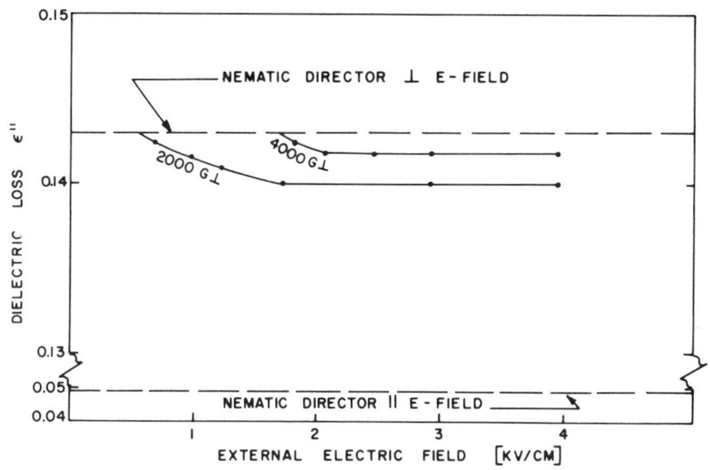

Figure 5. Dielectric loss at 24.5 GHz versus a 100 Hz external electric field for NOBA at T=120°C. The individual curves are for different magnetic fields perpendicular to the electric field.

Figure 5 shows the effect of a 100 Hz external electric field applied perpendicular to the magnetic fields. When a certain value of the electric field is reached the molecular alignment becomes less complete. This is consistent with the results shown in Fig. 4 in that a rotation of 90 degrees for the nematic directors (associated with various clusters) did not occur at high electric field intensities. It is also consistent with earlier results[7] involving positive conductivity and negative dielectric anisotropies, if we assume that the ordering mechanisms are associated with the conductivity and dielectric anisotropies.

Dielectric Anisotropy

A change in the dielectric anisotropy at 20 kHz is shown in Fig. 6 at the nematic-smectic C transition in NOBA.

Figure 6 shows the dielectric loss in NOBA when cooled in a 20 kHz electric field at 7000 v/cm. Measurements in a magnetic field of 10kG are also indicated for comparison, but the fields were not acting simultaneously. The preferred direction in the electric field for the long axes of the molecules changed from parallel to the electric field in the nematic to perpendicular in

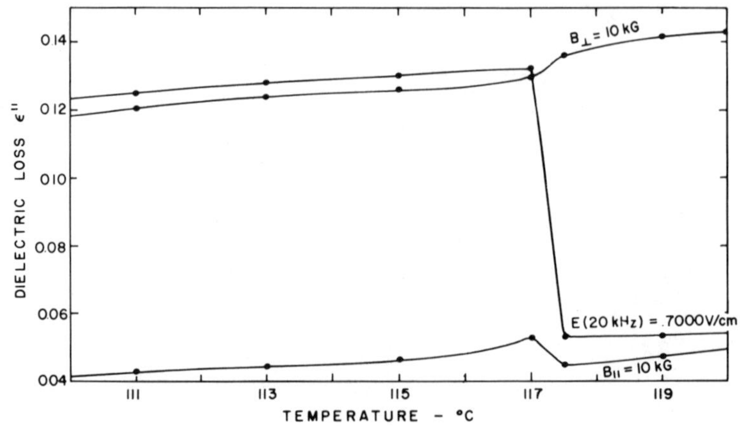

Figure 6. Dielectric loss at 24.5 GHz for NOBA as a function of temperature in the presence of a 20 kHz electric field at 7 kV/cm.

the smectic phase. This implies that the static dielectric anisotropy changes sign. This is a small change in the dielectric constant and could be associated with small changes in the percentage of pure monomer at the transition. A mechanism reported by Maier and Meier[11] and recently discussed by Martin, Meier and Saupe[12] may be needed to allow for a dispersion involving a small percentage of monomer. This involves a rotation about an axis perpendicular to the long molecular axis. The idea of a small change in the percentage of monomer or open dimer at a transition was used earlier[5] to explain a change in the dielectric loss at the nematic-isotropic liquid transition in NOBA.

The results shown in Fig. 7 imply a decrease in the dielectric anisotropy with a decrease in temperature. Figure 7 shows the dielectric loss at 24.5 GHz versus a 40 kHz electric field applied perpendicular to a 1 kG magnetic field. The change in the dielectric loss from its maximum to minimum value, shown for each curve, corresponds to a rotation of the nematic director from a direction perpendicular to parallel to the electric field. The maximum and minimum values are a function of temperature.[5] Figure 7 shows that a much higher electric field is required at 118°C than at 136°C to produce a rotation of the director. Since the order parameter[5,13] probably shows a small change, the dielectric anisotropy likely decreases with a decrease in temperature. The decrease in the dielectric anisotropy indicates that the

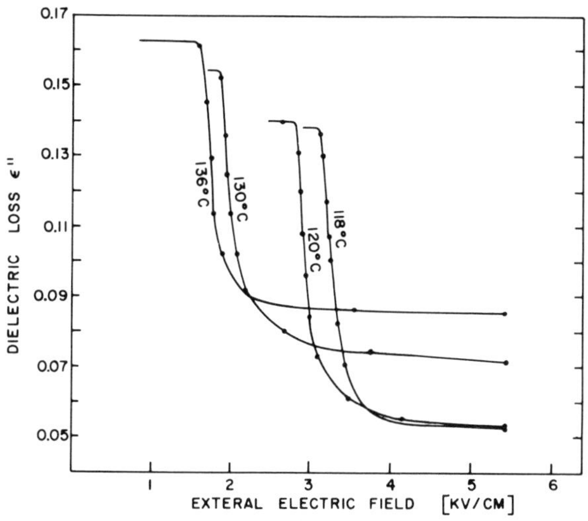

Figure 7. Dielectric loss at 24.5 GHz versus a 40 kHz external electric field for NOBA at different temperatures. A 1 kG magnetic field is perpendicular to the electric field.

mechanism responsible for the decrease in the nematic phase is probably associated with the change in the anisotropy at the transition of the smectic C phase. Although Fig. 7 clearly shows changes in the dielectric anisotropy with temperature, the changes are very small. It will be difficult to explain this change until more data are available involving low frequency dielectric measurements, but one might consider associating this change with a decrease in the percentage of monomer with a decrease in temperature. The component of the dipole moment of the monomer parallel to the long molecular axis will not be zero, therefore it can contribute to the static dielectric constant parallel to the nematic director. There will also be an additional contribution to the static dielectric constant in a direction perpendicular to the director, but if this contribution is less than in the parallel case the change in the percentage of monomer becomes a possible explanation. Work by Delodre and Cabane[13] supports the idea that the nematic phase in the p-alkoxybenzoic acids contains some monomer.

Molecular Alignment in the Smectic C

Phase Owing to Electric Fields

In order to study the frequency dependence of the molecular alignment due to an external electric field, the sample was cooled from the nematic to the smectic phase in the presence of magnetic and electric fields. Figure 8 shows the results. An 8 kG magnetic field was applied parallel to a 6500 V/cm external electric field at different frequencies. The upper and lower curves represent the molecular alignment in a 8 kG magnetic field (E = 0) applied perpendicular and parallel to the microwave electric field when the sample cooled from the nematic to the smectic phase. The 8 kG magnetic field is slightly less effective than higher fields in aligning the long axes of the molecules parallel to the field in the smectic phase. The results indicate there is a difference in alignment at different frequencies, namely, low audio frequency electric fields seem to have more effect in aligning the long molecular axes perpendicular to the electric field than frequencies above the audio region. This phenomenon is quite similar to the anomalous alignment in nematic which exhibit negative conductivity and negative dielectric anisotropies. However, the results in Fig. 4 and 8 show that, for the field intensities used in this work, the

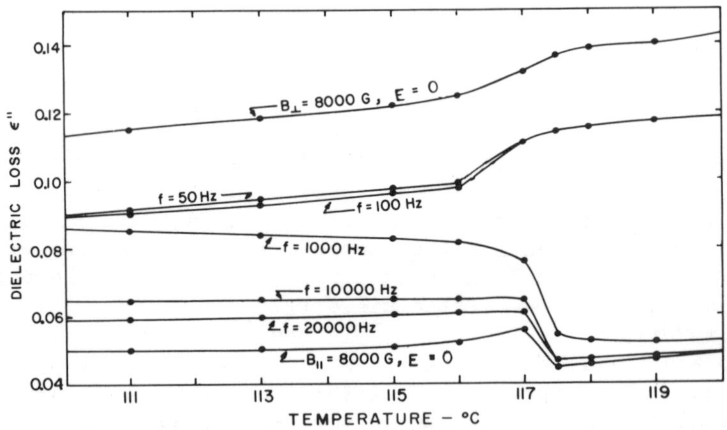

Figure 8. Dielectric loss at 24.5 GHz for NOBA as a function of temperature. A 6,500 V/cm electric field is parallel to a 8 kG magnetic field.

low frequency fields are more effective in the nematic than in the smectic C phase in producing an ordering with the long molecular axes preferring a direction perpendicular to the field. This is probably due to a tilt angle[14] in the smectic C phase. The conductivity is likely to be a maximum in the plane of the layered structure rather than always perpendicular to the direction preferred by the long molecular axes. The results in Fig. 8 do not give much detail, but they show that there is an effect due to the conductivity and that it is frequency dependent.

CONCLUSIONS

We believe that the lowest possible value of the ratio of the conductivities $\sigma_{\parallel}/\sigma_{\perp}$ corresponds to the most stable state in the nematic phase of NOBA. Experimentally, the best results can be obtained by cooling slowly from the isotropic phase. The results shown in Fig. 1 suggest that the sample stabilizes in about 6 hours at 134°C after melting but the results from Fig. 2 imply that the system was not in equilibrium. We must conclude that the ratio $\sigma_{\parallel}/\sigma_{\perp}$ for the most stable state is not known. It is possible that the length of time required to obtain stability would lead to enough decomposition to affect the ratio. A preliminary check on the presence of a magnetic field while stabilizing did not yield any positive results, but the effect of a magnet field must be further investigated.

In order to explain the temperature dependence of the conductivity anisotropy we assumed that at least two mechanisms must be considered. One mechanism is a cybotactic structure involving dimers and the other is polymerization involving hydrogen bonds. We must also consider the possibility of a small amount of monomer being present which we have used to explain changes in the static dielectric anisotropy. There is also the possibility that the nematic phase contains cybotactic groups that behave more like a smectic C than a smectic A phase and the tilt angle is temperature dependent. This would help explain some of the results shown in Fig. 2, but on the basis of the evidence available we are inclined to think of the clusters in the nematic phase as behaving more like a smectic A than a smectic C phase. When the results of Figures 4 and 5 are compared with results[15] on materials exhibiting positive conductivity and negative dielectric anisotropies, it appears that the conductivity is a maximum in the planes normal (or nearly normal) to the long molecular axes. This was also the case at higher temperatures. We are assuming that the conductivity is a maximum in the planes of a layered structure. Measurements of the viscosities, which are not presently available, are needed for a detailed discussion of the results shown in Figures 4 and 5.

Although the idea of polymerization may appear reasonable from a two dimensional point of view it becomes much more complicated in three dimensions. This may be helpful in explaining some of the results obtained in the nematic phase, but it presents difficulties when explaining the smectic C phase in NOBA, unless the structure simplifies in the smectic phase so that a two dimensional model is helpful. Preliminary measurements indicated that the structure in the smectic phase was dependent on the structure in the nematic phase prior to cooling. When a sample, exhibiting a positive conductivity anisotropy in the nematic phase, was cooled to the smectic and reheated to the nematic phase, it returned with a positive anisotropy. A sample with a negative conductivity anisotropy returned with a negative anisotropy after cooling and reheating. The results did not appear to depend on the length of time that the sample remained in the smectic phase.

In view of some of the conflicting reports concerning tilt angles in smectic C phases, more serious consideration should probably be given to the structure in the nematic prior to cooling to a smectic C phase. The effect of small amounts of impurities (ionic and nonionic) should be considered.

REFERENCES

* *Supported by U. S. Army Research Office, Durham, Grant No. DA-ARO-D-31-124-G135.

1. D. P. McLemore and E. F. Carr, J. Chem. Phys. 57, 3245 (1972).

2. A. De Vries, Mol. Cryst. Liq. Cryst. 10, 219 (1970).

3. F. Rondelez, Solid State Comm. 11, 1675 (1972).

4. E. F. Carr, Mol. Cryst. and Liq. Cryst. 13, 27 (1971).

5. L. S. Chou and E. F. Carr, Phys. Rev. A, 7, 1639 (1973).

6. L. S. Chou and E. F. Carr, Bull. Am. Phys. Soc. 18, 437(1973).

7. E. F. Carr and L. S. Chou, J. Appl. Phys. (to be published July 1973).

8. E. F. Carr, Adv. Chem. Ser. 63, 76 (1967).

9. G. W. Gray, Molecular Structure and the Properties of Liquid Crystals, Academic Press, Inc.(London).Ltd.,London (1962).

10. C. C. Gravatt and G. W. Brady, Mol. Cryst. and Liq. Cryst. 7, 355 (1969); T. W. Stinson and J. D. Litster, Phys. Rev. Lett. 25, 503 (1970); B. Chu, C. S. Bak. and F. L. Lin, Phys. Rev. Lett. 28, 1111 (1972).

11. W. Maier and G. Meier, Z. Naturforsch, A 16, 1200. (1961).

12. Anna J. Martin, G. Meier, and A. Saupe, Symposium of the Faraday Society, No. (5), 119 (1971).

13. B. Delodre and B. Cabane, Mol. Cryst. Liq. Cryst. 19,25 (1972).

14. T. R. Taylor, J. L. Fergason, and S. L. Arora, Phys. Rev. Lett. 24, 359 (1970).

15. E. F. Carr, Mol. Cryst. Liq. Cryst. 7, 253 (1969).

STABILITY OF MOLECULAR ORDER IN THE SMECTIC A PHASE OF A LIQUID CRYSTAL*

C.E. Tarr, R.M. Dennery†, and A.M. Fuller§

Department of Physics, University of Maine

Orono, Maine 04473

ABSTRACT

Ethyl-[p-(p-methoxybenzilidene)-amino]-cinnamate (EMC) was ordered in its smectic A phase by ordering in the nematic phase and cooling through the nematic-smectic A phase transition in the presence of a magnetic field. The direction and degree of order were determined using pulsed nuclear magnetic resonance techniques. The order was observed to change with time following a rotation of the magnetic field. The degree and stability of order produced were found to be functions of the concentration of small amounts of decomposition products. A model is discussed that satisfactorily accounts for the time dependence of the reordering.

INTRODUCTION

In a previous study[1] it was noted that the molecular order in the smectic A phase of ethyl-[p-(p-methoxybenzilidene)-amino]-cinnamate (EMC) could be disrupted by a magnetic field. This effect was not observed in very pure samples and appeared to be due to impurites, probably thermal decomposition products.

In the work reported here, fresh samples of EMC were partially decomposed by heating to the isotropic liquid phase for different lengths of time. The samples were then ordered by cooling from the nematic phase to the smectic A phase in the presence of a moderate magnetic field. The magnet was then rotated, and the change with time of the molecular order was observed using pulsed N.M.R. techniques.

Two models for the reorientation mechanism were considered. In both models it was assumed that the decomposition products disrupt the molecular order in small regions of the sample, permitting the molecules in these regions to reorient in the magnetic field. Each of these models is discussed below and compared to the experimental data.

EXPERIMENTAL METHODS

Commercial[2] EMC was purified by repeated recrystallization from ethanol. Samples of freshly recrystallized material showed no order instabilities when heated to the smectic A phase. If such samples were first heated to the isotropic liquid phase and maintained at that temperature for a short time, however, the molecular order in the smectic A phase was observed to be unstable. This effect appeared to be due to the presence of small amounts of decomposition products produced during the time the sample was maintained in the isotropic liquid phase. In order to test this hypothesis, a series of freshly recrystallized samples was sealed and heated to a temperature of 165°C, well above the nematic- isotropic liquid transition temperature of 138 °C. The samples were maintained at 165 °C for times ranging from 1 to 120 minutes and then quickly cooled to room temperature. The time spent in the isotropic (normal) liquid phase will be referred to as the "normal time" (N.T.) below.

The stability of molecular order in the smectic A phase was determined using pulsed N.M.R. techniques in the following way. First, the sample was heated into the nematic phase and ordered in a 7 kG magnetic field. The ordered sample was then cooled in the magnetic field to 112 °C (the middle of the smectic A temperature range) and allowed to stand undisturbed for an additional "ordering time" (O.T.). The magnetic field was then rotated quickly through an angle Θ and the free induction decay (F.I.D.) following a 90° pulse was recorded. The shape of the F.I.D. was observed to change with time as the molecules reordered in the magnetic field.

The apparent time constant for the reordering is relatively short (a few hundred seconds) so that some measure of the order was needed that could be acquired and recorded in a short time. Several parameters seemed obvious choices. For example, the moments of the resonance line are obtainable from the F.I.D.[3], and the second moment M_2 is directly related to the relative orientation of the molecular axes with respect to the magnetic field[4], so that a knowledge of M_2 as a function of time would be particularly useful. Unfortunately, even by making use of a signal averaging computer, the short reordering time prohibits making determinations of M_2 from the F.I.D. because of the time required to output a digitized representation of the F.I.D.

A less accurate measure of the order may be obtained directly from the F.I.D. without requiring a detailed numerical analysis. If one approximates the F.I.D. by an exponential, then the time constant of this exponential yields an effective spin-spin relaxation time T_2. Since $T_2 \propto 1/\sqrt{M_2}$, similar information is obtained. This method requires significantly less time for data acquisition than does the moment method, and was used to obtain the data discussed below.

Standard pulsed N.M.R. apparatus[5-6] was used, and all measurements were made at 30 MHz. The F.I.D. signal following a 90° pulse was averaged on a signal averaging computer for 64 repetitions in 8 seconds and a mild numerical smoothing performed to reduce high frequency noise. Thus, each data point represents an average of the F.I.D. over eight seconds. The resulting average signal was displayed on an oscilloscope and photographed for later analysis. The sample was heated by forced convection in a closed probe[7] and the sample temperature was electronically regulated to better than ±0.2 °C. The maximum temperature gradient over the sample volume was ±0.2 °C.

MODELS FOR MOLECULAR REORDERING

Let us assume that the decomposition products mix with the pure material in the sample in such a way as to lower the nematic-smectic A transition temperature for a portion of the sample. In this way it is possible to disrupt the molecular order to varying degrees within the sample. Let us further assume that the partially decomposed sample may be described as a mixture of three components as follows: a fraction n_s composed of material of sufficient purity to maintain stable smectic A order; a fraction n_o composed of poorly ordered material which reorders in a magnetic field in a time short compared to that required for a measurement of T_2; and, finally, a fraction n_f composed of material of intermediate stability that may reorder slowly in comparison to the fraction n_o.

Now, let us further assume that the free induction decay of each of these fractions is separately describable by a single exponential. Thus, if we assume that each constituent fraction is sufficiently large (so that interactions between molecules in different fractions may be neglected) then we may approximate the F.I.D. of the composite sample as:

$$f(\tau,t) \simeq e^{-\tau/T_2^e(t)} \simeq n_o e^{-\tau/T_2^o} + n_s e^{-\tau/T_2^s} + n_f e^{-\tau/T_2^f(t)}, \quad (1)$$

where: τ is the time along the F.I.D., t is the reordering time (i.e. time measured from the rotation of the magnetic field), T_2^o is the effective T_2 that would be observed if the entire sample were ordered with the molecular axes parallel to the magnetic field, T_2^s is the T_2 that would be observed if the sample were pure and the order stable (i.e. well ordered with the molecular axes parallel to the origional ordering direction), $T_2^f(t)$ is the T_2 that would be observed if the entire sample were composed of the unstably ordered component associated with n_f, and $T_2^e(t)$ is the value of T_2 measured experimentally at time t.

Clearly these approximations for the effective T_2 are generally rather poor for arbitrary τ. If we note, however, that the evaluation of $T_2^e(t)$ requires only a knowledge of $f(\tau,t)$ for $\tau \simeq T_2^e(t)$, then the approximations, though crude, provide unique and self consistent values of the parameters that we seek to extract from the data.

Now, if we denote the value of $T_2^e(t)$ measured at very short ordering times by $T_2^e(0)$ and evaluate Eq. 1 for $\tau = T_2^e(0)$, noting that $T_2^f(0) = T_2^s$, we obtain:

$$n_o = \frac{e^{-1} - e^{-T_2^e(0)/T_2^s}}{e^{-T_2^e(0)/T_2^o} - e^{-T_2^e(0)/T_2^s}} \quad . \tag{2}$$

Further, evaluating Eq. 1 at $\tau = T_2^e(t)$ for long reordering times denoted as $T_2^e(\infty)$, and noting that $T_2^f(\infty) = T_2^o$, then we obtain:

$$n_s = \frac{e^{-1} - e^{-T_2^e(\infty)/T_2^o}}{e^{-T_2^e(\infty)/T_2^s} - e^{-T_2^e(\infty)/T_2^o}} \quad , \tag{3}$$

and, finally, we obtain n_f from:

$$n_o + n_s + n_f = 1 \quad . \tag{4}$$

Stochastic Process

If we assume that the reordering takes place via a process in which small portions of the origional fraction n_f reorient in a very short time from the original ordering direction to a direction parallel to the magnetic field, and further that this process is describable by a single exponential time constant, then $f(\tau,t)$ becomes:

$$f(\tau,t) = (n_s + n_f e^{-t/T_r})e^{-\tau/T_2^s} + [n_o + (1 - e^{-t/T_r})n_f]e^{-\tau/T_2^o}, \tag{5}$$

where T_r is the characteristic time constant for the process.

Thus, using the stochastic model and evaluating $f(\tau,t)$ at $\tau = T_2^e(t)$, we obtain a time dependence that is given by:

$$e^{-t/T_r} = \frac{e^{-1} - n_s e^{-T_2^e(t)/T_2^s} + (1 - n_s)e^{-T_2^e(t)/T_2^o}}{n_f(e^{-T_2^e(t)/T_2^s} - e^{-T_2^e(t)/T_2^o})} \quad . \tag{6}$$

Viscous Damping

Alternatively, let us assume that the fraction n_f reorients in the magnetic field in a continuous manner that can be described by the time dependence of an angle ϕ between the average direction of the molecular axes and the magnetic field. Further, assume that ϕ is decreased by a restoring torque exerted by the magnetic field on the magnetic moment of the partially aligned molecules in n_f, and that the motion is viscously damped. If the inertial term associated with the motion is small compared to the torque and damping terms, then the motion can be approximated by:

$$\alpha\frac{d\phi}{dt} = -\beta\sin\phi \text{ , so that:} \tag{7}$$

$$e^{-t/T_r} = \tan\frac{\phi}{2}/\tan\frac{\theta}{2} \quad . \tag{8}$$

If we approximate the angular dependence of $T_2^f(t)$ to include the dipolar broadening and intrinsic broadening of the line, then $T_2^f(t)$ is given by:

$$\frac{1}{T_2^f(t)} = A|3\cos^2\phi - 1| + B \quad . \tag{9}$$

From this it follows that:

$$\phi = \cos^{-1}\left[\frac{1}{3}\left(\frac{2\left(\frac{T_2^s}{T_2^f(t)} - 1\right)}{\frac{T_2^s}{T_2^o} - 1} + 1\right)\right]^{1/2} , \tag{10}$$

where $T_2^f(t)$ is obtained from $T_2^e(t)$ by evaluating Eq.1 at $\tau = T_2^e(t)$ and is given by:

$$T_2^f(t) = \frac{T_2^e(t)}{\ln\left[\dfrac{n_f}{e^{-1} - n_o e^{-T_2^e(t)/T_2^o} - n_s e^{-T_2^e(t)/T_2^s}}\right]} \, . \qquad (11)$$

The time dependence that should be obeyed by $T_2^e(t)$ for the viscous damping model is then given by substituting Eq. 10 into Eq. 8.

RESULTS AND CONCLUSIONS

A number of freshly recrystallized samples were studied as a function of normal time and ordering time. In all cases a rotation angle $\theta = 54.7°$ was chosen to maximize the variation in T_2. The results of typical measurements on two samples having rather different reordering times are shown in Figs. 1 and 2. The error bars shown are ± 10% and are somewhat conservative.

The T_2 data from each sample were least squares fitted to Eqs.6 and 8 to yield reordering time constants, T_r, for the stochastic and viscous damping models. The fits to the viscous damping model were quite good, and, in all cases, were better than the fits to the stochastic model. Finally, values obtained from the data including the least squares values for T_r were used to generate $f(\tau,t)$ for each sample using Eq. 1. A computer iteration was then performed on the $f(\tau,t)$'s obtained to "measure" T_2 using the same criteria used in the experimental determinations of T_2. The solid lines in Figs. 1 and 2 show the values obtained from the computer measurements, and, thus, show the dependence of $T_2(t)$ predicted by the viscous damping model. The model depends critically upon the initial effective angle of the molecules in n_f so that small departures from θ are quite noticeable, making the agreement with experiment worst for small t. The agreement between the measured values of T_2 and those predicted from the time dependence calculated for the viscous damping model seem sufficiently good to lend support to this model.

Figure 3 shows the results of a series of measurements in which the normal time was varied while the ordering time was held constant. These data show that T_r decreases with increasing decomposition, as expected.

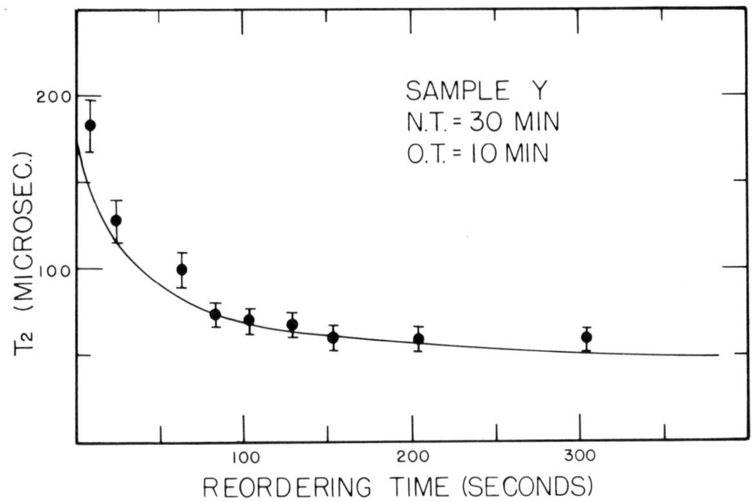

Figure 1. T_2 vs. reordering time for a sample maintained in the normal liquid phase for 30 minutes and ordered in the smectic A phase for 10 minutes. The solid line is the time dependence derived using the viscous damping model.

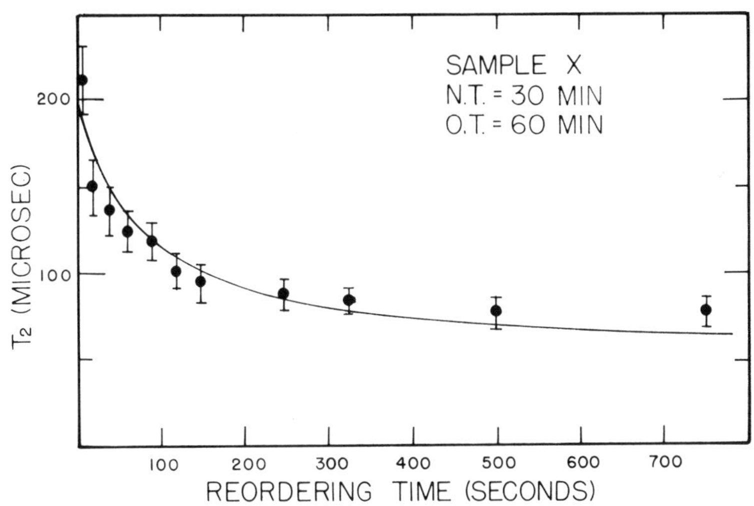

Figure 2. T_2 vs. reordering time for a sample maintained in the normal liquid phase for 30 minutes and ordered in the smectic A phase for 60 minutes. The solid line is the time dependence derived using the viscous damping model.

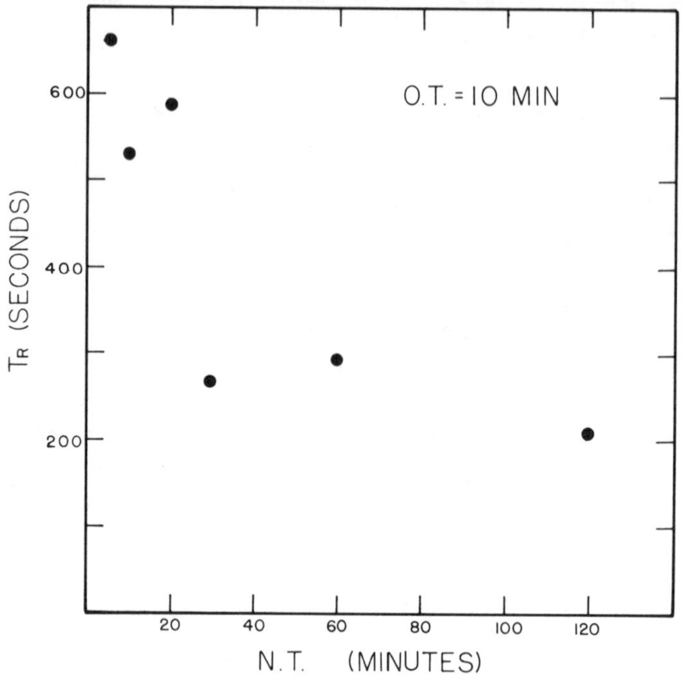

Figure 3. Characteristic reordering time vs. time in the normal liquid phase for a series of samples each ordered for 10 minutes in the smectic A phase.

Measurements on a series of samples each heated in the normal liquid phase for the same time and allowed to order in the smectic A phase for different periods of time indicated that T_r increases with increasing ordering time for a fixed impurity concentration. This result is not altogther surprising, as one possible interpretation may be that the molecules in the partially ordered fraction n_f continue to aggregate in a magnetic field to form larger smectic planes. This would have the effect of increasing the viscous damping term in Eq. 7 and, thus, would lead to an increase in T_r.

Because it was necessary to begin each measurement with a fresh sample, and because of the large number of possible combinations of experimental parameters, the results presented here are necessarily of a preliminary nature. We are presently extending the measurements on partially decomposed samples as well as making measurements on samples that are selectively doped with impurities.

To conclude, our observations appear to be adequately described by assuming that decomposition products mix with pure material in the smectic A phase to produce a mixture of three components: regions displaying stable smectic A order, regions in which the order is sufficiently disrupted to allow the molecules to reorder in a magnetic field in a very short time, and, finally, regions in which the order is partially disrupted and in which the molecules reorder relatively slowly in a magnetic field. Further the data appear to support a reordering process in which the aligning torque due to an applied magnetic field is opposed by viscous damping.

ACKNOWLEDGEMENT

The authors wish to thank Mr. Michael E. Field for his assistance in the data analyses.

FOOTNOTES AND REFERENCES

* Work supported in part by the National Science Foundation .

† Present address: Naval Weapons Laboratory, Dahlgren, Virginia.

§ Present address: Dept. of Chemistry, University of Maine, Orono, Maine.

1. C.E. Tarr and A.M. Fuller, Fourth International Liquid Crystal Conference, Kent State University, August 1972 (to appear J. Mol. Cryst. and Liquid Cryst.)

2. Eastman Kodak Co., Eastman Organic Chemicals, Rochester, N.Y.

3. I.J. Lowe and R.E. Norberg, Phys. Rev. 107, 46 (1968).

4. G.H. Brown, J.W. Doane, and V.D. Neff, Critical Reviews in Solid State Sciences 1, 344 ff. (1970).

5. I.J. Lowe and C.E. Tarr, J. Sci. Instr. 1, 320 and 604 (1968).

6. C.E. Tarr and M.A. Nickerson, J. Sci. Instr. 5, 328 (1972).

7. A.M. Fuller, Ph.D. Thesis, University of Maine,1973 (unpublished).

MOLECULAR DIFFUSION IN THE NEMATIC AND SMECTIC C PHASE OF 4-4'-DI-n-HEPTYLOXYAZOXYBENZENE (HOAB)

J.A. Murphy, J.W. Doane, and D.L. Fishel

Kent State University, Kent, Ohio 44242

Earlier measurements of the diffusion constant of the impurity molecule tetramethylsilane (TMS) when dissolved in a smectic A and smectic B liquid crystal have shown this quantity to be strongly anisotropic.[1] That is, the diffusion constant in a direction perpendicular to the long molecular axis, $D_\perp$, (diffusion within the smectic layers) is a much larger quantity than that in a direction parallel to the long molecular axis, $D_{||}$, with a typical value for the ratio $D_{||}/D_\perp \sim 0.1$. In the earlier work it was not possible, however, to obtain a measure of this ratio in the nematic phase. In this paper we report such a measurement as well as a measurement of the anisotropy in the smectic C phase. Much to our surprise, the ratio $D_{||}/D_\perp$ in the nematic phase varies strongly with temperature, exhibiting strong pretransition effects over most of the nematic temperature range.

The measurements were made using pulsed nmr techniques[2] in a method described earlier.[1] Basically, in this method the diffusing molecules are labeled by polarizing some of their nuclear spins by use of a pulse of radio frequency. This is done on a time scale of a few microseconds.

After labeling, a large magnetic field gradient is applied across the sample which has the effect of sectioning the liquid crystal. If a molecule diffuses in the direction of the gradient, its nuclear spins lose their coherence with the other spins and this coherence cannot be restored. After a few milliseconds (10 milliseconds in this case) observation pulses are applied. The coherence of those spins which have not diffused in the direction of the applied gradient is restored and those which have diffused is not. The result is observed in the form of an echo; the strength of the echo giving a measure of the diffusion constant. If the amplitude of the echo is A(O) when the pulsed gradient is not applied and A(G) when it is applied, the ratio A(G)/A(O) is related to the diffusion constant, D, by the expression[3]:

$$\ln[A(G)/A(O)] = -\gamma^2 D\delta^2(\Delta - 1/3\delta)G^2$$

Where γ is the nuclear gyromagnetic ratio, G is the magnitude of the linear gradient, Δ is the time between the leading edges of the gradient pulses and δ is the gradient pulse width. In earlier measurements the direction of the gradient was fixed in the direction of the large uniform magnetic field; this uniform field being necessary for the nmr observation and the uniform alignment of the liquid crystal sample. In this case the anisotropy could only be measured in the smectic A or B phase since the sample could be oriented in the gradient without distortion from the large magnetic field. In the nematic phase, however, the molecular director always remains parallel to the field. Therefore, in this phase, it is necessary to change the direction of the gradient[4] relative to the direction of the uniform magnetic field in order to measure the anisotropy. We might add that this is also true for the smectic C phase since the molecules have the ability to reorient within the layers in this particular smectic phase.

Figure 1 shows the results of the measurement of $D_{||}$ and $D_{\perp}$ of 10 mol% of TMS dissolved in the liquid crystalline phases of 4-4'-di-n-heplyloxyazoxybenzene (HOAB). In the isotropic phase where it is impossible to distinguish between the TMS and HOAB molecules, there is no expected anisotropy. In the nematic phase on the other hand, two unexpected results appear. First is that the anisotropy becomes larger as one approaches the smectic C phase. This implies the onset of short range smectic order. In fact, at the smectic phase transition there is little discontinuity in the diffusion constants. Using radioactive tracers Yun and Frederickson[5] and using nmr methods, Blinc, Zupancic and Pirs[6] observed very little anisotropy, $D_{||}/D_{\perp} \sim 1.3$, in the nematic phase of PAA and MBBA. However, de Vries[7] has shown the presence of "cybotactic groups" in materials such as HOAB which exhibit the smectic C phase. The presence of this short range smectic order could alter this measurement in HOAB as we observe here. An observation similar to ours has recently been made in the electrical conductivity as measured by Rondelez.[8] Also, it may be that TMS shows a greater anisotropy in general than the self diffusion of the host liquid crystal molecules. A second interesting result is the anomolous peak in $D_{||}$ in the nematic phase. The origin of this peak at this point is not altogether clear, however, it is believed that it may have to do with the lifetime of the short range smectic order or "cybotactic groups". The time scale of the diffusion measurement is ~10 milliseconds which is approximately the value of Δ. If the life time of the fluctuations' short range smectic order is larger than this, then the diffusion will have the characteristic of a smectic. If on the other hand the life time is short compared to

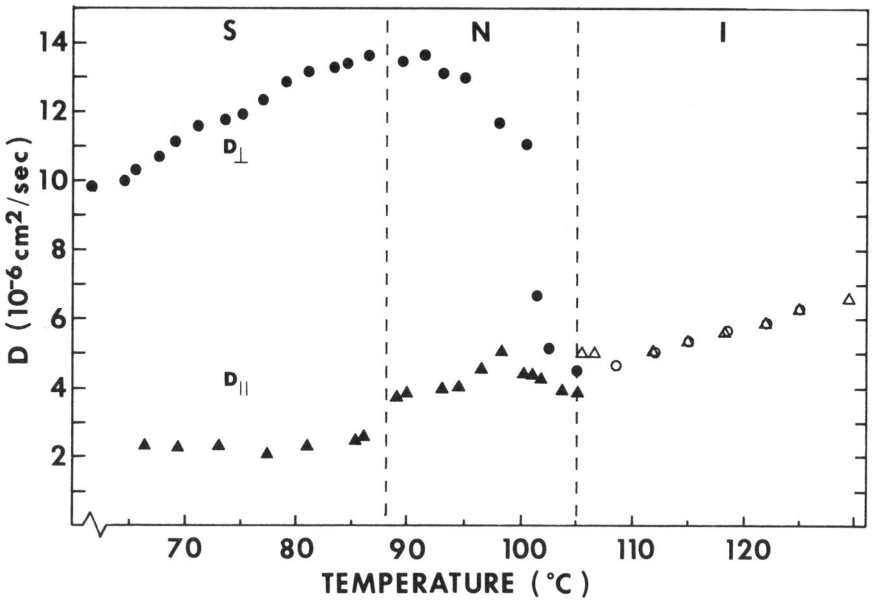

Fig. 1. Temperature dependence of the anisotropic diffusion constant in the isotropic (i), nematic (n), and smectic (S) phases of TMS doped HOAB. The dashed lines indicate the transition temperature.

10 ms, then the diffusion may appear nematic like. The anomaly may be that regime where the life time is near 10 ms, the time of the measurement. Why the anomally only appears is $D_{||}$ may be due to shape of the cybotactic groups. More study is in progress on this speculation.

Finally, it should be mentioned that the phase transition temperatures are somewhat surpressed from that in pure HOAB due to the dissolved TMS.

[1] J.A. Murphy, J.W. Doane, Y.Y. Hue and D.L. Fishel (to appear in Mol. Cryst. and Liq. Cryst.)
[2] H.Y. Carr, E.M. Purcell, Phys. Rev. 94, 630(1954).
[3] E.O. Stejskal, J.E. Tanner, J. Chem. Phys. 42, 288(1965).
[4] R.S. Parker, I. Zupancic and J. Pers (to be published).
[5] C. Yun and A.G. Frederickson, Mol. Cryst. and Liq. Cryst. 12, 73 (1970).
[6] R. Blinc, J. Pirs and I. Zupancic, Phys. Rev. Lett. 30, 546 (1973).
[7] A de Vries, Mol. Cryst. and Liq. Cryst. 10, 31 (1970); 10, 219 (1970) and 11, 361 (1970).
[8] F. Rondelez, Solid State Comm. 11, 1675 (1972).

DIFFUSION IN ORIENTED LAMELLAR PHASES BY PULSED NMR*

Mingjien Chien,[+] B.A. Smith,[‡] E.T. Samulski,[‡‡]

and C. G. Wade

Department of Chemistry, University of Texas at Austin

Introduction

The bilayer structure (lamellar phase) of lipid-water and soap-water systems is the subject of considerable research because of the known occurrence of bilayer structures in biological membranes.(1) The nuclear magnetic resonance (NMR) line widths in this phase are very broad due to the incomplete motional averaging of the static dipolar interactions. This has until recently precluded the use of NMR spin echo techniques to measure diffusion because spin echo occurrence requires the removal of the static dipolar interactions. DeVries and Berendsen (2) have shown by optical and NMR measurements that the potassium oleate (KO)-water lamellar phase can be oriented between glass plates with the optical axis (along the hydrocarbon chains) perpendicular to the glass plates. They demonstrated that the line width varied as $(3\cos^2\Omega - 1)$ where Ω is the angle between the optical axis and the magnetic field. This provided a conclusive demonstration that the broad lines arise from incompletely averaged dipolar interactions which have residual static interactions along the chain direction. In particular, they noted that the lines got very narrow at the so-called "magic angle" where $\Omega = 54°44'$ and the term $(3\cos^2\Omega - 1)$ (and hence the dipolar interactions) becomes very small.

As we demonstrated in an earlier paper, (3) this orientational minimization of the dipolar interaction permits the observation of a pulsed NMR Carr-Purcell spin echo (CPSE) for the KO protons in the lamellar phase of KO-D_2O. The echo can be seen only at the magic angle and its observation affords, in principle, the opportunity to measure diffusion in these oriented systems using well established pulsed gradient NMR techniques.(4) Of particular interest in membrane dynamics is the diffusion of the oleate molecules parallel and perpendicular to the bilayer and the diffusion of H_2O in the channels between the polar head groups. By utilizing the fact that the NMR experiment measures diffusion in the direction of an applied gradient, we are able to study a variety of phenomena by varying the gradient direction with respect to the aligned sample optical axis; the latter is fixed at 54°44' to the static magnetic field. The small values of the diffusion coefficient, D, expected for the oleate ($D_{\parallel} \approx 10^{-8}$ cm^2/sec for lecithin in lecithin-water systems as measured by McConnell and co-workers (5) using ESR spin label techniques) place stringent requirements on the pulsed gradient apparatus. We report here some initial results of our efforts to study diffusion in these systems.

Nuclear magnetic relaxation studies also provide information on molecular motions. A unique property of liquid crystals is that the spin-lattice relaxation time (T_1) is always longer than the spin-spin relaxation time (T_2). This indicates the presence of slow motions which contribute to T_2 more effectively than to T_1. Several types of motions have been suggested to be present in the lamellar phase. They are translational diffusion,(6,7) isomeric rotation and angular displacements of chains which are a consequence of the rotations.(8) These motions are all based on the behavior of single molecules. In nematic liquid crystals, another important motional phenomenon is known (9): the collective mode fluctuations of ordered molecules resulting from the cooperative effects. This dominates the relaxation of nematogen protons in certain cases.(9) For lamellar phases the collective motions have almost been completely neglected, yet there is no reason to doubt the existence of cooperativity between molecules and this may become very important when the molecules are all aligned in the same direction as in our experiments. Since the contribution to the nuclear magnetic relaxation from each type of motion yields different frequency dependence, it is in principle possible to tell which type of motion dominates by studying the frequency dependence of relaxation.

We have measured the spin-lattice relaxation time in the rotating frame ($T_{1\rho}$) as a function of the frequency ω_1 ($=\gamma H_1$, where H_1 is the intensity of the RF field and γ is the gyromagnetic ratio) for both oriented and unoriented samples. We have also measured the spin-lattice and, where possible, the spin-spin relaxation times at a fixed frequency (30 MHz).

Experimental

The sample preparation (KO-D_2O and KO-H_2O, both 75% by weight KO) has been described in a previous paper.(3) The compounds were not degassed but the short relaxation times encountered indicated that dissolved oxygen is not a dominant contributor to T_1. Relaxation times were measured with a Ventron pulsed NMR spectrometer (30 MHz) with a 90° pulse length of about 3 μsec. The techniques were similar to those previously reported except that a Varian twelve inch V2100A electromagnet was used instead of the six inch magnet. Space limitations required the removal of the magnet homogeneity coils. T_1 was measured by the saturation-τ-90 sequence. The Meiboom-Gill modification of the CPSE sequence (90-τ-180-τ-echo) was used to measure T_2, while $T_{1\rho}$ was determined with conventional spin lock techniques.

Translational diffusion measurements were carried out by the pulsed magnetic field gradient method. This technique uses the basic CPSE sequence and adds a square gradient pulse between the 90° and 180° pulses and an identical gradient pulse between the 180° pulse and the echo. The gradient and pulsing system, constructed in this laboratory used a nickel-cadmium battery of 20 ampere hour capacity to provide the gradient current source. The rise time of the gradient pulse was ~5 μsec. Water and glycerin were used to calibrate the gradient.

Signal averaging was again used to improve the signal to noise. All experiments were performed at ambient temperature.

Results and Discussion

Water molecules in the lamellar phase are much more mobile than the KO molecules and consequently the water protons have a longer T_2 than KO protons. Thus it is possible to measure H_2O diffusion separately from the KO protons in the oriented samples by either orienting away from the magic angle (no KO echo) or by using a long τ value. In the unoriented (i.e., not aligned between glass plates) samples, only the H_2O echoes could be observed. In the unoriented sample, uncertainties as to the possible boundaries of the microscopic structure probably preclude the extraction of D of water. The (probably invalid) assumption of unbounded diffusion (eqn. 1) leads to a value of $D \approx 2.5 \pm 0.5 \times 10^{-6}$ cm^2/sec. This compares with the value of $D \approx 2.3 \times 10^{-5}$ cm^2/sec for pure H_2O. For unrestricted media, the spin echo obeys equation 1 in a pulsed gradient experiment (4)

$$\ln R = -\gamma^2 g^2 \delta^2 (\Delta - 1/3\,\delta) D \qquad (1)$$

where R is the factor by which diffusion reduces the amplitude of the spin echo, γ is the magnetogyric ratio, g is the gradient (Gauss/cm), δ is the width of the gradient pulse and Δ is the leading edge separation of the gradient pulses.

Theories of diffusion in bounded media are known.(10) Stejskal (10) has presented a theory applicable to H_2O diffusion in the channels of the oriented KO-H_2O sample; that is, diffusion occurring between two flat boundaries "infinite" in extent. His results indicate that diffusion parallel to the boundary should obey eqn (1) but the expression for diffusion perpendicular to the boundaries includes the effects of the boundary separation. In the case of the KO-H_2O system, estimates of the "boundary" (polar head group) separation (10-20 Å), g, δ, Δ, and D indicate that we should <u>not</u> expect attenuation of the echo for the gradient perpendicular to the bilayers. Physically, this means the H_2O molecule bounces rapidly between the bilayers but does not diffuse to a significantly different value of magnetic field during the experiment. Our value of D for H_2O parallel to the bilayers is $5 \pm 1 \times 10^{-6}$ cm^2/sec, an order of magnitude less than that for pure H_2O.

We have attempted to measure oleate diffusion in the KO-D_2O system but have been unsuccessful due to instabilities in the gradient system. This method is very sensitive to imbalances in the gradient pulses and the imbalances increase as g increases. The small D expected (5) and the short value of T_2 combine to demand a large g. Our efforts in this area continue.

Relaxation Measurements

In these systems, the spin-lattice relaxation time may vary along the free induction decay (FID) because in the oriented samples the FID is highly anisotropic with respect to the optic axis-magnetic field angle. The value of the oleate proton (KO-D_2O) T_1 measured at the initial portion of the decay ($\sim$15 μsec after the 90° pulse), however, is independent of orientation and in fact is the same as that measured in the unoriented samples, $\approx$240 msec. This portion of the FID is that which is least anisotropic. For the KO-D_2O aligned samples oriented at the magic angle, the oleate $T_2 \approx 13.5$ msec. A very small spin echo was observable in the unoriented samples, perhaps arising from freely rotating terminal methyl groups, or from inhomogeneous portions of the sample or from lamellar domains fortuitously arranged at the magic angle. The echo decay was non-exponential with time constants ranging from 0.9 to 12. msec.

$T_{1\rho}$ for the oleate protons was measured over a frequency range of 40 KHz to 400 KHz in unoriented samples and in oriented samples aligned at the magic angle. The relaxation was non-exponential in all cases and the analysis is at present incomplete. However, the $T_{1\rho}$ values vary with frequency, approaching T_1 as frequency increases. It is hoped that further analysis will yield information on the details of molecular dynamics in the lamellar phase.

Conclusions

Lamellar phases oriented between glass plates provide a unique system to study molecular dynamics using NMR techniques.

The oleate proton and H_2O proton effects can be separately studied by using KO-D_2O solutions or by utilizing the angular variation of the oleate spin echo and FID. Water diffusion parallel to the bilayers has been measured; presumably such techniques could be extended to the study of the diffusion of ions as well.

Studies of the frequency and orientational dependence of T_1, $T_{1\rho}$ and, where possible, T_2 are expected to yield data on the molecular dynamics of model membrane bilayer systems.

Acknowledgements

The authors are grateful to Cecil Dybowski for assistance and discussions throughout this study and to Roberta Matthews for sample preparation.

REFERENCES

* Research supported in part by the Robert A. Welch Foundation, the National Institutes of Health (Grant HE-12528) and the Biomedical Sciences Support Grant of the University of Texas.

+ Robert A. Welch Foundation Pre-doctoral Fellow. Present address: Department of Pharmacology, Albert Einstein College of Medicine.

‡ Robert A. Welch Fellow (Undergraduate). Present address: Department of Chemistry, Harvard University.

‡‡ Robert A. Welch Postdoctoral Fellow; present address: Institute of Materials Science, University of Connecticut.

1. S.J. Singer and G.L. Nicolson, Science 175, 720 (1972).
2. J.J. DeVries and H.J.C. Berendsen, Nature 221, 1139 (1969).
3. E.T. Samulski, B.A. Smith, and C.G. Wade, Chem. Phys. Lett. 20, 167 (1973).
4. E.O. Stejskal and J.E. Tanner, J. Chem. Phys. 42, 288 (1961).
5. C.J. Scandella, P. Devaux and H.M. McConnell, Proc. Natl. Acad. Sci. U.S. 69, 2056 (1972).
6. A.G. Lee, N.J.M. Birdsall, and J.C. Metcalfe, Biochem. 12, 1650 (1973).
7. J. Chavolin and P. Rigny, J. Chem. Phys. 58, 3999 (1973).
8. A.F. Horwitz, W.J. Horsley and M.P. Klein, Proc. Natl. Acad. Sci. U.S. 69, 590 (1969).
9. P.Pincus, Solid State Comm. 7, 415 (1969); J.W. Doane, R. Blinc, and M. Vilfan, Solid State Comm. 11, 1073 (1972).
10. E.O. Stejskal, Adv. in Mol. Relax. Processes 3, 27 (1972) and references therein.

ROTATIONAL DIFFUSION IN THE NEMATIC PHASE: PART I

K. S. Chu, B. L. Richards,

D. S. Moroi, and W. M. Franklin

Department of Physics and Liquid Crystal Institute

Kent State University, Kent Ohio 44242

ABSTRACT

A theoretical expression for the rotational diffusion coefficient in the nematic phase is derived in this part of the work. The calculations are based on the time dependent angular distribution function which is obtained by solving a Vlasov-type equation including the effect of molecular collisions. A perturbation technique of the Bhatnager-Gross-Krook (BGK) model is used for the collision term and for the linearization of the equation.

INTRODUCTION

Molecular dynamics in liquid crystal systems have been investigated extensively in recent years. The determination of the diffusion coefficients is a primary interest in the study of this field.

In order to derive an expression for the rotational diffusion coefficient, we introduce a single relaxation time serving as an adjustable parameter in the problem. As a phenomenonological result of Vlasov's equation[1-2] we assume that the driving torque originates from the self-consistent molecular field due to the nematic potential. By adopting BGK's perturbation technique[3] we calculate the rotational diffusion at the lowest order of approximation.

THEORY

The Hamiltonian of a single molecule at equilibrium in a nematic liquid crystal has the form

$$H = \frac{m}{2}\vec{v}^2 + \frac{1}{2I}\left(P_\theta^2 + P_\phi^2/\sin^2\theta\right) + U(\theta, S) \tag{1}$$

where m is the mass of the molecule, I is the moment of inertia of the molecule about an axis perpendicular to the director, and

$$U(\theta, S) = -\frac{AS}{V^2}\left(\frac{3}{2}\cos^2\theta - 1\right) \tag{2}$$

is the nematic potential. $S = \left\langle \frac{3}{2}\cos^2\theta - 1 \right\rangle$ is Saupe's order[4] parameter which describes the degree of orientational order of the molecules about their preferred direction of alignment. For S=0 it represents an isotropic fluid. For s=1 it corresponds to the highest nematic order. In general S has a value between zero and unity. A is a characteristic constant of the substance. V is the molar volume. θ is the angle between the directors of the molecules and their preferred direction of orientation.

From equation (1) we have the following equations of motions which are in the absence of molecular collisions.

$$\begin{aligned} \dot{P}_\theta &= -\frac{\partial U}{\partial \theta} + \frac{P_\phi^2 \cos\theta}{I \sin^3\theta} \\ \dot{P}_\phi &= -\frac{\partial H}{\partial \phi} = 0 \qquad \text{(uniaxial symmetry)} \\ P_\phi &= \text{Constant} \\ \dot{\theta} &= P_\theta / I \\ \dot{\phi} &= P_\phi / I\sin^2\theta \end{aligned} \tag{3}$$

Consider a nematic system at equilibrium state. Its angular distribution function, provided the translational and rotational motions are independent, can be written in the form

$$f_0(\theta, P_\theta, P_\phi) = \frac{1}{2\pi IKT} \exp\left\{-\frac{1}{2IKT}\left(P_\theta^2 + \frac{P_\phi^2}{\sin^2\theta}\right) - \frac{U(\theta,S)}{KT}\right\} \tag{4}$$

in which $\vec{r}$ and ϕ disappear from the distribution function as a result of random distribution of the center of mass of the molecules and the uniaxial symmetry about the preferred direction.

Now suppose this system is slightly deviated from its equilibrium state, then we can express the time dependent distribution function as

$$f(\theta, P_\theta, P_\phi, t) = f_0(\theta, P_\theta, P_\phi)\left\{1 + g(\theta, P_\theta, P_\phi; t)\right\} \tag{5}$$

where $|g| \ll 1$ is a fraction of perturbation to the equilibrium distribution. This one particle distribution function we expect to obey a Vlasov-type equation which includes the effect of molecular collisions

$$\frac{\partial f}{\partial t} + \frac{P_\theta}{I}\frac{\partial f}{\partial \theta} + \left(-\frac{\partial u}{\partial \theta} + \frac{P_\phi^2 \cos\theta}{I \sin^3\theta}\right)\frac{\partial f}{\partial P_\theta} = D_c f \tag{6}$$

where $D_c f$ represents the rate at which molecules enter and leave an element volume in the phase space as a result of molecular collisions.

Since we consider only a small deviation from equilibrium, we can take the approximation

$$D_c f \cong -\frac{g f_0 (e-1)}{e\tau} \tag{7}$$

where τ is the relaxation time during which the perturbed part of distribution function is reduced to $1/e$ of its initial value. This is an adjustable parameter characteristic of the system and can be obtained experimentally.

In order to find the conditional equation for g, we substitute equations (5) and (7) into equation (6). Noting that $df_0/dt = 0$ we obtain

$$\frac{\partial g}{\partial t} + \frac{P_\theta}{I}\frac{\partial g}{\partial \theta} + \left(-\frac{\partial u}{\partial \theta} + \frac{P_\phi^2 \cos\theta}{I \sin^3\theta}\right)\frac{\partial g}{\partial P_\theta} = -\frac{g(e-1)}{e\tau} \tag{8}$$

As a physical consequence of our basic assumptions we assume

$$g = T(t) F(\theta, P_\theta, P_\phi) \tag{9}$$

Equation (8) separates into two equations,

$$\frac{1}{T}\frac{\partial T}{\partial t} = -\left\{\lambda + \frac{e-1}{e\tau}\right\} \tag{10}$$

$$\frac{P_\theta}{IF}\frac{\partial F}{\partial \theta} + \frac{1}{F}\left(-\frac{\partial u}{\partial \theta} + \frac{P_\phi^2 \cos\theta}{I \sin^3\theta}\right)\frac{\partial F}{\partial P_\theta} = \lambda \tag{11}$$

where λ is a separation constant. The solution of equation (10) is

$$T = B e^{-\lambda t - \frac{e-1}{e\tau} t} \tag{12}$$

and B is the integration constant.

Utilizing the definition of τ as $t \to \tau$ it can be shown that

$$\lambda = \frac{1}{e\tau} \tag{13}$$

Now we turn to solve equation (11) which is a non-homogeneous and quasi-linear partial differential equation. Its subsiduary differential equation is

$$\frac{d\theta}{P_\theta/I} = \frac{dP_\theta}{-\partial U/\partial\theta + P_\phi^2\cos\theta/I\sin^3\theta} = \frac{dF}{\lambda F} \tag{14}$$

Solving the above equation in the phase space we obtain two characteristic curves,

$$F_1 = U + \frac{1}{2I}(P_\theta^2 + P_\phi^2/\sin^2\theta) = C_1$$

$$F_2 = \ln F/I\lambda + V\left\{2V^2U + \frac{V^2}{I}(P_\theta^2 + P_\phi^2/\sin^2\theta) + AS\right\}^{-\frac{1}{2}} \times$$

$$\left\{\Phi\left(\sin^{-1}\sqrt{1 - \frac{3AS\cos^2\theta}{2V^2U + \frac{V^2}{I}(P_\theta^2 + \frac{P_\phi^2}{\sin^2\theta}) - 2AS}}, \frac{2V^2U + \frac{V^2}{I}(P_\theta^2 + \frac{P_\phi^2}{\sin^2\theta}) - 2AS}{2V^2U + \frac{V^2}{I}(P_\theta^2 + \frac{P_\phi^2}{\sin^2\theta}) + AS}\right)\right. \tag{15}$$

$$\left. - \Phi'\left(\sin^{-1}\sqrt{1 - \frac{3AS}{2V^2U + \frac{V^2}{I}(P_\theta^2 + \frac{P_\phi^2}{\sin^2\theta}) - 2AS}}, \frac{2V^2U + \frac{V^2}{I}(P_\theta^2 + \frac{P_\phi^2}{\sin^2\theta}) - 2AS}{2V^2U + \frac{V^2}{I}(P_\theta^2 + \frac{P_\phi^2}{\sin^2\theta}) + AS}\right)\right\}$$

Where Φ and Φ' are elliptical integrals of the first kind. The Jacobian of Functions F_1 and F_2 satisfies the following condition

$$\frac{\partial(F_1, F_2)}{\partial(\theta, P_\theta)} = 0 \tag{16}$$

Thus the general solution of equation (11) satisfies

$$F_2 = \chi(F_1) \tag{17}$$

Where $\chi(F_1)$ is an arbitrary function of F_1.
For the purposes of mathematical simplication and physical significance, we choose the solution of the form

$$F_2 = -\frac{1}{KT}F_1 \tag{18}$$

This gives the desired solution of equation (11)

$$F = \exp\left\{-\frac{I\lambda}{KT}\left[\frac{P_\theta^2 + \frac{P_\phi^2}{\sin^2\theta}}{2I} + U\right] + V\left[2V^2U + \frac{V^2}{I}(P_\theta^2 + \frac{P_\phi^2}{\sin^2\theta}) + AS\right]^{-\frac{1}{2}}(\Phi - \Phi')\right\} \tag{19}$$

So

$$g = BFe^{-t/\tau} \tag{20}$$

Where the integration constant B is determined from the boundary condition at t = 0, and it gives the measurement of the initial perturbation of the system. then the solution to equation (6) is

$$f = f_0\left\{1 + BFe^{-t/\tau}\right\} \tag{21}$$

Note that as $t \to \infty$, $f \to f_0$, the Maxwell-Boltzman distribution.

To calculate the rotational diffusion coefficient, we begin with the evaluation of the ensemble average of $\theta(t)$ in the phase space

$$\langle \theta(t) \rangle = \frac{\int \theta f \, dP_\theta \, d\Omega}{\int f \, dP_\theta \, d\Omega} \tag{22}$$

Where the solid angle $d\Omega = 2\pi \sin\theta \, d\theta$. Substitute f from equation (21) and we obtain

$$\langle \theta(t) \rangle = \frac{\alpha_1 + B\alpha_2 e^{-t/\tau}}{\alpha_3 + B\alpha_4 e^{-t/\tau}} \tag{23}$$

Where

$$\alpha_1 = 2\pi \int \theta f_0 \, dP_\theta \sin\theta \, d\theta \qquad \alpha_2 = 2\pi \int \theta F f_0 \, dP_\theta \sin\theta \, d\theta$$
$$\alpha_3 = 2\pi \int f_0 \sin\theta \, dP_\theta \, d\theta \qquad \alpha_4 = 2\pi \int F f_0 \sin\theta \, d\theta \, dP_\theta \tag{24}$$

Then the average angular velocity of the non-equilibrium system at any time t, using the commutation relation between the time derivative operator and ensemble average operator, is

$$\langle \omega(t) \rangle = \frac{d}{dt}\langle \theta(t) \rangle = \frac{B(\alpha_4\alpha_1 - \alpha_2\alpha_3)e^{-t/\tau}}{\tau(\alpha_3 + B\alpha_4 e^{-t/\tau})^2} \tag{25}$$

The diffusion coefficient in the angular space can be defined as

$$D \equiv \int_0^\infty \langle \omega(0)\omega(t) \rangle_{t=\infty} dt \tag{26}$$

Since the system is only deviated slightly from the equilibrium state ($t = \infty$), a good approximation can be taken as

$$D \cong \int_0^\infty \langle \omega(0)\omega(t) \rangle_{t=t} dt \tag{27}$$

We divide the angular velocity at any time into an average value and a random fluctuating part, denoted by a prime

$$\omega(0) = \langle \omega(0) \rangle + \omega'(0) \quad ; \quad \omega(t) = \langle \omega(t) \rangle + \omega'(t) \tag{28}$$

Then equation (27) becomes

$$D \cong \int_0^\infty \left\{ \langle \omega(0) \rangle \langle \omega(t) \rangle + \langle \omega'(0) \rangle \langle \omega(t) \rangle + \langle \omega(0) \rangle \langle \omega'(t) \rangle + \langle \omega'(0)\omega'(t) \rangle \right\} dt$$

The last three terms vanish because of the uncorrelation of the random part and the average value part, and the zero value of the ensemble average of the random angular velocity. So

$$D \cong \langle \omega(0) \rangle \langle \theta(t) \rangle \Big|_0^\infty \tag{29}$$

Utilizing equation (23) and (25), this becomes

$$D = \frac{B}{\tau} \frac{(\alpha_4\alpha_1 - \alpha_2\alpha_3)^2}{\alpha_3(\alpha_3 + B\alpha_4)^3} \tag{30}$$

Adopting the conventional way, we define the adjustable parameter τ as

$$\tau \equiv \frac{\xi_0}{nKT} \; ; \quad n \equiv \frac{1}{B} \tag{31}$$

Where ξ_0 is the rotational friction constant of the isotropic fluid. As an example, for a fluid composed of polar molecules, n = 2.

Finally we obtain a theoretical expression for the rotational diffusion coefficient in a nematic liquid crystal

$$D = \frac{KT}{\xi_0} \left\{ \frac{(\alpha_4\alpha_1 - \alpha_2\alpha_3)^2}{\alpha_3(\alpha_3 + B\alpha_4)^3} \right\} \tag{32}$$

Due to the nematic field, the number in the brackets gives the correction of the rotational friction constant, and thus the rotational diffusion coefficient.

CONCLUSION

In this paper we just finished the formalism part of the work as we indicated as part I. The actual numerical evaluations of the numbers $\alpha_1, \alpha_2, \alpha_3, \alpha_4$, and the elliptical functions and are being programmed for the computer calculations. As we see from the formalism, certain approximations have to be taken in a real liquid crystalsystem in order toobtain possible results. For instance, we can take $\text{Sin}\,\theta \sim \theta$ as a fairly accurate approximation in the calculations. Since the thermal fluctuation angle between the director of a nematic liquid crystal molecule and its preferred direction of alignment is less than 20 degrees,[5] the error at even the larger angle such as 20 degree caused by this approximation is only about one percent difference. As another example, the approximation we took for the value of P_ϕ which has been treated as a constant in our derivation is based on the equalpartition theory. The time average of P_ϕ^2 therefore equals to $2(1-S)IKT/3$ The reason why we treat P_ϕ as a conserved quantity in the theory as well as P_θ is that we neglect the contribution due to the random fluctuating torque. This is again concluded from the small deviation assumption of the distribution function. The perturbed part of the Hamiltonian due to the effect of molecular collisions is therefore negligible compared to the equilibrium Hamiltonian. The higher terms in the elliptical integrals are also truncated in the numerical calculations. In the second part of this work we will concentrat on the numerical calculations and the comparison with experienmental results. the generalization of this theory can be made by using a nematic potential including the short-range order effect, by considering the translational and rotational correlation, and by treating the friction coefficient as a positional dependent quantity. A continued work of this paper will be published or presented elsewhere.

ACKNOWLEDGMENT

This work has benefited very substantially from discussions with Dr. Kenji K. Kobayashi.

REFERENCES

1. A. A. Vlasov, Many-Particale Theory, Moscow-Leningrad, 1950, trans. AEC-tr-3406, Washington, D.C., 1959. Earlier paper: A. A. Vlasov, J. Exp. Theoret. Phys. (USSR), 8, 291(1938)
2. Kenji K. Kobayashi, Wilbur M. Franklin, and David S. Moroi, Phys. Rev., A 7, 1781(1973).
3. P. L. Bhatnager, E. P. Gross, and M. Krook, Phys. Rev., 94, 511(1954).
4. W. Maier and A. Saupe, Zeitschrift fur Naturforschung 14A, 511(1954).
5. G. W. Gray, "Molecular Structure and Properties of Liquid Crystals", (Academic Press Inc., N. Y., N. Y., 1969).

POSSIBLE PHASE DIAGRAMS FOR MIXTURES OF 'POSITIVE' AND 'NEGATIVE' NEMATIC LIQUID CRYSTALS*

Richard Alben

Dept. of Engineering & Applied Science, Yale University

New Haven, Connecticut 06520

Common nematogenic materials consist of rod-like molecules whose longest axes tend to be parallel in the nematic phase. In addition to such 'positive' nematics, it is reasonable to expect that there might be 'negative' nematics with planar molecules aligned with their shortest axes parallel.[1,2] We have studied a theoretical model which indicates that, if indeed such 'negative' nematics exist, then mixing them with suitable 'positive' nematics would produce some remarkable effects.[3,4] We here describe our most significant findings. The details of our calculation may be found in Refs. 3 and 4.

In Fig. 1 we show a calculated phase diagram for a mixture of rods of length 5 units mixed with effectively spherical impurities. As expected, such impurities lower the nematic-isotropic transition temperature and also lead to a region of phase coexistence. (See Appendix.) However, the detailed calculations[4] show that they do not have any significant effect on the discontinuities in various properties at the first-order phase transition.

In Fig. 2 we show a calculated phase diagram for the more interesting case where the rods are mixed with flat plates 3.5 units on a side and 1.0 units thick. In the model, which is based on steric interactions between molecules, the planar molecules tend to avoid configurations in which they are perpendicular to neighboring rods because of mutual excluded volume effects. The plates thus participate in the order in the liquid crystal phase and their presence does not necessarily depress, and indeed, in

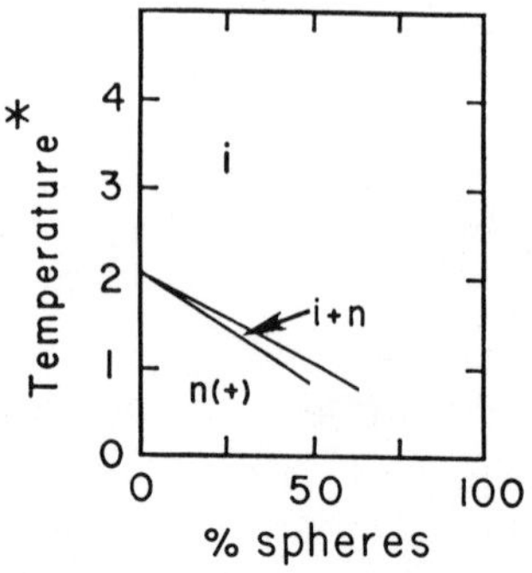

Figure 1. Calculated phase diagram of a mixture of rods of length 5 units with non-ordering 'spherical' particles. Temperature* is proportional to temperature divided by pressure. The isotropic liquid region is denoted by i, the rod-like nematic region is denoted n(+) and the phase coexistence region is denoted i+n. Spherical impurities strongly depress the nematic isotropic transition temperature but do not change the first order character of the transition.

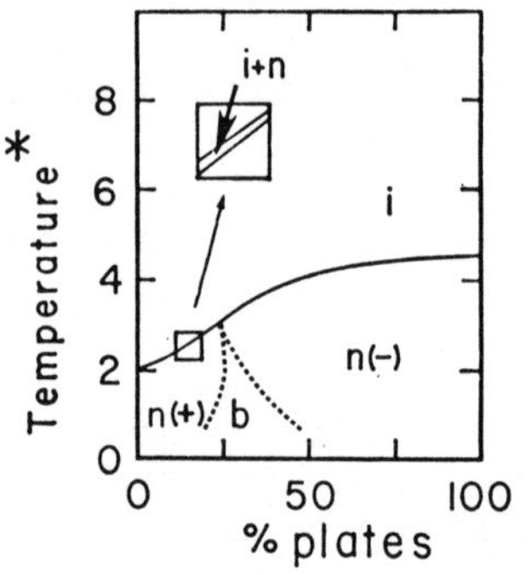

Figure 2. Calculated phase diagram of a mixture of rods of length 5 units with plates 3.5 units on a side and 1 unit in thickness. The plate-like nematic phase is denoted n(-), the biaxial phase is denoted b, and other symbols are defined in the caption to Fig. 1. First-order phase transition boundaries are indicated by solid lines and second-order boundaries by dotted lines. First-order transitions are generally associated with phase coexistence regions as shown in the insert. Plate-like impurities participate in the liquid-crystal order. The interaction of rods and plates leads to the appearance of the biaxial phase and the special critical point where the second-order lines meet the nematic isotropic phase boundary.

this case, increases the nematic-isotropic transition temperature. As the concentration of planar molecules is increased, the discontinuities at the first-order nematic-isotropic transition are found to decrease until, at a special critical point, the transition becomes second order. This point occurs where the boundary of a non-uniaxial (i.e., biaxial) phase intersects the nematic-isotropic phase boundary. For higher concentrations of planar molecules, a negative nematic phase appears.

Our model is of the mean-field type and in many respects is only approximate. Predictions of order parameter behavior near the phase transitions and especially the precise manner in which the phase transition lines come together at the special critical point are certainly not to be taken at face value. Nonetheless, there are significant aspects of our results which can serve as a guide in a systematic study of liquid crystal mixtures involving plate-like molecules.

Firstly, the model is not limited to steric interactions. Anisotropic dispersion forces[5] lead to the same conclusions, provided that they are such that 'positive' molecules tend to be parallel to one another, 'negative' molecules tend to be parallel to one another and the axes of the 'positive' molecules tend to lie parallel to the planes of the 'negative' molecules. Of course, not all molecules which are flat in shape will have the proper dispersion interactions. For example, many aromatic ring complexes, which have perpendicular configurations in their crystalline forms, are not as likely to show the effects described as are certain phosphorins. Secondly, it should be noted that it is not possible to continuously change a 'positive' nematic to a 'negative' nematic without encountering some sort of phase transition. Thus, if indeed negative nematics exist, then it is likely that some sort of novel behavior would be manifested by mixtures of them with positive nematics. Finally, it is most likely that plate-like molecules which depress the transition temperature of nematics least participate most strongly in the liquid crystal order and are thus most likely to change the character of the phase transition.

Helpful discussions with Professor James R. McColl are gratefully acknowledged.

APPENDIX

The relation between the change in the transition temperature due to an impurity and the associated region of phase coexistence has been given by deKock.[6] The relation can be expressed as follows: We let $\bar{T} = (T_1 + T_2)/2$, where T_1 is the lowest temperature for which some isotropic phase appears and T_2 is the highest temperature for which some liquid crystal phase remains. Also we let $\delta = T_2 - T_1$ and x be the mole fraction of impurity. We then have for small x

$$\frac{d\delta}{dx} = 2\left(\sqrt{1 + \left[\frac{d\bar{T}}{dx}\frac{L}{RT^2}\right]^2} - 1\right) \Big/ \frac{L}{RT^2}$$

$$\approx \frac{d\bar{T}}{dx}\left[\frac{d\bar{T}}{dx}\frac{L}{RT^2}\right] ,$$

where L is the latent heat of the pure nematic-isotropic transition, R is the gas constant and for the second line we assume that $\frac{d\bar{T}}{dx}\frac{L}{RT^2}$ is small. From various results[7,8] for $\frac{d\bar{T}}{dx}$ and a typical value for L,[9] we find that .01 mole fraction of a non-ordering impurity will change the average transition temperature by 2K and lead to a phase coexistence region about 0.15K in width. A careful distinction between phase separation effects and true critical effects is important in the studies which we have suggested.

REFERENCES

*Work supported under ONR Contract No. N00014-67-A-0097-0013.

1. L. Onsager, Ann. N.Y. Acad. Sci. 51, 627 (1948).

2. J.F. Dreyer in Liquid Crystals and Ordered Fluids (Plenum Press, 1970), p. 311.

3. R. Alben, Phys. Rev. Letters 30, 778 (1973).

4. C.S. Shih and R. Alben, J. Chem. Phys. 57, 3055 (1972); R. Alben (submitted to J. Chem. Phys.).

5. M.J. Frieser, Phys. Rev. Letters 24, 1041 (1970).

6. A.C. deKock, Z. Phys. Chem. (Leipzig) 48, 129 (1904).

7. A. Sussman, Appl. Phys. Letters 21, 126 (1972).

8. P.E. Cladis, J. Rault, and J.-P. Burger, Mol. Cryst. Liquid Cryst. 13, 1 (1971).

9. H. Arnold, Z. Phys. Chem (DDR) 226, 146 (1964).

VIBRATIONAL SPECTRA OF LIQUID CRYSTALS. VIII. INFRARED SPECTROSCOPIC MEASUREMENTS OF ORDER IN NEMATICS AND NEMATIC SOLUTIONS

Bernard J. Bulkin, Terry Kennelly and Wai Bong Lok

Hunter College of the City University of New York

New York, New York 10021

In an early study of the infrared spectrum of a nematic phase, Neff et. al. (1) used infrared dichroism to examine the effect of an electric field on the intensity of the CN stretching vibration in 4-methoxybenzylidene-4'-cyanoaniline. The dichroic ratios were used to obtain values of the order parameter as a function of field strength.

In this paper, we report results of infrared spectroscopic investigations of 4-methyoxybenzylidene-4'-n, butylaniline (MBBA) and 4-ethoxybenzylidene-4'-n-butylaniline (EBBA) aligned in homogeneous and homeotropic configurations. In addition, we present the results of a study of solutions in EBBA of benzonitrile and 4-biphenylcarbonitrile (I), which indicate the effect of molecular chain length on alignment by a nematic phase. Finally, results for a twisted nematic phase are presented. These give evidence for the presence of infrared optical activity, a phenomenon which has recently been reported for cholesteric phases as well (2,3).

BACKGROUND

It is possible to determine the order parameter from the

$$S = \frac{3\langle\cos^2\theta\rangle - 1}{2}$$

spectrum. For our purposes, in this paper, S (θ) is a measure of the deviation of the average long avis of a solute molecule, e.g.

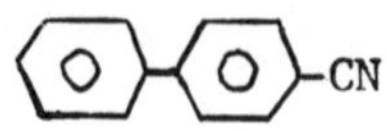

(I)

from the director of the nematic phase. The measurement and calculation are straightforward if polarized infrared radiation is available. The molecules are aligned in a homogeneous configuration by rubbing; spectra are taken with polarizer set to pass light with E parallel to and perpendicular to the director. The dichroic ratio, R, so measured

$$R = I_{\perp}/I_{\parallel}$$

is related to θ by (1)

$$R = 2\cot^2\theta.$$

Thus with a polarizer one can determine S from the homogeneous alignment alone. If no polarizer is used, it is still possible to determine S from the spectra of homogeneous and homeotropically aligned samples. Consider a molecule such as I. Those infrared bands with transitions along the long axis ($M_z \neq 0$, A_1 symmetry) should have absorbances equal to

$$\frac{A \text{ (homeotropic)}}{A \text{ (homogeneous}} \cong \tan^2\theta$$

For modes which have their transition moments perpendicular to the long axis (M_x, $M_y \neq 0$, B_1 or B_2 symmetry) the corresponding ratio is predicted to be

$$\frac{A \text{ (homeotropic)}}{A \text{ (homogeneous)}} \cong \frac{1 + \cos^2\theta}{1 + \sin^2\theta}$$

assuming that the molecules rotate freely about the long axis. Qualitatively, it is interesting to note that B_1 and B_2 modes increase in intensity in the homeotropic orientation, while A_1 modes decrease.

We believe that it preferable to measure the order parameter using a solute, rather than from the intensities of the molecular vibrations of the nematic molecule itself. This is because of two problems with the latter method:1- None of the easily factored group vibrations of a nematic molecule have transition moments exactly on the long axis;2- The nematic molecules all have C_1 symmetry, hence mixing in of off axis modes can occur (i.e. the internal vibrations may be inaccurate representations of the normal modes). Since the magnitude of such ininteraction force constants is unknown it must be assumed to be zero in most cases. With high symmetry solutes, such as those used herein, these problems are avoided. Nonetheless, it will be seen that the intensity measurements on the nematic molecules can provide useful data other than order parameters.

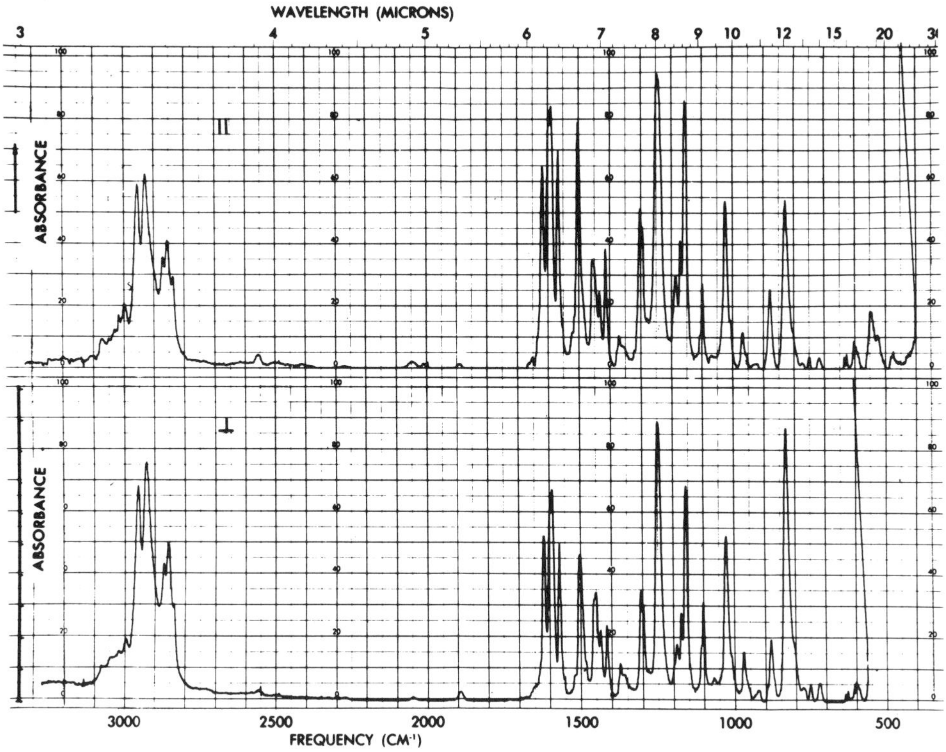

Fig. 1. Infrared spectra of homogeneous (labeled //) and homeotropically (labeled ⊥) aligned MBBA taken with no polarizer in the beam.

INFRARED SPECTRA OF MBBA

Figure 1 shows the infrared spectrum of homogeneous (labeled //) and homeotropically (labeled ⊥) aligned samples of MBBA in the nematic state. (These spectra show the mid-infrared region. The far infrared has been reported previously (4).). These spectra were taken with unpolarized light. It is clear that there are a number of relative intensity changes in these spectra. Qualitatively, one sees that, as expected, intensity ratios greater and less than one are observed between the two spectra. Although the molecule has low symmetry, it is reasonable to say that those bands which have greater intensity in the homogeneous orientation have transition moments along the long axis. Conversely, those which are enhanced in the homeotropic orientation have transition moments perpendicular to the long axis. The actual attenuation observed will depend upon the angles and mixing of modes, as discussed above.

Qualitatively we see that these ideas are confirmed. Consider the C-H stretching region (2800-3100 cm^{-1}). The n-butyl chain extends out along the long axis. The C-H bonds of the methylene groups are thus oriented primarily perpendicular to the long axis. The symmetric and asymmetric stretching vibrations of the CH_2 groups occur at 2850 and 2930 cm^{-1}. It is clear that these modes have greater intensity in the homeotropic orientation. The CH_3 stretching vibrations occur at 2960 (asym) and 2870 (sym) cm^{-1}. Although the bands are strongly overlapped, it appears that the 2960 cm^{-1} has greater intensity in the homeotropic orientation while the 2870 cm^{-1} band has about equal intensity in both spectra. This is in accord with qualitative predictions.

In the middle frequency region (1720-400 cm^{-1}) one again can pick out the transition moments of the bands. Here these become quite useful in assigning the absorption frequencies to group vibrations. Qualitative arguments predict, and the measurements show, that the ratios observed should cluster about several values. In this case one observes ratios of ca. 0.7, 1.0 and 1.4. The clustering about three values is proof that considerable local symmetry exists in the molecule, as it does in most complex molecules. The fact that the range is only from 0.7 to 1.4 reflects the amount of molecular motion present with respect to the director. A perfectly aligned material would show a range from 0 to 2.

Turn now to the assignments of these bands. Recently, these have been attempted by Heger and Mercier (5) who reasoned from the Raman spectrum of MBBA and infrared/Raman spectra of p-n butylaniline as well as p-methoxybenzaldehyde. These spectra are at relatively low resolution, and no depolarization ratios of the isotropic liquids are reported in support of the assignments. Our assignments are based mainly on qualitative group frequency arguments (6) and the intensities in Fig. 1.

In the 1550-1650 cm^{-1} region, there are four main bands at 1620, 1598, 1592 and 1568 cm^{-1}. All are polarized along the long axis. We assign these to C=N and C=C (aromatic) stretching. These internal coordinates are undoubtedly mixed, but based upon the spectra of other substituted benzenes it appears that the highest frequency mode of this group has the largest C=N component while the other three are primarily phenyl modes. This is partially in conflict with ref. 5, where the 1620 cm^{-1} band is assigned to a vibration associated with the $O-CH_3$ group (just which vibration is unclear). Those authors assign a band at

1270 cm^{-1} to the C=N valence vibration, but this is too low based on analogous compounds. The strong infrared band at 1500 cm^{-1} is again common to aromatic compounds, which almost always show such a band at this position. This is also a C=C stretching mode. It is not assigned by Heger.

The broad doublet at 1450 and 1460 cm^{-1} is due to CH_3 and CH_2 deformation vibrations. The polarization is hard to measure here due to overlapping of several bands, probably changing in opposite directions as discussed above for C-H stretching modes. The overall intensity is slightly greater in the homeotropic spectrum. A study of partially deuterated MBBA, in progress, should resolve some of the questions in both the stretching and deformation regions.

In the balance of the fingerprint region assignments become considerably more difficult. Bands at 1157 and 1248 cm^{-1} are undoubtedly due to normal modes which are principally methyl-oxygen and phenyl-oxygen stretching vibrations. Both show polarization behavior characteristic of modes along the long axis, which is qualitatively what is expected. Of note in the lower frequency region is the polarization of the band at 830 cm^{-1}. This is an out-of-plane motion of the phenyl ring. As such it is predicted to be more intense in homeotropic alignment, and this is the case. Related to this are the two characteristic in plane C-H deformation modes which are well known for para substituted aromatics. These are the two particularly sharp bands at 1175 and 1100 cm^{-1}. Despite overlapping, one can see that these have opposite polarization; clearly they belong to different symmetry species, with sufficiently little coupling so that local symmetry is being maintained. All of these assignments are rather standard, we believe, but differ in details from those of ref. 5. It appears that infrared spectroscopy of oriented nematic samples is useful for marking these assignments. Further work, including Raman studies, is in progress on this aspect of the problem.

INFRARED SPECTRA OF NEMATIC SOLUTIONS

To measure the order parameter via the infrared spectrum, we have examined solutions of 4 biphenyl carbonitrile and benzonitrile in EBBA. Though saturated, the solutions were quite dilute, so that the only band of the nitrile which could be seen clearly, unoverlapped by EBBA bands, was the CN stretching vibration at 2225 cm^{-1}.

Figure 2 shows the spectra of the CN stretching vibration of I in EBBA. Three sets of spectra are shown. In (a) the band is shown using unpolarized infrared light in the homogeneous and homeotropic configuration. Some polarization is seen. In (b)

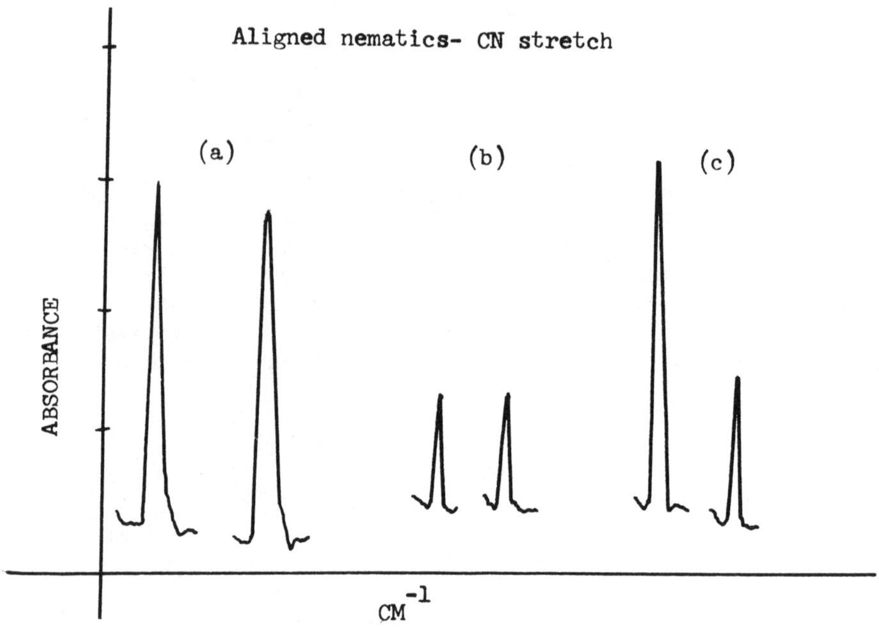

Fig. 2. Infrared band at 2225 cm^{-1} of I 1. to r. a) unpolarized light, homogeneous and homeotropic samples; b) polarizer in place, x and y polarizations, homeotropic sample; c) polarizer in place, x (along director) and y(normal to director) polarizations, homogeneous sample.

the sample is in the homeotropic orientation, and the infrared light is polarized. The two spectra, taken at orthogonal positions of the polarizer, show no difference in intensity. Finally, in (c) the spectra are taken in the homogeneous orientation using polarized light. Here we clearly see the expected attenuation in the absorbance when the polarizer is perpendicular to the director.

Using these data, the order parameter S for the solute can be calculated in two ways, either using the homogeneous alignment data alone or in combination with the homeotropic intensity. The first method yields a value of S=.30 at 40°C. The second method in which homogeneous and homeotropic intensities are ratioed, gives S= .40 for the same solution. It is not clear to us why this difference occurs and in particular, whether or not it reflects a higher order in the homeotropic orientation than in the homogeneous.

One other point of note regarding Figure 2b is the lack of measurable polarization of the CN stretching vibration in the homeotropic alignment. This shows that the molecules tilt off axis from the homeotropic director with cylindrical symmetry (i.e. no preferred direction of disorders).

The order parameter has been measured for the EBBA solutions of I as a function of sample thickness. We have found that it is, within experimental error, independent of thickness between 15 and 50 microns when only the homogeneous orientation is used, but that S decreases with decreasing thickness when the homogeneous and homeotropic intensities are ratioed. This implies that the thinner homeotropic films are less well ordered than the thicker. The decrease is from .40 to .29 as thickness varies from 50 to 15 microns.

Preliminary spectroscopic measurements have been carried out on solutions of benzonitrile in EBBA, aligned homogeneously and homeotropically. These results indicate that the order parameter for such solutions is equal (.40) to that for solutions of I in EBBA (all at $40^{o}C$). Further work is in progress on this and related systems.

TWISTED NEMATIC PHASE

When two plates are rubbed so as to give homogeneous alignment, and the cell is constructed with the rubbing directions perpendicular to one another, a twisted nematic phase (pseudo-cholesteric) results. The infrared spectra of I - EBBA solutions have been examined in the twisted nematic state using polarized infrared radiation.

In the absence of other effects one might expect an equal distribution of angles in the sample plane for such phases, hence no polarization. This is not the case.

The twisted nematic spectra show polarization in both the solute and EBBA bands. This is illustrated with the CN stretching region for I and the 900-1000 cm^{-1} region for the EBBA. In the latter region, strong bands appear at 915 and 970 cm^{-1}. As seen in Figure 3, which is a series of spectra analogous to those of Fig. 2 for the ordinary nematic phase, those bands have opposite polarization. This is of interest when one examines the twisted nematic phase spectra.

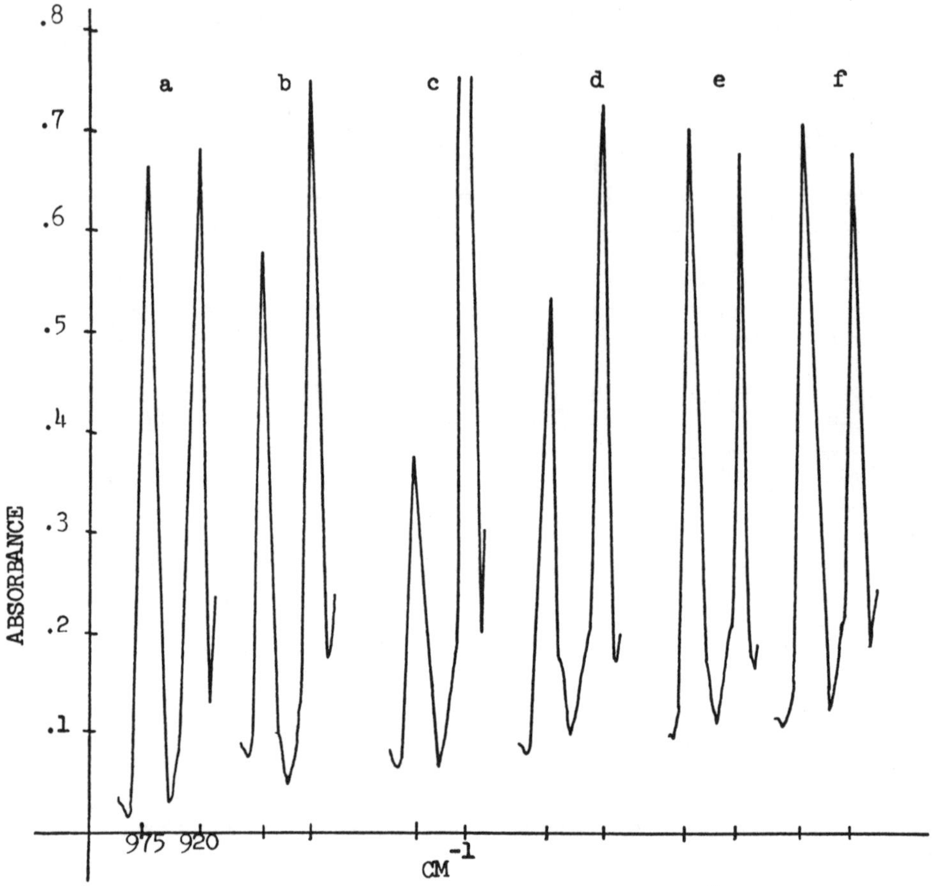

Fig. 3. Infrared bands of EBBA in the 900-1000 cm^{-1} region. a) unpolarized light, homogeneous; b) unpolarized light, homeotropic; c) x polarization (along director), homogeneous; d) y polarization (normal to director), homogeneous; e) x polarization, homeotropic; f) y polarization, homeotropic.

In Fig. 4 the spectra of the twisted nematic phase are shown in the CN stretching region. Two sets of spectra are shown, each exhibiting considerable polarization with opposite sign. These two spectra represent opposite helical senses. The sign of the polarization is thus determined by the handedness of the helix for this configuration. Fig. 5 shows a plot of the dichroic difference ($A_{90^o} - A_{0^o}$) for the twisted nematic phase. This illustrates clearly the different polarizations of the bands observed in 900-1000 cm^{-1} region of the spectrum.

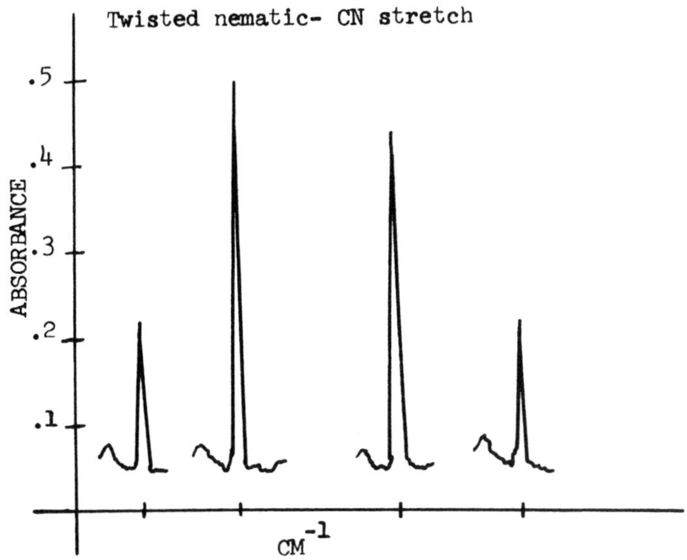

Fig. 4. Infrared band at 2225 cm^{-1} of I in EBBA in a twisted nematic sample. from l. to r. the first two spectra are x and y polarizations for one helical sense, the second two spectra are x and y polarizations for the opposite helical sense.

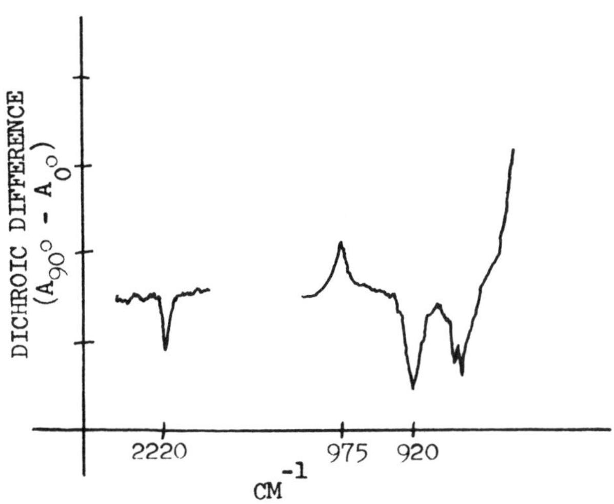

Fig. 5. Dichroic difference spectra , showing the difference in absorbance of two polarizations, for solutions of I inEBBA in a twisted nematic phase.The high frequency band is due to I, the lower group to EBBA.

We believe that the difference in absorbance shown here arises from the chiral environment in the sample which is caused by the twist. This is another demonstration of infrared optical activity in liquid crystals, a phenomenon which has been recently seen in cholesteric phases by Chabay (2) and Schrader (3). The fact that one can observe this for optically inactive molecules such as I and EBBA shows that molecular motion is sufficiently restricted in this phase as to maintain the chiral environment for a molecular vibration. The system is a potentially good one for investigating the parameters affecting the sign and magnitude of the infrared optical activity.

EXPERIMENTAL

All chemicals were purchased from Eastern Organics and purified as necessary according to standard methods. Transition temperatures of the nematic materials agreed well with literature values.

Infrared spectra were obtained using Perkin-Elmer Model 180 spectrometer run at ca. 2 cm^{-1} resolution. For polarization methods to yield standard dichroic ratios, a polarizer was placed in the recombined beam after the absorber, as is usual. The dichroic difference was also recorded for several samples. This was accomplished by placing the sample in the common beam at a focus in the source housing, and placing polarizers oriented at right angles to each other in the "sample" and "reference" beam positions respectively. It can be readily shown that this configuration yields a signal at the detector which is proportional to $A_{90°} - A_{0°}$. This configuration was of particular use for measurements on the twisted nematic phase. A Perkin-Elmer Model 521 infrared spectrometer was also used for measurements without a polarizer.

Homogeneous alignment on KBr windows was achieved by rubbing with cat's fur. Homeotropic alignment was done by treatment of the plates with lecithin, according to the method suggested by Dreyer. Two treated windows were separated by teflon spaces for varying thickness.

ACKNOWLEDGMENTS

This work was supported by a grant from the U.S. Army Research Office-Durham. We thank Dr. Robert Hannah of the Perkin-Elmer Corp for use of the P-E Model 180 and for his assistance with measurements made on the Model 180.

REFERENCES

1. V.D. Neff, L.W. Gulrich, and G.H. Brown, in G.H. Brown, G.J. Dienes and M.M. Labes, eds., Liquid Crystals, Gordon & Breach, New York, 1966, pp. 21-35.

2. I. Chabay, Chem. Phys. Lett., 17, 283 (1972).

3. B. Schrader and E-H Korte, Angew. Chem. Internat. Ed., 11, 226 (1972).

4. B.J. Bulkin and W.B. Lok, J. Phys. Chem., 77, 326 (1973).

5. J.P. Hegar and R. Mercier, Helv. Phys. Acta., 45, 886 (1972).

6. See for example, L.J. Bellamy, The Infrared Spectra of Complex Molecules, 2nd ed., Methuen, London, 1958.

CHANGES IN THERMODYNAMIC AND OPTICAL PROPERTIES ASSOCIATED WITH MESOMORPHIC TRANSITIONS

J. R. Flick, A. S. Marshall, and S.E.B. Petrie

Research Laboratories, Eastman Kodak Company

Rochester, New York 14650

ABSTRACT

In parallel thermal and optical studies of the phase transitions in some liquid-crystal-forming compounds, it has been observed that the magnitude of the enthalpy change associated with small alterations in structural order is not an indication of the extent of the change in optical behavior to be anticipated. Of the multiple phase transformations observed in N,N'-bis(4-n-octyloxybenzylidene)-1,4-phenylenediamine, the $S2$-S_3 transition, which is accompanied by the smallest heat of transition, involves a marked difference in the characteristic textures of the smectic phases. The structural changes associated with some of the other S-S transitions in this compound, however, result in subtle textural changes that are difficult to detect, although the corresponding enthalpy changes are substantial. In 4-n-octyloxyphenyl 4-(4-n-octyloxybenzoyloxy)benzoate, the enthalpy change associated with the S_A-S_C transition at 145°C is so small that it would not be detected normally in dsc studies. This transition can be detected optically, however, by observing the change in the characteristic texture. Although the thermal effect in N,N'-terephthalylidenebis(4-n-butylaniline) is slightly greater, essentially similar results are obtained. It is advisable, therefore, to use several methods in studying the phase transitions in liquid-crystal-forming compounds in order to ensure that characterization of the compound is as complete as possible.

INTRODUCTION

Transitions involving structural changes between mesophases in liquid-crystal-forming compounds are normally detected by observing the changes in optical behavior[1-8] and/or the thermal effects[7-9] associated with the alterations in structure. It has been recognized for several years that an exclusively optical approach frequently yields ambiguous results.[7] Consequently, there has been an increasing tendency to rely on sensitive thermal methods for the detection of phase transitions involving subtle changes in order.

Furthermore, calorimetry provides quantitative information on the partitioning of order among the various phases and on mesophase stability.[1-4,7,9] Since the fraction of total transitional entropy that is associated with the mesophase-isotropic liquid transition is generally characteristic of the basic mesophase type,[1-4,7,9] such quantitative information can be used for identifying and classifying mesophase types.

For some compounds, however, transition entropies involving smectic phases have been found to be atypical.[10,11] Thus, although detection may be made thermally, identification of mesophases is ultimately made on the basis of their textures, as observed under a polarizing microscope,[5,6] and their x-ray patterns.[6,12,13]

In thermal and optical studies of phase transitions in liquid-crystal compounds, it has been found generally that any structural change that resulted in a change in optical behavior is accompanied by a detectable thermal effect. No detectable enthalpy change has been reported for the "blue phase"[1,8,14,15] that has been observed to form initially in some cholesteric-forming compounds on cooling from the isotropic liquid, but these phase changes, observed optically, are irreversible and are not well understood.

In parallel optical and thermal studies of reversible phase transitions between smectic phases in certain compounds, however, we have found that some of the transitions involve marked alterations in characteristic textures, but very small enthalpy changes. The results suggest that for some compounds, the transitions between mesophases that are accompanied by distinctive optical changes would not be detected normally in dsc studies. Extremely small latent heats of transition from smectic phases to the smectic A phase have also been reported by Smith, et al.[16]

The optical and thermal data are given here for the phase transitions, especially the smectic C to smectic A transitions, in N,N'-terephthalylidenebis-(4-n-butylaniline)[17] (TBBA) and in 4-n- octyloxyphenyl 4-(4-n-octyloxybenzoyloxy)benzoate (PBOB).[18] In addition to the three enantiotropic smectic phases reported previously for TBBA,[17] two monotropic smectic phases have been detected thermally. Because of the multiplicity of mesophases observed in N,N'-bis(4-n-octyloxybenzylidene)-1,4-phenylenediamine (BPD),[19] a comparison of optical and thermal data for this compound was also undertaken.

EXPERIMENTAL

The preparations of TBBA,[17] PBOB,[18] and BPD[19] have been described elsewhere. Satisfactory elemental analyses were obtained for the compounds used in these studies.

For the investigation of the thermal behavior, a Perkin-Elmer DSC-2 differential scanning calorimeter, which has a sensitivity range of 0.01 mcal sec^{-1} $in.^{-1}$ to 0.2 mcal sec^{-1} $in.^{-1}$, was employed. A programming rate of 5°C min^{-1} was used, and the instrument sensitivity scales selected are indicated on the figures of the thermograms. The transition temperatures were determined to $\pm$0.5°C from the extrapolated onset of the endotherms, or exotherms in cooling cycles, associated with the transitions, i.e., from the temperature corresponding to the intersection of the linear extrapolation of the leading edge of the endotherm with the baseline. The estimated error in the determination of the heats of transition is 5%.

For the optical studies, microscope slides of the various liquid-crystalline materials were prepared by melting a small quantity of the material between a slide and a cover glass. The thickness of the liquid-crystal layer required in order to see both color changes and texture changes was achieved by varying the amount of material used.

The hot stage used to control the temperature of the slide was a Mettler FP52 Microfurnace in conjunction with a Mettler FP1 Temperature Control and Measurement Unit. The phase transitions and textures

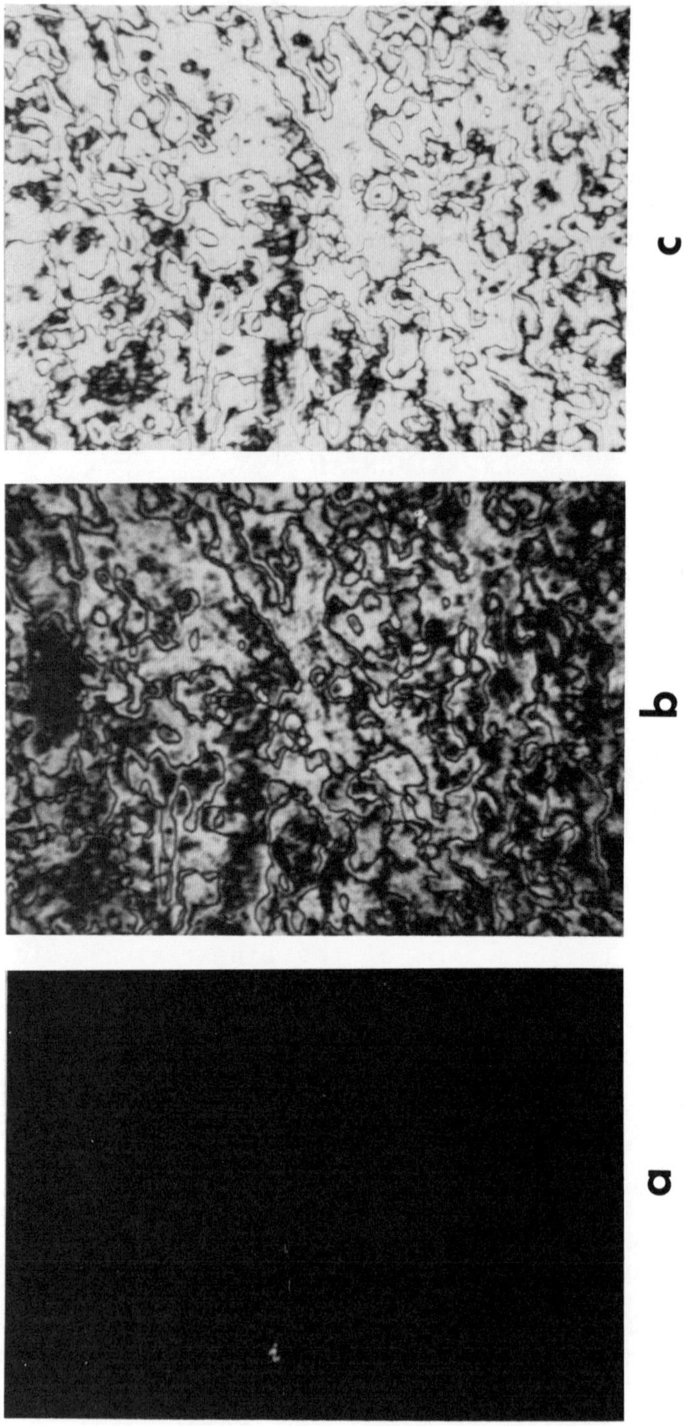

Fig. 1 Photomicrographs of PBOB between crossed polarizers, magnification ca. 100X: (a) 150°C, (b) 140°C, and (c) 90°C.

in both heating and cooling cycles were observed with a Zeiss Universal polarizing microscope. Photomicrographs were taken of the various phases with a Zeiss C-35 35-mm camera attached to the microscope.

RESULTS AND DISCUSSION

4-n-Octyloxyphenyl 4-(4-n-Octyloxybenzoyloxy)benzoate (PBOB)

It has been established previously[18] that PBOB forms a smectic phase on the loss of crystalline structure at 84°C, a nematic phase at 163°C, and an isotropic liquid phase at 189°C. The thermograms obtained with a Du Pont 900 Differential Thermal Analyzer support these observations.[20]

As a result of our attempts to identify the smectic phase on the basis of its optical texture, it was discovered that an alteration in mesophase structure, which had not been detected thermally, occurred at 145°C. The change in optical properties at this transition was well defined. The dark, essentially textureless field, Fig. 1a, observed for PBOB layers between crossed polarizers over the temperature range between 145°C and 189°C suggests that the molecules spontaneously adopted homeotropic alignment, i.e., that they spontaneously aligned normal to the slide surfaces. Presumably, the associated mesophase is smectic A. The mesophase that forms on melting and persists up to 145°C appears to be smectic C. The smectic schlieren texture, illustrated in Fig. 1b, is typical of smectic C phases in sections less than 0.1 mm in thickness.[5]

The birefringence of the smectic C phase was found to be temperature dependent. A much brighter field, as illustrated in Fig. 1c, was observed at 90°C for PBOB between crossed polarizers. This result is consistent with a postulate of a temperature-dependent tilt angle, such as that reported for TBBA.[17]

Further thermal studies of PBOB were undertaken, and at a sensitivity of 0.02 mcal sec^{-1} $in.^{-1}$, a small, barely detectable baseline shift was observed at 145°C. The dsc thermogram obtained at 5°C min^{-1} is illustrated in Fig. 2. A well-defined, measurable latent heat is not associated with the smectic C to smectic A (S_C-S_A) transition. Also, the enthalpy change of 0.07 kcal mol^{-1} observed for the smectic A to nematic (S_A-N)

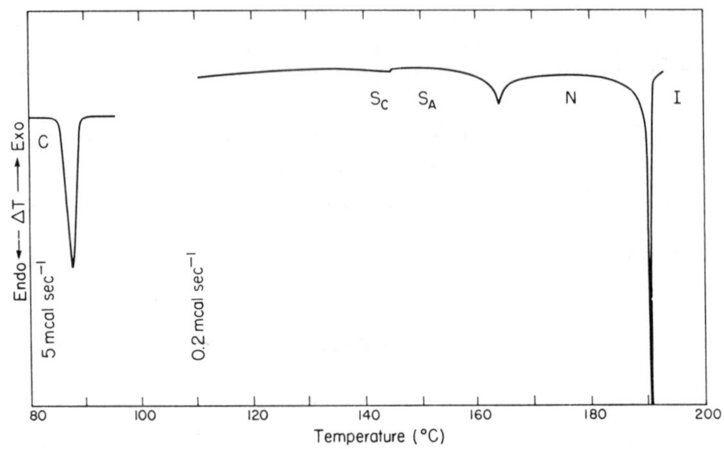

Fig. 2 DSC thermogram for PBOB on heating at 5°C min^{-1}.

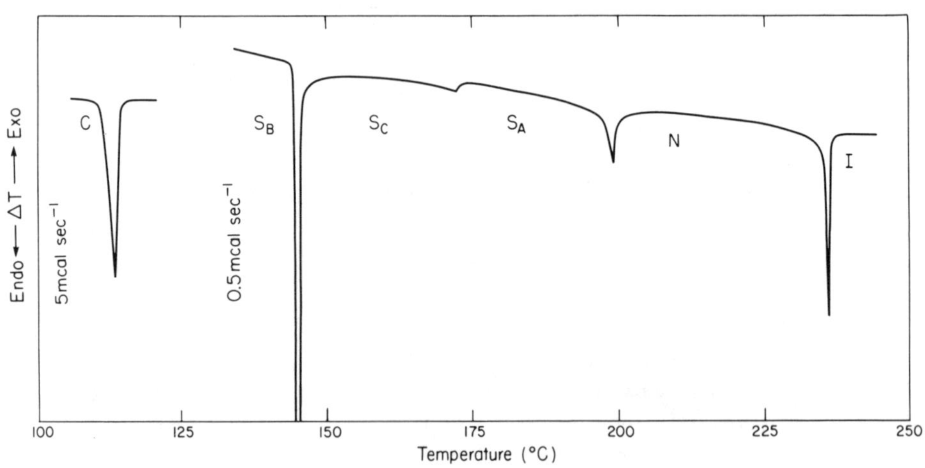

Fig. 3 DSC thermogram for TBBA on heating at 5°C min^{-1}.

transition is an order of magnitude smaller than that normally measured for S_A-N transitions.[1-3,9]

The gradualness of the S_C-S_A transition found in the dsc thermogram is consistent with the temperature dependence of the birefringence observed in the optical studies. These observations suggest that the smectic C phase exhibited by PBOB has a temperature-dependent tilt angle. Furthermore, the temperature dependence of the specific heat through the S_C-S_A transition suggests that this transition may be second order.[21]

N,N'-Terephthalylidenebis(4-*n*-butylaniline) (TBBA)

In order to identify better the characteristics of mesomorphic transitions involving little, or possibly no, latent heat, changes in thermal and optical properties associated with the phase transitions exhibited by TBBA were explored in greater detail than those reported previously.[17] In prior studies of TBBA,[17] it has been established that this compound forms a highly ordered[22] smectic B phase on the loss of crystalline structure at 113.0°C,(actually x-ray studies suggest that the mesophase is smectic H),[22] a smectic C phase at 144.1°C, a smectic A phase at 172.5°C, a nematic phase at 199.6°C, and the isotropic liquid at 236.5°C.

Qualitatively, there is a strong similarity between the dsc curve for PBOB, illustrated in Fig. 2, and that for TBBA, illustrated in Fig. 3, over the temperature ranges involving the S_C-S_A, the S_A-N, and the nematic to isotropic liquid (N-I) transitions. Quantitatively, however, the enthalpy changes measured for PBOB are significantly less than those found for TBBA for the S_C-S_A and S_A-N transitions. The latent heats measured for the N-I transitions exhibited by these compounds, however, are similar, and typical of those found generally for this phase change.[1-3,9] In Tables I and II, the latent heats of transition measured for PBOB and TBBA are given. A comparison of the dsc thermograms indicates that the baseline shift associated with the S_C-S_A transition in PBOB is about 20 percent of that observed in TBBA, and, therefore, considerably more difficult to detect thermally.

Similarly to the observations for PBOB, the change in texture detected optically at the S_C-S_A transition in TBBA is well defined, as indicated in the photomicrographs of the textures of the S_A and S_C phases

TABLE I.

Transition Enthalpies and Entropies for PBOB

Transition from Preceding Phase to	Temperature, °C	Enthalpy kcal mol^{-1}	Entropy cal mol^{-1} $°K^{-1}$
S_C	87.0	11.9	33.1
S_A	145.0		
N	163.0	0.07	0.16
I	189.5	0.40	0.87

TABLE II.

Trasition Enthalpies and Entropies for TBBA

Transition from Preceding Phase to	Temperature, °C	Enthalpy kcal mol^{-1}	Entropy cal mol^{-1} $°K^{-1}$
Crystal - S_B	112.5	4.2	10.9
S_5			
S_4	80.0*	0.07_5	0.21
S_B	89.5	0.26	0.72
S_C	144.0	0.88	2.1
S_A	172.5		
N	198.5	0.16	0.35
I	235.5	0.21	0.41

* The S_4-S_5 phase transition in TBBA on cooling occurred at 75.5°C

reproduced in Fig. 4. Although regions of homeotropic alignment apparently formed in the liquid-crystal layer, some of the characteristic features of the S_A phase are evident in Fig. 4a.[1-7] An increase in the overall brightness of the field observed with crossed polarizers of the S_C phase at lower temperatures is apparent in Fig. 4c. This gradual change in the optical properties of the S_C phase with temperature is consistent with the observation of a temperature-dependent tilt angle in the S_C phase formed by TBBA.[17] The tilt angle changed from 0°C at the S_A-S_C transition to approximately 26° at the S_C-S_B transition.

In TBBA on cooling, two monotropic phases, which were not reported in the previous thermal study of this liquid-crystalline compound,[17] have been detected thermally. The two phases are labelled S_4 and S_5 in order of decreasing temperature in the dsc thermogram of the cooling cycle reproduced in Fig. 5. Both of these phases are believed to be smectic, since (1) they exhibited the mosaic texture typical of highly ordered smectic phases,[1-7] such as that observed in Fig. 6c for the S_B phase, and (2) the temperatures at which they formed in cooling cycles were found to be independent of cooling rate over rates ranging from 5 to 20°C min^{-1}. The entropy change associated with the S_B-S_4 transition suggests that the S_4 phase is significantly more ordered than the S_B phase, which has been found by x-ray studies to be highly ordered in that each smectic plane has hexagonal order.[22]

It is obvious from the dsc thermogram illustrated in Fig. 5 that the S_B-S_4 and S_4-S_5 transitions can be easily detected thermally. The differences between the textures exhibited by the S_B, S_4, and S_5 phases, on the other hand, are small, as illustrated by the photomicrographs given in Fig. 6. The absence of an observable change in optical properties between mesophases has been reported by Arora, et al. for a number of 4-n-alkoxybenzylidene-4'-aminopropionphenones.[23]

N,N'-Bis(4-n-octyloxybenzylidene)-1,4-phenylenediamine (BPD)

As indicated by the multiple endotherms in the dsc trace of BPD illustrated in Fig. 7, order is lost in a number of discrete steps in this compound on heating from the crystalline state. It has been established that a nematic and five smectic phases, S_1 to S_5 in order of decreasing temperature, are exhibited by BPD.[19]

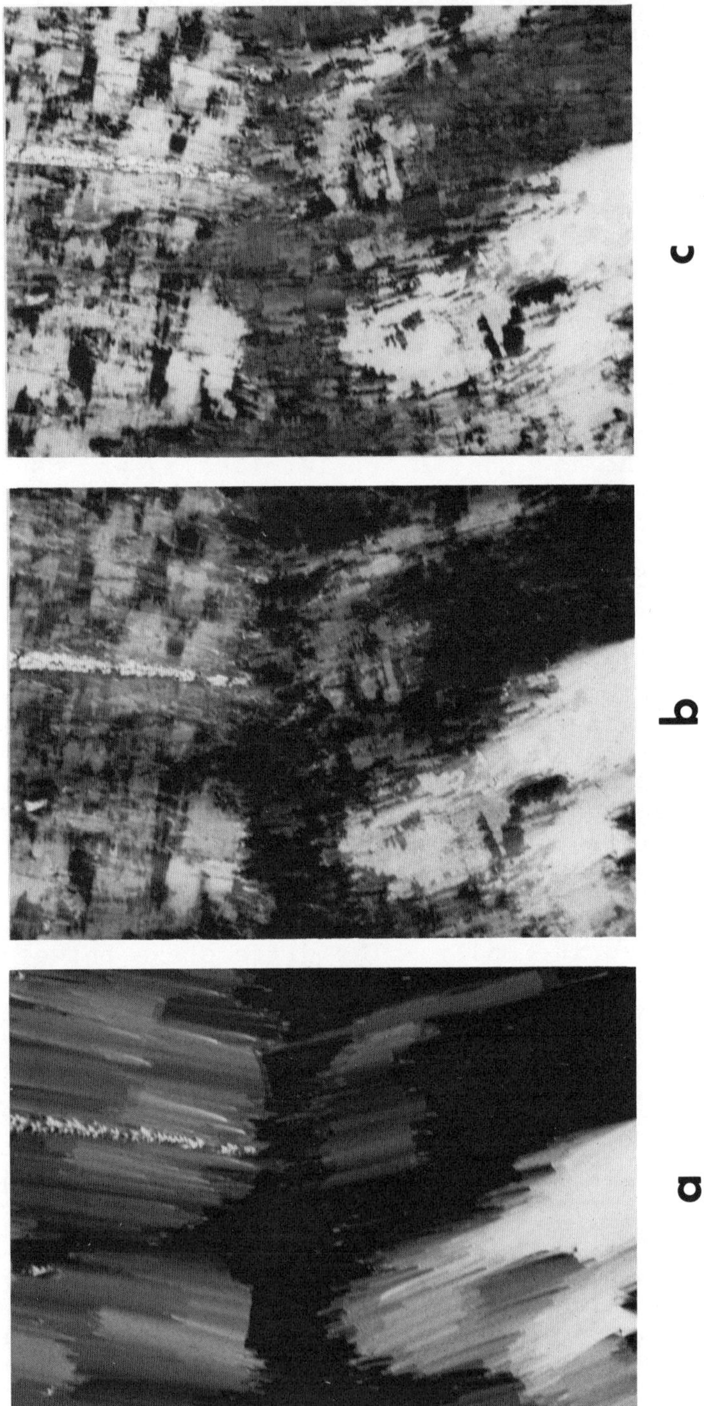

Fig. 4 Photomicrographs of TBBA between crossed polarizers, magnification ca. 100X: (a) 195°C, (b) 165°C, and (c) 145°C.

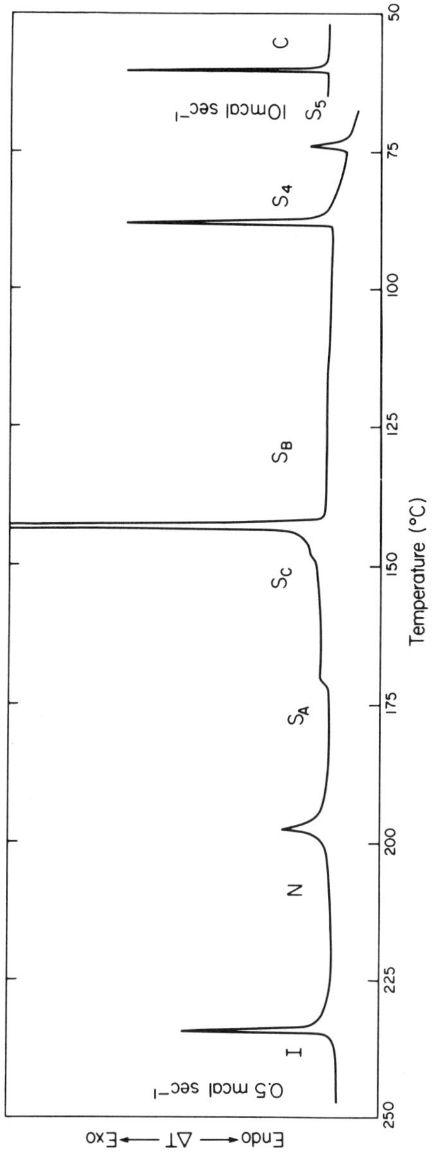

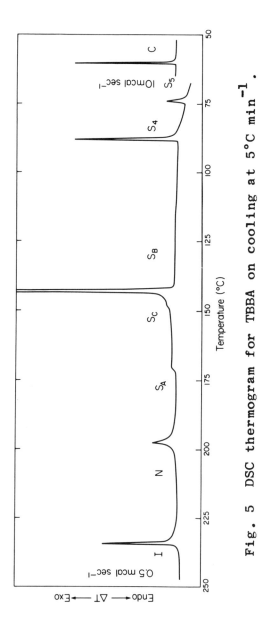

Fig. 5 DSC thermogram for TBBA on cooling at 5°C min^{-1}.

Fig. 6 Photomicrographs of TBBA between crossed polarizers, magnification ca. 100X: (a) 137°C, (b) 80°C, and (c) 70°C.

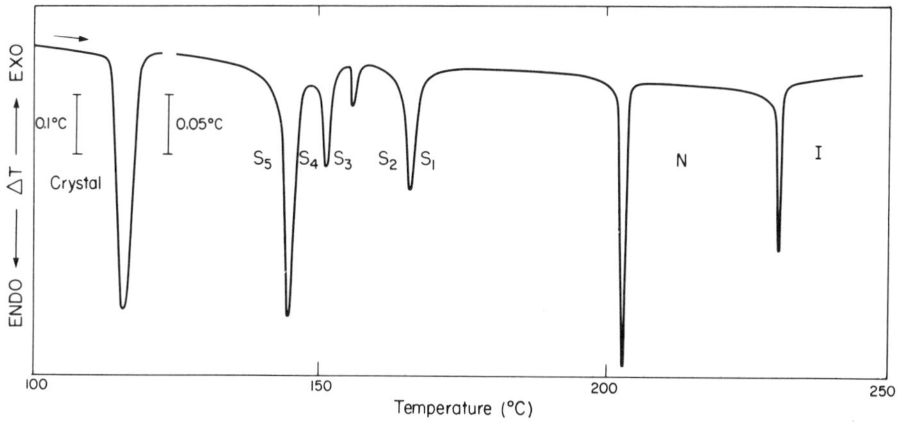

Fig. 7 DSC thermogram for BPD on heating at 5°C min^{-1}.

TABLE III.

Transition Enthalpies and Entropies for BPD

Transition from Preceding Phase to	Temperature, °C	Enthalpy kcal mol^{-1}	Entropy cal mol^{-1} $°K^{-1}$
S_5	115	4.2	10.9
S_4	142	1.6	3.9
S_3	149	0.50	1.2
S_2	155	0.10	0.23
S_1	164	0.76	1.7
N	203	0.85	1.8
I	231	0.54	1.1

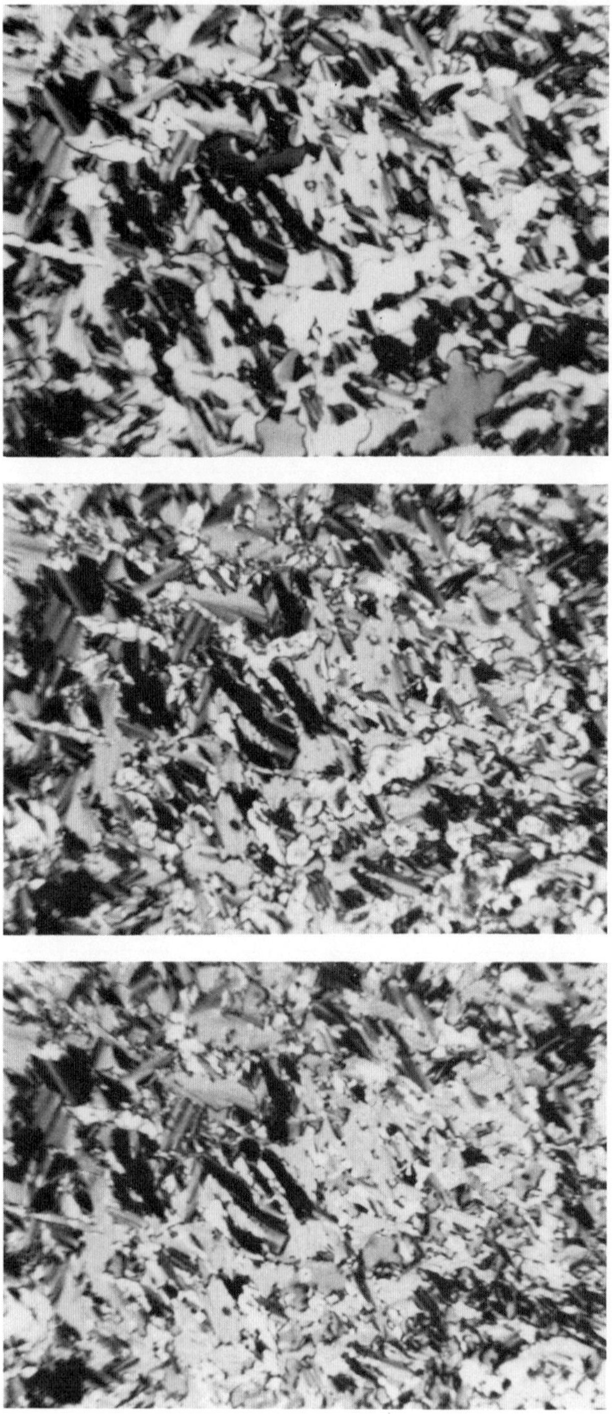

a

b

c

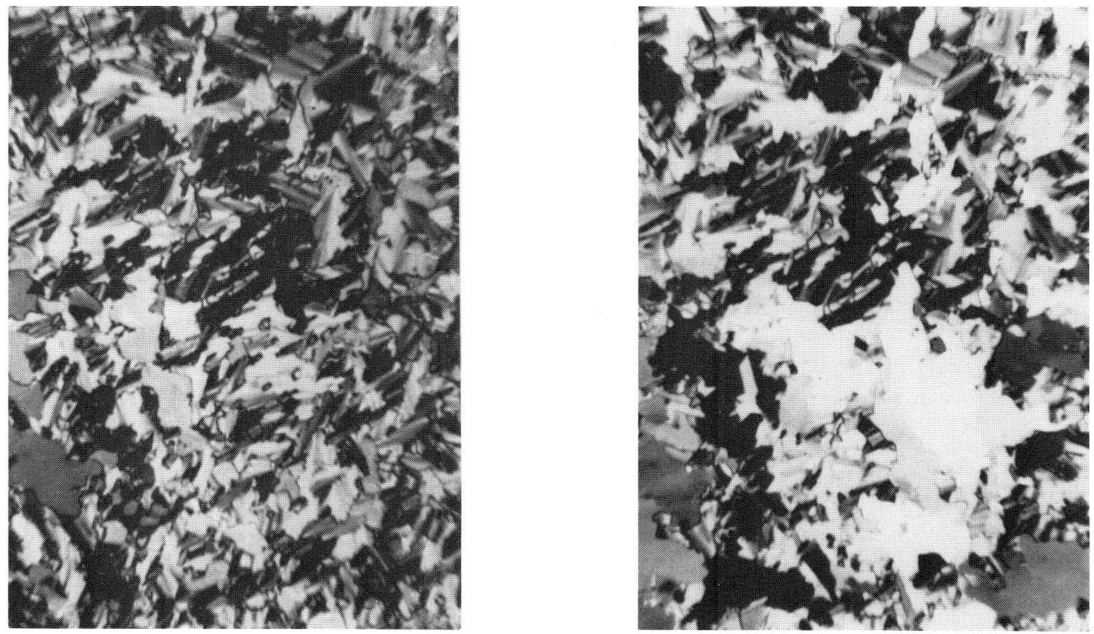

d e

Fig. 8 Photomicrographs of BPD between crossed polarizers, magnification ca. 100X: (a) 175°C, (b) 160°C, (c) 172°C, (d) 146°C, and (e) 120°C.

The thermodynamic data determined in our studies are given in Table III.

Although a valid classification of the various smectic phases in BPD cannot be made on the basis of the characteristic textures illustrated in Fig. 8, it is evident that all of the phase changes can be detected optically as well as thermally. The S_2-S_3 transition, which involves a very small latent heat, the smallest latent heat of all the phase transitions observed in BPD, is readily detected optically. The adjacent S_3-S_4 transition, however, involves a change in the overall brightness of the field obtained with crossed polarizers, rather than a well-defined change in textures, and is, therefore, more difficult to determine optically. Because of the substantially larger enthalpy change associated with this transition, as given in Table III, it is well characterized thermally.

CONCLUSIONS

From parallel optical and thermal studies of phase transitions in certain liquid-crystal-forming compounds, in particular PBOB and TBBA, it is evident that some phase transitions could easily be passed through undetected if only the thermal or optical method of detection was employed. It is advisable, therefore, to use several methods for detecting the phase transitions between phases in liquid-crystal-forming compounds, in order to ensure that characterization of such compounds is as complete as possible.

The preliminary studies presented here of the thermal and optical behavior of a smectic C phase having a temperature-dependent tilt angle suggest that this phenomenon may be more common than it has been considered to be previously. Furthermore, the observations indicate that such mesophases have characteristic thermal and optical behavior. We are exploring further the possibility that S_C-S_A transitions involving S_C phases having this property are second order.

ACKNOWLEDGMENTS

The authors are indebted to J. P. Van Meter, of the Eastman Kodak Research Laboratories, for synthesizing the liquid-crystal compounds used in these studies. For technical assistance in the preparation of the photographs, the authors wish to express their gratitude to R. T. Klingbiel and B. M. Spinell, also of these Laboratories.

REFERENCES

1. G. W. Gray, "Molecular Structure and the Properties of Liquid Crystals," Academic Press, New York, N.Y., 1962, chapt. 2.

2. G. H. Brown and W. G. Shaw, Chem. Rev., **57**, 1049 (1957).

3. G. H. Brown, J. W. Doane, and V. D. Neff, "A Review of the Structure and Physical Properties of Liquid Crystals," CRC Press, Cleveland, Ohio, 1971.

4. A. Saupe, Angew. Chem. Int. Ed. (English), **7**, 97 (1968).

5. H. Sackmann and D. Demus, Mol. Cryst., **2**, 81 (1966).

6. H. Sackmann and D. Demus, Fortschr. Chem. Forsch., **12**, 349 (1969).

7. R. D. Ennulat, Mol. Cryst. Liq. Cryst., **3**, 405 (1967).

8. D. Coates, K. J Harrison, and G. W. Gray, paper presented at the Fourth International Liquid Crystal Conference, Kent State University, Kent, Ohio, 1972, Abstract No. 109.

9. R. S. Porter, E. M. Barrall, and J. F. Johnson, Accounts Chem. Res., **2**, 53 (1969).

10. J. P. Van Meter and S.E.B. Petrie, paper presented at the 162nd National Meeting of the Amer. Chem. Soc., Washington, D.C., 1971, Abstract No. Phys. 177.

11. A. DeVries and D. L. Fishel, Mol. Cryst. Liq. Cryst., **16**, 311 (1972).

12. A. DeVries, Mol. Cryst. Liq. Cryst., **10**, 31 (1970).

13. A. DeVries, paper presented at the Fourth International Liquid Crystal Conference, Kent State University, Kent, Ohio, 1972, Abstract No. 152.

14. G. W. Gray, J. Chem. Soc., 3733 (1956).

15. T. Asuda and R. S. Stein, paper presented at the Fourth International Liquid Crystal Conference, Kent State University, Kent, Ohio, 1972, Abstract No. 41.

16. G. W. Smith, Z G. Gardlund, and R. J. Curtis, Mol. Cryst. Liq. Cryst., **19**, 327 (1973).

17. T. R. Taylor, S. L. Arora, and J. L. Fergason, Phys. Rev. Lett., **25**, 722 (1970).

18. J. P. Van Meter and B. H. Klanderman, Mol. Cryst. Liq. Cryst., in press.

19. S. L. Arora, T. R. Taylor, J. L. Fergason, and A. Saupe, J. Amer. Chem. Soc., **91**, 3671 (1969).

20. A. S. Marshall, J. P. Van Meter, and S.E.B. Petrie, paper presented at the Fourth International Liquid Crystal Conference, Kent State University, Kent, Ohio, 1972, Abstract No. 115.

21. W. L. McMillan, Phys. Rev. A, **4**, 1238 (1971).

22. A. M. Levelut and M. Lambert, C. R. Acad. Sci. Paris, Series B, **272**, 1018 (1971).

23. S. L. Arora, T. R. Taylor, and J. L. Fergason in "Liquid Crystals and Ordered Fluids," J. F. Johnson and R. S. Porter, Eds., Plenum Press, New York, 1970, p. 321.

KINETICS OF FIELD ALIGNMENT AND ELASTIC RELAXATION IN TWISTED NEMATIC LIQUID CRYSTALS

T. S. Chang, P. E. Greene and E. E. Loebner

Hewlett-Packard Laboratories, Palo Alto

California 94304

INTRODUCTION

It has been shown by Schadt and Helfrich[1] that the orientation pattern of nematic liquid crystals can be made to correspond to the planar texture of cholesteric liquid crystals if the molecular alignment is parallel to the walls but differs in direction at the two surfaces. Such an arrangement is known as a twisted nematic. A 90° twist leads to a 90° rotation of linearly polarized light travelling normally to the orienting surfaces provided the direction of polarization is collinear with the aligned nematic molecules at either of the two surfaces. These authors have further shown that the application of a sufficiently high electric field between the surfaces destroys the molecular twist in the liquid crystal bulk and eliminates most of the rotation of the polarized light provided the material has a positive dielectric anisotropy[1]. Upon removal of the field, the initial orientation pattern is re-established.

The Schadt-Helfrich electro-optical effect is becoming increasingly important for light modulated display applications. One of the limiting parameters of such displays is their response time. The purpose of this work is to study the kinetics of both the field induced alignment and the elastic relaxation-reorientation in order to provide guiding principles for the selection of superior materials for the above applications.

THEORETICAL

Our theoretical considerations derive from the work of Jakeman and Raynes[2], who offered a differential equation for the internal rotation of the molecular director under the action of an electric field during an electrically induced cholesteric to nematic change and who obtained expressions for the decay and rise times of this effect. Noting the similarity of their effect to that of Schadt-Helfrich, they suggested that similar relations should apply to both. The Jakeman-Raynes model balances three forces: electrical moment proportional to the dielectric anisotropy, elastic restoring forces dependent on the elastic moduli and viscous drag. The decay and rise times are expressed as follows:

$$\tau_D = \eta L^2/\tilde{k}\pi^2 \tag{1}$$

$$\tau_R = \eta \bigg/ \left\{ (\varepsilon_{11}-\varepsilon_{\perp})\frac{E^2}{4\pi} - \tilde{k}\pi^2/L^2 \right\} \tag{2}$$

where η is the average viscosity, $\tilde{k}$ is an effective elastic constant, to be discussed, E is the applied electric field strength and L is thickness of the cell.

We now show that the threshold voltage can be derived by equating the denominator of Equation (2) to zero since it gives the limit of an infinitely long rise time:

$$V_{th} = E_{th}L = \pi \left[\frac{4\pi\tilde{k}}{\varepsilon_{11}-\varepsilon_{\perp}} \right]^{1/2} \tag{3}$$

This is in agreement with determinations of Leslie[3] and Helfrich[4] if we assume the identity of:

$$\tilde{k} = \left[k_{11} + (k_{33}-2k_{22}) \right] /4 \tag{4}$$

where k_{11}, k_{22}, k_{33} are the splay, twist and bend elastic constants, respectively. Their threshold voltage expressions were obtained by equating the dielectric energy gained by the field induced deformation to the elastic energy required to destroy the initial twisted configuration and replace it by transition regions of splay and bend.

The above-referred equation of motion of Jakeman and Raynes[2] reduces to:

$$\frac{\partial\phi}{\partial t} = \frac{\tilde{k}}{\eta}\frac{\partial^2\phi}{\partial x^2} \qquad (5)$$

in the absence of the applied field, with x the distance measured in the direction perpendicular to the wall. Equation (5) can be interpreted as a kinetic diffusion equation, where the diffusing species is the molecular director diffusing solely along the x direction. This is in agreement with the suggestion of Williams[5] that the re-orientational elastic wave of nematics propagates from the walls at a rate controlled by diffusion kinetics. However, in his case, the negative material was aligned homeotropically. Thus, the diffusion coefficient for this case can be found to be:

$$D = \frac{\tilde{k}}{\eta} = \frac{k_{11} + (k_{33} - 2k_{22})}{4\eta} \qquad (6)$$

EXPERIMENTAL

Three positive nematic mixtures were studied: Mixture A (Eastman Kodak mixture number 11900), Mixture B (Princeton Organic LC360) and Mixture C (Merck ZLI319). Surface alignments were obtained by a process similar to Janning's[6] where films of surface energy higher than the surface tension of nematics were evaporated at a shallow angle of incidence onto the cell walls. Homogeneous alignment was thus obtained[7].

A polarized He-Ne laser was used as the light source for the measurement of optical response times. The cells were oriented with surfaces perpendicular to the laser beam which was in the line of sight of a photodetector. A pair of crossed polarizers were placed with one on each side of the cell. Square wave voltages of 100 Hz were used to drive the cells.

Three response times were measured. The rise and decay times (τ_R and τ_D), are the times needed for a 90% change in optical transmission following the application and removal of the voltage, respectively. The starting portion of rise time needed for a 10% change is called the delay time.

The dependence of τ_D on cell thickness, as expressed by Equation (1), was tested using cells of various thicknesses. The other materials' parameters were established using a single thickness cell for each of the three nematic mixtures as shown in Table I.

TABLE I

CELL THICKNESSES AND DRIVING VOLTAGES FOR STUDY OF $\tilde{k}$, η, V_{th} AND TEMPERATURE VARIATION OF RESPONSE TIMES

	Cell Thickness (μm)	Vrms (Volts, 100 Hz Square Wave)
Mixture A	20	8
Mixture B	17	8
Mixture C	19	6

RESULTS AND DISCUSSION

Figure I gives a typical example of a response time measurement. Oscillatary ringing is normally observed during decay. This complication is more pronounced for thicker cells and for operation at lower temperature. This pehnomenon is not yet clearly understood.

Figure II shows τ_D versus cell thickness for all the three mixtures. It can be seen that the τ_D dependence on the cell thickness predicted by Equation (1) is well satisfied. The diffusion constants of these three mixtures have been calculated using Equation (1). They are shown in Table II.

Figure I Optical response of a twisted nematic cell filled with mixture B at 23°C, the upper trace is the output of a photodiode; the lower trace is the 100 Hz square wave applied voltage signal (Vrms = 8 volts); cell thickness = 17μm; time scale = 200msec./div.; delay time = 100msec.; rise time = 180msec. and decay time = 700msec.

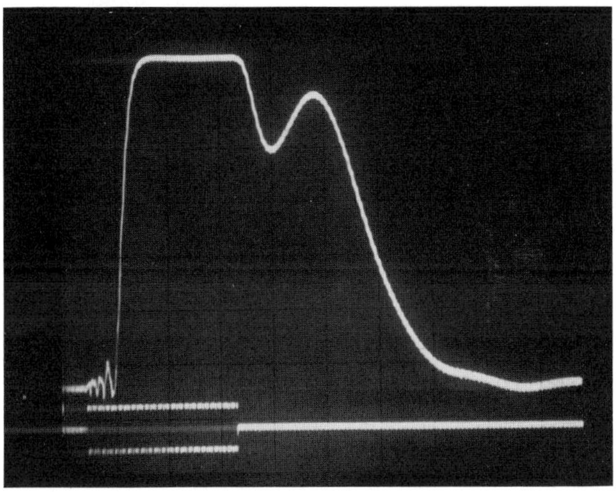

TABLE II

MEASURED DIFFUSION CONSTANTS AND CALCULATED ELASTIC CONSTANTS, VISCOSITIES AND THRESHOLD VOLTAGES FROM ROOM TEMPERATURE RESPONSE TIME DATA OF THREE POSITIVE NEMATIC MIXTURES; THE MEASURED THRESHOLD VOLTAGES (V_{th}) ARE SHOWN ON THE LAST COLUMN.

	D (cm^2/sec.)	k (dyne)	η (poise)	$\pi\left[\frac{4\pi k}{\varepsilon_{\parallel}-\varepsilon_{\perp}}\right]^{1/2}$ (Volts)	V_{th} (Volts)
Mixture A	3.2×10^{-7}	2.5×10^{-6}	8.0	4.1	3.5-4 (8)
Mixture B	4.2×10^{-7}	2.2×10^{-6}	5.2	3.6	3.0 (9)
Mixture C	1.3×10^{-7}	4.2×10^{-7}	3.3	0.8	1.1 (10)

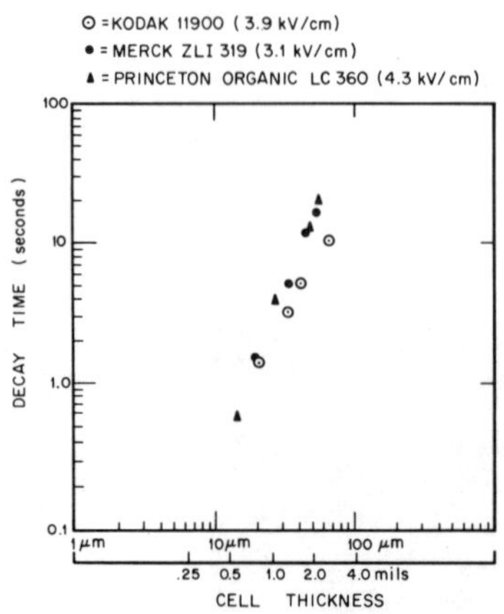

Figure II Decay time vs cell thickness for three positive mixtures; applied field strength is kept constant.

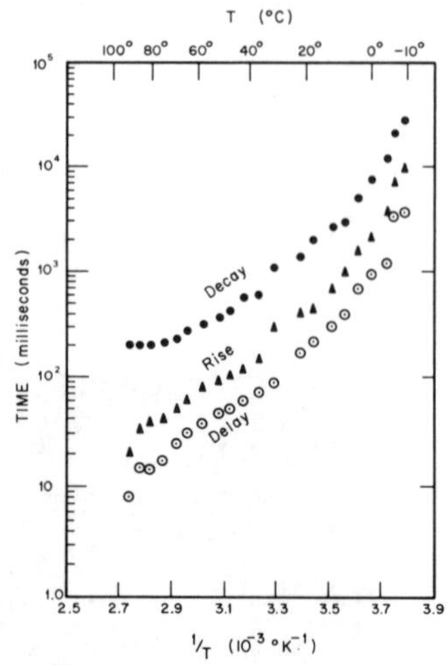

Figure III Response times of Eastman Kodak 11900 mixture vs temperature; the cell thickness is 20 μm.

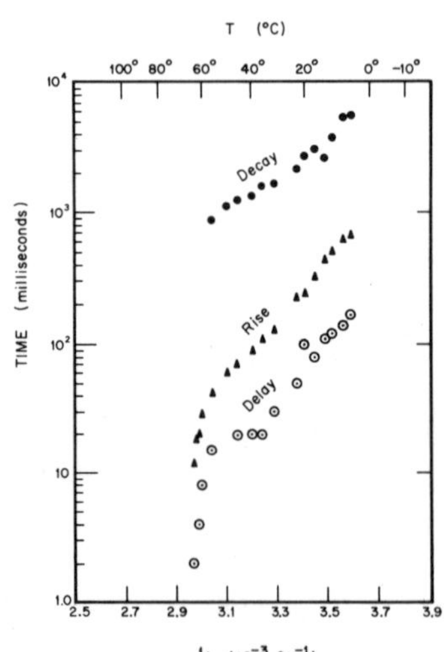

Figure IV Response times of Princeton Organics LC360 mixture vs temperature; the cell thickness is 17 μm.

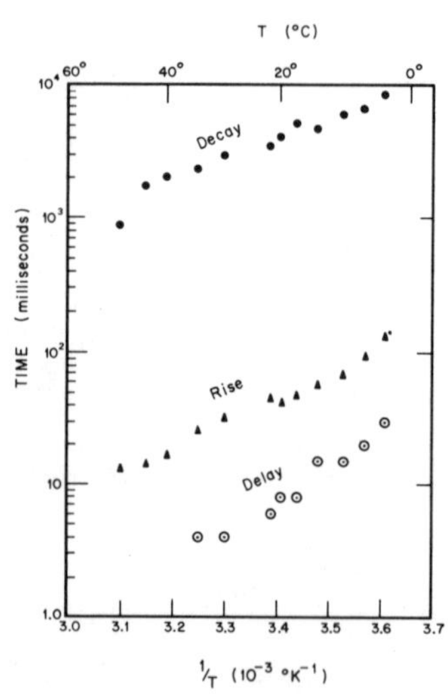

Figure V Response times of Merck ZLI319 mixture vs temperature; the cell thickness is 19 μm.

The effective elastic constants and average viscosities calculated from rise and decay responses using Equation (1) and (2), are also shown in Table II. These values are used to calculate the threshold voltage following Equation (3). They are compared with the measured values in Table II. In spite of the good agreement between calculated and measured V_{th}, an independent verification of the $\tilde{k}$ and η values would be useful to validate their indirect deviations from τ_D and τ_R.

The temperature variations of the response times are shown in Figures III, IV and V for the mixtures A, B and C, respectively. It should be noted that deviations occur near room temperature in these plots, which we attribute to our experimental procedure combined with slight permanent changes in the response times following approach or passage through the nematic-isotropic transition temperature. The temperature dependence of delay time was found to follow closely that of τ_R.

These data are plotted in semi-log vs 1/T fashion in order to discover a possible exponential dependence on 1/T. The data in Figure IV and V show that it is quite reasonable to assume a single value for the slope of each curve, especially within the mesomorphic ranges removed sufficiently from the transition temperatures. For Mixture A, at least two slopes are needed to fit an exponential model. The activation energies, derived from the data, are defined as follows:

$$\tau_j(T) = \tau_{jo} e^{E_j/kT} \tag{7}$$

where j = D, R and τ_{jo} is a constant.

The results are tabulated in Table III.

In our case, E_R is mainly related to the activation energy associated with viscosity because the applied voltage is many times above threshold. Consequently, the first term in the denominator of Equation (2) is much larger than the second term. Furthermore, the dielectric anistropy does not vary significantly with temperature outside of either transition range. This viscosity activation energy is a measure of short-range intermolecular force against alignment in the field or re-orientation after the removal of the field and has been discussed by others.[11]

TABLE III

THE ACTIVATION ENERGIES ASSOCIATED WITH THE DECAY AND RISE RESPONSES OF THREE POSITIVE NEMATIC MIXTURES IN TWISTED STRUCTURE.

	E_D (eV)	E_R (eV)
Mixture A	.63 (-10°∿23°C) .28 (23°∿95°C)	.67 (-10°∿23°C) .30 (23°∿95°C)
Mixture B	.27	.43
Mixture C	.26	.38

E_D depends on both η and $\tilde{k}$. The temperature behavior of all three elastic constants can be assumed to be exponential as we deduce to be the case for (PAA) using data presented in Reference(12). temperature dependence derives from squaring the long-range oder parameter, which is related to the cosine of the deviation angle of the molecular director from its aligned directions. This elasticity activation energy measures the long-range intermolecular elastic energy that restores the orientational pattern to its original twisted configuration. Following Equation (1), E_D is the difference of the viscosity and elasticity activation energies. Therefore, it is the effective potential barrier height, caused by intermolecular forces hindering the re-orientation of molecules to their original twisted pattern. The diffusion coefficient, D, should also have the same activation energy E_D. As expected E_D is found to be smaller than E_R for all three mixtures.

In conclusion, the re-orientation of positive nematic liquid crystals to the twisted structure is found to have a diffusion type kinetics. Available models derived from the continuum theory are used to calculate various material parameters from time response data and are found to be self-consistent. Various activation energies of different physical origins are also evaluated from the temperature variation of response times. We are grateful to Mr. Don Bradbury for supplying the cells, Mr. Arthur Yeap for carrying out some of the measurements and Dr. W. Helfrich for providing us his manuscript prior to publication.

REFERENCES

1. M. Schadt and W. Helfrich, Appl. Phys. Lett. 18, 127 (1971)

2. E. Jakeman and E. P. Raynes, Phys. Lett. 39A, 69 (1972)

3. F. M. Leslie, Mol. Cryst. and Liq. Cryst. 12, 57 (1970)

4. W. Helfrich, to be published in Mol. Cryst. and Liq. Cryst. (1973)

5. G. H. Heilmeier, L. A. Zanoni and L. A. Barton, Proc. IEEE 56, 1162 (1968)

6. J. L. Janning, Appl. Phys. Lett. 21, 173 (1972)

7. L. T. Creagh and A. R. Kmetz, Fourth International Liquid Crystal Conference, Kent, Ohio (1972)

8. Kodak publication NO.JJ-14 and earlier released data (1973)

9. Princeton Organics news release on field effect liquid crystal (Feb. 1973)

10. Dr. B. Hample (private communication)

11. A. Sussman, IEEE Trans. PHP 8, 24 (1972)

12. R. G. Priest, Mol. Cryst. and Liq. Cryst. 17, 129 (1972)

STUDIES ON THE MOLECULAR ARRANGEMENT IN LIQUID CRYSTALS BY POLARIZATION OF FLUORESCENCE

S. Sakagami, A. Takase, M. Nakamizo and H. Kakiyama

National Industrial Research Institute of Kyushu

Tosu-shi, Saga-ken 841, Japan

ABSTRACT

Molecular arrangements in liquid crystals have been studied by the method of fluorescence polarization. Fluorescence molecules with a highly optical anisotropy, dispersed homogeneously in a liquid crystalline material are used as a probe to detect the molecular alignment in the liquid crystal. The angular distribution of the polarized components of the fluorescence gives useful information on the extent as well as the types of molecular orientation. Examples of the application of this method are given for the determination of the molecular arrangement in liquid crystals at the phase transitions from the isotropic liquid state through mesomorphic state to the crystalline state. The results obtained for the nematic state of para-n-octyloxybenzoic acid suggest that the molecules can be regarded as adopting almost perfect parallel alignment over regions much larger than the molecular dimensions. The observed angular distribution patterns of the polarized components of fluorescence can be well explained by assuming a molecular arrangement of uniaxial prolate ellipsoid for the nematic state. Discussion is briefly given to the angular distribution patterns for the smectic and crystalline states.

INTRODUCTION

Liquid crystalline states are well characterized by a nearly parallel alignment of their molecules over regions much larger than the molecular dimensions while in a liquid state. The parallel alignment of the nematic state can be relatively easily obtained by rubbing the surfaces of glass plates or applying electric and magnetic fields. Heilmeier et al. have shown in their studies on guest-host interactions in nematic liquid crystals that the orientation of a guest dichroic dye is controlled by the orientation of its nematic host in an applied field (1). This implies that the orientation of the guest molecules is closely related to the orientation of the liquid crystalline molecules and, hence, that the guest molecules can be possibly used as a probe for the detection of the extent and types of molecular orientation present in liquid crystals. A similar technique has been employed by Nishijima et al. in their studies on molecular orientations in polymer solids, which utilizes a homogeneously dispersed fluorescent compound (2). In both studies, the added guest compounds are long, rod-like molecules which possess a highly optical anisotropy. These works show that the absorption and fluorescence intensities of the guest molecules are strongly dependent on the molecular orientation of the host matrix with respect to the electric vector of the incident light.

In this paper, the fluorescence method of determining orientation in the liquid crystalline states are described with particular applications to some liquid crystalline materials.

THEORETICAL

When a molecule absorbs ultraviolet (UV) or visible light, it is electronically excited from the ground state (S_o) to the excited singlet states (S_1, S_2....). The reverse electronic transition from the absorption process, from S_1 downwards to S_o, is usually accompanied with emission of fluorescence radiation (Figure 1).

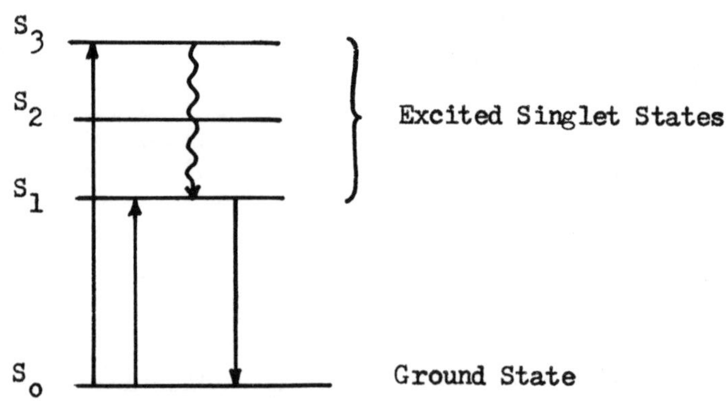

Figure 1. Schematic energy level diagram showing the electronic transitions of fluorescence molecules.

According to the classical theory of electronic transitions, the molecule can be treated as an anisotropic (linear) or partially anisotropic oscillator. The polarization of light absorption and fluorescence emission by molecules has been explained by assuming the vibrations of the linear oscillators induced along the fixed axes within the molecules. Many of fluorescent organic molecules are more or less optically anisotropic. For these complicated polyatomic molecules, the simple oscillator model is still the only possible one to explain the polarization phenomena of their absorption and fluorescence spectra. The most general model of a fluorescent molecule consists of an absorbing oscillator A and an emitting oscillator F which have different frequencies but are coupled to each other so that light absorbed in A is re-emitted by F (3, 4). In general, A and F will coincide if the absorption and emission processes correspond to the same electronic transition, e.g., $S_1 \longleftrightarrow S_0$ in Figure 1. For highly anisotropic fluorescent molecules, it can be assumed that the directions of these oscillators coincide with a long molecular axis (M) for the electronic transition between the ground state and the first excited state. The maximum probability of light absorption can be obtained if the absorbing oscillator A lies parallel to the electric vector (P_1) of incident beam. The molecules of intermediate orientation are excited with the maximum probability multiplied by $\cos^2\alpha$, where α is the angle between the absorbing oscillator A lying parallel to the longer molecular axis (M) and the exciting electric vector (P_1), as shown in Figure 2. The fluorescence light from each excited molecule can be looked upon as consisting of a plane-polarized light whose electric vector coincides with the direction of

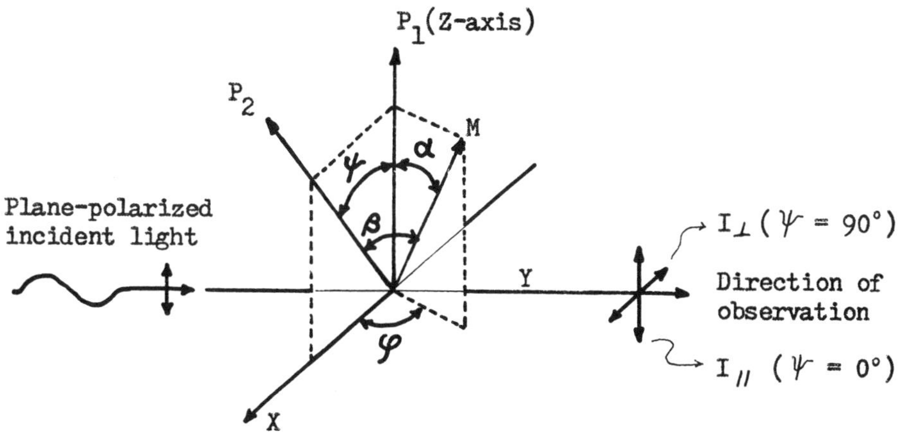

Figure 2. Optical coordinate system for the observation of the angular distribution of the polarized components of fluorescence.

the longer molecular axis (M). If the fluorescence from the excited molecule is observed in a direction perpendicular to M, the maximum intensity of the polarized component of fluorescence can be obtained when the vibration direction (P_2) of an analyzer coincides with that of the emitting oscillator F, e.g., the longer molecular axis (M). Generally the intensity of the polarized component of fluorescence is given by,

$$I_{\alpha\beta} = K\Phi \cos^2\alpha \cos^2\beta \ , \tag{1}$$

where K is the probability of the electronic transition ($S_1 \leftarrow S_0$), Φ is the fluorescence yield and β is the angle between M and P_2, as shown in figure 2.

If a sample of liquid crystal in thin layer between glass plates is placed on the X-Z plane in Figure 2 and the fluorescence from the sample is observed in the direction of the Y axis with the exciting electric vector (P_1) fixed along the Z axis, the polarized components of fluorescence intensity, $I_{//}$ and $I_{\perp}$, are measured when the direction of the electric vector of the analyzer (P_2) is parallel ($\varphi = 0°$) and perpendicular ($\varphi = 90°$) to P_1, respectively. The angular distributions of both the polarized components of fluorescence are obtained by rotating the sample within the X-Z plane. When the fluorescence molecules are dispersed in a medium with a certain type of molecular orientation, the intensities of both the polarized components of fluorescence can be generally expressed by the following equations,

$$I_{//} = (K\Phi/2\pi)\int_0^{2\pi}\int_0^{\pi/2} N(\omega,\varphi)\, M_x^2\, M_x^2 \sin\omega\, d\omega\, d\varphi \ , \tag{2}$$

and

$$I_{\perp} = (K\Phi/2\pi)\int_0^{2\pi}\int_0^{\pi/2} N(\omega,\varphi)\, M_x^2\, M_z^2 \sin\omega\, d\omega\, d\varphi \ , \tag{3}$$

where $N(\omega, \varphi)$ is an angular distribution function of the molecular axes of the fluorescence molecules in the orientation coordinate system (a, b, c), as shown in Figure 3 (II), and M_x and M_z are the components of M in the directions of the X and Z axes in the optical coordinate system in Figure 3 (I), respectively. These components, M_x, M_y and M_z in the optical system can be easily obtained from the corresponding components, M_a, M_b and M_c, in the orientation coordinate system using a transformation matrix (T), as expressed in Eqs. (4) and (5). The extent as well as the types of molecular orientation in the system are obtained from measurements of the angular dependency of polarized fluorescence components, $I_{//}$ and $I_{\perp}$, and of the degree of polarization of fluorescence (P), which is defined by Eq. (6).

$$\begin{pmatrix} M_x \\ M_y \\ M_z \end{pmatrix} = (T) \begin{pmatrix} M_a \\ M_b \\ M_c \end{pmatrix} \tag{4}$$

$$(T) = \begin{pmatrix} \cos\theta \sin\gamma, & -\sin\theta, & \cos\theta\cos\gamma \\ \sin\theta\sin\gamma, & \cos\theta, & \sin\theta\cos\gamma \\ -\cos\gamma, & 0, & \sin\gamma \end{pmatrix} \tag{5}$$

$$P = (I_{//} - I_{\perp})/(I_{//} + I_{\perp}) \tag{6}$$

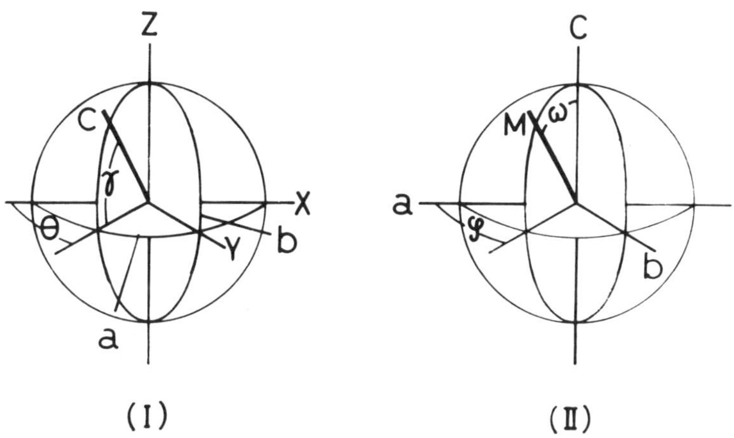

Figure 3. Coordinate systems: (I) optical coordinate system and (II) orientation coordinate system. (see ref. 5)

EXPERIMENTAL

The optical system used in this study is schematically illustrated in Figure 4. The light of the 365 mμ wavelength was separated from a mercury lamp using a monochromator (M) and was used as an exciting light; it illuminated fluorescent molecules dispersed in liquid crystals after passing through a polarizer (P). The wavelength of the exciting light coincides with that of the maximum of the principal absorption peak, which corresponds to the electronic transition between the ground state and the first excited singlet state of the fluorescent molecules. The polarized components of the fluorescence intensity are measured by a photomultiplier (PM) through an analyzer (A) and a cut-off filter (CF) and, then, recorded as the sample is being rotated through 360° on the rotating stage. The analyzer can be also rotated so that the direction of the electric vector of the fluorescent light lies either parallel ($I_{//}$) or perpendicular ($I_{\perp}$) to that of the exciting light. An ordinary polarizing microscope was modified as the optical system for this study. The observed sample area is about 100 μ in diameter.

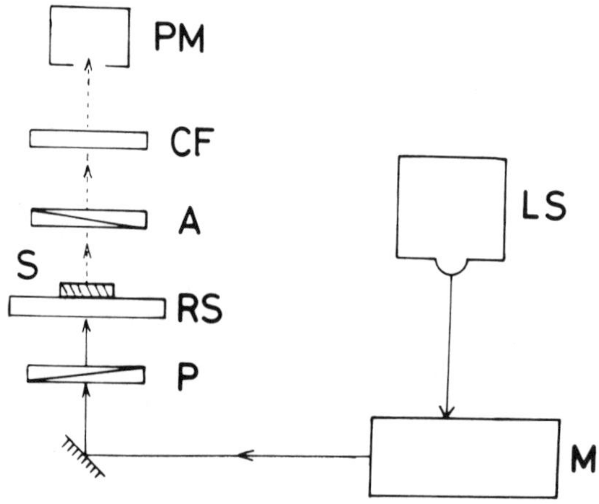

Figure 4. Schematic diagram of the apparatus. LS: light source (Hg lamp), M: monochromator, P: polarizer, RS: rotating stage, S: sample, A: analyzer, CF: cut-off filter, PM: photomultiplier.

The liquid crystalline materials used in this study are listed as follows along with the types of mesophase and the phase transition temperatures.

p-n-Amyloxybenzoic acid(AOBA), $CH_3(CH_2)_4OC_6H_4COOH$

$$C \xleftrightarrow{124^{\circ}C} N \xleftrightarrow{151^{\circ}C} I$$

p-n-Hexyloxybenzoic acid (HOBA), $CH_3(CH_2)_5OC_6H_4COOH$

$$C \xleftrightarrow{105^\circ C} N \xleftrightarrow{153^\circ C} I$$

p-n-Octyloxybenzoic acid (OOBA), $CH_3(CH_2)_7OC_6H_4COOH$

$$C_2 \xleftrightarrow{68^\circ C} C_1 \xleftrightarrow{101^\circ C} Sm(C) \xleftrightarrow{108^\circ C} N \xleftrightarrow{145^\circ C} I$$

Butyl-p-(p'-ethoxyphenoxycarbonyl)-phenyl carbonate (BEPC)
$CH_3(CH_2)_3OCOOC_6H_4COOC_6H_4OC_2H_5$

$$C \xleftrightarrow{55^\circ C} N \xleftrightarrow{82.4^\circ C} I$$

Cholesteryl-n-nonanoate (CN), $CH_3(CH_2)_7COOC_{27}H_{45}$

$$C \xleftrightarrow{74^\circ C} Sm(A) \xleftrightarrow{81^\circ C} Ch \xleftrightarrow{93^\circ C} I$$

Here C represents a crystalline phase, Ch cholesteric phase, I isotropic liquid phase, N nematic phase, and Sm(A) and Sm(C) smectic phases of A and C types, respectively.

The fluorescent compound used in this study was 1,6-diphenyl-1,3,5-hexatriene, C_6H_5-$(CH{=}CH)_3$-C_6H_5(DH). It emits a strong fluorescence covering the spectral region from 4000 A through an intensity maximum around 4500 A to 5500 A. This fluorescent molecule can be regarded as a nearly perfect linear oscillator in view of its molecular symmetry. A highly optical anisotropy of this molecule was confirmed from a limiting value of the degree of polarization (P), which approaches 0.5 at the extrapolation of $T/\eta = 0$ when the degree of polarization (P) of the fluorescence is measured in various solvent with a wide range of viscosities (η) at absolute temperature (T) (3).

A mixture of the liquid crystal compound and fluorescent substances of about 10^{-3} wt % was dissolved in chloroform in order to obtain a uniformly-dispersed system, and then the solvent was removed by evaporation at room temperature under reduced pressure. The liquid crystal state was attained by heating this mixture to the isotropic state of the liquid crystal between a glass slide and a cover slip on a microscopic stage with a heating block, and then by slowly cooling it to the mesomorphic state. The sample thickness was approximately 5μ. The aligned nematic state was obtained by the well-known method of rubbing the glass surface.

The effect of the addition of fluorescent compound was examined on the optical textures and phase transition temperatures of liquid crystalline materials by means of a polarizing microscope and a differential thermal analyzer over temperature ranges covering all phases from solid to isotropic liquid. Figure 5 shows the effect of fluorescent diphenylhexatriene (DH) on the phase transition temperatures of p-n-octyloxybenzoic acid (OOBA). As can be seen in Figure 5, there is no appreciable change in the plase transition temperatures at concentrations of the fluorescent

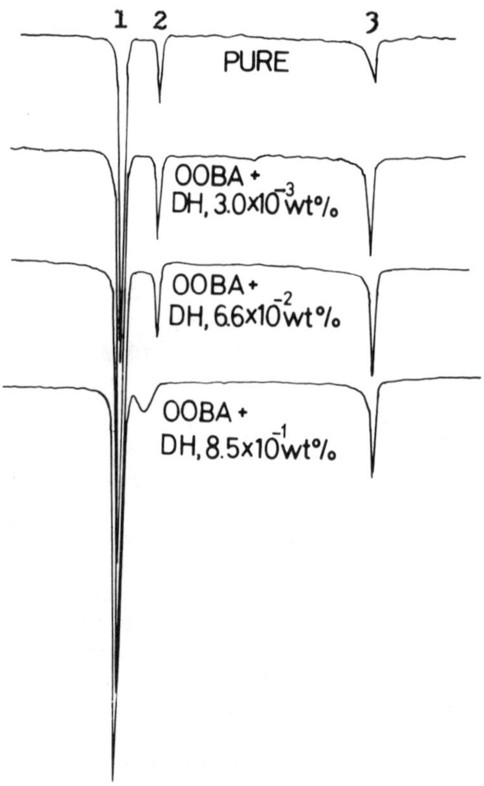

Figure 5. Differential thermograms of p-n-octyloxybenzoic acid (OOBA) containing diphenyl hexatriene (DH). Phase transition temperatures; 1, C_1 ⟷ Sm(C) at 101°C. 2, Sm(C) ⟷ N at 108°C. 3, N ⟷ I at 145°C.

compound (DH) up to 6.6 x 10^{-2} wt %. At 8.5 x 10^{-1} wt % of DH, the transition temperature between the smectic (C-type) and nematic phases decreased from 108°C to 104°C while the others remain unchanged. This was also confirmed by microscopic observations on the optical textures of OOBA in the liquid crystalline state. In this study the concentration of DH in the liquid crystals was at most 10^{-3} wt %.

RESULTS AND DISCUSSION

The polarized components of the fluorescence intensity, $I_{//}$ and $I_{\perp}$, obtained for OOBA at each state, are shown in Figure 6 as a function of the angle of sample rotation (γ).

As is evident from Figure 6, both the components, $I_{//}$ and $I_{\perp}$, for the isotropic state are independent of the angle of rotation (γ). This can be easily understood by assuming three-dimensional random orientation of the fluorescent molecules, as is expected for the isotropic state. In the case of a completely random distribution, in which $N(\omega, \varphi) = 1$, the intensities of the polarized components of fluorescence are obtained by integrating Eqs. (2) and (3) over all molecules and given by the following Equations:

$$I_{//} = (3/15)\,K\Phi, \tag{7}$$

and

$$I_{\perp} = (1/15)\,K\Phi. \tag{8}$$

These equations indicate that the polarized components of the fluorescence intensity are independent of the angle of sample rotation (γ) and that the intensity ratio of the components, $I_{//}/I_{\perp}$, is equal to 3. Therefore, the degree of polarization (P) defined by Eq. (6) is 0.5. The observed intensity ratio for the isotropic state of OOBA is about 2, which is smaller than the ratio, 3, theoretically expected for the isotropic liquid with randomly distributed molecular axes, whose directions are maintained during the lifetime of excitation. The reasons for this discrepancy are connected with the partial depolarization of fluorescence due to the Brownian rotational motion and the internal rotation of the fluorescent molecules during the lifetime of the excited state. Such motions of the molecules reduce the anisotropy of the spacial distribution of excited molecules, and consequently, also reduce the degree of polarization of the fluorescence. It is apparent that these motions are strongly dependent on the viscosity and temperature of the medium. However, the angular distribution of the polarized components of fluorescence at the isotropic state indicates the perfectly random orientation of the fluorescent molecules, that is, the OOBA molecules, as is clearly shown in Figure 6.

At the phase transition from the isotropic liquid to the nematic state, both $I_{\parallel}$ and $I_{\perp}$ drastically change, as is illustrated in Figure 6, and their patterns show no significant change

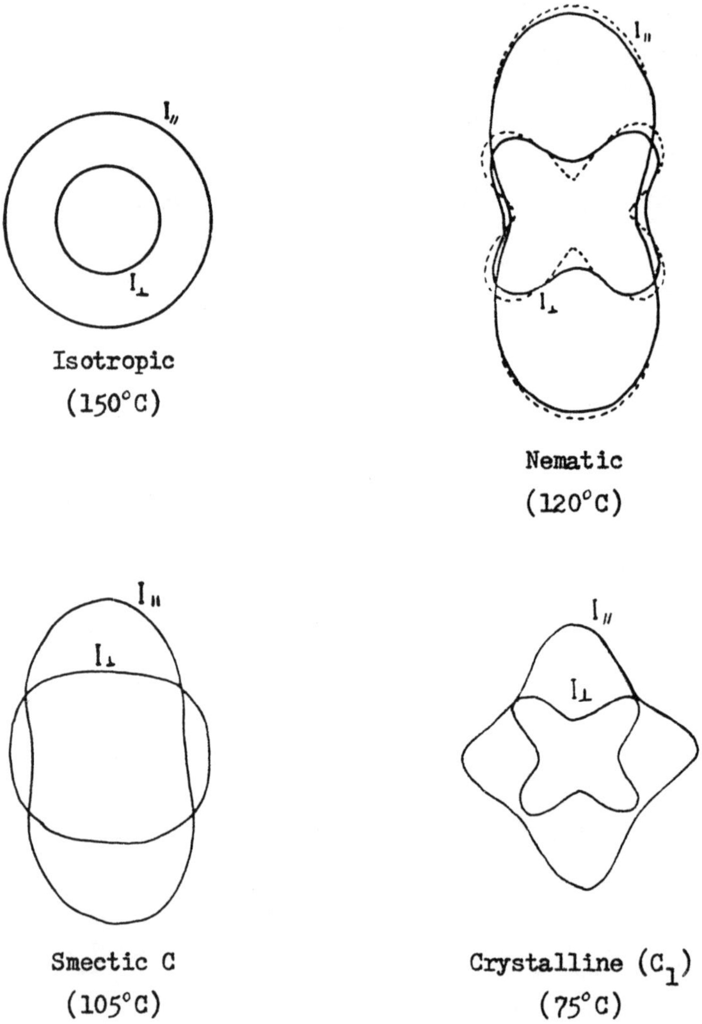

Figure 6. Angular distribution patterns of the polarized components of fluorescence for each state of OOBA.

with the variation in temperature within the nematic state, except that the fluorescence intensity decreased slightly with decreasing temperature. As is evident from Figure 6, both of the polarized components have a maximum intensity at a definite angle of γ. This suggests a partially uniaxial molecular orientation along a preferred axis. This is consistent with the molecular orientation which has been confirmed in the nematic state by means of microscopic and other physical methods.

The observed angular dependencies of both $I_{/\!/}$ and $I_{\perp}$ for the nematic state can be theoretically calculated using Eqs. (2) and (3) if the angular distribution function, $N(\omega, \varphi)$, is known of the long axes of fluorescent molecules. It seems reasonable to assume that the molecular orientation of the nematic state can be approximately expressed by a uniaxial prolate ellipsoid of ratation because of the partially parallel orientation of the nematic molecules. The distribution function for the prolate ellipsoid of rotation, as shown in Figure 7, can be given by:

$$N(\omega, \varphi) \propto (\lambda^2 \sin^2\omega + \cos^2\omega)^{-3/2}, \qquad (9)$$

where λ is the axial ratio, C/A. The angular distribution of $I_{/\!/}$ and $I_{\perp}$ can be straightforwardly calculated by substituting Eq. (9) into Eqs. (2) and (3).

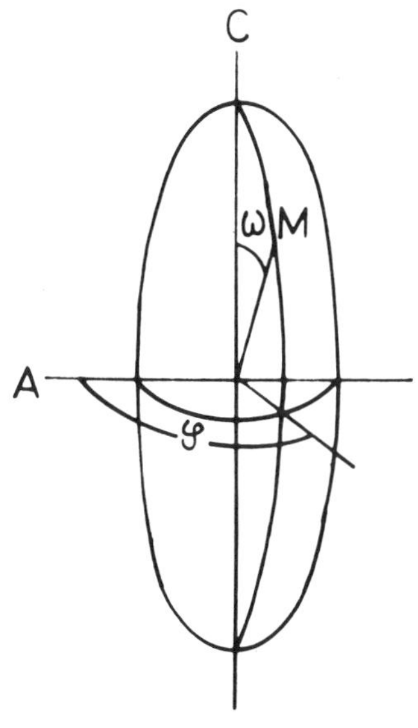

Figure 7. Schematic representation of the distribution function $N(\omega, \varphi)$ for the prolate ellispoid of rotation.

The final equations representing $I_{//}$ and $I_{\perp}$ include λ, Γ (angle of sample rotation) and θ (angle between a preferred orientation axis and the X-Z plane of the optical coordinate system as shown in Figure 3) as parameters. Calculations were carried out on the angular intensity distribution of both the polarized fluorescent components for various values of λ. Results of the calculated intensities at $\lambda = 5$ are shown by the dotted line for each component in Figure 6. These results agree well with the observed angular distributions of $I_{//}$ and $I_{\perp}$. This indicates distinctly that the nematic state have a more or less complete orientation of the molecules with their long axes parallel and that the molecular orientation can be approximately represented by the uniaxial prolate ellipsoid of rotation with the axial ratio of about 5. This is the results obtained from the aligned nematic state covering the area of about 100 μ in diameter. It is expected that an almost perfect alignment of the nematic molecules along a preferred axis would be obtained if the measurements of the angular distribution of polarized fluorescent components are made on a area much smaller than that examined in this study. However, the fact that the axial ratio is nearly equal to 5 indicates a nearly parallel alignment over regions much larger than the molecular dimensions.

OOBA exhibits another mesomorphic state, i.e., smectic state of C type, at temperatures between 102°C and 108°C in addition to the nematic state. It is well-known that the smectic C state has such a layer structure that the long molecular axis is tilted to a layer and shows optical properties characteristic of biaxial crystals (6 - 10). The observed angular distributions of $I_{//}$ and $I_{\perp}$ for the smectic C state of OOBA are strikingly different from those for the nematic state, as is shown in Figure 6. Both $I_{//}$ and $I_{\perp}$ are very little dependent on the angle of rotation (Γ). Microscopic observations on OOBA in the liquid crystalline state indicated that its smectic C state exhibits the smectic schlieren textures (11, 12), while the nematic state gives the patterns characteristic of the so-called aligned thin layers due to surface action (13). This suggests that the smectic C state can not be aligned by rubbing glass surfaces, in contrast to the nematic state. The observed patterns for the smectic C state of OOBA show that the molecular orientation is slightly uniaxial and that the directions of the long molecular axes vary successively around a preferred orientation axis. This can not be explained by using the tilted smectic structure with optically properties of biaxial crystals. For the molecular arrangement of the smectic C state, another structural model has been suggested by Saupe (14) and supported by Sakagami, et al. (15). This is a twisted smectic structure, in which the molecules are arranged in layers and have their long axes successively twisted with respect to the normal layer. If this is the case, the observed $I_{//}$ and $I_{\perp}$, which are substantially independent of the rotation angle, are reasonable, since the long molecular axes in the twisted smectic layers are likely to be essentially distributed at random over the whole region where the polarization measurement of the fluorescence from the system is made.

The polarized fluorescence intensity decreases at the phase transition from the smectic C to the crystalline state of OOBA. The angular dependencies of both polarized components are significantly different from those observed for the nematic and smectic C states, as is evident from Figure 6. These patterns of the crystalline state indicate definitely a nearly perfect biaxial molecular orientation with an axial ratio equal to about 1 and with the axes perpendicular to each other.

Other nematic compounds like AOBA, HOBA and BEPC yield nearly the same fluorescence patterns as those observed for the nematic state of OOBA. Mesomorphic compounds, which exhibit cholesteric state, are also studied by using the fluorescence polarization method. Figure 8 shows the angular distribution patterns of the polarized components of fluorescence obtained from each state of cholesteryl-n-nonanoate (CN) containing a small amount of the fluorescent compound (DH). The cholesteric state gives the patterns similar to the isotropic state. It can be explained by considering that the major axes of cholesteric molecules are arranged parallel to one another within a layer but twisted successively from layer to layer.

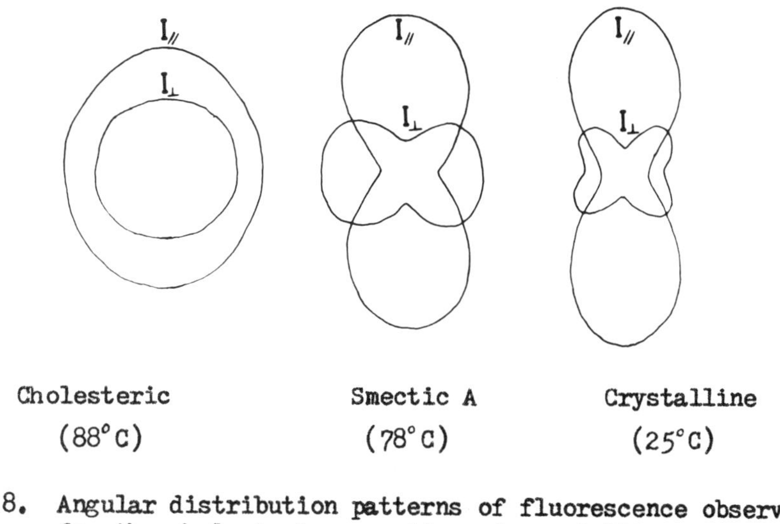

Figure 8. Angular distribution patterns of fluorescence observed for the cholesteric, smectic and crystalline states of cholesteryl-n-nonanoate.

The preferred orientation axes of cholesteric molecules are uniformly distributed in all directions perpendicular to the negative optic axis of the cholesteric state when viewed from the direction parallel to the optic axis.

Cholesteryl-n-nonanoate adopts a smectic A state at temperature between 74°C and 81°C. The observed angular distributions of the polarized components of fluorescence in Figure 8 indicate a partial uniaxial alignment of the molecules, in which a preferred orientation axis is slightly tilted to the glass plates. It seems that this does not coincide with the structure of smectic A state, in which the molecules are arranged completely parallel to the preferred axis and perpendicular to the layer planes. The crystalline state of cholesteryl-n-nonanoate gives angular distribution patterns characteristic of a perfect parallel alignment of the molecules along a preferred axis, which lies parallel to the glass plates.

It is evident that the angular distribution patterns of polarized fluorescent components give useful information on the molecular orientation in liquid crystals and can be employed as a promising tool to characterize ordered fluids. It was assumed throughout this study that the molecular axes of the fluorescent molecules always align parallel to those of the liquid crystalline molecules and that the lifetime of the fluorescent molecule in the first excited singlet state is much shorter than its relaxation time of rotation, because of a high viscosity of the mesophase. The observed angular distribution patterns of fluorescence are closely related to the molecular orientation in liquid crystals. This study is still in progress to find the influence of the type of fluorescent molecule on the fluorescence patterns which are obtained from the liquid crystalline matrices.

The authors wish to thank Professor Yasunori Nishijima of Kyoto University for his valuable suggestions.

REFERENCES

1. G. H. Heilmeier, J. A. Castellano and L. A. Zanoni, Mol. Cryst. Liquid Cryst., 8, 293 (1969)
G. H. Heilmeier and L. A. Zanoni, Appl. Phys. Letters, 13, 91 (1968)
2. Y. Nishijima, Y. Onogi and T. Asai, J. Polymer Sci., Part C, No. 15, 237 (1966)
Y. Nishijima, ibid., Part C, No. 31, 353 (1970)
3. P. Pringsheim, "Fluorescence and Phosphorescence", Interscience, New York (1949)

4. P. P. Feofilov, "The Physical Basis of Polarized Emission", Consultants Bureau, New York (1961)
5. Y. Nishijima, Ber. Bunsen-gesellschaft, 74, 778 (1970)
6. G. R. Luckhurst and F. Sundholm, Mol. Phys., 21, 349 (1971)
7. S. Diele, P. Brand, and H. Sackmann, Mol. Cryst. Liquid Cryst., 16, 105 (1972)
8. J. G. Chistyakov and W. M. Chaikowsky, ibid., 7, 269 (1969)
9. T. R. Taylor, J. L. Fergason, and S. L. Arora, Phys. Rev. Lett., 24, 359 (1970)
10. T. R. Taylor, S. L. Arora, and J. L. Fergason, ibid., 25, 722 (1970)
11. H. Sackmann and D. Demus, Mol. Cryst., 2, 81 (1966)
12. E. F. Carr, Mol. Cryst. Liquid Cryst., 13, 27 (1971)
13. A. Saupe, Mol. Cryst. Liquid Cryst., 16, 87 (1972)
14. A. Saupe, Mol. Cryst. Liquid Cryst., 7, 59 (1969)
15. S. Sakagami, A. Takase, M. Nakamizo, and H. Kakiyama, Mol. Cryst. Liquid Cryst., 19, 303 (1973).

MOLECULAR ORDER AND MOLECULAR THEORIES OF LIQUID CRYSTALS

W. L. McMillan

Dept. of Physics and Materials Research Laboratory

University of Illinois, Urbana, Illinois

I want to review for you the molecular theories of liquid crystals which have been developed recently. You are all familiar with the Maier-Saupe (1) theory of the nematic phase. Kobayashi (2) and I (3) independently formulated a theory of the smectic A phase which has proved quite useful. This year I (4) worked out a molecular theory of the smectic C phase and my student, Bob Meyer, (5) has extended that theory to the smectic B and H phases.

Let me first tell you what a molecular theory is and then review some of the results. In order to do a molecular theory of a particular phase one must first understand the molecular order in that phase. These liquid crystal phases are characterized by the positional and orientational order of the molecules as described in Table I.

	Isotropic liquid	Nematic	Smectic A	Smectic C	Smectic B	Smectic H
long axes parallel?	no	yes	yes	yes	yes	yes
smectic planes?	no	no	yes	yes	yes	yes
dipoles alligned?	no	no	no	yes	no	yes
2-D hexagonal lattice?	no	no	no	no	yes	yes

Table I. Molecular order in the various liquid crystal phases.

Next one must choose the intermolecular potential. The calculational methods which we will use are not very sophisticated and we cannot handle hard core potentials. Thus we are unable to treat realistic intermolecular potentials and the potentials are usually chosen for mathematical convenience and are unrelated to the chemical structure of the molecules. It should be obvious that the more realistic potential one is able to use the more meaningful one's results are likely to be. Only in the smectic C case where we use the molecular dipole-dipole interaction is the connection with the chemical structure established. Having settled on a potential we now use classical statistical mechanics with the self-consistent field approximation to calculate the properties of our model. In fact we derive self-consistency equations for the order parameters and these equations must usually be solved on a computer. The information that we get out of such a calculation is a) the phase diagram, that is the transition temperatures between the various phases as a function of the parameters of the potential, b) the thermodynamic properties, the entropy and specific heat as a function of temperature and c) the temperature dependence of the order parameters.

Now for the results. The first model I want to discuss encompasses the smectic A, nematic and isotropic liquid phases. The intermolecular potential was chosen for convenience to be

$$U_{12} = \frac{V_o\, e^{-(r_{12}/r_o)^2}}{N r_o^3 \pi^{3/2}} \left(\frac{3}{2}\cos^2\theta_{12} - \frac{1}{2}\right) \qquad (1)$$

where r_{12} is the distance between molecular centers and θ_{12} is the angle between the long molecular axes. V_o is the strength of the potential and r_o is the range. The two order parameters are the Maier-Saupe orientational order parameter

$$S = \langle\, 3/2 \cos^2\theta - 1/2 \,\rangle \qquad (2)$$

which indicates how well alligned the long axes are and the smectic order parameter

$$\tau = \langle\, \cos\,(2\pi z/d) \,\rangle \qquad (3)$$

which indicates the strength of the smectic planes. Here θ is the angle between the long axis of a molecule and the z axis and d is the smectic interplanar spacing. In Figure 1 I show a sketch of the phase diagram of transition temperatures versus molecular length. The light line indicates a second order phase transition where the smectic order parameter goes continuously to zero and there is no latent heat. The heavy lines indicate first order phase transitions where the order parameters drop discontinuously to zero and there is a latent heat.

The temperature dependence of the order parameters are sketched in Figure 2 for the second order case.

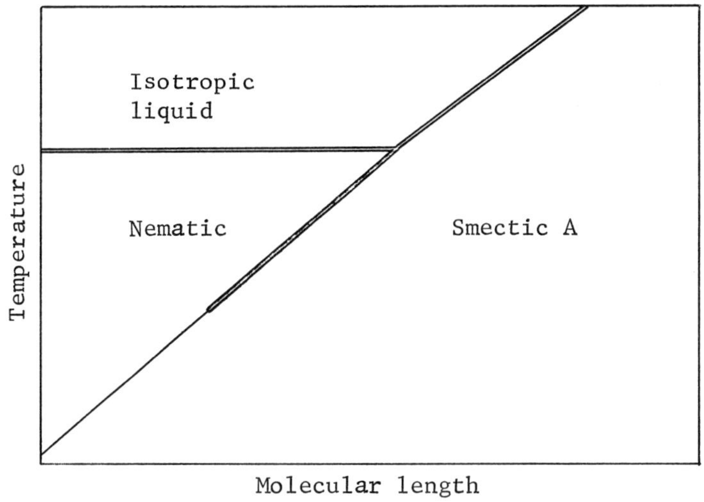

Figure 1. Sketch of the phase diagram of the smectic A model.

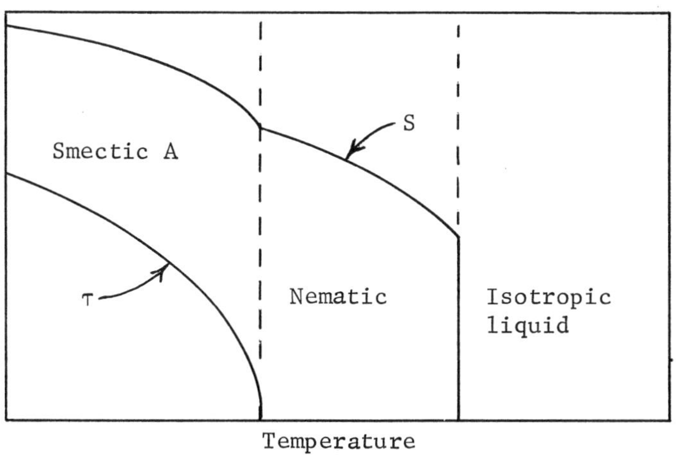

Figure 2. Sketch of the temperature dependence of the order parameters in the smectic A model.

One of the more interesting predictions of the theory is the existence of the second order region of the smectic A-nematic phase transition. This second order region has been confirmed experimentally and is under intensive study at the moment. When one bases one's theory on an ad hoc potential such as Eq. 1 it is important to have rigorous experimental tests of the theory. The Maier-Saupe order parameter can be measured with NMR (6) and the smectic order parameter can be measured by Bragg scattering of X-rays (7). Both of these measurements as well as latent heat measurements (8) show that the theory is qualitatively correct and in some regions quantitatively correct.

The second model assumes that the smectic A order is well established and that the long axes are parallel and the molecules sit on planes. The model examines the motions of molecules in one smectic plane and encompasses the four smectic phases. The potential is chosen to be the dipole-dipole interaction of the molecular electric dipoles plus a soft core repulsion. Liquid crystal molecules usually have two or three dipoles associated with amide, ether or ester bonds. For example, terephthal-bis-butylaniline has two oppositely directed dipoles above and below the molecular center and approximately perpendicular to long axis. According to the model the dipoles are alligned in the smectic C and H phases and the arrangement of the dipoles is shown in Figure 3. There is no net dipole moment and these phases are not ferroelectric. If the dipoles are alligned the molecules will tilt over to maximize the dipole-dipole interaction as sketched. This accounts for the tilt in the smectic C and H phases. The molecules rotate freely about the long axis and the dipoles are not alligned in the smectic A and B phases. The phase diagram is sketched in Figure 4.

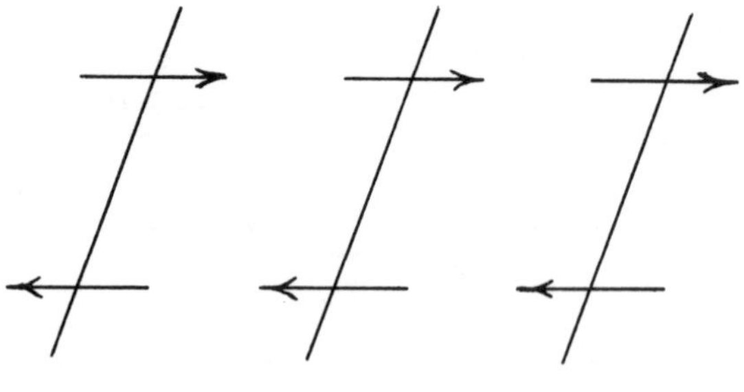

Figure 3. Sketch of the arrangement of dipoles in the smectic C phase.

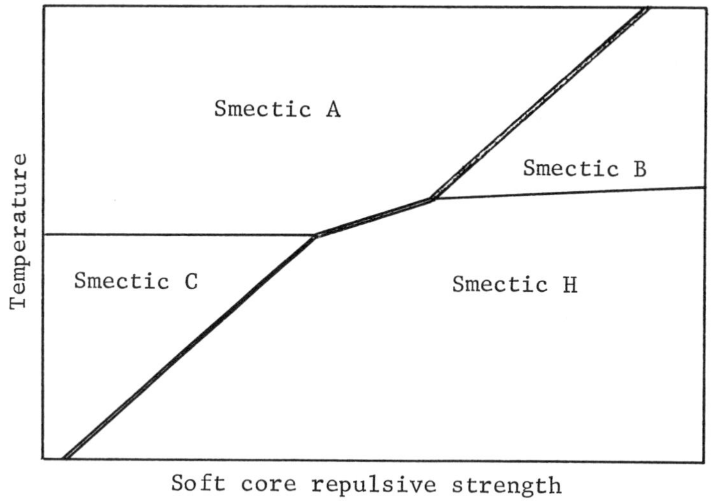

Figure 4. Sketch of the phase diagram of the second model.

The smectic A-smectic C and the smectic B-smectic H transitions are second order. The theory is brand new and the experimental testing is not yet under way. It has not even been confirmed experimentally that the dipoles are alligned in the smectic C and H phases.

In conclusion, I have tried to sketch for you the development of one aspect of the study of liquid crystals, the molecular theory approach. This approach might have been a sterile mathematical exercise; in fact it has been very fruitful and we are fortunate that our simple theories have yielded so much insight into liquid crystals. We can look forward to the next stage of experimental study and to the development of more quantitative theories.

REFERENCES

1. W. Maier and A. Saupe, Z. Naturforsch. 13A 564 (1958); 14A 882 (1959); 15A, 287 (1960).

2. K. K. Kobayashi, Phys. Letters 31A, 125 (1970); J. Phys. Soc. Japan 29 101 (1970).

3. W. L. McMillan, Phys. Rev. A4, 1238 (1971).

4. W. L. McMillan, Phys. Rev., in press.

5. R. J. Meyer and W. L. McMillan, to be published.

6. J. W. Doane, R. S. Parker, B. Cvikl, D. L. Johnson and D. L. Fishel, to be published.

7. W. L. McMillan, Phys. Rev. A6, 936 (1972).

8. P. Durek, J. Baturic-Rubcic, S. Marcelja, and J. W. Doane, to be published.

This work was supported in part by U.S. Army grant DAHC-15-73-G-10.

A NEW LYOTROPIC NEMATIC MESOPHASE

Robert C. Long, Jr. and J. H. Goldstein

Chemistry Department, Emory University

Atlanta, Georgia 30322

INTRODUCTION

Several years ago Lawson and Flautt reported a new type of nematic phase suitable for use as an ordering solvent in NMR spectroscopy. This lyotropic phase is a quaternary system composed of sodium decyl sulfate, decanol, sodium sulfate and water, with corresponding solvent capabilities not generally exhibited by thermotropic liquid crystals. In addition it possesses the property, of special interest in NMR spectroscopy, that its ordering is not disturbed when it is spun in a non-solenoidal magnetic field. A number of investigators have employed this ordering medium for the determination of molecular structures and orientations of various solutes using NMR methods applicable to anisotropic media.[1]

The availability to NMR spectroscopists of additional lyotropic nematic solvents would be potentially very advantageous for several reasons. For example, the orientation of solutes would be expected to vary with the medium, particularly in the case of hydrophilic molecules, to a greater extent than is possible simply through minor composition changes in a given quaternary system. Not only are such variations inherently interesting, but also they can in certain cases lead to elimination of ambiguities in values of dipolar couplings which in turn complicate the problem of structure determination. There are, in addition, the obvious advantages arising from differences in temperature ranges, pH and solvent characteristics, whenever these do in fact occur.

With these considerations in mind, we have now prepared a new quaternary nematic system based on potassium laurate as the amphiphile and containing, in addition, decanol, potassium chloride and D_2O. The temperature range over which the nematic phase is stable, as well as the texture of the system, have been determined by thermal microscopy. The ordering of the water component has been followed by observing the quadrupolar splitting in the magnetic resonance spectrum of D_2O.

The use in NMR spectroscopy of a nematic solvent whose amphiphile contains the carboxylate head group leads to some interesting consequences. Previous work here with the sodium decyl sulfate system has shown that while planar lipophilic solutes in this medium generally orient with their planes perpendicular to the applied magnetic field, the orientation of hydrophilic solutes is more variable and unpredictable.[2] It is possible that the behavior of this latter class of solutes is related to their greater sensitivity to conditions at the interface between the polar head group of the amphiphile and the surrounding aqueous material. Under such circumstances the transition from sulfate to carboxylate head groups would be expected to perceptibly affect the extent and the nature of orientation of at least some hydrophilic solutes but not, in general, of lipophilic solutes. Several examples are described of NMR studies carried out in the laurate system which serve to illustrate the extent to which the above considerations are borne out.

EXPERIMENTAL

The lauric acid used was material of guaranteed purity supplied by Aldrich Chemical Co. Decanol was distilled, its hydroxyl proton exchanged by shaking with D_2O and the resulting material dried and redistilled. The potassium chloride was dried and stored under vacuum. Potassium laurate was prepared from lauric acid by neutralization with KOH and was recrystallized from hot anhydrous ethanol. The liquid crystal was prepared from the components following the technique described by Black, Lawson and Flautt. The resulting material was homogeneous, of a stringy consistency and flowed easily. Storage of the material for as long as three months did not lead to phase separation.

Most of the microscopic examination of the laurate system was carried out using a small sample holder, expecially designed to minimize water loss (and subsequent compositional changes) and to fit into a Mettler FP-52 hot stage linked to a programmable temperature controller.

For NMR studies of dissolved solutes the samples were made up to approximately 1% w/w, and were then conditioned in the magnetic field of the spectrometer for a number of hours before use. During the spectroscopic studies, temperature was controlled by placing a thermocouple as close to the sample as possible, and both the thermocouple and the spectrometer temperature controller were calibrated against the platinum resistance thermometer of the Mettler FP-52. Temperatures of the samples were maintained to within $\pm$ 1.0°C. All spectra were taken on a Bruker HFX-90 spectrometer.

CHARACTERIZATION OF THE PHASE

Potassium laurate-water systems possess the advantage that their neat and middle phases can be observed at room temperature. The same phases are present when KCl is added to the system.[4] As pointed out by Rosevear, mesophase systems exhibiting nematic properties are sometimes found on the higher-water and higher-electrolyte side of the middle phase.[5] We originally prepared mixtures of potassium laurate (KL), potassium chloride (KCl) and deuterium oxide (D_2O) of composition 34.6%, 3.0% and 64.4%, respectively. The resulting solution was brittle and would not flow under the influence of gravity, characteristics possessed by middle soap phases.[6] It was found that addition of decanol decreased the viscosity of the mixture. Examination of this material with the polarizing microscope revealed the "schlieren" or "threaded" microscopic texture. This texture differentiates the nematic from the neat, middle, and isotropic phases.[5] A typical photomicrograph is shown in Fig. 1.

Samples were prepared in the cell arrangement described previously (4.0% KCl, 30% KL, 60% D_2O and 6.0%

Figure 1. "Threaded texture" of the laurate mesophase.

decanol, by weight) and the microscopic texture examined as a function of temperature. The characteristic nematic texture extends over the range of temperature from ∿283°K to 342°K. At 342°K the mesophase separates into two phases, clearly visible in the absence of the analyzer. Insertion of the analyzer reveals "Batonnets" and in some cases "oily streaks" similar to those observed by Rosevear for the middle phase dispersed in isotropic liquid.[6] Lowering the cell temperature rapidly to 298°K brings on the formation of droplets of nematic phase exhibiting distorted extinction crosses. These grow in size and coalesce, giving the characteristic "sinuous" nematic texture. If the material is left undisturbed in the nematic state the field of view becomes dark. The nematic pattern can be regenerated by placing the cell in a magnetic field or by the application of slight pressure to the cell surface. The nematic threads appear sharp if the sample is viewed immediately after removal from the magnetic field. As time progresses the threads broaden leading to dark areas which eventually increase in size until the field is dark. Increasing the temperature hastens this process. At 297 ± 1°K the pD of the laurate phase was determined to be 11.3 as compared with a pD value of 7.2 measured for the decyl sulfate system.

DEUTERON MAGNETIC RESONANCE SPECTRA

The deuteron spectrum of the laurate mesophase, when first placed in the magnetic field, exhibits the "powder-type" structure previously reported in other ordered systems.[7,8] The frequency separation of the two principal peaks of this pattern is given, to first order, by the expression $\Delta\nu_Q = (3/4h)e^2qQ$, where e^2qQ is the quadrupole coupling constant of the deuteron, under the assumption of an axially symmetric electric field gradient at the deuteron. More refined treatments have been described but are not required for the purpose of this study.[9] On remaining in the field, the spectral pattern is transformed into a well-defined doublet corresponding to a uniform ordering of the axis of the electric field gradients relative to the magnetic field direction. The broad doublet pattern is similar to that found in a crystalline powder with random orientation. In isotropic liquid media, however, the quadrupole coupling averages to zero and the pattern collapses to a single line.

Fig. 2A shows the deuteron spectrum of the nematic laurate phase, taken immediately after the sample is placed in the magnetic field. The dissymetry in the pattern reveals the progress of ordering by the magnetic field during the course of a single spectral scan. After several minutes of non-spinning orientation, the sharper doublet of Fig. 2B is obtained, indicating the existence of uniform orientation in the phase.

Rotation of the sample through 90° leads to reappearance of the powder pattern. If the sample is then spun, the sharp doublet quickly reappears and persists with further spinning. This behavior is identical in all respects with that observed in the decyl sulfate system.[3] The doublet separation (for a sample of 4% KCl, 30%KL, 60% D_2O and 6% decanol) was ∿450 Hz, quite similar to the value obtained in the decyl sulfate system.[3]

Deuteron spectra have been observed by Charvolin and Rigny at 90°C for the neat (lamellar) and middle (hexagonal) phases of the binary system composed of potassium laurate and D_2O.[8] In both cases the spectra consist of a quadrupolar doublet centered around a central D_2O peak, indicating the presence of two forms of water, exchanging at a rate which is slow compared to $1/\Delta\nu_Q$. Significantly, the nematic phase of the new quaternary system does not show the central peak below about 336°K.

The values of $\Delta\nu_Q$ for the binary system ranged from $\sim 1.5 \times 10^3$ to 3.0×10^3 Hz (neat) and $\sim 0.6 \times 10^3$ to 1.0×10^3 Hz (middle). The corresponding doublet separation in D_2O ice is $\sim 2 \times 10^5$ Hz,[10] indicating, at least qualitatively, that the quadrupolar splitting in all the laurate systems has been considerably reduced by motion of the D_2O relative to the constraints of the medium.

Fig. 3 shows the variation of $\Delta\nu_Q$ for the quaternary system with temperature. The doublet separation is roughly constant up to 336°K at which point the central D_2O peak appears, indicating the presence of the isotropic phase.

The temperature at which the central peak is first observed, 336°K, is 6° below the temperature at which the system is observed through microscopy to separate

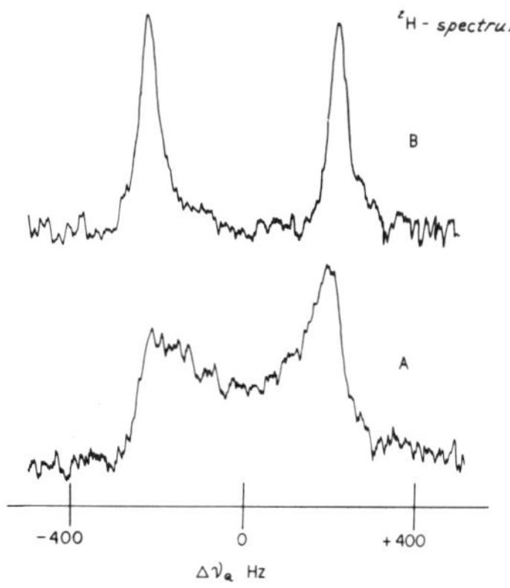

Figure 2. Deuterium spectrum of the D_2O in the potassium laurate mesophase. A) immediately after placing in the magnetic field; B) several minutes later (non-spinning).

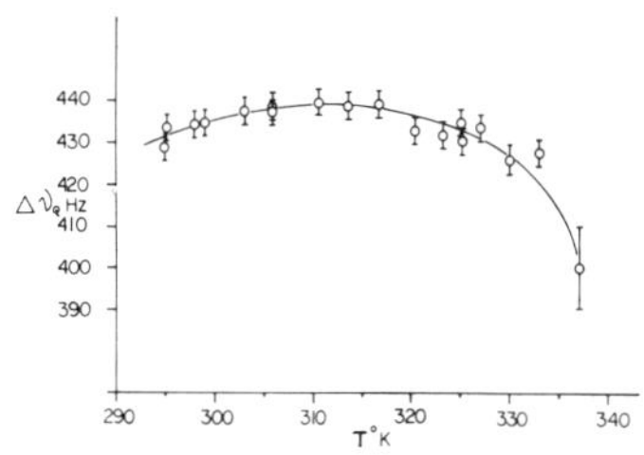

Figure 3. Plot of $\Delta\nu_Q$ vs T for the laurate mesophase.

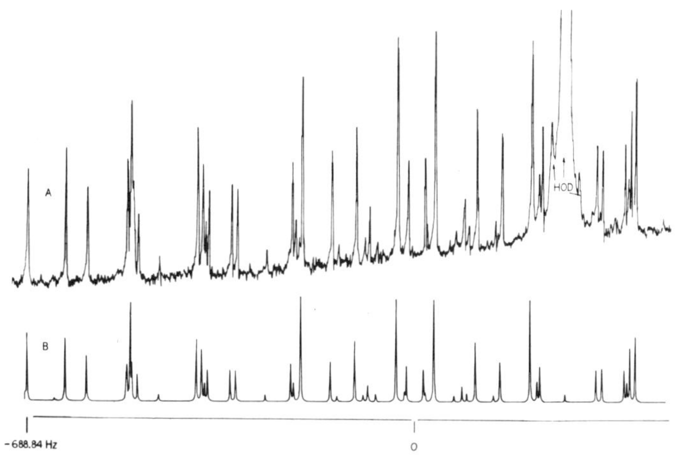

Figure 4. Experimental and calculated spectrum of benzene in the laurate mesophase (128 scans, ∿.5% w/w, 295°K).

into two distinct phases. Analogous observations have been reported for other lyotropic systems.[7] It would appear that the behavior of the system above 336°K reflects the presence of two phases, one isotropic and the other anisotropic, with D_2O exchange between the two occurring at a slow rate.

PROTON SPECTRA OF SOLUTES

Figs. 4 and 5 show the proton spectra of benzene (∿ .5% w/w, 128 accumulations) and pyridazine (1.1% w/w, single scan), both taken at 295°K. Fig. 5 also shows the single-scan pyridazine spectrum obtained in the decyl sulfate system (0.68% w/w). These spectra illustrate the resolution obtainable for both hydrophilic and lipophilic materials.

In the case of benzene the spectral parameters which reproduce the observed spectrum to an rms error of .22 Hz are: J_{ortho} = 7.53 $\pm$ 0.02 Hz, D_{ortho} = 365.22 $\pm$ 0.01 Hz; J_{meta} = 1.36 $\pm$ 0.02 Hz, D_{meta} = 70.93 $\pm$ 0.01 Hz; J_{para} = .86 $\pm$ 0.02 Hz, D_{para} = 46.20 $\pm$ 0.01 Hz.

The calculated spectrum is shown in the lower portion of Fig. 4. The dipolar couplings determined by Black, Lawson and Flautt[11] for benzene partially oriented in the decyl sulfate phase are D_{ortho} = 330.80 $\pm$ 0.03 Hz, D_{meta} = 64.51 $\pm$ 0.03 Hz, D_{para} = 42.10 $\pm$ 0.04 Hz. The ratios of the dipolar couplings are 1: 0.1942: 0.1265 and 1: 0.1950: 0.1273 for the laurate and decyl sulfate phases, respectively. These results are in good agreement with the ratios calculated using hexagonal geometry, 1: 0.1925: 0.1250. Assuming C-C and C-H bond lengths of 1.39 Å and 1.08 Å, respectively. The motional constant, $C_{3z^2-r^2}$, for benzene partially oriented in the decyl sulfate and laurate phases is + .09283 and + .10250, respectively. The relative orientation of the C_6 symmetry axis to the magnetic field direction can be expressed by the probability function introduced by Snyder,[12] which for the two phases takes the explicit form shown below:

$P(\theta,\emptyset) = 0.0713 + 0.0248 \cos^2\theta$ (decyl sulfate phase)

$P(\theta,\emptyset) = 0.0705 + 0.0274 \cos^2\theta$ (laurate phase)

where $P(\theta,\emptyset)$ is the probability per unit solid angle that the C_6 axis and the magnetic field are at a rel-

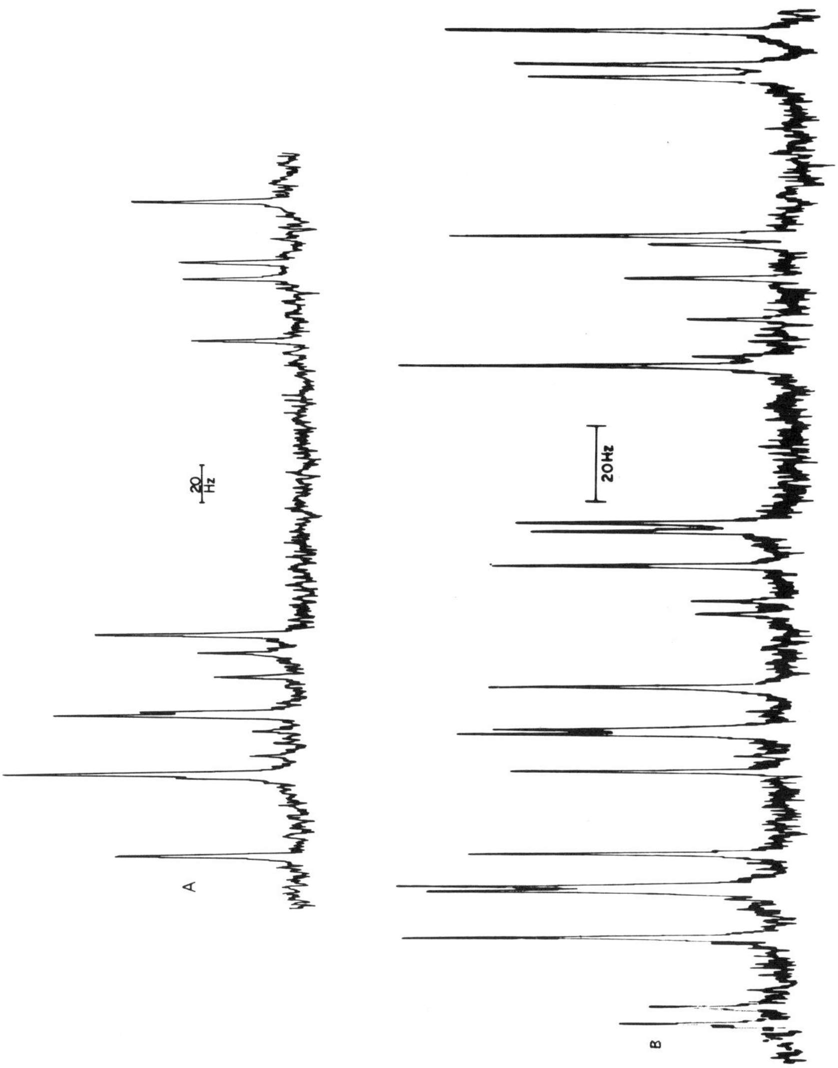

Figure 5. Spectra of pyridazine obtained in the decyl sulfate A) and laurate (B) mesophases (∿1.% w/w; single scan).

ative orientation $\theta, \emptyset$. The orientation of maximum probability for benzene in both mesophases occurs at $\theta=0$ (independent of $\emptyset$), corresponding to an applied field direction perpendicular to the molecular plane.

This result is obtained for both phases despite the fact that the molar ratio of D_2O to surfactant is somewhat different in the two systems, 18.7 and 23.8, respectively, for the decyl sulfate and laurate phases. No significant change in the orientation was found on lowering the concentration of benzene from 1.6 to .5%, w/w, although an improvement in resolution was noted.

For more hydrophilic solutes greater variations in solute ordering are observed. In the case of ethyleneimine, for example, the dipolar couplings obtained in the laurate phase at 298°K are $\mp$ 400.44 $\pm$ .10 Hz (gem), $\mp$ 8.41 $\pm$ .05 Hz (cis) and $\mp$ 31.90 $\pm$ .06 Hz (trans). The corresponding values obtained by Gazzard[13] at 309°K are $\mp$ 726.1 $\pm$ .6 Hz (gem), $\mp$ 2.9 Hz (cis) and $\mp$ 53.43 Hz (trans). These values indicate a significant difference in the degree of ordering.

The ratios of the <u>cis</u> to <u>gem</u> proton-proton distance is 1.364 $\pm$ .005 and 1.363 $\pm$ .001 for the decyl sulfate and laurate mesophases, respectively. Rapid exchange of the nitrogen-bound hydrogen with D_2O in the mesophase results in effective C_{2v} symmetry on the NMR time scale. The values r_{cis}/r_{gem} are intermediate between the ratio for <u>syn</u> and <u>anti</u> cases, 1.3572 and 1.3847, respectively, as determined by Bak and Skaarup[14] from microwave data.

Methanol exhibits a triplet pattern as a result of the anisotropic motion in the laurate system. The splitting is $\Delta\nu = 3/2|D_{HH}|$.[12] The value of $|D_{HH}|$ is 388.86 Hz (298°K) for the laurate phase as compared with 254.9 Hz (298°K) in the decyl sulfate phase.

We have previously reported the results of studies on a series of planar molecules partially oriented in the decyl sulfate phase. Among these were furan, thiophane, pyrazine, pyrimidine, and pyridazine.[2,15] For the more lipophilic molecules in this group, furan and thiophene, the plane of the ring orients in a similar manner to that for benzene. If the optical axis of the liquid crystal is perpendicular to the applied field direction, this corresponds to an orientation of the molecular planes along the liquid crystal optical axis. For planar molecules this is normally the case in thermotropic media. For the diazines in the lyotropic phase the results were somewhat more variable. The striking example was pyridazine in which the molecular plane was preferentially perpendicular to the optical axis.[2]

A re-examination of the orientations of furan, thiophene and the diazines in the laurate phase has revealed some interesing differences.

In the case of furan and thiophene the molecular planes still tend to be perpendicular to the magnetic field direction and parallel to the optical axis of the mesophase, analogous to benzene. The results for the diazines are different, as is shown in Table 1. Here the motional constants used in Snyder's notation have been converted to order parameters according to the well known relations[12]:

Table 1

S Values for several Diazines in the Decyl Sulfate (DS) and Potassium Laurate (PL) Mesophases

	(DS)[a]	(PL)
Pyrazine		
S_{xx}	+0.01835	+0.04077
S_{yy}	-0.01182	-0.00106
S_{zz}	-0.00653	-0.03971
Pyridazine		
S_{xx}	+0.01901	-0.00224
S_{yy}	-0.01768	-0.00661
S_{zz}	-0.00132	+0.00885
Pyrimidine		
S_{xx}	+0.00451	-0.01887
S_{yy}	+0.00910	+0.02138
S_{zz}	-0.01361	-0.00251

[a] Calculated from motional constant given in ref. (2).

$$C_{3z^2-r^2} = 5^{1/2}\ S_{zz}\ 2^{-1}\ (3\cos^2\alpha-1)$$

$$C_{x^2-y^2} = 5^{1/2}\ 3^{-1/2}\ (S_{xx} - S_{yy})\ 2^{-1}\ (3\cos^2\alpha-1)$$

where α is the angle between the magnetic field direction and the optical axis of the mesophase. Since the optical axis is perpendicular to the applied field direction (as would appear from the ability to spin the sample without destroying the orientation) $3\cos^2\alpha-1=-1$. The order parameters are compared in Table 1 with those obtained in the decyl sulfate.[2]

The difference in orientation of the three diazines in the two lyotropic media (decyl sulfate and laurate) can be conveniently described in terms of ratios of the order parameters as shown in the tabulation below:

Table 2

Distance Ratios for Hydrogen Atoms in the Diazines

Pyrazine[a]	(PL phase)	(DS phase)[d]	Thermotropic[e]
(r_{14}/r_{12})	$1.643\pm.002$	$1.651\pm.004$	$1.66\pm.02$
Pyrimidine[b]			
(r_{12}/r_{23})	$1.651\pm.005$	$1.642\pm.011$	$1.62\pm.01$
(r_{13}/r_{23})	$1.940\mp.005$	$1.930\mp.009$	$1.90\mp.02$
(r_{24}/r_{23})	$1.701\mp.001$	$1.701\mp.005$	$1.70\bar{6}\pm.004$
Pyridazine[c]			
(r_{12}/r_{23})	$1.000\pm.008$	$.983\pm.015$	$.988\pm.010$
(r_{13}/r_{23})	$1.701\mp.004$	$1.692\mp.008$	$1.693\mp.007$
(r_{14}/r_{23})	$1.898\mp.003$	$1.897\mp.004$	$1.890\mp.004$

a. H_1 and H_2 are ortho, H_1 and H_3 para to each other.
b. H_1 is between the two N atoms; H_1 and H_3 are para to each other.
c. The H atoms are numbered in cyclic order; H_1 is adjacent to N.
d. Ref. (2).
e. Thermotropic values for pyrazine, pyrimidine, and pyridazine from refs. 1, 16 and 17, respectively.

	S_{yy}/S_{xx} (DS)	S_{yy}/S_{xx} (PL)	S_{zz}/S_{xx} (DS)	S_{zz}/S_{xx} (PL)
pyrazine	-0.644	-0.026	-0.356	-0.974
pyridazine	-0.930	+2.951	-0.069	-3.951
pyrimidine	+2.018	-1.133	-3.018	+0.1330

(in pyrazine z is perpendicular to the plane and x is along the N-N axis. In the other two cases x is perpendicular to the plane and z bisects the N-N axis.)

From the above values it is apparent that significant differences in orientation occur for all three solutes in going from one phase to the other. The most striking effect occurs in the case of pyridazine (see also values in Table 1). Here the molecular plane tends to align parallel to the liquid crystal optical axis for the laurate phase but perpendicular to it in the decyl sulfate phase.

Table 2 summarizes the structures of the diazines determined in the laurate phase, the decyl sulfate phase and corresponding thermotropic values reported in the literature.

Acknowledgment. The authors wish to thank Dr. F. B. Rosevear and Dr. T. J. Flautt, of Proctor and Gamble Co., for helpful discussions which led to the choice of laurate as a possible amphiphile. We also thank Kathryn R. Long for help in the experimental work. This work was supported by the National Institutes of Health.

REFERENCES

1. P. Diehl and C. L. Khetrapal, NMR Basic Principles and Progress, Vol. 1, P. Diehl, E. Fluck and R. Kasfeld, Ed. (Springer-Verlag, New York, N. Y. 1969).
2. R. C. Long, Jr. and J. H. Goldstein, paper presented at Fourth International Liquid Crystal Conference, Kent, Ohio (1972); Mol. Cryst. and Liq. Cryst., in press.
3. P. J. Black, K. D. Lawson and T. J. Flautt, Mol. Cryst. and Liq. Cryst., 7, 201 (1969).
4. J. W. McBain and M. C. Field, J. Phys. Chem., 30, 1545 (1926).
5. F. B. Rosevear, J. Soc. Cosmetic Chemists, 19, 581-594 (1968).

6. F. B. Rosevear, J. Amer. Oil Chemists' Soc., 31, 628 (1954).
7. K. D. Lawson and T. J. Flautt, J. Phys. Chem., 72, 2066 (1968).
8. J. Charvolin and P. Rigny, J. de Physique, Colloque C 4, 76 (1969).
9. E. A. Lucken, Nuclear Quadrupole Coupling Constants (Academic Press, New York, 1969).
10. P. Waldstein, W. Rubideau, and J. A. Jackson, J. Chem. Phys., 41, 3407 (1964).
11. P. J. Black, K. D. Lawson and T. J. Flautt, J. Chem. Phys., 50, 542 (1969).
12. L. C. Snyder, J. Chem. Phys., 43 , 4041 (1965).
13. I. J. Gazzard, Mol. Phys., 25, 469 (1973).
14. B. Bak and S. Skaarup, J. Mol. Structure, 10, 385 (1971).
15. R. C. Long, Jr., S. L. Baughcum, and J. H. Goldstein, J. Mag. Res., 1, 253 (1972).
16. C. L. Khetrapal, A. V. Patankar, P. Diehl, Org. Mag. Res., 2, 405 (1970).
17. E. E. Burnell and C. W. DeLange, Mol. Phys., 16, 95 (1969).

NMR STUDIES OF THE INTERACTION BETWEEN SODIUM IONS AND ANIONIC SURFACTANTS IN SOME AMPHIPHILE-WATER SYSTEMS

H. Gustavsson, G. Lindblom, B. Lindman, N.-O. Persson, and H. Wennerström

Division of Physical Chemistry, The Lund Institute of Technology, Chemical Center, Lund, Sweden

INTRODUCTION

An elucidation of the interaction mechanism between small ions and ionic amphiphiles in lyotropic mesophases is of considerable chemical interest since biological membranes have been shown to have several properties in common with liquid crystals formed by ionic amphiphiles and water.[1] Furthermore, it has been observed that phase equilibria[2] and aggregate shapes[2,3] depend markedly on the nature of the counter-ion in surfactant systems; and, therefore, a better understanding of the conditions for occurrence of different types of lyotropic mesophases may be expected by probing into the ion binding in solutions containing ionic amphiphiles. Additionally, studies of the ion binding in colloidal systems may prove valuable in connection with theoretical approaches to the ion binding in polyelectrolyte solutions.

In order to obtain information on the binding of small ions in surfactant systems we have started systematic investigations of different surfactant systems using a variety of NMR methods. Owing to its negligible perturbation of the system being studied, NMR is attractive especially in connection with biological systems for which the same experimental methods are applicable as those outlined in the present paper.

The present study is mainly concerned with the sodium ion binding in lamellar mesophases formed by an anionic surfactant, water and a second amphiphile. The NMR parameters employed are alkali ion quadrupole splittings, alkali ion transverse and longitudinal relaxation rates, counter-ion chemical shifts and water deuteron and ammonium nitrogen quadrupole splittings. For some of the methods employed the investigations are not yet completed and more extensive reports will be given at a later date.

EXPERIMENTAL

The alkali ion quadrupole splittings were obtained by means of a Varian V-4200 wide-line spectrometer as described previously.[4] The ^{23}Na NMR line widths at half height of the central peak in quadrupole-split spectra were obtained with the side-band technique as described in ref. 5 with the modification that a flux stabilizer was introduced to improve the magnetic field stability. The magnetic field inhomogeneity amounted to at most 1.5 μT in our ^{23}Na NMR studies. No correction for inhomogeneity broadening or for broadening due to overmodulation or saturation has been applied. The broadening due to the two latter effects amounts to 1-2%.

^{23}Na chemical shifts were determined at 15.82 MHz using a Varian V-4200 wide-line spectrometer. In order to improve homogeneity and stability shim coils, a flux stabilizer and a frequency stabilizer were employed. External reference solutions were inserted coaxially in the sample tubes in the case of the amorphous isotropic solutions whereas for the liquid crystalline solutions the reference solutions were outside the sample. The spectra were swept in both increasing and decreasing field directions to reduce the effect from magnetic field drift. The difference in bulk susceptibility between sample and reference was estimated to give an error smaller than the precision of the experimental shifts (±0.15 ppm for the isotropic solutions). A positive chemical shift corresponds to a shift to lower field.

^{23}Na longitudinal relaxation times were determined by means of a Bruker BK-322s pulse spectrometer as described in ref. 6.

Deuteron NMR spectra were obtained as described in ref. 7. ^{14}N NMR spectra were recorded analogously at a magnetic field strength of 1.40 T.

If not otherwise stated all the measurements were performed at 27±2°C and with randomly oriented samples.

Sample preparation was achieved as described previously[4,7]. The compounds employed are abbreviated as follows: sodium n-octanoate (NaC_8), n-decanol ($C_{10}OH$), n-octanoic acid (C_8OOH), sodium n-octylsulphate (NaC_8SO_4), sodium n-octylsulphonate (NaC_8SO_3), ammonium n-octanoate (NH_4C_8) and sodium sulpho-di(2-ethylhexyl) succinic ester (Aerosol OT).

EXPERIMENTAL RESULTS

For the lamellar mesophases in the two ternary systems, sodium octanoate-decanol-heavy water and sodium octanoate-octanoic acid-heavy water, (for phase diagrams see ref. 8), we investigated the ^{23}Na NMR line width and the ^{23}Na first order quadrupole splitting as a function of the sample composition. The data obtained are illustrated in Figs. 1 and 2 where plots of the line width, $\Delta\nu_{1/2}$ and the quadrupole splitting, Δ, against R_{ionic}, the molar ratio between ionic amphiphile and total amphiphile content, at two constant molar fractions of water ($x(D_2O)$) are shown. It may be inferred from these figures that:

a) For both these systems both the line width and the quadrupole splitting decreases with increasing water content.

b) At a constant mole fraction of water (0.77 for the NaC_8-C_8OOH-D_2O system) both the line width and the quadrupole splitting are almost constant up to a value of $R_{ionic} \simeq 0.50$. For larger R_{ionic} values the line width increases with increasing R_{ionic} whereas the quadrupole splitting decreases monotonically. An analogous behaviour is displayed by the NaC_8-$C_{10}OH$-D_2O system. Here the critical value of R_{ionic} is slightly lower.

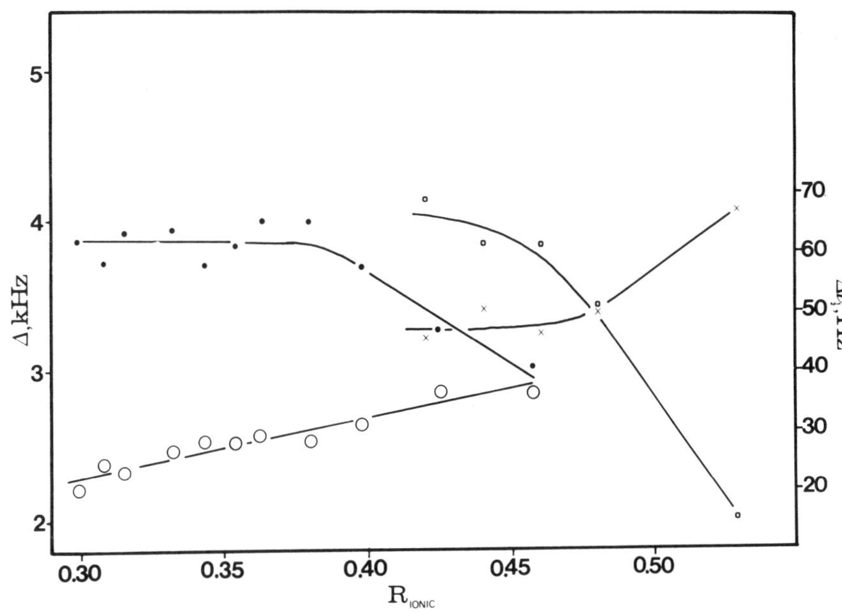

Fig. 1. ^{23}Na first order quadrupole splittings and line widths for powdered lamellar mesophase samples composed of sodium octanoate, decanol and heavy water obtained at 27°C. The splitting, Δ (in kHz), is given as a function of R_{ionic} for $x(D_2O)=0.77$ (□) and $x(D_2O)=0.83$ (•). The line width data, $\Delta\nu_{1/2}$ (in Hz), are shown at $x(D_2O)=0.77$ (X) and $x(D_2O)=0.83$ (o).

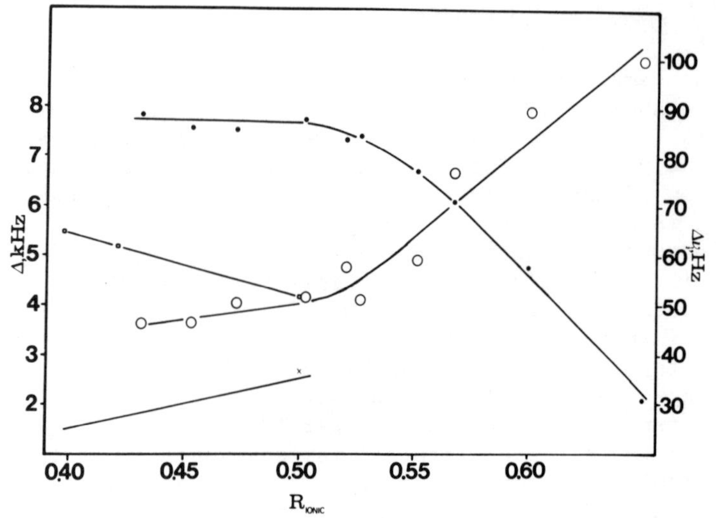

Fig. 2. ^{23}Na first-order quadrupole splittings (Δ, in kHz) and line widths ($\Delta\nu_{1/2}$, in Hz) for powdered lamellar mesophase samples composed of sodium octanoate, octanoic acid and heavy water obtained at 27°C. Quadrupole splittings are denoted by • at $x(D_2O)=0.77$ and by □ at $x(D_2O)=0.83$. Line widths are denoted by o at $x(D_2O)=0.77$ and by X at $x(D_2O)=0.83$.

c) At a constant mole fraction of water (0.83), the ^{23}Na line width increases slowly with increasing R_{ionic} for both the systems considered. The ^{23}Na quadrupole splitting decreases with increasing R_{ionic} for the NaC_8-C_8OOH-D_2O system. For the NaC_8-$C_{10}OH$-D_2O system the quadrupole splitting is essentially constant up to R_{ionic} 0.38 and then it decreases markedly as R_{ionic} is increased.

In order to investigate how the ^{23}Na quadrupole splitting depends on the end-group of the surfactant, exploratory studies were performed with the lamellar phases of the systems sodium octylsulphonate-decanol-D_2O, sodium octylsulphate-decanol-D_2O and Aerosol OT-H_2O. The ^{23}Na quadrupole splitting is for these systems much larger than when sodium octanoate is the surfactant. From Fig. 3 it may be seen that the splitting increases in the sequence octylsulphonate < octylsulphate < Aerosol OT, but is rather independent of the water content in the respective systems. In Fig. 3 also the deuteron quadrupole splitting has been included for the Aerosol OT-D_2O system. It can be seen that there is a strong and

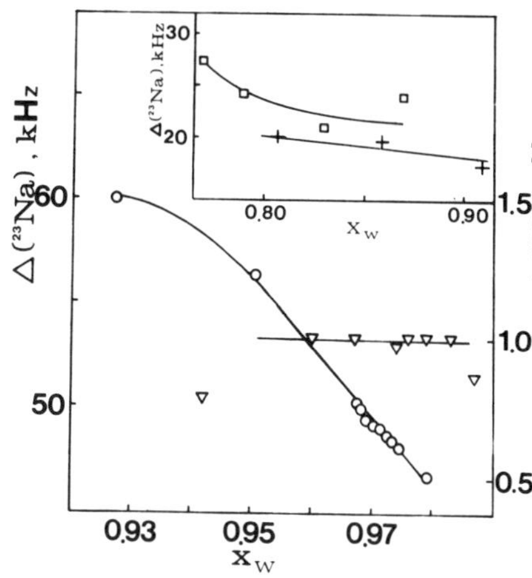

Fig. 3. ^{23}Na and ^{2}H first-order quadrupole splittings for some powdered lamellar mesophase samples as a function of water (x_w).
^{23}Na splittings for the Aerosol OT-H_2O system (▽). The splittings are referred to the left hand ordinate axis.
^{2}H splittings for the Aerosol OT-D_2O system (o). The splittings are referred to the right hand ordinate axis.
The ^{23}Na splittings of the systems NaC_8SO_4-$C_{10}OH$-D_2O (□) and NaC_8SO_3-$C_{10}OH$-D_2O (+) at R_{ionic}=0.31 are shown in the upper right corner.

regular increase in the deuteron splitting as the water content is decreased.

For reasons of interpretation (cf. below) the ^{23}Na line width studies were complemented with some pulsed NMR experiments. It was found, for example, that the ^{23}Na longitudinal relaxation time, T_1, for a lamellar sample in the NaC_8-$C_{10}OH$-H_2O system with R_{ionic}=0.41 and $x(H_2O)$=0.80 is 11.7 ms and for a reversed hexagonal sample in the same system with R_{ionic}=0.256 and $x(H_2O)$=0.63 the T_1 is 3.5 ms. For a lamellar sample in the NaC_8-C_8OOH-D_2O system with R_{ionic}=0.60 and $x(D_2O)$=0.78 T_1 amounts to 3.9 ms.

To complement the ^{23}Na quadrupole splitting studies, we have investigated lamellar mesophase samples with a nonionic surfactant (nonylphenoldecaethylene glycolether) instead of an ionic one and with sodium chloride added. With the non-ionic surfactant no ^{23}Na quadrupole splitting was observed.

In our previous investigations[3,4,9-11] only monoatomic ions were investigated by the quadrupole splitting method. In order to investigate the possible extension of the method to other types of ions, the ^{14}N NMR spectra of samples composed of NH_4C_8, $C_{10}OH$ and D_2O were recorded. Since the phase diagram of this ternary system has not been determined, the sample compositions were chosen to be well within the borders of the lamellar phase of the NaC_8-$C_{10}OH$-H_2O system. A typical ^{14}N NMR spectrum is shown in Fig. 4 and, as expected for a nucleus with a spin quantum number I=1 in an anisotropic environment, the signal is split into two component signals. As can be seen in Table I the magnitude of the ^{14}N splitting varies only moderately with the sample composition. The same observation applies to the deuteron splittings which are included for comparison in Table I.

An approach which has previously not been applied to the study of ion binding in liquid crystals is to investigate the NMR chemical shifts. Recently we demonstrated the feasibility of this method for studying counter-ion binding in micellar solutions.[12] Thus it was shown with aqueous NaC_8 solutions that at the critical micelle concentration the signal starts to move rapidly towards lower fields with increasing soap concentration. In order to obtain information on how the ^{23}Na chemical shift depends on the surfactant endpgroup we have now also investigated aqueous solutions of sodium n-octylsulphate. As may be seen in Fig. 5 the same general behaviour is found as with NaC_8 with the interesting difference that on micelle formation the signal shifts in the opposite

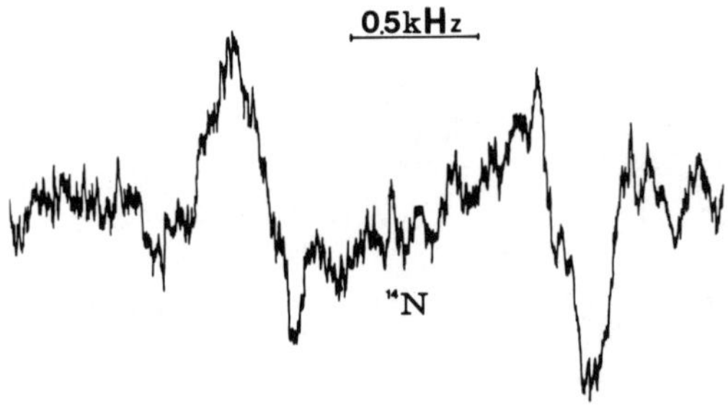

Fig. 4. ^{14}N NMR spectrum of a powdered mesophase sample composed of 11.0 mole-% ammonium octanoate, 11.9 mole-% decanol and 77.1 mole-% heavy water. Temperature: 27°C.

Table I. ^{14}N and ^{2}H quadrupole splittings of powdered mesophase samples composed of ammonium octanoate, decanol and heavy water. Temperature: 27°C.

Sample composition		Quadrupole splittings (kHz)	
R_{ionic}	$x(D_2O)$	^{14}N	^{2}H
0.480	0.77	1.33	0.60
0.500	0.77	1.30	0.51
0.521	0.77	1.25	0.48
0.400	0.80	1.59	0.65
0.454	0.80	1.34	0.54
0.500	0.80	1.19	0.47
0.519	0.80	1.41	0.46
0.420	0.81	1.42	0.62
0.454	0.81	1.29	0.52
0.454	0.83	1.55	0.49
0.480	0.83	1.04	0.48

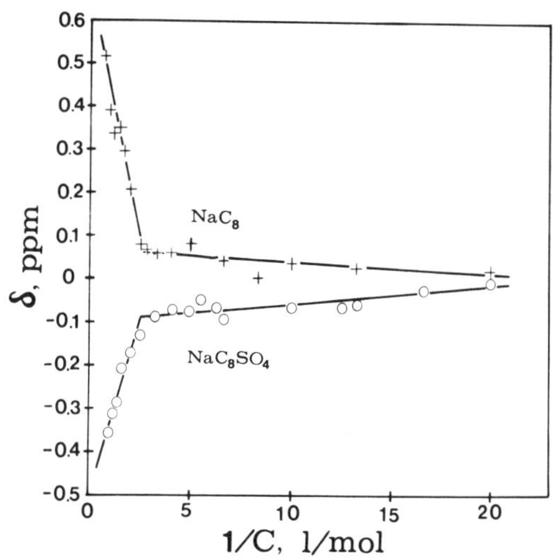

Fig. 5. ^{23}Na chemical shifts (in ppm) for amorphous isotropic solutions of sodium octanoate (X) and sodium octylsulphate (O) as a function of the inverse soap concentration (cf. ref. 12). A positive δ denotes a shift to lower field.

Table II. ^{23}Na chemical shifts for some lamellar mesophase samples. 0.5 M NaCl was used as an external reference. A positive shift corresponds to a shift to lower magnetic field. Compositions are given in mole-%.

Sample composition	Chemical shift (ppm)
83.0% D_2O, 5.1% NaC_8, 11.9% $C_{10}OH$	0.73
83.0% D_2O, 6.0% NaC_8, 11.0% $C_{10}OH$	0.82
83.0% D_2O, 7.2% NaC_8, 9.8% $C_{10}OH$	0.92
87.9% D_2O, 3.3% NaC_8, 8.8% $C_{10}OH$	0.33
77.5% D_2O, 9.9% NaC_8, 12.6% $C_{10}OH$	1.17
76.9% D_2O, 9.8% NaC_8SO_4, 13.3% $C_{10}OH$	-1.89
83.0% D_2O, 5.1% NaC_8SO_4, 11.9% $C_{10}OH$	-0.81
87.2% D_2O, 4.0% NaC_8SO_4, 8.8% $C_{10}OH$	-0.56

directions for the two soaps. Counter-ion chemical shift studies of lyotropic mesophases have recently been started and some preliminary observations are given in Table II. For the NaC_8-$C_{10}OH$-D_2O-system the chemical shift increases with increasing R_{ionic} and with decreasing $x(D_2O)$. The dependence of chemical shift on surfactant end-group is qualitatively the same for the mesophases as for the micellar solutions. Compared to our studies on isotropic solutions[12] the error is somewhat larger for the mesophases since it is difficult to find an optimal location of the external reference solution due to the procedure for preparing the mesophase samples.

Deuteron quadrupole splittings for the NH_4C_8-$C_{10}OH$-D_2O and Aerosol OT-D_2O systems are presented above. Some additional deuteron NMR results which are utilized in the discussion will be presented in full elsewhere.[13]

QUADRUPOLE EFFECTS OF SPIN-3/2 NUCLEI

The counter-ion quadrupole relaxation and quadrupole splitting studies were concerned with $^{23}Na^+$, which has the spin quantum number I=3/2. In order to have a basis for the interpretation of the experimental data it is appropriate to briefly describe the quadrupole effects for spin 3/2 particles. (For general treatments of the quadrupole interactions on which our approach is based see refs. 14 and 15.) We consider an assembly of quadrupolar nuclei in a strong external magnetic field. The hamiltonian, in frequency units, for such a system can be written

$$H = \omega_o(1-\sigma)I_z + \sum_{q',q=-2}^{2} (-1)^q V_{-q}A_{q'}D^{(2)}_{q'q}(\alpha,\beta,\gamma) \qquad (1)$$

where ω_o is the Larmor precession frequency and σ the magnetic shielding tensor. V_{-q} is a component of the irreducible electric field gradient tensor in the principal axes system at the nucleus, $A_{q'}$ is a component of a second rank irreducible spin operator in the laboratory coordinate system and $D^{(2)}_{q'q}(\alpha,\beta,\gamma)$ is a second rank Wigner rotation matrix element, where the Euler angles α, β and γ define transofrmations between the laboratory-fixed and the molecule-fixed coordinate systems.

The hamiltonian in Eq. (1) can often be separated into a time-independent and a time-dependent part

$$H = H_o + H_1(t)$$

where $$H_o = \omega_o(1-\sigma)I_z + \sum_{q,q'} (-1)^q V_{-q}A_{q'} \overline{D^{(2)}_{q'q}} \qquad (2)$$

and $$H_1(t) = \sum_{q,q'} (-1)^q V_{-q}A_{q'}(D^{(2)}_{q'q}(t) - \overline{D^{(2)}_{q'q}}) \qquad (3)$$

For an isotropic solution $\overline{D^{(2)}_{q'q}}=0$, _i.e._ the mean value of the quadrupole hamiltonian is zero, which implies that the quadrupole interaction only gives relaxation effects. In anisotropic fluids, like lamellar liquid crystals, the $\overline{D^{(2)}_{q'q}}$'s are finite. The time averages are taken over a time long compared to the inverse quadrupole splitting, $(\nu_Q \overline{D^{(2)}_{00}})^{-1}$, where

$\nu_Q = \frac{3V_0eQ}{I(2I-1)h} = \frac{3e^2qQ}{2I(2I-1)h}$. Here eQ is the quadrupole moment of the nucleus, $V_0=1/2$ eq, eq is the main component of the electric field gradient tensor in the principal coordinate system. H_o determines the positions and intensities of the signals in the NMR spectrum whereas $H_1(t)$ describes their line widths and the relaxation behaviour of the spin system.

If the quadrupole interaction is small compared to ω_o a first order perturbation calculation shows that only the spin operator $A_o=1/3(3I_z^2-I^2)$ in Eq. (2) contributes to the energy eigen-values. The difference in frequency between the transitions $m=-3/2 \rightarrow m=-1/2$

and m=3/2→m=1/2 (m being the magnetic quantum number), for a spin 3/2 nucleus being

$$\Delta_{\Omega}^{(1)} = |p \cdot \nu_Q \cdot S(3\cos^2\Omega - 1)| \qquad (4)$$

$\Delta_{\Omega}^{(1)}$ is the first-order quadrupole splitting, S is the order parameter characterizing the degree of orientation of the electric field gradient tensor and is simply related to the $\overline{D_{q'q}^{(2)}}$'s.[11] Ω is the angle between the external magnetic field and the direction of highest symmetry in the microcrystallites. (Here we are using the concept "microcrystallite" in a wide sense.) p is the fraction of nuclei located in an anisotropic environment. $\Delta_{\Omega}^{(1)}$ gives the splitting for a sample where all the microcrystallites are aligned in the same direction. For a powdered sample where all orientations of Ω are present the splitting is obtained by an integration over all the equally probable values of cos Ω. This gives

$$\Delta_{p}^{(1)} = |p \cdot \nu_Q \cdot S| \qquad (5)$$

provided it can be assumed that the nuclei are not allowed to diffuse between regions having different Ω's in times small compared to the inverse of the difference of the quadrupole splittings in the regions considered. If such an exchange between different "microcrystallites" exists, then a partial or total reduction of the quadrupole splitting results. This has also been observed experimentally.[4,11] Such an exchange process could also give rise to relaxation effects.

Using Eq. (3) for the time-dependent hamiltonian $H_1(t)$ the relaxation times can be calculated to be

$$\frac{1}{T_1} = \frac{1}{T_2} = \frac{3}{40}\left(\frac{e^2Qq}{\hbar}\right)^2 \frac{2I+3}{I^2(2I-1)} \cdot \tau_c \qquad (6)$$

in the extreme narrowing limit ($\omega_o\tau_c \ll 1$). Here τ_c is the correlation time describing the time-dependence of the electric field gradient tensor and T_1 and T_2 are the longitudinal and transverse relaxation times, respectively. Provided the line width is determined by quadrupole relaxation then, at extreme narrowing conditions,

$$\Delta\nu = K \cdot q^2 \cdot \tau_c \qquad (7)$$

where K is a constant which depends on the properties of the nucleus studied.

For a system where the nuclei exchange rapidly between different sites i, i.e. when exchange proceeds much more rapidly than the difference in relaxation rate or quadrupole splitting[x] between the sites, then the observable quantities are

$$\Delta_{obs} = \sum p_i \Delta_i \qquad (8)$$

and

$$\Delta\nu_{obs} = \sum p_i \Delta\nu_i \qquad (9)$$

Here p_i is the probability of finding the nucleus at site i.

Arguments for the applicability of Eqs. (6), (7) and (9) to surfactant systems have been given previously.[3,16] A confirmation of the applicability of the extreme narrowing condition was obtained for some mesophase samples. Thus it was found that the observed T_1 values are equal to the T_2 values calculated from the width of the central peak in the NMR spectrum.

DISCUSSION

From Figs. 1 and 2 it can be seen that the $^{23}Na^+$ transverse relaxation rate decreases with increasing water content for the two lamellar mesophases investigated. This may be ascribed to a lessened interaction of the counter-ions with the charged lamellar surfaces; but whether the variation of line width is a result of variations in the fraction of the counter-ions bound, of variations in the intrinsic relaxation rates or of both is difficult to ascertain.

Counter-ion translational diffusion coefficients have recently been employed to obtain information on the degree of counter-ion binding to micelles in aqueous sodium octanoate solutions.[17] From the diffusion coefficients the number of counter-ions diffusing with the micelles was calculated. It was found that the ratio between counter-ions and soap anions in the micelles is about 0.60 just above the critical micelle concentration and that this figure increases to about 0.7 at the highest concentrations. These figures may not be directly applicable in the interpretation of quadrupole relaxation data. However, since the alterations in both the diffusion coefficients and the quadrupole relaxation rates on soap aggregation are due to changes in the mobilities (which may or may not be of different kinds) of the counter-ions or their immediate environment, there is some justification for assuming that the

---

[x] In connection with the splittings only exchange between different sites in a given microcrystallite is considered here. Exchange between different microcrystallites was treated above.

same counter-ion association degrees appear in both types of experiments. By studying the counter-ion diffusion coefficients in lyotropic mesophases the p_i's in Eq. (9) could then be obtained.

Diffusion experiments on liquid crystals are more difficult to perform than for amorphous isotropic solutions and the data needed are presently not available. Self-diffusion coefficients may be determined by pulsed NMR rather independently of the macroscopic properties of the samples, but in the present situation it can be envisaged that the presence of static quadrupole effects (especially if they are of second-order) will cause problems. However, the stimulated echo method recently used for the elimination of static dipolar couplings in lyotropic mesophases[18] may constitute a solution to this problem. Studies of the kind indicated have been initiated. Lacking direct information on the p_i's of the liquid crystalline samples we may use the observation that on going from a concentrated micellar solution to a liquid crystalline phase, with which it may be in equilibrium, the changes in counter-ion quadrupole relaxation rates are small.[19] It is reasonable to assume then that changes in the p_i's are also small (cf. Eq. 9). On the basis of these assumptions the $^{23}Na^+$ quadrupole relaxation rates in the lamellar mesophase samples in the NaC_8-$C_{10}OH$-D_2O and NaC_8-C_8OOH-D_2O systems having the highest water contents, may be interpreted as due to about 70% of the sodium ions being bound to the charged lamellar surfaces whereas the rest should have intrinsic relaxation rates comparable to those observed in dilute aqueous solutions. It follows then from the magnitude of the observed variations in the ^{23}Na line widths that they can not be interpreted solely in terms of increases in the fraction of bound counter-ions as $x(D_2O)$ is decreased or as R_{ionic} is increased but that changes in the relaxation rates of the bound counter-ions must occur when the sample composition is varied. A possible interpretation of our relaxation data is therefore that in both of the systems considered there is an enforced interaction between the counter-ions and the lamellae as the water mole fraction is decreased and as the ratio between ionic and non-ionic amphiphile is increased. It is interesting to note from Figs. 1 and 2 that at low water contents the increase in line width starts to be considerable at a rather well defined R_{ionic} close to 0.5. This may be due to some type of complex formation between the ionic and non-ionic amphiphiles in the lamellae.

Our previous studies of counter-ion quadrupole relaxation in surfactant systems[19,20] strongly indicate that water motion is causing relaxation. This conclusion is based on a comparison of the relaxation and the energy of activation of the relaxation process in different phases. Substitution of heavy water for ordinary water gave an increase in line width by about 20%. This is consistent with the conclusion above. A possible interpretation of the observed increase in ^{23}Na line width when $R_{ionic} \simeq 0.5$ is one in terms of a retardation of water molecular motion. This interpretation is consistent with water deuteron quadrupole splitting studies.[13] Calculations from X-ray data also show that the thickness of the water layers in this region falls to small values, 8-10Å, and that the whole water content can be regarded as "bound" to the lamellar surfaces.[2]

Interpretation of ^{23}Na quadrupole splittings for powdered samples is complicated by the presence of exchange effects.[4] If the dependence of the splittings on sample composition is due to exchange phenomena it can be seen that as R_{ionic} is increased above ca. 0.5 the exchange of counter-ions between domains with different orientations of the lamellae with respect to the magnetic field increases considerably. Since the required information on the microcrystallites is not available it is not possible to decide whether these observations should be attributed to a decrease in microcrystallite size (or increase in lamellar curvature) or to an increased counter-ion translational mobility. It is significant that exchange between different microcrystallites is not observed in deuteron NMR spectra. This indicates that the observed effect is not due to a marked microcrystallite size reduction.

Studies of the ^{23}Na quadrupole splittings, and their temperature dependence, for oriented samples are being undertaken but have yet not given an unambiguous solution to the problem of interpretation. Preliminary observations indicate, however, that at high R_{ionic} values for the NaC_8-$C_{10}OH$-D_2O and NaC_8-C_8OOH-D_2O systems the splitting for an oriented sample with $\Omega=90^o$ may be considerably larger than the splitting for a powdered sample.

It is interesting to note that ^{14}N NMR may conveniently be applied to study counter-ion binding in liquid crystals by the quadrupole splitting method (Fig. 4). It should be noted that over a wide range of concentrations both the ^{14}N and the ^{2}H static quadrupole interactions are nearly independent of sample composition in the NH_4C_8-$C_{10}OH$-D_2O system. No attempt has yet been made to distinguish the effects of the different locations of the deuterons on the deuteron splittings.

It was noted above that the sodium ion exchange between different microcrystallites is rapid enough to partially or completely eliminate the static quadrupole interactions. Since for the same samples we observe no exchange effect on the water deuteron quadrupole splitting, it can be inferred that this exchange process occurs more rapidly for the sodium ions than for the water molecules. The information about the translational mobility of ions from quadrupole splittings can have important implications for biological systems. For example, it is possible that lateral transport of ions is an important process in nerve conduction along the axon membrane.

Applying the present method on biological model membrane studies might reveal which factors are important in facilitating or hindering ion lateral diffusion along a membrane surface.

One possibility for obtaining information on the mode of interaction between the counter-ions and the surfactant is to investigate how the interaction is modified with variations in the surfactant end-group. It is interesting to note that several of the NMR parameters considered are quite different for alkanoates and alkylsulphates. Thus the following observations were made:
a) The ^{23}Na NMR line widths observed for lamellar mesophases samples at comparable compositions are considerably larger for the $NaC_8-C_{10}OH-D_2O$ than for the $NaC_8SO_4-C_{10}OH-D_2O$ and $NaC_8SO_3-C_{10}OH-D_2O$ systems.
b) The ^{23}Na quadrupole splittings are generally an order of magnitude larger for the $NaC_8SO_3-C_{10}OH-D_2O$ and $NaC_8SO_4-C_{10}OH-D_2O$ systems than for the $NaC_8-C_{10}OH-D_2O$ system. It should be noted that a difference between the two types of end-groups is also observed for oriented samples. (The fact that we observed no splitting when a non-ionic surfactant was used makes it reasonable to ascribe the splittings to interactions of the counter-ions with the surfactant end-groups.)
c) The positions of ^{23}Na NMR signal shifts in different directions on micelle formation for sodium octylsulphate and sodium octanoate (Fig. 5). This marked difference in chemical shift pertains also to lamellar mesophase samples (Table II).

All the observations suggest that the mode of counter-ion binding in surfactant systems is strongly dependent on surfactant end-group. Detailed considerations of this will be deferred to a later date when studies on other end-groups are also completed. It is tempting, however, to ascribe the effects observed to differences in end-group basicity which would affect the interaction between the surfactant end-group and the water of counter-ion hydration.

The lamellar phase of the Aerosol OT-water system was included in this study to provide information on the end-group effects but also with the hope of contributing to the elucidation of the peculiar irregularity observed with other experimental techniques. Thus in the region of 35-42% by weight of Aerosol OT Rogers and Winsor[21] observed a change in sign of the birefringence and Fontell[22] observed a discontinuity in the X-ray spacing. From Fig. 3 it can be inferred that both the deuteron and the ^{23}Na static quadrupole interactions vary regularly over the whole range of stability of the lamellar phase. According to these data no change in either the water or the counter-ion binding can be detected in the region where a structural change has been deduced from other types of measurements. However, since the water deuteron splitting are dominated by shortrange interactions between the lamellae and the water molecules, we can not rule out alterations in the long-range order of the water lamellae.

ACKNOWLEDGEMENTS

We are grateful to Dr. Krister Fontell for proposing the study of the Aerosol OT system and for helpful discussions. Mr. Sven Andersson is thanked for technical assistance. Dr. T. Bull kindly made a linguistic revision of the manuscript.

REFERENCES

1. D. Chapman, in "Membranes and Ion Transport" (E.E. Bittar Ed.) vol. 1, Wiley-interscience, London 1970, p. 23.

2. P. Ekwall, I. Danielsson and P. Stenius, in MTP International Review of Science, Physical Chemistry, Series one, vol. 7, Butterworths 1972, p. 97.

3. G. Lindblom, B. Lindman and L. Mandell, J. Colloid Interface Sci. 1973, 42, 400.

4. G. Lindblom and B. Lindman, Mol.Cryst. Liquid Cryst. in press.

5. B. Lindman and P. Ekwall, Kolloid-Z. Z. Polym. 1969, 234, 1115.

6. J. Andrasko, I. Lindqvist and T.E. Bull, Chemica Scripta 1972, 2, 93.

7. N.-O. Persson, H. Wennerström and B. Lindman, Acta Chem.Scand., in press.

8. P. Ekwall, L. Mandell and K. Fontell, Mol.Cryst. Liquid Cryst. 1969, 8, 157.

9. G. Lindblom, H. Wennerström and B. Lindman, Chem.Phys.Lett. 1971, 8, 849.

10. G. Lindblom, Acta Chem.Scand. 1971, 25, 2767.

11. G. Lindblom, Acta Chem.Scand. 1972, 26, 1745.

12. H. Gustavsson and B. Lindman, Chem.Commun. 1973, 93.

13. N.-O. Persson and B. Lindman, to be published.

14. M.H. Cohen and F. Reif, Sol. State Phys. 1957, 5, 321.

15. A. Abragam, "The Principles of Nuclear Magnetism", Clarendon Press, London 1961.

16. G. Lindblom and B. Lindman, Mol.Cryst. Liquid Cryst. 1971, 14, 49.

17. B. Lindman and B. Brun, J. Colloid Interface Sci. 1973, 42, 388.

18. R.T. Roberts, Nature 1973, 242, 348.

19. G. Lindblom and B. Lindman, Proc.Intern.Congr. Surface Active Agents, 6th Zürich 1972, in press.

20. H. Wennerström, G. Lindblom and B. Lindman, to be published.

21. J. Rogers, and P.A. Winsor, Nature 1967, 216, 477.

22. K. Fontell, J. Colloid Interface Sci., in press.

THE DEPENDENCE OF SOME PROPERTIES OF AQUEOUS LIQUID CRYSTALLINE PHASES ON THEIR WATER CONTENT

Per Ekwall

Stockholm, Sweden

The number of different phases is rather large in the aqueous systems of amphiphilic substances; this holds especially for the ternary systems which besides the amphiphile contain another organic component possessing lipophilic or amphiphilic properties. As always when it is the question of equilibria between different phases at constant temperature and pressure the conditions for equilibrium between the phases is the equality of the activities of the various components; aqueous phases in equilibrium with each other should thus have the same water activity. Fig. 1 presents the equilibrium vapour pressures for on the one hand the isotropic aqueous solution L_1, and on the other hand the isotropic alcoholic solution L_2, or various liquid crystalline phases (designated B,C,D and E) respectively in the ternary system of sodium caprylate, decan-1-ol and water at 25°C (1).

For ternary systems, where phases sometimes exist over very extended concentration regions, it is of importance to know the direction of the tie-lines through the two-phase zones that separate the homogeneous phases and thereby of the points on the phase boundaries that are in mutual equilibrium; in spite of the widely different contents of water they are characterized by the same vapour pressure. Along the boundary of a particular phase the composition of the phase and its water activity vary; as illustrated in Figure 1 these changes differ widely from phase to another. When one proceeds from the phase boundary inside the homogeneous phase region the composition and the water activity also will change. These changes will be dealt with more in detail below.

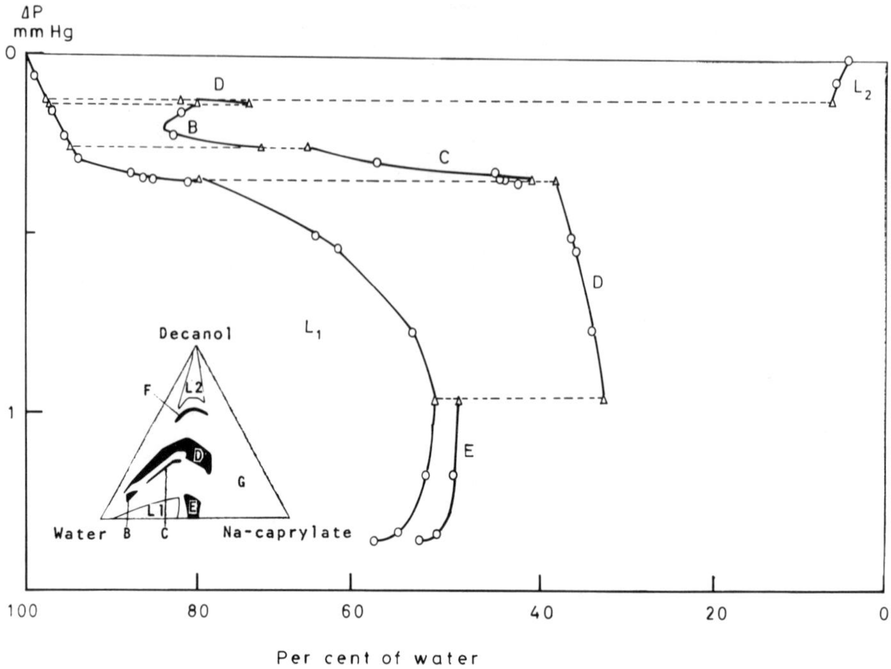

Fig. 1. The equilibrium water vapour pressure between the aqueous solution L_1 and the decanolic solution L_2 and various mesomorphous phases B,C,D and E, respectively, in the system sodium caprylate - decan-1-ol - water at 25°.

The full-drawn curve to the left gives the depression of the vapour pressure ΔP at the boundary of the aqueous solution and the full-drawn curves to the right the depression of the vapour pressure ΔP at the boundary of the phases L_2, D,B,C,D and E. The dashed horisontal lines join together the points on the phase boundaries, Δ, that are in equilibrium with each other in three-phase zones.

The insert shows a schematic triangular phase diagram of the system.

Generally speaking one can thus state that an aqueous phase continues to take up water until its water vapour pressure has risen to the value of the neighbouring phase with which it is in equilibrium. The manner in which the water vapour pressure increases with the content of water inside the phase depends on the state of the water - that is, in which way it is influenced or engaged by the amphiphilic substance. The variation in the latter respect is rather large in various types of liquid crystalline phases and also in phases of the same type in different amphiphilic systems.

Already J.W. McBain observed that the uptake of water in "soap boiler's neat soap" varies in different amphiphile systems. He distinguished between an "expanding" and a "non-expanding" type (2). His observations have later been confirmed.

If the coherent double layers of amphiphile molecules are uninfluenced of the water added to a lamellar liquid crystalline system, e.g. the neat phase (designated D), the water will cause a typical one-dimensional swelling with the result that the slope of log d vs log ($1/\phi_a$) will be unity (d = the fundamental X-ray repeat distance and ϕ_a = the volume fraction of the amphiphilic substance). This slope has been obtained for the neat phase of the expanding type (e.g. various monoglycerides, diethyleneglycol monolaurate, the dialkylsulphosuccinates Aerosol OT and Aerosol MA, Fig. 2 a). In these neat phases the thickness of the amphiphilic layers and the interfacial area per polar group are almost independent of the water content (Fig. 2 b); that is to say no additional water seems to be incorporated between the molecules of the amphiphile layers, all water being intercalated between the layers.

According to the few measurements existing of the water vapour pressure of such phases, e.g. in the water-rich parts of the neat phase of the dialkylsulphosuccinate Aerosol OT - water system, the water vapour pressure decreases very slowly with decreasing content of water from the value obtained for saturated (1.3 %, 25°) aqueous solution of Aerosol OT (Fig. 3). This suggests that at least part of the water in that section of the mesophase region is in unbound state, which is understandable as it is the question of as large contents of water as 250 to 50 moles per mole of amphiphile.

The uptake of water produces in contrast no swelling of the neat phase of many typical ionic association colloids (Fig. 4 a). The addition of water results in a reduction of the thickness of the double layers and a rise in the calculated interfacial area per polar group (Fig. 4 b and c). The water molecules seem to penetrate the amphiphile layers forcing their molecules apart

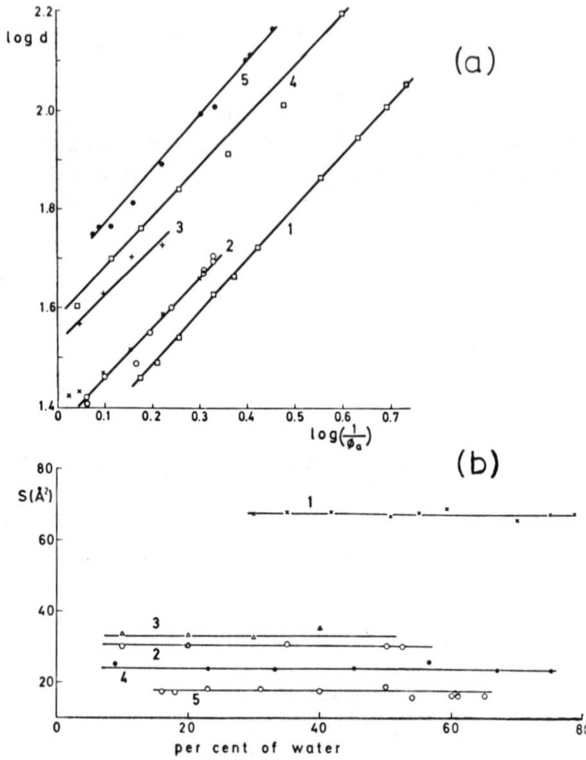

Fig. 2. Neat phase D in binary systems of swelling amphiphiles and water. (a) Variation of the fundamental repeat distance d and the volume fraction of amphiphilic substance, $\emptyset_a$. The slopes of the log d vs log $(1/\emptyset_a)$-curves are all about unity. (b) Variation of the interfacial area per polar group, S, with the water content. 1: Aerosol OT - water, 20°, after ref. 3; 2: Monocaprylin - water, o - 20°, x - 25°, after refs. 4 and 5; 3: Monopalmitin - water, 65°, after ref. 5; 4: Monolaurin - water, 29°, after ref. 5; 5: Ethylene glycolmonolaurate - water, 29°, after ref. 2.

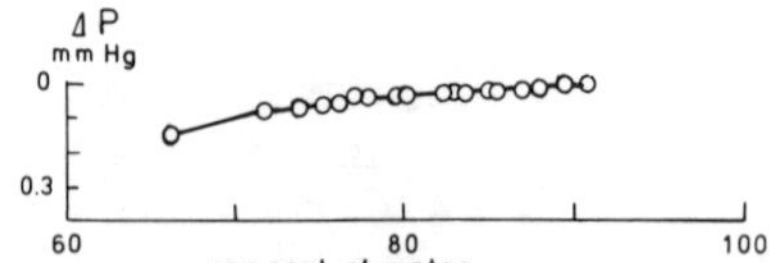

Fig. 3. The system Aerosol OT - water, 25°. The depression of the water vapour pressure ΔP of the neat phase D <u>vs</u> the water content, after ref. 6.

and reducing simultaneously the thickness of these layers so that the phase as a whole is non-expanding. The increase in the interfacial area per polar group seems to be independent of the chain length of the amphiphile and only dependent on the type of the polar group and on the temperature (7). Even at the lowest water contents this area is as large as 26 Å^2 so that the hydrated counter-ions can be situated between the ionized end groups of the amphiphile molecules. The large increase in the interfacial area accompanying the uptake affords a proof that most of the water, too, is interposed between the groups. This seems to be connected with the large water-binding capacity of the ions and ionized groups. The maximum amount of water that the neat phases of the alkali soaps are able to take up would seem not to exceed that which can be bound by the alkali ions by ion-dipole attraction and by the carboxylate groups by hydrogen bonds. The maximum values of the interfacial area per polar group are of the order of 40 -46 Å^2, the charge density thus remaining rather high.

When a long-chain alkanol is solubilized in the neat phase of these ionic amphiphiles the uptake of water by the phase changes in nature; as the alkanol content is increased the water uptake gradually approaches that of the neat phase of the above mentioned swelling amphiphiles - that is to say, the additional water will finally be taken up under one-dimensional swelling. As a result of the insertion of alcohol molecules between the amphiphilic molecules of the double layers hydrated counter-ions are expelled and the mean interfacial area per polar group is reduced. As the amount of alkanol incorporated is increased the slope of the curve of log d <u>vs</u> log ($1/\phi_a$) increases from zero for the pure surfactant to unity (Fig. 5). This means that the alcohol causes an increasingly larger fraction of the additional water to remain outside the amphiphile layers as intercalated separate water layers. Ultimately no additional water can be incorporated between the polar groups at the interfaces. The mean interfacial area per polar group has now decreased to 24 - 25 Å^2. This value is unaffected by any further addition of water or any further increase in the alcohol to soap ratio (8,9).

At this limit the properties of the neat phase are modified in many other respects. E.g. the capacity of the phase for incorporating water increases abruptly (Fig. 6 c) with the result that the neat phase region in the triangular diagram projects as long narrow salient towards the water apex; furthermore, the rheological behaviour of the phase changes (Fig. 6 e and f). The reason for these changes has been sought in the fact that up to a limit all the water of the phase is engaged by the polar groups of the amphiphile, whereas above the limit the additional water, which is intercalated between the double layers under

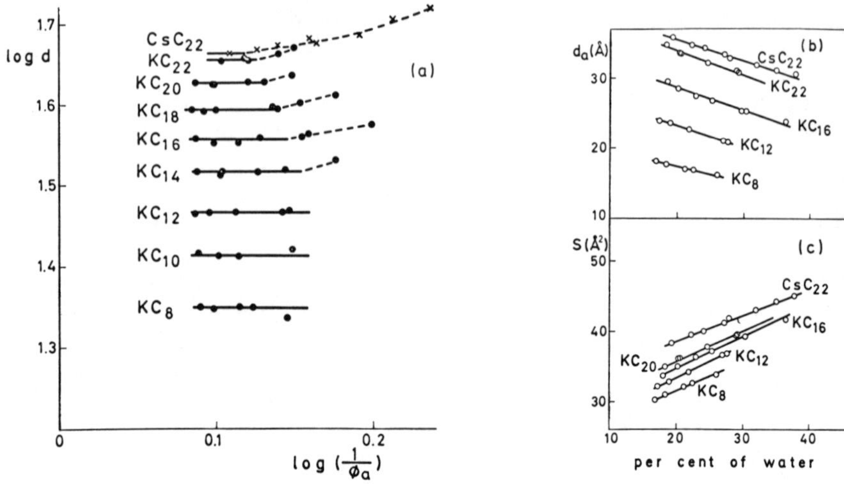

Fig. 4. Neat phase D in binary systems of fatty acid alkali soaps and water at 86°. According to measurements of Gallot and Skoulios (7).
(a) Variation of the fundamental interplanar repeat distance d with volume fraction of amphiphilic substance, ϕ_a. The slopes of the curves log d vs log $(1/\phi_a)$ are almost all zero.
(b) Variation of the thickness of the double amphiphile layers with water content.
(c) Variation of the interfacial area per polar group with water content.

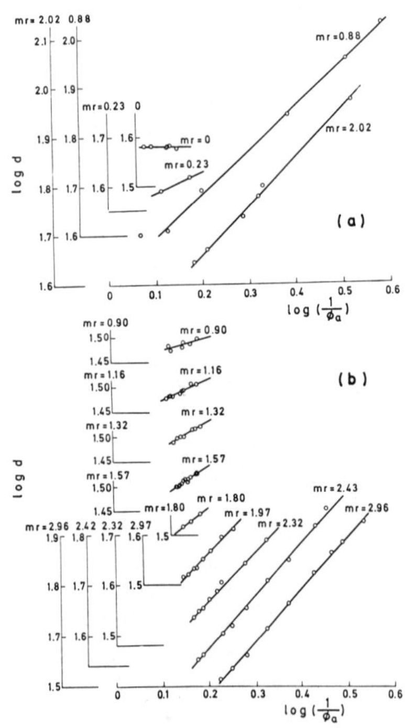

Fig. 5. Neat phase D in ternary systems of fatty acid alkali soaps, alcohol and water at 20°. Variation of the fundamental interplanar repeat distance d with volume fraction of amphiphilic substance ϕ_a (soap + alcohol) at various molar ratios between alcohol and soap. The slope of curves log d vs log $(1/\phi_a)$ increases to unity with the molar ratio.
(a) potassium oleate, decan-1-ol and water(8); (b) sodium caprylate, decan-1-ol and water (9).

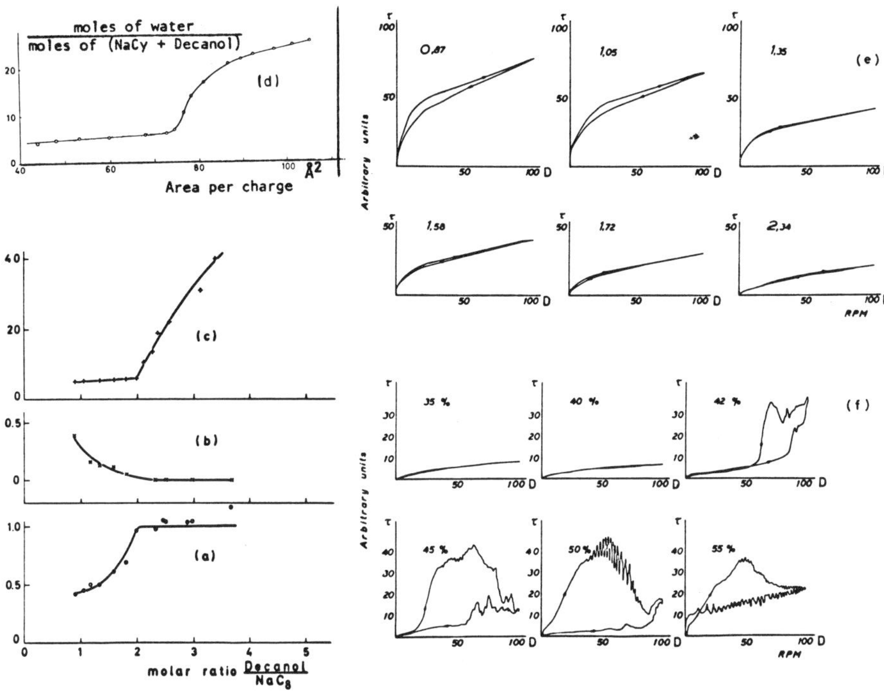

Fig. 6. Some properties of the neat phase D in the system sodium caprylate, decan-1-ol and water, 20°.

(a) Variation of the slope of the curve log d vs log $(1/\phi_a)$ with the molar ratio decanol/ sodium caprylate (9).
(b) Variation of the slope of the curve log S vs log Z with the molar ratio decanol / sodium caprylate. S = the interfacial area per polar group, Z = the molar ratio water / sodium caprylate(9).
(c) Variation of the maximum water content of the neat phase with the molar ratio decanol / sodium caprylate (9).
(d) Variation of the maximum water content along the lower boundary of the neat phase region with the charge density (interfacial area per carboxylate group) of the amphiphile/water interface(9,10).
(e) and (f) rheograms of the neat phase at various molar ratios between alcohol and soap and at various water contents. τ-D curves obtained with a Ferranti-Shirley viscometer (τ =shear stress in arbitrary units; D = shear rate, RPM), (e) weigth ratio water/soap = 45/55; molar ratio decanol/soap 0.87 - 2.34, (f) molar ratio decanol/soap 2.45; water content 35 - 55 % (11).

one-dimensional swelling, seems to behave as free bulk water. Support for this opinion has been obtained by measurements of the water vapour pressure in the neat phase of the system sodium caprylate, decan-1-ol and water.

The dependence of the vapour pressure of the water content differs for different parts of the mesophase regions; while the vapour pressure decreases rapidly with decrease of the water content in the water-poor parts of the phase, it is rather unaffected in the water-rich parts where it is close to that of pure water or dilute aqueous solutions (Fig. 7). The dividing line between these two regions is about there where one may conceive the water-binding capacity of the hydrophilic groups to be exhausted (ca 6 moles of water per mole of sodium ions, 5 moles per mole of ionized carboxylate groups and 3 moles per mole of alcohol groups, Fig. 8), this limit coincides with that where other properties of the mesophase are changed.

The sudden increase in the capacity for incorporating water has been ascribed the liberation of counter-ions of the soap as a result of the decrease in the charge density at the interface of the double layers from a value of about one charge per 37 $Å^2$ to one per 72 - 75 $Å^2$ (Fig. 6 d). This liberation creates the conditions for a Donnan distribution of counter-ions which in turn enables increasing amount of water to be retained until the vapour pressure reaches the value characteristic of the more water-rich phase in equilibrium with the neat phase.

While the extent of the neat-phase region towards higher water contents in the alcohol- and water-rich part of the region thus seems to be caused by the Donnan distribution of liberated counter-ions, in other parts containing lesser amounts of alcohol its existence seems to be determined by the maximal water-binding capacity of the polar groups. Towards the lowest water contents, however, the extension of the region may be limited by the facts that the alkali ions of the soap must be hydrated and that the interfacial area per polar group cannot decrease below a value 23 - 24 $Å^2$. In respect of high alcohol contents the region of existence seems to be limited by the fact that the amphiphilic layers are not able to solubilize more than about 4 moles of alcohol per mole of soap (12).

In the alkali soap system the above type of neat phase, with its large capacity for taking up water, is formed by solubilizing paraffin chain alcohols from decan-1-ol to pentan-1-ol. The alcoholic neat phases of alkyl sulphates and sulphonates are of the same type. This holds, with some modifications, also for the alcoholic neat phases of the alkylammonium halides. In all these cases the interaction between the polar groups of the

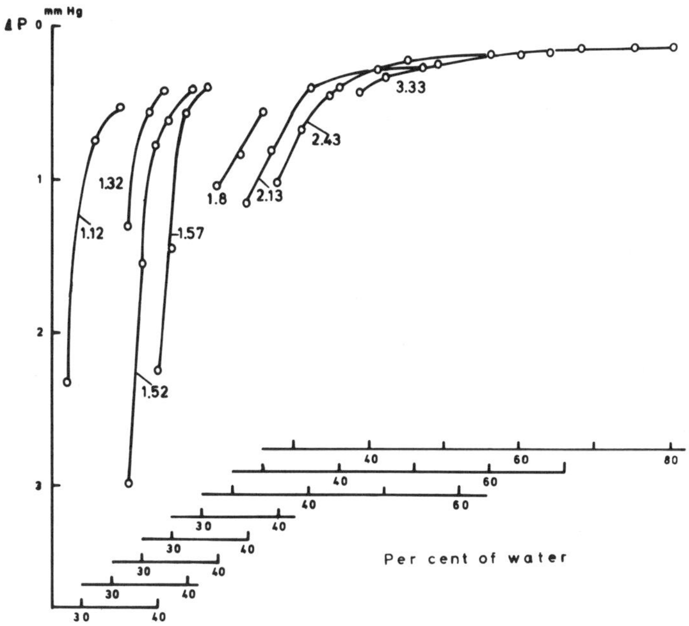

Fig. 7. The water vapour pressure of the neat phase D in the system sodium caprylate, decan-1-ol and water at 25°.
The depression of the vapour pressure, ΔP, vs the water content at various molar ratios of decanol/sodium caprylate (1).

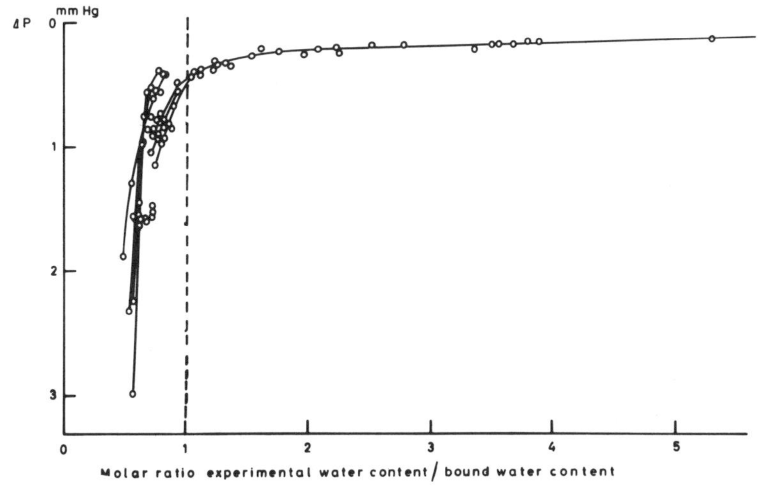

Fig. 8. The water vapour pressure of the neat phase D in the system sodium caprylate, decan-1-ol and water at 25°.
The depression of the vapour pressure, ΔP, vs the water content expressed as the molar ratio q of the experimental content of water to the estimated maximum content of bound water (1).

amphiphile and alcohol seems to be the critical factor for the formation of the actual type of neat phase and especially for its stabilization at high water content. The importance of this interaction is supported by the fact that the same type of neat phase appears also in systems of soap, fatty acid and water, where a corresponding interaction between the carboxylic group of the fatty acid and the carboxylate group of the soap takes place. On the other hand, this type of neat phase has not been encountered in ternary systems where such an interaction between solubilizate and the ionic association colloid is lacking; when the solubilizate is a hydrocarbon the region of the neat phase seems not to extend towards higher water contents than that can be conceived bound by the polar groups of the amphiphile. The same applies when the solubilizate is a compound with a weak polar group, such as a methyl ester, nitrile or aldehyde.

Another lamellar mesophase consisting of double amphiphile layers is the so-called "mucous woven phase" (designated B). This phase has so far not been observed in binary systems of amphiphile and water but only in ternary systems where in addition to a typical association colloid and water there is another amphiphilic compound such as a long-chain alkanol or a fatty acid. It appears in the part of the system containing much water and its water content exceeds considerably the maximum that can be bound by the polar groups. In the direction of increasing water content the mesophase is in equilibrium with the dilute aqueous solutions of the amphiphile between c.m.c. and l.a.c.

The water uptake of mesophase B takes place under one-dimensional swelling, the curve log d vs log $(1/\phi_a)$ being about unity (Fig. 9 a); all the additional water, thus, is intercalated between the amphiphilic layers. The interfacial area per polar group is constant at 25 - 27 Å^2 throughout the phase region (Fig. 9 b). The charge density is low - one charge per 68 - 83 Å^2. Conditions exist thus for the liberation of counter-ions from the amphiphile layers and the commencement of a Donnan distribution that enables the phase to retain large amounts of unbound water. Consequently the water vapour pressure is high throughout the phase (Fig. 9 c); it rises slowly with the water content from a value that coincides with that of the points on the borders of phases D and C, which are in equilibrium with the B-phase until it reaches the value of the aqueous solution region L_1, between c.m.c. and l.a.c. with which it is in equilibrium at the border towards higher water contents.

The "white phase" or "square phase type 1" (designated C), another mesophase that up to now only has been encountered in ternary systems of typical association colloids together with another amphiphilic compound such as a long-chain alcohol or a fatty acid, occurs also in the water-rich part of the system in

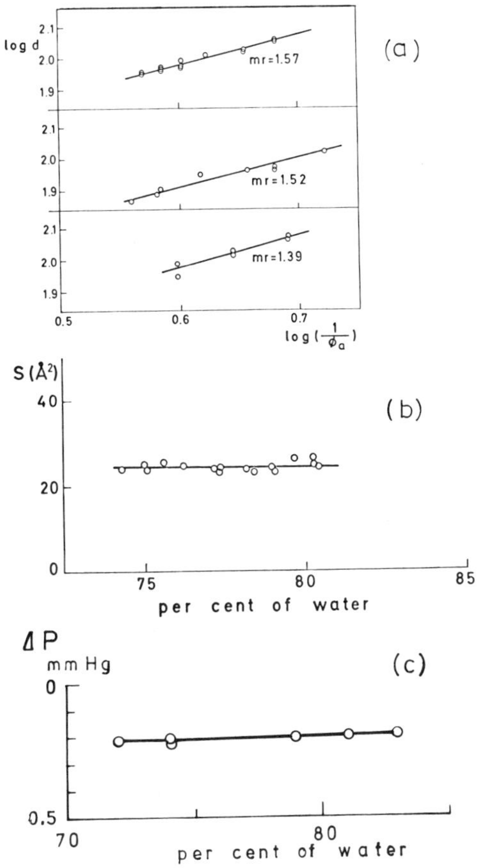

Fig. 9. Mucous woven phase B in the system sodium caprylate, decan-1-ol and water.

(a) Variation of the fundamental interplanar repeat distance d with the volume fraction of amphiphilic substance, $\emptyset_a$. The slope of the curves log d vs log $(1/\emptyset_a)$ is about unity, 20° (9).
(b) Variation of the interfacial area per polar group S with the water content, 20° (9).
(c) Variation of the depression of the water vapour pressure ΔP with the water content, 25° (1).

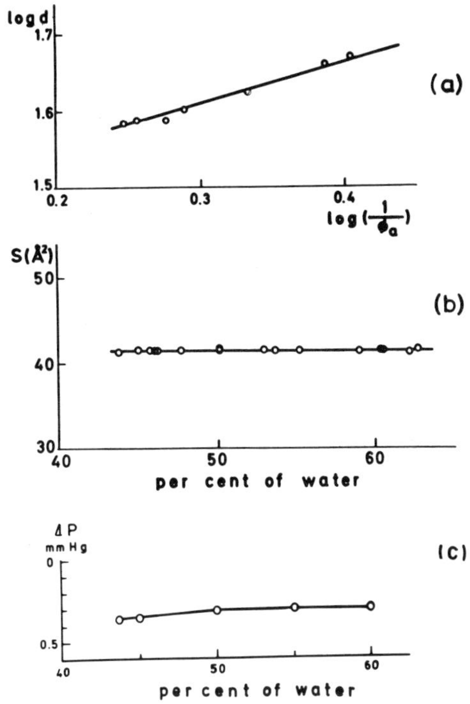

Fig. 10. Square phase of type 1, C, in the system sodium caprylate, decan-1-ol and water. Molar ratio decanol/sodium caprylate 1.57 -1.58.

(a) Variation of the fundamental interplanar repeat distance d with the volume fraction of amphiphilic substance, $\emptyset_a$. The slope of the curve log d vs log $(1/\emptyset_a)$ is about 0.5, 20° (9).
(b) Variation of the interfacial area per polar group S with the water content, 20° (9).
(c) Variation of the depression of the water vapour pressure ΔP with the water content, 25° (1).

the area between the regions for the neat phase D and isotropic aqueous solution L_1, between c.m.c. and 2nd c.m.c. Its content of water is throughout so high that all of it cannot be bound by the polar groups. This is mirrored by its rather high water vapour pressure which only slowly decreases as the water content is decreased (Fig. 10 c); the values for the vapour pressure are close to those of the neighbouring parts of the aqueous solution region.

The curves for log d vs log $(1/\phi_a)$ have a slope of about 0.5 (Fig. 10 a); that is consistent with the supposed structure of long rod-shaped amphiphile aggregates with constant cross-section in a square array and with all additional water intercalated between the rods. The mean interfacial area per polar group is constant (about 38 Å^2 in the system sodium caprylate, decan-1-ol and water, Fig. 10 b) and the charge density is low (one charge to 76 - 83 Å^2). Also in this case there are thus conditions for the liberation of counter-ions from the amphiphilic layers and the originating of a Donnan distribution that enables the phase to retain large amounts of unbound water.

The common middle phase of type 1 (designated E) with its two-dimensional hexagonal structure consists as known of long mutually parallell rods in hexagonal array, the rods consisting of more or less radially arranged amphiphilic molecules with their polar groups located in the interfaces with the surrounding water continuum. The cross-section of the rod aggregates is circular but will obtain a hexagonal shape when the volume fraction of the hydrated amphiphile exceeds o.91; the slope of log d_p vs log $(1/\phi_a)$ (d_p = the parameter of the hexagonal structure) is often somewhat below 0.5 and tends in many systems to become still smaller as the water content is diminished (Fig. 11a) As a consequence there is a slow decrease in the interfacial area per polar group with the water content; the rate is independent of the chain length but varies with the type of the polar groups and the temperature (Fig. 11 b). Even at the lowest water contents the calculated values of the interfacial area per polar group is so high (39 - 40 Å^2) for alkali soaps that the alkali ions and the water molecules may well be located between the carboxylate groups at the surface of the rod-shaped aggregates. There must be a minimum value for the interfacial area per polar group, that increases with the chain length of the soap. The minimum amount of water of the middle soap is probably determined by this fact; the space between the end groups in the interfaces must remain completely occupied by hydrated counter-ions and water molecules if the structure in question is to be maintained. Accordingly, the region of existence of the middle phase is displaced towards higher water contents with increasing chain length of the amphiphile when the polar end group is the

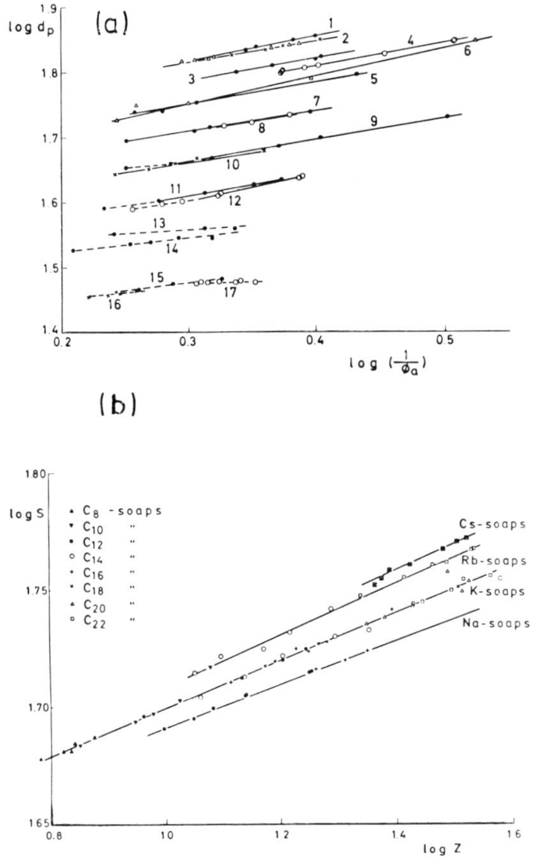

Fig. 11. Middle phase type 1, E, in binary soap water systems.

(a) Variation of the lattice parameter d_p with the volume fraction of amphiphilic substance.
1: KC_{22}, 86°; 2: RbC_{22}, 86°; 3: KC_{20}, 86°: 4: NaC_{18}, 86°; 5: KC_{18}, 86°; 6: $KC_{18:1}$, 20°; 7: KC_{16}, 86°; 8: NaC_{16}, 86°; 9: KC_{14}, 86°; 10: RbC_{14}, 86°; 11: KC_{12}, 86°; 12: NaC_{12}, 86°; 13: KC_{10}, 20°; 14: KC_{10}, 86°; 15: KC_8, 20°; 16: KC_8, 86°; 17: NaC_8, 20°. Values at 86° according to measurements of Gallot and Skoulios, ref. 7, values at 20° according to ref. 9.

(b) Variation of the interfacial area S per polar group with the water content Z(in moles per mole of soap), 86°, After measurements of Gallot and Skoulios (7).

same. On the other hand for amphiphiles of the same type and same chain length the region is displaced towards lower water contents when the size of the end group is increased.

It seems probable that the increase of the value of the area per polar group with increasing water content is due to the fact that part of the additional water, too, is located between the carboxylate groups at the interface. The factors that restrict this uptake of water and thus the extent of the middle phase region towards higher water contents are still unknown. The interfacial area per polar group rises here to about 60 $Å^2$. The charge density in the interface has thus fallen to a level where it is no longer certain that all counter-ions remain bound at this surface. The possibility that in some of these phases, too, a Donnan distribution of liberated ions can contribute to the uptake of water, cannot be ruled out.

The middle phase of most systems is able to incorporate various additional lipophilic and amphiphilic compounds without imparing the basic structure. The capacity for such an incorporation differs with the character of the additive.

For n-alkanols sparingly soluble in water the maximum amount that the middle phase of a given alkali soap is able to solubilize is independent of the chain length of the alcohol; for sodium caprylate at 20° it is 0.33 moles of alcohol (decan-1-ol to pentan-1-ol) per mole of soap (12). This has been regarded as proof that the interaction between the carboxylate group of the soap and the hydroxyl group of the alcohol determines the number of alcohol molecules that can be inserted between the soap molecules of the rod-shaped aggregates. For other surfactants similar ratios have been recorded. The calculated mean interfacial area per polar group decreases rapidly and proportionally to the increase in the ratio of alcohol to soap (Fig. 12). This decrease is due not only to the smaller space requirements of the alcohol group and the interaction between the polar groups, but obviously primarily to expulsion of the hydrated alkali ions that were interposed between the carboxylate groups.

Only few measurements of the water vapour pressure of the normal middle phase seem to have been performed; they are from the system sodium caprylate, decan-1-ol and water at 25° and refers furthermore to specimens at the boundaries of the phase. Already at the boundary towards concentrated aqueous solution L_1 the vapour pressure depression is rather large (Fig. 13 a) and the measurements show that the vapour pressure rapidly decreases with increasing soap content (fig. 13 b). Of interest is that the water vapour pressure increases with the decanol content; a parallell phenomenon to the case when alcohol is solubilized

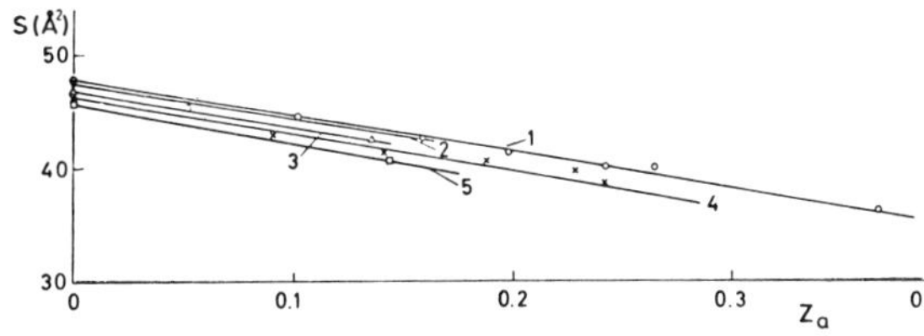

Fig. 12. Middle phase of type 1, E, in ternary sodium caprylate - water systems containing a solubilized alkanol, 20°. Variation of the interfacial area S per polar group with the alkanol content(in moles per mole of soap, Z_a) (9,12).
1: weigth ratio water/soap 54.0/46.0; solubilization of decan-1-ol. 2: 55.3/46.7; solubilization of hexan-1-ol. 3: 51.0/49.0; solubilization of decan-1-ol. 4: 49.4/50.6; solubilization of octan-1-ol.

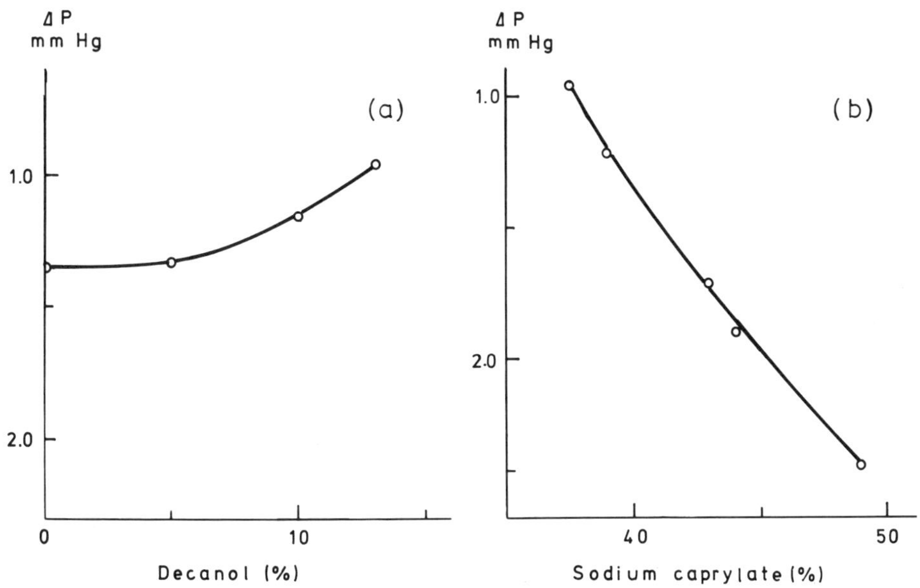

Fig. 13. Middle phase of type 1, E, of the system sodium caprylate, decan-1-ol and water. Water vapour pressure at 25°(1).
(a) Depression of the vapour pressure, ΔP, vs decanol content of the phase at the boundary against the aqueous solution L_1.
(b) Depression of the vapour pressure, ΔP, vs sodium caprylate content of the phase at the boundary against the neat phase D.

by the micelles of the aqueous solution (Fig. 13 a). This depends obviously on the extrusion of hydrated counter-ions from the interfaces of polar groups of the amphiphile aggregates.

The <u>middle phase of type 2</u>, the reversed middle phase (designated F, H_2 or M_2) occurs in one, two or multi-component systems of many amphiphiles in the water-poor parts of the system. It is often in equilibrium on the one hand with isotropic solution L_2 of the amphiphile in an organic solvent, the solution containing micelles of the reversed type, and on the other hand with lamellar neat phase D. It possess the same basic structure as middle phase of type 1 but the rods are now composed of amphiphile molecules whose polar groups are oriented towards the center while the hydrocarbon parts of the molecules are oriented outwards forming a hydrocarbon continuum through the structure. The experimental observations are still so few and the molecular structure so diverse, that it is difficult to define, in quantitative terms, a connection between the properties of the phase and its content of water.

Measurements of the water vapour pressure are almost completely lacking, but one may dare the prediction that the dependence of the vapour pressure on the water content basically is the same as in the L_2 solutions containing reversed micelles with solubilized water. Measurements by ourself and other have shown that the vapour pressure in such solutions decreases rapidly with the water content from a high value in the water-rich regions or at the phase boundary towards the solution L_1 and the neat phase when one proceeds in the L_2 phase region (Fig. 14 a and b).

When the reversed middle phase takes up water, this will be incorporated by the polar core of the rod aggregates, whereby the cross-section of both the core and the rods will increase. The phase can often in such a way take up large amounts of water; this holds especially for systems containing besides the amphiphile another organic component, a hydrocarbon, a n-alkanol or a fatty acid. Bonds between the amphiphile molecules seem often to stabilize the aggregates and thus render the incorporation of large amounts of water possible.

The lattice parameter d_p is rather unaffected by small additions of water, while with increasing amounts d_p increases at a faster rate (Fig. 15, log d_p <u>vs</u> log $(1/\phi_w)$, ϕ_w = volume fraction of water). This suggests that the first water molecules penetrate between the polar groups and are bound to them. The interfacial area per polar group in most systems increases rapidly for the first additions but slower for the subsequent additions (Fig. 15 b); sometimes the increase ceases completely after a certain addition of water.

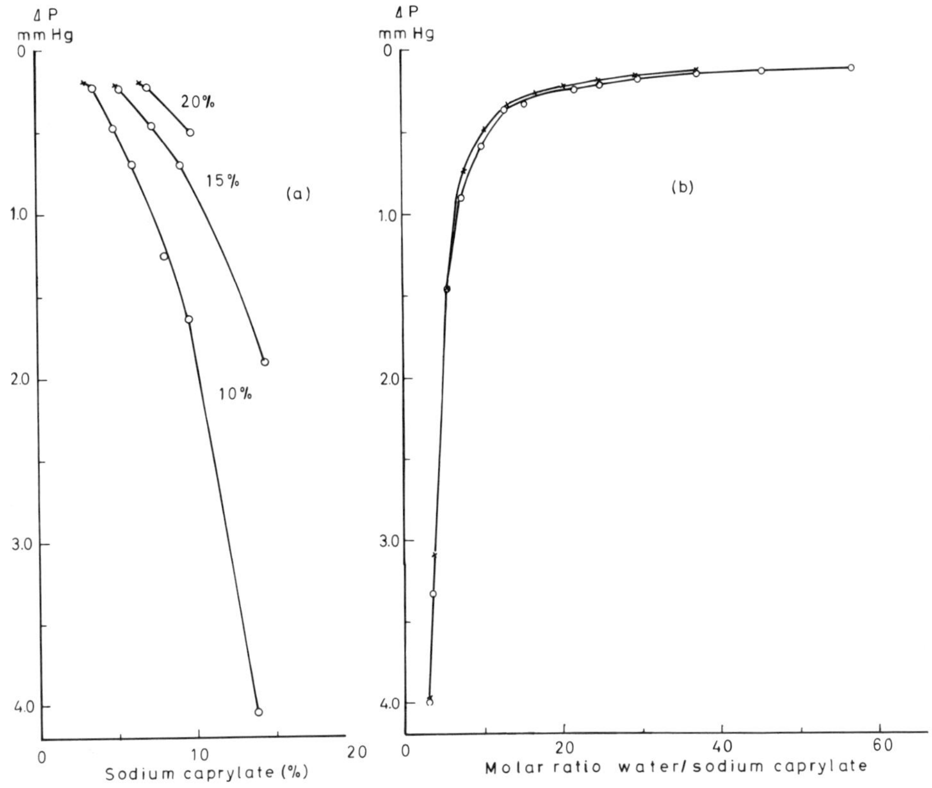

Fig. 14. Water vapour pressure in L_2-solutions containing micelles of the reversed type, 25° (1)

(a) Alcoholic solution L_2 of the system sodium caprylate, decan-1-ol and water. Depression of the vapour pressure, ΔP, vs sodium caprylate concentration; series with constant water content, 10, 15 and 20 %. X - values at the boundary of the solution phase.

(b) Fatty acid solution L_2 of the system sodium caprylate, caprylic acid and water. Depression of the vapour pressure, ΔP, vs the water content (in moles per mole of soap). Series with constant ratio of caprylic acid to sodium caprylate.

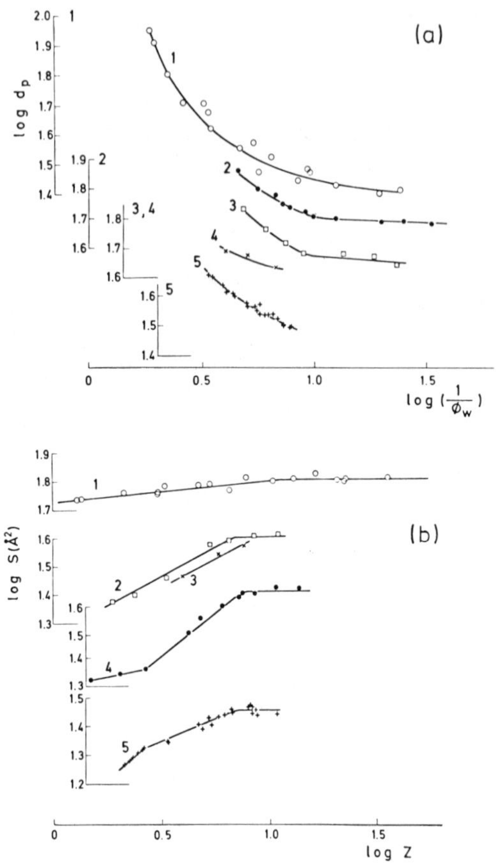

Fig. 15. Middle phase of type 2, F, in systems of amphiphiles and water,

(a) variation of the lattice parameter d_p with volume fraction of water, ϕ_w; log d_p vs log $(1/\phi_w)$.

(b) Variation in the interfacial area per polar group, S, with the water content (in moles of water per mole of amphiphile, Z); log S vs log Z.

1: Aerosol OT - p-xylene - water, 20°, after ref. 3;

2: phosphatidyletanolamine - water, 55°, after refs. 13,14;

3: human brain lipids - water, 37°, after refs. 13, 15;

4: monobehenin- water, 83°, after ref. 5;

5: sodium caprylate - decan-1-ol - water, 20°, after ref. 9.

The upper limit for the water solubilization seems in some systems to be determined by the maximum water-binding capacity of the polar groups while in other systems this capacity can be exceeded. An example of the latter is provided by the reversed middle phase of the Aerosol OT system containing hydrocarbon, a system which has a long salient towards high water contents. In this phase the interfacial area per polar group becomes independent of the water content and the charge density at the interface approaches values (one charge to 64 $Å^2$) that should allow the liberation of counter-ions and a mechanism for the uptake of water regulated by the Donnan distribution (3).

In spite of the swelling of the core of the rods through the solubilization of water the thickness of the hydrocarbon sheath around the rod-aggregates remains fairly constant throughout the regions of existence of the phase; it is slightly less than the length of the hydrocarbon chains of the amphiphile. The molecules of an added hydrocarbon will be chiefly located between the rods and possibly also within the outermost part of the hydrocarbon sheath. On the other hand if the added component is an n-alkanol or a fatty acid only a portion of it is located between the rods, the remainder being solubilized in the rod aggregates with the polar groups directed towards the core. The latter amount seems to be regulated by the interaction between the polar groups of the two organic components (3,9).

One seems thus to be able to distinguish between two main types for the uptake of water by liquid crystalline phases of amphiphiles. For one type the water, more or less unaffected by the polar groups of the amphiphile, is intercalated between the amphiphile aggregates (lamellar phases and mesophases of type 1) or in their core (mesophases of type 2). For the other type the water is in one or another manner influenced by or bound to the polar groups of the amphiphile and at least to some extent is built in between them. In the first type the value of the interfacial area per polar group remains unaffected by changes in the water content and the water vapour pressure of the phase is only little affected by this content and does not differ much from that of pure water or dilute aqueous solutions. In the latter type the interfacial area per polar group changes with the water content simultaneously as the water vapour pressure of the phase decreases rather rapidly with decreasing content of water. Probably there exist intermediate forms, too, between these extreme types, because of the rather large variation in the interaction between water and the polar groups of the amphiphiles; e.g. there are ion-dipole effects, hydrogen bonds and influences on the water structure.

An important task for future research should be to study

the water uptake of lyotropic mesophases more close and systematically and the state of the water in the phases.

Often the transition from one phase to another is accompanied by an abrupt change in the value of the interfacial area per polar group in the amphiphile aggregates; in direction towards higher water contents, as a rule, the area increases. This does not imply, however, that it generally would be such an increase of the area at increased contents of water that induces the phase transition - a conclusion which seems to be the purport of a theory launched by Winsor (16). The main factors that determine the location of the boundaries for the region of existence of a phase and thus its transition to another phase seem to be its ability or unability to incorporate water. The binding of water to the polar groups, the Donnan distribution of free counter-ions that makes the uptake of unbound water possible, bonds between the amphiphile molecules, that stabilize water-rich aggregates seem to be factors of greater importance than the value of the interfacial area per polar group.

As we have seen above in many phases the interfacial area per polar group will not at all be influenced by the increase in the water content. The capacity of the phase to incorporate water is not exhausted and the phase continues to take up water until its water vapour pressure obtains a value that coincides with that of a neighbouring phase and an equilibrium between the phases and thus the transition from one phase to another becomes possible. In these cases Donnan equilibria and stabilizing of the amphiphile aggregates by mutual bonds play a role.

In other phases the interfacial area per polar group continues to increase with the water content until the phase boundary is reached. The factor that determines the location of the phase boundary, however, seems not to be the value of the mentioned area but the exhaustion of the capacity of the polar groups to bind water and the lack of some other factor that would enable the retaining of further amounts of water; therefore, the water vapour pressure rapidly rises to a value that enables an equilibrium with and the transition to another phase.

In other cases again, e.g. at decreasing water contents the cause for a phase transition may be the unability of a phase to exist with a lesser content of water than that needed for complete hydration of the counter-ions of the amphiphile or of that needed to fill up a characteristic space between the polar groups required of geometric reasons in the original structure. Also further reasons are known which regulate the location of the boundaries of a particular phase and the transition from one phase to another.

References

1. Ekwall,P. and Mandell, L., Unpublished measurements
2. Marsden, S.S. and McBain, J.W., J. Phys. Coll. Chem. 52, 110 (1948)
3. Ekwall, P., Mandell: L. and Fontell, K., J. Coll. Interf. Sci. 35, 215 (1969)
4. Ekwall, P., Mandell, L. and Fontell, K., to be published
5. Larsson, K., Z. physik. Chemie N.F. 56,173 (1967)
6. Fontell, K., J. Coll. Interf. Sci. in press
7. Gallot, B. and Skoulios, A.E., Kolloid-Z. 208 37 (1968)
8. Ekwall, P, Mandell, L. and Fontell, K., J. Coll. Interf. Sci. 31, 508, 530 (1969)
9. Fontell, K., Mandell, L., Lehtinen, H. and Ekwall, P., Acta Polytechnica Scand.. Chem. Met. Series 74,III 1968
10. Mandell,L., Fontell,K. and Ekwall, P., Am.Chem. Soc. "Advances in Chemistry Series"63 89 (1967)
11. Solyom, P. and Ekwall,P., Rheologica Acta 8,316 (1969)
12. Mandell, L., to be published
13. Luzzati, V., "Biological Membranes", ed. D. Chapman, Academic Press 1968 p. 71
14. Reiss-Husson, F., J. Molec. Biol. 25, 363 (1967)
15. Luzzati, V. and Husson, F., J. Coll. Biol. 12 207 (1962)
16. Winsor, P.A., Chem. Rev. 68 ,1 (1968)

DIELECTRIC RELAXATION IN LIPID BILAYER MEMBRANES

S. Takashima and H.P. Schwan

Department of Bioengineering, Moore School of Electrical Engineering, University of Pennsylvania Philadelphia, Pennsylvania, 19174

Introduction

Takashima, Schwan and Tasaki(1) observed that the membrane capacitance of squid axon undergoes a dielectric relaxation in the proximity of 10KHz. They found, after careful analyses of data(vide infra) that the widely accepted membrane capacity $1 uF/cm^2$ of nerve axon membrane(2) decreased to approximately $0.6\ uF/cm^2$ between 2 and 50KHz. The origin of the frequency dependent membrane capacitance is still unknown and the present research is a further attempt to identify the component which undergoes the dispersion in the low frequency region.

The membrane capacitance and resistance of lipid bilayers have been investigated by many workers(3,4,5,6,7,8) and it is generally agreed that the membrane capacitance and resistance of lipid bilayers do not undergo a dielectric dispersion. The dielectric behavior of axon membrane is, therefore, substantially different from those of artificial lipid bilayer membranes. The possible reasons for the difference may be in the fact that lipid bilayer membranes used thus far are not functional, i.e., they do not contain active elements such as EIM, alamethicin or valinomycin which are known to evoke ionic currents upon stimulation(9,10).

There is another important difference between lipid bilayer membranes and biological membranes. It is now establishedthat biological membranes contain proteins as well as lipids while artificial bilayer membranes usually contain only lipids. As well known, protein molecules are dipolar ions which have a dipole moment approximately 500-1000 DU(1 Debye Unit=10^{-18}esu cgs) (11,12,13). Hence if one applies an electric field to a suspension of protein molecules in aqueous media or in membrane matrices, these protein molecules respond to the field by rotating themselves in the entirety along the electric field vector. The dielectric relaxation of globular proteins in aqueous solution is usually observed in the proximity of 1MHz. The dispersion of axon membrane capacitance was observed around 10KHz and is considerably lower than the relaxation frequencies of globular proteins in aqueous solution. However, there are indications that membrane proteins are elongated rather than spherical(14) and moreover, the viscosity of the membrane matrix is much higher than that of aqueous media. Under these circumstances, it is not really surprising that the relaxation frequency of proteins is shifted considerably to a low frequency region.

These possible causes for the frequency dependent membrane capacitance do not exist in the artificial bilayer membranes used by other investigators. In an attempt to further investigate the possible cause of the frequency dependent capacitance in excitable membranes, we decided to use artificial lipid bilayer membranes, though do not contain proteins, containing an active element, alamethicin. Alamethicin is known to evoke a time dependent ionic current in lipid bilayers, which resembles ionic currents in nerve membranes.

Experiment

The experimental procedures for the preparation of lipid bilayer membranes are already described by Mueller and Rudin(10) and the methods used in this research are exactly the same as before. The phospholipids used for the measurements of electrical properties are oxidized cholesterol, sphingomyelin and egg-lecithin. Lipids are dissolved in n-octane and the solution is smeared with a small brush over an orifice of an area 0.25 mm^2 drilled on the side wall of a small teflon cup. The formation of membrane can be observed with a low power microscope with a reflected light. The thinning of the membrane and the formation of bilayer can be confirmed with the blackening of the membrane in a reflected beam. Also the formation of bilayer results in the stabilization of the membrane capacity and the limiting value is approximately 0.6-0.7 uF/cm^2.

The electrical admittance of the membrane is measured with a Wayne-Kerr wide frequency bridge B-221 which covers between 100Hz and 100KHz. The major experimental errors arise from the mechanical and electrical instability of the membrane. Immediately after the formation of membranes, they have a multilayer structure and the capacitance is relatively low. The membranes begin to thin and the bilayer structure is reached within a few minutes and the limiting capacitance value 0.6-0.7 uF/cm^2 is construed as the electrical capacity of bilayer membranes. The stability of membranes depends upon the composition of lipids and also depends on the amount of alamethicin. In the presence of alamethicin, membranes are more unstable electrically and mechanically. Frequently, membranes break down before the completion of measurement. In these cases, the wide frequency measurement can be completed only with the use of more than one membrane with sufficient frequency overlap between different membranes. The possible discrepancies in the values with different membranes are adjusted simply by shifting one of the curves so that the ends of both curves match with each other.

The bathing solution is usually 0.1 mol KCl solution unless otherwise stated. As mentioned later, the conductivity of the bathing solution has a profound effect on the frequency dependence of the measured capacitance(4). As shown in Fig.1, the conductivity of the bathing solution constitutes a series resistance to the membrane admittance and thus the measured admittance includes the series resistance as well as the membrane admittance. Hence, the measured capacitance and resistance do not neccessarily represent the true membrane admittance. In order to determine the membrane admittance, the measured capacitance and conductance must be corrected for the series resistance. As will be shown later, the measured capacitance and conductance show a marked frequency dependence. However, this is partly or totally due to the presence of the series resistance as can be seen from the following equations.

$$C = C_{\infty} + \frac{C_o - C_{\infty}}{1 + (wT)^2} \qquad (1)$$

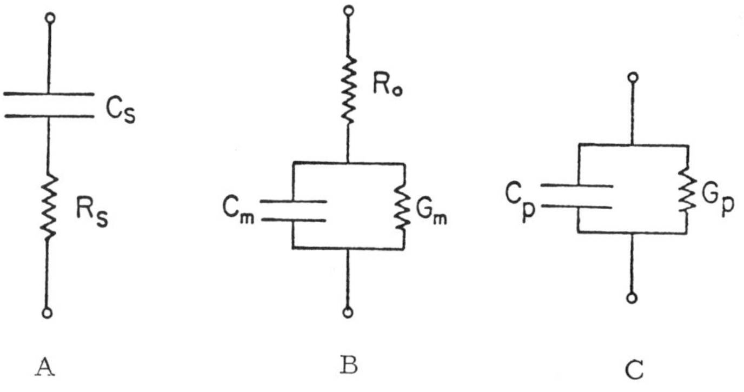

Fig.1. (A) Impedance representation of circuit(2).
(B) Membrane equivalent circuit including series resistance R_o.
(C) Admittance representation of circuit(2).

and

$$G = G_o + \frac{(G_\infty - G_o)(wT)^2}{1 + (wT)^2} \qquad (2)$$

where C and G are capacitance and conductance at given frequencies, C_o and C_∞ are the low and high frequency capacitances, G_o and G_∞ are the conductances at low and high frequencies. These limiting values are given, from the equivalent circuit shown in Fig.1, by:

$$C_o = \frac{C_m}{1 + R_oG_m} \qquad C_\infty = 0 \qquad (3)$$

$$G_o = \frac{G_m}{1 + R_oG_m} \qquad G_\infty = 1/R_o \qquad (4)$$

where R_o is the series resistance (= $R_1 + R_3$). The time constant, which is equal to the inverse of the center frequency, is:

$$T = \frac{C_m}{G_m + 1/R_o} \qquad (5)$$

These two equations (1) and (2) are derived using the equivalent circuit in Fig.1 without any assumption as to the frequency dependence of C_m and G_m. It is obvious from these equations that the measured capacitance C and conductance G will undergo a dispersion even if C_m and G_m are frequency independent. Therefore, in order to study the frequency dependence of C_m and G_m, the measured capacitance and conductance must be corrected for the series resistance R_o.

The equivalent circuit of the membrane admittance with the series resistance is given by Fig.1B in terms of C_m, G_m and R_o. The quantities measured with the bridge are represented by Fig.1A in terms of C_p and R_p. These admittance quantities can be easily transformed into impedance quantities as shown in Fig. 1C in terms of R_s and X_s.

$$-X_s = \frac{C_mR_m^2w}{1 + w^2C_m^2R_m^2} = \frac{wC_pR_p^2}{1 + w^2C_p^2R_p^2} \qquad (6)$$

$$R_s = R_o + \frac{R_m}{1 + w^2C_m^2R_m^2} = \frac{C_pR_p^2}{1 + w^2C_p^2R_p^2} \qquad (7)$$

If one plots $-X_s$ against R_s, there will be an arc plot(15,16). Of importance is the fact that the reactance X_s is unaffected at all by the presence of the series resistance. The presence of the series resistance merely shifts the locus along the resistance axis by the magnitude of the series resistance without changing the value of the reactance. Hence, by locating the intersection between the arc and the abscissa, we can determine the value of series resistance which in turn can be used in eqs. (6) and (7) to calculate the membrane capacitance C_m and conductance G_m.

In addition to the impedance measurement, voltage clamp experiments are performed to establish the current-voltage relationship for the lipid bilayer membranes used in this experiment. Lipid bilayer membranes have a very high resistivities and usualy permit only a capacitive spike current if a step voltage is applied. However, if the membranes are doped with alamethicin, these membranes become more permeable to ionic species and begin to permit time dependent ionic currents. The ionic currents have a very long rise time, which depends upon the composition of the membrane and the amount of alamethicin. Alamethicin is usually added in the bathing solution in the cup. Alamethicin molecules, apparently, diffuse into the membrane upon the applicarion of an electrical field. The current-voltage relationship for oxidized cholesterol in the presence of 10^{-7} mol alamethicin is shown in Fig.2. As shown, the I-V curve is highly non-linear except a narrow region between -30 and +60mv on both sides of the origin. As mentioned before, we are interested in the linear electrical properties of membranes and the I-V curve shown in Fig.2 indicates that the applied potential for the impedance measurements must be limited between -30 and + 30 mv in order to avoid a non-linearity.

The electrodes are silver-silverchloride plates placed on both sides of the orifice and the flow of currents is limited by the size of the orifice rather than the area of the electrodes. Hence, the measured capacitance and conductance are converted to capacity and conductivity/cm^2 by dividing the measured values by the area of the orifice. The temperature of measurements is 22-25° and controlled by ambient air.

Results

The capacitances of an oxidized cholesterol membrane bathed in KCl solutions at various concentrations are shown in Fig.3. As discussed earlier, the frequency profile of the measured capacitance of membranes depends strongly on the conductivity of the bathing solution. The center frequency of the dispersion is inversely proportional to the value of series resistance(see eq. (5)). Namely, the center frequency of the dispersion shifts toward high frequencies when the KCl concentration is increased as eq.(5) predicts.

The measured admittance quantities are converted into the impedance quantities R and X and by plotting them against each other, we obtain the impedance plot. Three impedance plots shown

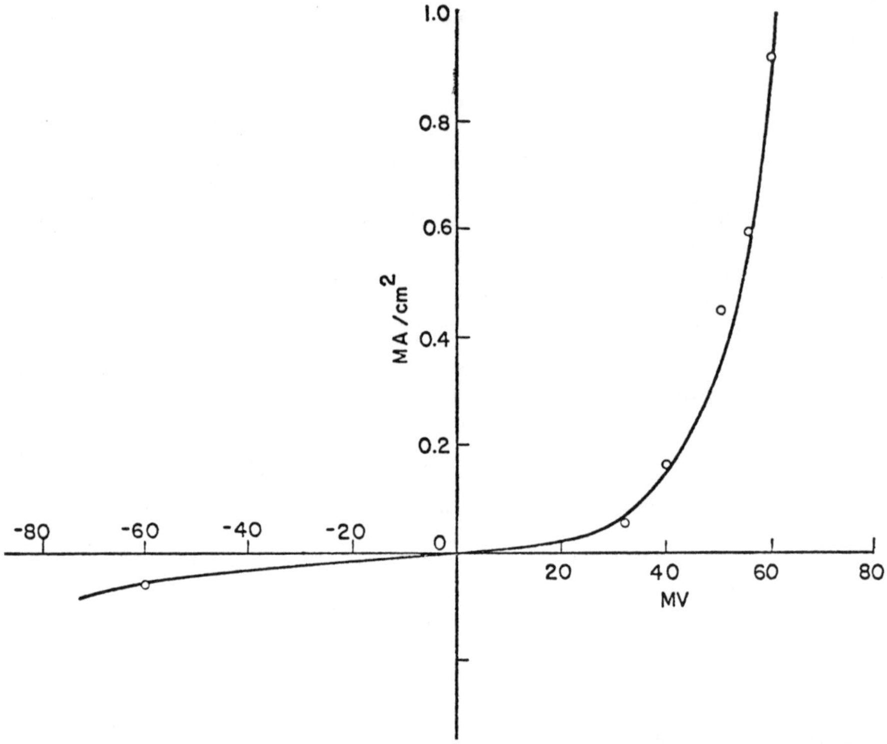

Fig.2. Current-voltage curve of oxidized cholesterol membrane in the presence of alamethicin(10^{-7}mol/l). The ordinate is the current in mA/cm^2 and the abscissa is the potential in mV. Alamethicin is added inside the cup.

in Fig.4 are obtained from the three curves shown in Fig.3. The impedance plots are linear within the experimental error and almost vertical to the abscissa at high frequencies. The intersections of these plots with the abscissa gives the values of series resistances for each KCl solution. From Fig.4, we obtain the values of series resistances, 480, 1250 and 5500 ohms for curves 3, 2 and 1 in Fig.3. These values are used in eqs.(6) and (7) to calculate the membrane capacitance and conductance. The calculated membrane capacitances are shown in Fig.3 by dotted lines. As seen, the calculated membrane capacitances are practically independent of frequency for all these KCl solutions. Hence we can conclude that the membrane capacitance and conductance, though not shown in the figure, of artificial lipid bilayer membranes, without alamethicin, is independent of frequency. This is the same conclusion reached by Hanai, Haydon and Taylor(4) and also by Schwan and Thompson(3). The same measurements and the method of analysis are extended to other lipid membranes such as egg-lecithin and sphingomyelin and we found consistently

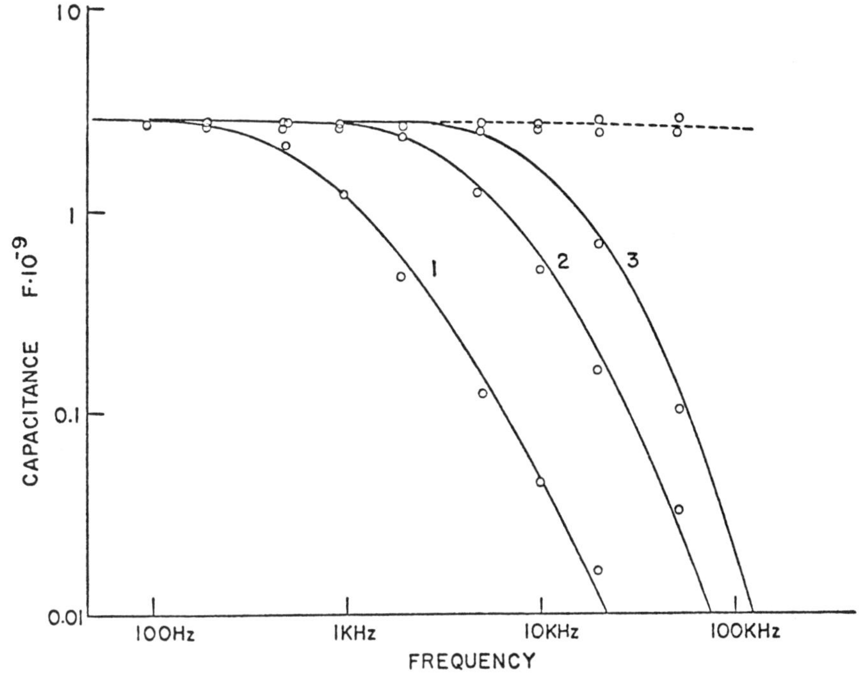

Fig.3. Measured capacitances of oxidized cholesterol membrane in KCl solutions. Solid curves 1,2 and 3 are obtained in KCl solutions with resistances 480, 1250 and 5500 ohms. Dotted lines with open circles are after correction for series resistance R_o.

that the membrane capacitances are independent of frequency.

The membrane capacitance can be converted to membrane capacity using the area of the orifice 0.25 mm^2. Although the membrane capacity varies, to some extent, from one membrane to another, we obtain, as an average, a value 0.7 uF/cm^2 for various lipid bilayer membranes. This value is consistent with the values reported by other investigators and is smaller than the membrane capacity of nerve axon membrane, i.e., $1uF/cm^2$.

As shown in Fig.2, alamethicin makes bilayer membranes permeable to ionic species and permits ionic currents through these membranes. From the linear portion of the I-V curve shown in Fig.2, we can calculate the membrane resistivity to be $2x10^6$ ohms cm^2. The I-V curve also indicates the limits of linearity of the membrane admittance in the presence of alamethicin. This means that the field strength of the a.c. potential used for the bridge measurements must not exceed these limits and should be constricted within 30 mv in order to avoid the non-linearity. In all admittance measurements in this experiment, the strength of the a.c. field is limited to 10mv peak to peak

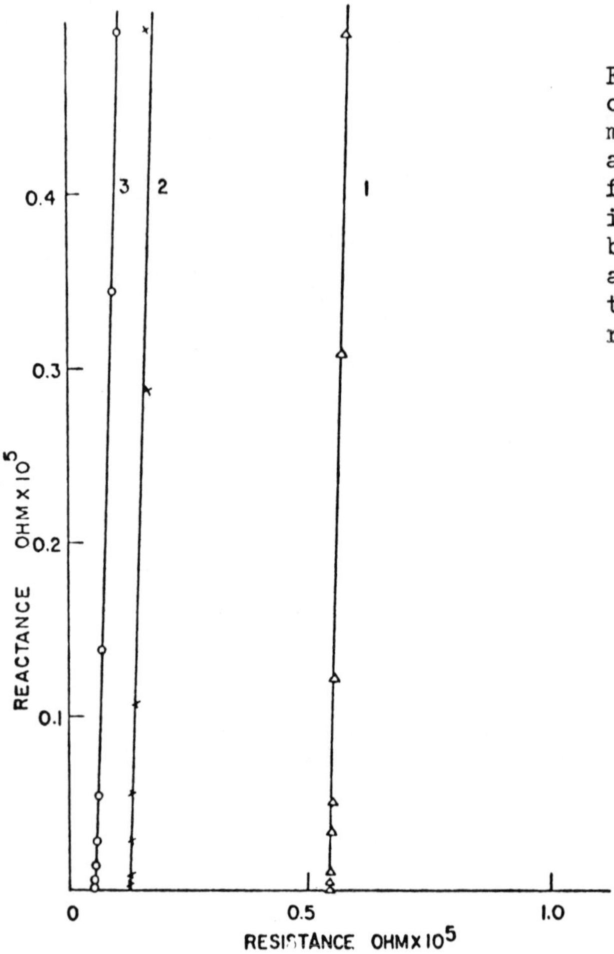

Fig.4. Impedance loci of oxidized cholesterol membrane. Curves 1,2 and 3 are calculated from corresponding curves in Fig.3. Intersections between impedance loci and the abscissa give the values of series resistances.

and this ensures the linearity of the measurement.

The membrane capacitance of oxidized cholesterol bilayers and its frequency profiles are shown in Fig.5 with and without alamethicin. The apparent difference in the static membrane capacitances between two curves is merely due to the difficulty of obtaining the same membrane capacity consistently and the difference should not be construed as the effect of alamethicin. Of importance in this figure is that the membrane capacitance is still frequency independent even with alamethicin. This indicates that even in the presence of active elements and possible ionic currents, the overall structure of the membrane matrices does not undergo a significant change. To reinforce this

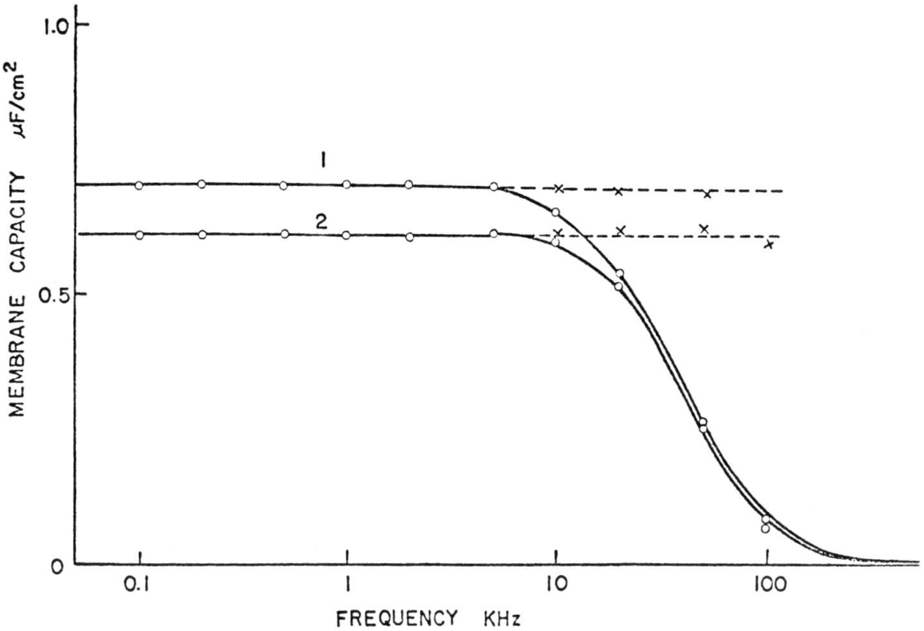

Fig.5. Capacitances of oxidized cholesterol membrane with and without alamethicin. Curve 1, with alamethicin and curve 2, without. Crosses and dotted lines are after correction for series resistances.

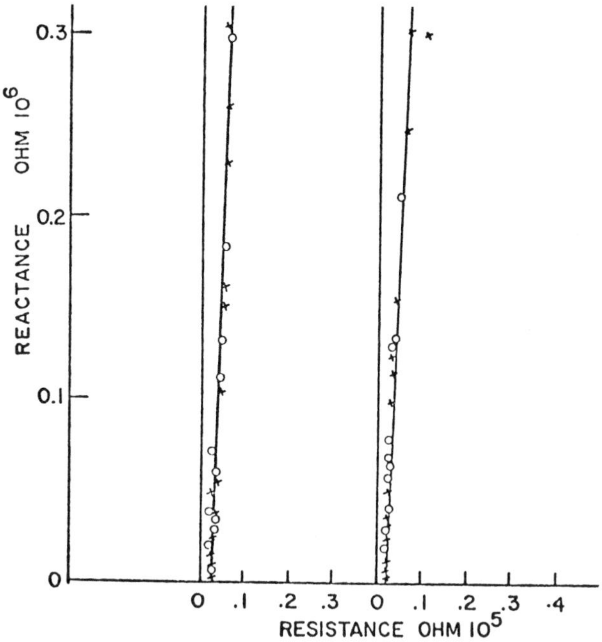

Fig.6. Impedance loci of oxidized cholesterol(1) and egg-lecithin(2) membranes. Circles are in the presence of alamethicin and crosses are without alamethicin.

conclusion, the impedance plots of two lipid bilayer membranes are shown in Fig.6. As shown in this figure, the impedance plots of oxidized choesterol and egg-lecithin membranes are quite unaffected by the addition of alamethicin. These two figures suggest that the addition of active elements creates a structural change in the membrane which is limited only to the small region around alamethicin molecules.

Discussion

As discussed, the capacitance of artificial lipid bilayer membranes is virtually independent of frequency indicating the lack of dipolar relaxation in these membranes. The presence of an active element, alamethicin, had been expected to give rise to relaxation phenomena in the membrane. This anticipation was based on the possible cooperativity of the membrane structure, that is, the creation of active sites by the presence of alamethicin would have a long range effect on the membrane structure. However, the results described above indicate contrary and the presence of active elements induces only limited changes in the membrane properties.

Another reason for this expectation is based on the theoretical consideration by Cole(17). As mentioned before, the presence of alamethicin evokes a time dependent ionic current which resembles the outward current in axon membrane. Eliminating the terms for inward current and leakage current, the Hodgkin-Huxley equation(18) becomes:

$$I = C_m\, dV/dt + g_k n^4 (V - V_k) \qquad (8)$$

where V_k is the Nernst potential, n is a parameter whose numerical value ranges from 0 to 1 and g_k is the conductivity related to the outward current. If the current is perturbed by a sinusoidal field, the additional current can be calculated by expanding the H-H equation in the Taylor series. The Laplace transformation of eq.(8) after the Taylor expansion and retaining only the first order term, gives rise to an expression for the admittance in the frequency domain:

$$G(p) = C_p + g_k n^4 + 4 g_k n^3 V \frac{dn/dV}{1 + p\tau_n} \qquad (9)$$

where p is the Laplace transform operator(jw). The first two terms represent the high frequency conductance and the third term represents the complex conductance. The numerical calculation of the membrane conductance and capacitance was performed by Majer(19) for axon membrane using somewhat different derivation. Majer's calculation indicates that the membrane capacity due to time dependent ionic current is of the order of 10^{-9} farad per cm^2 for squid axon. Thus the possible frequency dependent membrane capacitance evoked by the ionic currents is too small to be detected by the conventional bridge method, when it is superimposed with a much larger frequency independent capacitance. The magnitude of ionic current in the artificial membranes is approximately the same as those in squid axon and Majer's calculation can be applied to the present case without major revision. From these considerations, the possible frequency dependent membrane capacitance evoked by alamethicin is not likely to be detected under the condition of the present experiment.

The authors are indebted to Dr.P.Mueller. Without his guidance, this research was not possible. This work was supported by a grant NIH-HE-01253.

Reference

1) S.Takashima, H.P.Schwan and I.Tasaki, Abstract, Annual Meeting of Biophysical Society, Paper SaAM-E2, Toronto, Canada, 1972.
2) H.P.Schwan, Advances Biol.Med.Phsics,5,147,1957.
3) H.P.Schwan and T.E.Thompson, Abstract, Annual Meeting of Biophysical Soc., Paper MemB, Boston, Mass, 1966.
4) T.Hanai, D.A.Haydon and J.Taylor, Proc.Roy.Soc.London,A, 281,377,1964.
5) S.Ohki, Biophys.J.,9,1195,1969.
6) S.Ohki, J.Colloid and Interface Sci.,30,413,1969.
7) S.H.White, Biophys.J., 10,1127,1970.
8) H.G.L.Coster and R.Simmons, Biochim. Biophys. Acta, 203, 17, 1970.
9) G.Starl, B.Ketterer, B.Benz and P.Lauger, Biophys.J.,11,981, 1971.
10) P.Muller and D.O.Rudin, Nature, 217,713, 1968.
11) J.L.Oncley, Protein, Amino Acids and Peptides, Ed.E.J.Cohn and J.T.Edsall, Reinhold, New York, 1943.
12) S.Takashima, Physical Principles and Techniques of Protein Chemistry, Part A, Ed.S.J.Leach, Academic Press, N.Y.1969.
13) J.Wyman, Biophysical Chemistry, Vol.1, Chapter 6, Academic Press, New York, 1958.
14) S.J.Singer and G.L.Nicolson, Science, 175, 720, 1972.
15) K.S.Cole, J.Gen.Physiology, 12, 29, 1928.
16) K.S.Cole, J.Gen.Physiology, 12, 37, 1928.
17) K.S.Cole, Membranes, Ions and Impulses, University of California Press, Berkeley, 1968.
18) A.L.Hodgkin and A.F.Huxley, J. Physiology, 117, 500,1952b.
19) B.Majer, Thesis, University of Pennsylvania, 1973.

NUMERICAL COMPUTATIONS FOR THE FLOW OF LIQUID CRYSTALS

Bruce A. Finlayson

Department of Chemical Engineering

University of Washington, Seattle, Washington 98195

We report here calculations for average properties - viscosity, thermal conductivity, and dielectric constant - when a nematic liquid crystal is undergoing flow in three situations: Poiseuille flow in a capillary, and Couette and Poiseuille flow between flat plates including the presence of a magnetic field. By presenting the results in dimensionless form we are able to derive "universal" curves for these average properties in their dependence on the appropriate driving forces: pressure drop, surface velocity and magnetic field. The "universal" curves are nearly the same for p,p'-dimethoxy-azoxybenzene (PAA) and p'methoxybenzylidene-p-n-butylaniline (MBBA) even though their respective viscosities differ markedly.

EQUATIONS

The three physical situations studied are illustrated in Figure 1. For Poiseuille flow between flat plates the equations governing the director orientation θ(y') are given by Leslie [1].

$$f'(\theta) \frac{d^2\theta}{dy'^2} + \frac{1}{2} \frac{df'}{d\theta} \left(\frac{d\theta}{dy'}\right)^2 + \frac{1}{2} \frac{ay'}{g'(\theta)} [\lambda_1 + \lambda_2 \cos 2\theta]$$

$$+ \frac{\Delta\chi}{2} H^2 \sin 2\theta = 0 \qquad (1)$$

where $f'(\theta) = k_{11} \cos^2\theta + k_{33} \sin^2\theta$

$$2g'(\theta) = 2\mu_1 \sin^2\theta\cos^2\theta + (\mu_5-\mu_2)\sin^2\theta + (\mu_6+\mu_3)\cos^2\theta + \mu_4$$

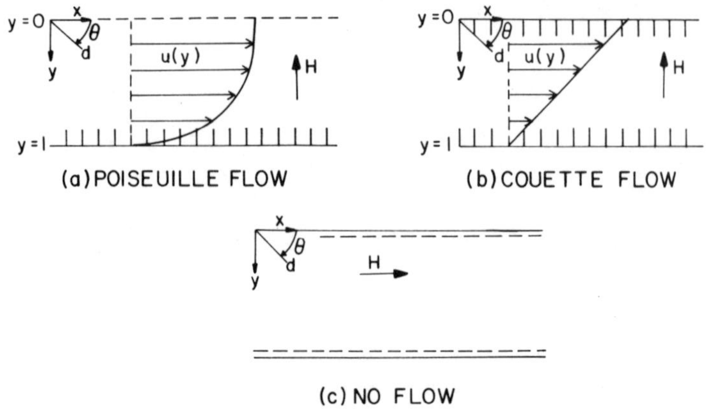

Figure 1. Flow Cases

and y' is the dimensional distance from the midpoint between the two planes, $a = (P_L - P_o)/L < 0$ is the pressure drop per unit length, H is the magnetic field, k_{11} and k_{33} are the splay and bend elastic constants, μ_1 through μ_6 are viscosity coefficients, $\lambda_1 = \mu_2 - \mu_3$ and $\lambda_2 = \mu_5 - \mu_6$, and $\Delta\chi = \chi_a - \chi_t$, the anisotropy of magnetic susceptibility. The velocity is governed by

$$g'(\theta)\ \frac{du'}{dy'} = ay' \qquad (2)$$

where u' is the dimensional velocity.

The key result of this paper depends on making these equations dimensionless. We do that by defining

$$f(\theta) = \cos^2\theta + \frac{k_{33}}{k_{11}} \sin^2\theta$$

$$g(\theta) = \frac{1}{\mu_4} g'(\theta) = \frac{\mu_1}{\mu_4} \sin^2\theta\cos^2\theta + \frac{\mu_5-\mu_2}{2\mu_4} \sin^2\theta + \frac{\mu_6+\mu_3}{2\mu_4} \cos^2\theta + \frac{1}{2}$$

$$y = y'/b,\ u = \mu_4 u'/(ab^2)$$

where b is the half-thickness for Poiseuille flow between flat plates. Inserting these definitions into equations (1) and (2) yields, after rearrangement,

$$\frac{d^2\theta}{dy^2} + \frac{1}{2f(\theta)}\frac{df}{d\theta}\left(\frac{d\theta}{dy}\right)^2 + \frac{A_2 y}{f(\theta)g(\theta)}\left[1 + \frac{\lambda_2}{\lambda_1}\cos 2\theta\right] + \frac{A_3}{f(\theta)}\sin 2\theta = 0 \tag{3}$$

$$\frac{du}{dy} = \frac{y}{g(\theta)} \tag{4}$$

where

$$A_2 \equiv \frac{ab^3\lambda_1}{2k_{11}\mu_4}, \quad A_3 \equiv \frac{\Delta\chi H^2 b^2}{2k_{11}}$$

b = half-thickness.

The equations are to be solved subject to the boundary conditions.

$$\frac{d\theta}{dy}(0) = 0, \ \theta(1) = \theta_{wl}, \ u(1) = 0 \tag{5}$$

Here we consider only the case $\theta_{wl} = 90^o$ (i.e. the director is perpendicular to the wall). After solving equations (3) - (5) for $\theta(y)$ and $u(y)$ we define the following average properties

$$\mu_{avg} = - 1/(3\ u_{avg})$$

$$\frac{1}{k_{avg}} = \int_o^1 \frac{dy}{k(\theta(y))}$$

$$\frac{1}{\varepsilon_{avg}} = \int_o^1 \frac{dy}{\varepsilon(\theta(y))}$$

where

$$k(\theta) = 1 + \frac{k_a - k_t}{k_t}\sin^2\theta$$

$$\varepsilon(\theta) = 1 + \frac{\varepsilon_a - \varepsilon_t}{\varepsilon_t}\sin^2\theta$$

are the dimensionless thermal conductivity and dielectric constant functions, and $k_{avg} = k'_{avg}/k_t$, $\varepsilon_{avg} = \varepsilon'_{avg}/\varepsilon_t$. The primed terms have dimensions, as do the principle values k_t, k_a, the transverse and axial thermal conductivity. The average viscosity is defined in such a way that we recover the viscosity for a Newtonian fluid.

Now it is clear that the average thermal conductivity, for example, depends on $(k_a - k_t)/k_t$ as well as the distribution of $\theta(y)$. The solution for $\theta(y)$ depends primarily on the parameters A_2 and A_3, and secondarily on the ratios λ_2/λ_1 and those appearing in the definitions of $f(\theta)$ and $g(\theta)$. The results of this paper demonstrate that the predominate influence of the parameters is through A_2 and A_3.

For Couette flow between flat plates the dimensionless equations are

$$\frac{d^2\theta}{dy^2} + \frac{1}{2f(\theta)} \frac{df}{d\theta} \left(\frac{d\theta}{dy}\right)^2 + \frac{A_2}{f(\theta)g(\theta)} \left[1 + \frac{\lambda_2}{\lambda_1} \cos 2\theta\right] + \frac{A_3}{f(\theta)} \sin 2\theta = 0$$

$$\frac{du}{dy} = \frac{1}{g(\theta)}$$

$$\theta(0) = \theta_{wo} \qquad \theta(1) = \theta_{w1} \qquad u(1) = 0$$

The definition of A_3 remains the same except that now b is the total distance between the flat plates. In addition the following definitions apply for Couette flow.

$$A_2 \equiv \frac{c\lambda_1 b^2}{2\mu_4 k_{11}} \quad , \ b = \text{total thickness}$$

$$\mu_{avg} = -1/u(0)$$

$$u = \mu_4 u'/(cb)$$

The parameter c is the stress exerted on the moving surface, and the velocity of the surface is $V = c\ b\ u(0)/\mu_4$. Here we consider only the case $\theta_{wo} = \theta_{w1} = 90^o$, except when $A_2 = 0$ and we take $\theta_{wo} = \theta_{w1} = 0^o$.

Table I. Physical Constants

	PAA	MBBA
k_{33}/k_{11}	1.894	1.228
μ_1/μ_4	0.632	0.078
$(\mu_5-\mu_2)/2\mu_4$	0.853	0.744
$(\mu_6 + \mu_3)/2\mu_4$	- 0.184	- 0.214
λ_2/λ_1	- 1.052	- 1.031
$(k_a-k_t)/k_t$	0.461	0.640
$(\varepsilon_a-\varepsilon_t)/\varepsilon_t$	- 0.033	- 0.059

In the equation for $f(\theta)$ it is clear that when $\theta \sim 0°$, $f \sim 1$, while for $\theta \sim 90°$, $f \sim 1 + k_{33}/k_{11}$. Thus we expect that when the wall orientation θ is 90°, then k_{33} will be a better parameter to non-dimensionalize f' with than k_{11}. Thus we also take

$$B_2 = \frac{A_2}{k_{33}/k_{11}} = \frac{ab^3\lambda_1}{2k_{33}\mu_4} \quad \text{or} \quad \frac{c\lambda_1 b^2}{2\mu_4 k_{33}}$$

The equations were solved numerically, using a new technique, orthogonal collocation on finite elements, which will be described elsewhere.

PARAMETERS

Calculations are reported for PAA at 122°C and MBBA at 25°C. The ratios of parameters needed are listed in Table I. The elastic constants for PAA are from Gruler [2]; for MBBA they are from Haller [3]. The viscosities for PAA are the same as used in Tseng, *et al.* [4], while the MBBA values are from Gähwiller [5]. The thermal conductivity values are from Longley-Cook and Kessler [6] and Pieranski, *et al.* [7], and the dielectric constants are from Maier [8] and Gerritsma, *et al.* [9]. The ratios of parameters are not too different for the two fluids even though the original values differ markedly. For example μ_4 is 6.8 centipoise for PAA and 83.2 centipoise for MBBA.

To present the results we define the following functions:

$$\psi_\mu = \frac{\mu_{avg} - \mu(\theta_o)}{\mu_{avg}(A_2{=}0) - \mu(\theta_o)} \tag{6}$$

$$\phi_k = \frac{1 + \frac{\Delta k}{k_t}}{\Delta k/k_t} \left(1 - \frac{1}{k_{avg}}\right) \tag{7}$$

where $-\lambda_1/\lambda_2 = \cos 2\theta_o$, $\mu(\theta_o)$ is the value of $g(\theta)$ for $\theta = \theta_o$, μ_{avg} $(A_2 = 0)$ is the value of μ_{avg} when $A_2 = 0$. μ_{avg} $(A_2 = 0)$ is the same as $\mu_{avg}(\theta = \frac{\pi}{2})$ except for Poiseuille flow in a capillary. Similar definitions apply to ϕ_ε, χ_k and χ_ε. The purpose of these definitions is to normalize the μ_{avg}, k_{avg}, and ε_{avg} so that they vary between 0 and 1.

RESULTS

First consider Poiseuille flow through a cylindrical capillary tube. Calculations have been done for PAA previously [4]*, and were presented as a graph of μ'_{avg} versus Q/R, where Q is the flow rate and R is the capillary radius. Since $a = -\mu'_{avg} 8Q/\pi R^4$ by definition (note $a < 0$),

$$A_2 = \frac{aR^3\lambda_1}{2k_{11}\mu_4} = \frac{\mu'_{avg}(-\lambda_1)}{k_{11}\mu_4}\frac{4Q}{\pi R}$$

and we can think of A_2 as a dimensionless Q/R. Similar calculations were done for MBBA. If plotted as μ'_{avg} versus Q/R we would obtain two widely different curves for the two fluids, due to the large differences in viscosity parameters. If we use Eq. (6), together with the calculated results at $A_2 \to 0$ of μ'_{avg} = 8.88 centipoise for PAA and 95.9 centipoise for MBBA, and plot ψ_μ as a function of $B_2 = a R^3 \lambda_1/(2k_{33}\mu_4)$, since k_{33} is the relevant parameter when the wall orientation is 90°, we obtain the results in Figure 2.

---

*The previous calculations use slightly different values of k_{11} and k_{33} than are reported in Table I here.

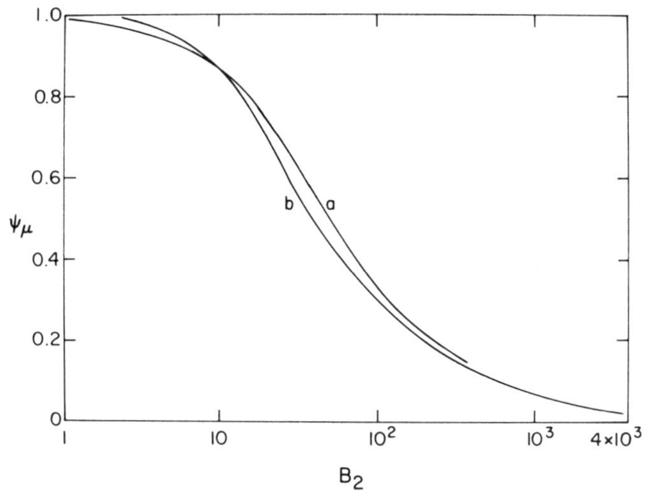

Figure 2. Viscosity Variation with Pressure Drop, Flow in Capillary (MBBA: a ; PAA = b)

The curves are very nearly the same, illustrating the major thesis of this paper: the dimensionless average properties are nearly independent of the material. It is apparent that to achieve flow orientation it is necessary to have $B_2 \geq 600$ to have $\psi < 0.1$.

Next consider case (c) in Figure 1. Previous work has shown that the thermal conductivity function ϕ_k will depend on H/H_c and k_{33}/k_{11} [7]. The parameter A_3 is another form of H/H_c and the solution is

$$\theta = 0 \text{ for } A_3 \leq \frac{\pi^2}{2}$$

The b in the definition of A_3 here is the total thickness between the plates. Figure 3 illustrates the results for ϕ_k and ϕ_ε for the two fluids. Pieranski, et al. [7] showed that ϕ_k should be linear for $H \sim H_c$, as is the case. For high magnetic fields ($A_3 \geq 30$) the k_{avg} and ε_{avg} are linear functions of $1/\sqrt{A_3}$. For $\phi \geq 0.9$ it is necessary that $A_3 \geq 200 - 360$ (or $H/H_c \geq 40 - 70$) depending on whether one is interested in ε or k and depending on the fluid.

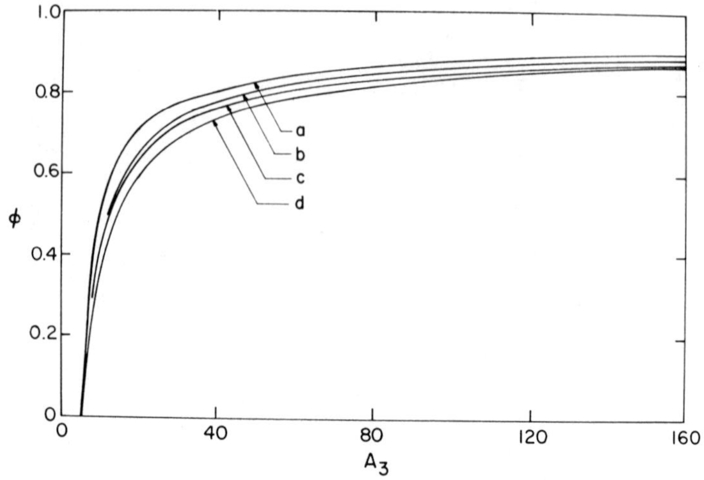

Figure 3. Thermal Conductivity and Dielectric Constant Variation with Magnetic Field (MBBA: a-k, c-ε; PAA: b-k, d-ε)

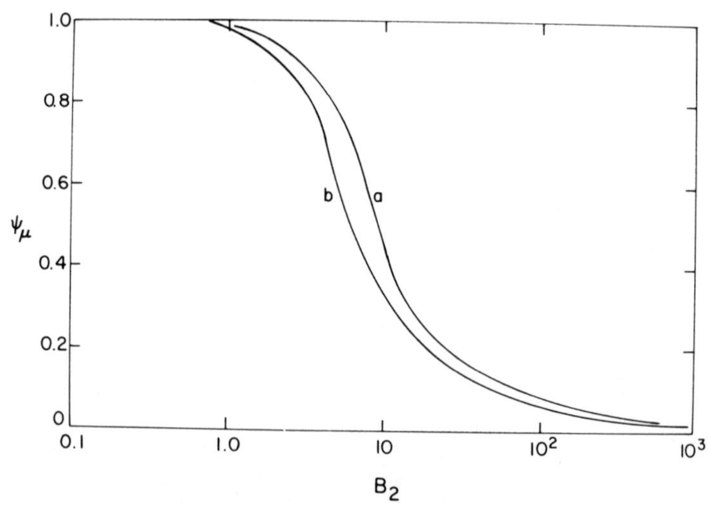

Figure 4. Viscosity Variation with Velocity for Couette Flow (PAA-a; MBBA-b)

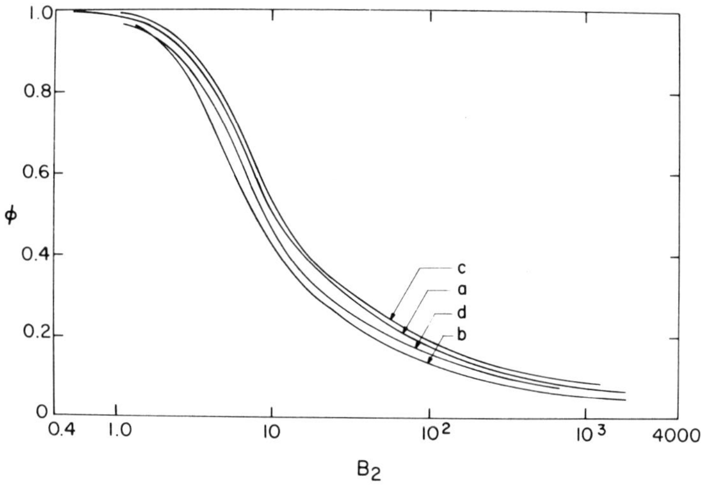

Figure 5. Thermal Conductivity and Dielectric Constant Variation with Velocity for Couette Flow (MBBA = a-k, b-ε; PAA = c-k, d-ε)

Figure 4 gives results for ψ_μ as a function of B_2 for Couette flow. Again the two curves are almost the same and it is necessary for $B_2 \geq 70$ before flow orientation occurs (defined arbitrarily as when $\phi_\mu < 0.1$). However, even though the viscosity has decreased to within 10% of its ultimate value for $B_2 \geq 70$, the thermal conductivity and dielectric constant have not, as shown in Figure 5. It is necessary for $B_2 \geq 300$-500 before those values are within 10% of their asymptotic values. For a value of $B_2 \sim 1$ the director has essentially the wall orientation ($\theta = 90^o$) throughout the fluid. The values of B_2 needed for ψ or ϕ to be 0.9, 0.5 or 0.1 are listed in Table II. Note in Figure 5 that the four curves, ϕ_k and ϕ_ε for the two fluids, are similar - again showing that "universal" curves of this type are possible provided the proper dimensionless groups are used.

Figure 6 illustrates the results for Poiseuille flow between flat plates. Here b is the half-thickness. A value of $B_2 > 120$ is necessary before flow orientation occurs.

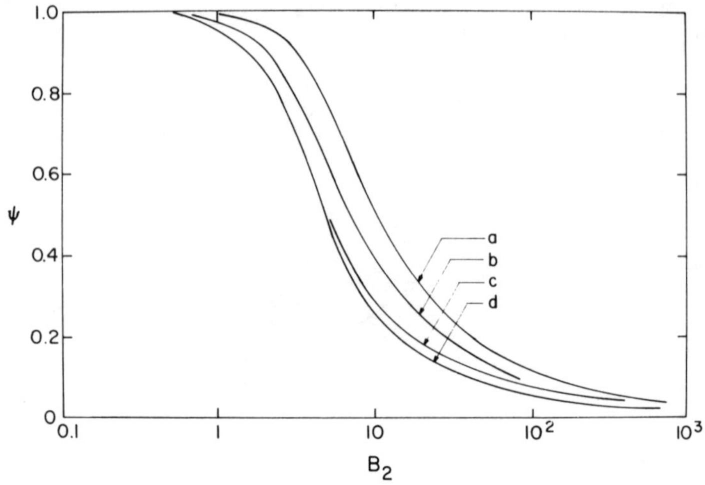

Figure 6. Viscosity, Thermal Conductivity and Dielectric Constant Variation with Pressure Drop for Poiseuille Flow (PAA: a-μ, c-ε, d-k; MBBA: b-μ. The curves of ε and k for MBBA are not shown. They are very close to but slightly below the respective curves for PAA.)

Table II. Limiting Values of Driving Force to Achieve ψ or φ of 0.9, 0.5 or 0.1

Type of flow	Parameter	90%	50%	10%
Poiseuille flow in capillary	B_2	8	40-50	600
Poiseuille flow between flat plates	μ B_2	2-3	7-10	120
	k or ε B_2	1.5	5	40-60
Couette flow between flat plates	μ B_2	2-3	6-9	70
	k or ε B_2	2-3	7-12	300-500
No flow, with magnetic field	A_3	200-360	9-15	6

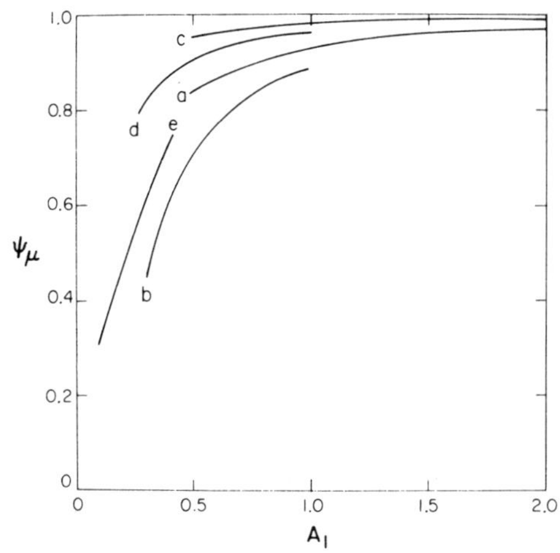

Figure 7. Viscosity Variation with both Pressure gradient and Magnetic Field for Poiseuille Flow Between Flat Plates (MBBA: a-A_2 = 10, b-A_2 = 20; PAA: c-A_2 = 10, d-A_2 = 20, e-A_2 = 50)

Solutions are also possible when both A_2 and A_3 are non-zero. Figure 7 presents results when A_2 covers the range from 10 to 50 and A_3 is from 5 to 20 (b = half-thickness, or from 20 to 80 when b = whole thickness).

Figures 6 and 3 show that this covers a range of strong (but not complete) flow orientation and strong (but not complete) magnetic field orientation. For the case illustrated in Figure 7 (corresponding to case a in Figure 1, with $\theta_{w1} = 90^o$) the flow and magnetic effects are competing with each other. The dimensionless group

$$A_1 = \frac{A_3}{A_2} = \frac{\Delta\chi\ H^2\ \mu_4}{b\ a\ \lambda_1}\ ,\quad b = \text{half-thickness}$$

measures the relative importance of the magnetic effect to the flow effect. The results in Figure 7 confirm that for high velocity (high A_2) the value of A_1 must be large before the magnetic field dominates the orientation.

The figures can be used to determine if wall effects are important in experiments. For example, Marinin and Zwetkoff [10] measured the dielectric constant of PAA flowing in a parallel plate capacitor in the presence of a magnetic field. H varies from 500 to 3000 gauss and for this particular situation then A_3 varies from 5.4 to 190. Figure 3 shows that wall effects will be important at the smaller range of A_3. The experimental range of u'_{avg} is 0.17 to 2.12 cm/sec. These values correspond to values of B_2 from 100 to 1000. Figure 6 shows that these values of B_2 are sufficient to achieve almost complete flow orientation. Values of A_1 are then below 0.5, and Figure 7 suggests that much larger values of magnetic field would be needed to overcome the flow orientation. This was observed in the experiment.

In the thermal conductivity experiments of Fisher and Fredrickson [11], for Couette flow a shear rate of 30 sec^{-1} was necessary to achieve flow orientation. The thermal conductivity anisotropy deduced from that experiment is opposite to that measured by Longley-Cook and Kessler [6], but it is still of interest to compare the value of B_2 for the k_{avg} to change 50% of the total possible change. The experiments gave a shear rate of 1.4 sec^{-1} which corresponds to $B_2 = 48$, whereas from Figure 5 we get $B_2 = 9$.

SUMMARY

The dimensionless graphs presented here enable easy predictions of the importance of flow orientation and magnetic field orientation in diverse flow situations. The "universal" curves are computed for PAA and MBBA. While no claim is made that all nematic liquid crystals would follow similar curves, it is shown that the dimensionless average properties depend primarily on the driving forces A_2 (pressure drop) and A_3 (surface velocity) and only secondarily on the ratios of elastic constants, and viscosity parameters, and thermal conductivity and dielectric constant anisotropies.

ACKNOWLEDGMENT

This research was supported by the National Science Foundation Grant No. GK-12517. The author thanks Mr. H. C. Tseng for doing the computations reported in Figure 2.

REFERENCES

1. Leslie, F. M., Arch. Ration. Mech. Anal. 28, 265 (1968).

2. Gruler, H., Z. Naturforsch. 28A, 474 (1973).

3. Haller, I., J. Chem. Phys. 57, 1400 (1972).

4. Tseng, H. C., Silver, D. L. and Finlayson, B. A., Phys. Fluids 15, 1213 (1972).

5. Gähwiller, Ch., Phys. Lett. 36A, 311 (1971).

6. Longley-Cook, M. and Kessler, J. O., private communication.

7. Pieranski, P., Brochard, F. and Guyon, F., J. Phys. (Paris) 33, 681 (1972).

8. Maier, W., Z. Naturforsch. 16A, 470 (1961).

9. Gerritsma, C. J.,DeJeu, W. H. and Van Zanter, P., Phys. Lett. 36A, 389 (1971).

10. Marinin, W. and Zwetkoff, W., Acta Physicochim. URSS 11, 837 (1939).

11. Fisher, J. and Fredrickson, A. G. Mol. Cryst. Liq. Cryst. 6, 255 (1969).

THE INVESTIGATION OF LIPID-WATER SYSTEMS PART 5

INFRA RED SPECTRA OF MESOPHASES

M. P. McDONALD, L. D. R. WILFORD

DEPARTMENT OF CHEMISTRY AND BIOLOGY

THE POLYTECHNIC, SHEFFIELD, ENGLAND

INTRODUCTION

There have been relatively few infra red studies of the liquid crystal (l.c.) state and most of these have been concerned with thermotropic mesophases (1). Because of the ease of orientation of nematic mesophases it has been possible to estimate the degree of order in these phases by infra red spectroscopy and the results have been seen to compare favourably with those obtained by other methods (2).

The only reported infra red measurements on lyotropic mesophases are those of Bulkin (3,4) on phospholipid-water mixtures. He showed that changes in the methylene deformation band at 1470 cm^{-1} can be used to detect phase transitions in these systems in which both lamellar and hexagonal l.c. phases occur.

In this laboratory broad line nmr has been used to investigate the structure of the solid and l.c. phases formed in monoglyceride-water systems (5,6,7). The effects of changing temperature and water content on the degree of order in the hydrocarbon and water regions of the lamellar mesophase in the 1-mono-octanoin(MG8)-water system have been observed through measurement of the proton resonance splittings of samples ordered between glass slides. A parallel investigation of the structure of ordered lyotropic mesophases has been carried out using infra red spectroscopy.

In this paper we report the phase diagram of the 1-mono-undecanoin(MG11)-D_2O system and infra red measurements on ordered mesophases in this system and on the polymorphic forms of anhydrous MG11. The MG11/D_2O system was chosen because of the large composition and temperature range of stability of its lamellar l.c. phase.

EXPERIMENTAL

Racemic MG11 was prepared by Malkin's method (8), using di-isopropyl ether in the extraction to reduce the loss of mono-glyceride into the aqueous layer. After recrystallization the material was found to be at least 99% pure by thin layer chromatography (Mpt 56°C). The D_2O used was Koch-Light 99.7% grade.

The phase diagram was determined by visual observation of samples on glass slides or in sealed tubes using the polarizing microscope and also by differential scanning calorimetry (d.s.c.) of samples in sealed pans using the Du Pont Thermal Analyzer. The points around and below 0°C were determined by d.s.c. only. Samples were made up by warming weighed mixtures to the temperature at which they formed isotropic solutions and then cooling them. Care was taken to reduce to a minimum the contact with the atmosphere of deuterium containing samples.

Because of the supercooling which occurs in monoglyceride-water systems it was not possible to determine the transition temperatures on cooling runs. All samples were therefore cooled to -50°C and allowed to warm up at a programmed rate of 5°C min^{-1} to record the thermogram. The transition temperatures were taken to be the onset of the peaks on the thermogram.

The infrared spectra were obtained using a Grubb Parsons Spectromaster with Beckman-RIIC variable temperature cells. Temperature of the samples was controlled to within ± 3°C. Polarized spectra were obtained using a silver chloride stack in the common beam so that the E vector of the radiation was parallel to the slit. For the observation of decoupled ν (OH) bands it is necessary to reduce the OH/OD ratio in the sample to ca. 0.05(9). To achieve this the lipids were shaken with approximately 15% by weight of D_2O at temperatures in the isotropic solution region and the mixtures cooled to the l.c. state. They were then freeze dried using liquid nitrogen, mixed with more D_2O, and the process repeated a further six times.

Ordered samples of l.c. phases were prepared by smearing a thin layer of material on one of the silica plates of the spectrometer cell and pressing the second plate gently down on to the l.c. layer.

The extent of the ordering along an axis normal to the silica plates was checked on the polarizing microscope and in most cases a completely dark field was obtained.

RESULTS AND DISCUSSION

Phase Study

The main features of the MG11/D_2O phase diagram (Fig.1) are similar to those of the MG12/H_2O system (5,10) and the MG10/H_2O system (11). It has been possible to detect the two phase region between N and FI visually but not by calorimetry. The N $\rightarrow$ FI transition temperature increases to a maximum value of 98°C. at 0.88 mole fraction of D_2O. The transitions from solid to gel and from gel to FI or N are clearly seen on the d.s.c. traces. The latter transition temperature shows a gradual decrease with increasing D_2O concentration except for a region near 0.5 mole fraction of D_2O where a significant hump occurs. This is again reminiscent of the MG12/H_2O system where a similar peak was attributed to the existence of a monohydrate (5). Larsson commented that crystalline hydrates cannot exist in the gel phase since the water in that phase is disordered (11). If as Larsson states the gel phase has the same structure as the α-crystal form of monoglycerides it is possible that at certain water contents the structure may have an optimum stability as reflected in a higher melting point. This sort of thing occurs in the α-phases of long chain alcohols which take up to one quarter mole of water with an increase in melting point of up to 3°C (12).

Other evidence of monoglyceride-water association appears in the low temperature d.s.c. results. Samples containing less than 0.47 mole fraction of D_2O were homogenized as isotropic fluid and then cooled to temperatures below -50°C in the calorimeter. No transition which could be associated with D_2O freezing or melting was observed on either cooling or subsequent heating runs. However, just above 0.47 mole fraction of D_2O a peak was observed at -11.5^{o}C which with increasing water content increases in size and moves steadily up to + 3°C at 0.93 mole fraction of D_2O as shown in Fig.1. In this laboratory the only other monoglyceride-D_2O systems investigated at low temperatures have been the MG8/D_2O and mono-olein/D_2O systems and in both these cases the D_2O melting transitions have all occurred between 0°C and 3°C. In the MG8/D_2O system the D_2O melting peak was visible in mixtures containing 0.3 mole fraction of D_2O. It is therefore considered that some specific lipid-water interaction is occurring at low temperatures in the MG11/D_2O system which probably occurs also in the MG12/H_2O system but not in MG8/D_2O or mono-olein/D_2O.

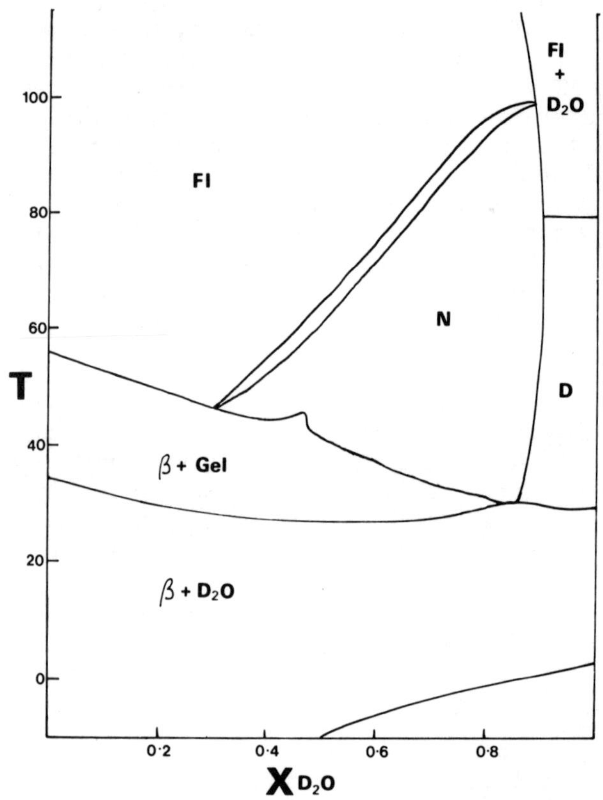

Fig.1. 1-Mono-undecanoin-deuterium oxide phase diagram

ν(OH) bands

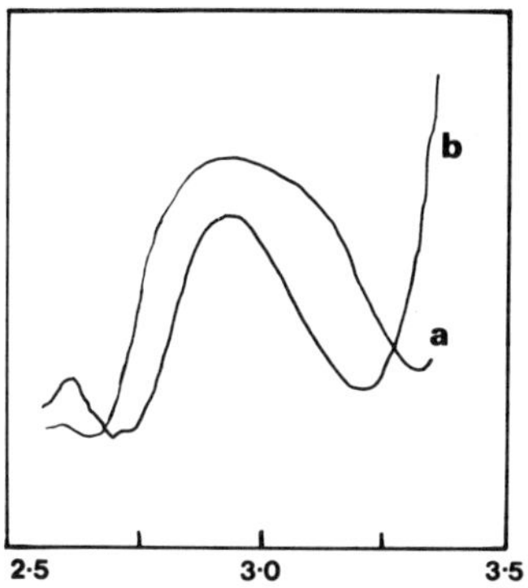

Fig.2. ν(OH) bands: a - coupled, b - decoupled. Wavelength 2.5 - 3.5μ.

The ν(OH) band is extremely broad in samples of MG11 containing ca. 0.5 mole fraction of H_2O with a maximum at 3410 cm^{-1} and with a pronounced shoulder at 3300 cm^{-1}. The half width of the band, $\Delta\nu\frac{1}{2}$, is greater than 400 cm^{-1} and the band remains unchanged both in position and half width by increase in concentration of water. When most of the hydroxyl protons have been replaced by deuterium the amount of coupling between neighbouring bonds is much reduced and the band is narrower as shown in Fig.2. ($\Delta\nu\frac{1}{2}$ = 270 cm^{-1}). The decoupled band is asymmetrical at low water contents and becomes more symmetrical as the water content increases until it is almost identical with the decoupled ν(OH) band in liquid water.

When the spectra of ordered samples of l.c. were recorded using polarized radiation no significant dichroisms were obtained by rotating the samples through 90° about the axis of the spectrometer beam.

Changes in the spectra were observed however when the sample was turned through 45° about an axis perpendicular to the spectrometer beam. This procedure has been carried out satisfactorily with thin silica plates but with silver chloride plates the transmission was too greatly reduced when they were used at an angle of 45° to the beam. It has only been possible therefore to obtain spectra in the 3000 cm^{-1} region since the silica plates do not transmit satisfactorily below this frequency.

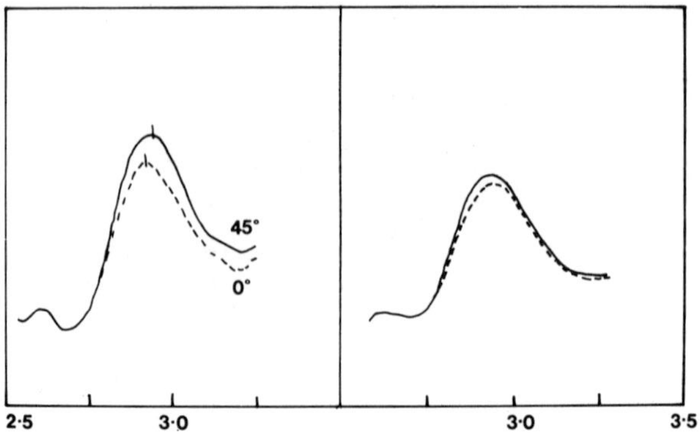

Fig.3. Decoupled ν(OH) bands in DMG11 containing 0.53 and 0.86 mole fraction of D_2O at 0° (V=0) and 45° (V=45) to the beam. Wavelength 2.5 - 3.5 μ.

Fig.3 shows the changes in the decoupled ν(OH) band in deuterated MG11 (DMG11) with concentration of D_2O at 45°C and the frequencies of the bands are given in Table 1.

Table 1

X_{D_2O}	ν(OH) at V=0	ν(OH) at V=45
0.53	3418	3406
0.86	3416	3413

As well as a decrease in frequency of the ν(OH) band at V=45 there is a marked increase in its intensity in the sample containing 0.53 mole fraction of D_2O. The differences in both frequency and intensity are both negligible in the sample containing 0.86 mole fraction of D_2O. An exactly similar set of data has been obtained from measurements on the lamellar phase in the DMG8/D_2O system. The experiments were repeated with a very thin film of water between the plates and no dichroic effects were observable.

On heating a sample of DMG11 containing 0.73 mole fraction of D_2O from 45°C to 75°C the dichroic effect disappeared almost completely.

The asymmetry of the unpolarized ν(OH) band at low water contents and the dichroic effects in the polarized spectra are both due to the structure within the water layer of the mesophase. The presence of such a structure has already been detected by nmr observations on ordered samples (7). The arrangement of monoglyceride molecules in a lamellar structure with their polar groups in contact with layers of water molecules will mean that there is a greater component of the transition moment of the ν(OH) vibration of these groups and of water molecules hydrogen bonded to them in a direction perpendicular to the layer.

It is known that hydrogen bonding between alcoholic hydroxyl groups and water molecules is stronger than that between the hydroxyl groups or water molecules alone (13) and one would therefore expect the stretching vibration of the OH groups involved in such bonding to occur at lower frequencies.

Thus the low frequency shoulder on the unpolarized spectra at low water contents may arise from these vibrations and the spectra at V=45 should represent the interaction of a greater proportion of vertically oriented ν(OH) components of low frequency than that at V=0.

When the concentration of D_2O in the system increases it leads to an increase in the thickness of the water layer but not in the number of molecules instantaneously hydrogen bonded to the monoglyceride OH groups. One would therefore expect the relative intensity of the low frequency vibration to diminish and in fact both the asymmetry of the unpolarized spectra and the differences between the two polarized spectra decrease with increasing water content. The similar effects produced by an increase of temperature are due to a greater relative decrease in the numbers of the most strongly associated species which give rise to the low frequency vibration. This is a quite general effect of temperature on hydrogen bonding equilibria.

ν(C=O) bands

At present there is no detailed knowledge on the hydrogen bonding structure in 1-monoglycerides in the solid, liquid or liquid crystalline states. In his X-ray analysis of two β forms of MG-18 and a β' form of 1-mono-11-bromoundecanoin Larsson (14) did not propose a hydrogen bonding scheme for the head groups but concluded that the hydrogen bonding occurred probably between hydroxyl groups only. Chapman (15) related the frequencies of the single bands he observed in different polymorphs of MG-18 and MG-16 to the degree of involvement of the carbonyl group in hydrogen bonding.

Table 2. ν(OH) and ν(C=O) band positions in 1-mono-undecanoin

Phase	Temp.	ν(OH)	ν(C=O)
Liquid	65°C	3448	1739,1731
α	30°C	3361	1735,1717
β'	20°C	3408,3302	1737,1729
β	30°C	3292,3253	1739,1731

In Table 2 we have recorded the positions of the peaks or shoulders of the ν(OH) and ν(C=O) bands in MG11 and in Fig.4 we show the actual bands for the three solid phases.

The occurrence of two ν(C=O) bands usually indicates an involvement of the carbonyl group in hydrogen bonding, the lower frequency component arising from the bonded carbonyl. Unfortunately there are a number of different hydrogen bonding possibilities in the case of the monoglyceride molecule since the carbonyl may bond with either of its own two hydroxyl groups or either of those on adjoining molecules.

The spectrum of 0.005M MG8 in carbon tetrachloride contains two well resolved carbonyl bands of equal intensity at 1754 cm^{-1} and 1731 cm^{-1} and in the ν(OH) region a number of closely spaced monomer peaks around 3600 cm^{-1} and a broader band of medium intensity centred at 3523 cm^{-1}.

There are three possible explanations for this latter band. It may be due to intra-molecular bonding between the two OH groups but when the groups are on adjacent carbon atoms the association band usually occurs at higher frequencies (16). There is also a possibility of bonding between OH and the ether oxygen of the carboxyl group but in a recorded example of this type of bond the intensity of the corresponding ν(OH) band was very low (17). Finally there may be hydrogen bonding between the carbonyl and one of the OH groups and this seems quite feasible since such associations have been found to give rise to ν(OH) bands near 3550 cm^{-1} (17,18) and bonded carbonyl bands between 1705 and 1740 cm^{-1} (17-22).

Further information has been obtained from nmr spectra of MG8 in deuterated dimethyl sulphoxide (d_6-DMSO) (23). The primary and secondary OH resonances can be distinguished in this solvent by their spin splitting multiplicity and they occur at δ_p=4.6ppm and δ_s=4.85ppm from TMS. Chapman and King found for propan-1,2,diol in this solvent that δ_p=4.45 ppm and δ_s=4.38 ppm (24) and it may be argued that the change in order of δ values arises from the interaction of the secondary OH with the carbonyl group in the monoglyceride solution.

There are thus some grounds for supposing that monoglycerides in solution may adopt the configuration shown below

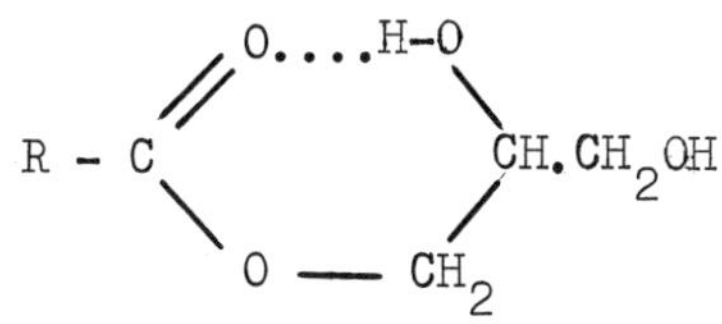

If this configuration persists in the solid and liquid state it may account for the fact that the low frequency carbonyl band is always strong as can be seen from Fig.4.

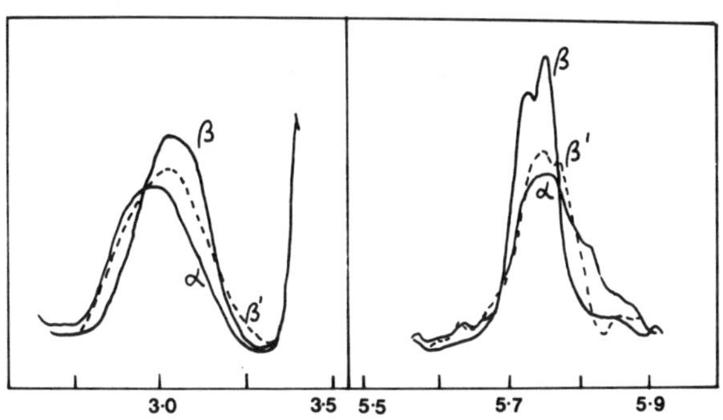

Fig.4. ν(OH) and ν(C=O) bands in anhydrous polymorphic forms of MG11.
Wavelength 2.5-3.5μ and 5.5-5.9μ.

The fact that there are differences in relative intensity of the two peaks suggests that the carbonyl group must get involved in other associations. It is even possible that both components of the ν(C=O) band in the solid and liquid states arise from hydrogen bonded carbonyl groups since their frequencies are significantly lower than that of the 1754 cm^{-1} band in dilute carbon tetrachloride solution.

The differences in relative intensity of the components of the ν(C=O) band in the α, β' and β phases are seen to be small but significant. The component at 1717 cm^{-1} in the α-phase suggests that an extra degree of hydrogen bonding becomes possible in this phase.

In the lamellar l.c. phase the ν(OH) band is located around 3400 cm^{-1} and the carbonyl doublet at 1744 cm^{-1} and 1733 cm^{-1} at all mole fractions of D_2O as shown in Fig.5.

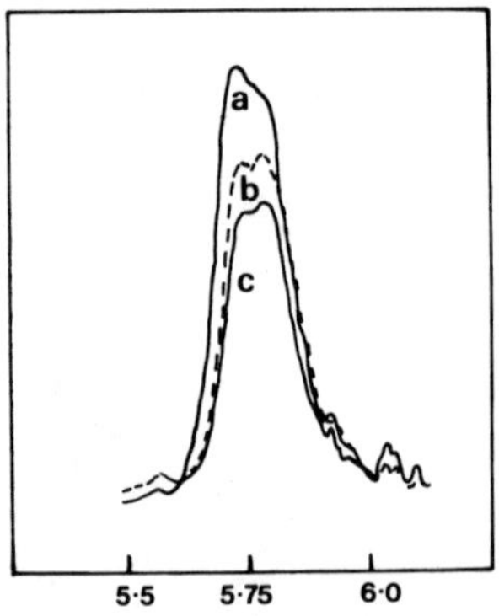

Fig.5. ν(C=O) bands in the lamellar l.c. phase of the DMG11/D_2O system.

a 0.53)
b 0.77) mole fraction D_2O
c 0.86)
Wavelength 5.5 - 6μ.

The relative intensity of the low frequency component is seen to increase with increasing D_2O content of the sample, and this sequence also occurs with increasing content of D_2O in the l.c. phase of the MG8/D_2O system. In the MG11 system increasing D_2O content causes increased melting point of the l.c. phase but in the MG8 system the melting point goes through a maximum at ca. 0.8 mole fraction of D_2O. Since the relative intensity of the low frequency component of the carbonyl band continues to increase above 0.8 mole fraction of D_2O it seems that the hydrogen bonding in which the carbonyl group is involved is not important to the stability of the lamellar phase.

References

1. S. Chandrasekhar, N.V. Madhusudana. App. Spec. Revs. 1973, 6, 193

2. A. Saupe, W. Maier, Z. Naturforsch. 1961, 16a, 816.

3. B.J. Bulkin, N. Krishnamachari. Biochem.Biophys.Acta 1970 211, 592

4. B.J. Bulkin, N. Krishnamachari. J. Amer. Chem. Soc. 1972, 94, 1109

5. A.S.C. Lawrence, M.P. McDonald. Mol. Crystals 1966, 1, 205

6. B. Ellis, A.S.C. Lawrence, M.P. McDonald, W.E. Peel, Liquid Crystals and Ordered Fluids, ed. J.F. Johnson and R.S. Porter (Plenum Press 1970) p.277

7. M.P. McDonald, W.E. Peel. Trans.Faraday Soc. 1971, 67, 890

8. T. Malkin, M.R. El Shurbagy. J. Chem. Soc. 1936, 1628

9. H.J. Hrowstowski, G.C. Pimentel. J. Chem. Phys. 1951, 19, 661

10. E.S. Lutton. J. Amer. Oil Chemists Soc. 1965, 42, 1068

11. K. Larsson. Zeit. Phys. Chem. 1967, 56, 173

12. A.S.C. Lawrence, M.A. Al-Mamun, M.P. McDonald. Trans. Faraday Soc. 1967. 63. 2789

13. N.N. Ugarova, L. Radic, I. Nemes. Russ. J. Phys. Chem. 1967, 41, 835

14. K. Larsson. Arkiv Kemi. 1964, 23, 35

15. D. Chapman. J. Chem. Soc. 1956, 55

16. L.P. Kuhn. J. Amer. Chem. Soc. 1952, 74, 2492

17. T.C. Bruice, T.H. Fife. J. Amer. Chem. Soc. 1962, 84, 1973

18. L.J. Bellamy, R.J. Pace. Spectrochim.Acta 1971, 27A, 705

19. R.E. Kagarise, K.B. Whetsel. Spectrochim.Acta. 1962, 18, 341

20. H. Minato. Bull.Chem. Soc. Japan. 1963, 36, 1020

21. W.K. Thompson, D.G. Hall. Trans.Faraday Soc. 1967, 63, 1553

22. S. Searles, M. Tamres, G.M. Barrow. J. Amer. Chem. Soc. 1953, 75, 71

23. W.E. Peel, Ph.D. Thesis, Sheffield Polytechnic (1972).

24. O.L. Chapman, R.W. King J. Amer. Chem. Soc. 1964, 86, 1256

OPTICAL PROPERTIES OF NEMATIC POLY-γ-BENZYL-L-GLUTAMATE

Donald B. DuPré and James R. Hammersmith

Department of Chemistry, University of Louisville

Louisville, Kentucky 40208

Poly-γ-benzyl-L-glutamate (PBLG) in certain solvents adopts an α-helical conformation and thus geometrically resembles a long, rigid rod with side chains containing benzyl radicals extending latterally outwards along the rod axis. As the refractive indices parallel and perpendicular to the helical axis are different, alignment of the macromolecule results in optical birefringence. The effect is dramatically seen by simply allowing a PBLG solution to flow between crossed polars. The rod shaped macromolecules tend to conform with hydrodynamic flow lines, producing a birefringent solution of remarkable color (streaming birefringence). The PBLG molecule itself has a positive dielectric and diamagnetic anisotropy, with the components of the dielectric and magnetic susceptibility tensor parallel to the long molecular axis larger than those perpendicular. In the presence of an electric or magnetic field individual PBLG molecules will also tend to align themselves, in this case such that their long axes are parallel to the direction of imposed field.

In certain solvents above a critical concentration of polymer (usually $\gtrsim$ 12% w/v), PBLG forms a lyotropic liquid crystal of the cholesteric structure. In the presence of sufficiently strong electric or magnetic fields liquid crystalline PBLG is known to undergo a mesomorphic phase transition from the natural cholesteric structure to one that resembles an aligned nematic. Presumably the applied field unwinds the helicoidal superstructure of the cholesteric phase and forms an oriented nematic liquid crystal with the rodlike PBLG molecules more or less parallel to the applied field. Indeed, the nematic stasis has been defined as a special case of the cholesteric, one in which the pitch of the helix is infinite. Electric or magnetic fields induce a pitch dilation in the liquid crystal and, at some critical field strength, the mesophase looses the high optical rotatory power and diffraction properties associated with cholesteric liquid crystals. The phenomenon of field induced phase transition may be termed electro- or magnetotropism, in analogy to the more usual lyotropic or thermotropic mesomorphic transitions which occur with changes in solvent concentration or temperature. Even in strong fields however the transition to nematic order of liquid crystal PBLG is slow, taking as long as 100 hours before equilibrium is reached. Electric fields have been shown to be more effective than magnetic fields in producing alignment of the molecular aggregates in this polymeric system. The field induced cholesteric to nematic transition of this liquid crystal has been followed spectroscopically by NMR[1-4] and IR dichroism,[5,6] and also by optical methods.[7] We have observed visually the slow cholesteric-nematic phase transition, as well as a more rapid macromolecular reorientation in the acquired nematic stasis, and a new critical orientation of nematic PBLG that results in unusual anisotropic optical effects. The phenomena are accounted for in terms of molecular order and dynamics of the associated polypeptide helices.

PBLG of molecular weight 310,000 was obtained from Pilot Chemicals, Inc. and used without further purification. Liquid crystal solutions of 13.7% (w/v) in millipore filtered chloroform were prepared in 1 cm x 1 cm x 3 cm spectrophometric cells which had been ultrasonically cleaned after a chromic acid rinse in acetic acid and high purity deionized water. It should be noted that PBLG samples in the studies cited above were of small volume, held in containers with large surface to sample contact. In an effort to reduce the influence of boundaries and surfaces, our samples were prepared in bulk.

Upon preparation PBLG solutions are non-uniform and striated. After several weeks of maturation, samples become uniformly bright but colorless when viewed between crossed polars. This is characteristic of the high optical rotatory power of lyotropic liquid crystals in the cholesteric state. The absence of brilliant iridescent color sometimes seen in light reflected from cholesteric solutions of PBLG is an indication that the helical pitch of the superstructure in our undisturbed samples is beyond the wavelength of light.[8]

After maturation our samples were subjected for an extended period of time to a strong magnetic field (12 kG) applied perpendicular to two of the cell wall faces. Visual observations were made with the sample placed between crossed polars, parallel and perpendicular to the field axis, in transmitted white light. When it was necessary to remove the cell from the magnet for field parallel observation, the sample was quickly returned to the same position between the poles. The magnetotropic cholesteric-nematic phase transition was visually evident in textural changes in the solution. Samples became clearer and more uniform. After several weeks residence time in the field observations were again made in transmitted white light with the sample placed between mutually perpendicular polars. Curious optical effects appeared which were dramatically different parallel and perpendicular to the field axis.

The view perpendicular to the field is one of regions of light and dark polygonal areas extending the height of the sample in a stable pattern. These blotches are of macroscopic dimensions covering areas as large as $.5cm^2$. Some have a slight green or pink coloration. There is no evidence of blotch boundaries when the sample is viewed in unpolarized light. When the polarizers are tilted slightly (± 10°) to the left and the right of the extinction position, a reversal of the light and dark regions occurs with the zone boundaries maintained. Careful observation revealed that the blotches are optically active wall domains of equal but opposite optical rotatory power that can be transformed from light to dark by equal clockwise or counterclockwise rotation of the polars from orthogonality. A similar effect has been observed in large drops of thermotropic liquid crystals placed horizontal to an applied magnetic field.[9,10] It is clear in observing our macroscopic samples that these optically active nematic blotches are purely a surface effect. It is evident that one is looking through an arrangement of blotched regions on the front surface of the cell to others on the rear.

More striking is the view between crossed polars parallel to the field axis which is marked by brilliant iridescent color (blues, greens, reds, orange browns). If the sample is thermally insulated between the pole faces the colors appear uniform over the width and height of the cell except at the meniscus boundary where irregular and complicated surface conditions result in a small swirl contortion of multi-color. The precise color seen is a very sensitive function of the angle of observation. When viewed directly along the optic axis the sample may be almost entirely extinguished between crossed polars. In our setup the PBLG samples were maintained in the magnet slightly below room temperature. If observations are made outside the magnet for more than a few seconds color uniformity is lost with evident pattern alterations proceeding inward from the warming cell surfaces. Deliberate thermal perturbation of a cell wall results in rapid color response.

In this critical condition liquid crystal PBLG was observed to be extraordinarily sensitive to thermal fluctuations. Even thermal gradients set up by air currents register in a color response. As our magnet is water cooled, a sample left exposed between the pole faces in the colder months is subject to a downwards air flow which cools the cell walls relative to the cell center. The result is convection currents which set up in the liquid crystal a regular, birefringent swirl pattern parallel to the magnetic field. The birefringent pattern extends outwards from a central oval which itself may be extinguished between the polars. About the central oval are smaller partially colored contortions reminiscent of the sky in Van Gogh's The Starry Night. The pattern along the field direction extends throughout the volume of the cell and is clearly more than just a surface boundary effect. When the polars are rotated the colors go over into one another continuously. This swirl or elliptical vortex pattern is also seen in natural light but the sample is otherwise colorless. In this critically aligned condition nematic PBLG when viewed parallel or near parallel to the major director is a very sensitive optical register to thermal fluctuations.

That this aligned liquid crystal may be extinguished when viewed directly along the transmission axis (parallel to applied magnetic field) is not surprising. The diamagnetic anisotropy of PBLG is such that individual molecules will tend to align themselves with their long axes parallel to the field. As the components of the refractive index perpendicular to the rod axis are equal one would expect optical extinction between crossed polars in perfectly aligned samples as in an uniaxial single crystal viewed along the c-axis. The brilliant colors observed though reminscent of those noted by others in undisturbed PBLG liquid crystal solutions[11] are seen only in transmitted polarized light parallel to the alignment axis. They are due therefore to optical retardation of preferential components of white light in the anisotropic sample. The presence of birefringent color in this bulk sample when the view is slightly off axis represents a recovery of the longitudinal component of the refractive index tensor of the macromolecule ($n_{//} > n_{\perp}$) and a fortunate phasal difference between the ordinary and extraordinary components of white light as they make their way through the 1 cm pathlength cell. The precise component of $\Delta n = n_{//} - n_{\perp}$ that one sees is dependent upon the orientation of the PBLG axes with respect to the viewer. These orientations in the critically aligned condition are apparently quite sensitive to disturbances such as thermal gradients. In observing the sample along and at slight angle to the field direction one goes from a condition of extinction of polarized light when the view is parallel to the nematic director to one in which an off-axis component of Δn and therefore color is seen. A slight change in this component (resulting from a change in viewing angle or thermal perturbation) is manifested by rapid change in color. Note should be made that if the macromolecular aggregates are not set up in proper orientation with the field (as in the case of insufficiently matured samples) such differences in overall refractivity of the sample would not be so extreme, and the resultant optical spectacle less vivid.

The field-parallel effect may be a property of the molecular size as well as liquid crystallinity as it is apparently not observed in other nematics composed of smaller molecules.

A macromolecular reorientation process on a much shorter time scale may be followed visually if the sample is rotated in the magnet while in its nematic stasis. When the liquid crystal is turned 90° in the field a rapid and continuous destruction of the pattern described above is seen to begin after 1-2 minutes. Birefringent regions become confused and diffuse with colors in the central region of the vortex persisting longer than those on the outer borders. The surface blotches then parallel to the new field axis disappear. The field is dark although not perfectly extinct with a salt and pepper appearance. Curious parallel striations run up and down the length of the cell. These striations are probably a result of cell surface preparation but may reflect a more fundamental property of the liquid crystal. They could result, for example, from lines of "disinclination" or optical discontinuity separating molecular aggregates that rotate to the left from those that rotate to the right in the re-establishment of alignment with the new field axis. Repeated 90° rotations in the field produce identical molecular reorientations which are essentially complete in 10-12 minutes. A reorientation process in PBLG on such a time scale has been followed by other workers[4] using NMR. We suspect that we are following the same macromolecular reorientation process optically. If the sample is allowed to remain in any one of these new orientations for another 3-4 days the perpendicular blotches and parallel birefringent vortices return. The general characteristics outlined above were always recovered.

A more detailed report of these observations along with color photographs will be published elsewhere.

Acknowledgement

The support of the Petroleum Research Fund administered by the American Chemical Society and Research Corporation is gratefully acknowledged.

Summary

Above a critical concentration in certain solvents poly-γ-benzyl-L-glutamate (PBLG) forms a lyotropic liquid crystal of the cholesteric structure. In the presence of sufficiently strong electric or magnetic fields this liquid crystal undergoes a mesomorphic phase transition to nematic order. Reported here are the results of our long term observations of bulk samples of liquid crystalline PBLG subjected to strong magnetic fields. We have observed visually several magnetohydrodynamic processes that occur on very different time scales: (a) the slow cholesteric-nematic phase transition, (b) a more rapid reorientation in the acquired nematic stasis, and (c) a new critical orientation of nematic PBLG with unusual anisotropic optical effects and high, but directional, thermal sensitivity. The phenomena are accounted for in terms of molecular order and dynamics of the associated polypeptide helices.

References

1. E. T. Samulski and A. V. Tobolsky, Macromolecules 1, 555 (1968).
2. S. Sobajima, J. Phys. Soc. Japan, 23, 1070 (1967).
3. M. Panar and W. D. Phillips, J. Amer. Chem. Soc., 90, 3880 (1968).
4. R. D. Orwoll and R. L. Vold, J. Amer. Chem. Soc., 93, 5335 (1971).
5. E. Iizuka, Biochem. Biophys. Acta 243, 1 (1971).
6. E. Iizuka and Y. Go, J. Phys. Soc. Japan, 31, 1205 (1971).
7. W. J. Toth and A. V. Tobolsky, Polymer Letters 8, 531 (1970).
8. C. Robinson, Trans. Faraday Soc. 52 571 (1956).
9. R. Williams, Phys. Rev. Letters 21, 342 (1968).
10. R. Williams, J. Chem. Phys. 50, 1324 (1969).
11. C. Robinson, Mol. Crystals Liq. Crystals 1, 467 (1966).

LIQUID CRYSTAL-ISOTROPIC PHASE EQUILIBRIA IN STIFF CHAIN POLYMERS

Wilmer G. Miller, Juey H. Rai and Elizabeth L. Wee

Department of Chemistry, University of Minnesota
Minneapolis, Minnesota 55455

INTRODUCTION

Most polymers when in solution exist as polymer random coils irrespective of the extent of intramolecular order in the solid state. Certain classes of macromolecules, however, do retain their intramolecular order when molecularly dispersed and behave as stiff chain or rodlike particles. Among these are synthetic polymers such as polyisocyanates[1] and polypeptides[2] and biological macromolecules such as DNA, tropocollagen[3] and tobacco mosaic virus.[4] In solution the concentration dependence of stiff and flexible chain polymers can be qualitatively different. In a good solvent flexible chain polymers mix in all proportions with little change in intramolecular or intermolecular order. As the polymer concentration is increased stiff chain polymers may undergo a phase transition from an isotropic to an anisotropic or liquid crystalline solution.[4-8] This phenomenon was predicted on the basis of molecular asymmetry alone by Onsager[9] and Isihara,[10] and later by Flory.[11,12] The Flory lattice model is applicable over the entire composition range and is particularly convenient for comparison with experimental studies. Earlier a partial temperature-composition phase diagram was published[13] for the two component system of α-helical poly-γ-benzyl-α,L-glutamate (PBLG) in dimethylformamide (DMF). We wish to report here an extension of these studies, a comparison of the phase boundaries determined by different experimental techniques, and a comparison of the experimental with the theoretical phase diagrams.

EXPERIMENTAL METHODS AND RESULTS

We have found no single experimental method applicable in determining the phase boundaries over the entire range of composition and temperature. On the basis of four techniques - isopiestic, hydrodynamic, nuclear magnetic resonance, and polarizing microscope - we have determined the phase diagram as shown in Figure 1 for PBLG with a molecular weight of 310,000 (M_w), as in the earlier study.[13] We turn now to the details of the determinations.

The high concentration phase boundary at temperatures below 20°C was determined by isopiestic measurements. Solid PBLG was placed on the pan of a quartz spring balance in a temperature controlled environment located in an isolatable section of a vacuum line. A large excess of a PBLG-DMF solution which was biphasic at the temperature of interest was introduced, the system briefly evacuated, and then isolated. The PBLG takes up solvent until it reaches the phase boundary at which point the solvent vapor pressure is the same as in the biphasic solution. The equilibrium nature of the measurements was checked by equilibration with biphasic solutions of differing mean composition. The results are shown in Figure 2.

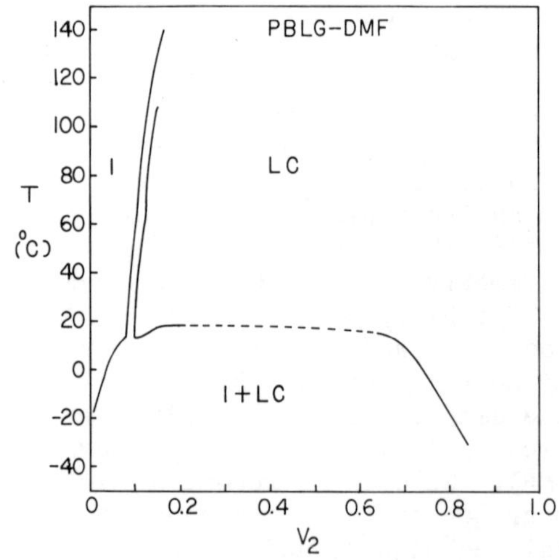

Figure 1. Phase diagram for polybenzylglutamate (M_w, 310,000) in dimethylformamide. Dashed line indicates area of insufficient data. Phases present were isotropic (I) or liquid crystalline (LC).

In the sections of the phase diagram previously published the line bounding the liquid crystal phase was determined by nuclear magnetic resonance. This was based on the fact that in the liquid crystal phase the solvent when near the ordered PBLG rods experiences an anisotropic environment. Consequently the proton direct dipolar interaction is not averaged to zero. When one crosses the phase boundary into the biphasic region the direct dipolar splitting collapses. We have noticed in a number of cases that the solvent proton spectra in the isotropic and biphasic region differ in that the ratio of the methyl doublet peak heights are reversed. In the figures to follow classification as isotropic or biphasic by nmr is merely a statement of relative methyl group peak heights.

Viscosity as a function of temperature, concentration and shear rate was determined with a Haake Rotovisco Couette type rotational viscometer. Typical results are shown in Figure 3. On the basis of correspondence to other data one is easily led to conclude that the reversal in the concentration dependence of the viscosity results from crossing the biphasic region. The viscosity

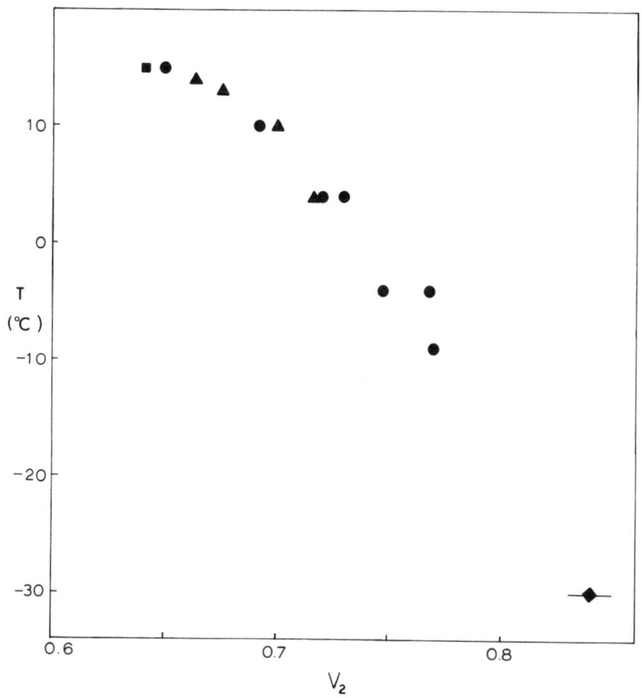

Figure 2. Phase boundary as determined by isopiestic measurements. Biphasic solutions used for equilibration had mean composition of 5(●), 7(◆), 13(▲) or 15(■) volume percent polymer.

drops as the system continues to acquire a higher proportion of the less viscous liquid crystal phase and then starts rising again in the usual manner after the pure liquid crystal phase is reached. In the isotropic region the viscosity curves at different temperatures form a family of parallel lines which may be shifted to fall on a common curve. This allows one to draw with confidence the lines as shown in Figure 3. The resulting maxima and minima may be taken as the position of the phase boundaries. The phase boundaries determined in this manner are shown in Figure 4. The points on the line bounding the liquid crystal phase are shear rate independent in contrast to those bounding the isotropic phase. The shear gradient tends to order the isotropic phase which is consistent with other hydrodynamic studies on helical polypeptides.[14] The shear independent phase boundary must lie somewhere between the lowest shear rate measured and the values obtained by linear extrapolation to zero shear rate.

Previously the line bounding the isotropic phase was located by polarizing microscope by determining the temperature at which light begins to be transmitted through crossed polars.[13] It was noted then that the phase diagram was thought to be reliable except below 9 volume percent. In this low concentration region the phase boundary has been re-examined by polarizing microscope measurements.

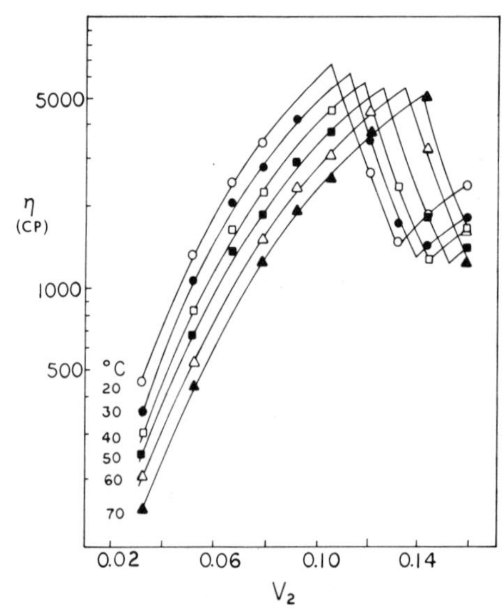

Figure 3. Concentration dependence of viscosity at the temperatures indicated. The shear rate was 38.8 sec^{-1}, calculated at the rotor surface.

The samples were sealed into clear rather than frosted nmr tubes, as was the case in the previous study, and a better temperature controlled microscope stage was employed. The results are shown in Figure 5.

COMPARISON OF EXPERIMENTAL RESULTS

The high concentration, low temperature phase boundary has been determined only by isopiestic measurements. Equilibration with biphasic solutions of different mean composition led to the same value (Figure 2). We thus feel that these represent equilibrium values for the sample employed. In earlier vapor sorption studies[15] it was found that the method of solid sample preparation did not effect the equilibrium vapor sorption. Other studies[16] show the side chains and solvent to be quite mobile at solvent compositions even lower than the equilibrium values reported here. We consequently felt that no special preparation of the solid polymer for isopiestic measurements was necessary.

A comparison of the various determinations in the narrow biphasic region is shown in Figure 6. As these data at the lower temperatures approach the area where the narrow biphasic region transforms into the wide biphasic region we have expanded the scale and included many determinations. The viscosity determined

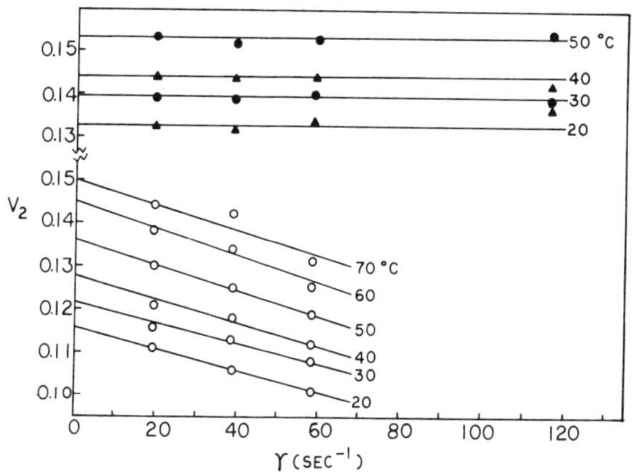

Figure 4. Viscosity determined phase boundaries as a function of shear rate. Open symbols - isotropic phase boundary; filled symbols - liquid crystal phase boundary.

lines seem open to little ambiguity except that the line bounding the isotropic phase in the absence of a shear gradient should lie somewhere between the two extremes shown. We will use the viscosity data as a standard for comparison. The polarizing microscope results are in substantial though not exact agreement with the viscosity results. When viewed on a scale comparable to Figure 1 the differences seem minor. The nmr data, however, have greater differences and need discussing. Just as the shear gradient is a nonspherically symmetric force field and can serve to align the highly asymmetric rodlike molecules, the magnetic field can serve a similar function. In fact the magnetic field untwists the cholesteric phase into a nematic one in the nmr procedure used to determine the phase boundary. Although we saw that a shear gradient moved the isotropic boundary to lower concentrations but not the liquid crystal boundary, we could argue that both boundaries might be shifted to lower concentrations in the magnetic field. There is a tendency in this direction though only the 10.5% solution offers serious discrepancy. The viscosity determined phase

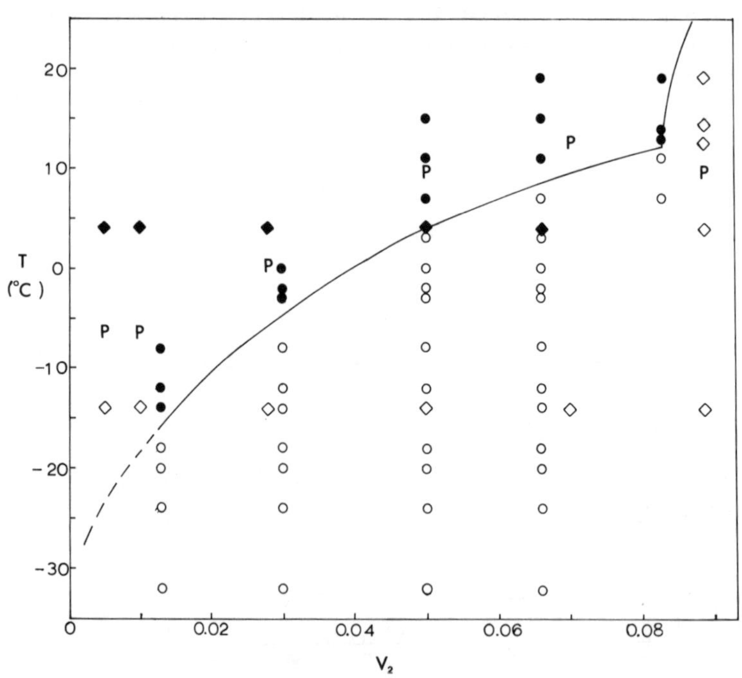

Figure 5. Low concentration phase boundary as determined by polarizing microscope (● - isotropic; ○ - containing ordered phase; P - previous determination[13]). Also shown are nmr results where spectra were classified as biphasic (◇) or isotropic (◆).

boundaries tend to extend smoothly into the microscope and nmr determined values at higher temperatures, which leads us to suggest this combination as the best representation of the narrow biphasic region.

As the narrow biphasic region joins the wide biphasic region the shape at the point of "contact" is important but a difficult one experimentally, particularly with respect to the line coming off to higher concentration. The viscosity is anomalous and difficult to interpret. The nmr results show clearly a positive slope with increasing concentration making identification manditory as a small biphasic region containing two liquid crystal phases of different composition. The isopiestic measurements are compatible with this interpretation. Even if the values are quantitatively incorrect due to the influence of the magnetic field, it is difficult to see how the 6° rise in going from 10 to 20 volume percent could be qualitatively incorrect. The area from 10 to 60 volume percent polymer, however, needs further study.

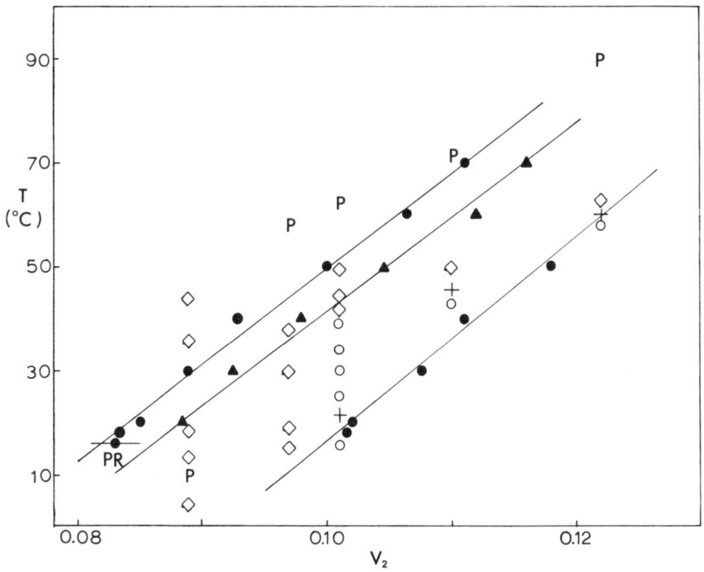

Figure 6. Comparison of phase boundary determinations in the narrow biphasic region. Viscosity - liquid crystal line (lower ●); isotropic line at 19.4 sec^{-1} shear rate (upper ●) or linear extrapolation to zero shear rate (▲). Polarizing microscope - earlier results[13] (P) or new determination (PR) for line bounding the isotropic phase. Nuclear magnetic resonance results[13] characterized as liquid crystal (○), biphasic (◇) or inbetween (+).

Finally the boundary lying to the left of the narrow biphasic region, shown in expanded scale in Figure 5, needs to be considered. The less precise microscope results reported previously are in substantial agreement with the newer ones. The joining of the narrow with the wide biphasic boundary seems rather well defined. Even the nmr classification is in generally good agreement with the microscope data. We would not be unduly concerned if it were not, as we know of no reason why the peak heights of the DMF methyl doublet should be reversed in the isotropic and biphasic region at all concentrations and temperatures.

COMPARISON OF THEORETICAL WITH EXPERIMENTAL DIAGRAMS

The isotropic-liquid crystal phase equilibria for rigid, impenetrable rods may be calculated[12] from equations 1-3

$$\ln(1-v_2)+(1-1/x)v_2+\chi v_2^2 = \ln(1-v_2^*)+(y-1)v_2^*/x+2/y+\chi v_2^{*2} \quad (1)$$

$$\ln v_2 + (x-1)v_2 - \ln x^2 + \chi x(1-v_2)^2 = \ln v_2^* + (y-1)v_2^* + 2-\ln y^2 + \chi x(1-v_2^*)^2 \quad (2)$$

$$v_2^* = [x/(x-y)][1-\exp(-2/y)] \quad (3)$$

where x is the rod axial ratio, y the equilibrium degree of disorientation in the liquid crystal phase, χ the polymer-solvent interaction parameter, and v_2 and v_2^* are the equilibrium volume fraction of polymer in the isotropic and liquid crystal phases, respectively. In addition there is a small region where equilibrium between two liquid crystal phases minimizes the free energy of the system. The phase diagram for an axial ratio of 150 is shown in Figure 7. This is slightly larger than the axial ratio of 135 calculated for the experimental polymer assuming a mean cross section of 15.5 A.

The similarities between the experimental and theoretical phase diagrams are striking. Both are characterized by a narrow biphasic region where the coexisting isotropic and liquid crystal phases differ only slightly in concentration, and then a sudden transition to a wide biphasic region wherein nearly pure solvent is in equilibrium with highly concentrated polymer. The narrow biphasic region and wide range stability of the ordered phase is a consequence of molecular asymmetry. The predicted transition to the wide biphasic region is a consequence of the catastrophic effect of unfavorable polymer-solvent interaction.

Close inspection of the two phase diagrams does reveal differences, the more obvious of which is the curvature of the narrow biphasic region to higher concentrations as the temperature is

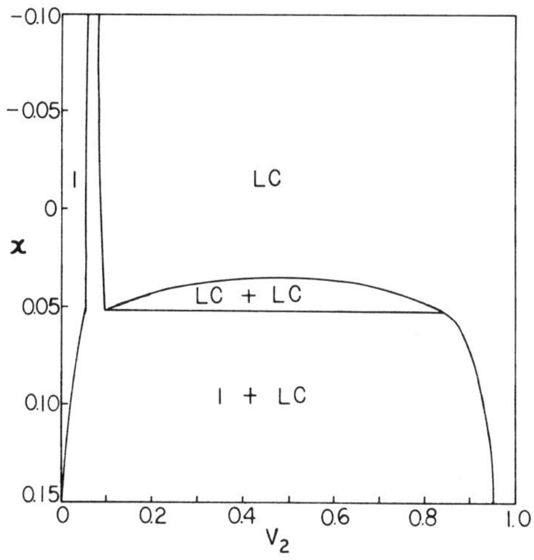

Figure 7. Lattice model phase diagram for rigid, impenetrable rods[12] of axial ratio 150.

increased. This is not unique to the PBLG-DMF system as polycarbobenzoxylysine in DMF has an identical behavior. Noting further that the theoretical narrow biphasic region moves toward higher concentrations as the axial ratio is reduced it seemed tempting to suggest that the apparent axial ratio of the experimental polymer decrease with increasing temperature. Extensive measurements in the temperature dependence of the intrinsic viscosity of the lysine polymer do indeed show increasing deviation from rigid behavior which should appear as a reduced effective axial ratio.[17] A brief temperature survey of the intrinsic viscosity of PBLG, as shown in Figure 8, reveals a similar effect. The theoretical narrow biphasic region is nearly independent of χ. The apparent axial ratio necessary to make the theoretical and experimental values agree was then determined assuming $\chi = 0$. The results are shown in Figure 9 where the lines bounding the isotropic phase were forced to agree. A corresponding determination using the line bounding the liquid crystal phase lies parallel but shifted to axial ratios about 25 units larger.[18] These results suggest that PBLG has some flexibility even at room temperature, which is not contrary to dilute solution studies.[19]

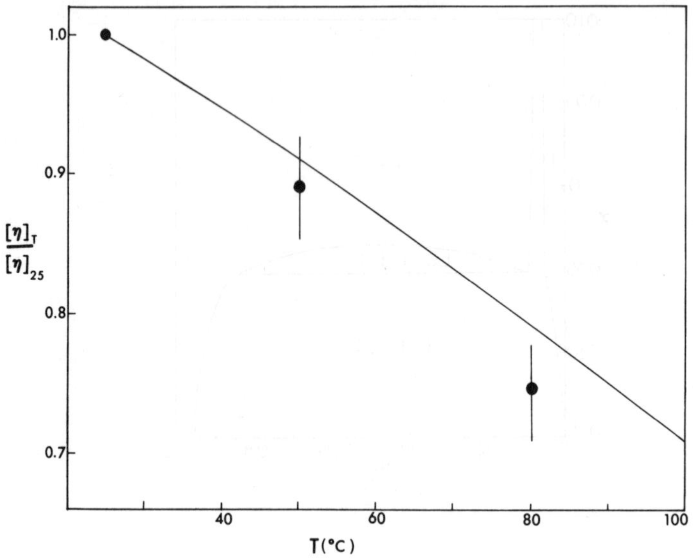

Figure 8. Comparison of the temperature dependence of the intrinsic viscosity of PBLG with polycarbobenzoxylysine[17] in DMF. Vertical lines - PBLG, M_W = 310,000; solid curve - polycarbobenzoxylysine, M_W = 690,000.

Inasmuch as rod flexibility is present and appears to affect the phase boundaries, it seemed of interest to compare the experimental results with the lattice model prediction for semiflexible rods. The isotropic-anisotropic equilibria may be calculated[11] from equations 4 and 5

$$\ln(1-v_2)+(1-1/x)v_2 + \chi v_2{}^2 = -\ln[1+v_2{}^*/x(1-v_2{}^*)] + \chi v_2{}^{*2} \tag{4}$$

$$\ln v_2 + (x-1)v_2 + (x-2)\ln(1-f)-\ln(z/2) + x\chi(1-v_2)^2 = -\ln[x^{-1} + (1-v_2{}^*)/v_2{}^*] + \chi x(1-v_2{}^*)^2 \tag{5}$$

where

$$f = 1/[1 + (z-2)^{-1} \exp(E/RT)] \tag{6}$$

Here f is the equilibrium degree of flexibility, z the lattice coordination number, E the bending energy for the "best" unit, and the other synbols are as in equations 1-3. Shown in Figure 10 is

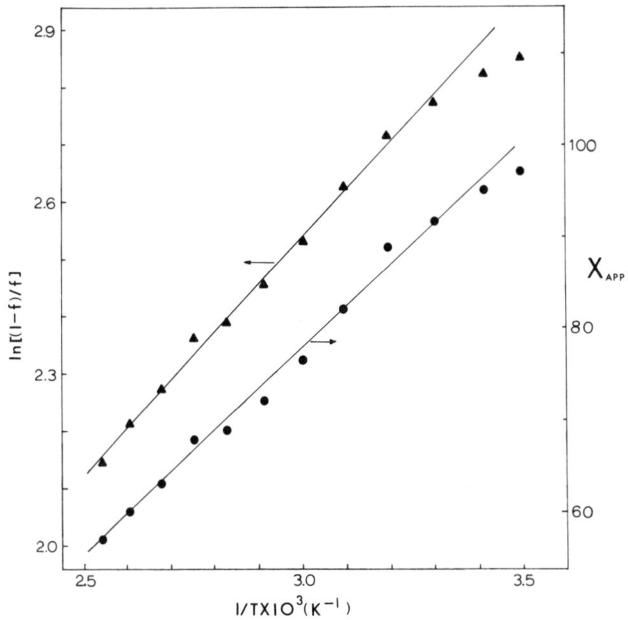

Figure 9. The flexibility of PBLG as deduced from the line bounding the isotropic phase. Triangles - determined from equations 4 and 5 with $\chi = 0$; circles - determined from equations 1-3 with $\chi = 0$.

the phase diagram for two values of the flexibility. We were surprised to find such a strong similarity between the phase diagram calculated with f vanishingly small and that for the rigid, impenetrable rod model. We had thought that the disorientation parameter played an essential role in predicting the narrow, parallel biphasic region yet equations 4 and 5 are predicated on the basis of complete order in anisotropic phase at all concentrations. Terms in the disordered phase obviously compensate. Ignoring this momentarily we notice that increasing the flexibility parameter shifts the narrow region to higher concentrations analogous to a decreased axial ratio in the rigid, impenetrable rod model. We consequently did the analogous procedure by finding the f necessary to make the lines bounding the isotropic region agree, again taking $\chi = 0$. We note from equation 6 that $\ln[(1-f)/f]$ should be a linear function of $1/T$. The results are shown in Figure 9 with $z = 12$. Different z values only shift the line slightly with E remaining constant and equal to 1.5-1.7 kcal. The f values run from about 0.06 at low temperature to 0.11 at high temperature. Analogous treatment of the line bounding the liquid crystal phase also yields a linear plot, but the f values run from 0.01 to 0.02. The rigid rod model

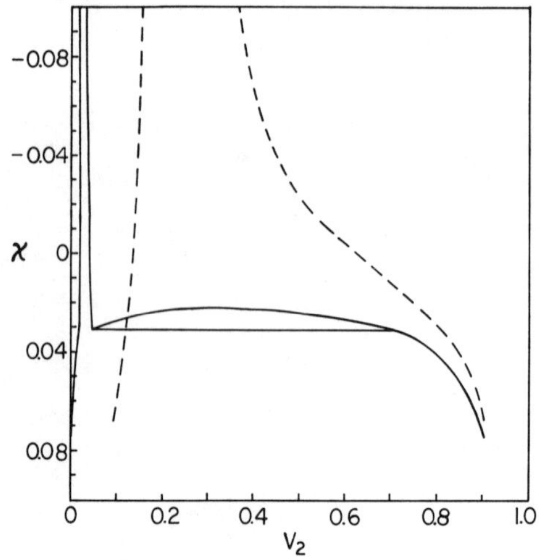

Figure 10. Effect of flexibility on phase equilibria for rods of axial ratio 150 calculated from equations 4 and 5 with f = 0 (solid lines) or f = 0.1 (dashed lines).

with adjustable axial ratio gives far more consistent values. In addition there is ample evidence to show that the helical molecules have a distribution of orientations in the liquid crystal phase.[20,21] Thus in spite of flexibility present in the molecular rods we feel the rigid rod model is a better representation.

The shift from the narrow to the broad biphasic equilibria is predicted to occur when polymer-solvent interaction becomes unfavorable. Ample experimental evidence indicates that such behavior exists in the PBLG-DMF system.[22,23] Finally we note that the transition to the broad biphasic region does not occur as suddenly as the theoretical rigid, impenetrable rod model predicts. We believe that this results from the effect of the long, flexible side chains attached to the molecular rods.[15,18]

This work was supported in part by the National Institutes of Health, U.S. Public Health Service.

REFERENCES

1. H. Yu, A. J. Bur and L. J. Fetters, J. Chem. Phys. 44, 2568 (1966).

2. P. Doty, J. H. Bradbury and A. M. Holtzer, J. Amer. Chem. Soc. 78, 947 (1956).

3. P. Doty and T. Nishihara in "Recent Advances in Gelatin and Glue Research", G. Stainsby, Ed., Pergamon Press, London, 1958.

4. G. Oster, J. Gen. Physiol. 33, 445 (1950).

5. F. C. Bawden and N. W. Price, Proc. Royal Sci. 123B, 274 (1937).

6. J. D. Bernal and I. Fankuchen, J. Gen. Physiol. 25, 111 (1941).

7. A. E. Elliott and E. J. Ambrose,Disc. Faraday Soc. 9, 246 (1950).

8. C. Robinson, Trans. Faraday Soc. 52, 571 (1956).

9. L. Onsager, Ann. N. Y. Acad. Sci. 51, 627 (1949).

10. A. Isihara, J. Chem. Phys. 19, 1142 (1951).

11. P. J. Flory, Proc. Roy. Soc. (London) A234, 60 (1956).

12. P. J. Flory, Proc. Roy. Soc. (London) A234, 73 (1956).

13. E. L. Wee and W. G. Miller, J. Phys. Chem. 75, 1446 (1971).

14. J. Hermans, Jr., J. Colloid Sci., 17, 638 (1962).

15. J. H. Rai and W. G. Miller, Macromolecules, 5, 45 (1972).

16. J. H. Rai, W. G. Miller and R. G. Bryant, Macromolecules, 6, 262 (1973).

17. G. L. Santee and W. G. Miller, to be submitted to Macromolecules.

18. W. G. Miller, C. C. Wu, E. L. Wee, G. L. Santee, J. H. Rai, and K. G. Goebel, 12th Prague Microsymposium on Macromolecules, August 20-23, 1973.

19. H. Fujita, A. Teramoto, K. Okita, T. Yamashita and S. Ikida, Biopolymers 4, 769, 781 (1966).

20. R. D. Orwoll and R. L. Vold, J. Amer. Chem. Soc. 93, 5335 (1971).

21. E. L. Wee and W. G. Miller, J. Phys. Chem. 77, 182 (1973).

22. K. D. Goebel and W. G. Miller, Macromolecules, 3, 64 (1970).

23. J. H. Rai and W. G. Miller, J. Phys. Chem. 76, 1081 (1972).

NUCLEAR MAGNETIC RESONANCE IN POLYPEPTIDE LIQUID CRYSTALS

William A. Hines and Edward T. Samulski

Departments of Physics and Chemistry and Institute of Materials Science, University of Connecticut, Storrs, Connecticut 06268

INTRODUCTION

Some twenty years ago it was found that synthetic polypeptides, $\left(\text{NH-CHR-CO}\right)_n$, can exist as rigid, rodlike α-helical molecules, in contrast with the random coil conformation adopted by most other synthetic polymers in solution. This observation with its implications in the study of protein structure stimulated a concentrated and sustained investigation of the dilute solution (1-5 wt.% polymer) properties of this class of polymers. In more concentrated polypeptide solutions (10-15 wt.% polymer), poly (γ-benzyl-L-glutamate) (PBLG; R = $CH_2CH_2COOCH_2C_6H_5$), forms a lyotropic liquid crystal. Robinson[1] extensively characterized the PBLG liquid crystal and found similarities between its supramolecular structure and the structure existing in thermotropic cholesteric liquid crystals.

The polymer concentrations at which the birefringent liquid crystalline phase initially appears (giving a two phase solution) and that at which the entire solution becomes a continuous birefringent liquid crystal are designated $(N_p/N_s)_A$ and $(N_p/N_s)_B$ respectively; N_p/N_s is the ratio of peptide residues to solvent molecules. The A and B points depend slightly on the particular solvent used but no correlation with solvent properties is apparent. On the other hand, in any particular solvent the A and B points are related to the degree of polymerization, n, of the synthetic polypeptide. The A and B points occur at lower polymer concentrations the higher the value of n.

These qualitative observations alone suggest that the rodlike shape of the polypeptide molecule and not specific polymer-solvent interactions are responsible for the formation of the polypeptide liquid crystal. Theoretical investigations of the possibility that phase separation or self-ordering should occur in solutions of rodlike solutes were carried out by Onsager[2], Flory[3] and more recently by Straley[4]. These extensions of the lattice theory of solutions, which incorporate the constraints involved in packing rodlike molecules with non-interacting solvent molecules, satisfactorily predict the concentrations at which the phase boundaries occur as a function of the axial ratio.

Herein we report measurements of the nuclear magnetic resonance spin-lattice relaxation time, T_1, of the solvent protons and the polypeptide protons in PBLG-CH_2Cl_2 and PBLG-CD_2Cl_2 solutions respectively.

EXPERIMENTAL

The polypeptide solutions were degassed using conventional freeze-pump-thaw cycles. The solvent proton T_1 was measured with a saturation-90° pulse sequence at 8 MHz and 30 MHz. The large differences between the T_1 for the solvent protons (4 to 28 sec) and T_1 for the polypeptide protons ($\sim$ 0.1 sec) enabled us to extract solvent T_1 values from relaxation data from PBLG-CH_2Cl_2 solutions. Our value for neat CH_2Cl_2, T_1 = 28 sec at 20 ^oC, is in good agreement with reported values (T_1 = 32 sec at 35 ^oC)[5]. The polypeptide proton T_1 was determined in PBLG-CD_2Cl_2 solutions using a 180°-90° pulse sequence at 5, 8, 15, 30 and 60 MHz. All measurements were carried out at 20 ^oC.

RESULTS

In this study, two PBLG molecular weights were used, M.W. = 550,000 and M.W. = 13,000 (Pilot Chemical Co. Lots G-131 and G-135 respectively). Assuming a rigid α-helical conformation for PBLG, these molecular weights correspond to axial ratios p = 151 and p = 3.56 respectively. Using a polypeptide diameter of 25 Å (obtained from a molecular model with a radially extending side-chain) and a length = 1.5(M.W./m.w.) Å, where m.w. is the PBLG peptide residue molecular weight (219) and 1.5 Å is the projection of a peptide residue along the helix axis. With these axial ratios and the lattice theory referred to above, the isotropic-two phase boundary (A point) occurs at N_p/N_s = 0.019 (0.355) and the two phase-liquid crystal boundary (B point) at N_p/N_s = 0.031 (0.375) for p = 151 (3.56).

Solvent Spin-lattice Relaxation

The solvent T_1 was measured over a range of N_p/N_s for both samples of PBLG, p = 3.56 and p = 151. In the latter samples, this range included isotropic, two phase and liquid crystalline PBLG – CH_2Cl_2 solutions; the solutions with p = 3.56 were in all cases isotropic, ordinary polymer solutions. In Fig. 1, the CH_2Cl_2 spin-lattice relaxation rate, $1/T_1$, is shown as a function of polypeptide concentration. The calculated phase boundaries for p = 151 are indicated. The position of the isotropic-two phase boundary, N_p/N_s = 0.019, is in good agreement with the previously

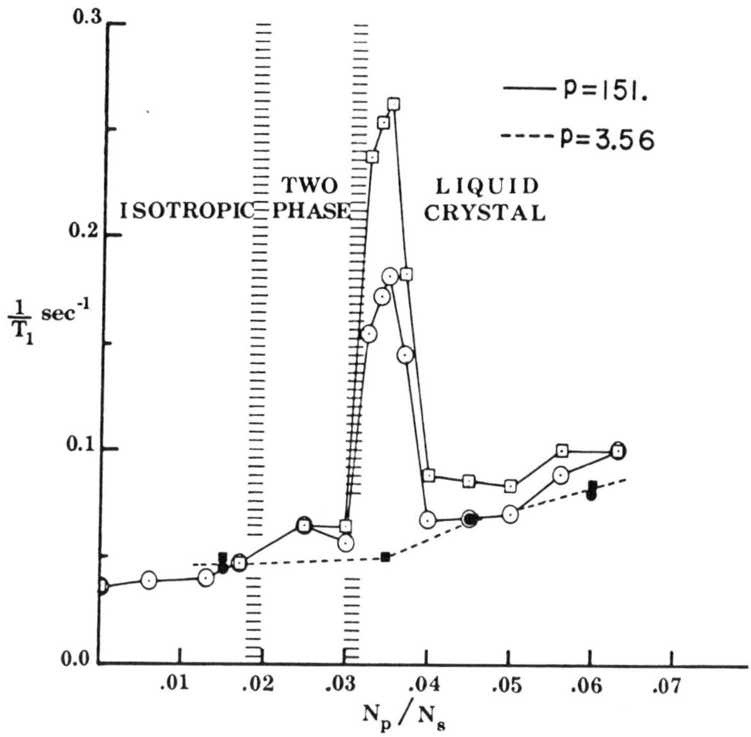

Fig. 1. CH_2Cl_2 proton spin-lattice relaxation rate ($1/T_1$) versus polypeptide concentration (N_p/N_s); open symbols correspond to the high molecular weight (p = 151) PBLG (⊡ = 8 MHz, ⊙ = 30 MHz) and closed symbols correspond to the low molecular weight (p = 3.56) PBLG (■ = 8 MHz, ● = 30Mhz).

reported dramatic change in solution viscosity which occurs on formation of the liquid crystalline phase[6].

The gradual monotonic increase in the solvent relaxation rate exhibited by the low molecular weight PBLG solutions with increasing polymer concentration (followed in the high molecular weight polypeptide solutions in the isotropic phase and later in the liquid crystal) is characteristic of ordinary isotropic polymer solutions and reflects increasing contributions to the relaxation rate from non-specific solvent-polymer interactions[7]. For the high molecular weight polypeptide, there is an anomalous increase in the solvent relaxation rate at the two phase-liquid crystal boundary. Concomitantly, $1/T_1$ becomes frequency dependent. $1/T_1$ for the low molecular weight polypeptide solutions is independent of frequency over the concentration range investigated. Consideration of the solvent relaxation data for the high molecular weight and low molecular weight polypeptide solutions separately, as a function of N_p/N_s, and together, at a single value of N_p/N_s, reveals a conspicuous absence of correlation between solvent relaxation rates and solution viscosities[6]. This indicates that unlike simple liquids, macroscopic polymer solution viscosity does not reflect microscopic dynamics.

Polypeptide Spin-lattice Relaxation

The polypeptide relaxation rate was determined in two liquid crystalline PBLG-CD_2Cl_2 solutions, $N_p/N_s = 0.035$ and $N_p/N_s = 0.060$ (p = 151, high molecular weight). $1/T_1$ of the polypeptide protons is shown as a function of the Larmor frequency, ν, in Fig. 2. For the higher frequencies, the data can be approximately described by the functional form $1/T_1 = A + B\nu^{-n}$, where A is independent of frequency ($A \approx 7.0\ sec^{-1}$; the high frequency asymptotic value of $1/T_1$) and n = 2. If the liquid crystalline solutions were left in the magnetic field, the observed proton nuclear magnetization free induction decay, or T_2^{eff} (the time for the magnetization to decay to 1/e of its initial value), changed with time ($T_2^{eff} \approx 300$ μsec initially and $T_2^{eff} \approx 200$ μsec after several hours; $\nu = 60$ MHz). After coming to equilibrium in the field, T_2^{eff} was influenced significantly by the relative angular orientation of the sample tube in the field. This phenomenon can be described by the following characteristics of polypeptide liquid crystals.

The application of a sufficiently strong magnetic field to these liquid crystals untwists the cholesteric structure and forms a macroscopic nematic structure with the nematic optic axis oriented parallel to the field[8]. The low degree of mobility of

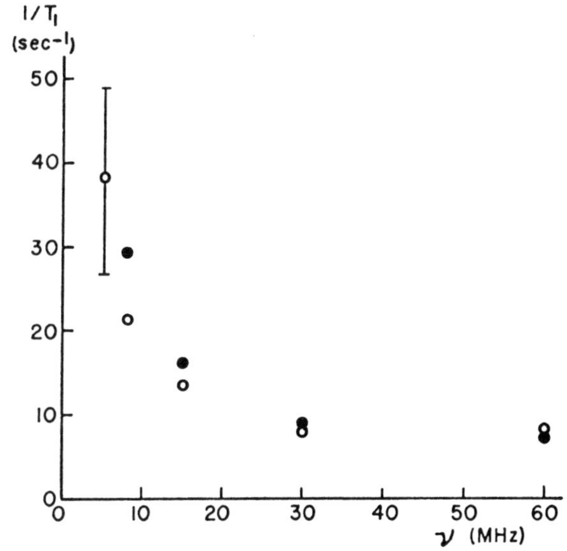

Fig. 2. PBLG proton spin-lattice relaxation rate ($1/T_1$) versus Larmor frequency (ν); open symbols correspond to N_p/N_s = 0.060 and closed symbols correspond to N_p/N_s = 0.035. of the magnetic field induced cholesteric-nematic transition.

the polypeptide molecules in the viscous solutions accounts for the rather long times (up to several hours) required for completion.

The static dipolar Hamiltonian (and T_2^{-1}) contains a multiplicative factor, $\langle 3\cos^2\theta - 1\rangle$, where θ is the angle between the internuclear vector and the magnetic field and $\langle\rangle$ indicates an average over molecular motion. In liquid crystals, this factor can be factored into several terms

$$\langle 3\cos^2\theta - 1\rangle = P_2(\cos\alpha)\cdot P_2(\cos\beta)\cdot(3\cos^2\Omega - 1) \qquad (1)$$

where $P_2(\cos\alpha)$ accounts for intramolecular reorientations of the internuclear vector, $P_2(\cos\beta)$ describes reorientations of the molecular axis relative to the liquid crystal optic axis arising, for example, from thermal fluctuations, and Ω is the angle between the liquid crystal optic axis and the magnetic field. Since the last term can be varied experimentally (the high viscosity of the polypeptide liquid crystal enables one to rotate the nematic optic axis away from the magnetic field direction without immediate reorientation of this axis), the free induction decay will exhibit angular dependence if T_2^{eff} is dominated by dipolar interactions. This appears to be the case because the free induction decay is a maximum, i.e. limited by the magnetic field inhomogeneity when $\Omega = 55^o\ 54'$ (from Eqn. 1, the dipolar interactions vanish for this value of Ω)[9].

DISCUSSION

Spin-lattice relaxation is frequency dependent if the correlation time for relavant molecular motion is the order of the Larmor frequency. Two general types of processes which fulfill this condition and could be operative in the polypeptide solutions are: (1) motion associated with the overall rotational tumbling of the macromolecule and (2) motions derived from structural fluctuations of the molecular arrangement in the liquid crystal. Some time ago, frequency dependent relaxation rates were reported for polypeptide protons in <u>isotropic</u> PBLG-$CDCl_3$ solutions[10]. In NMR studies of protein solutions, frequency dependent solvent relaxation is commonly found[11]. Both of these observations can be interpreted in terms of type (1) motional processes; slow tumbling of the macromolecule. The absence of frequency dependent solvent relaxation rates in isotropic PBLG-CH_2Cl_2 solutions and its abrupt appearance in the liquid crystalline solutions suggest that motional processes of type (2) are important in these systems. However, the observed frequency dependence of the polypeptide protons, $1/T_1 \sim \nu^{-2}$, is not in agreement with currently accepted models of structural fluctuations in thermotropic liquid crystals; long range collective fluctuations of the nematic axis modulate $P_2(\cos\beta)$ in Eqn. 1 to give frequency dependent relaxation rates of the form $1/T_1 \propto \nu^{-1/2}$ (see reference 12). Neither is the observations in polypeptide liquid crystals in agreement with another source of frequency dependent relaxation; namely slow translational diffusion, $1/T_1 \propto \nu^{1/2}$ (see reference 13). Consideration of both the slow macromolecule rotational motion together with the constraints imposed on this motion by the molecular arrangement in the polypeptide liquid crystal does yield a regime of correlation times in the radio frequency range. Below we outline a model calculation of these correlation times derived from the structural characteristics of the polypeptide liquid crystal.

X-ray studies of the polypeptide liquid crystal have shown that the rodlike molecules pack in a hexagonal arrangement as schematically illustrated in Fig. 3. It was found experimentally

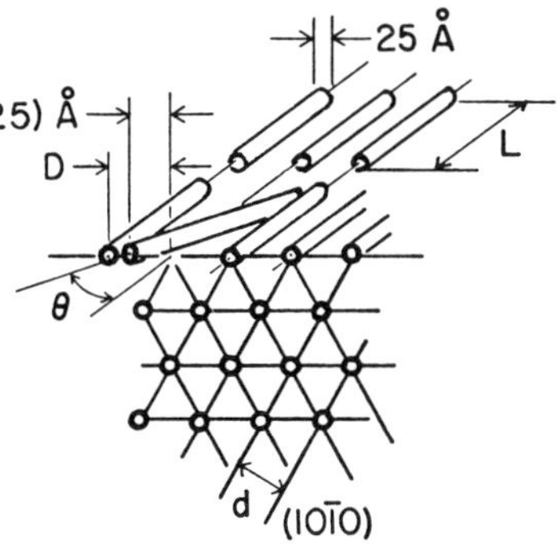

Fig. 3. Schematic representation of rodlike polypeptide molecules packed on a hexagonal net, where rod length L = (M.W./m.w.)(1.5 Å), rod diameter = 25 Å, d = spacing between (10$\bar{1}$0) planes, D = $2d/\sqrt{3}$ and $\theta = \sin^{-1}[(D - 25)/(L/2)]$.

that the concentration dependence of the distance between the (10$\bar{1}$0) planes, d, is given by

$$d = \left\{\frac{\sqrt{3}}{2} \cdot \frac{10^{24}}{N_o h} \cdot \frac{N_s}{N_p} \cdot \frac{(m.w.)_s}{\rho_s}\right\}^{1/2} \quad \text{in Å}, \qquad (2)$$

where N_o is Avogadro's number, h = 1.5 Å, N_p/N_s is the concentration, ρ_s is the density of the solvent and $(m.w.)_s$ is the solvent molecular weight[14]. On the hexagonal net, the long rodlike molecule is only allowed to execute a very small rotation about an axis through its center and perpendicular to its length before contacting a nearest neighbor. The rotary diffusion coefficient, D_r, of a rodlike particle in a medium of viscosity η can be obtained by a suitable modification of the Stokes or Einstein relation[15],

$$D_r = \frac{k_B T\,[\ln(2p) - \lambda]}{4\pi\eta a^3} \quad \text{in sec}^{-1} \qquad (3)$$

where a is one-half the rod length, p is the axial ratio and λ is a constant of order unity. The mean square angular displacement of the rod, $\overline{\theta^2}$, in a time t is given by the random walk formulation, $\overline{\theta^2} = 4D_r t$. Taking θ to be defined as shown in Fig. 3, the average time between polymer collisions for a single polymer molecule can be related to the concentration by

$$t = \left\{\frac{2d}{\sqrt{3}} - 25\right\}^2 \Bigg/ 4D_r a^2 \quad \text{in sec.} \tag{4}$$

Using this expression, we find that the helical polypeptide molecule in the liquid crystal executes oscillatory motion about its mean direction of orientation on a time scale comparable to the Larmor frequency. For example, using the high molecular weight polypeptide with $N_p/N_s = 0.035$ and $N_p/N_s = 0.060$ in Eqn. 4, we find that $1/t = 11.9$ MHz and 45.2 MHz respectively.

Classical theoretical description of nuclear spin-lattice relaxation due to a time dependent dipole-dipole interaction in terms of the spectral density of the dipolar fluctuations at a frequency ν are well known[15]. If it is assumed that the fluctuations can be described by an exponential correlation function, with a characteristic correlation time t_c, the functional dependence of the spectral density on t_c is fixed for dipolar relaxation

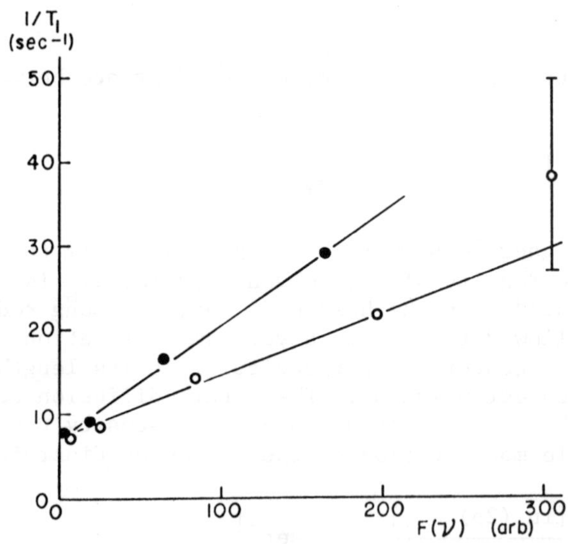

Fig. 4. PBLG proton spin-lattice relaxation rate ($1/T_1$) versus $F(\nu)$, the quantity contained in the brackets of Eqn. 5; open symbols correspond to $N_p/N_s = 0.060$ and closed symbols correspond to $N_p/N_s = 0.035$.

$$\frac{1}{T_1} = S\left[\frac{t_c}{1 + 4\pi^2\nu^2 t_c^{\ 2}} + \frac{4t_c}{1 + 16\pi^2\nu^2 t_c^{\ 2}}\right] \text{ in sec}^{-1}. \tag{5}$$

S is a constant containing the nuclear constants and is a measure of the strength of the relaxation interaction. The term in brackets, $F(\nu)$, is the correlation time dependence of the spectral densities.

The oscillations of the polypeptide molecule about the liquid crystal optic axis (the magnetic field) as described above should, through $P_2(\cos\beta)$ in Eqn. 1, contribute to polypeptide dipolar spin-lattice relaxation. In this case, a plot of $1/T_1$ versus $F(\nu)$ would be linear for the correct value of t_c. In Fig. 4, we have made such a plot of the polypeptide relaxation rate at the two concentrations, $N_p/N_s = 0.035$ and $N_p/N_s = 0.060$. The straight lines correspond to the best fits to the data and were obtained using values of $1/t_c = 60$ MHz and $1/t_c = 70$ MHz respectively. Although there appears to be an indication of a concentration dependence as the characteristic $1/t$ values for polypeptide fluctuations calculated from the model, a more thorough study of the frequency dependence of T_1 will be necessary in order to determine the validity of this model for motional processes in polypeptide liquid crystals.

REFERENCES

1. C. Robinson, Mol. Cryst. 1, 467 (1966) and ref. cited therin.
2. L. Onsager, Ann. N. Y. Acad. Sci. 56, 627 (1949).
3. P. J. Flory, Proc. Roy. Soc. A 243, 73 (1956).
4. J. P. Straley, Mol. Cryst. and Liq. Cryst., In Press.
5. W. G. Rothschild, Macromolecules 5, 37 (1972).
6. W. A. Hines and E. T. Samulski, Macromolecules, In Press.
7. J. E. Anderson, K-J. Liu and R. Ullman, Discuss. Faraday Soc. 49, 175 (1970).
8. S. Sobajima, J. Phys. Soc. Japan 23, 1070 (1967); M. Panar and W. D. Phillips, J. Amer. Chem. Soc. 90, 3380 (1968); E. T. Samulski and A. V. Tobolsky, Macromolecules 1, 555 (1968) and Biopolymers 10, 1013 (1971).
9. E. T. Samulski, B. A. Smith and C. G. Wade, Chem. Phys. Letters 20, 167 (1973).
10. R. Kimmich and F. Noack, Ber. Bun.-Gesellschaft. 75, 269 (1971).
11. S. H. Koenig and W. E. Schillinger, J. Biol. Chem. 244, 3283 (1969).
12. A. F. Martins, Phys. Rev. Letters 28, 289 (1972).

13. J. F. Harmon and B. H. Muller, Phys. Rev. 182, 400 (1969).
14. C. Robinson, J. C. Ward and R. B. Beevers, Discuss. Faraday Soc. 25, 29 (1958).
15. S. Broersma, J. Chem. Phys. 32, 1626, 1632 (1960).
16. A. Abragam, The Principles of Nuclear Magnetism, Chapter VIII, Oxford (1961).

MAGNETIC RELAXATION OF POLY-γ-BENZYL-L-GLUTAMATE SOLUTIONS IN DEUTEROCHLOROFORM

B. M. Fung and Thomas H. Martin

Department of Chemistry, University of Oklahoma
Norman, Oklahoma 73069

ABSTRACT

For liquid and liquid crystalline solutions of poly-γ-benzyl-L-glutamate (PBLG) in deuterochloroform, the spin-lattice relaxation time (T_1) of deuterium in $CDCl_3$ is strongly dependent upon temperature and concentration, while the proton T_1 of PBLG has no concentration dependence and only a small temperature dependence. The data are analyzed based upon a model in which highly associated and bulk solvent molecules are in chemical exchange, and its validity is discussed.

INTRODUCTION

Since the discovery by Robinson and co-workers (*13,14*) that poly-γ-benzyl-L-glutamate (PBLG) forms a lyotropic liquid crystal in many organic solvents, there has been a number of studies made on this system.

Lyotropic liquid crystalline PBLG solutions are cholesteric in nature, but in a strong magnetic or electric field the α-helices slowly align along the field and the phase changes into nematic. Sobajima (*19*) first reported the ordering of solvent molecules of PBLG solutions by observing dipolar splitting of CH_2Cl_2 and CH_2Br_2 in the nuclear magnetic resonance (NMR) spectrum. Since then, NMR spectra of liquid crystalline PBLG and poly-L-glutamic acid solutions have been extensively studied (*4-8,11,12,15-19*).

In addition to the continuous wave (CW) NMR method, magnetic relaxation also gives valuable information on the physical state and properties of many complex systems. Kimmich and Noack (*7*) studied the spin-lattice relaxation time (T_1) of protons in a solution of PBLG in deuterochloroform, and reported a considerable dependence of proton T_1 of PBLG on the resonance frequency. The system was apparently liquid crystalline in nature but was not identified as such by the authors.

We have performed a detailed study on the magnetic relaxation of proton and deuteron in solutions of PBLG in $CDCl_3$. The dependence of T_1 on concentration and temperature was measured, and the results are discussed in relation to the nature of the liquid crystalline solutions.

EXPERIMENTAL METHOD AND RESULTS

The liquid crystal samples were prepared by adding a weighed amount of PBLG to a weighed amount of $CDCl_3$ in a 10 mm NMR tube which was immediately sealed. De-gassing several samples did not affect the deuterium T_1; consequently all of them were sealed without de-gassing. The PBLG was obtained from Cyclo Chemicals and was reported to have a molecular weight of 200,000 to 400,000. The $CDCl_3$ was obtained from Stohler Isotopes and was used without further purification. All samples were centrifuged to insure complete dissolution.

In calculating a mole fraction or mole ratio we used the molecular weight of one benzyl glutamate residue of the polymer as one "mole."

The spin-lattice relaxation time (T_1) was measured by the standard 180°-t-90° technique (*3*). The logarithm of $M_\infty - M_t$, where M is the amplitude of the magnetization immediately following the 90° pulse, was plotted against t, the time lapse between the two pulses. In all cases studied, the plot was linear within experimental error, and T_1 was determined from the slope of the line.

The pulse NMR spectrometer was "home-made." The proton and deuteron resonances were measured at several frequencies ranging from 4.5-10.5 MHz, with a pulse width of about 5 μsec for a 90° pulse for proton, and a width of about 10 μsec for a 90° pulse for deuteron.

We found that T_1 of the protons in PBLG has no dependence on the concentration of the polypeptide, and very little dependence on temperature (Fig. 1). It has a considerable dependence on the resonance frequency as reported in ref. *7*. The values of T_1 are slightly lower in our work, the most likely reason being a difference in the molecular weight of the PBLG samples used. Since

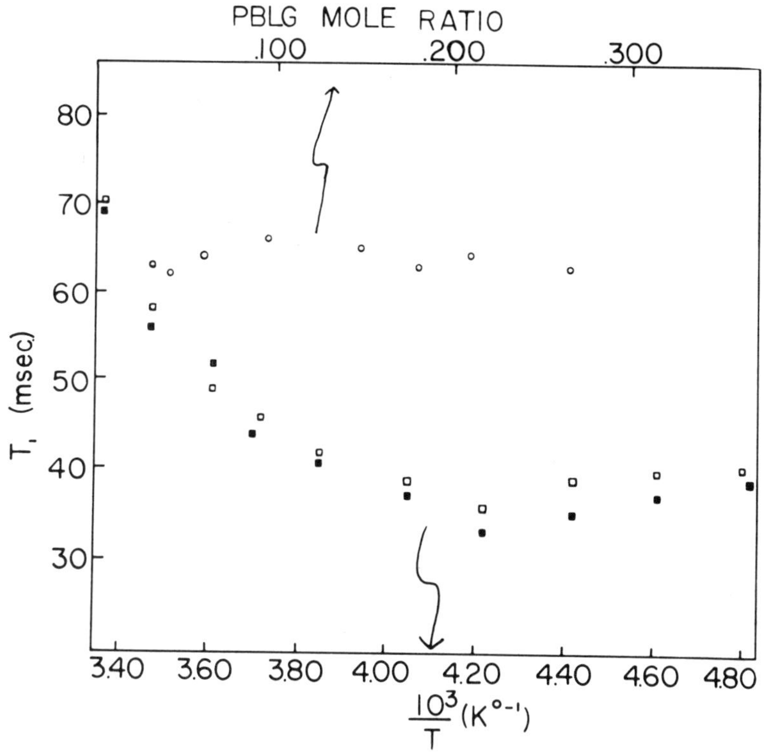

Figure 1. Spin-lattice relaxation time for proton of PBLG in solution with $CDCl_3$ of varying concentrations as a function of reciprocal temperature (bottom) and as a function of PBLG mole ratio (top). O– measurements made at 25°C and 10.5 MHz. PBLG mole ratio: □–.095; ■–.265.

the data in ref. 7 cover a wider range of frequency, the reader is referred to it for a more detailed presentation.

T_1 of the deuteron in $CDCl_3$ is strongly dependent upon temperature and the concentration of PBLG (Fig. 2). However, it did not show any dependence on the resonance frequency within experimental error in the frequency range accessible to our spectrometer (4.5-10.5 MHz for deuterium). T_1 for each sample stayed unchanged

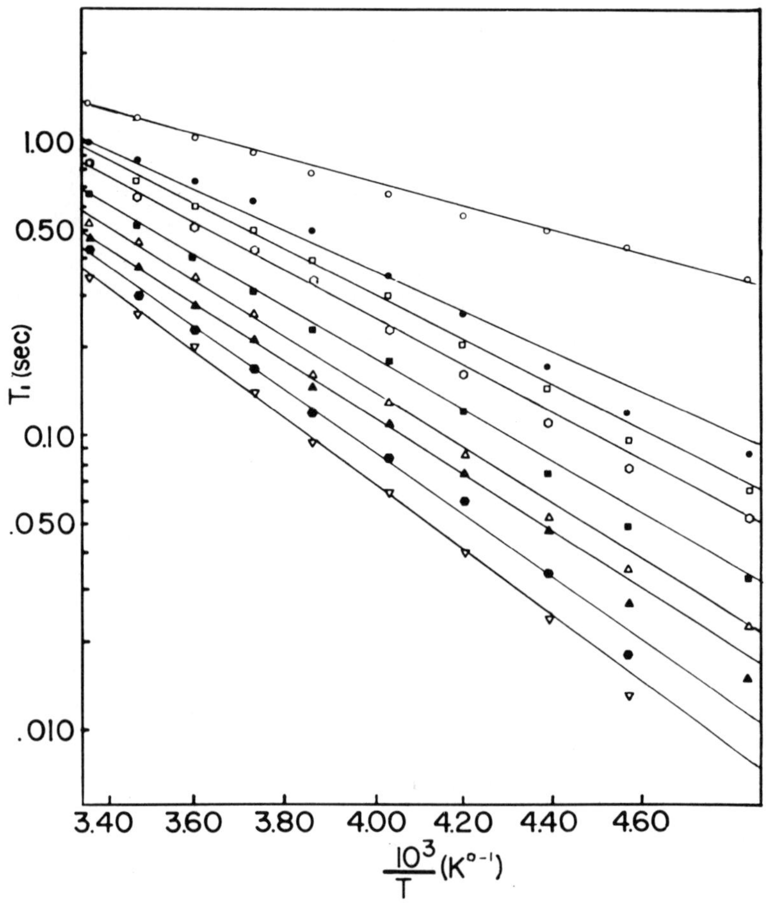

Figure 2. Spin-lattice relaxation rates at 10.5 MHz for deuterium of $CDCl_3$ in PBLG-$CDCl_3$ solutions of varying concentrations as a function of reciprocal temperature. PBLG mole ratio: ○ 0, ● .029, □ .037, ⬡ .061, ■ .095, △ .149, ▲ .180, ⬢ .210, ▽ .265. Points are experimental and lines are least square fits.

when the sample was aligned in the magnetic field overnight. There was no abrupt change in T_1 for both deuterium in $CDCl_3$ and proton in PBLG during the transition from liquid to liquid crystal as the concentration increased.

The spin-spin relaxation time (T_2) for deuterium in the liquid crystalline solutions cannot be determined without ambiguity for the following reasons. In the CW spectrum, even if the field inhomogeneity can be neglected, the line width is determined by T_2 and the distribution of the orientation factor due to the imperfect alignment of the polypeptide (*11*). When the pulse method is used, the Carr-Purcell-Meiboom-Gill spin-echo train (*2,10*) is non-exponential and dependent upon the pulse spacing because of the quadrupole splitting. However, the order of magnitude for T_2 can be estimated by using very small pulse spacing and extrapolate the data to zero pulse spacing. For example, a sample of 21 mole % PBLG in $CDCl_3$ has a value of $T_2 \sim 60$ msec at 25°C for deuterium. The corresponding value from the CW spectrum (taken on a Varian XL-100 spectrometer) is about 10 msec.

DISCUSSION

The lack of dependence of T_1 for proton in PBLG on its concentration indicates that the proton relaxation is intramolecular in nature. The very small dependence of proton T_1 on temperature indicates that the rotation of PBLG has a very small activation energy. Kimmich and Noack (*7*) suggested that the relaxation of PBLG and protein solutions may be mainly due to spin diffusion. We feel that this is highly improbable because the spin diffusion mechanism is important only for solid at low temperatures, at which there is no rotational motion (*1*). The dipolar relaxation for an ellipsoid (*20*) is adequate in describing the proton relaxation of PBLG solutions, and the treatment will be discussed elsewhere.

Our main interest is in the deuterium relaxation times for the solvent, $CDCl_3$. It is interesting that in PBLG solutions, small solvent molecules such as chloroform and dichloromethane in large quantities do have definitive orientations, as evidenced by the dipolar and quadrupolar splittings in the NMR spectra. Panar and Phillips (*12*) suggested that the solvent molecules exist in environments in which they are highly ordered by the solute (PBLG) and in environments remote from PBLG in which solvent ordering is non-existent, and the CW NMR spectra are time-averaged spectra. This is an attractive postulation and we are going to examine its validity in the light of our relaxation data.

The magnetic relaxation in multiple phase systems has been discussed in detail by Zimmerman and Brittin (*21*). For a two-phase system after a 90° pulse,

$$M_z = M_o(1-a_1e^{-\mu_1 t} + a_2e^{-\mu_2 t}), \qquad (1)$$

where M_z and M_o are the magnetization at time t and at equilibrium, respectively; a_1, a_2, μ_1 and μ_2 are parameters related to the exchange rate, populations, and relaxation rates in the two phases. In general, the decay of the magnetization is not a single exponential. Only under special conditions, the solution of the relaxation equation after a 180° pulse can be expressed as

$$M_z = M_o \ (1-2e^{-t/T_1}), \qquad (2)$$

and a single T_1 can be defined. Since in all our experimental measurements, log (M_o-M_z) for deuterium was always linear with respect to t within experimental error, equation (2) is applicable. We will discuss these special conditions in relation to the PBLG solutions. In the exchange model, it will be assumed that the solvent molecules that are closely associated with PBLG and are highly ordered have a spin-lattice relaxation time T_{1b}, and the isotropic solvent molecules have a spin-lattice relaxation time T_{1f} equal to that of the free solvent.

(1) When $T_{1b} \gtrsim T_{1f}$, the two exponentials in (1) are indistinguishable from each other regardless of the exchange rate, and (2) holds. This is not the case for the PBLG solutions, otherwise the deuterium T_1 would not change much with the concentration of PBLG.

(2) If the fraction of solvent molecules that is associated with the polypeptide (f) is very small, i.e. f<<1, equation (2) is also a good approximation, and

$$\frac{1}{T_1} = \frac{f}{\tau + T_{1b}} + \frac{1}{T_{1f}}, \qquad (3)$$

as shown by Luz and Meiboom (*9*). In (3), τ is the life time of the "bound" species and

$$f = p \cdot q, \qquad (4)$$

where p is the mole ratio of PBLG/$CDCl_3$, and q is the number of solvent molecules associated with each peptide unit in PBLG. In this case, the quantity $(1/T_1-1/T_{1f})/p$ should have no concentration dependence. We have plotted this quantity against the reciprocal temperature for deuterium in the PBLG-$CDCl_3$ solutions in Figure 3, the data in which indeed show no systematic dependence on concentration. The essential linear relation between $(1/T_1-1/T_{1f})/p$ and

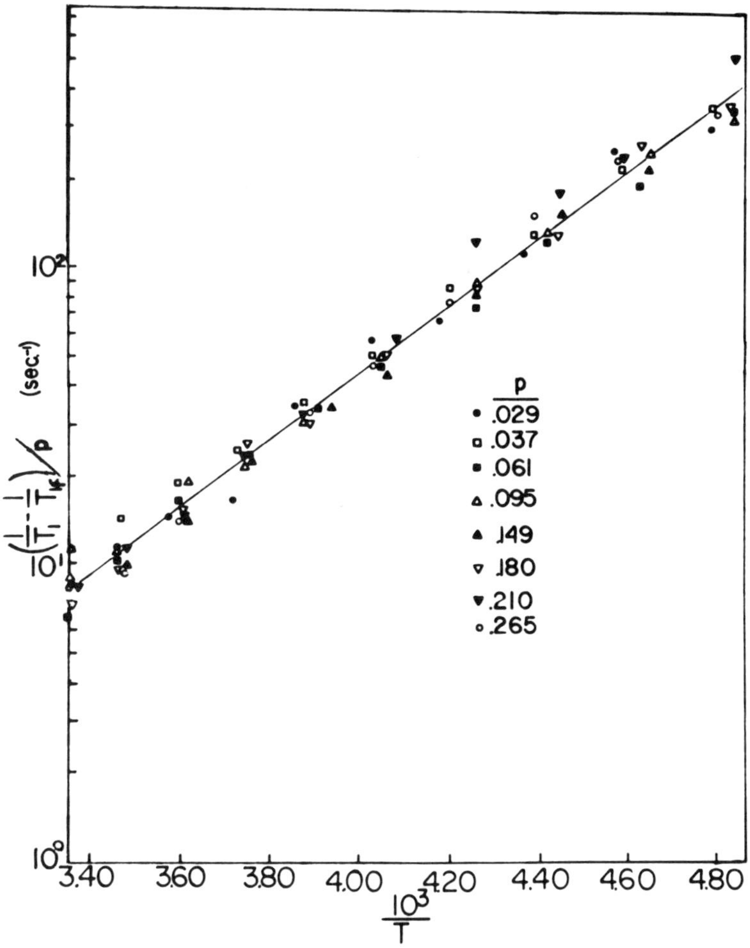

Figure 3. Plot of $(\frac{1}{T_1} - \frac{1}{T_{1f}})/p$ as a function of reciprocal temperature. This expression is described in the text. The points are experimental and the line is a least square fit.

1/T indicates that $\tau \gg T_{1b}$ and has the form $\tau = \tau_o \cdot e^{E_a/kT}$, where E_a is the activation energy of exchange, if eq. (3) is to hold. An alternative is that τ and T_{1b} have the same temperature dependence, which is not likely. A least square fit of the data in Figure 3 yields the value of E_a = 5.1 kcal/mole.

The rate of exchange, $1/\tau$, cannot be determined from Figure 3, because q is not known. On the other hand, one can estimate the order of magnitude for q by combining the relaxation and CW NMR data. If the observed peaks in the CW spectra are time-averaged signals between the isotropic and the highly ordered $CDCl_3$ molecules, the product $\pi\tau \cdot \Delta\nu_Q$ must be of the order of unity or less, otherwise a central peak would have been observed. For a solution with a mole ratio of PBLG/$CDCl_3$ = 0.265, $\Delta\nu_Q \sim 1.5$ kHz at room temperature. Since $q/\tau \sim 8.0$ at 25°C (Fig. 3), then $q < 1.7 \times 10^{-3}$. This means that there would be less than two $CDCl_3$ molecules that are closely associated with PBLG per 1,000 peptide units, which is not impossible but highly improbable.

(3) If the exchange between associated and "free" solvent molecules is very fast so that $\tau \ll T_{1b}$, T_{1f}, equation (2) also holds with (*21*)

$$\frac{1}{T_1} = \frac{f}{T_{1b}} + \frac{1-f}{T_{1f}} . \tag{5}$$

A rearrangement of (5) yields

$$\left(\frac{1}{T_1} - \frac{1}{T_{1f}}\right)/p = q \cdot \left(\frac{1}{T_{1b}} - \frac{1}{T_{1f}}\right). \tag{6}$$

The relaxation times of deuterium in the $CDCl_3$ molecules associated with PBLG can be adequately described by the relaxation of a quadrupolar nucleus in an ellipsoid with internal rotation (*20*):

$$\frac{1}{T_{1b}} = Q[J_1(w) + J_2(2w)], \tag{7}$$

$$\frac{1}{T_{2b}} = \frac{Q}{4}[J_o(o) + 10\, J_1(w) + J_2(2w)], \tag{8}$$

where
$$Q = \frac{9}{32}\left(1 + \frac{\eta^2}{3}\right)\left(\frac{e^2qQ}{\hbar}\right)^2, \tag{9}$$

and
$$J_i(w) = 2K_i\left[\frac{A\tau_A}{1+\omega^2\tau_A^2} + \frac{B\tau_B}{1+\omega^2\tau_B^2} + \frac{C\tau_C}{1+\omega^2\tau_C^2}\right] \tag{10}$$

are the Fourier intensities. In (9), e^2qQ/h is the quadrupole coupling constant, and η is the asymmetry parameter. In equation (10), K_i's are constants, A, B and C are coefficients related to the angle between the C-D bond and the long axis of the ellipsoid, and τ_A, τ_B, and τ_C are correlation times determined by the motion of the ellipsoid (PBLG) and the rotation of $CDCl_3$. The expression in (10) would take a more complicated form if the C-D bond does not form a constant angle with the axis of the α-helix during the rotation of the associated $CDCl_3$ molecules. Since the proton T_1 (and therefore the motion) of PBLG has only a very small temperature dependence (Fig. 1), the rather large temperature dependence of the deuterium T_1 must come from the rotation of the associated $CDCl_3$ molecules. A comparison of equation (6) and Figure (3) tends to indicate that the activation energy of the rotational motion is close to 5 kcal/mole. Actually, Figure (3) is disturbingly simple if equations (7)-(10) are adequate in describing the relaxation mechanism of T_{1b}.

The fact that $T_1=T_2$ for $CDCl_3$ and $T_1>T_2$ for the PBLG solutions can also be explained by (5), (7)-(10). Unfortunately, a quantitative calculation is not possible because T_2 cannot be accurately determined as discussed above, and the operating frequency range for deuterium (4.5-10.5 MHz) in our spectrometer is limited. (In this range the deuterium T_1 of $CDCl_3$ showed no variation, and the proton T_1 of PBLG changed by only 29%).

From the above discussion, it can be seen that the exchange model for liquid crystalline PBLG solutions can be substantiated by the relaxation data but the evidence is not conclusive. Further experiments are to be performed and other relaxation mechanisms (e.g. a continuous distribution of correlation time) are to be considered in order to understand the nature of the liquid crystalline PBLG solutions.

Acknowledgement. This work was supported by the National Science Foundation. We would like to thank Dr. Saul Meiboom for a helpful discussion.

REFERENCES

1. Abragam, A., "The Principles of Nuclear Magnetism," Oxford University Press (London, 1961).

2. Carr, H. Y., Purcell, E. M., Phys. Rev. *94*, 630 (1954).

3. Farrar, T. C., Becker, E. D., "Pulse and Fourier Transform NMR," Academic Press (New York, 1971).

4. Fung, B. M., Gerace, M. J., Gerace, L. S., J. Phys. Chem. *74*, 83 (1970).

5. Gerace, M. J., Fung, B. M., J. Chem. Phys. *53*, 2984 (1970).

6. Gill, D., Klein, M. P., Kotowycz, J. Amer. Chem. Soc. *90*, 6870 (1968).

7. Kimmich, R., Noack, F., Zeitschrift für Electrochemie, *75*, 269 (1971).

8. Klein, M. P., Gill, P., Kowtowycz, G., Chem. Phys. Letters *2*, 677 (1968).

9. Luz, Z., Meiboom, S., J. Chem. Phys. *40*, 2687 (1964).

10. Meiboom, S., Gill, D., Rev. Sci. Instrum. *29*, 688 (1958).

11. Orwoll, R. D., Vold, R. L., J. Amer. Chem. Soc. *93*, 5335 (1971).

12. Panar, M., Phillips, W., J. Amer. Chem. Soc. *90*, 3880 (1968).

13. Robinson, C., Trans. Faraday Soc. *52*, 571 (1956).

14. Robinson, C., Ward, J. C., Beevers, R. B., Discussions Faraday Soc. *25*, 29 (1958).

15. Samulski, E. T., Tobolsky, A. V., Mol. Cryst. Liquid Cryst. *7*, 433 (1969).

16. Samulski, E. T., Tobolsky, A. V., Liquid Crys. and Ordered Fluids. (ed.: J. F. Johnson and R. S. Porter, Plenum Press, New York, 1970).

17. Samulski, E. T., Tobolsky, A. V., Biopolymers *10*, 1013 (1971).

18. Samulski, E. T., Berendsen, H. J. C., J. Chem. Phys. *56*, 3920 (1972).

19. Sobajima, S., J. Phys. Soc. Jap. *23*, 1070 (1967).

20. Woessner, D. E., J. Chem. Phys. *37*, 647 (1962).

21. Zimmerman, J. R., Brittin, W. E., J. Phys. Chem. *61*, 1328 (1957).

POLYMERIZATION OF P-METHACRYLOYLOXYBENZOIC ACID AND METHACRYLIC ACID IN MESOMORPHIC N-ALKOXY-BENZOIC ACIDS

A. Blumstein*, R. Blumstein*, G. J. Murphy*, C. Wilson*, J. Billard**

*The Polymer Science Program, Dept. of Chemistry, Lowell Technological Institute, Lowell, Mass. USA.

**Laboratoire de Physique de la Matiere Condensee, 11 Place Marcellin Berthelot, Paris V France.

INTRODUCTION

Liquid crystals, which combine molecular order with molecular mobility, are of considerable interest as media for polymerization. Monomers which cannot be polymerized in the crystalline state easily polymerize in a mesophase because of molecular mobility. Orientation of the monomer molecules prior to polymerization can influence the kinetics of the reaction and the structure of the resulting polymer.

The synthesis and polymerization of several mesomorphic vinyl monomers has been described (1-6). Strzelecki and Liebert (7,8,9) have recently succeeded in "freezing" the long-range organization of several mesomorphic vinyl monomers polymerized in the presence of a mesomorphic crosslinking agent.

The polymerization of a non mesomorphic monomer within a mesomorphic solvent has also been reported. Amerik, Krentzel and al. (10,11) and Blumstein and al. (12) have polymerized a vinyl monomer, p-methacryloyloxybenzoic acid (MBA), admixed into a smectic or nematic p-n-alkoxybenzoic acid. The reactions were carried out with free radical initiators and N,N-dimethylformamide (DMF) was used to prepare reference polymers in an isotropic solvent.

The results presented here were obtained in the framework of a study of the possibility of topochemical control of free radical polymerization of non mesomorphic solvents. In attempting to interpret the kinetics of polymerization of MBA (10,11,13) and of methacrylic acid (13) within mesomorphic p-n-alkoxybenzoic acids we have found it necessary to establish the solubility of the monomers and of the polymers in the mesophase, the latter as they appear "in situ". The mesomorphic solvents were p-n-cetyloxybenzoic acid (CBA) in its smectic state and p-n-heptyloxybenzoic acid (HBA) in its nematic state. Results of differential scanning calorimetry (DSC) and thermal polarizing microscopy are presented. The molecular weight distributions of the polymers, established by gel permeation chromatography (GPC), are also discussed.

EXPERIMENTAL

Materials

Monomers: methacrylic acid (MA), purchased from Rohm and Haas, was freshly distilled prior to use; MBA was synthesized from p-hydroxybenzoic acid and methacryloyl chloride as described previously (14). Special care was exercised to avoid oligomer formation during recrystallization. Mp 182-183°C.

Saturated model compounds: isobutyric acid (IBA) was distilled prior to use and p-propionoxybenzoic acid (PBA) was synthesized from p-hydroxybenzoic acid and propionic anhydride as described previously (14). Mp: 192°C (DSC)

Reference solvents: o-ethoxybenzoic acid (EBA) and N,N-dimethylformamide (DMF) were distilled prior to use.

Mesomorphic solvents: p-n-heptyloxybenzoic acid (HBA), from Frinton Laboratories was recrystallized thrice; capillary Mp: 92°C,p-n-cetyloxybenzoic acid (CBA) was synthesized from p-hydroxybenzoic acid and cetyl iodide as described previously (14); capillary Mp: 99°C.

Phase diagrams

Binary mixtures of PBA-CBA and PBA-HBA were prepared by melting of accurately weighed out quantities of both components until complete miscibility has been achieved (150-180°C), cooling the melt and reducing the crystals to a fine powder. The approximate shape of the phase diagrams was established using the contact method (15). The accurate assessment of the diagrams was performed using spot checks on mixtures of known composition by means of thermal microscopy (Leitz Panphot microscope and Mettler heating stage) and DSC (Perkin Elmer Thermal Analyzer). We have used as temperature of transition the temperature corresponding to the intersection of the baseline with the line of early transition. These temperatures tended to be lower than that recorded by microscopy and capillary observations.

Solubility of isobutyric acid (IBA) in HBA and of the polymers poly(methacrylic acid) (PMA) and poly(methacryloyloxybenzoic acid) (PMBA) in the mesophase was observed in melting point capillaries and by thermal microscopy.

Polymerizations

Reference polymers: polymerizations in isotropic solvents were carried out with either benzoylperoxide (BPO) or t-butylperbenzoate (TBPB) as free radical initiators; MA was polymerized in solution in o-ethoxybenzoic acid (EBA) and MBA in solution in DMF, in thoroughly degassed polymerization tubes.

Polymers prepared in CBA or HBA: mixtures of NBA-HBA and MBA-CBA of known composition were finely divided (200 mesh sieve). The initiator was added in solution in ether, which was then driven off. Polymerization tubes were successively evacuated to 10^{-4}mm Hg and flushed with dry argon several times. In the case of MA-HBA systems the reaction mixtures were frozen with liquid nitrogen during degassing to avoid losses of the volatile monomer.

After polymerization unreacted MA and the solvent HBA were extracted with benzene (added with hydroquinone) and unreacted MBA and the solvent HBA or CBA with an ether-benzene mixture (added with hydroquinone).

Methylation of polymers: The polymers were esterified to their methyl esters with 1-methyl-3-p-tolyltriazene, as described elsewhere (16). The percentage of esterification can be estimated from the NMR spectrum of the polyester, by use of the planimeter using eq.:

$$\frac{\%\ \text{esterification}}{100} = \frac{\text{area O-CH}_3}{3/5\ (\text{area CH}_2 + \text{area } \alpha\text{-CH}_3)}.$$

The completeness of methylation was also checked by elementary analysis of the methylated polymers poly(methylmethacrylate) (PMMA) and poly(methylmethacryloyloxybenzoate) (PMMB).

Characterization of polymers: The NMR spectra were obtained with a Hitachi-Perkin Elmer 60 MHz instrument and a Waters Associates Model 200 Gel Permeation Chromatograph was used to establish the molecular weight distributions of the polymers. (solvent: tetrahydrofuran (THF) at 25°C).

RESULTS AND DISCUSSION

Solubility of the monomers in the mesophase

The dimers of p-n-alkoxybenzoic acids form a homologous series of mesomorphic compounds, beginning with n-propyloxybenzoic acid (17). In the present work HBA and CBA served as nematic and smectic solvent, respectively.

The monomer MBA does not polymerize in the crystalline state; at room temperature we have irradiated the monomer crystals with γ-rays without detecting any polymerization after application of a dose of 44 Megarad (0.66 Mgrd/hour). After similar irradiation at 100°C the degree of conversion to polymer was less than 1%, the polymerization probably induced by the presence of impurities and crystal defects. Upon melting or dissolution in a mesophase MBA polymerizes rapidly, so that reproducible phase transition temperatures are difficult to obtain for MBA-HBA or MBA-CBA systems on successive heating and cooling. Consequently, the saturated model compound PBA was used to establish the phase diagrams. A number of transition temperatures were verified by replacing PBA with MBA additioned with hydroquinone to retard polymerization (1% by weight with respect to monomer). Within the limits of experimental error the results obtained by both methods were similar.

Figure 1 shows, as an example, the thermograms of an equimolecular mixture PBA-CBA upon heating and cooling (18). The nematic mesophase appears clearly upon heating (126.5 - 131.5°C) and a strong supercooling effect is illustrated in fig. 1b (isot. ⟶ nem. transition observed at 122°C).

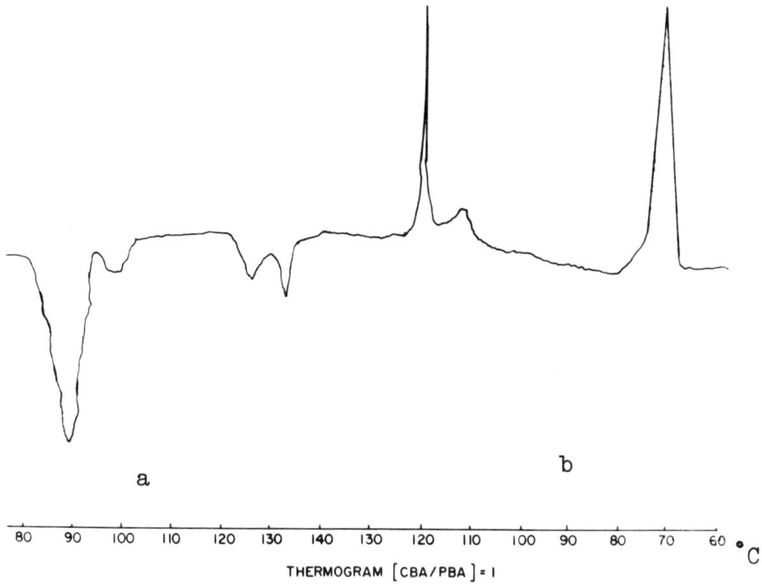

Fig. 1 DSC thermogram of an equimolecular mixture PBA-CBA

(a): heating; (b): cooling

Fig. 2 and 3 show phase diagrams for the systems PBA-CBA and PBA-HBA. The concentration of PBA is expressed in mole fraction and the symbols N,S,C and L designate nematic, smectic, crystal and isotropic liquid, respectively. A nematic region can be observed for the system PBA-CBA (18).

The solubility of monomer in the smectic mesophase of CBA does not exceed 0.2 (in mole fraction of monomer). The monomer is somewhat more soluble in the nematic mesophase of HBA, up to a maximum of approximately 0.3 (mole fraction of monomer).

In determining solubility of methacrylic acid in HBA the monomer was replaced by the saturated model compound IBA, in order to avoid complications due to polymerization upon heating and cooling. Table I shows the transition temperatures upon heating for systems IBA-HBA, observed in the melting point capillary (a) and under polarizing microscope (b).

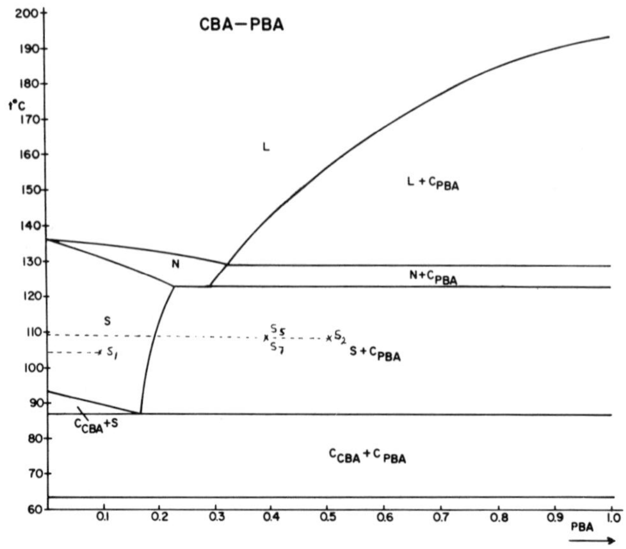

Fig 2 Phase diagram for the system PBA-CBA (see Table II for explanation of symbols S_1, S_2, S_5, S_7)

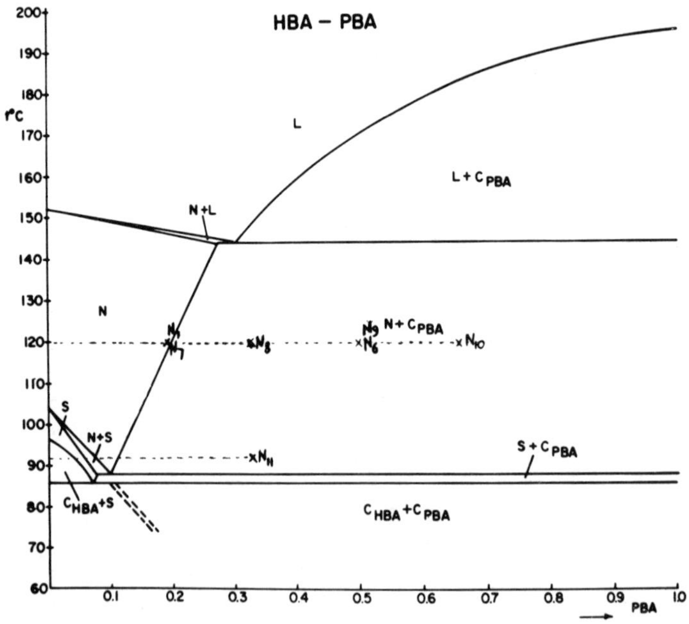

Fig. 3 Phase diagram for the system PBA-HBA (see Table II for explanation of symbols N_1, N_7, N_8, etc.)

Table I

Transition temperatures for Systems IBA-HBA

Conc. of IBA (mole fraction)	Sol → sm(°C) (a)	(b)	smec → nem(°C) (a)	(b)	nem → isot (°C) (a)	(b)
0.090	87.5	87	88.5	88.5	136	136.5
0.142	85	85	86.5	86.5	124	124
0.167	85	85	86	86	122.5	122.5
0.200	84.5	84.5	85.5	85.5	119.5	119.5
0.208			85.5	85.5	112	105
0.215	-	excess monomer			-	

A small smectic region could be observed. The transition nematic ⟶ isotropic occurs over an interval of temperatures. The temperatures recorded in table I are that at which the first isotropic droplet was observed.

A method for verifying the dissolution of IBA in the nematic mesophase of HBA consists of recording the proton magnetic resonance spectrum for a system of known composition at different temperatures. Fig. 4 shows the NMR spectrum obtained for a system IBA-HBA (mole fraction IBA: 0.167) at 110°C and 130°C. At 130°C the solution is isotropic (spectrum b), but at 110°C it is nematic (spectrum a). The spectrum of the solvent and solute, oriented by the magnetic field, is washed out in the background (19), except for one peak attributed to the rapidly exchanging carboxylic hydrogens of HBA.

Molecular weight distributions of the polymers

Table II summarizes the conditions of preparation of some PMBA's and gives their molecular weights, as measured by GPC. The symbols S, N and I refer to polymers prepared in CBA, HBA and DMF, respectively. The symbols $\overline{M}_n$ and $\overline{M}_w$ are the number and weight average molecular weights, respectively and $\overline{M}_w/\overline{M}_n$ the "heterogeneity ratio" of the molecular weight distribution. The molecular weights were computed from gel permeation chromatograms obtained for the methyl esters poly(methylmethacryloyloxybenzoates) (PMMB), as satisfactory GPC data would not be obtained directly from the PMBA's. Methylation was easily carried out with 1-methyl-3-p-tolyltriazene which we have found to be a more convenient methylating agent than the commonly used diazomethane (16). For PMMB we have established the

TABLE II

Poly(methacryloyloxybenzoic acids): conditions of preparation and molecular weight distributions

Sample	Monomer conc. (mole fraction)	Temp. of polym. (°C)	Initiator	Initiator conc. (% by weight with respect to monomer)	$\overline{M}_n$ ($\times 10^{-3}$)	$\overline{M}_w$ ($\times 10^{-3}$)	$\overline{M}_w/\overline{M}_n$
S-5	0.39	105	TBPB	0.5	273	711	2.6
S-7	0.39	105	BPO	0.5	220	600	2.7
N-6	0.50	120	BPO	0.05	413	1,097	2.7
N-11	0.33	92	BPO	1.0	351	1,026	2.9
I-1	0.20	109	TBPB	0.5	137	348	2.5
I-2	0.10	95	BPO	1.0	31	62	2.0
S-1	0.09	105	BPO	1.0	42	454	10.8
S-2	0.50	109	TBPB	0.5	14	72	5.2
N-1	0.20	120	BPO	0.5	194	563	2.9
N-7	0.20	120	TBPB	0.5	205	540	2.6
N-8	0.33	120	TBPB	0.5	165	406	2.6
N-9	0.50	120	TBPB	0.5	188	489	2.6
N-10	0.66	120	TBPB	0.5	152	453	2.9

following viscosity law in THF at 25°C:

$$[\eta]_{dl/g} = 1.04 \times 10^{-4}\ M^{0.65} \quad (M \text{ between } 2.1 \times 10^{6} \text{ and } 2.96 \times 10^{5}).$$

This was used for computing molecular weights from gel permeation chromatograms. Polymerization of MBA in HBA and CBA was carried out at monomer concentrations both below and above the saturation of the mesophase (see fig. 2 and 3). Within the interval of concentrations studied here polymerization is rapid and quasicomplete regardless of the amount of excess monomer crystal. According to Krentzel and Amerik (11) the equilibrium concentration of monomer is close to zero throughout the entire smectic or nematic range of temperatures. This is in sharp contrast with the increase in equilibrium concentration of monomer, from $[MBA]_e = 0.065$ at 85°C to $[MBA]_e = 0.508$ at 130°C, during polymerization in the isotropic solvent DMF (11). We have confirmed this point, as illustrated in fig. 5 which shows conversion to polymer versus time for some representative systems. For the N and S samples mentioned in table II the percentage of conversion to polymer is approximately 95%. For sample I-2 the conversion to polymer was stopped at about 30% and for sample I-1 polymerization was pushed to the limit of conversion (approximately 80%).

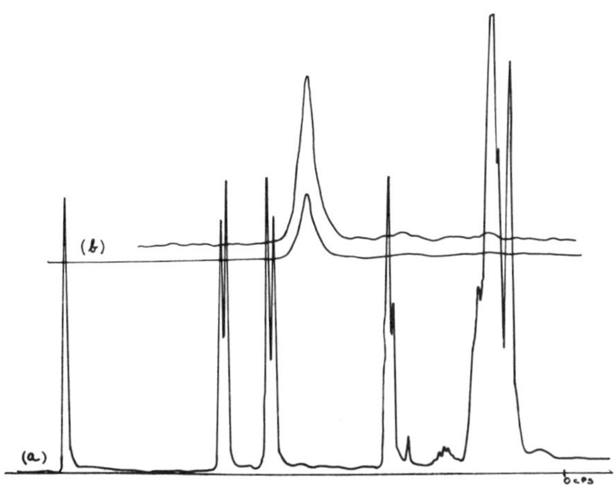

Fig. 4 NMR spectrum at 130°C (a) and 110°C (b) for a system IBA-HBA (mole fraction IBA: 0.167). Sweep width: 1200 cps; 60 MgHz instrument.

Samples S-5, S-7, N-6, N-11, I-1 and I-2 were prepared in thoroughly degassed polymerization tubes. The weight average molecular weights of polymers prepared in isotropic solution in DMF were approximately between 6×10^4 and 3.5×10^5. The value $\overline{M}_w/\overline{M}_n = 2.0$ observed for sample I-2 is typical of the type of molecular weight distribution obtained by free radical polymerization in an isotropic solvent. The distribution in sample I-1 is somewhat broader, as the polymerization was pushed to the limit of conversion.

The molecular weights of polymers prepared in the smectic and nematic solvents are much higher than the $\overline{M}_w$ and $\overline{M}_n$ of PMBA's prepared in DMF, and the molecular weight distributions are somewhat broadened. ($\overline{M}_w/\overline{M}_n$ = 2.5 - 2.9). It is interesting to observe the drastic influence of oxygen on the shape of the molecular weight distribution curve of polymers prepared in the smectic solvent: Sample S-1 for example was prepared by spreading the finely divided reaction mixture to form a layer of approximately 0.5 cm in thickness and polymerizing under a stream of dry commercial nitrogen. No further effort was made to exclude oxygen. Sample S-2 was

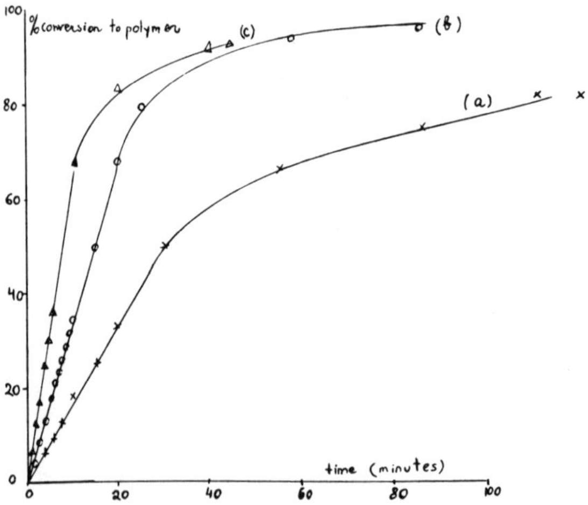

Fig. 5 Conversion to polymer as a function of time: (a) MBA in solution in DMF (b) MBA in CBA (c) MBA in HBA. Concentration monomer: 20% by weight; initiator: 0.1% by weight (with respect to monomer); temperature: 103°C.

prepared in a tube evacuated to 10^{-1} mm pressure and flushed twice with argon. The molecular weight distribution curve of samples S-1 and S-2 is binodal (see fig.6) with predominance of molecular weights of a few thousands and less on one hand and a few hundred thousands on the other. This appears to indicate that two different mechanisms of polymerization are involved.

Samples N-1, N-7, N-8, N-9 and N-10 have been prepared in the nematic solvent HBA under conditions similar to that described for samples S-1 and S-2. Although the values of molecular weights are depressed by the presence of small amounts of oxygen, the shapes of the molecular weight distribution curves are not altered. In samples N-7 through N-10 the mole fraction of monomer concentration was varied from 0.2 just below the saturation level of the mesophase, to 0.66. (see fig.3) Neither the molecular weights nor the width of the distribution appear to be significantly influenced by the presence of excess monomer crystal.

Table III summarizes the conditions of preparation of some PMA's and gives their molecular weights as measured by GPC in THF at 25°C. As in the case of PMBA's the polymers were methylated with triazene prior to GPC measurements. To compute molecular weights from GPC data, we have used the intrinsic viscosity law given by Provder and al. for PMMA in THF at 25°C (20).

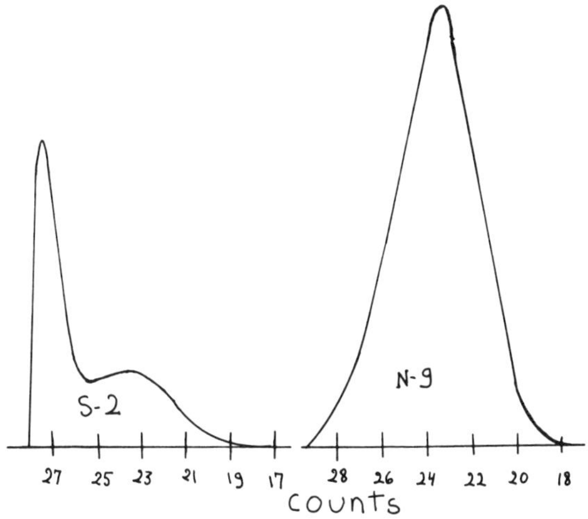

Fig. 6 Gel permeation chromatograms of samples S-2 and N-9.

$$[\eta]_{dl/g} = 1.04 \times 10^{-4} M^{0.697} \quad (M \geqslant 31,000) \quad (20).$$

Table III

Polymerization of MA and molecular weights of the polymers

Solvent	Temp.Polym. °C	% conversion	$\overline{M}_w(x10^{-3})$	$\overline{M}_n(x10^{-3})$	$\overline{M}_w/\overline{M}_n$
HBA	90	31	895	507	1.8
HBA	115	72	544	268	2.0
CBA	110	20	387	158	2.5
EBA	105	20	1,438	863	1.7
EBA	135	81	530	264	2.0

All polymerizations summarized in Table III were carried out in thoroughly degassed tubes at a monomer concentration 0.2 (mole fraction) below the saturation of the mesophase; the initiator used was TBPB (0.2% weight with respect to monomer). The molecular weights and distribution curves appear to be the same for PMA's prepared in the nematic solvent HBA and the isotropic solvent EBA.

Solubility of polymers in the mesophase

When the polymerization of MBA in HBA or CBA and of MA in HBA is observed under polarizing microscope the polymer can be seen to precipitate out of the mesophase. Polymer precipitation can also be observed in the melting point capillary. Insolubility of PMBA in the smectic phase of CBA is indirectly confirmed by the binodal distributions observed for samples such as S-1 or S-2. A tentative explanation for this type of gel permeation chromatograms is as follows: the polymerization is initiated in the mesophase, where the initiator is dissolved. The polymer is insoluble in the mesophase and the growing macroradical precipitates out in a coiled conformation after a certain number of addition steps. If a macroradical is deactivated by a chain transfer to solvent or oxygen, chain growth stops and polymers of low molecular weight are observed. At higher conversions, however, trapped macroradicals may continue to grow, the monomer being supplied through diffusion. The fact that no significant amounts of low molecular weight polymers were observed for samples such as N-7, N-8, N-9 or N-10 could well indicate that PMBA is more soluble in the nematic mesophase of HBA than it is in the smectic CBA. The conditions of chain growth may therefore be more favorable in the former solvent. This is supported by the fact that the polymerization rate is somewhat higher in the nematic solvent.

Although adequate knowledge of monomer-mesophase and polymer-mesophase compatibility appears to be essential for a clear interpretation of any results of polymerization a brief survey of the literature appears to yield but little information on the solubility of polymers within mesophases. Amerik and Krentzel (11) assume that PMBA actually stabilizes the mesophase. A similar view is reported by Hardy and al (21) in their discussion of the kinetics of polymerization of cetyl vinyl ether in the smectic state. These authors have observed an auto-acceleration of the reaction, which they explain by assuming that the polymer stabilizes the smectic mesophase. On the other hand Strzelecki and Liebert (7) report a separation of polymer from the nematic monomer N-p-acryloyloxybenzylidene-paracyanoaniline. Our microscopic observation shows that PMBA and PMA are insoluble in CBA and HBA.

As a rule, if we consider the free energy of mixing of a polymer with a smectic mesophase the entropy gain will be small. The situation is not unlike that of the mixing of two polymers. It is not unlogical to expect poor polymer solubility in liquid crystalline solvents. All other things taken equal solubility might be expected to decrease in the order:

isotropic ⟶ nematic ⟶ cholesteric ⟶ smectic

Insolubility of the polymer in the mesophase can adequately explain the virtual completeness of polymerization of MBA throughout the entire temperature interval of the mesophase. Amerik, Krentzel and al. (10, 11) explain this observation by assuming polymerization of ordered monomer sequences or microaggregates of monomer dissolved in the mesophase up to unspecified but high monomer concentrations. It is interesting, for instance, to cite their model for the structure of an equimolecular mixture of MBA and CBA: the mesophase is assumed to be formed of "dimers" MBA-CBA, a layer of MBA alternating with a layer of CBA. If such were the case each mole of solvent would be expected to dissolve at least one mole of monomer and this is well above the solubility limit illustrated in fig. 2 (Demixion of excess crystalline MBA was also confirmed by x-ray diffraction).

We assume that the mesophase is formed by wedging monomer molecules between solvent molecules and completeness of polymerization can then be simply explained by the fact that the polymer precipitates out, so that one can no longer consider monomer-polymer equilibria within the mesophase. When excess crystalline MBA is present at the onset of polymerization, it merely serves as a reservoir of monomer. This is indicated by the fact that crystalline MBA does not polymerize (although "plasticized" MBA polymerizes slowly) and also by the fact that presence of excess crystalline monomer does not influence the molecular weight distribution.

Acknowledgment: The authors wish to thank Mrs. Z. Gallot-Grubisic, Dr. A. Skoulios and Mr. J. Helfer of the C.R.M., Strasbourg, France for some x-ray, viscosity measurements and helpful discussions. Acknowledgment is made to the donors of the Petroleum Research Fund, administered by A.C.S. for partial support of this research.

BIBLIOGRAPHY

1. Y.B. Amerik, and B.A. Krentzel, J. Polymer Sci., C 16, 1383 (1967)

2. G. Hardy, F. Cser, A. Kallo, K. Nyitrai, G. Bodor and M. Lengyel, Acta Chim. Acad. Sci., Hung., 65, 287, (1970)

3. C.M. Paleos and M.M. Labes, Molecular Crystals and Liquid Crystals, 11, 385 (1970)

4. W.J. Toth and A.V. Tobolsky, J. Polymer Sci., B 8, 289 (1970)

5. A.C. de Visser, J. Feyen, K. de Groot and A.Bantjes, J. Polymer Sci., B 8, 805 (1970)

6. Y. Tanaka, S. Kabaya, Y. Shinura, A. Okada, Y. Kurihata and Y. Sakakibara, J. Polymer Sci., B, 110, 261 (1972)

7. L. Strzelecki and L. Liebert, Bull. Soc. Chim. France, 2, 597, (1973)

8. L. Liebert and L. Strzelecki, Bull. Soc. Chim. France, 2, 603, (1973)

9. L. Strzelecki and L. Liebert, Bull. Soc. Chim. France, 2, 605, (1973)

10. Y.B. Amerik, I.I. Kontstantinov, B.A. Krentzel and E.M. Malachaev, Vysokom. Soed., A, IX, 12, 1591 (1967)

11. B.A. Krentzel and Y.B. Amerik, Vysokom. Soed., A, XIII, 6, 1358 (1971)

12. A. Blumstein, N. Kitagawa and R. Blumstein, Molecular Crystals and Liquid Crystals, 12, 215 (1971)

13. R. Blumstein, A. Blumstein, G. Murphy and P. Ray, in preparation.

14. N. Kitagawa, M.Sc. Thesis, Lowell Technological Institute (1969)

15. L. Kofler, "Thermomikromethoden", Vg Chemie, Weinheim, (1954)

16. R. Blumstein, G.J. Murphy, A. Blumstein and A.C. Watterson, J. Polymer Sci., B, 11, 21 (1973)

17. G.W. Gray, "Molecular Structure and the Properties of Liquid Crystals", Academic Press, New York, 1962

18. A. Blumstein, J. Billard and R. Blumstein, Molecular Crystals and Liquid Crystals (in print)

19. A. Saupe and W. Maier, Z. Naturforshung, 19a, 172 (1964)

20. T. Provder, J.C. Woodbrey, J.H. Clark, Separation Sci., 6 (1), 101, (1971)

21. Gy. Hardy, K. Nyitrai, F. Cser, Gy. Cselik and I. Nagy, Europ. Polymer J., 5, 133 (1969)

THERMOTROPIC LIQUID CRYSTALS VI. THE PREPARATION AND MESOPHASE PROPERTIES OF ASYMMETRICALLY 4,4'-DISUBSTITUTED PHENYL BENZOATES

Mary E. Neubert, Leo T. Carlino, Richard D'Sidocky and D. L. Fishel

Liquid Crystal Institute and Chemistry Department
Kent State University, Kent, Ohio 44242

We have undertaken preparation of complete series of 4,4'-disubstituted phenyl benzoates, 1, and their thermal characterization to test the effect of terminal alkyl group branching at various positions with respect to the aromatic ring. We also wished to establish which of these compounds would be most useful in more detailed physical studies of structure and conformation[1-3]. Some members of these series for which substituents are normal alkyl and/or alkoxyl groups have been reported by other investigators[4,5].

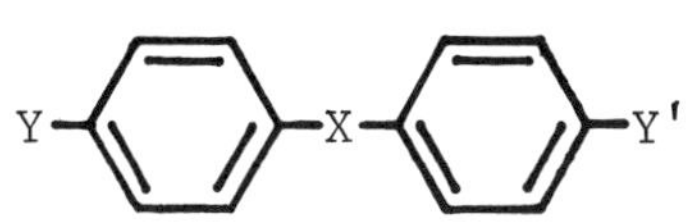

1, X= -O-C(=O)-

2, X= -CH=N-, -N=N-, -CH=CH-, -CH=NO-, -N=NO-, etc.

Studies of the effect of molecular structure changes on mesomorphic behavior have been inherent in recent synthetic efforts toward low temperature nematic compounds. These have usually been concerned with alteration of the nature of the central group or with control of relative conformation for the two aromatic rings in systems generally represented by 2.[6-12]

Changes in the terminal group structure have been usually limited to variation of electrical nature (substitution of nitrile, halogen, alkoxyl, acyl, etc.) or homologation within the normal chain series[13-16]. Although branched-chain compounds often melt at lower temperatures than the isomeric normal derivatives, few terminally branched compounds of type 2 have been reported to exhibit mesophase properties. Extensive work on such compounds has probably been inhibited because of the well established generalization that thermotropic liquid crystals have molecules that are long, rigid, rod-shaped and with no protuberances on the major axis. Only the work reported by Gray is a recent exception to this[17]; his studies have involved branching of the alkyl moiety in benzylidene derivatives of alkyl 4-aminocinnamates. Branching in these compounds is significantly removed from the aromatic ring system that delineates the principal molecular dimensions.

RESULTS:

The synthesis of precursor phenols and benzoic acid derivatives for this study followed well documented procedures with the exception of the preparation of 4-alkylbenzoyl chlorides. These latter compounds were obtained in one step from the corresponding alkylbenzenes by modified Friedel-Crafts acylation with oxalyl chloride. A detailed study of this reaction is reported elsewhere[18].

A two step synthesis of 4-alkoxyphenols by Williamson alkylation of p-benzyloxyphenol, then hydrogenolysis, has been reported by VanMeter and Klandermann[4]. In our hands this technique worked well; typical overall yields were ~70%. We are inclined, however, to favor the one step procedure we have outlined as somewhat less time consuming for small to medium size batches.

Thermal characterization of the 4,4'-disubstituted phenyl benzoates was accomplished by hot-stage microscopy and confirmed by DTA using well-established techniques. Thermal data are listed in Tables 1, 2, and 3 and are illustrated in plots versus carbon number, Figures 1 - 4.

We are in essential agreement with most data previously reported for several of the normal members of these series[4,5]. Discrepancies are footnoted in Table 1 for 4-methoxylphenyl 4'-n-butylbenzoate (Code≠: 10-4) and for compounds 40-4, 70-4 and 4-0$\overline{4}$. Generally the phenyl benzoate esters have relatively low melting points ($< 100^o$) and there is significant occurence of

---

≠ We have found it convenient to use a simple alphanumeric code for identification of 4,4'-disubstituted phenyl benzoates with terminal alkoxyl and alkyl groups. As an example: 4-n-propylphenyl 4'-sec-butylbenzoate is coded: $3\text{-}03^{1(1)}$; the standard script Arabic numerals refer to the length of the longest carbon chain as a 4- or 4'-substituent and superscript numbers refer to position and size (in parentheses) of a branched carbon chain on the main chain of carbon atoms. Oxygen position is indicated by "O"; the dash is the phenyl benzoate moiety.

FIGURE 1 NEMATIC-ISOTROPIC TRANSITION POINTS FOR 4,4' DISUBSTITUTED PHENYL BENZOATES

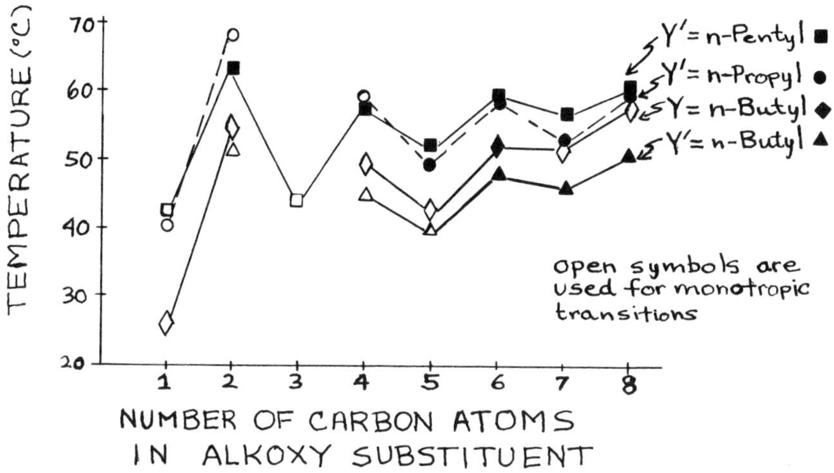

FIGURE 2 MELTING TEMPERATURES FOR 4,4'-DISUBSTITUTED PHENYLBENZOATES

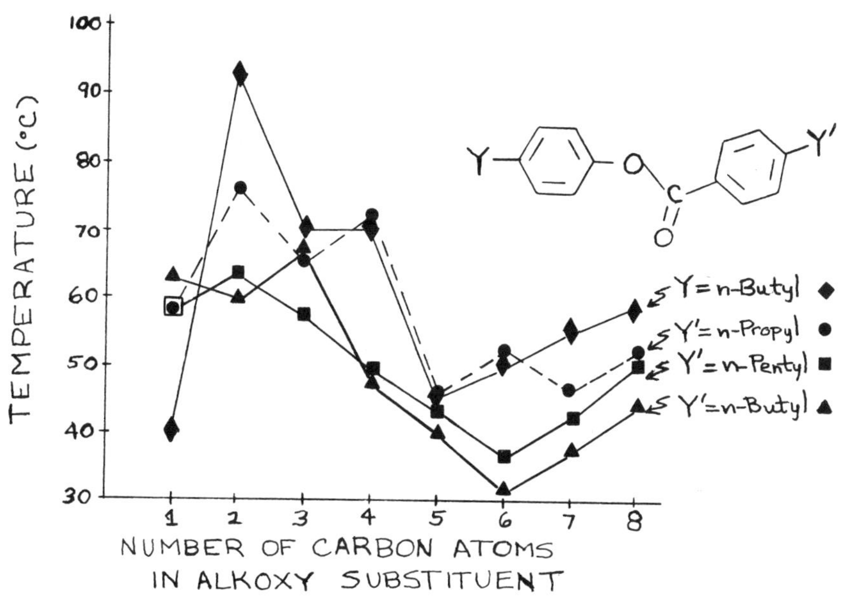

FIGURE 3 TRANSITION TEMPERATURES FOR 4-ALKOXY-4'-ISOBUTYLBENZOATES

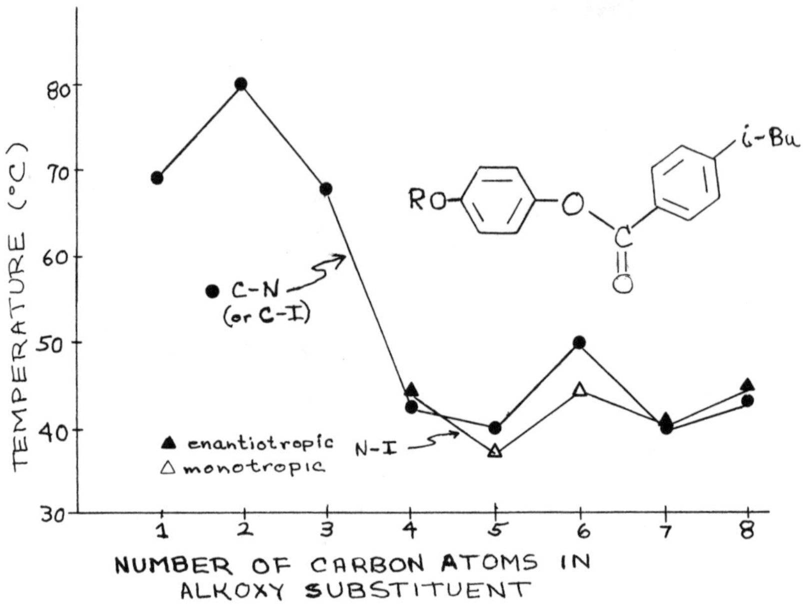

FIGURE 4 TRANSITION TEMPERATURES FOR 4-ISOBUTYLPHENYL-4'-ALKOXYBENZOATES

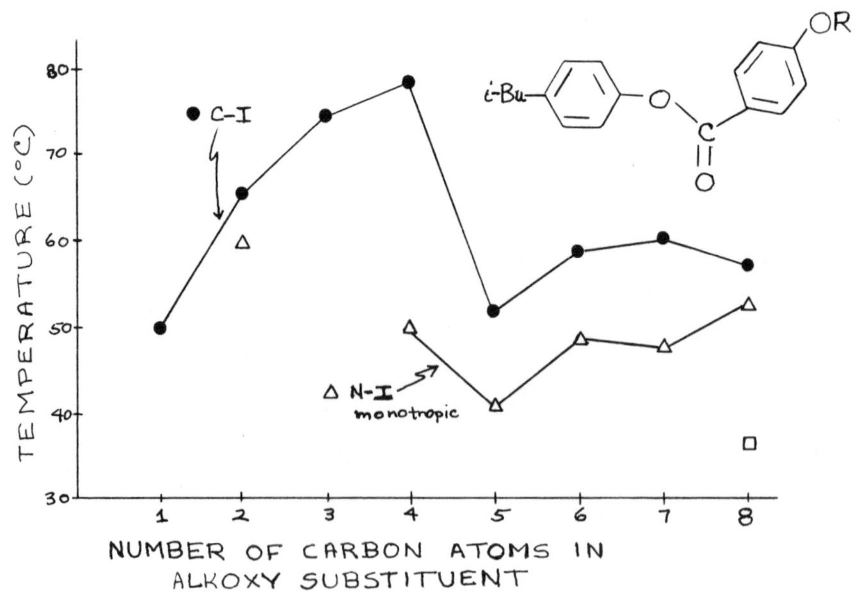

Table 1. Transition Temperatures for Normal 4,4'-Disubstituted Phenyl Benzoates, 1

Y	Y'	C[a]	N	I
10	3	(38.7)[b]	(40.4)	56.8-58
10	4	(34.8)	-c	61.7-62.6
10	5	(38.6)	(42.2)	57.0-57.9
4	01	(12)	(29)	39.9-40.8
20	3	(58.2)	(68.5)	75.0-76.1
20	4	(38.3)	(51.0)	58.0-59.7
20	5	(39.6)	61.2-63.0	63.4
4	02	(50.8)	(54.0)	91.5-92.7
30	3	(50.6)	-	63.7-65.1
30	4	(33.8)	-	64.7-65.8
30	5	(38.6)	(44.0)	56.0-57.3
4	03	(36.2)	-	69.3-69.9
40	3	(34.2)	(58.7-59.2)	70.3-72.3
	4[d]	(39.8)	(45.3)	46.3-47.3
	5	(30.5)	(48.4-49.6)	57.7
4	04[e]	(35.5)	(49.6)	69.2-70.2
50	3		44.9-45.5	49.9
	4	(28.6)	(39.2)	39.1-40.0
	5	(27.5)	(42.0-42.8)	51.8
4	05		(42.7)	43.9-44.8
60	3		50.8-52.3	58.5-59.1
	4	(14°)	28.9-31.6	48.4
	5		34.2-36.4	59.9
4	06	(43.2)	49.2-49.5	51.9
70	3	(27.5)	45.5-46.3	53.2
	4[b]	(28.2)	35.7-37.1	46.0
	5		40.5-42.2	57.4
4	07	(41.9)	(51.6)	54.1-54.8
80	3	(31.8)	50.2-51.7	58.8-59.4
	4	(24.8)	41.8-43.6	51.2
	5	(32.7)	49.0-49.8	60.6
4	08	(47.5)	(57.8) (50.8)[g]	57.7-58.1

a freezing temperature to crystalline phase
b all parentheses values are monotropic
c ref. 4 reports (23°)
d ref. 4 reports N(42); I, 44.
e ref. 5 reports N(50); I, 69.
f ref. 5 reports N; 35; I, 43.
g smectic B transition for 4-08

Table 2. Transition Temperatures for β-Methyl-Branched Alkyl 4,4'-Disubstituted Phenyl Benzoates, <u>1</u>

Y	Y'	C[a]	N	I
10	$3^{2(1)}$	(47.4)[b]	-	67.6-69.0
20	$3^{2(1)}$	(49.0)	-	78.9-79.7
30	$3^{2(1)}$	(28.1)	-	65.2-67.5
40	$3^{2(1)}$	-	40.1-42.1	43.7
50	$3^{2(1)}$		(36.7)	38.8-40.0
60	$3^{2(1)}$	(27.0)	(44.0)	48.0-49.9
70	$3^{2(1)}$		38.3-39.5	39.7
80	$3^{2(1)}$	(30.8)	41.2-43.2	44.4
$3^{2(1)}$	01			49.3-49.8
$3^{2(1)}$	02	(35.0)	(59.5)	64.5-65.5
$3^{2(1)}$	03	(42.8)		73.9-74.9
$3^{2(1)}$	04	(43.6)	(49.8)	76.8-78.2
$3^{2(1)}$	05	(29.9)	(40.7)	51.2-51.7
$3^{2(1)}$	06	(35.6)	(48.7)	57.8-58.5
$3^{2(1)}$	07	(32.3)	(47.5)	59.1-60.0
$3^{2(1)}$	08	(29.6)	(52.2)	55.9-57.0
			(36.0)[c]	

a freezing point
b all parentheses values are monotropic
c smectic transition for $3^{2(1)}$-08

nematic behavior. Monotropic nematic compounds are obtained for terminal groups of low carbon number whereas relatively short range enantiotropic behavior occurs for longer (>C_5) terminal groups.

Clearing points for all the normal series followed smoothly alternating curves which were closely parallel, even when alkoxy and alkyl groups were reversed for Y and Y' (Figure 1).

The astonishing feature of these data is the existence of nematic behavior for most of the isobutyl derivatives whether on the phenol or the benzoic acid moiety, Table 2. This is in light of the complete absence of mesomorphic properties, either smectic <u>or</u> nematic for any α-methyl derivative of either series. As can be seen from Table 3, crystal isotropic transition temperatures for the α-methyl compounds range from near room temperature (70-$3^{1(1)}$, 27.7-30.2°) to over 100° (20-$2^{1(1)}$, 105.2-106.6°). Phenyl benzoate derivatives with an isobutyl group as a substituent and which exhibit enantiotropic nematic behavior, do so over a narrow temperature range and at temperatures very close to 40°. The high-

Table 3. Crystal-Isotropic Transition Temperatures for α-Methyl Derivatives of 4,4'-Disubstituted Phenyl Benzoates, 1

Y	Y'	I≠	C*
10	$2^{1(1)}$	96.4-98.0	
20	$2^{1(1)}$	105.2-106.6	
30	$2^{1(1)}$	85.9-90.4	(54.4)
40	$2^{1(1)}$	79.3-81.2	(40.0)
50	$2^{1(1)}$	69.2-70.9	(40.1)
60	$2^{1(1)}$	53.0-54.0	(45.5)
70	$2^{1(1)}$	51.7-54.3	(33.3)
80	$2^{1(1)}$	47.8-50.3	(27.5)
10	$3^{1(1)}$	81.7-85.0	(51.4)
20	$3^{1(1)}$	81.0-85.0	(52.5)
30	$3^{1(1)}$	79.5-82.7	(51.1)
40	$3^{1(1)}$	56.9-58.3	(36.6)
50	$3^{1(1)}$	46.0-50.6	(31.1)
60	$3^{1(1)}$	44.9-46.8	
70	$3^{1(1)}$	27.7-30.2	
80	$3^{1(1)}$	33.0-34.3	
$3^{1(1)}$	01	34.3-35.9	(28.4)
$3^{1(1)}$	02	85.7-87.3	(79.0)
$3^{1(1)}$	03	45.3-47.8	(27.0)
$3^{1(1)}$	04	40.3-41.7	
$3^{1(1)}$	05	33.7-36.5	(30.3)
$3^{1(1)}$	06	46.0-49.3	
$3^{1(1)}$	07	39.6-40.5	
$3^{1(1)}$	08	40.3-42.0	

≠ Melting point of stable crystalline modification obtained from ethanol solution

* Freezing point of melt, undisturbed microscope slide preparation

est temperature at which monotropically stable nematic properties are observed for an isobutyl derivative is 59.5° ($3^{2(1)}$-02).

DISCUSSION:

Although synthetic studies toward understanding structure-phase transition relationships have been devoted to a molecular approach, it is clear that melting behavior must ultimately be related to crystal structure and only in a secondary way to molecular structure. Thus correlative regularity between molecular structure and transition temperatures are common for mesophase-isotropic transitions and even for mesophase-mesophase transitions but only rarely for crystal-to-higher-phase transitions[19]. It is, however, possible for the organic chemist to gain some insight into melting phenomena from data which may be related in a general way to molecular structure. Furthermore, significant recent work in thermodynamics involves a structural approach.

We have constructed what we feel are reasonable models for the most highly populated conformational shapes of the 4,4'-disubstituted phenyl benzoates in condensed phases and over the temperature range of interest. We have drawn on structural models proposed in the past as required for mesophase structure as well as on recent statistical-thermodynamics concepts and experimental measurements[20,21].

Basically, the phenyl benzoate molecule must probably be arranged with the 4- and 4'-substituents as far removed from one another as is possible. Co-planarity of the two benzene rings is not essential. In fact, there is most likely a twist angle between the aromatic planes of at least 30°; this twist is principally involved with the phenol moiety carbon-oxygen bond, the carbonyl group being in-plane and overlapping with the aromatic pi electrons. This allows a measure of stabilization including not only the carbonyl group but also the unshared pair overlap between the ether link and the other aromatic ring. Such an arrangement for the aromatic rings is consistent with stable mesophase properties by analogy with compounds including a number of other central groups[7,8,22-24].

An alkoxyl group is expected to enhance stability by its oxygen unshared-pair overlap with the associated benzene ring. This assumption and normal carbon-oxygen bond angles leads to a model with 30° between the plane of the aromatic ring and the axis of the alkoxyl chain in its most extended conformation. That such a conformation is unnecessary and even improbable may be derived from numerous studies of saturated hydrocarbon chain conformations[25,26]. For straight chain hydrocarbons beyond four carbons in length energy minima well below those thermally allowed at 320° K exist for twisted chain conformations. Thus, a significant population of molecules in condensed phases of n-alkoxybenzene derivatives might be expected to have the chain beyond the first four carbon atoms lying close to an axis coincident with one in the plane of the aromatic ring.

We also suggest that n-alkylbenzene derivatives will have conformations generally analogous to this. To the extent that one may extrapolate the findings of Ouellette et. al.[27] beyond three carbon chains, the most stable arrangement for the first carbon is such that an α-H lies in the plane of the benzene ring. Thus, to adopt the proposed near-linear arrangement, some kinking or twisting of the chain is necessary. Such an arrangement does, however, always imply inherent steric interference toward closest approach to one side of the aromatic ring due to the bulk provided by the second methylene group. When the second or beta-methylene group is methyl-branched (i.e. isobutyl derivatives) an effective block toward any significant interaction on one side of the contiguous aromatic ring is realized.

Now a simple explanation may be proposed for observations regarding potential nematic properties for isobutyl (β-methyl branched) vs sec-butyl (α -methyl branched) compounds. When melting occurs, in order for a nematic mesophase to develop it would appear from our data that significant π-π contact between aromatic rings of adjacent molecules must occur. One may imagine the development of "string" shaped molecular complexes if at least one side of both aromatic rings is free of steric interference and the long axes of all molecules concerned are approximately parallel. Each succeeding molecule is shifted along the long axis so that it has an exposed aromatic ring "face" for binding to the next molecule; the arrangement is roughly that of a row of dominoes which has collapsed. However, if one ring of any given molecule is unavailable for this form of molecular complex formation, long strings cannot develop and nematic behavior should not occur. This is the situation for all α -methyl substituted alkylbenzene derivatives in this study. There are no meso-stable arrangements for molecules with such structure and melting gives at once the random character of the isotropic melt.

This model for the secondary structure of a nematic mesophase allows a variety of more or less specifically oriented π-contact complexes for compounds with branches further removed from the aromatic ring. Lateral displacement of the aromatic rings may occur to give the configuration necessary for the greatest stability. Terminal alkyl and alkoxyl groups would contribute to the polarization forces which hold the complex together and which are associated primarily with the aromatic rings. Terminal groups would lie in interstitial spaces against the unoccupied face of the benzene ring on an adjacent molecule. Homologation beyond four or five carbon atom terminal groups would require extensive chain coiling if it were retained in this position; otherwise a chain would begin to interact with the terminal group on the neighboring molecule.

These concepts are directly analogous to structures proposed for other aromatic molecular complexes including even the more strongly bonded charge-transfer type as well as those depending only on polarization forces[28,29].

Our proposal differs from the secondary structure proposed by Pohl and Steinsträsser[30] for binary eutectics. We feel an extension of their model does not provide a structural requirement to explain why α-branched derivatives are not liquid crystalline whereas β-branched compounds are. It is conceivable that multiple π-π contact bonding (involving both faces of any given aromatic ring) would occur for those compounds which are enantiotropic; collapse of a more ordered crystalline lattice directly into such a structure could easily occur.

It is interesting to consider this structural model for a compound with a chiral terminal group. Assuming the most probable conformation is one with minimum non-bonded interactions, a β-methylbutyl alkyl group, for example, imparts a definite twist to a model of a string-like complex such as we have described. Incorporating many such helical structures into a close-packed bundle gives a model not unlike proposals for the gross structure of cholesteric materials[31].

EXPERIMENTAL

Transition temperatures were measured and phases identified for all materials by examination with a Leitz Ortholux polarizing microscope equipped with a calibrated Mettler FP2 hot stage. Initial heating rates of 10°/minute were used up to $<10^o$ of the expected transition, then heating rates of 2°/minute were employed through all observed transitions. Cooling rates were $\sim 2^o$/minute. Transitions below room temperature were obtained on a Reichert Thermopan polarizing microscope equipped with a cold stage. All enantiotropic esters and those for which successive transition temperatures were within 1° were checked with differential thermal analysis on DuPont DTA 900 or DTA 920 instruments. Analytical spectroscopic data were obtained for all precursor phenols and benzoic acid derivatives. Instruments were Perkin-Elmer Model 337 (infrared) Varian Associates Models A-60 or XL-100 (Pmr) and Associated Electronics Industries Model MS-12 (mass spectra) spectrometers. Elemental analyses were obtained for selected esters and were performed by Chemalytics Inc., Tempe, Arizona.

All esters were prepared from suitable 4-substituted benzoyl chlorides and 4-substituted phenols by a modified Schotten-Baumann reaction. Most of the precursors were synthesized; typical procedures are described.

4-n-Butylbenzoyl Chloride

A chlorocarbonylation reagent was prepared by adding oxalyl chloride (37.9g, 0.29m) to a stirred mixture of aluminum chloride (39.8g, 0.29m) in sym-tetrachloroethane (180 ml). This reagent was added to a stirred solution of n-butylbenzene (40g, 0.29m) in sym-tetrachloroethane (265) ml). After one hour the reaction mixture was poured into ice-water; the organic layer was separated immediately and extracted with cold 5% potassium hydroxide solution (2X), washed with water, dried over sodium sulfate, filtered and stripped (Rotovap). The residual liquid was distilled at reduced pressure to give a fraction boiling from 108.5-109°/1.2 torr (33.0g, 56.3%) lit.[32] bp 155-156°/26torr. This was essentially pure 4-n-butylbenzoyl chloride; ir 1735, 1770 cm^{-1}(COCl), molecular ion: m/e 196; hydrolysis gave 4-n-butylbenzoic acid mp 98-100°, clearing temperature, 113-114°. lit.[33]mp 99.5°, clearing point: 113°.

4-n-Pentyloxyphenol

A solution of potassium hydroxide (119g, 2.13m) in water (238 ml) was added over thirty minutes to a refluxing mixture of hydroquinone (220g, 2m) and 1-bromopentane (302g, 2m) in 240 ml ethanol. After refluxing an additional four hours the mixture was extracted with ether (∿1 l.). The organic layer was extracted with 10% potassium hydroxide (2X 500ml) and washed with water. The combined aqueous phase was acidified with conc. hydrochloric acid, cooled and the solid collected, washed and dried to give a crude mixture of hydroquinone and 4-n-pentyloxyphenol (total weight 165g.). This was chromatographed on silica gel (typical: 90 g on 200g, J.T. Baker ∿60-200 mesh, column dimensions 42X90 mm) using chloroform as elutant. A trace of diether was collected initially (R_f 0.685) followed by 4-n-pentyloxyphenol (R_f 0.164). All fractions were analyzed by tlc. Combined monoether fractions were recrystallized from hexane to give pure 4-n-pentyloxyphenol (109g, 30.2%) mp 51.5-53°; lit.[34]mp 49-50°. ir 3550, 3350 cm^{-1} (OH) molecular ion m/e= 180.

4-N-Alkyloxybenzoic acids were prepared from 4-hydroxybenzoic acid as described in the literature[35]. 4-n-Alkyloxybenzoyl chlorides were obtained from the acids by reaction with thionyl chloride, then were distilled at reduced pressure.

4-Alkylphenols were synthesized by either of two methods illustrated by the following preparations:

4-sec-Butylphenol

4-Benzyloxyacetophenone was prepared from 4-hydroxyacetophenone (39.8g, 0.29m), benzyl bromide (50.0g, 0.29m) and potassium carbonate (121.2g, 0.88m) by refluxing in methylethylketone (400 ml) for thirty-five hours. The resulting mixture was stripped, taken up in methylene chloride, washed with 10% sodium hydroxide and water then dried. After stripping, the product was obtained by recrystallization from cyclohexane giving 45.6g (69%) mp 87-88.5°. Mass analysis (70eV): m/e 226 (2.9,M+); m/e 181 (2.3, M^+-CH_3CHO); m/e 91 (100, $\emptyset CH_2^+$); m/e 77 (12.1,$\emptyset^+$).

2-(4-Benzyloxyphenyl)but-2-ene was prepared from 4-benzyloxyacetophenone (40g, 0.18m) by the addition of ethyl Grignard reagent (excess, 0.4m prepared fresh from ethyl bromide and magnesium) in anhydrous ether. The reaction mixture was treated with sat. ammonium chloride solution (200 ml) at room temperature. The organic layer was dried, and stripped to give a solid product mp 54-58° weighing 42.1g (~99%). Analysis (nmr, ir) indicated that spontaneous dehydration must have occurred during work up to give the 2-butene derivative directly.

4-sec-Butylphenol. The crude product from the previous reaction (37g) was charged with 10% palladium on charcoal catalyst (2g, Sargent 13902) and 250 ml absolute ethanol in a Parr hydrogenation apparatus. After fifteen hours at 55 psi/60° the reaction was stopped, catalyst was filtered and solvent stripped. The residue was distilled at reduced pressure bp 66°/0.3ton, to give 22.5g (85%) of 4-sec-butylphenol. Recrystallization from hexane gave mp 59-60° ir (CCl_4), OH, 3623 cm^{-1}(sharp); 3350 cm^{-1}(broad).

4-n-Butylphenol

4-(ketobutyl)phenol. Boron trifluoride was bubbled through a solution of phenol (94g 1m) and butyric acid (88.1g, 1m) in 500 ml carbon tetrachloride over three hours at reflux. Reflux was maintained three hours more, the mixture was let stand forty-eight hours, and 200 ml chloroform was added; this solution was extracted with sat. solium bicarbonate (500 ml), washed with water (2X100ml) and dried. Solvent was stripped and the residue distilled at reduced pressure (bp 137-140°/0.4 torr) to give 98g (59.4%) 4-(1-ketobutyl)phenol. Recrystallization (CCl_4) gave mp 92-93°, mass spectrum (70eV):m/e 164 (11.1 M^+);m/e 136 (8.4,M^+-CH_2=CH_2);m/e 121 (100,M^+-C_3H_7);m/e 93 (18.2,M^+-C_3H_7CHO);m/e 77 (2.6); m/e 65 (22.6).

4-n-Butylphenol was obtained by direct hydrogenation of the above keto derivative (30g) in a Parr hydrogenation apparatus in a manner analogous to preparation of 4-sec-butylphenol except that it was conducted at room temperature. Reduced pressure distillation gave 18.0g (65.6%) of product bp 75°/0.1 torr. Mass analysis (70eV): m/e 150 (14.9, M^+); m/e 107 (100,M^+-C_3H_7);m/e 77 (9.7); m/e 65 (3.6); m/e 51 (3.6).

4-n-Octyloxyphenyl 4'n-Propylbenzoate

A solution of 4-n-propylbenzoyl chloride (4g, 21.9mm) in methylene chloride (10ml) was added to a solution of 4-n-octyloxyphenol (4.87g, 21.9mm) and triethylamine (3ml) in methylene chloride (25ml). After fifteen minutes triethylamine hydrochloride was removed by washing with water. The organic layer was extracted with

5% potassium hydroxide (20ml), water (20ml), dried and stripped to give 7.5g (92.2%) crude ester. Recrystallization from abs. ethanol gave pure ester mp 48-54°(C-N), clearing point 60°(N-I); ir 1735 ($-CO_2-$)cm^{-1}.

Acknowledgements:

We are indebted to Dr. Larry Williams of the L.C.I. for some of the nuclear magnetic resonance spectra and to Dr. John T. S. Andrews for useful discussions concerning phase transition phenomena. This work was supported by the National Science Foundation under Grant GH-34164X.

REFERENCES:

1. A. deVries, Mol. Cryst. and Liq. Cryst., 11, 361 (1971).
2. J. W. Doane, R. S. Parker, B. Cvikl, D. L. Johnson and D. L. Fishel, Phys. Rev. Lett., 28, 1694 (1972)
3. D. L. Uhrich, R. A. Detjen and J. M. Wilson in "Mossbauer Effect Methodology", Vol 8, Plenum Press, N.Y. (1973).
4. J. P. VanMeter and B. H. Klandermann, Abstracts 4th International Liquid Crystal Conference, Kent, Ohio, Aug. 1972. No. 122.
5. R. Steinstrasser, Z..Naturforsch., 27b, 774 (1972).
6. R. Steinstrasser and L. Pohl, Z. Naturforsch. 26b, 577 (1971).
7. R. E. Rondeau, M. A. Berwick, R. N. Steppel and M. P. Serve', J. Am. Chem. Soc., 94, 1096 (1972).
8. W. R. Young, A. Aviram and R. J. Cox, J. Am. Chem. Soc., 94, 3976 (1972) and papers referenced therein.
9. W. R. Young, I. Haller and D. C. Green, J. Org. Chem., 37, 3707 (1972).
10. J. vanderVeen and A. H. Grohben, Mol. Cryst. and Liq. Cryst., 15, 239 (1971).
11. R. J. Cox, Mol. Cryst. and Liq. Cryst., 19, 111 (1972).
12. R. A. Champa, Mol. Cryst. and Liq. Cryst., 19, 233 (1973).
13. J. B. Flannery and W. Haas, J. Phys. Chem., 74, 3611 (1970).
14. H. J. Dietrich and E. L. Steiger, Mol. Cryst. and Liq. Cryst., 16, 263 (1972).
15. M. T. McCaffrey and J. A. Castellano, Mol. Cryst. and Liq. Cryst., 18, 209 (1972).
16. G. W. Smith, Z. G. Gardlund and R. J. Curtis, Mol. Cryst. and Liq. Cryst., 19, 327 (1973).
17. G. W. Gray and K. J. Harrison, Mol. Cryst. and Liq. Cryst., 13, 37 (1971).
18. M. E. Neubert, L. Carlino and D. L. Fishel, Abstracts, Fifth Northeast Regional American Chemical Society meeting, Rochester, New York, October 1973.

19. A. R. Ubbelohde, "Melting and Crystal Structure", Paragraph 1, 2 and Chapters 5 and 13, Oxford University Press, London, 1965.
20. G. W. Gray, "Molecular Structure and the Properties of Liquid Crystals", Academic Press, New York (1962).
21. R. Alben, Mol. Cryst. and Liq. Cryst., 13, 193 (1971).
22. E. Haselbach and E. Heilbronner, Helv. Chem-Acta, 51, 16 (1968).
23. I. Teucher, C. M. Paleos and M. M. Labes, Mol. Cryst. and Liq. Cryst., 11, 187 (1970).
24. H. B. Buergi and J. D. Dunitz, J. Chem. Soc. D 472 (1969).
25. V. S. Blasenbrey and W. Pechhold, Rheologica Acta, 6, 174 (1967).
26. R. A. Scott and H. A. Scheraga, J. Chem. Phys., 44, 3054 (1966).
27. R. J. Ouellette, B. K. Sinha, J. Stolfo, C. Levin and S. Williams J. Am. Chem. Soc., 92, 7145 (1970).
28. M. A. Slifkin, "Charge Transfer Interactions of Biomolecules", Academic Press, London, 1971.
29. R. Foster, "Organic Charge-Transfer Complexes, "Academic Press, London, 1969.
30. L. Pohl and R. Steinstrasser, Z. Naturforsch, 26b, 26 (1971).
31. G. H. Brown, Anal. Chem., 41, 26a (1969).
32. H. A. Fahim, J. Chem. Soc., 520 (1949).
33. C. Weygand and R. Gabler, Z. Physik Chem., B46, 270 (1940).
34. E. Klarman, L. W. Gatyas and V. A. Shternov, J. Am. Chem. Soc., 54, 298 (1932).
35. G. W. Gray and B. Jones, J. Chem. Soc., 4179 (1953).

EFFECTS OF CERTAIN CENTRAL GROUPS ON THE LIQUID CRYSTAL PROPERTIES OF DICARBOXYLIC ESTERS

Lawrence Verbit and Robert L. Tuggey

Department of Chemistry, State University of New York at Binghamton, Binghamton, N. Y. 13901

INTRODUCTION

At present there is no useful theory in the field of liquid crystals which allows the prediction of whether a given molecule will show, for example, a smectic or nematic mesophase and what the thermal range of the mesophase will be. All nematogenic compounds to date which are useful in electronic display devices contain *para*-alkoxy/alkyl-substituted benzene rings held together by a central group containing π-electron density. For the past few years we have been carrying out heuristic investigations on the effect of different central groups on liquid crystalline behavior.[1] In this paper we report some recent results.

RESULTS

We have prepared several symmetrical diesters of the general structure

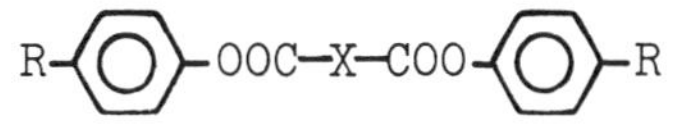

where the dicarboxylic acids used were fumaric, maleic, mesaconic (α-methylfumaric), and *trans*-1,4-cyclohexanedicarboxylic acid. The data are given in Table 1. Use of the substituent R = *n*-C_4H_9O allows comparison of our data with liquid crystal esters containing other central groups. These comparisons are shown in Table 2. Several esters of *trans*-1,4-cyclohexanedicarboxylic acid were also prepared which did not show mesomorphism. The data for these compounds are given in Table 3.

DISCUSSION

To facilitate comparison with the effects of various other central groups on mesomorphic behavior, we utilized the p-n-butoxyphenyl and p-n-pentylphenyl wing groups. This allows a representative para-alkoxy and para-alkyl group, respectively. The rationale used is that holding constant the wing groups allows differences in mesomorphic behavior caused chiefly by central groups to be delineated. Support for this approach comes from consideration of several series of mesogenic compounds containing both p-alkoxy- and p-alkylphenyl wing groups.[2] Plots of nematic-to-isotropic temperatures (clearing points) versus the number of atoms in the terminal groups give families of curves which are striking in their parallel nature, indicating that a given wing group exerts a consistent effect in the several series.

Table 1 shows that the fumaric acid esters, having the double bond trans, exhibit nematic mesophases. The di-p-n-butoxyphenyl fumarate, **1**, has an enantitropic mesophase some 32° broad while the corresponding pentyl ester, **2**, exhibits a monotropic nematic phase of considerably lower thermal stability. Introduction of an α-methyl substituent, compounds **3** and **4**, produces a steric effect which decreases the nematic thermal stability by some 50° for the p-butoxy ester. Corey-Pauling space-filling molecular models indicate that the fumarate esters may adopt a nearly coplanar arrangement and that introduction of an α-methyl group causes no additional steric interference to coplanarity. The increase in molecular breadth due to the spherically symmetrical α-methyl substituent probably causes an increase in the average intermolecular separation, resulting in the observed lower mesophase stability. Since intermolecular attractions are approximately proportional to the inverse sixth power of the distance between molecules, even such a small change as the introduction of an α-methyl group can have a profound effect on mesogenicity.

Large deviations from molecular colinearity are well known to lead to an absence of mesogenic properties.[1a,3] The maleic acid derivatives **5** and **6**, in which the double bond is cis were not mesomorphic.

Compounds **7**-**10** have the trans-1,4-cyclohexane ring as the central group. The p-methoxyphenyl derivative **7**, reported by Bacon and Brown,[4] has a nematic range of almost 100° and high nematic stability. The nematic stabilities of the butoxy and pentylphenyl esters in this series, **8** and **9**, are higher than the corresponding fumarate derivatives **1** and **2**, dramatically illustrating that *π-electron density in the central linkage is not necessary for mesomorphic behavior.*

TABLE 1. Transition Temperature Data for Some Dicarboxylic Esters[a]

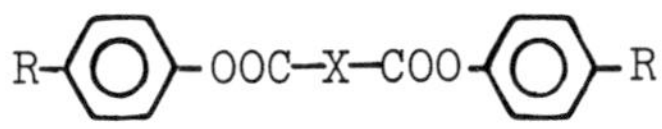

Compound No.	X	R	Phase transitions and temperatures[b]
1	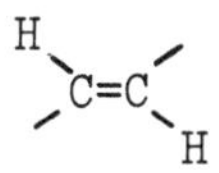	C_4H_9O	K 112 N 144 I
2		C_5H_{11}	K 86 I (85 N)
3	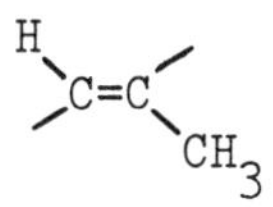	C_4H_9O	K 95 I (94 N)
4		C_5H_{11}	m.p. 68-69
5	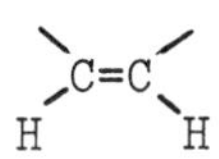	C_4H_9O	m.p. 57-58
6		C_5H_{11}	isotropic oil
7		CH_3O	K 143 N 242 I [c]
8		C_4H_9O	K 113 S 162 N 220 I
9		C_5H_{11}	K 101.5 S 135 N 161 I
10		$COOCH_2CH(CH_3)C_2H_5$	K 131 C 137 I

(a) All compounds gave satisfactory elemental analyses. (b) The linear notation is described in ref. 5. (c) Ref. 4.

The cyclohexanedicarboxylic esters 8 and 9 also exhibit smectic mesophases. Smectic mesophases have been suggested to become important when the lateral intermolecular attractions are relatively strong compared with the terminal attractions. *A priori*, one might expect that the polarizable π-electrons of the central double bond in a compound such as 1 would lead to greater lateral attractions than in the saturated cyclohexane analog 8, but the result contradicts this expectation.

Use of an optically active *primary*-amyloxycarbonyl terminal group in the cyclohexane ester 10, causes the occurrance of a twisted-nematic (cholesteric) mesophase but, interestingly enough,

TABLE 2. Transition Temperature Data for Some Mesogenic Esters with *p*-*n*-Butoxyphenyl Wing Groups

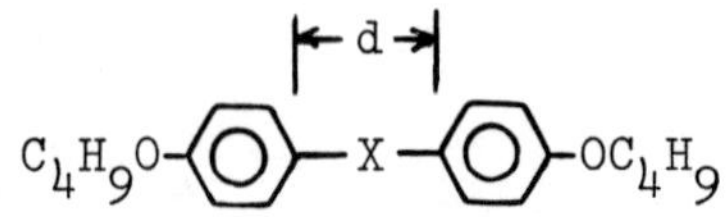

Compound No.	X	d, Å	Phase transitions and temperatures[a]	Reference
11	-COO-	3.5	K 87 N 92 I	b
3	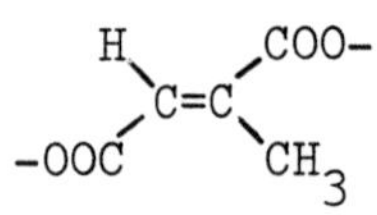	7.8	K 95 I (94 N)	c
12	-OOC-C≡C-COO-	8.15	K 91.5 N 99.5 I	d
1	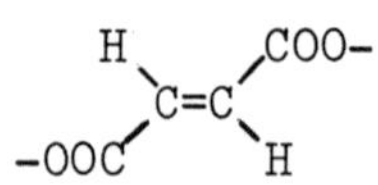	7.8	K 112 N 144 I	c
13		10.0	K 129 N 157 I	e
8		10.0	K 113 S 162 N 220 I	c
14	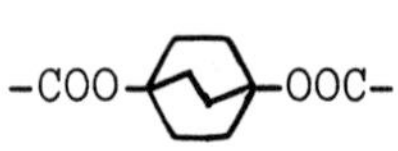	9.6	K 152 N 221 I	e
15	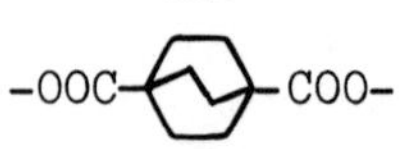	9.75	K 113 N 226 I	f
16	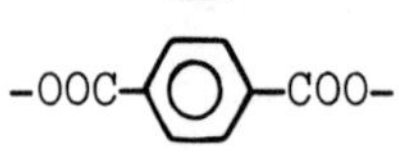	9.6	K 189 N 235 I	g
17	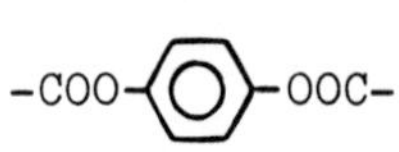	9.75	K 153 N 241 I	e
18	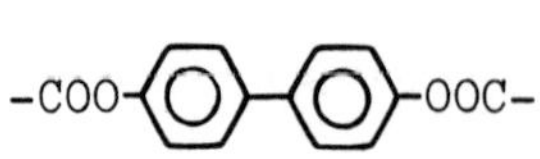	13.95	K 171 S 184 N 358 I	h

Footnotes for Table 2.

(a) The linear notation is described in ref. 5. (b) R. Steinsträsser, Z. Naturforsch., 27b, 774 (1972). (c) Present work. (d) L. Verbit and R. L. Tuggey, Mol. Cryst. & Liq. Cryst., 17, 49 (1972). (e) M. J. S. Dewar and R. S. Goldberg, J. Amer. Chem. Soc., 92, 1584 (1970). (f) R. S. Goldberg, Dissertation, University of Texas at Austin, 1969. (g) H. Kelker and B. Scheurle, J. Physique, 30-C4, 104 (1969). Dewar and Goldberg [J. Org. Chem., 35, 2711 (1970)] report K 183 N 229 I. (h) M. J. S. Dewar and J. P. Schroeder, J. Org. Chem., 30, 2296 (1965). The nematic phase is reported to decompose at 358° before changing to isotropic.

TABLE 3. Non-mesogenic Esters of trans-1,4-Cyclohexanedicarboxylic Acid Prepared in the Present Study [a]

R	n	m. p., °C
H	1	78-79
H	3	69-70
F	0	146-147
F	1	110-111
F	2	89-91
Cl	1	128-129
CH_3O	2	111-113
		134-135

(a) All compounds gave satisfactory elemental analyses.

no smectic phase is observed.

Table 2 compares data for a variety of central groups in esters having p-butoxyphenyl wing groups. The compounds are arranged in order of increasing clearing point. At this temperature the long-range intermolecular forces existing in the ordered liquid are destroyed by thermal energy effects and the mesophase becomes isotropic. The clearing point is thus a useful measure of the thermal stability of the mesophase. With respect to thermal stability, it does not matter whether the clearing point lies above or below the crystal-to-liquid point, i. e., whether the mesophase is enantiotropic or monotropic.

The structure of most nematic liquid crystals, particularly those useful in electro-optic devices, may be generalized as in structure A, below. The significant feature is the central unit which links the two para-substituted phenyl rings. These central groups generally contain polarizable π-electron density and are relatively rigid, thus serving to confer an overall rod-like shape to the molecule. Typical examples are benzene rings, double and triple bonds.

The esters in Table 2 show an interesting trend. Nematic mesophase stability is found to be approximately parallel to the size of the central group d, which is defined as follows:

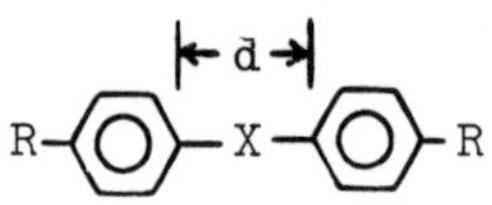

A

Measurements(Table 2) are taken from the projections of Dreiding models onto a plane surface.

The correlation between size of the central group and nematic stability appears only as a rough trend, due in part to the vastly differing electronic structures of the various central groups. What emerges as a factor of some significance in the contribution of the central group to mesophase stability, is the amount of electron density, both π and σ, and the total area (volume) over which it is dispersed. To a large extent, this is another way of viewing the molecular polarizability although it emphasizes the importance of σ- as well as π-electrons. There is no doubt that other factors such as rigidity, linearity, and width of the molecule are also important in contributing to mesophase stability.

The smallest central group in Table 2, the ester linkage, compound 11, exhibits the lowest clearing point. This is of interest in the search for materials which are nematic in the vicinity of room temperature. If one extrapolates our proposed correlation to the extreme of zero length of the central group then one would predict that the nematic clearing point for 4,4'-dibutoxybiphenyl will be less than about 90°. The compound has not been

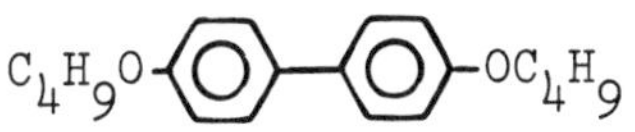

4,4'-Dibutoxybiphenyl

reported in the literature and it is in the process of being synthesized.

Several other esters of trans-1,4-cyclohexanedicarboxylic acid were prepared but none was mesogenic. The data are given in Table 3.

EXPERIMENTAL

Mesophases were identified using a 100-power AO Spencer polarizing microscope equipped with a variable temperature stage designed in these laboratories.[6] For mesogenic compounds, the temperatures reported are those at which the solid or mesophase has just disappeared. Temperatures are corrected. Microanalyses were performed by Galbraith Laboratories, Knoxville, Tenn.

The diesters were prepared by two routes. (1) The acid-catalyzed reaction of excess phenol with the diacid in benzene. Azeotroped water was separated in a Dean-Stark trap. (2) Reaction of the diacid chloride of trans-1,4-cyclohexanedicarboxylic acid with the alcohol or phenol in the presence of pyridine. Representative procedures are given below.

Di-p-n-Butoxyphenyl Fumarate (**1**). Fumaric acid is not soluble in benzene but was found to dissolve in an excess of the phenol used in the esterification. Hence, only enough benzene was added to azeotrope off the water formed during the reaction.

A mixture of p-n-butoxyphenol (Aldrich, 6.64 g, 0.04 mole), fumaric acid (Eastman, 1.16 g, 0.01 mole), and 5 ml of dry benzene were placed in a 50-ml round botton flask fitted with a Dean-Stark trap and reflux condenser. Conc. sulfuric acid (0.5 ml) was added and the reaction mixture refluxed for 48 hours. The reaction mixture was then cooled and 60 ml of ether added. The ether solution was washed twice with 25 ml of cold water, twice with 20 ml of 10% sodium bicarbonate solution, and once more with cold water. The solution was dried (sodium sulfate), filtered, and the ether evaporated to yield a yellow solid. Two recrystallizations from absolute ethanol afforded white crystals of **1**, 1.8 g, 44% yield, K 112 N 144 I.

Anal. Calc. for $C_{24}H_{28}O_6$: C, 69.89; H, 6.84; Found, C, 70.01; H, 6.88.

Di-p-n-pentylphenyl trans-1,4-Cyclohexanedicarboxylate (9). The diacid chloride was prepared by refluxing the diacid (Aldrich) in an excess of thionyl chloride for 6 hours, then distilling under vacuum. The diacid chloride was obtained as a clear, colorless liquid and was used immediately.

The diacid chloride (2.1 g, 0.01 mole) was dissolved in 5 ml of dry pyridine and a solution of p-n-pentylphenol (Eastman, 3.3 g, 0.02 mole) in 3 ml of pyridine was added. The solution was stirred at room temperature overnight, then diluted with 30 ml of ether. The ether solution was washed twice with 15 ml portions of cold water, then one with 10% HCl, twice with 10% sodium bicarbonate solution, followed by a cold water wash. The solution was stripped off on a rotary evaporator after being dried (magnesium sulfate). Two recrystallizations from mixed hexanes afforded white crystals of 9, 2.9 g, 62 % yield, K 101.5 S 135 N 161 I.

Anal. Calc. for $C_{30}H_{40}O_4$: C, 77.55; H, 8.68; Found; C, 77.68; H, 8.74.

ACKNOWLEDGEMENT

We thank the General Telephone and Electronics Laboratories for generous support of this research.

REFERENCES

1. (a) L. Verbit and R. L. Tuggey, Mol. Cryst. & Liq. Cryst., 17, 49 (1971); (b) L. Verbit and G. A. Lorenzo, 4th International Liquid Crystal Conference, Kent State University, August 1972, Mol. Cryst. & Liq. Cryst., in press.

2. J. van der Veen, W. H. de Jeu, A. H. Grobben, and J. Boven, ibid., 17, 291 (1972); W. R. Young, I. Haller, and A. Aviram, ibid., 15, 311 (1972).

3. G. W. Gray, "Molecular Structure and the Properties of Liquid Crystals," Academic Press, New York, N. Y., 1962, chapter 9.

4. W. E. Bacon and G. H. Brown, Mol. Cryst. & Liq. Cryst., 6, 155 (1969).

5. L. Verbit, ibid., 15, 89 (1971).

6. L. Verbit and T. R. Halbert, J. Chem. Educ., 48, 773 (1971).

RELATIONS OF TWO CONTINUUM THEORIES OF LIQUID CRYSTALS*

James D. Lee and A. Cemal Eringen

Department of Aerospace and Mechanical Sciences
Princeton University
School of Engineering and Applied Science
The George Washington University

For the thermomechanics of nematic and cholesteric liquid crystals, the Ericksen-Leslie theory and the theory based upon micropolar continuum mechanics are presented. The similarities and differences in basic laws of motion and constitutive equations between these two theories are discussed in detail. For illustrative purpose solutions of some special problems are given in order to illustrate special features of these theories.

I. INTRODUCTION

Since the discovery of liquid crystals by Reinitzer in 1888, the literature has registrated extensive entries both on experimental and theoretical grounds. In 1922 Friedel [1] classified liquid crystals into three major classes, namely, nematic, smectic, and cholesteric, according to molecular structures of the substances. Fairly complete surveys of the field were presented recently by Chistyakov [2] and Saupe [3]. Fergason [4] has discussed the molecular structures and optical properties of liquid crystals extensively. In regard to continuum theories of liquid crystals we mention the work of Oseen [5], Frank [6], Anzelius [7], Groupes d'Etudes des Cristaux Liquids [8], Stephen [9], Lubensky [10], Martin, et al [11], Forster, et al [12], Huang [13], Schmidt and Jahnig [14]. Complete thermomechanical theories of liquid crystals have been formulated by Ericksen [15-23], Leslie [24-27] and by Eringen and Lee [28-32]. In the present work, attention is focused on the comparison between the two distinct continuum theories of nematic and cholesteric liquid crystals, one of which was established by Ericksen and Leslie and the other by Eringen and

Lee. However we would like to mention that in 1964, Eringen [37] established the relationship of the balance laws of micropolar theory with those of anisotropic fluid theory of Ericksen [15] which in part, subsequently, constituted Ericksen-Leslie theory.

II. BASIC LAWS

The basic laws of motion of Ericksen-Leslie theory for liquid crystals are, [16,23,25,26]:

$$\dot{\rho} + \rho v_{i,i} = 0 \tag{A.1}$$

$$\sigma_{ji,j} + \rho F_i = \rho \dot{v}_i \tag{A.2}$$

$$\pi_{ji,j} + g_i + \rho_1 G_i = \rho_1 \dot{\omega}_i \tag{A.3}$$

$$\rho \dot{U} = \rho r - q_{i,i} + \sigma_{ji} A_{ij} + \pi_{ji} N_{ij} - g_i N_i \tag{A.4}$$

$$\sigma_{ij} - \pi_{ki} d_{j,k} + g_i d_j = \sigma_{ji} - \pi_{kj} d_{i,k} + g_j d_i \tag{A.5}$$

The basic laws of motion of micropolar theory established by Eringen and his coworker [33-36] are

$$\dot{\rho} + \rho v_{i,i} = 0 \tag{B.1}$$

$$\frac{d}{dt} i_{k\ell} + \varepsilon_{\ell mn} \nu_n i_{km} + \varepsilon_{kmn} \nu_n i_{m\ell} = 0 \tag{B.2}$$

$$t_{ji,j} + \rho f_i = \rho \dot{v}_i \tag{B.3}$$

$$m_{ji,j} + \varepsilon_{imn} t_{mn} + \rho \ell_i = \rho \dot{\sigma}_i \tag{B.4}$$

$$\rho \dot{\varepsilon} = \rho h + q_{i,i} + t_{ij}(v_{j,i} - \varepsilon_{ijk} \nu_k) + m_{ij} \nu_{j,i} \tag{B.5}$$

In equations (A.1 - B.5):

ρ	$\equiv$ mass density ,	v_i	$\equiv$ velocity vector
$\sigma_{ij} = t_{ij}$	$\equiv$ stress tensor ,	$F_i = f_i$	$\equiv$ body force
π_{ij}	$\equiv$ director stress tensor ,	g_i	$\equiv$ intrinsic director body force
d_i	$\equiv$ director ,	G_i	$\equiv$ external director body force
$r = h$	$\equiv$ heat source ,	q_i	$\equiv$ heat flux
$i_{k\ell} = i_{\ell k}$	$\equiv$ microinertia tensor,	m_{ij}	$\equiv$ moment stress tensor

$\ell_i \equiv$ body couple , $\varepsilon = U \equiv$ internal energy tensor

$\nu_i \equiv$ angular velocity vector , $\varepsilon_{ijk} \equiv$ alternating tensor

and

$$\omega_i \equiv \dot{d}_i \ , \qquad 2A_{ij} \equiv v_{i,j} + v_{j,i}$$
$$2\omega_{ij} \equiv v_{i,j} - v_{j,i} \ , \qquad N_i \equiv \omega_i + \omega_{ki}d_k \tag{1}$$
$$N_{ij} \equiv \omega_{i,j} + \omega_{ki}d_{k,j} \ , \qquad \dot{\sigma}_\ell \equiv \frac{d}{dt}[(i_{mm}\delta_{k\ell} - i_{\ell k})\nu_k]$$

A superposed dot represents material time derivative and an index following a comma indicates partial derivative with respect to spatial coordinates, e.g.,

$$v_k = \frac{dv_k}{dt} = \frac{\partial v_k}{\partial t} + v_{k,\ell}v_\ell, \quad v_{k,\ell} \equiv \frac{\partial v_k}{\partial x_\ell}, \quad x_{k,K} \equiv \frac{\partial x_k}{\partial X_K}$$

We also notice that q_i in Ericksen-Leslie theory is the outward heat flux while in Eringen's micropolar theory q_i is the inward heat flux.

In both theories A and B, the motion of the center of a material particle is represented by

$$x_k = x_k(X_K, t) \tag{2$_1$}$$

where x_k and X_K are respectively the spatial and material coordinates of the mass center of particle. In Ericksen-Leslie theory a director d_i with constant magnitude is introduced to describe the preferred direction inherent in liquid crystals. In micropolar theory the motion of the particle with respect to its center of mass is represented by

$$\xi_k = \chi_{kK}(\underline{X}, t)\Xi_K \tag{2$_2$}$$

with the constraints that χ_{kK} is an orthogonal matrix, i.e.,

$$\chi_{kK}\chi_{kL} = \delta_{KL} \tag{3}$$

To compare theory A with theory B, first of all we must establish the relationship between the orthogonal transformation χ_{kK} and the director. Suppose that in a special micropolar body, the relative position vector Ξ^*_K is selected to be of unit magnitude and parallel to the long axis of molecules locally, i.e.,

$$\Xi^*_K \Xi^*_K = 1 \ , \quad \underline{\Xi}^* \ || \quad \text{local axis of orientation}, \tag{4}$$

then

$$d_i = \xi_i^* \equiv \chi_{iK} \Xi_K^* \tag{5}$$

$$\xi_i^* \xi_i^* = \chi_{iK}\chi_{iL}\Xi_K^*\Xi_L^* = 1 \tag{6}$$

Equation (6) states that in such a body the assumption, $d_i d_i = 1$, of Ericksen-Leslie theory (cf. eqn.(1) of Ref.[23]) is satisfied.

For this special micropolar medium we can compare the balance laws of these theories. To this end we first note the following correspondence:

$$\begin{aligned} d_i &= \xi_i^* = \chi_{iK}\Xi_K^* \\ \rho_1 d_k d_\ell &= \rho i_{k\ell} \\ \varepsilon_{\ell mn} d_m \pi_{kn} &= m_{k\ell} \\ \varepsilon_{kmn} \rho_1 d_m G_n &= \rho \ell_k \end{aligned} \tag{7}$$

Now we proceed to show, under the assumption (4), and $(7)_2$ that

(1) The law for conservation of mass is identical: (A.1)=(B.1)

(2) The law for balance of linear momentum is identical: (A.2)=(B.3)

(3) The law for conservation of microinertia (B.2) is equivalent to

$$\dot{\rho}_1 + \rho_1 v_{i,i} = 0 \tag{8}$$

under the correspondence $(7)_1$ and $(7)_2$.

<u>Proof</u>:

$$\dot{d}_i \equiv \omega_i = \frac{d}{dt}(\chi_{iK})\Xi_K^* = -\varepsilon_{ijk}\nu_k d_j$$

$$\begin{aligned} \frac{d}{dt}(\rho i_{k\ell}) &= \dot{\rho} i_{k\ell} + \rho[-\varepsilon_{\ell mn}\nu_n i_{km} - \varepsilon_{kmn}\nu_n i_{m\ell}] \\ &= -v_{m,m}\rho i_{k\ell} + \frac{d}{dt}(\rho_1 d_k d_\ell) - \dot{\rho}_1 d_k d_\ell \end{aligned}$$

It follows that $\quad \dot{\rho}_1 + \rho_1 v_{i,i} = 0$.

This is why, in Ericksen-Leslie theory, incompressibility implies $\dot{\rho}_1 = 0$.

(4) The law for balance of angular momentum (B.4) is equivalent to the law for balance of director force (A.3) subjected to the condition (A.5) with the provision of the correspondence (7).

Proof: Multiply (A.3) by $\varepsilon_{k\ell i}d_\ell$ and we have

$$\rho_1\varepsilon_{k\ell i}d_\ell\ddot{d}_i = \rho_1\varepsilon_{k\ell i}d_\ell G_i + \varepsilon_{k\ell i}(d_\ell g_i - \pi_{ji}d_{\ell,j}) + (\varepsilon_{k\ell i}d_\ell\pi_{ji})_{,j} \tag{9}$$

We notice that $\rho_1\varepsilon_{k\ell i}d_\ell G_i = \rho\ell_k$ (cf. eqn.(7)$_4$ and p. 155 of Ref. [23]) and $\varepsilon_{k\ell i}d_\ell\pi_{ji} = m_{jk}$ and $\varepsilon_{k\ell i}(d_\ell g_i - \pi_{ji}d_{\ell,j}) = \varepsilon_{k\ell i}\sigma_{\ell i} = \varepsilon_{k\ell i}t_{\ell i}$ (cf. eqn.(A.5)). Using $\rho i_{k\ell} = \rho_1 d_k d_\ell$, it is straight forward to show that $\rho_1\varepsilon_{k\ell i}d_\ell\ddot{d}_i = \rho\dot{\sigma}_k$. Thus we complete the proof.

(5) The law for conservation of energy is identical: (A.4)≡(B.5) However in this case we must prove that

$$\sigma_{ji}A_{ij} + \pi_{ji}N_{ij} - g_i N_i = t_{ij}(\nu_{j,i} - \varepsilon_{ijk}\nu_k) + m_{ij}\nu_{j,i} \tag{10}$$

Proof: Substituting (1) into left hand side of eqn.(10) and using (A.5), we have

$$\sigma_{ji}A_{ij} + \pi_{ji}(\omega_{i,j} + \omega_{ki}d_{k,j}) - g_i(\omega_i + \omega_{ki}d_k)$$
$$= \sigma_{ji}v_{i,j} + \pi_{ji}\varepsilon_{\ell ik}(\nu_{k,j}d_\ell + \nu_k d_{\ell,j}) + \varepsilon_{ijk}\nu_k d_j g_i$$
$$= \sigma_{ji}v_{i,j} + m_{jk}\nu_{k,j} + \sigma_{ji}\varepsilon_{ijk}\nu_k$$
$$= t_{ij}(v_{j,i} - \varepsilon_{ijk}\nu_k) + m_{kj}\nu_{j,i}$$

Therefore, as far as balance laws are concerned, we have reached the following conclusion:

The balance laws in Ericksen-Leslie theory and in micropolar theory are essentially the same except that Ericksen and Leslie made an assumption about microinertia, i.e., $\rho i_{k\ell} = \rho_1 d_k d_\ell$, and therefore the law for conservation of microinertia is reduced to

$$\dot{\rho}_1 + \rho_1 v_{k,k} = 0$$

which in incompressible case is further reduced to

$$\dot{\rho}_1 = 0$$

However, it is important to note that, in Ericksen-Leslie theory by taking $\rho i_{k\ell}$ to be $\rho_1 d_k d_\ell$, the microinertia about the long axis of molecules has been taken to be zero. A matrix having the form of $\rho_1 d_k d_\ell$ can be diagonalized into a matrix with only one non-zero entry in the diagonal positions. Strictly speaking no such physical body can exist. Possibly in an approximate sense the particles in Ericksen-Leslie theory can be represented with a non-vanishing principal microinertia for a body consisting of thread-like elements. We also believe that the law of balance of angular momentum has more clear physical meaning than the law of balance of director force has.

III. CONSTITUTIVE THEORY

The essential differences between Ericksen-Leslie theory and Eringen-Lee theory about nematic and cholesteric liquid crystals are in the constitutive theory in which different concepts are adopted. In this section we discuss these differences. From now on, in the equations those terms with a single line underneath are only for the thermomechanical theory of cholesteric liquid crystals and those terms with a double line underneath are only for the non-heat-conducting theory of nematic liquid crystals. The remaining terms are common for both nematic and cholesteric liquid crystals.

In Ericksen-Leslie theory the independent constitutive variables are chosen to be [25,27]

$$\text{I.C.V.} = \{\rho, d_i, d_{i,j}, N_i, A_{ij}, T, \underline{T_{,i}}\} \tag{11}$$

while in the micropolar theory for liquid crystals they are chosen to be [28,32]

$$\text{I.C.V.} = \{\underline{\underline{\bar{\rho}^{1}}}, \underline{C_{KL}}, \dot{C}_{KL}, \Gamma_{KL}, \dot{\Gamma}_{KL}, T, \underline{T_{,K}}\} \tag{12}$$

where the Cosserat tensor, C_{KL}, and Wryness tensor, Γ_{KL}, are defined as follows:

$$C_{KL} \equiv x_{k,K}\chi_{kL}\ , \qquad \Gamma_{KL} \equiv \frac{1}{2}\, \varepsilon_{KNM}\chi_{kM}\chi_{kN,L} \tag{13}$$

We have

$$\begin{aligned} \dot{C}_{KL} &= (v_{\ell,k} - \varepsilon_{k\ell m}\nu_m)\, x_{k,K}\chi_{\ell L} \\ \dot{\Gamma}_{KL} &= \nu_{k,\ell}\, x_{\ell,L}\chi_{kK} \end{aligned} \tag{14}$$

We notice that

(1) All the independent constitutive variables in (11) and (12) are objective, i.e., they are invariant under arbitrary time-dependent rigid motions of the spatial frame of reference.

(2) In Ericksen-Leslie theory the assumption of incompressibility has further been made and that is why an unknown pressure (mathematically speaking a Langrange's multiplier) comes into the picture to replace ρ. Ericksen-Leslie theory could be generalized to include the constitutive variable N_{ij}. Without doing so the couple stress in this theory has no dissipative term.

(3) The results will be exactly the same if we replace (12) by

$$\text{I.C.V.} = \{\underline{\underline{\bar{\rho}^{1}}}, \underline{x_{k,K}}, \chi_{kK}, \chi_{kK,L}, \dot{C}_{KL}, \dot{\Gamma}_{KL}, T, \underline{T_{,K}}\} \tag{15}$$

Now we show the equivalency between the two sets of independent constitutive variables, (11) and (15), as follows:

$$\begin{aligned} d_i &\equiv \chi_{ik}\, \Xi^*_K \\ d_{i,j} &= \chi_{iK,L}\, X_{L,j}\, \Xi^*_K + \underline{\chi_{iK}\, X_{L,j}\, \Xi^*_{K,L}} \\ N_i &= (v_{j,i} - \varepsilon_{ijk}\nu_k - A_{ij}) d_j \\ \dot{C}_{KL} &= (v_{j,i} - \varepsilon_{ijk}\nu_k) x_{i,K} \chi_{jL} \\ N_{ij} &= (v_{m,i} - \varepsilon_{imn}\nu_n + A_{mi}) d_{m,j} - \varepsilon_{imn} d_m \nu_{n,j} \\ \dot{\Gamma}_{KL} &= \nu_{n,j}\, x_{j,L}\, \chi_{nK} \\ \underline{T_{,i}} &\underline{= T_{,K}\, X_{K,i}} \end{aligned} \tag{16}$$

From these we can see that in Ericksen-Leslie theory the generalized strains and rate of strains, d_i, $d_{i,j}$, N_i, A_{ij}, $T_{,i}$, are calculated with respect to the current (instantaneous) configuration and thus the theory is fluid-like. In other words the link between the deformed and the undeformed configurations is left out. Meanwhile in Eringen-Lee theory the existence of a set of reference states is assumed and the generalized strains and rate of strains, C_{KL}, Γ_{KL}, $\dot{C}_{KL}$, $\dot{\Gamma}_{KL}$, $T_{,K}$, are calculated with respect to the reference state. Thus Eringen-Lee theory is a nonlinear, micropolar generalization of the classical Kelvin-Voigt type of viscoelastic materials. This constitutes a fundamental conceptual difference between the two theories. However we would like to point out that the use of director d_i instead of the microrotation χ_{kK} implies that the rotation about the local axis of orientation cannot be represented. It must also be noted that the notions of "a certain motion cannot be represented," "a certain motion is prohibited," and "a certain motion is not important" are different in concept. The difference in the starting point leads to different consequences and we will discuss these points in the following sections.

IV. THE REFERENCE STATE

The constitutive equations for nematic and cholesteric liquid crystals proposed by Leslie [25,27] are

$$\begin{aligned} \sigma_{ij} = &- \rho\delta_{ij} - d_{k,j}[K_{22}d_{k,i} + K_{24}d_{i,k} + (K_{11} - K_{22} - K_{24})d_{m,m}\delta_{ki} \\ &+ (K_{33} - K_{22})d_i d_m d_{k,m} + K_2 d_\ell \varepsilon_{\ell ik}] + \underline{\alpha\varepsilon_{ik\rho}(d_\rho d_j)_{,k}} \\ &+ \mu_1 d_k d_\rho A_{k\rho} d_i d_j + \mu_2 d_i N_j + \mu_3 d_j N_i + \mu_4 A_{ij} \\ &+ \mu_5 d_i d_k A_{kj} + \mu_6 d_j d_k A_{ki} + \underline{d_\rho T_{,q}(\mu_7 \varepsilon_{j\rho q} d_i + \mu_8 \varepsilon_{i\rho q} d_j)} \end{aligned} \tag{17}$$

$$m_{ij} = \varepsilon_{j\ell k} d_\ell [K_{22} d_{k,i} + K_{24} d_{i,k} + (K_{11} - K_{22} - K_{24}) d_{m,m} \delta_{ki} + (K_{33} - K_{22}) d_i d_m d_{k,m}] + \underline{(\alpha - K_2)(\delta_{ij} - d_i d_j)} \tag{18}$$

$$q_i = \underline{C_1 T_{,i} + C_2 d_j T_{,j} d_i + C_3 \varepsilon_{ijk} d_j N_k + C_4 \varepsilon_{ijk} d_j A_{k\rho} d_\rho} \tag{19}$$

Equations (17-19) are highly nonlinear and the only linear terms are (notice that here we are not referring to the linearlization of Ericksen-Leslie theory, which is left for future study):

$$\begin{aligned} \overset{\circ}{\sigma}_{ij} &= -p\delta_{ij} + \mu_4 A_{ij} \\ \overset{\circ}{m}_{ij} &= \underline{(\alpha - K_2)\delta_{ij}} \\ \overset{\circ}{q}_i &= \underline{C_1 T_{,i}} \end{aligned} \tag{20}$$

We also notice that the director d_i does not appear alone in constitutive equations, except in the term $m^*_{ij} = (k_2 - \alpha) d_i d_j$ for cholesteric liquid crystals, and therefore Ericksen and Leslie use director as an "anisotropy indicator".

The constitutive equations for nematic and cholesteric liquid crystals obtained by Eringen and Lee [28,32] are

$$t_{k\ell} = \underline{\underline{\frac{\partial \psi_o}{\partial \rho^{-1}} \delta_{k\ell}}} + [\frac{\rho}{\rho_o}(\underline{-a_{KL}(T - T_o)} + \underline{A_{KLMN}(C_{MN} - \delta_{MN})} + C_{KLMN}\Gamma_{MN}) + a_{KLMN}\dot{C}_{MN} + \alpha_{KLMN}\dot{\Gamma}_{MN} + e_{KLM}T_{,M}] x_{k,K} \chi_{\ell L} \tag{21}$$

$$m_{k\ell} = [\frac{\rho}{\rho_o}[\underline{-b_{LK}(T - T_o)} + B_{LKMN}\Gamma_{MN} + \underline{C_{MNLK}(C_{MN} - \delta_{MN})}] + b_{LKMN}\dot{\Gamma}_{MN} + \underline{\beta_{LKMN}\dot{C}_{MN}} + \underline{f_{LKM}T_{,M}}] x_{k,K} \chi_{\ell L} \tag{22}$$

$$q_k = \underline{[h_{KL} T_{,L} + \gamma_{KLM}\dot{C}_{LM} + g_{KLM}\dot{\Gamma}_{LM}] x_{k,K}} \tag{23}$$

where a_{KL}, b_{KL}, h_{KL}, e_{KLM}, f_{KLM}, γ_{KLM}, g_{KLM}, A_{KLMN}, B_{KLMN}, C_{KLMN}, a_{KLMN}, b_{KLMN}, α_{KLMN}, and β_{KLMN} are material moduli and they are further restricted due to the material symmetry and physical considerations (cf. Ref.[28,32]). The material symmetry is characterized by the axis of orientation and axis of helix for nematic and cholesteric liquid crystals respectively. Constitutive equations (21-23) can be linearlized as follows:

$$t_{k\ell} = \underline{\underline{A(u_{m,m} - 1)\delta_{k\ell}}} - \underline{a_{k\ell}(T - T_o)} + \underline{A_{k\ell mn}(u_{n,m} - \varepsilon_{mn\rho}\phi_\rho)} + \underline{C_{k\ell mn}\phi_{m,n}} + a_{k\ell mn}(v_{n,m} - \varepsilon_{mn\rho}\nu_\rho) + \underline{\alpha_{k\ell mn}\nu_{m,n}} + \underline{e_{k\ell m}T_{,m}} \tag{24}$$

$$m_{k\ell} = \underline{-b_{\ell k}(T - T_o)} + B_{\ell kmn}\phi_{m,n} + \underline{C_{mn\ell k}(u_{n,m} - \varepsilon_{mn\rho}\phi_\rho)} + b_{\ell kmn}\nu_{m,n} + \underline{\beta_{\ell kmn}(v_{n,m} - \varepsilon_{mn\rho}\nu_\rho)} + \underline{f_{\ell km}T_{,m}} \tag{25}$$

$$q_k = \underline{h_{k\ell}T,_\ell} + \underline{\gamma_{k\ell m}(v_{m,\ell} - \varepsilon_{\ell mn}\nu_n)} + \underline{g_{k\ell m}T,_m} \tag{26}$$

where the new constants (e.g., $A_{k\ell mn}$) are related to their corresponding material moduli as follows, e.g.,

$$A_{k\ell mn} \equiv A_{KLMN}\ \delta_{kK}\ \delta_{\ell L}\ \delta_{mM}\ \delta_{nN} \tag{27}$$

with δ_{kK} being the direction cosine between the spatial coordinate and the material coordinate which is determined by the instantaneous axis of orientation and axis of helix of nematic and cholesteric liquid crystals, respectively. Now we may proceed to show the meaning of reference state. By definition reference state is only defined in the static case. Therefore, we consider the following static cases to define the reference states for liquid crystals.

Nematic Liquid Crystals

Ericksen-Leslie Theory.

$$d_{i,j} = 0 \quad \text{implies} \quad \sigma_{ij} = -p\delta_{ij}\ ,\ m_{ij} = 0 \tag{28}$$

This means that Ericksen and Leslie regard nematic liquid crystals as the kind of materials which assume any state that has an uniform alignment as reference state, in which moment stress vanishes and only an unknown pressure exists.

Eringen-Lee Theory.

$$\rho = \rho_o \text{ and } \Gamma_{KL} = 0 \text{ implies } t_{ij} = -\pi_o\delta_{ij} \text{ and } m_{ij} = 0 \tag{29}$$

This means that nematic liquid crystals assume any state that leaves density unchanged and rotation gradient vanished as reference state, in which moment stress vanishes and only a hydrostatic pressure exists. We can see that these two theories are essentially the same as far as the definition of reference state for nematic liquid crystals is concerned.

Cholesteric Liquid Crystals

Ericksen-Leslie Theory. Let

$$d_3 = T,_i = 0, \quad d_1 = \cos\theta, \quad d_2 = \sin\theta, \quad \theta = \Delta z \tag{30}$$

Substituting (30) into eqns. (17-19) and we obtain

$$\sigma_{33} = -p + \Delta(K_2 - K_{22}\Delta)$$
$$\sigma_{11} = -p + \alpha\Delta(\sin^2\theta - \cos^2\theta)$$

$$\begin{aligned}
\sigma_{22} &= -p - \alpha\Delta(\sin^2 - \cos^2) \\
\sigma_{12} &= \sigma_{21} = -2\alpha\Delta \sin\theta \cos\theta \\
m_{11} &= -K_{24}\Delta \sin^2\theta + (\alpha - K_2)(1 - \cos^2\theta) \\
m_{22} &= -K_{24}\Delta \cos^2\theta + (\alpha - K_2)(1 - \sin^2\theta) \\
m_{33} &= K_{22}\Delta + (\alpha - K_2) \\
m_{12} &= m_{21} = \sin\theta \cos\theta [K_{24}\Delta - (\alpha - K_2)] \\
\sigma_{23} &= \sigma_{32} = \sigma_{31} = \sigma_{13} = m_{23} = m_{32} = m_{31} = m_{13} = 0
\end{aligned} \tag{31}$$

Substituting eqns.(31) into balance laws of linear momentum and angular momentum, we realize that they are satisfied identically. Requiring $m_{33} = 0$, Leslie (p. 415, eqn.(3.16) of Ref. [27]) set $\Delta = (K_2 - \alpha)/K_{22}$. This means that Leslie regard cholesteric liquid crystals as materials which assume any state that leaves the helical structure (defined by axis of helix, in this case being x_3-axis, and twisting angle $\Delta = (K_2 - \alpha)/K_{22}$) unchanged and temperature gradient vanished as reference state, in which equations of motion are satisfied identically.

Eringen-Lee Theory. By setting $A_{1331} = A_{3113} = A_{3131} = C_{3131} = C_{3113} = 0$ (eqn.(3.23) of Ref. [32]), we state that cholesteric liquid crystals are materials which assume any state that leaves density, temperature, axis of helix, layered structure, and coefficient of optical activity unchanged as a reference state, in which both stress and moment stress vanish. However, some authors believe that there is no layered structure existing in cholesteric liquid crystals. If this is true, we let

$$x_{k,K} = \begin{bmatrix} 1 & 0 & 0 \\ 0 & 1 & 0 \\ \frac{du}{dx} & 0 & 1 \end{bmatrix}, \qquad \chi_{kK} = \begin{bmatrix} \cos\phi & \sin\phi & 0 \\ -\sin\phi & \cos\phi & 0 \\ 0 & 0 & 1 \end{bmatrix} \tag{32}$$

where $u = u(x)$ and $\phi = \Delta u$ (Δ is the coefficient of optical activity defined in Ref. [32]), and then we obtain the nonvanishing components of $t_{k\ell}$ and $m_{k\ell}$ as follows:

$$\begin{aligned}
t_{11} &= t_{22} = (A_{1111} + A_{1122})\cos\phi(\cos\phi - 1) + (A_{1212} - A_{1221})\sin^2\phi \\
t_{12} &= -t_{21} = -(A_{1111} + A_{1122})\sin\phi(\cos\phi - 1) + (A_{1212} - A_{1221})\cos\phi\sin\phi \\
t_{13} &= \frac{du}{dx}(A_{1313} + C_{1331}\Delta) \\
t_{31} &= \frac{du}{dx}[(A_{1111} + A_{1122})\cos\phi(\cos\phi - 1) + (A_{1212} - A_{1221})\sin^2\phi]
\end{aligned}$$

$$m_{11}=m_{22}=(C_{1111}+C_{1122})\cos\phi(\cos\phi-1)-(C_{1212}-C_{1221})\sin^2\phi$$
$$m_{12}=-m_{21}=-(C_{1111}+C_{1122})\sin\phi(\cos\phi-1)-(C_{1212}-C_{1221})\sin\phi\cos\phi$$
$$m_{13} = \frac{du}{dx}(B_{3131}\Delta+C_{1331})$$
$$m_{31} = \frac{du}{dx}[(C_{1111}+C_{1122})\cos\phi(\cos\phi-1)-(C_{1212}-C_{1221})\sin^2\phi] \tag{33}$$

By setting

$$A_{1111} + A_{1122} = 0, \qquad A_{1212} = A_{1221}$$
$$C_{1111} + C_{1122} = 0, \qquad C_{1212} = C_{1221} \tag{34}$$
$$A_{1313} = -C_{1331}\Delta = B_{3131}\Delta^2,$$

we may say the density, temperature, axis of helix, and coefficient of optical activity characterize the reference state of cholesteric liquid crystals. Thus we obtain essentially the same definition of reference state as proposed by Leslie.

V. SPECIAL PROBLEMS

In this section we investigate several special problems in order to show some characteristics of the two theories.

Rectilinear Shear Flow (Nematic Liquid Crystals)

For a shear flow between two parallel plates, $y = 0$ & $y = h$, one of which is stationary and the other is moving toward X-direction with a constant velocity v_o, we try the solutions in the following forms:

$$v_x = v(y), \quad d_x = \cos\theta, \quad d_y = \sin\theta, \quad \theta = \theta(t) \tag{35}$$

According to Ericksen-Leslie theory, we have for the equations of motion:

$$d^2v/dy^2 = 0$$
$$\rho_1\ddot{\theta}+(\mu_3-\mu_2)\dot{\theta}+\frac{1}{2}\frac{dv}{dy}[(\mu_5-\mu_6)(\sin^2\theta-\cos^2\theta)-(\mu_2-\mu_3)]=0 \tag{36}$$

with the boundary and initial conditions:

$$v(o) = 0, \quad v(h) = v_o$$
$$\theta(o) = \pi/2, \quad \dot{\theta}(o) = 0 \tag{37}$$

Therefore we obtain

$$v = v_o \frac{y}{h}$$
$$\rho_1 \ddot{\theta} + (\mu_3 - \mu_2)\dot{\theta} + \frac{1}{2}\frac{v_o}{h}[(\mu_5-\mu_6)(\sin^2\theta - \cos^2\theta) - (\mu_2-\mu_3)] = 0 \tag{38}$$

On the phase plane $(\theta,\dot{\theta})$ there is a singular point at

$$\dot{\theta} = 0$$
$$\theta = \frac{1}{2}\cos^{-1}[(\mu_3-\mu_2)/(\mu_5-\mu_6)] \ , \quad \text{if} \quad \mu_5-\mu_6 \geq 0$$
$$= \frac{\pi}{2} + \frac{1}{2}\cos^{-1}[(\mu_3-\mu_2)/(\mu_6-\mu_5)], \quad \text{if} \quad \mu_5-\mu_6 \leq 0 \tag{39}$$

and this singular point is stable branch point if $D \geq 0$, stable focus if $D < 0$, where D is defined as

$$D \equiv [(\mu_3-\mu_2)/\rho_1]^2 \ -4v_o[(\mu_5-\mu_6)^2 - (\mu_2-\mu_3)^2]^{\frac{1}{2}}/\rho_1 h \tag{40}$$

This implies that eventually the orientation θ, with respect to the flow direction will change to acquire the value given by eqn.(39). However, we believe that the singular point should be $\theta=\dot{\theta}=0$, namely, orientation should be parallel to flow direction eventually, (cf. Lee and Eringen [29]). For this to be true one must take

$$\mu_3-\mu_2 = \mu_5-\mu_6 \tag{41}$$

The requirement (41) is equivalent to setting $a_{1313} = a_{3113}$ in Eringen-Lee theory [29]. In our previous work [29] we obtained essentially the same set of equations of motion. However at that time we introduced the concept of instantaneous reference state which could be rewritten as follows:

<u>Instantaneous Reference State of Nematic Liquid Crystals</u>: If nematic liquid crystals undergo a dynamic process subject to the following conditions:

$$\rho = \rho_o, \quad \chi_{kK,L} = 0 \ , \tag{42}$$

then we take the current axis of orientation to characterize the instantaneous reference state so that the constitutive equations are reduced to

$$m_{ij}=0, \ t_{k\ell}=-p\delta_{k\ell}+a_{KLMN}(v_{n,m}-\varepsilon_{mnr}\nu_r)\delta_{kK}\delta_{\ell L}\delta_{mM}\delta_{nN} \tag{43}$$

where δ_{kK} is the direction cosine between the spatial reference frame and the instantaneous reference state.

Thus the implicit existance of the reference state and instantaneous reference state in the Ericksen-Leslie theory is equivalent to corresponding explicit concept of reference state introduced in Eringen-Lee theory.

Static Problem (Nematic Liquid Crystals)

Suppose a certain kind of nematic liquid crystals are placed in a container whose surface is specially treated (cf. Ref. [30]) such that the orientation of liquid crystals will be either parallel or perpendicular to the surface of the container, then an uniformly oriented situation cannot be expected. We would like to investigate the distribution of local orientations due to boundary effects in a static case. This problem has been solved by Lee and Eringen [30]. For Ericksen-Leslie theory, we try the solutions in the following forms:

$$d_x = \cos\theta, \quad d_y = \sin\theta, \quad d_z = 0, \quad \theta = \theta(x,y) \tag{44}$$

Substituting (44) into constitutive equations (17) & (18), we obtain the relevant nonvanishing components of σ_{ij} and m_{ij}

$$\begin{aligned} \sigma_{xx} &= -p + A\theta,_x \ , & \sigma_{yy} &= -p + B\theta,_y \\ \sigma_{xy} &= A\theta,_y \ , & \sigma_{yx} &= B\theta,_x \\ m_{xz} &= -A \ , & m_{yz} &= -B \end{aligned} \tag{45}$$

where

$$A \equiv g_{11}\sin\theta - g_{21}\cos\theta \ , \quad B \equiv g_{12}\sin\theta - g_{22}\cos\theta$$

$$\begin{aligned} g_{11} \equiv\ & -(K_{22}+K_{24})\sin\theta\ \theta,_x + (K_{11}-K_{22}-K_{24})(\cos\theta\ \theta,_y - \sin\theta\ \theta,_x) \\ & -(K_{33}-K_{22})\sin\theta\cos\theta(\cos\theta\ \theta,_x + \sin\theta\ \theta,_y) \\ g_{22} \equiv\ & (K_{22}+K_{24})\cos\theta\ \theta,_y + (K_{11}-K_{22}-K_{24})(\cos\theta\ \theta,_y - \sin\theta\ \theta,_x) \\ & +(K_{33}-K_{22})\sin\theta\cos\theta(\cos\theta\ \theta,_x + \sin\theta\ \theta,_y) \\ g_{12} \equiv\ & -K_{22}\sin\theta\ \theta,_y + K_{24}\cos\theta\ \theta,_x - (K_{33}-K_{22})\sin^2\theta(\cos\theta\ \theta,_x \\ & +\sin\theta\ \theta,_y) \\ g_{21} \equiv\ & K_{22}\cos\theta\ \theta,_x - K_{24}\sin\theta\ \theta,_y + (K_{33}-K_{22})\cos^2\theta(\cos\theta\ \theta,_x \\ & +\sin\theta\ \theta,_y) \end{aligned}$$

Therefore the equations of motion are

$$
\begin{aligned}
&-p,_x + (A\theta,_x),_x + (B\theta,_x),_y = 0 \\
&-p,_y + (A\theta,_y),_x + (B\theta,_y),_y = 0 \\
&-A,_x - B,_y + A\theta,_y - B\theta,_x = 0
\end{aligned}
\tag{46}
$$

We see that there are three equations in (46) and two unknowns θ and p. This is just an example and in general Ericksen-Leslie theory, in static case, has only three unknowns, p and d_i ($d_i d_i = 1$), but six equations (balance laws for linear and angular momenta). That is what we mean in static case, Ericksen-Leslie theory will end up with an over-determined system. Eringen-Lee theory, in static case, has six equations and six unknowns, i.e., three displacement components and three angles of microrotation. However we note that in Ref.[30] we have taken, in the deformed state, the average direction of local molecular axes as reference with respect to which the generalized strain measures are calculated, thus incorporated the elastic properties of the media.

Cholesteric Liquid Crystals Subject to Temperature Gradient

We investigate the following special problem: Cholesteric liquid crystals are placed between two parallel plates which are perpendicular to the axis of helix. Let the boundary conditions be

$$
\begin{aligned}
&T(z=0) = T_o , \qquad T(z=h) = T_1 \\
&m_{zz} = 0 \quad \text{at} \quad z = 0 \quad \text{and} \quad z = h
\end{aligned}
\tag{47}
$$

Now we try a set of solutions having the following form:

$$
\begin{aligned}
&d_x = \cos\theta , \quad d_y = \sin\theta , \quad d_z = 0 \\
&T = T(z) , \quad \theta = \theta(z,t)
\end{aligned}
\tag{48}
$$

Substituting (48) into eqns. (17-19), we obtain the relevant and nonvanishing components of $\sigma_{k\ell}$, $m_{k\ell}$, and q_k as follows:

$$
\begin{aligned}
\sigma_{xy} &= -2\alpha \frac{d\theta}{dz} \sin\theta \cos\theta + \dot{\theta}(\mu_2\cos^2\theta - \mu_3\sin^2\theta) \\
&\quad + T,_z(-\mu_7\cos^2\theta + \mu_8\sin^2\theta) \\
\sigma_{yx} &= -2\alpha \frac{d\theta}{dz} \sin\theta \cos\theta + \dot{\theta}(\mu_3\cos^2\theta - \mu_2\sin^2\theta) \\
&\quad + T,_z(-\mu_8\cos^2\theta + \mu_7\sin^2\theta)
\end{aligned}
$$

$$m_{zz} = K_{22}\frac{d\theta}{dz} + (\alpha - K_2)$$
$$q_z = C_1\, T,_z + C_3\,\dot{\theta} \qquad (49)$$

Substituting (49) into the heat equation (cf. eqn.(4.14) of Ref. [26]) and assuming no heat source, we obtain

$$\rho T\dot{S} = -C_1\, T,_{zz} - C_3\dot{\theta},_z + \alpha\dot{\theta},_z - (\mu_2 - \mu_3)\dot{\theta}^2 + (\mu_7 - \mu_8)T,_z\dot{\theta} \qquad (50)$$

Another equation of motion is

$$K_{22}\frac{d^2\theta}{dz^2} - (\mu_7 - \mu_8)T,_z + (\mu_2 - \mu_3)\dot{\theta} = \rho_1\ddot{\theta} \qquad (51)$$

Finally the solutions are (cf. Ref. [26]):

$$\dot{S} = 0$$
$$\theta = \theta_o + ((\mu_7 - \mu_8)(T_1 - T_o)/(\mu_2 - \mu_3)h)t + (K_2 - \alpha)z/K_{22} \qquad (52)$$
$$T = T_o + (T_1 - T_o)z/h$$

This shows that Ericksen-Leslie theory predicts that the temperature gradient induces a constant angular velocity

$$\omega \equiv \frac{d\theta}{dt} = \frac{\mu_7 - \mu_8}{\mu_2 - \mu_3}\,\frac{T_1 - T_o}{h}\ , \qquad (53)$$

but leaves the coefficient of optical activity unchanged, i.e.,

$$\Delta \equiv \frac{d\theta}{dz} = (K_2 - \alpha)/K_{22} \qquad (54)$$

And yet, according to the theory, the angular velocity of directors will not contribute to the increase of entropy. This problem demonstrates another one of the basic differences between Ericksen-Leslie theory and Eringen-Lee theory (cf. Ref. [32]). In other words, Eringen-Lee theory predicts that the temperature gradient induces, instead of a constant angular velocity, the changes of local density, orientation, and hence, accordingly, the coefficient of optical activity (cf. Eqn.(4.9) of Ref. [32]).

REFERENCES

* The present paper was partially supported by National Science Foundation. It owes its existance to referee's request during the review process of one of our previous papers on the subject.

1. G. Friedel, Ann. Physique 19, 273 (1922).
2. I. G. Chistyakov, Soviet Physics USPEKHI, V. 9, No. 4, 551 (1967).
3. A. Saupe, Angew. Chem. International Edit., V. 7, No. 2, 97 (1968).
4. J. L. Fergason, Sci. Am. 211, 76 (1964).
5. C. W. Oseen, Trans. Faraday Soc., 29, 883 (1933).
6. F. C. Frank, Discussions Faraday Soc., 25, 19 (1958).
7. A. Anzelius, Uppsala Univ. Arsskr. Mat. och Naturvet, 1 (1931).
8. Groupes d'Etudes des Cristaux Liquids, J. Chem. Phys., 51, 816 (1969).
9. M. Stephen, Phys. Rev. A2, 1558 (1970).
10. T. C. Lubensky, Phys. Rev. A2, 2497 (1970).
11. P. C. Martin, P. S. Pershan, and J. Swift, Phys. Rev. Lett., V. 25, No. 13, 844 (1970).
12. D. Forster et al, Phys. Rev. Lett., V. 26, 1016 (1971).
13. H. Huang, Phys. Rev. Lett., V. 26 1525 (1971).
14. H. Schmidt and J. Jahnig, Annals of Physics 71, 129 (1972).
15. J. L. Ericksen, Arch, Rat. Mech. Anal., 4, 231 (1960).
16. J. L. Ericksen, Trans. Soc. Rheol., 5, 23 (1961).
17. J. L. Ericksen, Arch. Rat. Mech. Anal., 9, 371 (1962).
18. J. L. Ericksen, Arch. Rat. Mech. Anal., 23, 266 (1966).
19. J. L. Ericksen, Phys. of Fluids, 9, 1205 (1966).
20. J. L. Ericksen, Quart. J. Mech. Appl. Math., 19, 455 (1966).
21. J. L. Ericksen, Appl. Mech. Rev., 20, 1029 (1967).
22. J. L. Ericksen, J. Fluid Mech., 27, 59 (1967).
23. J. L. Ericksen, Molecular Crystals & Liquid Crystals, 7, 153 (1969).
24. F. M. Leslie, Quart. J. Mech. Appl. Math., 19, 357 (1966).
25. F. M. Leslie, Arch. Rat. Mech. Anal., 28, 265 (1968).
26. F. M. Leslie, Proc. Roy. Soc. London A, 307, 359 (1968).
27. F. M. Leslie, Molecular Crystals & Liquid Crystals, 7, 407 (1969).
28. J. D. Lee and A. C. Eringen, J. Chem. Phys., 54, 5027 (1971).
29. J. D. Lee and A. C. Eringen, J. Chem. Phys., 55, 4504 (1971).
30. J. D. Lee and A. C. Eringen, J. Chem. Phys., 55, 4509 (1971).
31. J. D. Lee and A. C. Eringen, J. Chem. Phys., 58, 4203 (1973).
32. A. C. Eringen and J. D. Lee, Mechanics of Cholesteric Liquid Crystals, to be presented at the Third ACS Symposium on Ordered Fluids and Liquid Crystals. (1973).
33. A. C. Eringen and E. S. Suhubi, Int. J. Eng. Sci., 2, 189 (1964).
34. A. C. Eringen, Int. J. Eng. Sci., 2, 205 (1964).
35. A. C. Eringen, J. Math. Mech., 16, 1 (1966).
36. A. C. Eringen, J. Math. Mech., 15, 909 (1966).
37. A. C. Eringen, Proc. 11th Intern. Conf. Appl. Mech., 131, Springer-Verlag (1964).

STRUCTURE AND THERMAL CONDUCTIVITY OF SUPERCOOLED MBBA

J. O. Kessler and J. E. Lydon

Physics Department, University of Arizona, Tucson 85721, U.S.A. Astbury Department of Biophysics, University of Leeds, Leeds LS2 9JT, G.B.

Many nematic liquid crystals, when cooled to temperatures below the normal melting point, form a phase which is not the equilibrium crystalline solid. This behaviour has been shown in MBBA, (1) which will be the subject of the present paper.

If this non-equilibrium phase is a glass which has a spatial distribution of molecules very similar to that of the liquid crystal, but without the attendant molecular mobility, it may prove very useful in elucidating those aspects of the transport properties which do not depend on large scale molecular motion. Put another way, the hypothetical existence of such a glass opens the possibility that one part of a given transport property may be frozen out, permitting the measurement of the component remaining. For instance, one might be able to reduce greatly the ionic contribution to charge transport and thus measure any electronic contribution that might remain.

The present work is preliminary. Its objective is to demonstrate that the low temperature non-equilibrium phase of MBBA is in fact an amorphous (i.e. non-crystalline) one whose structure is closely related to that of the liquid crystal, and that at least one transport property can be measured and related to its high-temperature analog.

The transport property chosen was the thermal conductivity. The frozen-out part would be due to molecular transport, the remaining part, to lattice vibrations. The measurements in these preliminary experiments are not yet sufficiently accurate to permit the frozen-out and remaining parts to be distinguished.

However, the numbers show unambiguously that the anisotropy of of the liquid crystal phase continues into the quenched phase. This result, together with the structural information which demonstrates that the quenched phase is amorphous yet related to the equilibrium phases, achieves the stated objective.

STRUCTURE DETERMINATIONS

The material used in all these experiments was supplied by Eastman Chemicals. In the X-Ray and optical experiments a droplet was placed on a microscope slide and was plunged into liquid N_2. The slide and congealed specimen were then placed into the apparatus and examined in the appropriate temperature range.

A. X-Ray Diffraction

The X-Ray diffraction patterns of the quenched, crystalline and liquid crystalline forms of MBBA appeared in temperature sequence, as expected (1). As shown in Fig. 1, the crystalline phase gave a strong reflection at $2\Theta = 5.2^{o}$.

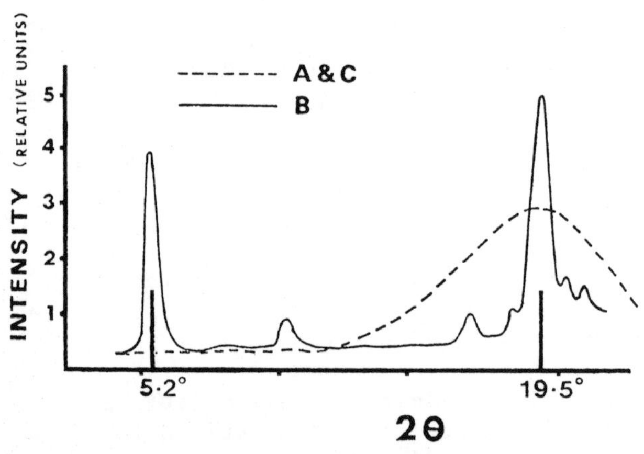

Fig. 1. X-ray diffraction intensity as a function of angle at three temperatures. The solid curve B was obtained with the crystalline intermediate-temperature phase. The dashed curve was obtained both with the high - temperature (liquid crystalline) phase C, and with the quenched phase A.

When a quenched sample at liquid N_2 temperature was placed in the diffractometer set at this angle, the observed intensity rose from an initial "background" value to a peak while the sample warmed and eventually crystallized. The intensity subsequently decreased to its initial value as the sample warmed further into the normal liquid crystalline phase. This effect, and the diffraction intensity in the region $2\Theta = 19.5^{\circ}$, which was studied similarly, are shown in Fig. 2. At the latter angle the crystalline form and the quenched phase both gave broad diffuse peaks of the same amplitude.

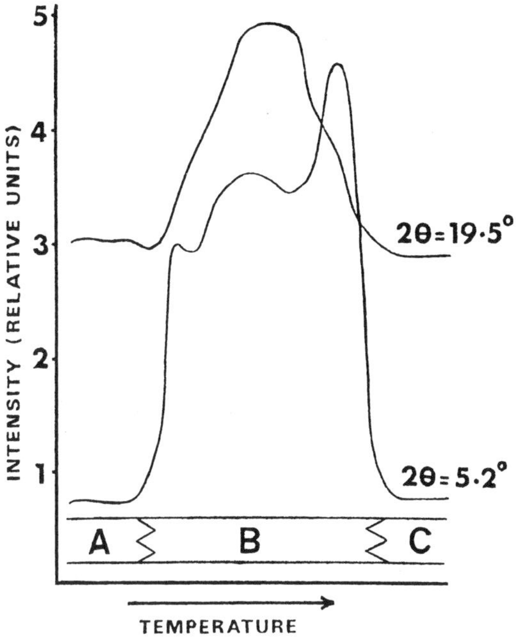

Fig. 2. X-Ray diffraction intensity as function of temperature, at two angles. The temperature regions A, B, C mark the presence of the quenched, crystalline, and liquid crystalline phases.

The coincidence of amplitude and angle of the diffuse peaks observed in the quenched and liquid crystalline phases (termed "A" and "C" in the figures) should be especially noted. This symmetry strongly suggests that the molecular arrangement of the liquid crystal is indeed frozen-in by the quenching process.

The mulitiple nature of the $2\Theta = 5.2^{\circ}$ peak, in the crystalline phase ("B"), was reproducible. It may indicate several phases or crystalline reorientation with warming. It should be noted that Mayer, Waluga and Janik inferred several intermediate phase transitions from their calorimetric data (1).

B. Optical Microscopy

If a liquid crystalline droplet is examined in a microscope one may observe the characteristic threads that give the nematic phase its name. These threads depend on the optical anisotropy of the material and on the local variations of the director axis. If one places a droplet of MBBA on a coverslip which is in turn placed on a cold piece of metal in a microscope stage, and if one examines the droplet as it cools from room temperature to somewhere near 100°K, one may observe the same threads all the way down in temperature. In the final stage the droplet feels quite solid when probed.

Unfortunately, icing problems prevented photographic documentation of this proof that liquid crystalline order is maintained in the quenched phase. However, pictures taken in transmitted white polarized light through the glass slide side of a MBBA drop and microscope slide arrangement do indicate the nature of the optical properties of the three phases. The micrographs are shown in Fig. 3. At low temperature the specimen is seen to be transparent but cracked, it is opaque and microcrystalline in the intermediate range, and again transparent at room temperature. The information in these micrographs therefore corroborates the phase identification discussed in the section on X-rays.

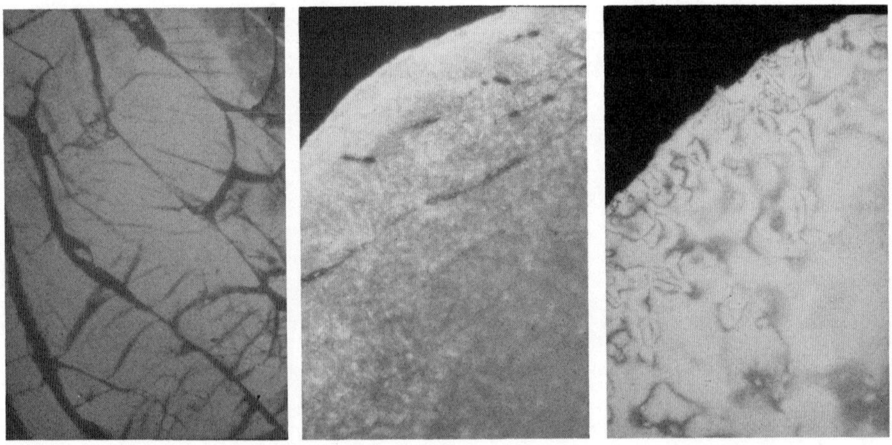

A B C

Fig. 3. Transmission micrographs of quenched (A), crystalline (B) and liquid crystalline (C) phases of MBBA. (After W. Kenchington)

C. Freeze Fracturing

The technique of freeze fracturing examines the nature of the cleavage interface of a congealed droplet of material. In our experiments a droplet of material is cooled in a bath of Freon 22 at about - $100^{\circ}C/sec$, to $123^{\circ}K$. It is then cleaved in vacuum by a steel blade at about $80^{\circ}K$. The interface is immediately shadowed and replicated with an evaporated Pt - C film. The replica is then examined in the electron microscope.

For freeze fracturing to be a useful technique for the study of liquid crystal phases two conditions must be satisfied. Firstly, the structure of the mesophase must be preserved in the quenching process and secondly the topography of the fracture surface should contain features recognizably characteristic of the mesophase. Previous studies carried out at Leeds (2) have shown that both of these conditions hold for the smectic and cholesteric phases. It was hoped that the characteristic threads might constitute recognizable features of freeze fractured nematics, but this did not prove to be the case. Fig. 4a, which is typical of the results obtained, shows no indication of two or three dimensional order. Fig. 4b shows for comparison the freeze fracture electron micrograph of a sample of silicone oil, which is presumably an isotropic glass. The similarity of the features of these fracture interfaces provides yet another demonstration that the quenched material is not microcrystalline.

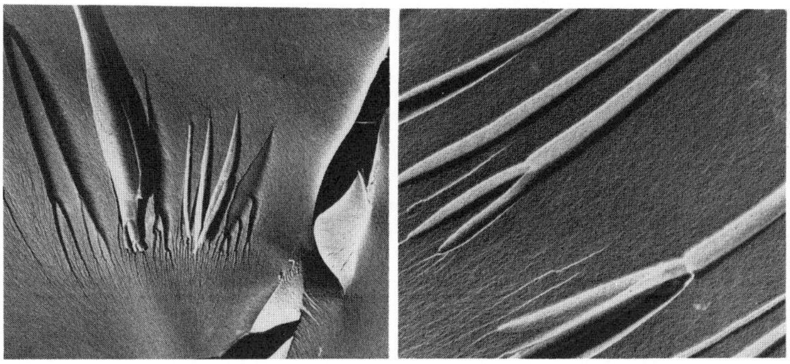

(a) (b)

Fig. 4. Electron micrographs of freeze fractured surface replicas. (a) is MBBA, shadowing in the direction from bottom left to top right; Magnification: 1mm in the illustration = 0.9 μm on the replica. (b) is Silicone oil, shadowing from top left to bottom right. 1mm ≈ 2μm. (Reproductions have been reduced to 27% of the original size.) (After A. Forge)

THERMAL CONDUCTIVITY MEASUREMENTS

A. Experimental

The apparatus consisted of two flat cells mounted rigidly at right angles to one another, each between two 1/16" copper plates, one of which acted as heat source, the other as heat sink. The walls of each cell were made out of 0.1 cm microscope slide glass. Copper platelets with 40 gauge copper-constantan thermocouples attached were mounted on the inside of each cell wall, in series with a 0.1cm thick square glass chip. The gap between the free interior faces of the copper platelets was 0.05 cm. These components, and the edges of the cells which were formed from 4mm glass tubing, were cemented together with a silastic compound. The MBBA was introduced into the cells with an eyedropper. Thermal contact between the glass cells and the heat source/sink was made through silicone vacuum grease. Thermocouples were also attached to the outside of the heat source/sink plates. Pairs of thermocouples were connected differentially, the output voltages being read on a Keithley digital VTVM and a strip chart recorder. These readings measured the temperature difference between the heat source and the first inside copper platelet, between that platelet and the second one, and between the second one and the heat sink. Steady state conditions were inferred from the stability of the signals.

In principle, the thermal conductivity of the sample between the inside platelets is proportional to the inverse of the steady state temperature difference between them. Thermal conductivity ratios are equal to temperature difference ratios, providing the total thermal flux is kept constant properly normalized and nothing changes in the apparatus.

The apparatus was located in a Dewar flask between the pole-pieces of an electromagnet. One of the cells was oriented with its face perpendicular to the magnetic field direction, the other was parallel. The apparatus could be turned to interchange the cells' roles for the purpose of eliminating some cell parameters.

Measurements of room temperature thermal conductivity were made to check the experimental conditions. It was found that the anisotropy attainable was developed at very low magnetic fields. Saturation was certainly complete at 7 kG, the field used eventually to align the MBBA before quenching.

The procedure used in obtaining the quenched phase was to place the MBBA-filled apparatus into the Dewar, with magnetic field applied. After 10 minutes in the field, liquid N_2 was poured over the apparatus. Cooling to 250°K took 5 sec and to 110°K, 40 sec. The temperature gradient across the inner platelets, i.e. across the MBBA specimen never exceeded 10°K. This fact, together with the rapidity of the process and the small gap size, implies that convection and concomitant flow alignment did not occur. It is not known whether the specimen remained in good contact with the copper platelets, but good contact and minimum cracking are implied by the reproducibility of the observed gradients.

After quenching, the field was turned off and thermal conductivity measurements at several heat fluxes were conducted with the heat sinks in contact with, or just above liquid N_2 level. The average temperature for each set of measurements on the quenched phase was kept between 90 and 200°K - more usually at the low end of the range.

Runs were made in various combinations of starting and measurement orientations (relative to Dewar and magnetic field). Systematic errors related to the difference between the measurement cells and differences arising from the orientation and thermal conductivity changes are thought to have been eliminated by the measurement and analysis procedure. Possible instrumental errors that could not be studied in the time available were heat flux paths in parallel with the specimens and distortions of the isotherms. Even small errors of this kind severely tend to reduce the measured anisotropy.

B. Results

The measured anisotropy is stated by (n-m)/m = a, where m is the thermal conductivity perpendicular to the magnetic alignment direction and n is the one parallel to it. It was found that at room temperature a = 0.23± .04, which is much lower than the published values (3, 4) of 0.44± .01 and 0.64± .04. The discrepancy suggests that either the specimen was very impure or that the instrumental errors mentioned above interfered. It is unlikely that the later published value is wrong.

The anisotropy of the quenched phase was found to be a = 0.30± .04. As before, the range quoted refers to the measurement series and does not include the possible effects of instrumental error. Since an instrumental error, if present, may be temperature-dependent, the high and low temperature anisotropies cannot as yet be directly compared, beyond the statement that they are similar.

Measurements were also performed on the thermal conductivity anisotropy of the intermediate microcrystalline phase, recooled to around $100^{\circ}K$. It was found that a = $0.00 \pm .05$, i.e. that the anisotropy vanishes in that case.

CONCLUSION AND SUMMARY

It has been shown by means of X-ray diffraction, optical microscopy and freeze fracturing that the nematic liquid crystal MBBA may be quenched from room temperature to a low-temperature solid phase with similar structural features. The thermal conductivity experiments, although of a preliminary nature, provide the additional very significant information that the characteristic macroscopic anisotropy, and its direction, is preserved during the quenching process. It has thereby been demonstrated that it is possible to produce anisotropic non-crystalline substances directly from an aligned liquid crystalline melt (5). Both practical and scientific applications may be expected.

It may also be inferred that the major thermal transport mechanism in the liquid depends on vibrational rather than translational motion of the molecules. Although this result was expected theoretically, it was put into some slight doubt by the well-known fact that molecular transport coefficients of liquid crystals are very anisotropic, whereas the ultrasonic velocity is not.

ACKNOWLEDGEMENTS

We wish to thank Mr. A. Forge, Dr. W. Kenchington and Mr. T. Croxon for help with the freeze fracturing technique, the optical microscopy and the thermal conductivity apparatus, respectively. One of us (JOK) would like to thank the University of Leeds Physics Department for its aid and hospitality, and the U.K. Science Research Council and the University of Arizona for their support.

REFERENCES

(1) J. A. Mayer, T. Waluga and J. A. Janik, Phys. Lett. 41A, 102 (1972), and private communication.

(2) J. E. Lydon and D. G. Robinson, Biochim et Biophys. Acta 260, 298 (1972).

(3) E. Guyon, P. Pieranski and F. Brochard, Comptes Rendus 273B, 486 (1971).

(4) P. Pieranski, F. Brochard and E. Guyon, to be published. We are indebted to Dr. Guyon for preprints of this work.

(5) L. Liebert and L. Strzelecki, Comptes Rendus 276C, 647, (1973), have described a similar effect that involves polymerization rather than quenching.

THE DIELECTRIC PROPERTIES OF NEMATIC MBBA IN THE PRESENCE OF ELECTRIC AND MAGNETIC FIELDS

P.G. Cummins, D.A. Dunmur and N.E. Jessup

Department of Chemistry
The University
Sheffield, S3 7HF
U.K.

1. INTRODUCTION

The dielectric properties of liquid crystalline mesophases have been the subject of numerous investigations,[1,2,3] and it has been observed that nematic mesophases become dielectrically anisotropic when subjected to electric or magnetic fields. This induced dielectric anisotropy is associated with a high degree of molecular alignment within the mesophase, although the precise molecular mechanism of the alignment process is still not known. The high degree of molecular alignment that may be induced in liquid crystalline mesophases by electric fields, magnetic fields or suitably treated glass surfaces forms the basis of many technological applications. These usually require thin films of the oriented mesophase, and so most dielectric studies have been made on such films. To avoid the surface effects associated with thin films, we have made dielectric measurements on bulk samples of nematic mesogens. In this paper we report measurements of the low frequency electric permittivity of bulk samples of p-methoxy-benzylidene-p-n-butyl aniline (MBBA) oriented by both static electric and magnetic fields. The dependence of the electric permittivity components on magnetic and electric field strength has also been determined.

Previous dielectric investigations of MBBA have been made on thin films in the presence of magnetic fields.[4,5] It was found that the material had a negative dielectric anisotropy, which is customarily associated with a molecular dipole moment lying across the long axis of the molecule.[6] The object of the present work was to provide more extensive experimental measurements in the hope that these will assist the molecular understanding of the dielectric properties of nematic mesophases.

2. EXPERIMENTAL

(i) Electric field measurements: A thermostatted cell was constructed from pyrex, having two pairs of stainless steel electrodes A & B arranged in the manner illustrated in fig.1. The volume of the cell was approximately 50 cm^3. To obtain measurements of the electric permittivity parallel to a static orienting field ($\varepsilon_{\parallel}$) the pair of electrodes A were connected to a capacitance bridge (Wayne Kerr type B331) operating at a frequency of 1592 Hz, and to a source of high voltage (Hewlett Packard model No. 6515A): see fig. 1. With the cell filled with MBBA the capacitance across the electrodes A was measured as a function of static electric field strength at various temperatures in the nematic range. To obtain values for $\varepsilon_{\parallel}$, the capacitance between the electrodes A was also measured using the same configuration as fig. 1, but with the cell containing standard dielectric liquids. The standard liquids were chosen to span the permittivity range of interest, and were cyclohexane, anisole and chlorobenzene; measurements with these liquids showed no dependence on static electric field strength. A plot of cell capacitance against liquid permittivity yielded a value for the cell constant, and hence capacitance measurements on the liquid

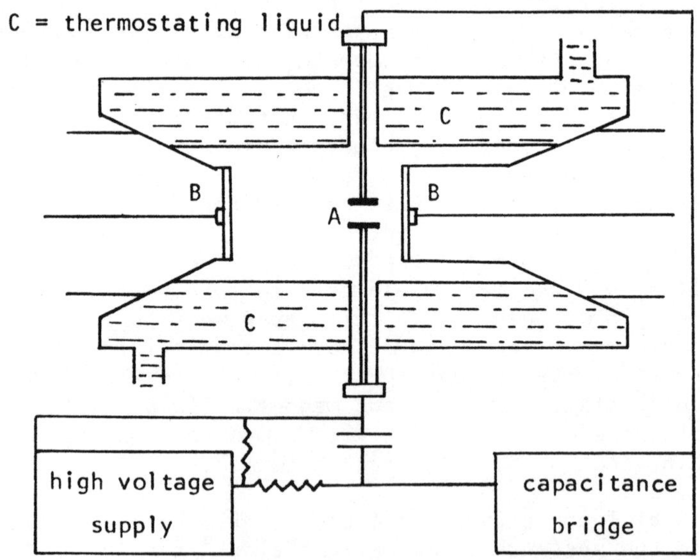

Figure 1. Permittivity cell for electric field measurements, approx. ½ scale.

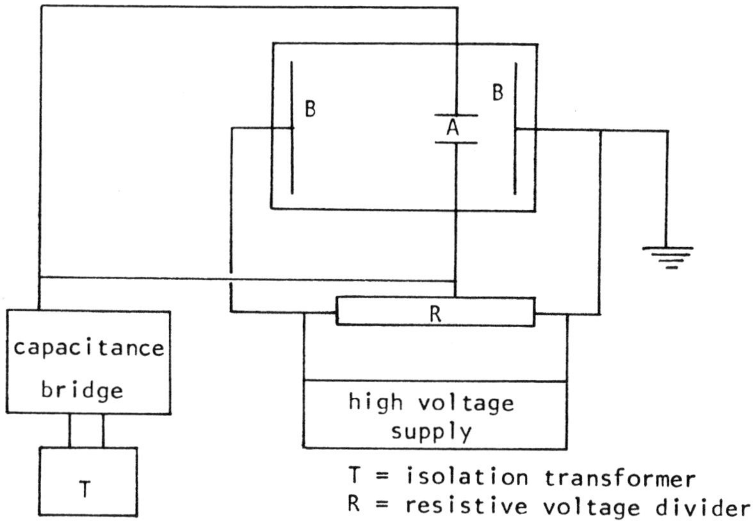

Figure 2. Experimental arrangement for $\varepsilon_\perp$ measurements in the presence of electric fields.

crystal could be directly converted to electric permittivities.

The electric permittivity component perpendicular to an orienting field ($\varepsilon_\perp$) was measured by using the pair of electrodes A as probe electrodes, and connecting the orienting electric field across the electrodes B, as illustrated in fig. 2. The probe electrodes were floated at a suitable d.c. potential to maintain a uniform orienting field between the electrodes B. The probe electrodes disturb the uniformity of the orienting field, and we have tried to minimise this effect by making them as small as possible. We believe that any effects due to field inhomogeneity are less than our quoted experimental uncertainties. Measurements of the capacitance between the probe electrodes were made by floating the capacitance bridge at the appropriate potential, and operating it by remote control. The cell was calibrated with standard dielectric liquids to obtain values for the permittivity component $\varepsilon_\perp$.

(ii) Magnetic field measurements: The two cells used to measure $\varepsilon_\parallel$ and $\varepsilon_\perp$ in the presence of a magnetic orienting field are illustrated in fig. 3. These cells were thermostatted and fitted between the pole pieces of an electro magnet (Mullard type EE1002). The cells' dimensions were such that the magnet could be operated with a pole gap of 5 cm. The magnetic field

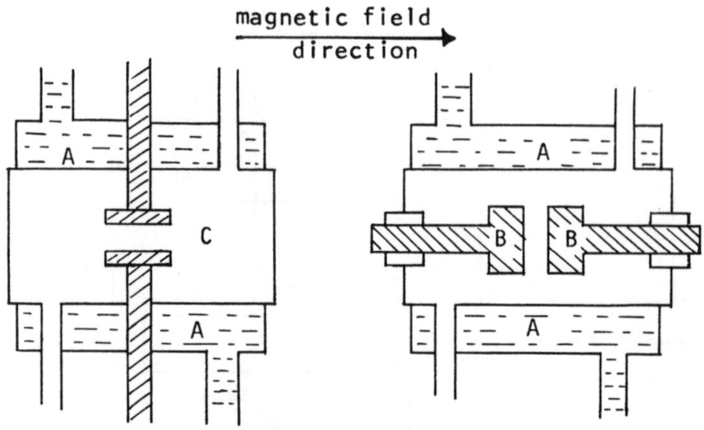

A = thermostating fluid. B = brass electrodes
C = stainless steel electrodes

Figure 3. Pyrex permittivity cells for magnetic field measurements, approx. ⅔ scale.

measured inside the cells filled with MBBA was found to be equal to the field strength in the absence of the cells. The capacitances of the cells filled with MBBA were measured as a function of magnetic field strength at various temperatures. Similar measurements were made on the cells filled with standard dielectric liquids to obtain the appropriate calibration constant.

The samples of MBBA were prepared and purified in our laboratory using standard techniques. Freshly purified samples had a clearing temperature of 47°C, but this fell rapidly on exposure to air to 44.5°C. The effect of continued application of large electric fields was to cause the clearing temperature to fall further. This was presumably a result of electrolytic decomposition.[7]

3. RESULTS.

Our results for the low frequency electric permittivity components of MBBA in the presence of static electric and magnetic orienting fields are presented in figs. 4 to 7 for various temperatures within the nematic range.

In common with other workers [5,8] we experienced difficulties concerning reproducibility of results from different samples of

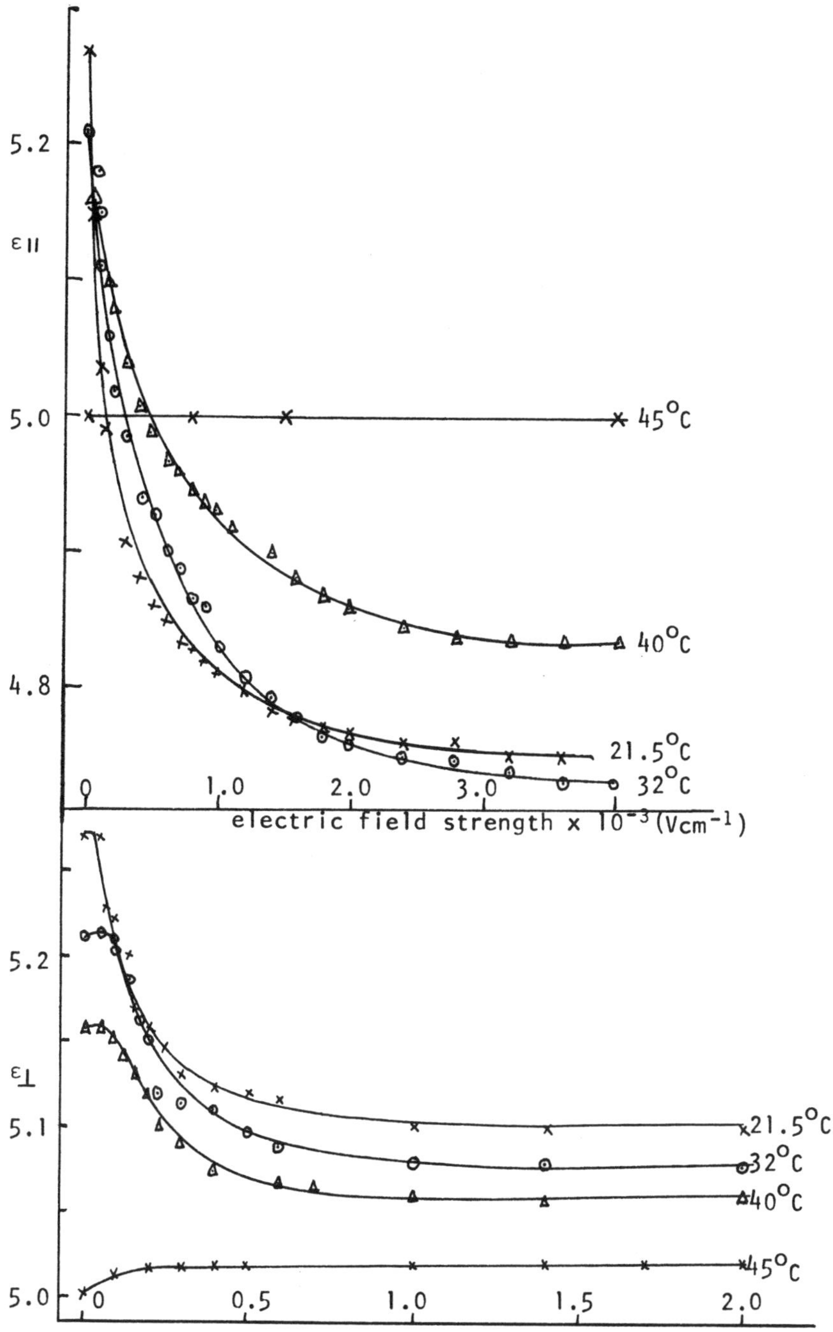

Figures 4 and 5. Permittivity components $\varepsilon_{||}$ and $\varepsilon_{\perp}$ as a function of electric field.

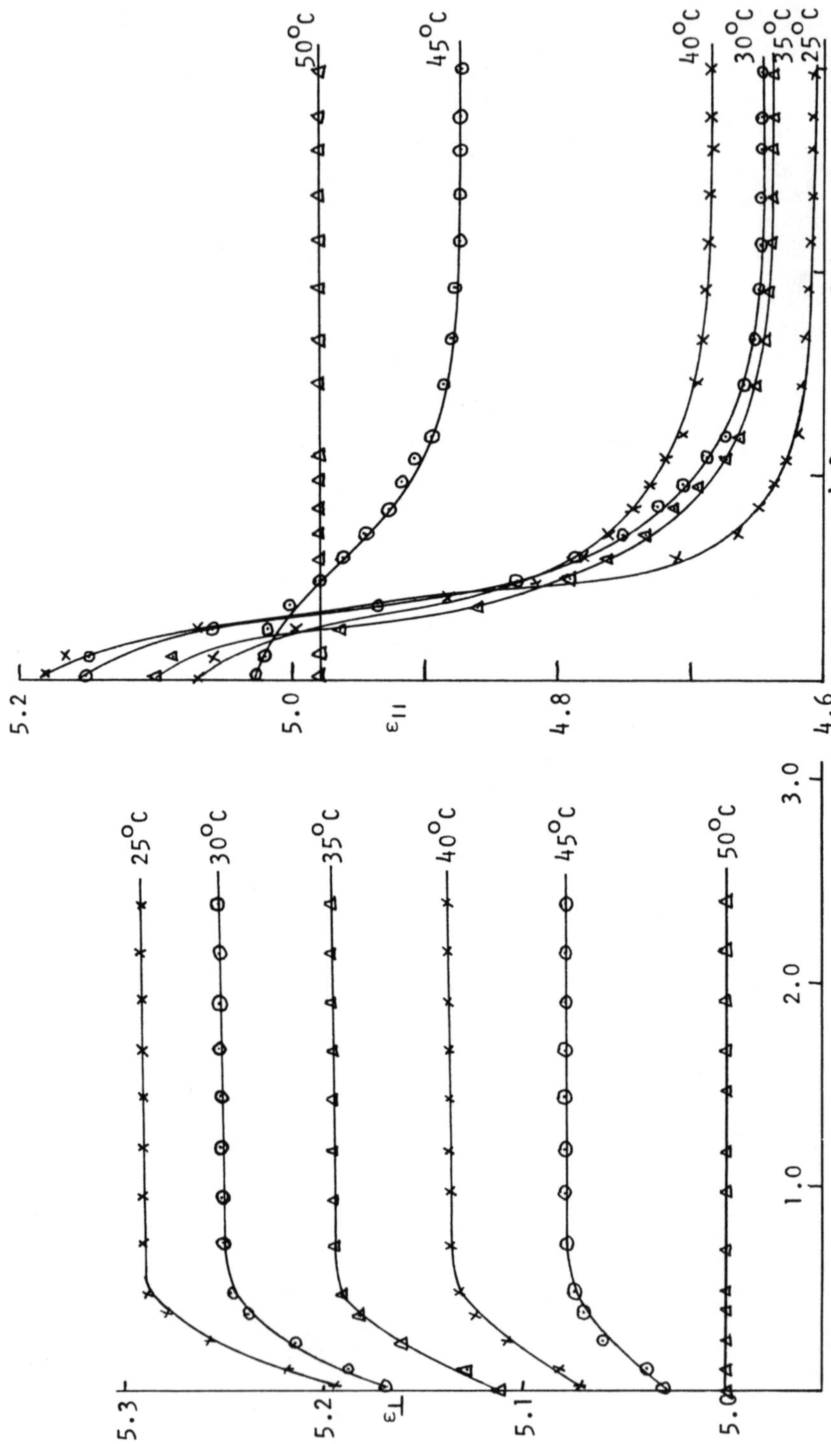

Figures 6 and 7. Permittivity components $\varepsilon_{\parallel}$ and $\varepsilon_{\perp}$ as a function of magnetic field strength in units of 10^3G.

MBBA. Consequently the accuracy of our permittivity measurements is estimated to be ± 0.1, although the precision of individual measurements on a particular sample is ± 0.02. The problems of irreproducibility were most marked in the case of the electric field measurements. All samples showed the qualitative behaviour illustrated in figs. 4 and 5, but the actual values of the permittivity varied with different samples by up to ± 0.07. We have attempted to correct for this problem by fixing the zero-field values of the permittivity components independently. These values were determined at various temperatures using a liquid permittivity cell (Wayne Kerr model D121), which had been provided with a thermostatting jacket. One important feature of this cell is the absence of electrodes in contact with the liquid, which is contained in a pyrex test tube. The arrangement of the external electrodes and the values of the voltages applied were such that there was no alignment of MBBA during the measurements. Our zero field results on nematic MBBA are given in fig. 8 as a function of temperature.

The zero-field permittivities measured in the magnetic field cells differ slightly from the values given in fig. 8. We have some evidence to suggest that there is a small degree of alignment between the electrodes in the absence of a magnetic field, and this would have the effect of reducing the measured

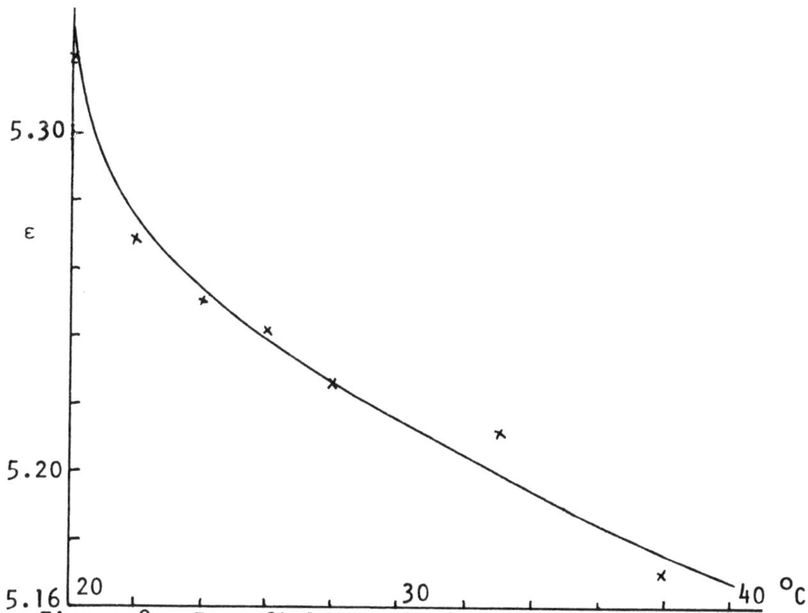

Figure 8. Zero-field permittivity of nematic MBBA as a function of temperature.

permittivity below the value corresponding to random orientation. There may have been a residual zero-field alignment in the electric field cells, but the accuracy of our measurements was not sufficiently high to detect it.

4. DISCUSSION

The orientation of nematic liquid crystals by electric or magnetic fields causes the electric permittivity to become anisotropic, and the oriented mesophase has the symmetry of a uniaxial crystal. The macroscopic electric susceptibility tensor $\chi_{\alpha\beta}$ is related to the electric permittivity by the expression:

$$\chi_{\alpha\beta} = (\varepsilon_{\alpha\beta} - \delta_{\alpha\beta})/4\pi$$

For an oriented fluid $\chi_{\alpha\beta}$ is diagonal, and has two independent components $\chi_{\parallel}$ and $\chi_{\perp}$, which refer to directions parallel and perpendicular to the orienting field respectively.

The analysing electric field E^a_α used to measure the permittivity is small, so that the measured electric susceptibility is given by:

$$\chi_{\alpha\beta} = \left(\frac{\partial P_\alpha}{\partial E^a_\beta}\right)_{E^a = 0}$$

The polarisation P_α is in general a function of the orienting field and of the analysing field E^a_α. It is convenient to consider the polarisation as arising from a distortion term P^D_α and an orientation term P^0_α; the latter is that part of the polarisation due to molecular reorientation. The degree of alignment of the liquid crystal may be measured by the order parameter

$$S = \overline{\tfrac{1}{2}(3\cos^2\theta - 1)}$$

where θ is the angle between the molecular long axis and the orientation direction. As $S \to 1$, corresponding to complete molecular alignment, the component of the orientation polarisation in the direction of the orienting field becomes independent of E^a, and hence does not contribute to $\chi_{\parallel}$. The distortion polarisation is linear in E^a, and is related to the molecular polarisability of the nematogen. It depends on the order parameter S both because of the intrinsic anisotropy in the molecular polarisability, and also because of the anisotropic local electric fields[9] that arise when $S \neq 0$. The distortion polarisation will contain non-linear contributions from an orienting electric field E^0, but the effect of a static magnetic field on the electric polarisability is expected to be small. We may therefore write:

$$P^D_\alpha = \alpha_{\alpha\beta}(E^o_\beta + E^a_\beta) + \tfrac{1}{2}\beta_{\alpha\beta\gamma}(E^o_\beta E^a_\gamma + E^a_\beta E^o_\gamma + E^o_\beta E^o_\gamma) + O(E^{o^2})$$

and it becomes clear that the distortion polarisation can be modified by an orienting electric field. The coefficients in the expression for P^D_α can be related to molecular polarisabilities and hyperpolarisabilities[10] if the local electric field is known in the oriented nematogen.

Our results indicate that $\varepsilon_\perp$ for MBBA increases to a limiting value in the presence of a magnetic orienting field, but decreases to a limiting value if orientation is achieved with a static electric field. There is no evidence for non-linear contributions to the distortion polarisation in this direction, since the limiting value of $\varepsilon_\perp$ is independent of further increase in the static electric field strength. The different behaviour for magnetic and electric fields must therefore be associated with the orientation polarisation term, and our conclusion is that reorientation of the molecules about their long axes is restricted more by an electric field than a magnetic field parallel to the long axis of the molecules. This is consistent with the view[11] that there is competition between flow alignment of the molecules[12] and dielectric alignment. In the case of MBBA the latter favours a molecular orientation such that the long axes of the molecules are perpendicular to the electric field. There is of course no such competition in the case of alignment by magnetic fields.

The results for the parallel permittivity component of samples oriented by an electric field show that even at relatively high field strengths $\varepsilon_{||}$ does not attain a limiting value. This could be a result of non-linear contributions to the distortion polarisation, although the effect appears to be small.

In table 1 we compare our values for the dielectric anisotropy of the fully aligned nematic phase of MBBA with those obtained by other workers. The agreement for magnetically oriented samples is satisfactory in view of the experimental difficulties mentioned. Unfortunately we do not have any values to compare with our electric field measurements, but we suggest that the dielectric anisotropy of samples oriented by electric fields is substantially less than the anisotropy measured for magnetically oriented samples.

We are engaged upon similar investigations of other nematic materials on the hope that such studies will further the molecular understanding of the dielectric properties of nematic liquid crystals.

Temperature °C	20	25	30	35	40	45	50
$\Delta\varepsilon$ electric field alignment (a)	-0.35	-0.36	-0.36	-0.33	-0.22	0.0	-
$\Delta\varepsilon$ magnetic field (b)	-	-0.69	-0.61	-0.56	-0.45	-0.21	0.0
(c)	-0.58	-0.53	-0.49	-0.44	-0.37	0.0	-

(a) and (b) - this work: the electric field values are interpolated from measurements at slightly different temperatures.

(c) - evaluated from Diguet et als ref. 4.

Table 1. dielectric anisotropy of MBBA at various temperatures.

Acknowledgements.

We are grateful to the Royal Society and the United Kingdom Science Research Council for financial assistance. One of us (P.G.C.) thanks the Science Research Council for the award of a Research Studentship. We also thank Mr. M.R. Manterfield for preparing and purifying the samples of MBBA.

References.

1. W. Kast, Ann. Physik. 73, 145 (1924)
2. G.H. Brown, J.W. Doane and V.D. Neff, CRC Critical Reviews of Solid State Physics 1, 303 (1970)
3. W. Maier and G. Meier, Z. Naturforsch. 16A, 470 (1961)
4. D. Diguet, F. Rondelez and G. Durand, C.R. Acad.Sc. Paris, 271B, 954 (1970)
5. F. Rondelez, D. Diguet and G. Durand, Mol.Cryst.Liq.Cryst. 15, 183 (1971)
6. G.W. Gray, Molecular Structure and Properties of Liquid Crystals, Academic Press, New York 1962 p.111
7. A. Denat, B. Gosse and J.P. Gosse, Chem.Phys.Lett. 18, 235 (1973)
8. E.F. Carr and C.R.K. Murty, Mol.Cryst.Liq.Cryst. 18, 369 (1972)
9. D.A. Dunmur, Chem.Phys.Lett. 10, 49 (1971)
10. A.D. Buckingham and B.J. Orr, Chem.Soc.Quart.Rev. 21, 195 (1967)
11. E.F. Carr, Mol.Cryst.Liq.Cryst. 7, 253 (1969)
12. W. Helfrich, J.Chem.Phys. 50, 100 (1969).

BULK VISCOSITIES OF MBBA FROM ULTRASONIC MEASUREMENTS

K. A. Kemp* and S. V. Letcher

Department of Physics, University of

Rhode Island, Kingston, R.I. 02881

The propagation of ultrasound in oriented nematic liquid crystals has been shown to be anisotropic.[1-6] The attenuation and, to a lesser extent, the velocity of sound depend on the direction of propagation relative to the preferred molecular orientation direction. The hydrodynamic theory of Forster, et al.[7] has produced an expression for the attenuation coefficient as a function of orientation angle, θ, which is given by

$$\frac{\alpha c^3}{4\pi^2 f^2} = \frac{1}{\rho}\left[\frac{\kappa_\perp(\gamma-1)}{C_p} + (\nu_2 + \nu_4)\right]\sin^2\theta$$

$$+\frac{1}{\rho}\left[\frac{\kappa_\parallel(\gamma-1)}{C_p} + (2\nu_1 + \nu_2 - \nu_4 + 2\nu_5)\right]\cos^2\theta$$

$$-\frac{1}{2\rho}\left[\nu_1 + \nu_2 - 2\nu_3\right]\sin^2 2\theta \qquad (1)$$

Here α is the amplitude attenuation coefficient at frequency f, such that the acoustic pressure would be given by $p(x,t) = p_0 \exp[i(kx - 2\pi ft) - \alpha x]$, c is the speed of sound, $\kappa_\perp$ and $\kappa_\parallel$ are thermal

---

*Present address: Naval Underwater Systems Center, Newport, R.I. 02840

conductivity coefficients, γ is the ratio of specific heats, ρ is the fluid density and ν_1 . . . ν_5 are viscosity coefficients. In liquid crystals, the thermal conductivity terms are usually negligible relative to the viscosity terms.

If the viscosities are independent of frequency, then α is proportional to the frequency squared. This is the behavior expected in any fluid in the absence of relaxation processes. We have previously reported that in para-azoxyanisole and in para-azoxyphenetole α/f^2 is apparently frequency independent between 5-15 MHz.[3] In MBBA, on the other hand, α/f^2 and the anisotropy of α/f^2 are frequency dependent in that frequency range.[1,5,6] It is the purpose of this paper to use the anisotropy data of Wetsel, et al.[5] and new data to be reported here to find the values and the frequency dependences of the absorption coefficient and of the viscosity coefficients of MBBA at 25°C.

The anisotropy of the absorption in MBBA as found by Wetsel, et al. is shown in Fig. 1. The quantity plotted on the vertical axis is proportional to the change in absorption as a function of orientation angle. All measurements are expressed relative to the unknown absorption at $\theta = 90°$. In order to use Eq. 1 to calculate the viscosity coefficients from the absorption data, it is necessary to know the actual values of $\alpha(\theta)$. We have, therefore, measured the absolute absorption coefficient at $\theta = 90°$ in the frequency range from 25-85 MHz. A pulse transmission technique was used with the sample located between two quartz delay lines. Because of the large sound absorption, the sample thickness was never greater than a few millimeters. Molecular orientation was achieved by rubbing the quartz surfaces and by the application of a reinforcing magnetic field of 800 Oe. The results are shown in the lower curve of Fig. 2 for $f \geq 25$ MHz. The upper curve of Fig. 2 (for $f \geq 25$ MHz) was obtained by adding to the lower curve the values of $[\alpha(0°) - \alpha(90°)]/f^2$ from Fig. 1. The data of Martinoty and Candau[8] in unoriented MBBA fall between the two curves as would be expected. Notice that both α/f^2 and its anisotropy increase at lower frequencies. Since a low frequency relaxation appears to be occurring, we have extended the frequency range by measuring $\alpha(\theta)$ at 12.5 and 7.5 MHz. This was done in a system that could be placed in an orienting field of 3000 Oe. The results for $\alpha(90°)/f^2$ and $\alpha(0°)/f^2$ are included in Fig. 2 and the anisotropy is shown in Fig. 3. The shape of the anisotropy is seen to be similar to the lower frequency curves of Fig. 1.

The combined data for $\alpha(\theta)/f^2$ from 7.5 to 85 MHz were used to calculate the ν_i values from Eq. 1. The number of parameters can, fortunately, be reduced by using independent values for ν_2 and ν_3. Gähwiller[9] measured the steady-flow viscosity and the flow alignment of MBBA, from which the zero frequency values of ν_2 and ν_3 could be

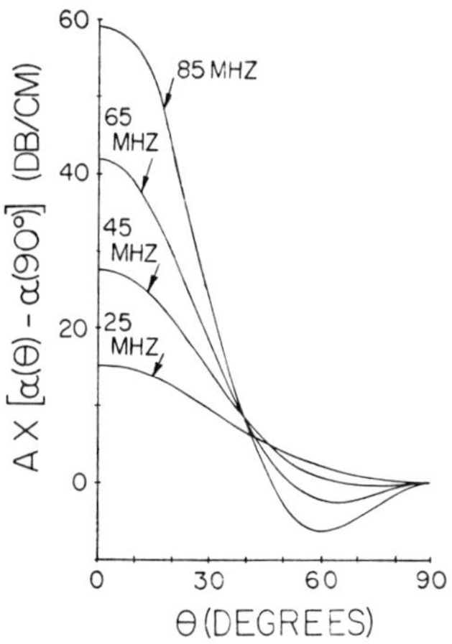

FIGURE 1. Absorption anisotropy in MBBA at 25°C after Wetsel, et al. (ref. 5).

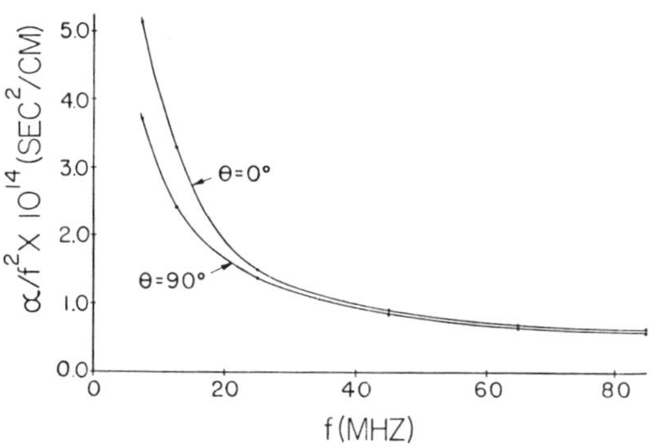

FIGURE 2. Frequency dependence of α/f^2 for $\theta = 90°$ and $\theta = 0$ in MBBA at 25°C. See text for the source of data.

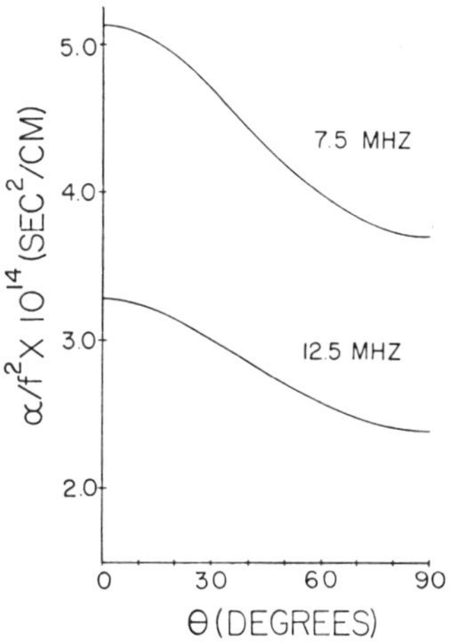

FIGURE 3. Anisotropy of α/f^2 at 7.5 and 12.5 MHz in MBBA at 25°C.

calculated. He obtained 41.6 cP and 23.8 cP for ν_2 and ν_3, respectively. Martinoty and Candau[10] measured the shear impedance of MBBA in the range 15-80 MHz, from which ν_2 and ν_3 could be obtained directly. They found ν_2 = 42 cP and ν_3 = 27 cP, with no frequency dependence. The values of Gahwiller and Martinoty and Candau agree within experimental accuracy and indicate that MBBA undergoes no shear viscoelastic relaxation below $\sim$ 100 MHz. Accordingly, for our calculations we have used the constant values of Martinoty and Candau for ν_2 and ν_3.

A least squares fit to the $\alpha(\theta)/f^2$ data at each frequency resulted in values for the three coefficients of the trigonometric functions of θ in Eq. 1. From these coefficients and the known values of ν_2 and ν_3, corresponding values of ν_1, ν_4 and ν_5 were obtained as shown in Fig. 4. Notice that ν_1 is obtained from the smallest coefficient in Eq. 1 and is considered to be the least reliable of the viscosity coefficients. Since ν_1 is a shear coefficient (it relates the dissipative part of the tensile stress to the tensile strain rate parallel to the molecular axis for an incompressible fluid), it would not be expected to exhibit relaxation at these low frequencies. The effect of ν_1 remaining constant at its high frequency limiting value over the entire frequency regime would not change ν_4 and would increase the value of ν_5 by less than 5% at low frequencies and less than 1% at high frequencies. Hence ν_4 and ν_5 are relatively insensitive to the uncertainty in ν_1.

The values of α/f^2, ν_4 and ν_5 can be fit to a relaxation equation of the form

$$g(f) = B + \frac{A}{1 + (2\pi f)^2\tau^2} , \qquad (2)$$

which involves a single relaxation time, τ. The values of A, B and τ are shown in Table I.

TABLE I

g(f)	B	A	$\tau \times 10^8$ sec
$\alpha(0)/f^2$	549×10^{-17} sec^2/cm	7353×10^{-17} sec^2/cm	1.65
$\alpha(90°)/f^2$	559×10^{-17} sec^2/cm	4790×10^{-17} sec^2/cm	1.56
ν_4	0.17 P	5.1 P	1.55
ν_5	0.01 P	6.1 P	1.58

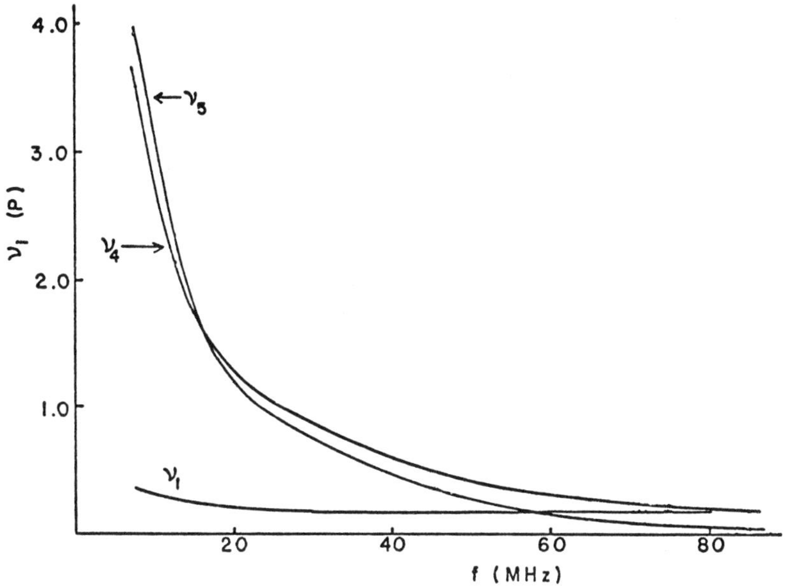

FIGURE 4. Frequency dependence of the viscosity coefficients ν_1, ν_4 and ν_5 of MBBA at 25°C. The shear coefficient ν_1 is essentially constant, while the bulk coefficients ν_4 and ν_5 exhibit a marked relaxation.

Notice that ν_4 and ν_5 have nearly identical relaxation times and that their high frequency limiting values reduce to the size of the shear viscosities or smaller. This latter observation indicates that any higher frequency relaxations are much smaller than the case reported here. The characteristic relaxation strengths and time are consistent with those at a thermal relaxation.[11] This is compatible with the work of Martinoty and Candau[8], who observed this relaxation in the isotropic liquid as well as in the unoriented nematic.

Measurements at lower frequencies would be useful to obtain better values for the relaxation times and strengths for α/f^2 and ν_i and, indeed, to determine if a single relaxation process is sufficient to fit the data. It should be noted that Natale and Commins[6] measure $[\alpha(0) - \alpha(90°)]/f^2$ at 3.5 MHz to be 0.85×10^{-14} sec^2/cm, which is about the size of our value at 12.5 MHz but smaller than our value at 7.5 MHz. If all data are correct, this would mean the curves of Fig. 2 begin to converge at low frequencies. Since no absolute absorption measurements exist at 3.5 MHz Natale and Commins' data are not included in Fig. 2.

This work was supported, in part, by the National Science Foundation and by the Naval Underwater Systems Center, Newport, R.I.

REFERENCES

1. A.E. Lord and M.M. Labes, Phys. Rev. Letters 25, 570 (1970).
2. E.D. Lieberman, J.D. Lee and F.C. Moon, Appl. Phys. Letters 18, 280 (1971).
3. K.A. Kemp and S.V. Letcher, Phys. Rev. Letters 27, 1634 (1971).
4. M.E. Mullen, B. Luthi and M.J. Stephen, Phys. Rev. Letters 28, 799 (1972).
5. G.C. Wetsel, Jr., R.S. Speer, B.A. Lowry and M.R. Woodard, J. Appl. Phys. 43, 1495 (1972).
6. G.G. Natale and D.E. Commins, Phys. Rev. Letters 28, 1439 (1972).
7. D. Forster, F.C. Lubensky, P.C. Martin, J. Swift and P.S. Pershan, Phys. Rev. Letters 26, 1016 (1971).
8. P. Martinoty and S. Candau, C. R. Acad. Sci., Ser. B 271, 107 (1970).
9. Ch. Gahwiller, Phys. Letters 36A, 311 (1971).
10. P. Martinoty and S. Candau, Mol. Cryst. and Liquid Cryst. 14, 293 (1971).
11. See, e.g., R.T. Beyer and S. V. Letcher, Physical Ultrasonics, Academic Press, New York (1969).

ORDER PARAMETERS AND CONFORMATION OF NEMATIC *p*-METHOXYBENZYLIDENE-*p*-n-BUTYLANILINE (MBBA) BY NMR STUDIES OF SOME SPECIFICALLY DEUTERATED DERIVATIVES

Y. S. Lee, Y. Y. Hsu and D. Dolphin

Department of Chemistry, Harvard University

Cambridge, Massachusetts 02138

INTRODUCTION

It is known that nematic liquids have a well defined degree of order.[1] The order parameter, S, which describes the fluctuation of the molecular axis from the direction of preferential orientation of the molecule, is given by

$$S = 1/2\ (3\cos^2\xi - 1) \qquad (1)$$

where ξ is the angle between the long axis of the molecule and the direction of its preferential orientation in the nematic phase. For complete order, $\cos^2\xi = 1$ and S=1 as in the case of a crystal, whereas for complete disorder, $\cos^2\xi = 1/3$ and S=0, representing an isotropic liquid. Thus the order parameter of a nematic will lie between 0 and 1 in a fluid. When a molecule is aligned in a magnetic field, H_o, each nuclear magnetic dipole will produce an additional field at neighboring nuclei, the component of which along the direction of H_o together with H_o will result in a total effective field[2]

$$H_{eff} = H_o \pm \alpha(3\cos^2\theta - 1) \qquad (2)$$

where α is an interaction field parameter and θ is the angle between H_o and the line joining the two interacting nuclei.

Equation (2) predicts a pair of resonance lines symmetrically disposed about the field value at which a single resonance line would occur in the absence of the additional field due to the neighboring nuclei. The separation, δH, of this doublet is thus

$$\delta H = 2\ |H_{eff} - H_o| = 2\alpha(3\cos^2\theta - 1) \quad (3)$$

For proton dipole-dipole interaction, $\alpha = 3/2\mu_H r_{H-H}^{-3}$, where μ_H is the proton nuclear moment (1.42×10^{-23} erg/gauss) and r_{H-H} is the distance between the two interacting protons. Thus, from equation (3), the separation of the doublet as a result of the magnetic dipole-dipole interaction of a pair of protons H_j and H_k held in a rigid orientation is

$$\delta H_{jk} = 3\ \mu_H r_{jk}^{-3}\ (3\cos^2\theta_{jk} - 1) \quad (4)$$

Molecules of a nematic liquid in a magnetic field of a few thousand gauss are aligned with their long axes approximately parallel to the field.[3] The angular-dependence term in equation (4) for a nematic liquid should be replaced by a mean value for the motions involved. Therefore,

$$\delta H_{jk} = 3\ \mu_H\ r_{jk}^{-3}\ \langle 3\cos^2\theta_{jk} - 1\rangle \quad (5)$$

This angular-dependence term can be evaluated by considering all possible ordering factors in the molecule with reference to the direction of the applied magnetic field.

Consider a molecule of MBBA-d_{17} lying arbitrarily with its long axis $\overrightarrow{OL}$ making an angle ξ with the preferred axis $\overrightarrow{OP}$ in a magnetic field H_o, as represented in Figure 1. The thermodynamic average term can be represented by a number of orientation terms:[4]

$$\langle 3\cos^2\theta_{jk} - 1\rangle = (3/2\cos^2\gamma - 1/2)\ (3/2\cos^2\phi - 1/2) \quad (6)$$
$$\times (3/2\cos^2\xi - 1/2)\ (3\cos^2\theta_o - 1)$$

where θ_o is the angle between the applied magnetic field and the preferred orientation of the molecule, γ is the angle between the

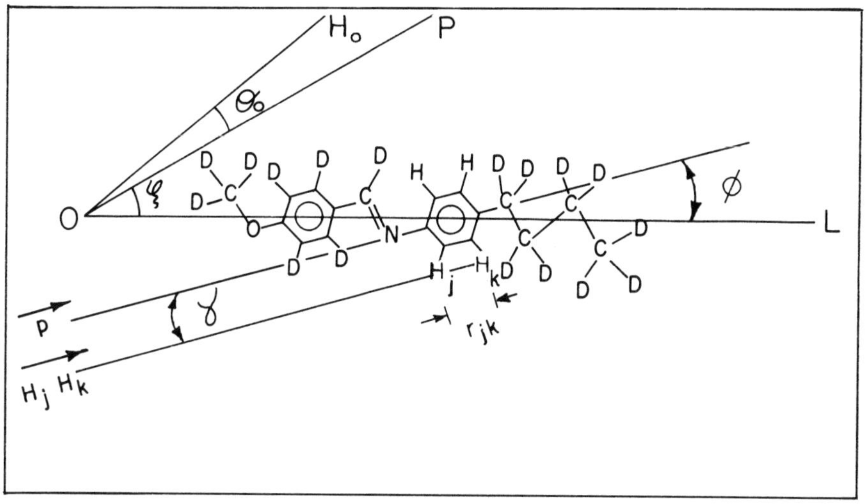

Figure 1. A Molecule of MBBA-d_{17} in the magnetic field. $\vec{OH_o}$: applied magnetic field; $\vec{OP}$: preferred orientation of the molecule; $\vec{OL}$: long axis of the molecule; $\vec{p}$: para axis of the benezene ring; $\vec{H_jH_k}$: direction of line joining two interacting protons H_j and H_k; r_{jk}: distance between two interacting protons.

para axis of the benzene ring and the line joining the two interacting protons H_j and H_k, ϕ is the angle between the para axis of the benzene ring and the long axis of the molecule.
By definition the angular term of ξ is the degree of order or the order parameter S. Such that:

$$S = (3/2\cos^2\xi - 1/2) \qquad (7)$$

In a magnetic field over 2000 gauss, the preferred orientation of the molecules of a nematic liquid is along the direction of magnetic field[4] and $\theta_o=0^o$. For the dipole-dipole interaction between the two protons H_j and H_k (Figure 1), $\gamma=0^o$. Therefore, in a strong magnetic field, equation (5) becomes

$$\delta H_{jk} = 4\alpha\ (3/2\ \cos^2\phi - 1/2)\ S \qquad (8)$$

If partially deuterated MBBA-d_{17} is considered, and taking[5] $r_{jk}=2.45\mathring{A}$ and $\phi = 10^o$, the order parameter for the nematic liquid can be expressed, from equation (8), as a linear function of the dipole-dipole splitting, δH_{jk}, of the interacting protons H_j and H_k; such that

$$S = \delta H_{jk}/5.503 \qquad (9)$$

RESULTS AND DISCUSSION

In the isotropic liquid state, the wide line NMR absorption spectra of both MBBA-d_0 (I) and MBBA-d_{17} (II) consist of two groups of side bands (Figure 2),

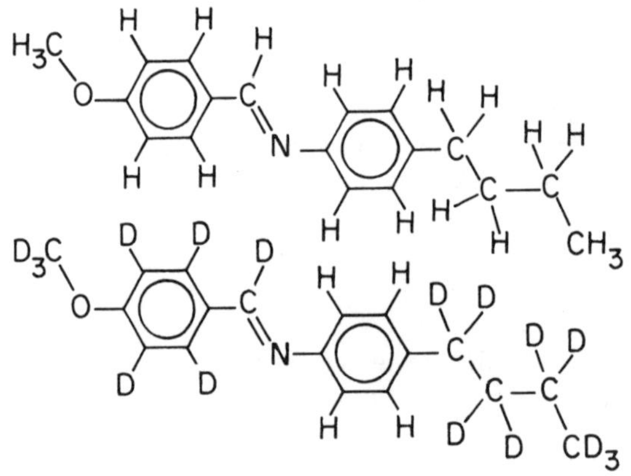

MBBA-d_0 (I)

MBBA-d_{17} (II)

with a separation of twice the modulation frequency[6] (2x14,000 Hz). A comparison of these spectra with the high resolution NMR absorption spectrum of MBBA-d_0 (Figure 3) allows each absorption peak to be identified. The weak signals on the high field side of the strong aromatic proton peak in the spectrum of MBBA-d_{17} are due to a trace of alkyl protons that have not been completely replaced by deuterons.

In the nematic phase, the recorded wide line NMR spectra of the two species of MBBA show magnetic dipolar splittings as shown in Figure 4. In this figure, only one of the two side bands is reported for each species of the nematogen. The other unreported side band is identical to it in shape but inverse in intensity, as in Figure 2. At all temperatures, both in the nematic and in the isotropic ranges the two side bands are separated by twice the modulation frequency.

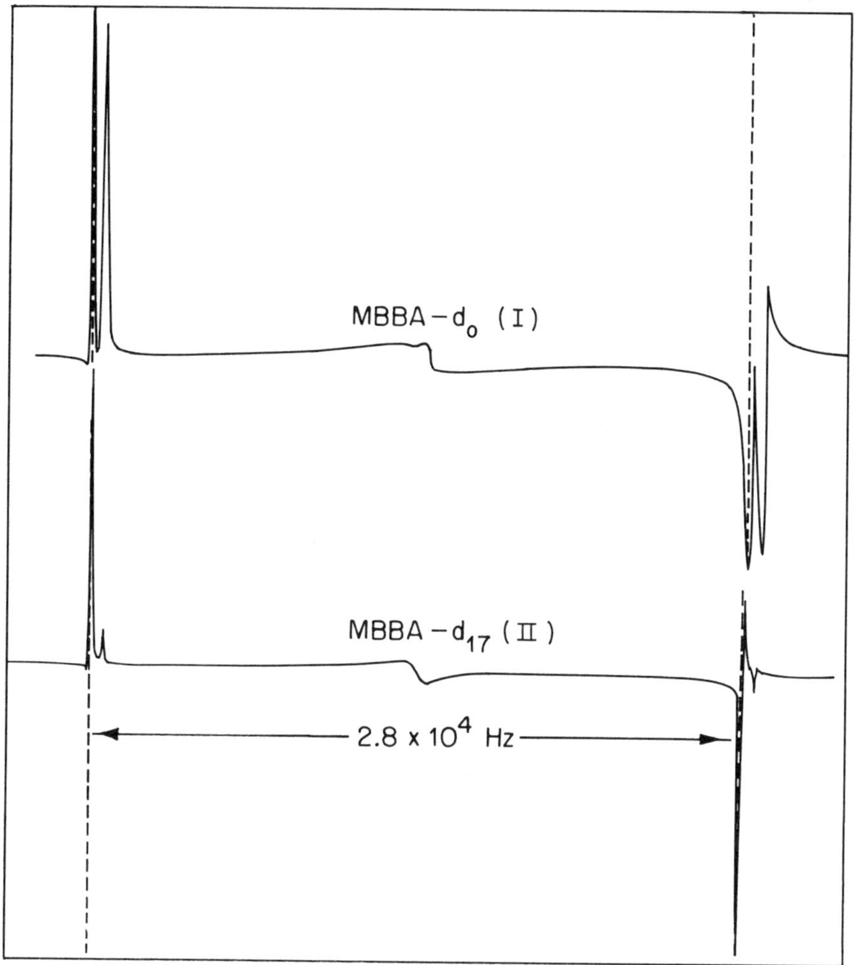

Figure 2. Wide line NMR absorption spectra of MBBA-d_o (I), MBBA-d_{17} (II) in the isotropic liquid state at 46°C, with a 14, 000 Hz. modulating frequency.

The spectrum of MBBA-d_{17} (Figure 4) consists of a well defined doublet which is assigned to the dipolar coupling of the ortho protons[7] on the aniline ring. It can be seen in Figure 4 that the spectrum of MBBA becomes more complicated when additional protons, other than these on the aniline ring, are present. The spectrum of MBBA-d_o has a large central peak assigned to the alkyl protons, and a rather broad symmetrical doublet, and the dipolar splitting of the adjacent aromatic protons become larger, due presumably to the contribution to the total field from these additional protons.

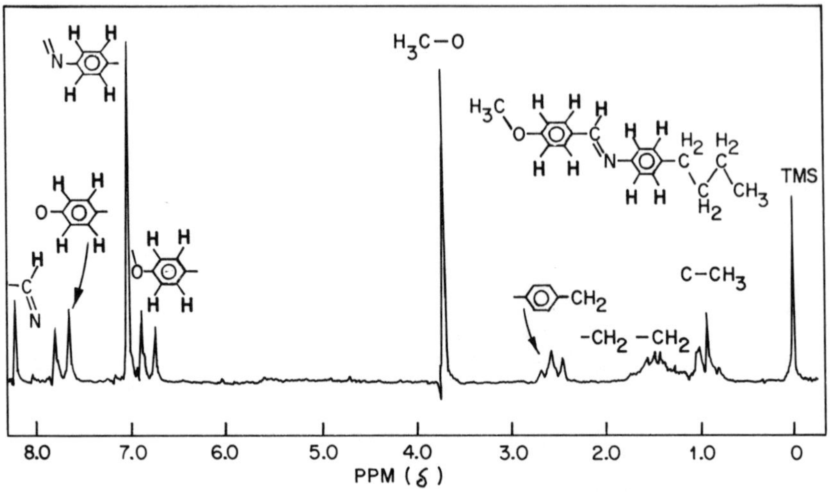

Figure 3. High resolution NMR absorption spectrum of MBBA-d_o (I) in CCl_4.

Thus it can be seen that the order parameter of MBBA calculated from the doublet separation obtained from non-deuterated molecules will yield a value which is larger than the true value, unless adjustment has been made to compensate for this discrepancy.

Separations of the doublet in the NMR absorption spectra of MBBA-d_{17} at different temperatures, together with the order parameters calculated according to equation (9), are shown in Figure 5. The separation of 726 Hz. between the $CHCl_3$ and TMS lines, under the same conditions, was used for the measurement of the dipolar splittings. The data of Watkins and Johnson[7] for non-deuterated MBBA (clearing point: 41°C) obtained by wide line NMR method seem to show considerable higher values for the order parameter than those reported in Figure 5 for MBBA-d_{17}, particularly if values of S of both compounds at temperatures equally below clearing points of the respective compounds are compared.

It has been shown[8] that changing the substitution pattern on the aniline ring does not affect the electronic absorption of the benzylidene ring of a Schiff's base, and at the same time substitution on the benzylidene ring does not affect that of the aniline ring. This suggests that two phenyl rings linked through a -CH=N- bridge are not conjugated and are therefore not coplanar. A similar conclusion has been reached by Anteunis et. al.[9] in their NMR conformational studies of cinnamaldehyde anils.

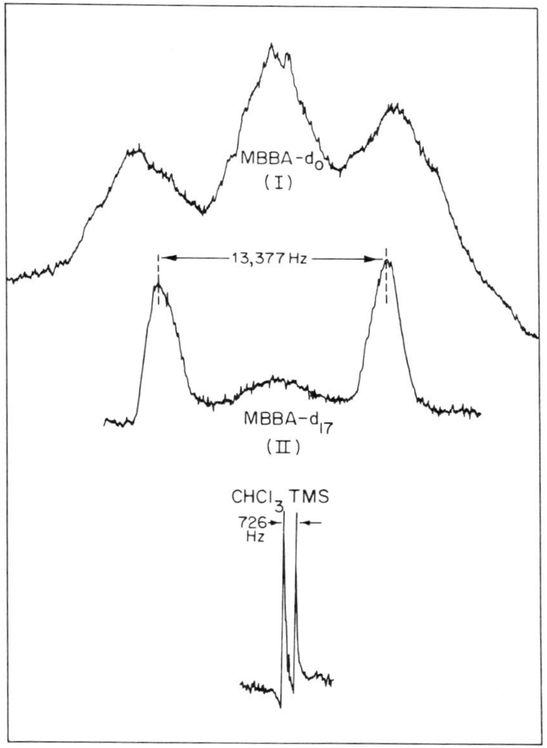

Figure 4. Wide line NMR absorption spectra of MBBA-d_o (I) and MBBA-d_{17} (II) in the nematic state at 30°C. (The separation of 726 Hz for the $CHCl_3$ and TMS signals was used to determine the dipolar separation of all species of MBBA.)

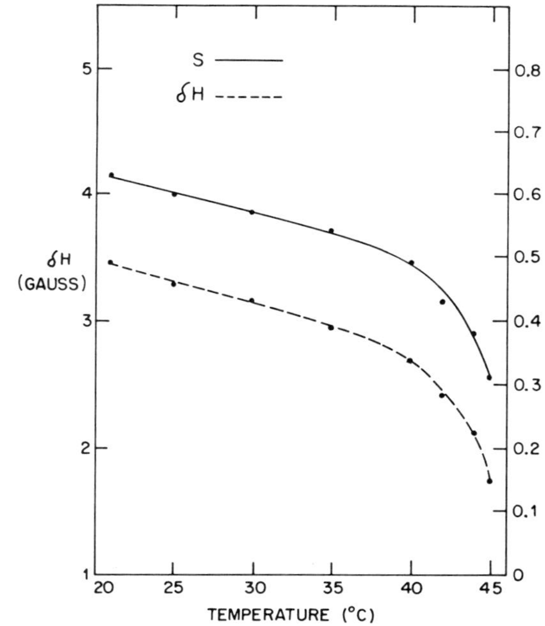

Figure 5. Dipole-dipole splittings and order parameters of MBBA from 21° to 45°C (1 gauss = 4260 Hz).

The conformation of the liquid crystalline Schiff's base telephthal-bis (4-aminofluorobenzene) (III) has been discussed by Bravo et. al.[10] From their data of second moments of the proton

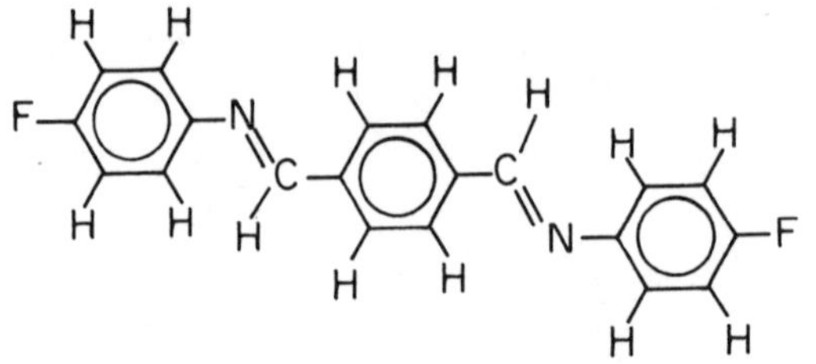

(III)

resonance they estimated that the plane containing the C-C=N-C linkage makes a preferred angle of about 20° relative to the plane of the terminal aromatic ring.

Burgi et. al.[11,12] have reported detailed structures of benzylidene-aniline and its para-substituted derivatives from crystallographic data. Similarities between the crystal reflectance spectra and the solution spectra of these compounds led these authors to conclude that the conformation of these molecules should not be too different in the crystalline and the free molecular states. They reported angles of between 40 and 55° for the twist of the aniline rings from the C-C=N-C plane for three different Schiff's bases.

Since rigid planar configurations and freely rotating terminal benzene rings are both unlikely in a Schiff's base,[10] a preferred angle between the C-C=N-C plane and the aniline ring is to be expected. It seems possible to determine the conformation of MBBA in its nematic state by observing the dipolar interaction between the methine proton and the aromatic protons. Depending upon the extent of twisting of the aniline ring from the linkage plane, the internuclear distance between the methine proton and the closest aromatic protons is comparable to that of the H_j and H_k protons on the aromatic ring. Preliminary observation of the wide line NMR spectrum of MBBA-d_{16} (IV) seems to reveal the dipolar interactions between the methine proton and the protons on the aniline ring.

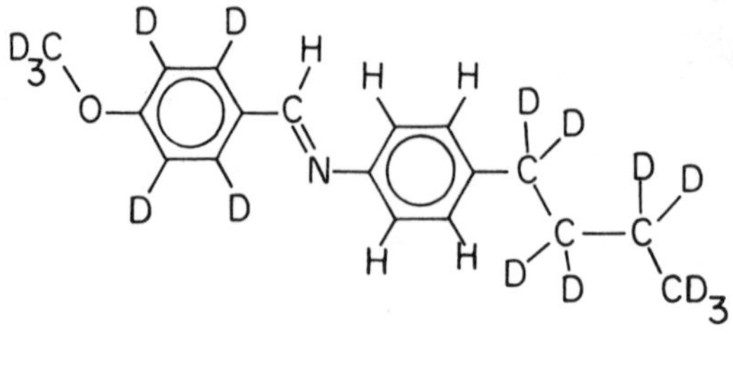

MBBA - d_{16} (IV)

The magnitude of this dipolar coupling seems to indicate that the aniline ring is twisting 47° away from the C-C=N-C linkage plane. However, other species of specifically deuterated MBBA are being synthesized in order to insure that only the dipolar interaction between the methine proton and an aromatic proton closest to it will be observed, so that the detailed conformation of nematic MBBA could be calculated.

EXPERIMENTAL DETAILS

A Varian HA-100 NMR spectrometer was used in this work. The spectra were recorded by sweeping the field, using a modulation frequency of 14,000 Hz. Under such conditions, two well separated side bands could be obtained.[6] The temperature of the probe was controlled by a Varian V6040 variable temperature controller. A methanol probe was used to measure the temperature by replacing the sample with it under identical instrumental conditions over the entire experimental temperature range (21° to 46°C). The stability of the temperature in the probe and the accuracy of the measured temperature were calibrated against a Cu-constantan thermocouple. The maximum fluctuation of the controlled temperatures was found to be less than 0.02°C and the methanol probe allowed the temperature to be measured better than 0.1°C.

A TMS-$CHCl_3$ probe, having separation of the TMS proton line and the $CHCl_3$ proton line of 726 Hz, was used to evaluate dipole-dipole splitting in the NMR absorption spectra of the nematic liquid.

In all cases spectra were recorded on non-spinning samples.

ACKNOWLEDGMENTS

We would like to thank Jane Carlton for assistance in preparation of some of the derivatives used in this work. The work was supported in part by the National Science Foundation under grants GH-33576 and GH-34401, and by the Advanced Research Projects Agency under grant DAHC-15-67-C-0219.

REFERENCES

1. A. Saupe, Angew. Chem. Internat. Edit., 7, 97 (1968).

2. G. E. Pake, J. Chem. Phys., 16, 327 (1948).

3. L. S. Ornstein and W. Kast, Trans Faraday Soc., 29, 93 (1933).

4. K. H. Weber, Ann. Physik, 3, 1 (1959).

5. J. C. Rowell, W. D. Phillips, L. R. Melby, and M. Panar, J. Chem. Phys., 43, 3442 (1965).

6. J. A. Pople, W. G. Schneider and H. J. Bernstein, "High-Resolution Nuclear Magnetic Resonance," McGraw-Hill Book Company, p. 74, 1959.

7. C. L. Watkins and C. S. Johnson, Jr., J. Phys. Chem., 75, No. 16, 2452 (1971).

8. M. Ashraf El-Bayoumi, M. El-Aasser and F. Abdel-Halim, J. Amer. Chem. Soc., 93, 586 (1971).

9. M. Anteunis and A. De Bruyn, J. Magnetic Resonance, 8, 7 (1972).

10. N. Bravo, J. W. Doane, S. L. Arora and J. L. Fergason, J. Chem. Phys., 50, No. 3, 1389 (1969).

11. H. B. Bürgi and J. D. Dunitz, J. Chem. Soc., D, 472 (1969).

12. H. B. Bürgi, J. D. Dunitz and C. Züst, Acta Cryst., B. 24, 463 (1968).

THE ANISOTROPIC ELECTRICAL CONDUCTIVITY OF MBBA CONTAINING ALKYL AMMONIUM HALIDES

Roger Chang

Science Center, Rockwell International

Thousand Oaks, California 91360

ABSTRACT

The anisotropic electrical conductivity of MBBA containing alkyl ammonium halide dopants in the 10^{-6} to 10^{-3} mole/liter concentration range was investigated. The conductivity data suggest the presence of associated cations and solvated anions. The anisotropic conductivity ratio, defined as $(\sigma_{\parallel}-\sigma_{\perp})/(1/3\sigma_{\parallel}+2/3\sigma_{\perp})$, increases with decreasing radius of the halide ion and is shown to be an important physical parameter determining the figure of merit of the liquid crystal composition used in alphanumeric display devices.

I. INTRODUCTION

We have reported on the anisotropic electrical conductivity of MBAA containing tetrabutyl ammonium tetraphenyl boride in a previous investigation.(1) The rapid advancement of "dynamic scattering" mode liquid crystal devices in a variety of alphanumeric display applications motivated us to continue investigating the problem in greater detail. "Dynamic scattering" of nematic liquid crystals having negative low-frequency dielectric anisotropy such as MBBA requires the presence of a sufficient amount of ions.(2,3) The cut-off frequency,(4) or the critical frequency (at a given applied voltage) above which "dynamic scattering" is quenched, is closely related to the ionic conductivity and is very temperature sensitive. It has been encountered in practice that a nematic liquid crystal mixture has a low nematic to crystalline transition temperature but fails to operate at these low temperatures due to a lack of sufficient ionic conductivity. Ionic dopants added to a nematic liquid crystal also lowers its order parameter and the nematic to isotropic transition temperature; the amount of lowering increases with increasing amounts of dopant present in the liquid crystal. The dopants must therefore be selectively chosen which have relatively high dissociation constant and produce at least one of the ion species having high mobility so that a minimum concentration of the dopant is used in order to avoid any appreciable lowering of the nematic to isotropic transition temperature. A good strategy might be the choice of a big and a small ion, the former giving high dissociation and the latter high mobility, provided that association does not occur. A detailed knowledge of the dopant molecular structure, degree of dissociation, and concentration and mobility of the dissociated ion species through careful studies and analyses of these parameters is essential. This is the first objective of the investigation.

Secondly, the continuum theory of the electrohydrodynamics of nematic liquid crystals, though still in its infancy, suggests that the anisotropic conductivity ratio is an important physical parameter determining the figure of merit of a given liquid crystal mixture for display device applications. A detailed investigation of the effect of various kinds and amounts of ionic dopants in the nematic liquid crystal host on the anisotropic electrical conductivity and alphanumeric display cell performance is the second objective of the investigation.

Thirdly, our results suggest that the "purity" of MBBA prepared by different methods varies and affects importantly the association behavior of a given ionic dopant. Time only permits a very brief discussion of this subject.

II. DESCRIPTION OF MBBA SOLVENTS AND DOPANTS

It has been our frustrating experience that the ionic conductivity of MBBA containing small concentrations (10^{-6} to 10^{-3} mole/liter) of ionic dopants having either small cations (such as the alkalis) or anions (such as the halides) is very sensitive to the "purity" of the MBBA solvent. It is difficult at present to define the "purity" of MBBA even qualitatively. Grossly speaking, there appear to be two principal kinds of impurities: the ionic and the non-ionic. For instance, MBBA prepared by distillation contains more ionic impurities than that obtained by zone-refining. The nematic to isotropic transition temperature of MBBA prepared in our laboratory by zone-refining is considerably lower than that prepared by distillation. Aging of MBBA causes both a gradual decrease of its electrical resistivity and nematic to isotropic transition temperature. The reasons for these disturbing influences are not thoroughly understood at present. Rather than belaboring the difficult task of characterizing the "purity" of MBBA from different sources we choose the easier task of identifying the sources of MBBA used in our conductivity measurements:

MBBA-Z-1 - Zone-refined in our laboratory, electrical resistivity 3.3×10^{10} ohm-cm, T_{NI} (nematic to isotropic transition) 44°C.

MBBA-EK - Purchased from Eastman Kodak, electrical resistivity 4×10^{9} ohm, T_{NI} 45°C.

MBBA-D-1 - Distilled in our laboratory, electrical resistivity 1×10^{10} ohm-cm, T_{NI} 45.2°C.

MBBA-D-2 - Distilled in our laboratory, electrical resistivity 1.1×10^{10} ohm-cm, T_{NI} 46°C.

The following ionic dopants were used in this study:

Tetra-alkyl ammonium halides } As purchased from Eastman
Dimethyl-dioctadecyl ammonium bromide } Kodac (reagent grade).

Tetrabutyl ammonium perchlorate - polarographic grade

III. PRESENTATION OF RESULTS

The variation of room-temperature angular average ionic conductivity of MBBA containing tetra-alkyl ammonium chloride in MBBA from different sources is illustrated by the σ-$\sqrt{C}$ plots shown in Figure 1, where σ is the angular average conductivity ($\sigma = 1/3\sigma_{\parallel} + 2/3\sigma_{\perp}$) and C is the dopant concentration in mole/liter. The following points are noted from a study of the results shown in Figure 1 in conjunction with the associated anisotropic conductivity data:

(a) When the σ-$\sqrt{C}$ plot is linear in the concentration range studied, the anisotropic conductivity ratio ($A_{\sigma} = (\sigma_{\parallel}-\sigma_{\perp})/(1/3\sigma_{\parallel} + 2/3\sigma_{\perp})$) is invariant with dopant concentration. This suggests that the ion species present are stable and do not change within the dopant concentration range investigated.

(b) When the σ-$\sqrt{C}$ plot is not linear, A_{σ} also changes with the dopant concentration. In these occasions it has always been observed that both $d\sigma/d\sqrt{C}$ (slope of the σ-$\sqrt{C}$ plot) and A_{σ} increase with increasing dopant concentration C. Typical σ and A_{σ} values for MBBA-EK containing various amounts of tetra-alkyl ammonium halides are illustrated in Table I.

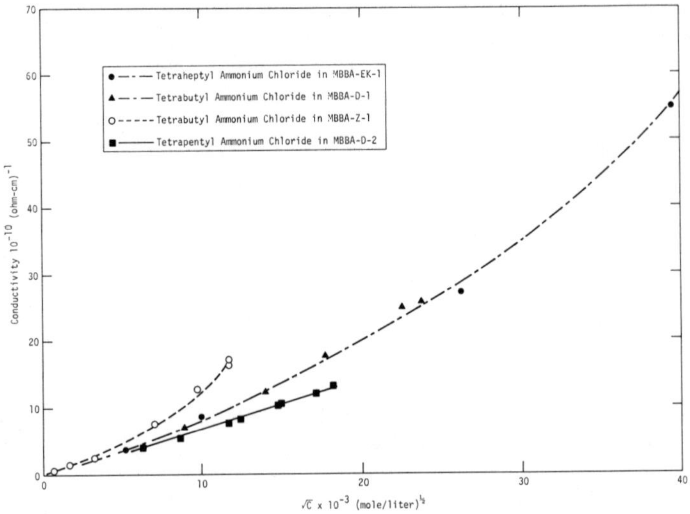

Fig. 1: Angular Average Ionic Conductivity of MBBA from Various Sources Containing Alkyl Ammonium Chloride versus Square-Root of Concentration of the Dopant.

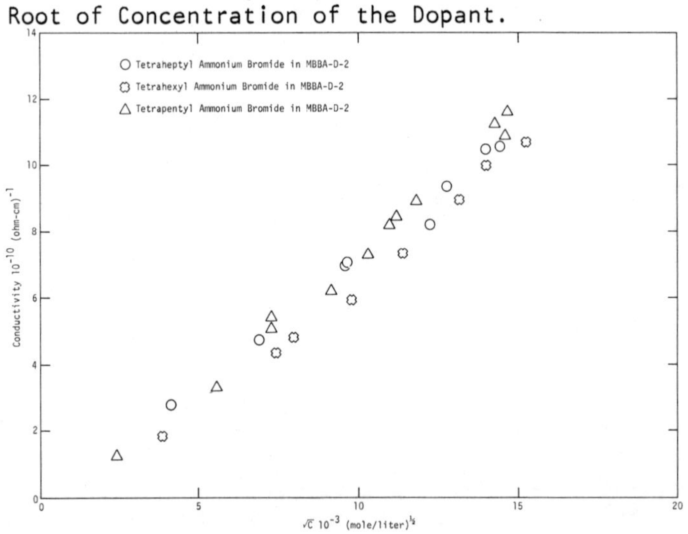

Fig. 2: Angular Average Ionic Conductivity of MBBA-D-2 versus Square-Root of Concentration of Alkyl Ammonium Bromide (Data from Table II, (A), (B) and (C)).

Fortunately one of the solvent source MBBA-D-2 yields linear σ-$\sqrt{C}$ plots for all the alkyl ammonium halides studied in the concentration range 10^{-5} to 10^{-3} mole/liter. The results are summarized in Table II. The σ-$\sqrt{C}$ plots are shown in Figures 2 and 3. Figure 2 indicates that for the tetra-alkyl ammonium bromides

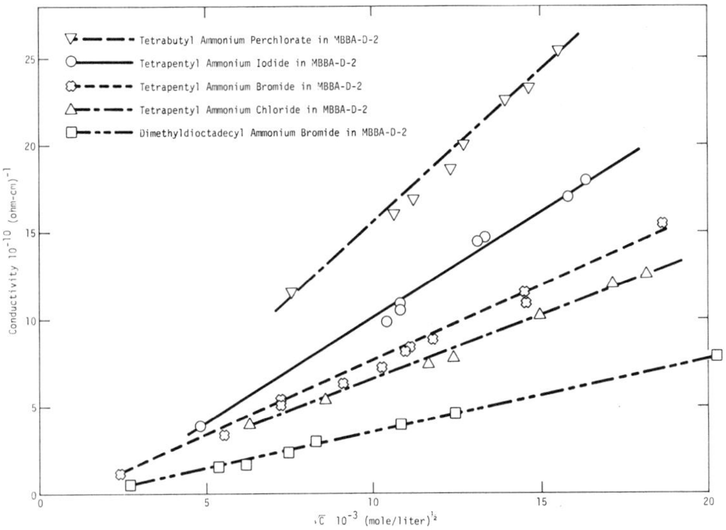

Fig. 3: Angular Average Ionic Conductivity of MBBA-D-2 versus Square-Root of Concentration of Tetrapentyl Ammonium Halides (Table II, (C), (D) and (E)), Tetrabutyl Ammonium Perchlorate (Table II, (F)), and Dimethyl-dioctadecyl Ammonium Bromide (Table II, (G)).

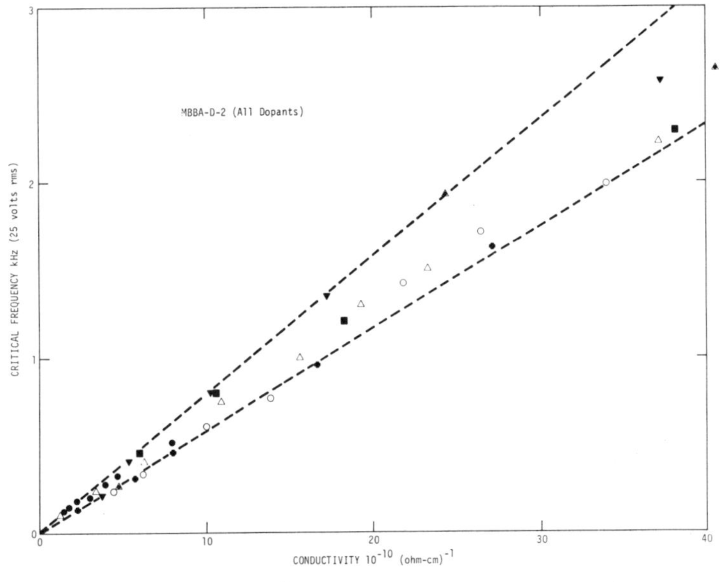

Fig. 4: Critical Frequency (excitation 25 volts rms) Beyond which "Dynamic Scattering" is Quenched versus the Angular Average Ionic Conductivity of Variously Doped MBBA-D-2.

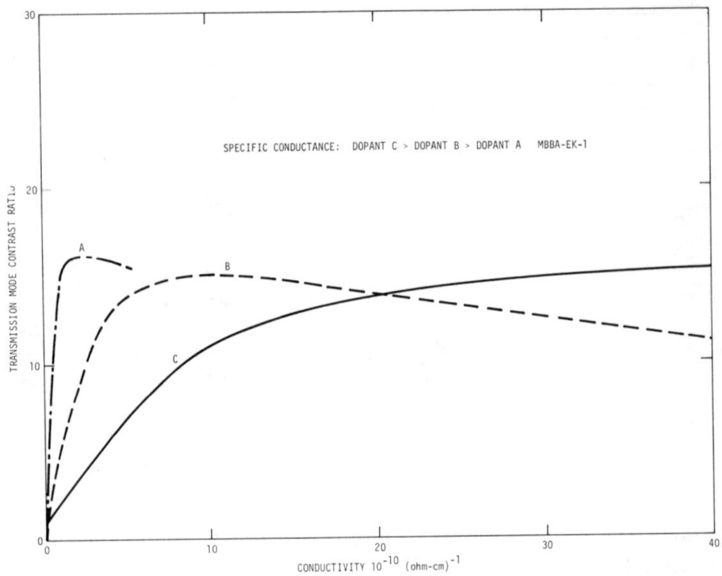

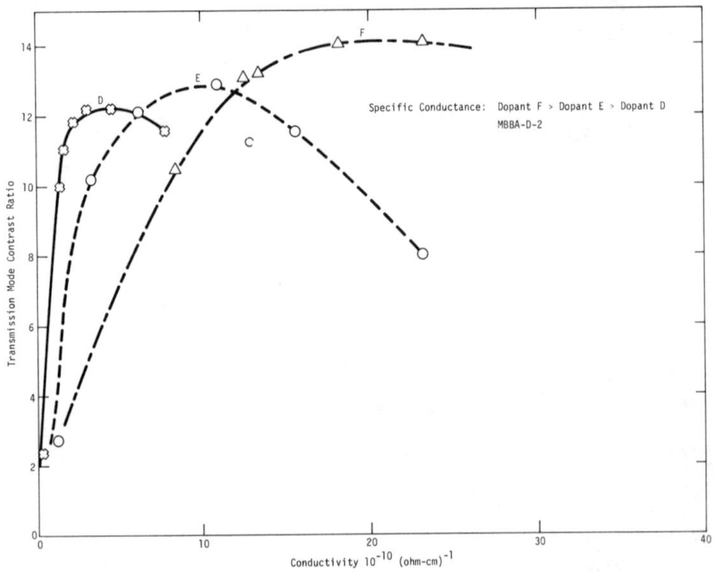

Fig. 5A: Transmission Mode Contrast Ratio (excitation at 25 volts rms, 60 Hz) of 19 micron-Cell Containing MBBA-EK having Various Specific Conductances.

Fig. 5B: Transmission Mode Contrast Ratio (excitation at 25 volts rms, 60 Hz) of 19 micron-Cell Containing MBBA-D-2 having Various Specific Conductances.

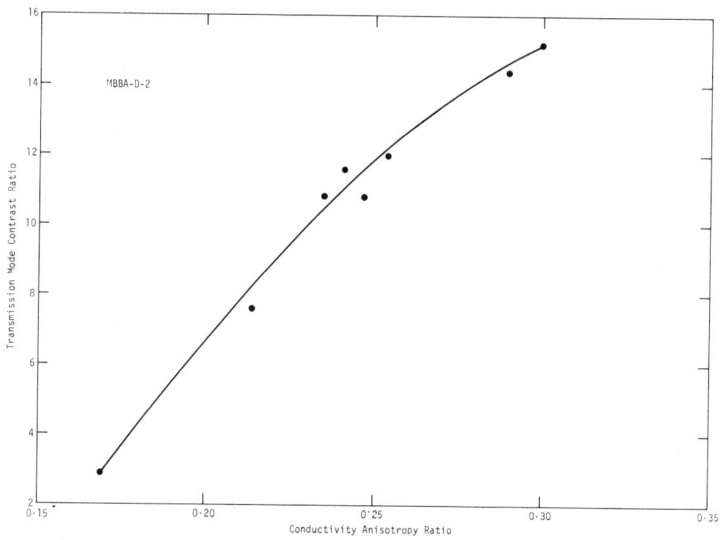

Fig. 6: Maximum Transmission Mode Contrast Ratio (excitation at 25 volts rms, 60 Hz) of 19 micron-Cells Filled with Doped MBBA-D-2 versus Anisotropic Conductivity Ratios of the Liquid Crystal.

$(C_nH_{2n+1})_4NBr$ the σ-$\sqrt{C}$ plots essentially coincide whether n equals 5, 6, or 7. On the other hand, for the alkyl ammonium halides the σ-$\sqrt{C}$ plots are sensitive to variations of the size of the halide ions.

All the conductivity data taken together, we find a linear correlation between the critical frequency and angular average conductivity in reasonable expectation of the continuum model formulated by de Gennes [4] and co-workers. The results for all the dopants studied in MBBA-D-2 are summarized in Figure 4.

We made cells (spaced with 3/4 mil Mylar) with these MBBA compositions and determined the transmission mode contrast ratio (TMCR, defined as the ratio of transmitted intensity of collimated monochromatic light at 5000°A without and with the application of 25 volts rms at 60 Hz) of these cells. When TMCR is plotted versus the angular average conductivity for a given dopant, the characteristics of various dopants are clear. These are illustrated in Figures 5A and 5B for, respectively, the MBBA-EK and the MBBA-D-2 solvents. The maximum TMCR (peak values in Figure 5B) is plotted versus the anisotropic conductivity ratio for the various alkyl ammonium halides in the MBBA-D-2 solvent (where the ion species are stable and the anisotropic conductivity ratio is invariant with dopant concentration) in Figure 6. The importance of the parameter A_σ in determining the figure of merit of a given liquid crystal composition is clearly brought out in Figure 6. Quantitative interpretation of these observations must await a parallel develop-

TABLE I. Anisotropic Electrical Conductivity (295.2°K) Data of MBBA-EK Containing Alkyl Ammonium Halides

(A) - Tetrapentyl Ammonium Bromide

C*	σ*	A_σ*
9.3	4.1	0.27
33.2	6.5	0.26
114.7	11.7	0.26
242.5	18.1	0.30
492.7	28.2	0.32
660.4	40.2	0.33
852.4	45.6	0.37

(B) - Tetra-iso-pentyl Ammonium Iodide

C*	σ*	A_σ*
3.4	2.3	0.17
19.4	5.7	0.21
39.2	8.1	0.25
147.3	16.7	0.23
330.3	27.1	0.27
571.2	46.5	0.29

(C) - Dimethyl-dioctadecyl Ammonium Bromide

C*	σ*	A_σ*
2.1	0.6	0.18
4.4	0.6	0.27
12.2	1.2	0.27
19.6	2.2	0.29
36.7	2.9	0.32
89.3	4.9	0.34

(D) - Tetraheptyl Ammonium Bromide

C*	σ*	A_σ*
19.4	5.3	0.22
70.9	10.3	0.30
217.5	17.2	0.31
649.1	37.2	0.32

* C   Concentration in 10^{-6} mole/liter

σ   Angular average conductivity $(1/3\sigma_{\parallel} + 2/3\sigma_{\perp})$ in $10^{-10}(\text{ohm-cm})^{-1}$

A_σ Anisotropic conductivity ratio $(\sigma_{\parallel} - \sigma_{\perp})/(1/3\sigma_{\parallel} + 2/3\sigma_{\perp})$

ment of the electrohydrodynamics of nematic liquid crystals in the "dynamic scattering" regime.

IV. DISCUSSION

It is beyond the scope of this investigation to discuss the conductivity data of doped MBBA solvents where the σ-$\sqrt{C}$ plots are not linear. It is believed that a prerequisite of such discussion is the capability to characterize the "purity" of MBBA prepared by various methods. Since the MBBA-D-2 solvent yields for all the alkyl ammonium halides: (a) conductivity reproducible to less than 10 percent standard deviation from statistical analyses, (b) σ-$\sqrt{C}$ plots linear, and (c) A_σ invariant with dopant concentration C in the concentration range 10^{-5} to 10^{-3} mole/liter investigated, further discussions are limited to the latter. The procedures to obtain the various parameters are identical to those reported in our first paper.[1] Comparison of the conductivity data reported here for the alkyl ammonium halides and our previously published data for tetrabutyl ammonium tetraphenyl boride strongly suggests that the alkyl ammonium halides are associated into complexes. From a purely speculative point of view we consider the trimer

$$A_3X_3 \rightleftharpoons A_2X^+ + AX_2^-, \quad K = \frac{(C_{A_2X^+})(C_{AX_2^-})}{(C_{A_3X_3})} \tag{1}$$

and the three-dimensional complex

$$A_9X_9 \rightleftharpoons A_5X_4^+ + A_4X_5^-, \quad K = \frac{(C_{A_5X_4^+})(C_{A_4X_5^-})}{(C_{A_9X_9})} \tag{2}$$

where A^+ is the alkyl ammonium ion and X^- is the halide ion. The A_2X^+ and AX_2^- are the trimmers, while the $A_5X_4^+$ and $A_4X_5^-$ ions may be visualized as face-centered tetrahedra. We treat the trimer association first and consider equation (1) as the sum of the following reactions:

$$A_3X_3 \rightleftharpoons 3AX \qquad K_d = \frac{C^3_{AX}}{C_{A_3X_3}} \tag{3}$$

$$AX \rightleftharpoons A^+ + X^- \qquad K_i = \frac{(C_{A^+})(C_{X^-})}{C_{AX}} \tag{4}$$

$$AX + A^+ \rightleftharpoons A_2X^+ \qquad K_{ac} = \frac{C_{A_2X^+}}{(C_{AX})(C_{A^+})} \tag{5}$$

$$AX + X^- \rightleftharpoons AX_2^- \qquad K_{aa} = \frac{C_{AX_2^-}}{(C_{AX})(C_{X^-})} \tag{6}$$

where C_i is the concentration in mole/liter of the i molecular or ion species and K_d, K_i, K_{ac}, K_{aa} are, respectively, the dissocation ($A_3X_3 \rightleftharpoons 3AX$), ionization ($AX \rightleftharpoons A^+ + X^-$) and ion-molecular association ($AX + A^+ \rightleftharpoons A_2X^+$, $AX + X^- \rightleftharpoons AX_2^-$) equilibrium constants. Combining equations (3), (4), (5) and (6) we find,

$$A_3X_3 \rightleftharpoons A_2X^+ + AX_2^- \qquad K = K_d K_i K_{ac} K_{aa} = \frac{(C_{A_2X^+})(C_{AX_2^-})}{(C_{A_3X_3})} \tag{7}$$

which is identical to equation (1).

Let C be the molal concentration of AX which is obtainable directly by weighing. A mass balance requires

$$C = 3C_{A_3X_3} + C_{AX} + \frac{1}{2}(C_{A^+} + C_{X^-}) + \frac{3}{2}(C_{A_2X^+} + C_{AX_2^-})$$

or

$$C = \frac{3}{K}(C_{A_2X^+})(C_{AX_2^-}) + \frac{1}{K_i}(C_{A^+})(C_X) + \frac{1}{2}(C_{A^+} + C_{X^-}) + \frac{3}{2}(C_{A_2X^+} + C_{AX_2^-}) \tag{8}$$

If we assume, in expectation of the experimental observations, the following,

$$\frac{1}{K_i}(C_{A^+})(C_{X^-}) << \frac{3}{K}(C_{A_2X^+})(C_{AX_2^-})$$

and

$$\frac{1}{2}(C_{A^+} + C_{X^-}) << \frac{3}{2}(C_{A_2X^+} + C_{AX_2^-}) \tag{9}*$$

---

*Footnote: Equation (9) is presumably satisfied when $C_{A^+} << C_{A_2X^+}$ and $C_{X^-} << C_{AX_2^-}$.

equation (8) becomes

$$C = \frac{3}{K}(C_{A_2X^+})(C_{AX_2^-}) + \frac{3}{2}(C_{A_2X^+} + C_{AX_2^-})$$
$$C_{A_2X^+} = C_{AX_2^-} \quad . \tag{10}$$

We define the following specific conductances,

$$\Lambda_{A_2X^+} = \lim_{C\to 0} \frac{1000\ \sigma_{A_2X^+}}{C_{A_2X^+}}$$
$$\Lambda_{AX_2^-} = \lim_{C\to 0} \frac{1000\ \sigma_{AX_2^-}}{C_{AX_2^-}} \tag{11}$$
$$\Lambda = \Lambda_{A_2X^+} + \Lambda_{AX_2^-}$$

Combination of equation (10) and (11) yields,

$$\frac{C}{3} = \frac{10^3}{\Lambda}\sigma + \frac{10^6}{K\Lambda^2}\sigma^2 \tag{12}$$

Equation (12) is cast in the same form as equation (7) of reference (1) when C is replaced by C/3. It can be easily shown that if the compex is of form A_nX_n which dissociates into the

$$A_{\frac{n+1}{2}}X_{\frac{n-1}{2}}^{+} \text{ and } A_{\frac{n-1}{2}}X_{\frac{n+1}{2}}^{-} \text{ ions,}$$

the same expression (12) results when C/3 is replaced by C/n. The computed specific conductances Λ and dissociation constants K assuming either of the complexes A_3X_3 and A_9X_9 (see equations (1) and (2)) from the experimental data of Figure 3 and Table II are listed in Table III for the various dopants in MBBA-D-2. Comparing with a specific conductance of 0.4 and dissociation constant of 5.7×10^{-6} for MBBA containing tetrabutyl ammonium (A^+) tetraphenyl boride (B^-) ions, it appears likely that the alkyl ammonium halides could be present as the complex A_9X_9 in MBBA-D-2 solvent.

TABLE II. Anisotropic Electrical Conductivity (295.2°K) Data of MBBA-D-2 Containing Alkyl Ammonium Halides

(A) Tetraheptyl Ammonium Bromide

C*	σ*	A_σ*
17.3	2.79	0.227
47.5	4.70	0.230
92.8	6.96	0.249
93.9	6.99	0.226
149.6	8.20	0.242
163.4	9.35	0.237
195.7	10.42	0.226
205.7	10.57	0.245
		(0.235 ± 0.010)

(B) Tetrahexyl Ammonium Bromide

C*	σ*	A_σ*
15.1	1.80	0.244
55.2	4.33	0.258
63.8	4.83	0.250
95.4	5.92	0.231
129.1	7.34	0.227
173.2	8.94	0.243
196.2	9.98	0.241
233.5	10.69	0.234
		(0.241 ± 0.010)

(C) Tetrapentyl Ammonium Bromide

C*	σ*	A_σ*
53.0	5.43	0.252
53.0	5.07	0.257
105.9	7.30	0.231
120.4	8.19	0.243
124.2	8.44	0.277
139.6	8.93	0.233
203.6	11.27	0.238
212.0	11.62	0.246
		(0.247 ± 0.015)

(D) Tetrapentyl Ammonium Chloride

C*	σ*	A_σ*
39.8	4.08	0.326
74.6	5.41	0.303
136.1	7.56	0.281
154.4	7.83	0.276
216.8	10.09	0.293
225.1	10.32	0.287
294.6	12.02	0.274
329.6	12.67	0.277
		(0.290 ± 0.017)

(E) Tetrapentyl Ammonium Iodide

C*	σ*	A_σ*
23.7	3.95	0.217
109.0	9.83	0.207
117.0	10.87	0.213
164.5	12.42	0.226
199.4	14.46	0.212
206.3	14.73	0.220
251.5	17.01	0.203
268.1	16.99	0.217
		0.214 ± 0.007

(F) Tetrabutyl Ammonium Perchlorate

C*	σ*	A_σ*
58.1	11.6	0.147
114.0	16.0	0.170
126.7	17.1	0.165
152.3	18.6	0.182
162.1	20.1	0.173
193.8	22.6	0.171
215.2	23.4	0.163
241.4	25.5	0.179
		0.169 ± 0.011

TABLE II. (Continued)

(G) - Dimethyl-Dioctadecyl Ammonium Bromide

C*	σ*	A_σ*
29.0	0.46	0.257
38.2	1.43	0.254
55.7	1.69	0.297
69.5	2.31	0.265
116.9	3.02	0.281
157.0	3.96	0.258
416.4	4.66	0.246
986.9	7.93	0.234
		(0.254 ± 0.021)

* C Concentration in 10^{-6} mole/liter

σ Angular Average Conductivity in 10^{-10} $(\text{ohm-cm})^{-1}$

A_σ Anisotropic Conductivity Ratio $(\sigma_{\parallel}-\sigma_{\perp})/(1/3\sigma_{\parallel} + 2/3\sigma_{\perp})$

The assumed stability of the tetrahedral $A_5X_4^+$ and $A_4X_5^-$ ions may have some theoretical justifications. Although we do not know exactly the size of the alkyl ammonium cations, they are of the order of 4 to 5Å.[5] The ionic radii of the halide ions are about 2 Å (Cl^- 1.81 Å, Br^- 1.95 Å, I^- 2.5 Å).[6] The radius ratio X^-/A^+ is about 0.4 to 0.5. The radius ratio cation/oxygen below which a tetrahedral corrdination is stable for the oxides is 0.414.[6] It is possible that in MBBA as solvent, a halide ion might like to be surrounded by four alkyl ammonium ions in the tetrahedral coordination. This, aided by electrostatic charge compensation, might well lead to the formation of tetrahedral $A_5X_4^+$ and $A_4X_5^-$ ions.

It is very difficult to establish the size of the complexes unambiguously without further painstaking investigation. The entropy and free energy changes associated with the formation of $A_5X_4^+$ and $A_4X_5^-$ ions appear to be too large to warrant serious consideration. Furthermore, these ions are nearly symmetrical and will not suffice to explain the significant and monotonic variation of the anisotropic conductivity ratio with the size of the halide ion

TABLE III. Calculated Λ and K Values (295.2°K) for Various Alkyl Ammonium Halide Dopants in MBBA-D-2 Assuming A_3X_3 or A_9X_9 Formation

	$(A_3X_3 \rightleftharpoons A_2X^+ + AX_2^-)$		$(A_9X_9 \rightleftharpoons A_5X_4^+ + A_4X_5^-)$	
Dopant	Λ	K	Λ	K
Tetrapentyl ammonium chloride	0.073	3.2×10^{-6}	0.218	1.1×10^{-6}
Tetrapentyl ammonium bromide	0.078	3.7×10^{-6}	0.232	1.2×10^{-6}
Tetrapentyl ammonium iodide	0.083	6.3×10^{-6}	0.248	2.1×10^{-6}
Tetrabutyl ammonium perchlorate	0.123	6.6×10^{-6}	0.367	2.2×10^{-6}

(Table II and Fig. 6). It is likely that the halide ions may be solvated. A perhaps more realistic hypothesis would be the association of the cation A^+ with the dopant molecule AX and the solvation of the anion X^-:

$$A^+ + AX \rightleftharpoons A_2X^+ \tag{13}$$

$$X^- + S\ (\text{solvent}) \rightleftharpoons XS^- \tag{14}$$

These reactions will not only lower very substantially the mobility, but also offers a plausible clue in regard to the sensitive variation of the anisotropic conductivity ratio with the size of the halide ion. Any quantitative treatment of the experimental data in light of these reactions is difficult at present and becomes more difficult to extend such treatment to the variance of the dopant behavior with MBBA solvents prepared by different methods (Table I and Fig. 1). It appears more important at this stage to discover the factors which lead to the variance of dopant conductivity behavior with solvent "purity". Until this is done, any effort directed to unfold the complex phenomena of dopant ion association dissociation, solvation, etc., seems fruitless and unwarranted.

ACKNOWLEDGEMENTS

Freeman B. Jones, Jr., supplied the zone-refined and distilled MBBA. The purified tetrabutyl ammonium perchlorate was furnished by Ira B. Goldberg. The author wishes also to thank John M. Richardson for helpful discussions.

REFERENCES

(1) Roger Chang and John M. Richardson, paper submitted to Molecular Crystals and Liquid Crystals.

(2) W. Helfrich, J. Chem. Phys. 51, 4092 (1969).

(3) G. H. Heilmeier, L. A. Zanoni and L. A. Barton, Proc. IEEE 56, 1162 (1968).

(4) Orsay Liquid Crystal Group, Phys. Rev. Letters 25, 1642 (1970).

(5) H. Falkenhagen and W. Ebeling, "Ionic Interactions", Edited by S. Petrucci, Academic Press, 1971, Volume 1, Chapter 1.

(6) L. Pauling, "The Nature of Chemical Bond", Cornell University Press, 1960, Chapter X.

CONTINUUM THEORY OF CHOLESTERIC LIQUID CRYSTALS

A. Cemal Eringen and James D. Lee

Princeton University, Princeton, NJ 08540

George Washington University, Washington, D.C. 20006

Abstract

Based on the theory of micropolar viscoelasticity a continuum theory of cholesteric liquid crystals is presented. The balance laws of motion are given and a constitutive theory is derived and restricted by the second law of thermodynamics. Material symmetry restrictions are obtained by means of physical considerations. The theory includes thermomechanical effects of dissipation. The axis of ellipse and coefficient of optical activity are defined to characterize the helical structure of cholesteric liquid crystals. Analytically it is shown that in general the coefficient of optical activity depends on the temperature variations, deformation and mechanical stresses. Several special cases of practical importance are studied in detail. A coupling of longitudinal and twist waves along the axis of helix is investigated theoretically and corresponding experiments are suggested.

1. INTRODUCTION

Liquid crystal is the general name for a certain class of organic substances which has an independent thermodynamic state, called liquid crystalline state or mesomorphous phase, separated from the normal liquid and from the crystalline state by first-order transition points, $[T_1, T_2]$, which are defined by latent heat of transition.[1] Liquid crystals in the mesomorphous phase are ordered on a molecular level hence they are anisotropic in some of their properties (a characteristic normally found only in solid crystals), yet possess mechanical properties of liquids. Molecules that form mesomorphous phases are more or less elongated. Based upon the molecular structures, liquid crystals are conventionally divided into three classes, namely, nematic, smectic, and cholesteric by Friedel.[2] In this work we focus our attention on cholesteric liquid crystals. Figure 1 shows the schematic structures of cholesteric substance.

The molecules in cholesteric liquid crystals are arranged in layers, with each layer, the parallel alignment of molecules is reminiscent of the nematic phase with the long axes of the molecules parallel to the plane of the layers. However, the direction of the long axes of molecules in each layer is displaced slightly from the corresponding direction in adjacent layers. The displacement is cumulative through successive layers, so that the overall displacement traces out a helical path.[3] In order to make the morphology clear, we define the following terminologies:

Axis of Molecular Layer: Since within each layer the molecular axes are parallel to each other, we may define a vector, called axis of molecular layer, to be parallel to the molecular axes and for each layer it is associated with a directional angle θ with respect to some fixed coordinate system.

Axis of Helix: The axis of helix, or optical axis, is the axis perpendicular to the layers, conventionally called X_3-axis, or Z-axis.

Coefficient of Optical Activity: The coefficient of optical activity, Δ, is defined to be the ratio of directional difference between the two axes of two molecular layers to the distance (along the axis of helix between these two layers, i.e.,

$$\Delta \equiv \frac{\theta(Z_1) - \theta(Z_2)}{Z_1 - Z_2} \tag{1.1}$$

In the undeformed state, Δ is a constant and equal to the limiting value

$$\Delta = \frac{d\theta}{dZ} \tag{1.2}$$

If Δ is positive, the substance is said to possess a right-hand helix, otherwise it possesses a left-hand helix.

As in smectic substances, a layer of cholesteric crystals could slide over neighboring layers without hindrance, and, hence, these substances behave as sort of two-dimensional fluids. The axis of helix, layered structure, and coefficient of optical activity characterize the reference state of cholesteric liquid crystals.

Besides the phenomenon of birefringence, cholesteric crystals possess a number of peculiar optical properties due to the unique helical structure of these substances. One of thes is the optical activity: When the plane of polarization of a linearly polarized light strikes perpendicularly to the molecular layers, it is rotated successfully through an angle proportional to the thickness of the transmitting material. The other is circular dichroism, which occurs when ordinary white light is directed at a cholesteric phase splitting into two components, right and left circularly polarized light. Depending on the material, one of these is transmitted and the other is reflected. It is this property that gives iridescent color for cholesteric substances.[3] Also because of the delicately balanced molecular structure, marked changes (in such optical properties as birefringence, circular dichroism, optical activity and color) can be produced in response to subtle variations of temperature, mechanical stresses, chemical environment and electromagnetic radiation, etc.[3]

Since the discovery of liquid crystals by Reinitzer in 1888, the literature has registered extensive entries both on experimental and theoretical grounds. A comprehensive review of the earlier works is to be found in Brown and Shaw.[4] Fergason[3] discussed the molecular structures and optical properties, and Porter et al.[5] summarized the rheological properties of liquid crystals. Fairly complete surveys of the field were presented by Chistyakov[6] and Saupe.[7] DeGennes[8,9] and Meyer[10] suggested possible magnetic effects on cholesteric liquid crystals. Selawry et al.[11] used cholesteric substances for thermographic measurement of skin temperature of the human body to locate veins, arteries, and other internal tissues.

In regard to continuum theories of cholesteric liquid crystals, we have the work of Leslie[12,13] who proposed constitutive equations for cholesteric liquid crystals based on a set of balance laws derived by Ericksen.[14] In Ericksen-Leslie's theory, the essential point is the introduction of a unit vector, $\underline{d}$, denoting the preferred direction inherent in liquid crystals.

The present work is, however, based on the theory of micropolar media introduced by Eringen[15-18]. The basic ideas of micropolar theory and the success of applicability to nematic liquid crystals have been discussed by Lee and Eringen.[19] As far as balance laws are concerned, one can show that the law of "balance of director force" proposed by Ericksen is equivalent to the law of "balance of angular momentum" of micropolar theory if we identify $\rho_1 d_k d_\ell$ with the microinertia tensor $\rho i_{k\ell}$ of Eringen, and $\varepsilon_{jk\ell} d_k \pi_{i\ell}$ by his moment stress m_{ij}. However, the essential difference between the present work and those of Ericksen-Leslie is in constitutive equations. We consider the microrotation as an independent constitutive variable instead of a director. Otherwise the rotation about the director cannot be expressed. Moreover the constitutive equations proposed by Leslie are highly nonlinear, and in the <u>static case</u> it seems that they end up with an overdetermined system.

The purpose of this work is to formulate a set of constitutive equations based on micropolar theory. Afterwards we solve some sample problems with the hope to explain certain optical properties of these substances on a mechanical background. The physical model we have in mind is to regard the cholesteric substance as a continuous medium made up of groups of molecules, called particles, bound together by molecular forces as a unity in such a way that they can perform microrotations about their centers of mass and, of course, macromotions as well. Mathematically these two motions can be expressed by:

Macromotion

(1.3) $$x_k = x_k(\underset{\sim}{X}, t), \qquad \det(x_{k,K}) \neq 0$$

Microrotation

(1.4) $$\xi_k = \chi_{kK}(\underset{\sim}{X}, t)\Xi_k, \qquad (\chi_{kK})^{-1} = (\chi_{kK})^T$$

where $\underset{\sim}{x}$ and $\underset{\sim}{X}$ are respectively the position vectors of centers of mass of a particle in deformed and undeformed states and ξ_k and Ξ_K are respectively the position vectors of points in deformed and undeformed particles with respect to $\underset{\sim}{x}$ and $\underset{\sim}{X}$. The macromotion (1.3) describes how the center of mass moves, and the microrotation (1.4) describes the local orientation. Since the particle size is finite, at each point $\underset{\sim}{x}$ and $\underset{\sim}{X}$ not only a mass density ρ and ρ_0 but also an inertia tensor $i_{k\ell}$ and I_{KL} are defined for deformed and undeformed states. Because of the helical symmetry of cholesteric liquid crystals in the undeformed state, I_{KL} is expected to have the following form

(1.5) $$I_{KL} = \begin{vmatrix} I\cos^2\Theta + J\sin^2\Theta & (I-J)\sin\Theta\cos\Theta & 0 \\ (I-J)\sin\Theta\cos\Theta & I\sin^2\Theta + J\cos^2\Theta & 0 \\ 0 & 0 & K \end{vmatrix}$$

In Section II we present the basic laws of motion and kinematical relations. Thermodynamically admissible nonlinear constitutive equations are developped and linearized in Section III. Restrictions arising from the material symmetry and physical consideration are discussed, and independent material moduli for cholesteric liquid crystals are obtained. The change of optical activity due to temperature variations, deformation, and pressure are presented in Section IV. In Section V, we investigate the propagations of twist waves and the coupling of these waves with the longitudinal waves.

II. LAWS OF MOTION

The basic laws of motion of a micropolar continuum are[15-18]

<u>Conservation of Mass</u>

(2.1) $$\dot{\rho} + \rho v_{k,k} = 0$$

<u>Conservation of Microinertia</u>

(2.2) $$i_{k\ell} = I_{KL}\chi_{kK}\chi_{\ell L}$$

<u>Balance of Linear Momentum</u>

(2.3) $$t_{k\ell,k} + \rho(f_\ell - \dot{v}_\ell) = 0$$

Balance of Angular Momentum

(2.4) $$m_{k\ell,k} + \varepsilon_{\ell mn} t_{mn} + \rho(\ell_\ell - \dot{\sigma}_\ell) = 0$$

Conservation of Energy

(2.5) $$\rho\dot{\varepsilon} = t_{k\ell}(v_{\ell,k} - \varepsilon_{k\ell m}\nu_m) + m_{k\ell}\nu_{\ell,k} + q_{k,k} + \rho h$$

Entropy Inequality

(2.6) $$\rho\dot{\eta} - (q_k/T)_{,k} - \rho h/T \geq 0$$

Inequality (2.6) is postulated to be valid for all independent processes. In equations (2.1) to (2.5) and inequality (2.6):

$\rho \equiv$ mass density

$i_{k\ell} = i_{\ell k} \equiv$ microinertia tensor in deformed state

$I_{KL} = I_{LK} \equiv$ microinertia tensor in undeformed state

$t_{k\ell} \equiv$ stress tensor

$m_{k\ell} \equiv$ moment stress

$\varepsilon \equiv$ internal energy

$q_k \equiv$ heat flux

$\eta \equiv$ entropy

$v_k \equiv$ velocity

$f_k \equiv$ body force

$\ell_k \equiv$ body couple

$\nu_k \equiv$ angular velocity

$h \equiv$ heat source

$T \equiv$ absolute temperature

and σ_ℓ is the spin vector defined by

(2.7) $$\sigma_\ell \equiv \frac{D}{Dt}[(i_{mm}\delta_{k\ell} - i_{\ell k})\nu_k]$$

where $\delta_{k\ell}$ is the Kronecker delta and $\varepsilon_{k\ell m}$ is the permutation tensor. We assume that the thermomechanical process is simple, ie., we take the entropy influx and entropy source to be the heat influx and heat source divided by absolute temperature respectively.

Throughout this paper we employ rectangular coordinate systems x_k, (k=1, 2, 3) and X_K, (K = 1, 2, 3) and assume summation convention over repeated indices. Indices following a comma represent partial differentiation, and a superposed dot or D/Dt indicates material differentiation, e.g.,

$$v_{k,\ell} \equiv \partial v_k/\partial x_\ell, \; x_{k,K} \equiv \partial x_k/\partial X_K, \; \dot{v}_k \equiv Dv_k/Dt \equiv \partial v_k/\partial t + v_{k,\ell}v_\ell$$

Equations (2.1) to (2.6) constitute a set of fourteen euqations involving thirty-seven unknowns since $\underset{\sim}{f}$, $\underset{\sim}{\ell}$, and h are considered to be prescribed. In order to have a complete theory we need a set of twenty-three constitutive equations for ε, η, $t_{k\ell}$, $m_{k\ell}$, and q_k. To this end we choose the motions, temperature, and their by-

products as independent constitutive variables.
Kinematical Relations:

$$v_k \equiv \dot{x}_k, \quad \dot{v}_k = \partial v_k/\partial t + v_{k,\ell} v_\ell$$

$$\nu_k \equiv -\frac{1}{2} \varepsilon_{k\ell m} \dot{\chi}_{\ell K} \chi_{mK}$$

$$C_{KL} \equiv x_{k,K} \chi_{kL}, \quad \Gamma_{KL} \equiv \frac{1}{2} \varepsilon_{KMN} \chi_{kM,L} \chi_{kN}$$

$$\dot{C}_{KL} = (v_{\ell,k} - \varepsilon_{k\ell m} \nu_m) x_{k,K} \chi_{\ell L}$$

$$\dot{\Gamma}_{KL} = \nu_{k,\ell}\, x_{\ell,L} \chi_{kK}$$

The Cosserat deformation tensor C_{KL}, Wryness tensor, Γ_{KL}, and their material derivatives $\dot{C}_{KL}$ and $\dot{\Gamma}_{KL}$, together with temperature gradient, are chosen to be independent constitutive variables which can be shown to be objective, i.e., invariant under time-dependent rigid motions of the spatial frame of reference.

III. CONSTITUTIVE EQUATIONS

Eliminating h between (2.5) and (2.6) we obtain the following entropy inequality which generalizes that of Clausius-Duhem

$$(3.1) \qquad -\rho(\dot{\psi} + \dot{T}\eta) + t_{k\ell}(v_{\ell,k} - \varepsilon_{k\ell m}\nu_m) + m_{k\ell}\, \nu_{\ell,k} + \frac{1}{T} q_k T_{,k} \geq 0$$

where ψ is the free energy defined by

$$(3.2) \qquad \psi \equiv \varepsilon - T\eta$$

Motivated by the facts that liquid crystals have the properties resembling those of fluids and solids, we assume that cholesteric liquid crystals have the constitutive equations reminiscent of micropolar viscoelastic solids generalizing the classical Kelvin Voigt model, Eringen[20-22].

$$\psi = \psi(\underset{\sim}{C}, \dot{\underset{\sim}{C}}, \underset{\sim}{\Gamma}, \dot{\underset{\sim}{\Gamma}}, T, \nabla T)$$

$$\eta = \eta(\underset{\sim}{C}, \dot{\underset{\sim}{C}}, \underset{\sim}{\Gamma}, \dot{\underset{\sim}{\Gamma}}, T, \nabla T)$$

$$(3.3) \qquad t_{k\ell} = \frac{\rho}{\rho_o} T_{KL}(\underset{\sim}{C}, \dot{\underset{\sim}{C}}, \underset{\sim}{\Gamma}, \dot{\underset{\sim}{\Gamma}}, T, \nabla T) x_{k,K} \chi_{\ell L}$$

$$m_{k\ell} = \frac{\rho}{\rho_o} M_{LK}(\underset{\sim}{C}, \dot{\underset{\sim}{C}}, \underset{\sim}{\Gamma}, \dot{\underset{\sim}{\Gamma}}, T, \nabla T) x_{k,K} \chi_{\ell L}$$

$$q_k = \frac{\rho}{\rho_o} Q_K(\underset{\sim}{C}, \dot{\underset{\sim}{C}}, \underset{\sim}{\Gamma}, \dot{\underset{\sim}{\Gamma}}, T, \nabla T) x_{k,K}$$

We now briefly discuss several important features of these equations. First, the same set of independent objective variables are present in all equations (equipresence). Second, they satisfy the

the axiom of objectivity, i.e., they are invariant under arbitrary time-dependent rigid motions of the spatial reference frame. Third we have assumed that cholesteric liquid crystals are homogeneous macroscopically. This simply means, that, macroscopically, the axis of helix appears as an axis of rotational symmetry. But we emphasize that in cholesteric liquid crystals there is no mirror symmetry.

Substituting (3.3) into (3.1) we obtain

$$-\rho_o[(\frac{\partial\psi}{\partial T}+\eta)\dot{T}+\frac{\partial\psi}{\partial T_{,K}}\dot{T}_{,K}+\frac{\partial\psi}{\partial C_{KL}}\dot{C}_{KL}+\frac{\partial\psi}{\partial \dot{C}_{KL}}\ddot{C}_{KL}+\frac{\partial\psi}{\partial \Gamma_{KL}}\dot{\Gamma}_{KL}$$

(3.4)

$$+\frac{\partial\psi}{\partial\dot{\Gamma}_{KL}}\ddot{\Gamma}_{KL}]+T_{KL}\dot{C}_{KL}+M_{KL}\dot{\Gamma}_{KL}+\frac{1}{T}Q_K T_{,K}\geq 0$$

This inequality is postulated to be valid for all the independent variations of T, $\dot{T}$, ∇T, $\nabla\dot{T}$, $\dot{\tilde{C}}$, $\ddot{\tilde{C}}$, $\dot{\tilde{\Gamma}}$, and $\ddot{\tilde{\Gamma}}$. Since (3.4) is linear in $\dot{T}$, $\nabla\dot{T}$, $\ddot{\tilde{C}}$, and $\ddot{\tilde{\Gamma}}$, it cannot be maintained for all values of these quantities unless their coefficients vanish, i.e.,

(3.5)
$$\eta = -\frac{\partial\psi}{\partial T}$$

$$\psi = \psi(\tilde{C}, \tilde{\Gamma}, T)$$

Thus, (3.4) is reduced to

(3.6) $$_dT_{KL}\dot{C}_{KL}+{_dM_{KL}}\dot{\Gamma}_{KL}+\frac{1}{T}Q_K T_{,K}\geq 0$$

where

(3.7)
$$_dT_{KL}\equiv T_{KL}-{_eT_{KL}}\equiv T_{KL}-\rho_o\frac{\partial\psi}{\partial C_{KL}}$$

$$_dM_{KL}\equiv M_{KL}-{_eM_{KL}}\equiv M_{KL}-\rho_o\frac{\partial\psi}{\partial \Gamma_{KL}}$$

Further if $_d\tilde{T}$, $_d\tilde{M}$, and $\tilde{Q}$ are continuous functions of $\tilde{C}$, $\tilde{\Gamma}$, and ∇T, then from (3.6) it follows that

(3.8) $\dot{\tilde{C}} = \dot{\tilde{\Gamma}} = \nabla T = 0$ implies $_d\tilde{T} = {_d\tilde{M}} = \tilde{Q} = 0$

We therefore proved:

Theorem: The general form of constitutive equations (3.3), of cholesteric liquid crystals, is thermodynamically admissible if and only if (3.5) to (3.8) are satisfied.

In order to obtain a set of linear constitutive equations, we expand $W \equiv \rho_o\psi$ as follows:

$$W = W_o + A_{KL}\,\phi_{KL} + B_{KL}\,\Gamma_{KL} + \frac{1}{2} A_{KLMN}\,\phi_{KL}\,\phi_{MN} + \frac{1}{2} B_{KLMN}\,\Gamma_{KL}\Gamma_{MN} + C_{KLMN}\,\phi_{KL}\,\Gamma_{MN} \tag{3.9}$$

where

$$\phi_{KL} \equiv C_{KL} - \delta_{KL} \tag{3.10}$$

and W_o, A_{KL}, B_{KL}, A_{KLMN}, B_{KLMN}, and C_{KLMN} are material moduli, and they are functions of temperature only. The elastic part of T_{KL} and M_{KL} are then obtained as

$$(3.11)\quad \begin{aligned} {}_eT_{KL} &\equiv \frac{\partial W}{\partial C_{KL}} = A_{KL} + A_{KLMN}\,\phi_{MN} + C_{KLMN}\,\Gamma_{MN} \\ {}_eM_{KL} &\equiv \frac{\partial W}{\partial \Gamma_{KL}} = B_{KL} + B_{KLMN}\,\Gamma_{MN} + C_{MNKL}\,\phi_{MN} \end{aligned}$$

We further write linear equations for ${}_d\tilde{T}$, ${}_d\tilde{M}$, and $\tilde{Q}$ in terms of their arguments $\tilde{C}$, $\dot{\tilde{C}}$, $\tilde{\Gamma}$, $\dot{\tilde{\Gamma}}$, and ∇T. Using (3.6) these read

$$(3.12)\quad \begin{aligned} \frac{\rho}{\rho_o}\,{}_dT_{KL} &= a_{KLMN}\dot{C}_{MN} + \alpha_{KLMN}\,\dot{\Gamma}_{MN} + e_{KLM}\,T_{,M} \\ \frac{\rho}{\rho_o}\,{}_dM_{KL} &= b_{KLMN}\dot{\Gamma}_{MN} + \beta_{KLMN}\,\dot{C}_{MN} + f_{KLM}\,T_{,M} \\ \frac{\rho}{\rho_o}\,Q_K &= h_{KL}\,T_{,L} + \gamma_{KLM}\,\dot{C}_{LM} + g_{KLM}\,\dot{\Gamma}_{LM} \end{aligned}$$

where the material moduli $\tilde{a}$, $\tilde{b}$, $\tilde{h}$, $\tilde{\alpha}$, $\tilde{\beta}$, $\tilde{\gamma}$, $\tilde{e}$, $\tilde{f}$, and g are functions of temperature only. The equation of heat conduction follows from the conservation law for energy and it reads:

$$\rho T\dot{\eta} = \rho h + q_{k,k} + \frac{\rho}{\rho_o}\left({}_dT_{KL}\,\dot{C}_{KL} + {}_dM_{KL}\,\dot{\Gamma}_{KL}\right) \tag{3.13}$$

For a linear theory, we further assume the strain, rotation, and the temperature variation from a reference temperature T_o are small. i.e.,

$$(3.14)\quad \begin{aligned} &T = T_o + \bar{T}, \qquad |\bar{T}| \ll T_o \\ &x_{k,K} \simeq \delta_{kK} + u_{k,\ell}\,\delta_{\ell K} \\ &\chi_{kK} \simeq \delta_{kK} - \varepsilon_{k\ell m}\,\phi_m\,\delta_{\ell K} \end{aligned}$$

where u_k is the displacement vector and ϕ_k is the angle of rotation about the x_k axis. Correspondingly, we have

(3.15)
$$W_o = S_o - \rho_o \eta_o \bar{T} - \frac{\rho_o \gamma}{2T_o} \bar{T}^2$$
$$A_{KL} = -a_{KL}\bar{T}, \qquad B_{KL} = -b_{KL}\bar{T}$$

where S_o, η_o, γ, a_{KL}, b_{KL}, and all other material moduli are functions of reference temperature T_o only. Finally we obtain the equations of motion and constitutive equations in the following linear spatial forms

$$t_{k\ell,k} + \rho f_\ell = \rho \ddot{u}_\ell$$
$$m_{k\ell,k} + \varepsilon_{\ell mn} t_{mn} + \rho \ell_\ell = \rho \dot{\sigma}_\ell$$
$$-\rho_o \gamma \dot{\bar{T}} - T_o [a_{k\ell}(\dot{u}_{\ell,k} - \varepsilon_{k\ell m} \dot{\phi}_m) + b_{k\ell} \dot{\phi}_{k,\ell}] + q_{k,k} + \rho_o h = 0$$

(3.16)
$$t_{k\ell} = -a_{k\ell}\bar{T} + A_{k\ell mn}(u_{n,m} - \varepsilon_{mnp} \phi_p) + C_{k\ell mn} \phi_{m,n}$$
$$+ a_{k\ell mn}(\dot{u}_{n,m} - \varepsilon_{mnp} \dot{\phi}_p) + \alpha_{k\ell mn}\dot{\phi}_{m,n} + e_{k\ell m} \bar{T}_{,m}$$
$$m_{k\ell} = -b_{\ell k}\bar{T} + B_{\ell kmn}\phi_{m,n} + C_{mn\ell k} (u_{n,m} - \varepsilon_{mnp}\phi_p)$$
$$+ b_{\ell kmn}\dot{\phi}_{m,n} + \beta_{\ell kmn}(\dot{u}_{n,m} - \varepsilon_{mnp}\dot{\phi}_p) + f_{\ell km} \bar{T}_{,m}$$
$$Q_k = h_{k\ell}\bar{T}_{,\ell} + \gamma_{k\ell m}(\dot{u}_{m,\ell} - \varepsilon_{\ell mp}\dot{\phi}_p) + q_{k\ell m} \dot{\phi}_{\ell,m}$$

where the new constants (e.g., $A_{k\ell mn}$) are related to their corresponding material moduli as follows, e.g.,

(3.17)
$$A_{k\ell mn} = A_{KLMN} \delta_{kK} \delta_{\ell L} \delta_{mM} \delta_{nN}$$

The material moduli appearing in the constitutive equations may be further restricted because of the symmetry conditions. The material symmetry may be characterized by a group of orthogonal transformations $\{S_{KL}\}$. Accordingly, the response functions must be form-invariant under all material coordinate transformations of the form

(3.18)
$$X^*_K = S_{KL} X_L$$

where

$$\delta_{KM} = S_{KL} S_{ML} = S_{LK} S_{KL}$$

The material moduli should obey the transformation rules of the form

$$h_{KL} = S_{PK}S_{QL}h_{PQ}$$

(3.19) $$\gamma_{KLM} = S_{PK}S_{QL}S_{RM}\gamma_{PQR}$$

$$A_{KLMN} = S_{PK}S_{QL}S_{RM}S_{SN}A_{PQRS}$$

Similar rules hold for all second, third, and fourth-order tensors. Since the cholesteric liquic crystals possess an uniaxial symmetry about their axis of helix, the orthogonal transformation $\{S_{KL}\}$ which characterizes this symmetry must have the form:

(3.20)
$$S_{KL} = \begin{vmatrix} \cos\alpha & \sin\alpha & 0 \\ -\sin\alpha & \cos\alpha & 0 \\ 0 & 0 & 1 \end{vmatrix}$$

Substituting (3.20) into (3.19) we find that:
(1) $h_{11} = h_{22}$, and h_{33} are the nonvanishing components of h_{KL}, and the same statement holds for A_{KL} and B_{KL}.

(2) $\gamma_{123} = -\gamma_{213}$, $\gamma_{132} = -\gamma_{231}$ and $\gamma_{312} = -\gamma_{321}$ are the nonvanishing components of γ_{KLM}, and the same statement holds for e_{KLM}, f_{KLM}, and g_{KLM}.

(3) $A_{1111} = A_{2222}$, $A_{1122} = A_{2211}$, $A_{1133} = A_{3311} = A_{2233} = A_{3322}$, A_{3333}, $A_{1313} = A_{2323}$, $A_{3131} = A_{3232}$, $A_{3113} = A_{1331} = A_{3223} = A_{2332}$, $A_{1212} = A_{2121}$, and $A_{1221} = A_{2112} = A_{1111} - A_{1122} - A_{1212}$ are the nonvanishing components for A_{KLMN} and the same statement holds for B_{KLMN}.

(4) $a_{1111} = a_{2222}$, $a_{1122} = a_{2211}$, $a_{1133} = a_{2233}$, $a_{3311} = a_{3322}$, a_{3333}, $a_{1313} = a_{2323}$, $a_{3131} = a_{3232}$, $a_{1331} = a_{2332}$, $a_{3113} = a_{3223}$, $a_{1212} = a_{2121}$, and $a_{1221} = a_{2112} = a_{1111} - a_{1122} - a_{1212}$

are the nonvanishing components for a_{KLMN}, and the same statement holds for b_{KLMN}, c_{KLMN}, α_{KLMN}, and β_{KLMN}.

Further restrictions shall be made upon the material moduli based on physical considerations. In this regard, we recall that, in the cholesteric phase, temperature is high enough to break the bonds between layers. The following example should serve to clarify this point.

Static Shear Deformation

Consider a static shear deformation specified by:

(3.21)
$$x_{kK} = \begin{vmatrix} 1 & 0 & S \\ 0 & 1 & 0 \\ 0 & 0 & 1 \end{vmatrix}, \qquad \chi_{kK} = \begin{vmatrix} 1 & 0 & 0 \\ 0 & 1 & 0 \\ 0 & 0 & 1 \end{vmatrix}$$

Substituting (3.21) into (3.11) and (3.3), with $T = T_o$, we obtain the nonvanishing components of $t_{k\ell}$, $m_{k\ell}$ as follows:

(3.22)
$$t_{11} = A_{3131} S^2, \quad t_{13} = A_{1331} S, \quad t_{31} = A_{3131} S$$
$$m_{11} = \mathcal{C}_{3113} S^2, \quad m_{13} = \mathcal{C}_{3131} S, \quad m_{31} = \mathcal{C}_{3113} S$$

This type of shear deformation does not alter the density, temperature, axis of helix, layered structure, and coefficient of optical activity of the cholesteric liquid crystals. Hence an elastic recovery is not expected. Therefore we require

(3.23)
$$A_{1331} = A_{3113} = A_{3131} = \mathcal{C}_{3131} = \mathcal{C}_{3113} = 0$$

This is what is meant by the statement that <u>cholesteric liquid crystals are materials which assume any state that leaves density, temperature, axis of helix, layered structure, and coefficient of optical activity unchanged as a reference state, in which both stress tensor and moment stress vanish</u>.

If we perform a static shear deformation specified by

$$x_{k,K} = \begin{vmatrix} 1 & 0 & 0 \\ 0 & 1 & 0 \\ S & 0 & 1 \end{vmatrix}, \qquad \chi_{kK} = \begin{vmatrix} 1 & 0 & 0 \\ 0 & 1 & 0 \\ 0 & 0 & 1 \end{vmatrix}$$

then we obtain the nonvanishing components of $t_{k\ell}$ and $m_{k\ell}$ as

(3.24)
$$t_{13} = A_{1313} S, \qquad t_{33} = A_{1313} S^2$$
$$m_{33} = \mathcal{C}_{1331} S^2, \quad m_{13} = \mathcal{C}_{1331} S, \quad m_{31} = \mathcal{C}_{1313} S$$

We emphasize that the material moduli $\mathcal{C}_{1313}$, $\mathcal{C}_{1331}$, and, especially A_{1313}, do not vanish, since, otherwise, in such a deformation, elements of one layer would move into positions in neighboring layers thus destroying the layered structure. If such a new state is to be considered as a reference state with vanishing stress tensor and moment stress, then the layered structure of cholesteric substances is meaningless.

In the following, we study another important feature of these constitutive equations.

<u>Case 1</u>: Let a static shear deformation take place in the plane of cholesteric layer represented by

(3.25)
$$x_{k,K} = \begin{vmatrix} 1 & S & 0 \\ 0 & 1 & 0 \\ 0 & 0 & 1 \end{vmatrix}, \qquad \chi_{kK} = \begin{vmatrix} 1 & 0 & 0 \\ 0 & 1 & 0 \\ 0 & 0 & 1 \end{vmatrix}$$

The nonvanishing components of the stress and couple stress tensor are calculated through (3.16)

(3.26)
$$t_{11} = A_{1212}S^2, \quad t_{12} = A_{1221}S, \quad t_{21} = A_{1212}S$$
$$m_{11} = C_{1221}S^2, \quad m_{12} = C_{1212}S, \quad m_{21} = C_{1221}S$$

For $S \equiv$ const., all equations of motion are satisfied except the following

(3.27)
$$(A_{1212} - A_{1221})S = 0$$

This simply means if $A_{1212} = A_{1221}$, then this kind of shear deformation can be maintained by imposing appropriate boundary conditions.

Case 2: Let a uniform microrotation take place about the axis of helix represented by

(3.28)
$$x_{k,K} = \begin{vmatrix} 1 & 0 & 0 \\ 0 & 1 & 0 \\ 0 & 0 & 1 \end{vmatrix}, \qquad \chi_{kK} = \begin{vmatrix} \cos\phi & \sin\phi & 0 \\ -\sin\phi & \cos\phi & 0 \\ 0 & 0 & 1 \end{vmatrix}$$

In this case the nonvanishing components of $t_{k\ell}$ and $m_{k\ell}$ are found to be

(3.29)
$$t_{11} = t_{22} = (A_{1111} + A_{1122})\cos\phi(\cos\phi - 1) + (A_{1212} - A_{1221})\sin^2\phi$$
$$t_{12} = -t_{21} = -(A_{1111} + A_{1122})\sin\phi(\cos\phi - 1) + (A_{1212} - A_{1221})\sin\phi\cos\phi$$
$$m_{11} = m_{22} = (C_{1111} + C_{1122})\cos\phi(\cos\phi - 1) - (C_{1212} - C_{1221})\sin^2\phi$$
$$m_{12} = -m_{21} = (C_{1111} + C_{1122})\sin\phi(\cos\phi - 1) + (C_{1212} - C_{1221})\sin\phi\cos\phi$$

If the angle of rotation, ϕ, is small, then we have

$$t_{12} = -t_{21} = (A_{1212} - A_{1221})\phi$$

$$m_{12} = -m_{21} = (\mathcal{C}_{1212} - \mathcal{C}_{1221})\phi$$

Again, if $A_{1212} = A_{1221}$, then this kind of microrotation can be maintained by imposing appropriate boundary conditions. The question as to whether $A_{1212} = A_{1221}$ or not should be answered by experiment. This is discussed in Section V.

IV. CHANGE OF OPTICAL ACTIVITY

In this section we investigate the change in optical activity due to temperature variations and motions. Consider a static problem in which the strains and microrotations are small so that the linear constitutive equations apply. Let the displacement, microrotation, and temperature fields be represented by

(4.1) $$u_1 = u_2 = \phi_1 = \phi_2 = 0, \quad u_3 \equiv u(z), \ \phi_3 \equiv \phi(z), \ \bar{T} \equiv T(z)$$

Substituting (4.1) into $(3.16)_4$ to $(3.16)_6$, we find the nonvanishing components of $t_{k\ell}$, $m_{k\ell}$, and q_k are given by

(4.2) $$\begin{aligned}
t_{11} &= t_{22} = -a_{11}T + A_{1133}u_{,3} + \mathcal{C}_{1133}\,\phi_{,3} \\
t_{33} &= -a_{33}T + A_{3333}u_{,3} + \mathcal{C}_{3333}\,\phi_{,3} \\
t_{12} &= -t_{21} = -(A_{1212} - A_{1221})\phi + e_{123}\,T_{,3} \\
m_{11} &= m_{22} = -b_{11}T + B_{1133}\,\phi_{,3} + \mathcal{C}_{3311}\,u_{,3} \\
m_{33} &= -b_{33}T + B_{3333}\phi_{,3} + \mathcal{C}_{3333}\,u_{,3} \\
m_{12} &= -m_{21} = (\mathcal{C}_{1212} - C_{1221})\phi + f_{213}\,T_{,3} \\
q_3 &= h_{33}\,T_{,3}
\end{aligned}$$

Substituting (4.2) into equations of motion with vanishing body force, body couple, and heat source, we find that all components of these equations are satisfied identically except

(4.3) $$\begin{aligned}
&-aT_{,3} + A\,u_{,33} + \ \phi_{,33} = 0 \\
&-bT_{,3} + C\,u_{,33} + B\,\phi_{,33} - D\phi + eT_{,3} = 0 \\
&T_{,33} = 0
\end{aligned}$$

where

$$A \equiv A_{3333},\ B \equiv B_{3333}, \mathcal{C} \equiv \mathcal{C}_{3333}$$

$$D \equiv 2(A_{1212} - A_{1221}),\ a \equiv a_{33}\ ,\ b \equiv b_{33}\ ,\ e \equiv 2e_{123}$$

The general solutions of (4.3) are

$$\begin{aligned} T &= c_1 z + c_2 \\ u &= a_1 + a_2 z + \frac{1}{2}(h_1/A)z^2 + a_3 e^{kz} + a_4 e^{-kz} \\ \phi &= \frac{Ch_1 - Ah_2}{AD} - \frac{A}{C}(a_3 e^{kz} + a_4 e^{-kz}) \end{aligned} \tag{4.4}$$

where

$$h_1 \equiv ac_1,\quad h_2 \equiv (b-e)c_1,\qquad k^2 \equiv AD/(AB-C^2)$$

and a_i's are constants to be determined by boundary conditions.

The coefficients of optical activity in the undeformed and deformed states are defined respectively by:

$$\begin{aligned} \Delta &\equiv \lim_{Z_1 \to Z_2} \frac{\Theta(Z_1) - \Theta(Z_2)}{Z_1 - Z_2} \\ \delta &\equiv \lim_{z_1 \to z_2} \frac{\theta(z_1) - \theta(z_2)}{z_1 - z_2} \end{aligned} \tag{4.5}$$

Since

$$\theta(z) \simeq \Theta(Z) + \phi(z), \qquad z \simeq Z + u(z)$$

We obtain

$$\delta = \Delta(1 - \frac{du}{dz}) + \frac{d\phi}{dz} \tag{4.6}$$

Now calculate the change of optical activity for the following special cases:

Case 1: Uniform increase (decrease) of temperature in a slab. Consider the case in which boundary tractions vanish. Thus, the boundary conditions are

$$T = \overset{0}{T}\ ,\quad t_{33} = m_{33} = 0, \qquad @\ z = 0,\ L$$

Equations (4.4), in this case, become

$$\begin{aligned} u &= \overset{0}{T}\,[az/A + f(z)] \\ \phi &= -\overset{0}{T}\ Af(z)/C \\ \delta &= \Delta(1 - \frac{a}{A}\overset{0}{T}) - \overset{0}{T}\ (\Delta + \frac{A}{C})\frac{df}{dz} \end{aligned} \tag{4.7}$$

where

$$f(z) \equiv \frac{C(Ca - Ab)}{2Ak(Ab-C^2)\sinh(kL)}[(1-e^{-kL})e^{kz} + (1-e^{kL})e^{-kz}]$$

An interesting case occurs where Ca = Ab, for which

(4.8)
$$u = T^0 az/A, \quad \phi = 0$$
$$\delta = \Delta(1 - \frac{a}{A}T^0)$$

This means there is a homogeneous expansion, and the optical activity decreases as the temperature increases[3]. We require that the temperature variation is in the temperature range for cholesteric mesomorphous phase.

Case 2: Temperature gradient in the z-direction.
For vanishing boundary tractions on the faces of the slab we have

$$t_{33} = m_{33} = 0, \qquad T = -T^0/2, \qquad @\ z = 0$$
$$t_{33} = m_{33} = 0, \qquad T = T^0/2, \qquad @\ z = L$$

Then we obtain

(4.9)
$$T = T^0(\frac{z}{L} - \frac{1}{2})$$
$$U = T^0[\frac{az}{2A}(\frac{z}{L} - 1) + g(z)]$$
$$\phi = T^0[\frac{Ca - Ab + Ae}{ADL} - \frac{A}{C}g(z)]$$
$$\delta = \Delta[1 - T^0\frac{a}{A}(\frac{z}{L} - \frac{1}{2})] - T^0(\Delta + \frac{A}{C})\frac{dg}{dz}$$

where

$$g(z) \equiv \frac{C(Ca - Ab)}{4Ak\ (AB - C^2)\sinh(kL)}[(1+e^{-kL})e^{kz} + (1+e^{kL})e^{-kz}]$$

If Ca = Ab, then we have

(4.10)
$$u = U = T^0\frac{az}{2A}(\frac{z}{L} - \frac{1}{2}), \qquad \phi = T^0\frac{e}{DL}$$
$$\delta = \Delta[1 - T^0\frac{a}{A}(\frac{z}{L} - \frac{1}{2})]$$

Case 3: Isothermal slab subjected to pressure.

For vanishing surface couple the boundary conditions are

$$T = m_{33} = 0, \qquad t_{33} = -p, \qquad @z = 0,L$$

The solution is now given by

$$T = 0$$

$$u = U = -p[\frac{z}{A} + h(z)]$$

$$\phi = p\frac{A}{C} h(z)$$

$$\delta = \Delta(1 + \frac{p}{A}) + p(\Delta + \frac{A}{C})\frac{dh}{dz}$$

where

$$h(z) \equiv \frac{C^2}{2Ak(AB-C^2)\sinh\ (kL)}[(1-e^{-kL})e^{kz} + (1-e^{kL})e^{-kz}]$$

In the case where the cholesteric layer is very thin, i.e., $kL \ll 1$ we have

$$\delta \simeq \Delta(1 + \frac{Bp}{AB-C^2}) + \frac{Cp}{AB-C} \tag{4.12}$$

Case 4: Displacement prescribed at the boundary.

Let the boundary conditions be

$$T = m_{33} = 0, \qquad U = 0, \qquad @\ z = 0$$

$$T = m_{33} = 0, \qquad U = -U_0, \qquad @\ z = L$$

Then we have

$$T = 0$$

$$U = \gamma[2-2\cosh\ (kL) - \frac{AD}{kC^2}\sinh\ (kL)z - (1-e^{-kL})e^{kz} - (1-e^{kL})e^{-kz}]$$

$$\phi = \frac{A}{C}\gamma[(1 - e^{-kL})e^{kz} + (1 - e^{kL})e^{-kz}]$$

$$\delta = \Delta[1 + \frac{2AD\gamma}{kC^2}\sinh(kL)] - \gamma g(z)\ (\Delta + \frac{A}{C})$$

where

$$\gamma \equiv \frac{U_0}{4\cosh(kL) - 4 + 2ADL\sinh(kL)/kC^2}$$

These examples show that the coefficient of optical activity (which characterizes the helical structure of cholesteric liquid crystals) changes in response to temperature variations, deformation, and mechanical stresses. This provides the necessary reasoning on the mechanical origin of the marked changes in optical properties when cholesteric substances are subjected to external disturbances.

V. TWIST WAVES

Here we investigate the wave propagation in a cholesteric substance along the axis of helix. For simplicity we neglect the thermal coupling and all the dissipative terms. The plane wave solutions are of the form:

$$U_3(z, t) \equiv U(x, t) = \bar{U} \exp (i\omega t - iKz)$$

(5.1)

$$\phi (z, t) \equiv \phi(z, t) = \frac{\bar{\phi}}{\sqrt{1 + J}} \exp (i\omega t - iKz)$$

and all other components of displacement and microrotational field vanish. The equations of motion are

$$A\, U_{,33} + C\, \phi_{,33} = \rho \ddot{U}$$

(5.2)

$$B\, \phi_{,33} + C U_{,33} - D\phi = \rho (I + J)\, \ddot{\phi}$$

where A, B, C, and D are defined as in the previous section. Substituting (5.1) into(5.2), we obtain

(5.3)
$$\begin{vmatrix} AK^2 - \rho\omega^2 & CK^2/(I+J)^{1/2} \\ CK^2/(I+J)^{1/2} & \frac{1}{I+J}(BK^2+D) - \rho\omega^2 \end{vmatrix} \begin{vmatrix} \bar{U} \\ \bar{\phi} \end{vmatrix} = 0$$

Equations (5.3) imply the following results

(1) The longitudinal and twist waves are coupled, a unique property for cholesteric liquid crystals.
(2) There are two modes propagating with the following dispersion relations:

(5.4) $$\rho\omega^2 = \frac{1}{2}[AK^2+\frac{1}{I+J}(BK^2+D)\pm\{[AK^2-\frac{1}{I+J}(BK^2+D)]^2+4C^2K^4/(I+J)\}^{1/2}]$$

One of these two modes is an acoustic mode and the other is an optical mode with a cut-off frequency

(5.5) $$\omega_c = [D\rho^{-1} (I+J)^{-1}]^{1/2}$$

The stress potential in this problem is given by

$$\Sigma \equiv \rho_o\psi - S_o = A_{KLMN}\phi_{KL}\phi_{MN} + \frac{1}{2} B_{KLMN}\Gamma_{KL}\Gamma_{MN}$$

(5.6)

$$+ C_{KLMN}\phi_{KL}\Gamma_{MN}$$

$$= A(U_{,3})^2 + B(\phi_{,3})^2 + 2C(U_{,3})(\phi_{,3}) + D\phi^2$$

This form is positive definite if and only if

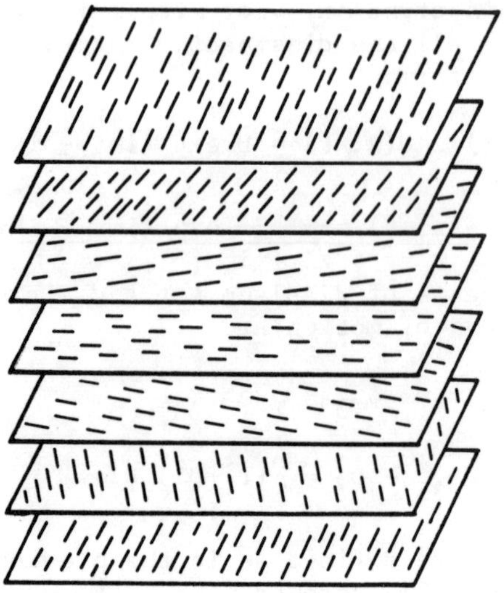

Fig. 1 Schematic Structure of Cholesteric Liquid Crystals

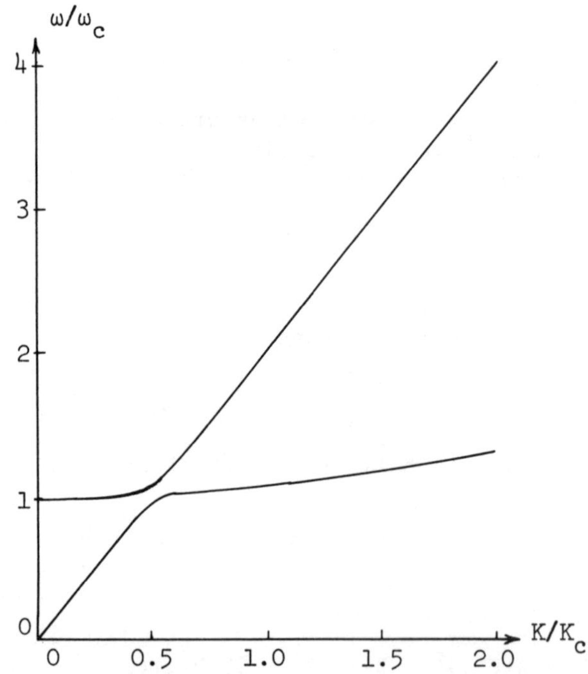

Fig. 2 Dispersion Relation of Twist Wave in Cholesteric Liquid Cryst[als]
$K_c^2 \equiv 4D/A(I+J)$, $B = A(I+J)/16$, $C = A(I+J)^{\frac{1}{2}}/8$

(5.7) $$A \geq 0, B \geq 0, D \geq 0, AB - C^2 \geq 0$$

Because of (5.7), one may verify that two modes exist. A plot of the dispersion relations is given in Fig. 2 for illustrative purpose.

This analysis suggests the following significant experiment: Generate a longitudinal wave along the axis of helix. Observe that when the frequency is below the cut-off frequency only one mode propagates and when the frequency is above the cut-off frequency, there are two modes propagating with different phase velocities. If such a wave propagation experiment is performed, one may clearly demonstrate the coupling between longitudinal and twisting waves.

From (5.4) one can notice that if D = 0, i.e., $A_{1212} = A_{1221}$, then both of the two modes are acoustic and nondispersive with the phase velocities

(5.8) $$v_{phase} = \{\frac{1}{2\rho} \{ A + \frac{B}{I+J} \pm [(A - \frac{B}{I+J})^2 + 4 \frac{C^2}{I+J}]^{1/2}\}\}^{1/2}$$

If this is indeed the case for cholesteric liquid crystals, then the shear deformation and the uniform microrotation shown in Section 3, (Case 1 and 2), can be maintained simply by imposing appropriate boundary conditions.

REFERENCES

1. R. S. Porter, E. M. Barrall II, and J. F. Johnson, J. Chem. Phys., V. 45, No. 5, 1452 (1966).

2. G. Friedel, Ann. Physique 19, 273 (1922).

3. J. L. Fergason, Sci. Am. V. 211, No. 2, 77 (1964).

4. G. H. Brown and W. G. Shaw, Chem. Rev., 57, 1049 (1957).

5. R. S. Porter and J. F. Johnson, Rheology IV (New York): Academic Press, 1967, edited by F. R. Eirich), p. 317.

6. I. G. Chistyakov, Soviet Physics USPEKHI, V. 9, No. 4, 551 (1967).

7. A. Saupe, Angew. Chem. International Edit. V. 7, No. 2, 97 (1968).

8. P. G. DeGennes, Solid State Communications V. 6, 163 (1968).

9. P. G. DeGennes, Molecular Crystals and Liquid Crystals, V. 7, 325 (1969).

10. R. B. Meyer, Applied Phys. Lett., 12, 281 (1968).

11. O. S. Selawry, H. S. Selawry, and J. F. Holland, Liquid Crystals (New York: Gordon and Breach, 1966, edited by Brown, Dienes, and Labes), p. 175.

12. F. M. Leslie, Proc. Roy. Soc. A., 307, 359 (1968).

13. F. M. Leslie, Molecular Crystals and Liquid Crystals, 7, 407 (1969).

14. J. L. Ericksen, Trans. Soc. Rheol., 5, 23 (1961).

15. A. C. Eringen and E. S. Suhubi, Int. J. Engng. Sci., 2, 189 (1964).

16. A. C. Eringen, J. Math. & Mech., 15, 909 (1966).

17. A. C. Eringen, J. Math. & Mech. 16, 1 (1966).

18. A. C. Eringen, Foundations of Micropolar Thermoelasticity, International Center for Mechanical Sciences, Udine, Italy, 1970.

19. J. D. Lee and A. C. Eringen, J. Chem. Phys., V. 54, No. 12, 5027 (1971).

20. A. C. Eringen, Mechanics of Continua (New York: John Wiley & Sons, 1967).

21. A. C. Eringen, Int. J. Engng. Sci., 5, 191 (1967).

22. A. C. Eringen, Proceedings of the Eleventh International Congress of the Eleventh International Congress of Applied Mechanics (held in 1964, Munich, Germany), edited by H. Gortler, Springer-Verlag (1966) 131-138.

This work was supported by the National Science Foundation.

CHIRALITY IN MIXED NEMATIC AND CHOLESTERIC LIQUID CRYSTALS

N. Oron, K. Ko, L. J. Yu and M. M. Labes

Department of Chemistry, Temple University, Phila.,

Pa. 19122

Abstract

A mechanism is discussed explaining the different types of pitch-concentration dependences encountered in chiral mesophases. Addition of a low concentration of a chiral compound to a nematic produces a mesophase in which pitch is initially inversely proportional to concentration, but as the concentration increases, three different types of behavior occur: (a) the pitch slowly saturates as it approaches that of the pure chiral (cholesteric) compound; (b) the pitch decreases to a minimum pitch value (MPV) and then increases again to that of the pure cholesteric; (c) the pitch decreases to a MPV, increases to infinity and then decreases again to that of the pure cholesteric. This type of "compensation" which usually occurs only when a right-handed and left-handed cholesteric are mixed, is reported here for a mixture of cholesteryl-2(2-ethoxyethoxy) ethyl carbonate and the nematic N-p-methoxybenzylidene-p-n-butylaniline. Thus, starting from a pure cholesteric, pitch may either increase or decrease upon adding a non-chiral solute. The common mechanism involves the induced circular dichroism (ICD) in the non-chiral compound caused by the helical arrangement of the chiral mesophase, which may be of the same or opposite sign as the ICD. The pitch of the pure cholesteric compound, the structure of the non-chiral compound and its affinity to ICD are important in determining which type of spectral behavior is observed.

In a previous publication (1), an experimental method was described and data were provided to show improved response times and intensity-ratios for conical helical perturbation of a ternary cholesteric mixture consisting of: cholesteryl chloride (CC) 1.1 parts; cholesteryl nonanoate (CN) 0.9 parts; and cholesteryl oleyl carbonate (COC) 2.0 parts by weight, when doped with MBBA (N-p-methoxybenzylidene-p-n-butylaniline). In the course of this work, a rather peculiar spectral behavior was observed. It was found that by adding 5-30% of MBBA to the mixture, a blue shift and bandwidth broadening in the transmission spectra occur. The response time is reduced by a factor of four and the intensity ratio is increased by a factor of seven, accompanied by a wavelength shift towards the blue. By increasing the percentage of MBBA from 30% to 50%, the blue mixture is shifted back to red, going through a minimum pitch value (MPV) at $\sim$25-30%. The trend of bandwidth broadening continues. The peak wavelength λ is known to be directly proportional to the pitch Z and the refractive index n, and inversely proportional to the twist angle θ.

In recent publications a similar spectral behavior was reported by Nakagiri et al, in a mixture of cholesteryl propionate (CP) and MBBA (2), and in a mixture of CP and p-azoxyanisole (PAA) (3). However, the MPV was not observed in a mixture of CP and p-cyanobenzylidene-p'-phenetidine (CBP) (3) where at high concentrations of CP a saturation is observed instead. A similar linear saturation was also observed (4) in mixtures of active and racemic p-alkoxybenzal-p-(β-methylbutyl) anilines, which behave thermodynamically as a one component chiral nematic liquid crystal.

It has previously been recognized (5) that the addition of a cholesteric material to a nematic produces a cholesteric mesophase, and that the pitch of this mesophase is inversely proportional to the concentration of the optically active molecule in the low concentration range. By extending the molecular statistical theory of Maier and Saupe (6), the dispersion interaction energy between two molecules in adjacent planes with a finite twist angle has been calculated (7) taking into account the dipole-dipole and the dipole-quadropole interactions. This theory explains the twisting power of optically active solute molecules in a nematic solvent and also the concentration dependence of the induced twist angle at low concentrations.

To explain their results and MPV's at higher concentrations, Nakagiri et al (2,3) further extend this theory, and by calculating the second order perturbation energy, introducing the quadropole-quadropole interaction as well, show that the dependence of pitch Z on concentration C of the optically active molecule is of the form:

$$Z \sim (\alpha/\beta)1/C + (\gamma/\beta)C \qquad (1)$$

where α is related to the anisotropy of the molecular polarizability; β, γ are related to the anisotropy of the molecule. At high concentrations C the second term in (1) will be dominant, resulting in increase of pitch with concentration.

Recently some additional results were reported by Saeva and Wysocki (8) on a mixture of CC and MBBA. They are shown on curve (c) of Fig. 1 in comparison to some other results previously mentioned. In view of these results, it is clear that the theories suggested to explain the pitch dependence are incomplete, since they do not take into consideration the sign of the circular dichroism due to the helical arrangement which is induced on the non-chiralic molecule by the cholesteric liquid crystal and the affinity of the non-chiralic molecule to induced circular dichroism (ICD).

Based on investigations of the circular dichroism of helically arranged molecules in cholesteric phases, Sackmann and Voss (9) established rules concerning the interrelation between the induced helical arrangement on the non-chiralic solute material and the cholesteric solvent. The solute molecules orient in the liquid crystal to an appreciable extent. An elongated molecule orients with its long molecular axis preferentially parallel to the local optical axis of the cholesteric molecule, which is represented as being composed of stacked layers of two-dimensional nematic liquid crystals. Therefore such solute molecules exhibit the same helical arrangement, on a macroscopic scale, as the solvent molecule. The sign of the "structural" circular dichroism depends on the polarization direction of the electronic transition of the solute molecule. If the absorption band of an elongated molecule is polarized paralell to the long molecular axis, the sign of the circular-dichroism band of the solute is opposite to the sign of the circular-dichroism resulting from the pitch of the cholesteric phase. If the transition moment is parallel to a short axis of the solute, the signs of both circularly-dichroic components will be the same. In light of this rule, the different reported variations of wavelength (or pitch) with concentration can be understood as follows.

<u>Case A</u>: The affinity of the nematic solute to ICD is weak. The result in this case is independent of the sign of the ICD band of the non-chiralic material (which can be equal or opposite in sign to the circular-dichroism resulting from the pitch of the cholesteric).

Example: In the mixture of CP and CBP shown in curve (a) of Fig. 1, CP, which forms a left-handed (LH) helix, will induce a weak right-handed (RH) stacking in the CBP solute. At low concentration of cholesteric, both cholesteric and nematic-induced components of the system act to gradually reduce the infinite pitch of the nematic. On increasing the wt. % of cholesteric, the much stronger influence of the cholesteric dominates, resulting in gradual saturation of the wavelength to match the natural pitch of

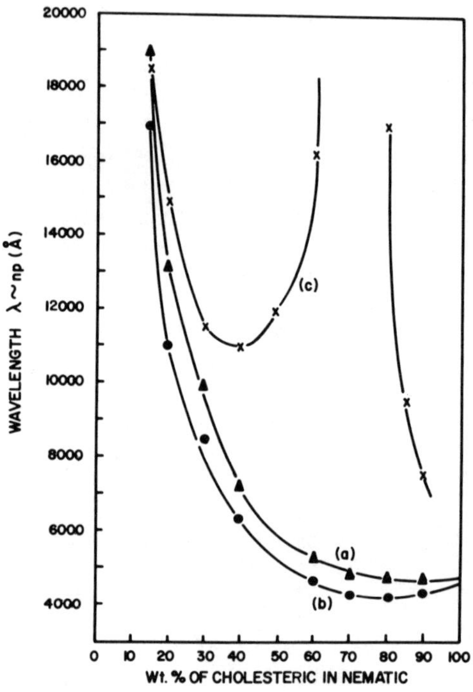

Figure 1: Wavelength of maximum reflection of various weight percent of cholesterics in nematic dopants.

(a) CP in CBP (see reference 3)

(b) CP in MBBA (see reference 2)

(c) CC in MBBA (see reference 8)

the pure cholesteryl propionate.

Case B: The affinity of the non-chiralic material to ICD is strong and the ICD band of the non-chiralic material is of the same sign as that resulting from the pitch of the cholesteric.

Example: Curve (b) of Fig. 1 relates to a mixture of CP and MBBA (2), where a MPV is reached at appr. 70 wt. % of CP. CP, which forms a LH helix, will induce LH stacking in the MBBA solute. When chirality is induced, both cholesteric and nematic-induced components of the system act to gradually reduce the infinite pitch of the nematic. At high concentrations of cholesteric - low concentrations of nematic - the contributions of the induced chirality of the nematic to the final formation of the pitch is weakened. The lowest pitch is reached at appr. 70 wt. % of CP. Increasing the wt. % of CP will result in reduction of the contribution of the chiral-nematic to the mixture and the pitch will gradually increase to match that of the original cholesteric component, thus forming the MPV in the pitch/concentration curve. The occurrence of the MPV in the pitch/concentration curve and the steepness of the slope will, thus, depend on the affinity of the nematic dopant to ICD and on the final pitch of the cholesteric component. An even clearer example demonstrating the former part of this statement was shown for a mixture of CP:PAA (3). Apparently PAA has a stronger affinity to ICD and it does reach a lower MPV.

Case C: The affinity of the non-chiralic material to ICD is strong and the sign of the ICD band of the non-chiralic material is opposite to the sign of the circular dichroism resulting from the pitch of the cholesteric.

Example: In the mixture of CC:MBBA (9), shown on curve (c) of Fig. 1, CC, which forms a right-handed helix (RH), will induce a left-handed (LH) stacking on the MBBA solute. At high concentrations of MBBA a LH chirality is induced, gradually reducing the infinite pitch of the nematic.

As soon as chirality is induced, the system will behave as a binary cholesteric system with one LH and one RH component. In such a system a compensation process (10) is expected to occur. For this particular mixture, the LH component is dominant up to 40 wt. % of CC. From this point and up, the RH component's influence increases gradually. The mixture is still LH but the pitch increases until, at about 70 wt. % of CC, the system is totally compensated and a nematic mesophase results. Further increase of the wt. % of CC allows the mixture to become RH, resulting in a gradual pitch decrease. At very low concentrations of the non-chiralic material the ICD is of no influence any more.

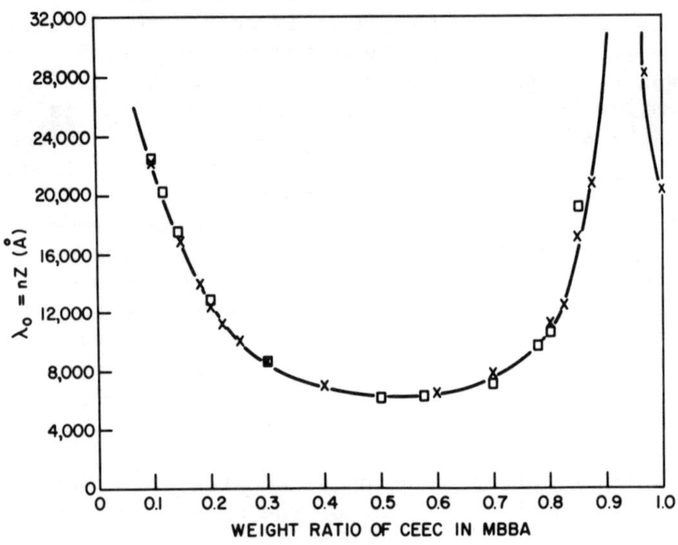

Figure 2: Wavelength of maximum reflection of various weight percent of CEEC in MBBA.

× Commercial CEEC

□ Purified CEEC

In order to re-confirm the behavior in this case, we carried out a similar experiment with a mixture of cholesteryl-2-(2-ethoxy-ethoxy) ethyl carbonate (CEEC) and MBBA. CEEC was used as obtained and purified by a method described previously (11). Samples, 12.7μ in thickness, were placed in a special cell (12) that could be fitted into the compartment of a Cary 14 Spectrometer. The quantity measured was the wavelength of maximum reflection $\lambda_o = nZ$, where n is the average refractive index. Determination of $\lambda_o = 28{,}000$ Å for the 98% CEEC in MBBA was made via infrared transmission measurements as previously described (13). All measurements were done up to 10°C below the cholesteric-isotropic transition point. The dependence of λ_o vs temperature was found to be negligible. CEEC is known to have a very long pitch (14) in the pure cholesteric mesophase and a RH rotatory power under certain conditions (15); MBBA will accept a LH ICD.

The results are given in Fig. 2 which shows the wavelength λ_o vs the concentration of CEEC in MBBA. The results for the commercial and pure CEEC are the same; it was verified that the helical sense is LH throughout the 0-85 wt. % concentration range and RH for higher concentrations. A pronounced turning point is reached at appr. 50 wt. % of CEEC followed by a steep increase to infinity representing the nematic compensated state, followed by a decrease towards the original pitch of CEEC. A similar behavior might be expected for a system of a LH cholesteric inducing RH chirality on a nematic dopant with strong affinity to ICD, but no experimental results have as yet been reported.

In conclusion, it should be mentioned that there is a totally different approach to viewing the mechanism of pitch-concentration relationships. There are several experimental results which indicate that a cholesteric material may change its effective chirality in different environments. This can be pictured as follows: a solvent phase may influence the orientation of the side chain at the 3-position in a cholesteric molecule; since the effective twisting power of the molecule is dependent on the configuration of this side chain (14), it may change its handedness in mixtures with other mesogens. For example, it has been observed that cholesteryl-2-(2-butoxyethoxy) ethyl carbonate (15) and cholesteryl iodide (16) change their chirality in different solvents. An experimental distinction between this mechanism and that of ICD has as yet to be made.

Acknowledgment: This work was supported by the National Science Foundation under Grant No. GP 25988 and the National Aeronautics and Space Administration under Grant No. NGR 39-012-024.

References

1. N. Oron, L. J. Yu and M. M. Labes, Mol. Cryst. Liq. Cryst. 21, 333 (1973).
2. T. Nakagiri, H. Kodama and K. K. Kobayashi, Phys. Rev. Letters 27, 564 (1971).
3. T. Nakagiri, H. Kodama and K. K. Kobayashi, Presented at the 4th International Liquid Crystal Conference, Kent State University, Kent, Ohio, August 1972.
4. D. Dolphin, Z. Muljiani, J. Cheng and R. B. Meyer, J. Chem. Phys. 58, 413 (1973).
5. R. Cano, Bull. Soc. Fr. Mineral Cristallogr. 90, 333 (1967); R. Cano and P. Chatelain, C. R. Acad. Sci. Ser. B, 253, 1815 (1961); G. Durand, L. Leger, F. Rondelez and M. Veyssie, Phys. Rev. Letters 22, 227 (1969).
6. W. Maier and A. Saupe, Z. Naturforsch 14a, 882 (1959).
7. W. J. A. Goossens, Mol. Cryst. Liq. Cryst. 12, 237 (1971).
8. F. D. Saeva and J. J. Wysocki, J. Amer. Chem. Soc. 93, 5928 (1971).
9. E. Sackmann and J. Voss, Chem. Phys. Letters 14, 528 (1972).
10. H. Baessler and M. M. Labes, Phys. Rev. Letters 21, 1791 (1968); H. Baessler and M. M. Labes, J. Chem. Phys. 51, 1846 (1969).
11. H. Baessler, P. A. G. Malya, W. R. Nes and M. M. Labes, Mol. Cryst. Liq. Cryst. 6, 329 (1970).
12. I. Teucher, K. Ko and M. M. Labes, J. Chem. Phys. 56, 3308 (1972).
13. H. Baessler and M. M. Labes, Mol. Cryst. Liq. Cryst. 6, 419, (1970).
14. K. Ko, I. Teucher and M. M. Labes, Mol. Cryst. Liq. Cryst. 22, 203 (1973).
15. L. B. Leder, Chem. Phys. Letters 6, 285 (1970).
16. J. E. Adams and W. E. L. Haas, Mol. Cryst. Liq. Cryst. 15, 27 (1971).

CYLINDRICALLY SYMMETRIC TEXTURES IN MESOPHASES OF CHOLESTERYL ESTERS*

Fraser P. Price and Chan S. Bak

University of Massachusetts

Amherst, Massachusetts 01002

ABSTRACT

We have produced cylindrically symmetric single crystal textures of the cholesteric and smectic phases of cholesteryl esters by containing them in glass capillaries of smaller than 50 μ radius and inducing shear either by air-pressure changes from one end of the capillary or by volume changes on phase transformations of the sample itself. In the cholesteric phase, when the long axes of the molecules at the capillary interface are aligned parallel to the tube axis, the optic (twist) axes of the weakly negative uniaxial indicatrix are all radially oriented. In the smectic phase, when the molecules are aligned perpendicular to the substrate, the optic axes of the somewhat stronger positive uniaxial indicatrix are again radial. With the molecules aligned parallel to the tube axis, the optic axes of the smectic phase are everywhere parallel to the tube axis. These cylindrically symmetric orientations allow determination of the birefringence of the mesophase. The axial ratios of the refractive index ellipsoids are close to unity, for example, 1.014 and 1.028 respectively for the cholesteric and smectic phases of cholesteryl nonanoate. So far as we know these birefringences have been measured for the first time from cylindrical textures.

INTRODUCTION

The experimental determination of the optical birefringence in liquid crystals (LC) has been carried out (1,2) conventionally by

*This work was supported by Grant No. HL13188 from the National Institutes of Health.

observing the optical behavior of a thin LC film supported between two glass plates. The orientation of molecules on the substrate surface is produced typically by rubbing the substrate, thus producing a uniformly oriented LC. In cholesterics one encounters the problem of matching the thickness of the LC film to a multiple of the helical pitch length, otherwise the orienting effects of the substrate alter the helical structure. This problem is more difficult to overcome when the pitch changes continuously during changes of temperature. If the sample thickness is too large a multiple of the pitch length, then a question arises as to how uniform sample one has under study, i.e., what portion of the observed effects is due to the influence of the container walls (substrates) and what portion is intrinsic to the sample itself.

We report here a new method we have developed to determine the LC birefringence by observing a cylindrically symmetric texture produced in capillaries. The orientation of the molecules is predetermined at the inside wall of the bore according to wall treatment and left free in the core region. With cholesteryl nonanoate we observed three different cylindrically symmetric textures depending on the mesophase and the condition of the inside wall of the bore. These uniform textures were readily seen under a polarizing microscope and exhibited numerous strips of color bands aligned parallel to the tube axis. The wavelength and position of the color bands are well explained by our theoretical calculation and permit determination of the optical birefringence. The obtained birefringences are compared with the measurements done by other investigators using different methods.

THEORY

In what follows we recognize three different cylindrically symmetric textures:

Texture 1. This texture, formed in the cholesteric phase with the long axes of the molecules (directors) aligned parallel to the substrate interface, is shown in Fig. 1(a) and 1(b). The texture can be envisioned as a set of coaxial cylindrical surfaces. In each surface the directors are parallel to each other (nematic type) and make a given angle with the cylinder axis. However in going from one surface to another separated by distance Δr, that angle changes by $\Delta r/p \cdot 2\pi$ where p is the pitch length of the cholesteric helix. In the outermost cylinder that angle is zero (the directors are parallel to the tube axis). Thus in traversing the set of cylinders in a radial direction the directors rotate in a manner demanded by the cholesteric structure. Within a given cylindrical surface the direction of the directors traces out helical paths with a given pitch length p'. However, in going from one surface to another, p' undergoes continuous and repeated changes unlike the fixed pitch (p) of the cholesteric helix. For example, p' is

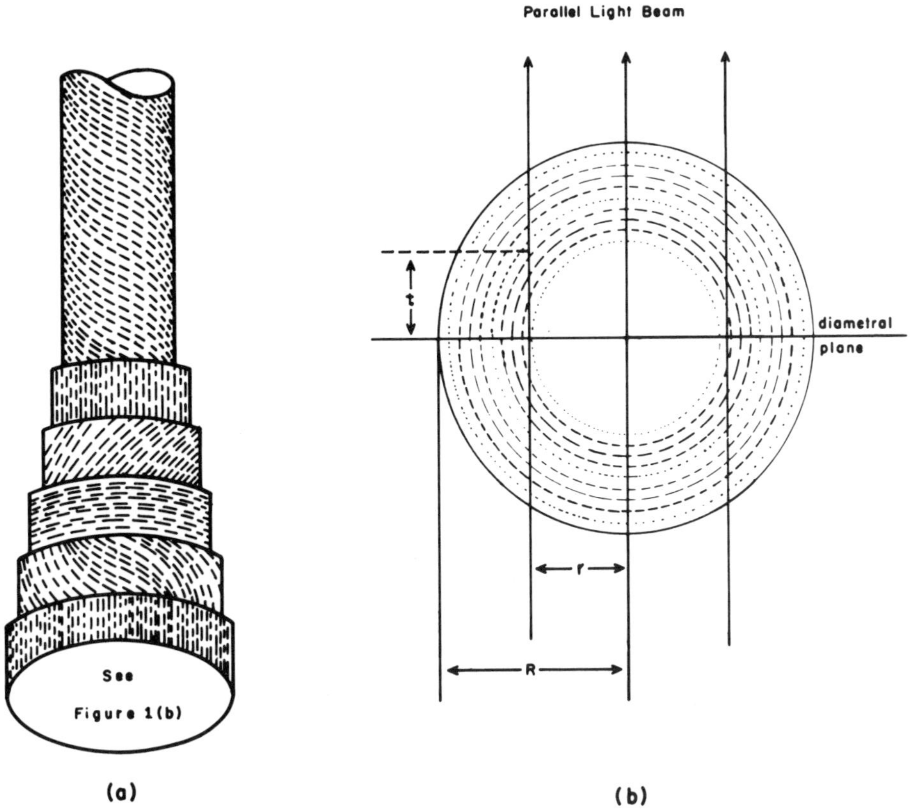

Figure 1. Cholesteric phase. Texture 1: (a) Schematic drawing showing director direction in coaxial layers, (b) cross-section defining terms of equations. In both cases radial pitch length exaggerated.

zero in those surfaces where the directors are perpendicular to the tube axis and infinity in those surfaces where the directors are parallel to the tube axis. We assume that the director is identifiable with the high polarizability direction within the cholesteryl ester molecule and that the polarizabilities transverse to this director are all equal. Then, since $p \ll R$, every small domain within this texture is characterized by a negative uniaxial indicatrix with the optic axis radially oriented. Then a parallel light beam incident on the tube as shown in Fig. 1(b) experiences a changing birefringence along the beam path. For a light beam traversing at the distance r from the tube axis, the indices of refraction ($n_{||}$ and $n_{\perp}$) parallel and perpendicular to the tube axis are:

$$n_{||} = n_o$$

$$n_{\perp} = n_o \left\{ \frac{t^2 + r^2}{t^2 + \left(\frac{n_o}{n_e}\right)^2 r^2} \right\}^{1/2}$$

where t is the distance from the diametral plane (see Fig. 1(b)). The retardation difference accumulated along the beam path for this light is

$$\int (n_{||} - n_{\perp})\, dt$$

where the integration is over the beam path. Provided $n_o/n_e - 1 \ll 1$ and retaining up to the third term in the power series of $n_{\perp}$, we have

$$\int (n_{||} - n_{\perp})\, dt = n_o R (r/R) \frac{n_{oe}^2 - 1}{n_{oe}(\frac{r}{R})} \left[\left\{ 1 + \frac{1}{8} \cdot \frac{n_{oe}^2 - 1}{n_{oe}^2} \right\} \tan^{-1} \frac{\{1 - (\frac{r}{R})^2\}^{1/2}}{n_{oe}(\frac{r}{R})} + \frac{1}{8}(\frac{r}{R}) \cdot \frac{n_{oe}^2 - 1}{n_{oe}} \frac{\{1 - (\frac{r}{R})^2\}^{1/2}}{(n_{oe}^2 - 1)(\frac{r}{R})^2 + 1} \right] \quad (1)$$

where R is the radius of the capillary bore and $n_{oe} = n_o/n_e$. From Eq. (1) it is expected that at the center line (r = 0) and at the edge of the tube (r = R) there is no retardation difference. Thus between crossed polars in the microscope the tube center and edges will appear as dark bands. In the intermediate range $0 < r < R$ there will appear strips of color bands aligned parallel to the tube axis. We characterize the overall birefringence of the specimen as positive in that the axial refractive index exceeds the radial.

The arrays of directors in this texture may be characterized in terms of Frank's elastic constants (3) such as for bend, twist, etc. The bend strain, for example, is zero in those cylindrical surfaces where directors lie parallel to the tube axis. However, in those surfaces where directors are perpendicular to the tube axis, the bend strain is nonzero and increases as the radius of the cylinder diminishes. Thus in going from the outer surface to the core region of the tube, the bend energy changes repeatedly from zero to increasing maxima but the surface energy (or the influence of the substrate-mesophase interface that tends to produce texture 1) decreases. If the intrinsic elastic constant for bend of the material is not small enough compared with the other elastic constants (splay, twist, etc.), then it is possible for the structure to assume other orientations to ensure the minimum energy of the system. Then the bend energy may fluctuate less than in texture 1 or may change monotonically in the capillary. Thus if both the bend energy and surface energy are large, the helix axes are not oriented radially but are tilted in some other direction to minimize the effect of the large bend energy. These tilted helix axes were indeed observed in cholesteryl hexanoate contained in a capillary of bore radius 18 μ. If the bend energy is small but the surface energy is large, then the texture 1 is formed in the capillary within the resolving power of the microscope. Cholesteryl nonanoate produces this texture right up to the core. If the surface energy is small, a focal conic texture is formed except very near the wall.

Texture 2. When the directors in smectic phase are aligned parallel to the tube axis, the overall alignment inside the capillary consists of a series of planar layers which lie perpendicular to the tube axis as shown in Fig. 2(a). In each layer the optic axis of a positive uniaxial

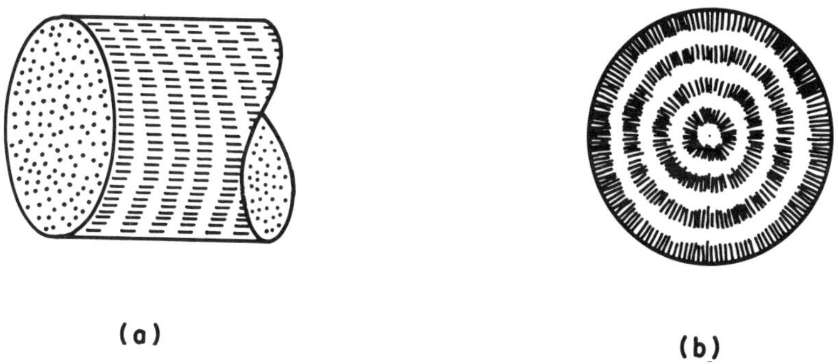

Figure 2. Smectic phase. (a) Texture 2, (b) Texture 3.

indicatrix is oriented perpendicular to the plane (parallel to the tube axis). The retardation difference is

$$\int (n_{\parallel} - n_{\perp})\,dt = 2n_o R(1-n_{oe})\{1-(\tfrac{r}{R})^2\}^{1/2} \tag{2}$$

It is noted from Eq. (2) that unlike texture 1, a broad color band appears at the center line (r = 0). The bands become narrower symmetrically on both sides of the center line. The overall birefringence of the specimen is positive in the sense described above.

Texture 3. This texture is formed in the smectic phase with the directors aligned perpendicular to the substrate surface as shown in Fig. (2b). The texture consists of a set of coaxial cylindrical layers. In every layers the optic axes are oriented radially. The retardation difference in this texture is

$$\int (n_{\perp} - n_{\parallel})\,dt = n_o R(\tfrac{r}{R})\,\frac{1-n_{oe}^2}{n_{oe}} \cdot \left[\left\{1 - \frac{1}{8}\cdot\frac{1-n_{oe}^2}{n_{oe}}\right\} \tan^{-1}\frac{\{1-(\frac{r}{R})^2\}^{1/2}}{n_{oe}(\frac{r}{R})} - \frac{1}{8}(\tfrac{r}{R})\cdot\frac{1-n_{oe}^2}{n_{oe}}\,\frac{\{1-(\frac{r}{R})^2\}^{1/2}}{(n_{oe}^2-1)(\frac{r}{R})^2+1}\right] \tag{3}$$

The expected color bands are similar to those from texture 1 except that texture 3 reveals more numerous bands for a given bore radius due to the larger retardation difference. The overall birefringence of the specimen is negative since the radial refractive index exceeds the axial. It must be noted that here it is the splay energy rather than the bend energy (texture 1) which becomes larger in the core region and which could cause the optic axis to be tilted from the radial direction. The four cholesteryl esters we studied (nonanoate, laurate, myristate and stearate) produced a uniform texture 3 up to the core region. This indicates that the esters we studied have a small elastic splay constant.

EXPERIMENTAL

The cholesteryl nonanoate used was obtained from Eastman Chemical Co. and purified by recrystallizing from n-pentanol, washed in methanol and dried under vacuum for several days until no further decrease in weight was noted. The pyrex glass capillaries were drawn to several different bore radii in the range between 10 ~ 50 μ. Cholesteryl nonanoate was introduced into one end of the capillary and the other end was sealed with epoxy. The sample was axially sheared repeatedly by changing the air pressure in the capillary by heating or cooling the sealed side of the capillary. With this procedure textures 1 and 2 were obtained in the cholesteric and smectic phases respectively. To obtain texture 3 a dilute solution of hexadecyltrimethyl amonium bromide (HTAB) (4) in toluene (about 1% by volume) was first announced into the capillaries and the solvent was removed by evaporating under vacuum. This process leaves a thin film of HTAB on the inside wall. These treated capillaries readily produced a well-oriented texture 3 upon the small shear resulting from the volume change on transition from the cholesteric to the smectic phase. When the phase changes from smectic to cholesteric, the texture 3 becomes focal conic in these treated capillaries. When the capillaries produce uniform textures, as expected from the calculation for textures 1, 2 and 3, numerous color bands appear under a polarizing microscope. The capillaries filled with nonanoate were embedded in epoxy resin supported between two glass plates. This minimizes the distortion of parallel light beam by the cylindrical glass tube walls. With the tube axis rotated at 45° from the crossed polars' direction of the microscope, color pictures were taken for several different bore radii. The overall birefringence was determined with the aid of a quartz wedge. The wavelength and relative position of color bands in the pictures were measured with a traveling microscope. The data were fitted to the computer-calculated theoretical color bands by assuming various values of n_o/n_e, combined with the value of the average refractive index measured with a temperature-controlled Abbe refractometer.

RESULTS AND DISCUSSION

Figures 3 and 4 show the observed and predicted color bands for texture 1 produced in the cholesteric phase of the nonanoate at 79.0^{o}C (4.4^{o}C above the cholesteric-smectic transition point). For two capillaries of radii 42 and 28 μ, the observed wavelength and position of color bands are in good agreement with the predictions for the values n_o/n_e=1.014 and n (average)=1.49. It was found that the retardation difference was very sensitive to the value n_o/n_e yielding an uncertainty of only $\pm$0.002. On the other hand the measurements are insensitive to n (average) permitting about 10% of error. Our obtained values n_o/n_e and n compare favorably with the measurements near the cholesteric-smectic transition point made by Bottcher and Graber (1) using different methods. However their measured n_o/n_e decreases monotonically as the temperature approaches cholesteric-isotropic transition point while our observation indicates that n_o/n_e remains same within the uncertainty of $\pm$ 0.002. This discrepancy, we believe, is attributable to differences in the sample preparation. When the samples are supported between two glass plates as prepared by Bottcher and Graber, the pitch change resulting from a temperature change would alter the helical structure since the substrate holds the adjacent molecules in the same relative orientation regardless of pitch length. The consequent change in birefringence will then represent that of cholesteric structure perturbed by the boundary conditions of the substrate. However the macroscopic

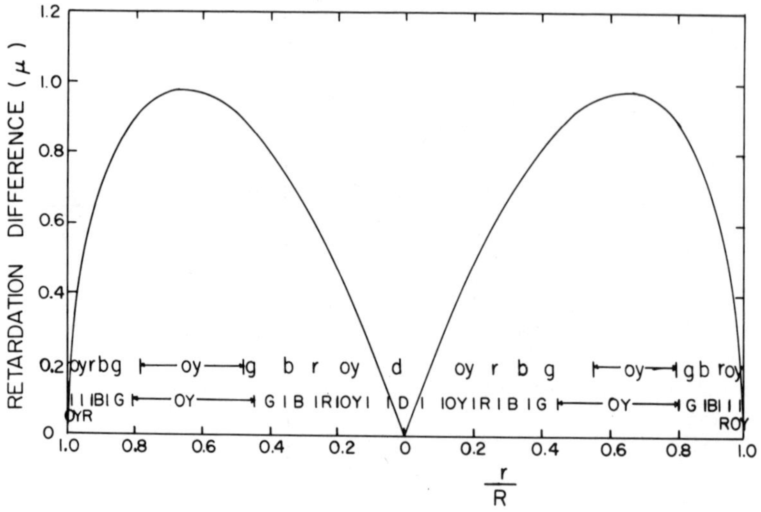

Figure 3. Plot of retardation vs r/R for texture 1 of cholesteryl nonanoate (n_{oe} = 1.014). Also noted are color band ranges calculated (upper case letters) and observed (lower case letters). D = dark, OY = orange yellow, R = red, B = blue, G = green.

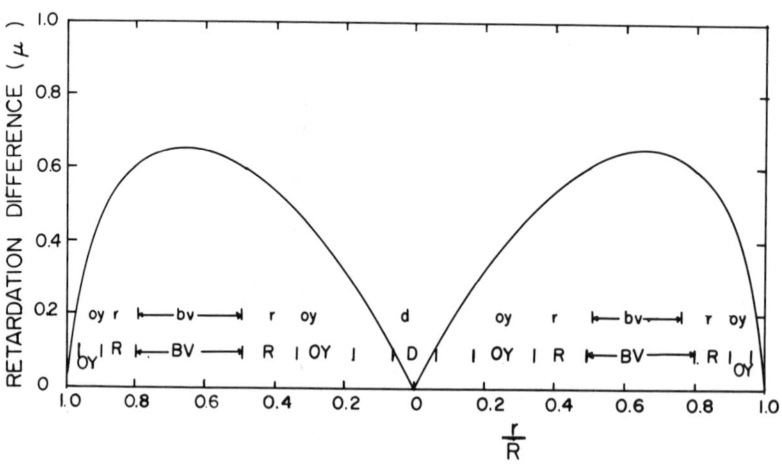

Figure 4. Plot of retardation vs r/R for texture 1 of cholesteryl nonanoate in a 28 μ capillary with n = 1.014. Also noted are calculated and observed ranges of r/R as shown in Figure 3.

birefringence (negative uniaxial) must not be changed by a pitch change alone unless it is accompanied by a change in density or order parameter (5). One may expect that the density change in an anisotropic material could be anisotropic. Thus in a cholesteric the linear thermal expansion in the plane of the nematic-type layer may differ from the linear expansion between the layers. This will result in a change in n_o/n_e. However this effect is not normally expected except in a small temperature range (typically 0.5°C) (6) in the vicinity of the phase transition point. The density change reflects approximately the order parameter change, and a study in cholesteryl nonanoate (7) indicates that the density changes rapidly only in a small temperature range ($\simeq$ 0.5°C) near the phase transition point. Thus the negligible change in n_o/n_e that we observed in the cholesteric phase is to be expected. Our data from textures 2 and 3 in the smectic phase of cholesteryl nonanoate yield n_e/n_o = 1.026 and 1.029 respectively. These are in good agreement with the value obtained by Bottcher and Graber (1). It is also interesting that the axial ratios of 1.028 and 1.014 for the positive uniaxial smectic indicatrix and the negative unaxial cholesteric indicatrix respectively are self consistent, if it is assumed that internal field effects are comparable in the smectic and the cholesteric phases and if the polarizability ellipsoid of the cholesteryl nonanoate molecule is unaxial positive with axes in the ratio $1:1:\sqrt{1.028}$. Then the axial ratios of the indicatrices of the smectic and the cholestric phases will be respectively 1:1:1.028 and 1.014: 1.014:1 (1.014 = (1.028 + 1.000)/2).

In conclusion the cylindrically symmetric textures we produced in capillaries provide a new and reliable method for the study of optical birefringence in liquid crystals. It is also useful in studying the orienting effects of substrates on liquid crystals because the uniformly oriented mesophase near the wall of a capillary is readily seen under the microscope. Further studies on the orienting effects of various substrates on liquid crystals have been done in our laboratory and will be reported.

REFERENCES

1. B. Bottcher and G. Graber, Mol. Cryst. Liq. Cryst. 14, 1 (1971).
2. G. Pelzl and H. Sackmann, Mol. Cryst. Liq. Cryst. 15, 75 (1971).
3. F. C. Frank, Discuss. Faraday Soc., 25, 19 (1958).
4. C. Williams, P. Pieranski and P. G. Cladis, Phys. Rev. Lett., 29, 90 (1972).
5. P. G. deGennes, Mol. Cryst. Liq. Cryst. 12, 193 (1971).
6. F. P. Price and J. H. Wendorff, Mol. Cryst. Liq. Cryst., in press.
7. F. P. Price and J. H. Wendorff, J. Phys. Chem. 76, 276 (1972).

INDUCED ROTARY POWER IN TERNARY MIXTURES OF LIQUID CRYSTALS

James Adams, Gary Dir and Werner Haas

Xerox Corporation, Webster, New York 14580

ABSTRACT

We have observed that a ternary mixture comprising one nematic component and a compensated binary cholesteric results in an inversion wavelength (λ_o) in the near infrared for a broad range of nematic ratios. In particular, the system consisting of N - (p - methoxybenzylidene) - p - butylaniline, cholesteryl nonanoate and cholesteryl chloride was studied over a wide compositional region. Measured pitch values indicate a large deviation from the linear additive law. An interpretation lies in the possibility that both cholesteric species interact with the nematic, but with different strengths. Consequently, the nematic exhibits some net rotary power. A determination of coupling constants will be presented.

I INTRODUCTION

Friedel[1] observed, in 1922, that mixtures of cholesterics with opposite chirality exhibited the chirality of the dominant component. An intrinsically left-handed (L.H.) species, when mixed with an intrinsically right handed (R.H.) species, resulted in a helical pitch value which approached infinity at some specific ingredient ratio, presumably a point at which the two opposite twisting tendencies compensated. A special case of that situation involves the mixture of a nematic, with no twisting tendency, and a cholesteric, the result of which is a cholesteric.

This system has been treated quantitatively by Cano[2,3,4], who showed that pitch versus composition could be calculated quite accurately if it was assumed that the nematic was passive. The more general case of two cholesterics was first treated quantitatively in 1969[5] and later extended to three and four component systems[6]. It was found that pitch dependence on composition was consistent with the viewpoint that each constituent has an intrinsic twisting tendency or effective rotary power (ERP) independent of its concentration or the ERP of other molecular species present, and, furthermore, that the final specific rotation (inverse pitch), θ was linear in both the ERP's and the amounts of ingredients. A surprisingly large number of known systems fall into this pattern. It is also true that non-mesomorphic species, when mixed with certain liquid crystals, play an active role (i.e., can be assigned an ERP different from zero) in helix formation[7]. On the other hand, several systems fail to obey the linear law. For example, cholesteryl iodide assumes the chirality of its partner in binary[8] mixtures (weak effect) and certain combinations of cholesterics and nematics have been observed to deviate so far from prediction that the approach appears not to be relevant to those mixtures[9,10]. We have recently shown[11] that these systems and, in fact, all known binary systems, can be treated effectively by including first order interaction terms, i.e., the ERP of species A may depend (linearly) on the concentration of species B, etc. In particular, the N - (p - methoxybenzylidene) - p - butylaniline (MBBA) cholesteryl chloride (CC) system is highly consistent with this model. It is the purpose of this communication to extend the argument to a three component system.

II THEORY

The linear additive law for a binary mixture is given by Eq.(1)

$$\theta = \alpha\theta_A + (1 - \alpha)\theta_B, \tag{1}$$

where α is the fraction by weight of ingredient A and θ_A is its effective rotary power. Assuming θ_A is independent of α, etc., a first order interaction would modify Eq.(1) according to

$$\theta = \alpha\theta_A + (1 - \alpha)\theta_B + \alpha(1-\alpha)K(A,B), \tag{2}$$

where K (A,B) is a measure of the interaction between A & B. We now extend (2) to a three component system comprising two cholesterics with θ_A and θ_B, respectively, and one nematic ($\theta_C = 0$). To first order, it is possible to rule out significant interaction between the cholesterics by choosing two cholesterics having the property that in binary mixtures (1) is obeyed.* A specific example of such a system consists of CC and cholesteryl nonanoate (CN) which is highly linear down to around 40% by weight CC at which point the CN rich mixture has a tendency to convert to the smectic mesophase (which it does at around 20% CC at room temperature[6]). If β represents the fraction by weight of the nematic in the total fraction mixture and α represents the fraction of CC in the CC/CN mixture, then

$$\theta = \alpha(1 - \beta)\theta_A + (1 - \alpha)(1 - \beta)\theta_B + \alpha(1 - \beta)\beta K(A,C) + (1 - \alpha)(1 - \beta)\beta K(B,C). \quad (3)$$

We will show in the results section, that (3) fits the ternary system described over a substantial compositional range.

III EXPERIMENT

The cholesterics were received from Eastman Kodak and were recrystalized from ethyl alcohol. The MBBA was synthesized by W. Werley (in our laboratories) via a Schiff base reflux reaction and purified by high vacuum distillation. Water content was determined to be less than 100 ppm by gas chromatography. Ingredients were weighed, placed together in a small vial, heated above the isotropic point and thoroughly mixed. Upon cooling, a small amount of material was deposited on a glass slide and caused to adopt the Grandjean texture by shearing with a second slide. The pitch (p) was then measured by determining the inversion wavelength. The inversion wavelength, $\lambda_o = 2np$,† corresponds to the wavelength of minimum transmission, as long as this value is far from the intrinsic absorption edge (∿4000Å for MBBA) and far from any vibrational modes. For most of the wavelengths measured, this was the case; however, as samples approached the blue, the peak corresponding to the reflection of one sense of circular polarized light[12] rides on top of the intrinsic absorption. This is a good example of a symmetric and an asymmetric broadening function combining to produce an apparent peak which is, in fact, shifted from the peak corresponding to the singularity in the symmetric function[13].

---

* First order, in this context, means simply that a strong interaction between the cholesterics is not induced by the presence of the nematic. Obviously, these higher order terms can be included, however, as in the present case to a large degree, they are not required for a fit.

† In the present work, we will not attempt to determine the dependence of index of refraction, n, on composition and wavelength. It is assumed that λ_o data are pitch data. The index variations are small in comparison to the range of λ_o's.

A Cary Spectrophotometer Model 14 was used to determine λ_o. Reproducibility was around 2% as long as data were taken promptly after film creation. Samples (films) exposed to air for periods of hours showed larger shifts (as large as 10%), presumably due to the adsorption of water vapor and other impurities. All data reported were taken within a few minutes after sample preparation. Shelf-life, on the other hand, was no problem and throughout the duration of the experiment ($\approx$6 weeks), previous results were reproduced (to within $\approx$2%), as long as data were taken on freshly prepared films.

IV RESULTS

The experimental data are shown in Figure 1. For convenience, 1/2np is plotted versus β with α as a parameter. In a plot of this nature, systems following (1) appear as a straight line and deviations indicate interactions. The higher curvature in the higher α regime indicates that the influence of the CC is larger as will be determined quantitatively later. A striking feature of this data* is found in the $\alpha = 0.6$ region. The cholesterics are essentially compensated, and treating them as a single ingredient with ERP = 0, infinite pitch when mixed with a nematic is naively expected. In fact, as can be seen, films in this range are almost in the visible. Evidently, the CC and CN influence the MBBA substantially differently. To determine this influence, an ERP (θ_C) for the MBBA for each of the data points in Figure 1 were calculated. The results are shown in Figure 2, where θ_C is plotted as a function of α for various β's. Clearly, for small β, the influence of the nematic increases linearly as the cholesteric fraction is shifted towards higher CC content. However, below 0.6, curvature is apparent and at $\beta = 0.1$, curvature is so large that the entire treatment breaks down. Evidently, in cholesteric rich mixtures, the linear viewpoint is not adequate. To determine K(A,C) and K(B,C), it is necessary to extrapolate the data in Figure 2 to both $\alpha = 0$ and $\alpha = 1.0$. This is done in Figure 3 for that part of the data set compatible with (3). We find $K(A,C) = 1.9 \times 10^5\ cm^{-1}$ and $K(B,C) = 1.1 \times 10^5\ cm^{-1}$. The influence of the CC is approximately twice that of the CN. Furthermore, the CC induces a chirality in the MBBA which is opposite to its own (i.e., induces L.H. behavior) whereas the CN induces its own chirality (L.H.).

---

* It was our experimental attempt to create a close-to-infinite pitch material with approximately equal amounts of cholesteric and nematic which stimulated this study.

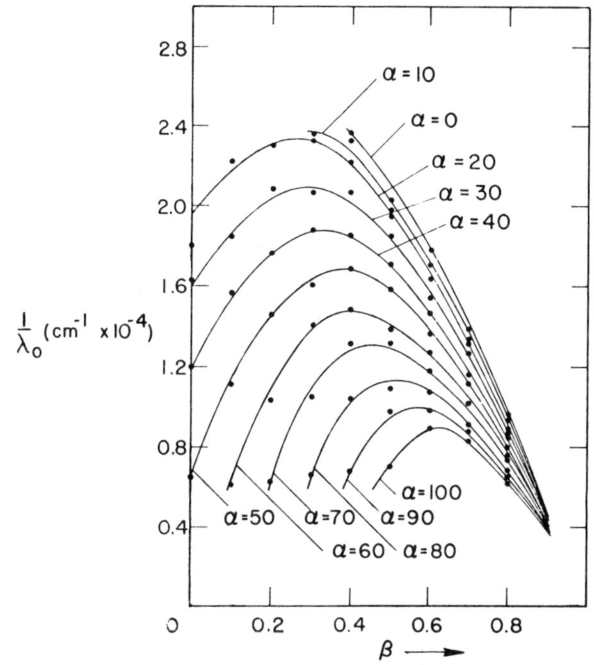

Figure 1. Measured values of $1/\lambda_o$ versus β for various values of α. Lines connect points of equivalent α.

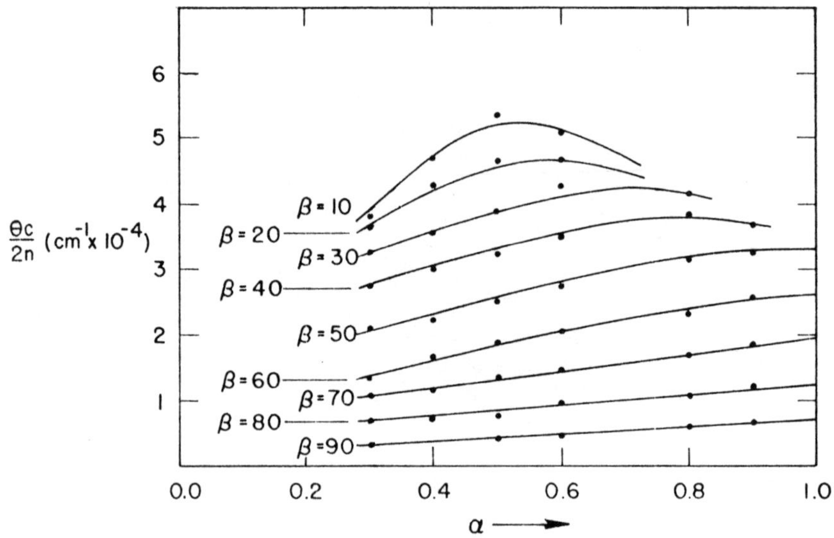

Figure 2. Calculated ERP of nematic as a function of α. Lines connect points of equivalent β.

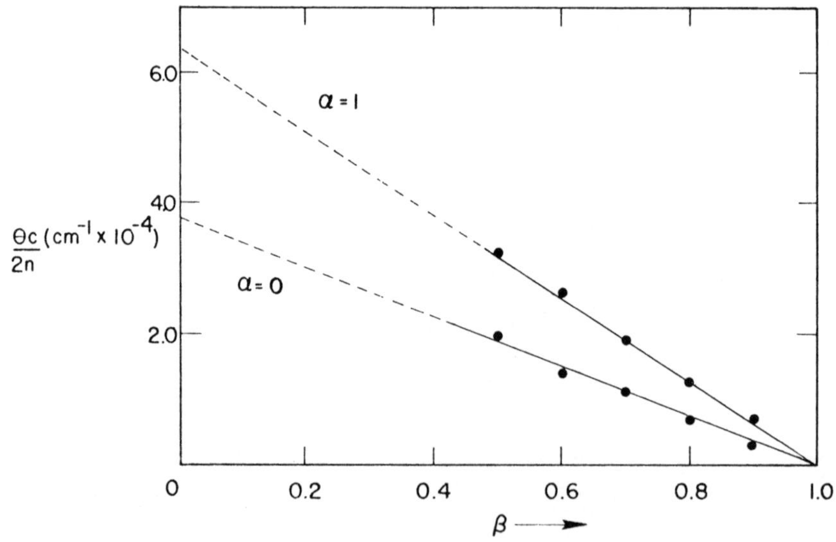

Figure 3. Calculated ERP of nematic in pure CC (α = 1) and pure CN (α = 0) as a function of β.

V CONCLUSIONS

We have shown that in a nematic rich ternary mixture of two cholesterics and one nematic, pitch versus composition data is consistent with the viewpoint that each cholesteric induces a twisting tendency in the nematic and that tendency is linear in the amount of each cholesteric present. This description is clearly not a unique one, however, it is often convenient to be able to express pitch versus composition in a simple functional form for the purpose of calculating pitch sensitivity[13] to temperature, vapor, etc. The approach taken was at best semi-empirical and we do not wish to imply that significant structural conclusions can be drawn. On the other hand, (3) does fit a large data set with only four parameters and, hopefully, will be a useful tool for predicting system response.

ACKNOWLEDGMENTS

The authors thank Ivo Gates for technical assistance.

BIBLIOGRAPHY

1. G. Friedel, Ann. Phys. (Paris) 18 (2), 274 (1922).

2. R. Cano and P. Chatelain, Comptes Rendus 253, 1815 (1961).

3. R. Cano, These, Montpellier, enregistree C.N.R.S. N° 1130, (1966).

4. R. Cano, Bull. Soc. Fr. Mineral. Cryst. 90, 333 (1967).

5. J.E. Adams, W. Haas and J. Wysocki, Bull. Am. Phys. Soc. 14, 6 (1969).

6. J.E. Adams, W. Haas and J. Wysocki, "Liquid Crystals and Ordered Fluids," Plenum Press, New York (1970).

7. J.E. Adams and L.B. Leder, Chem. Phys. Letters 6, 90 (1970).

8. J.E. Adams and W. Haas, Mol. Cryst. and Liquid Cryst. 15, 27 (1971).

9. T. Nakagiri, H. Kodama and K.K. Kobayashi, Phys. Rev. Letters 27, 564 (1971).

10. F. Saeva and J. Wysocki, Am. Chem. Soc. J. 93, 5928 (1971).

11. J.E. Adams and W. Haas, submitted to Mol. Cryst. and Liquid Cryst.

12. J. Fergason, Mol. Cryst. 1, 293 (1966).

13. J.E. Adams, B.F. Williams and R.R. Hewitt, Phys. Rev. 151, 238 (1966).

14. J.E. Adams and W. Haas, Mol. Cryst. and Liquid Cryst. 16, 33 (1972).

CONDUCTIVITY DIFFERENCES IN THE CHOLESTERIC TEXTURES

Gary Dir, James Adams and Werner Haas

XEROX CORPORATION

Webster, New York 14580

ABSTRACT

Both the focal conic and Grandjean textures can be characterized by a conductivity tensor with uniaxial symmetry. We have observed that the diagonal element corresponding to the conductivity in the special direction is ≈ 20% larger in the focal conic case. The film studied was a mixture of 80% N-(p-methoxybenzylidene)-p-butylaniline and 20% cholesteryl oleyl carbonate. The transformation to the focal conic texture was produced by the application of a dc voltage and the reverse transformation was induced by an ac voltage. At sufficiently low current levels, current does not perturb structure (at least to the extent that the optical and electrical properties are not altered). At higher levels, the measuring stimulus converts the film to an intermediate state and both textures approach the same current level since they are converted from their original symmetries into the same state. The conductivity differences observed are consistent with conductivity ratio data for the nematic component and the molecular distributions in the two textures. The influence of resistivity and temperature on the focal conic/Grandjean conductivity ratio is shown.

INTRODUCTION

The application of an electric field across a thin

nematic layer produces a current in that direction, the magnitude of which depends on the angular distribution of molecular axes. There are two limiting cases: molecules parallel to the field (to within the order parameter) and molecules perpendicular to the field. The components of the conductivity tensor for these two cases are called $\sigma_{\parallel}$ and $\sigma_{\perp}$, respectively. (If the field is applied in the Z direction, $\sigma_{\parallel}$ determines the current flow in the Z direction when the molecules are aligned in the Z direction, etc.) As early as 1914[1], measurements of this nature revealed a substantial anistropy in para-azoxyanisole. In the case of nematics, the orientation of the molecules can be controlled by external fields[2], surface treatments[3], and modified, of course, by temperature. In this study, we report an experimental determination of conductivity differences in the two cholesteric textures(comprising mainly nematic molecules) and show that these values are consistent with measured values of $\sigma_{\parallel}$ and $\sigma_{\perp}$ and the molecular distributions ascribed to the textures.

THEORY

The cholesteric state is characterized by local helical symmetry. The molecules lie in planes and within one plane there is a preferential alignment. This special direction rotates from plane to plane producing a helical structure. To describe the Grandjean and focal conic textures it is convenient to introduce a vector, we call the pitch vector, the direction of which corresponds to the local direction of the helical axis and the magnitude of which depends on the local pitch value. The Grandjean texture is characterized by a homogeneous pitch vector field, the pitch vector being everywhere approximately normal to the substrate. In the focal conic texture, the pitch vector is essentially parallel to the substrate and macroscopically randomly oriented in that plane.[4] However, over a space of a few microns the pitch vector field can be quite homogeneous, often making sharp changes of direction at the boundaries of uniform regions, giving rise to a domain-like structure. This molecular distribution is not isotropic. The appropriate conductivity tensors are shown in Table I (See next page). By convention, the field is always applied in the Z direction. The homeotropic case corresponds to molecules aligned parallel to Z and the homeogenous case corresponds to molecules aligned parallel to substrate in the (arbitrary) X direction.

HOMOGENEOUS

$$\begin{pmatrix} \sigma_{\parallel} & 0 & 0 \\ 0 & \sigma_{\perp} & 0 \\ 0 & 0 & \sigma_{\perp} \end{pmatrix}$$

HOMEOTROPIC

$$\begin{pmatrix} \sigma_{\perp} & 0 & 0 \\ 0 & \sigma_{\perp} & 0 \\ 0 & 0 & \sigma_{\parallel} \end{pmatrix}$$

GRANDJEAN

$$\begin{pmatrix} \left(\frac{\sigma_{\perp}+\sigma_{\parallel}}{2}\right) & 0 & 0 \\ 0 & \left(\frac{\sigma_{\perp}+\sigma_{\parallel}}{2}\right) & 0 \\ 0 & 0 & \sigma_{\perp} \end{pmatrix}$$

FOCAL CONIC

$$\begin{pmatrix} \left(\frac{3\sigma_{\perp}+\sigma_{\parallel}}{4}\right) & 0 & 0 \\ 0 & \left(\frac{3\sigma_{\perp}+\sigma_{\parallel}}{4}\right) & 0 \\ 0 & 0 & \left(\frac{\sigma_{\perp}+\sigma_{\parallel}}{2}\right) \end{pmatrix}$$

ISOTROPIC

$$\begin{pmatrix} \left(\frac{2\sigma_{\perp}+\sigma_{\parallel}}{3}\right) & 0 & 0 \\ 0 & \left(\frac{2\sigma_{\perp}+\sigma_{\parallel}}{3}\right) & 0 \\ 0 & 0 & \left(\frac{2\sigma_{\perp}+\sigma_{\parallel}}{3}\right) \end{pmatrix}$$

TABLE I

EXPERIMENTAL

A cholesteric-nematic mixture of 80 wt. % N-(p-methoxybenzylidene)-p-butylanilene (MBBA) and 20 wt. % cholesteryl oleyl carbonate (COC) was used throughout the experiments. This mixture was chosen because of its texture forming and retention properties. The pitch of the mixture was determined to be approximately 0.5 micron. The MBBA was synthesized by W. Werley of our laboratories, using a Schiff base reflux reaction. Vacuum distillation was used to purify the material. The COC was purchased from Eastman Kodak and used as received. The conductivity of the MBBA-COC mixture was controlled by doping with the ionizable salt hexadecl-trimethyl ammonium chloride.

Samples were made using the conventional sandwich layer structure consisting of glass-liquid crystal-glass. Transparent conductive indium oxide electrodes were coated on the glass surfaces contacting the liquid crystal. Cell thickness was controlled by use of 1/2-mil Tedlar spacers.

Samples were converted from Grandjean to focal conic by the application of a voltage substantially above the turbulent scattering threshold for a period of a few seconds.[5] After voltage removal, the sample relaxed into the focal conic state. Samples were converted from focal conic to Grandjean via the application of an ac field of the appropriate frequency and amplitude.[6] To insure that results were independent of the mechanism of texture conversion, the focal conic was also produced by heating the sample to a temperature above isotropic and allowing it to cool. Identical results were obtained. Values were independent of the polarity of the applied voltage. Conductivity ratios were also measured in two samples at one volt peak ac and found to agree with low voltage dc results.

The dc conductivities of the Grandjean and focal conic textures were determined via current-voltage measurements. The experimental apparatus consisted of a regulated dc power supply in electrical series with the test sample and a current sampling resistor. A Leitz ortholux polarizing microscope was used to observe optical behavior which was correlated with the I-V measurements. Sample temperature was regulated by use of a Mettler FP2 microthermal controller. This instrument has an accuracy of 0.2°C. A mixture of 2-propanol and solid carbon dioxide was used to lower temperatures below ambient values when necessary. All measurements (except for temperature dependence) were performed at 28°C.

RESULTS

In the low voltage regime, both σ_G ($\sigma_G \equiv \sigma_\perp$) and $\sigma_F[\sigma_F \equiv (\sigma_\parallel + \sigma_\perp)/2]$ the ZZ components of the Grandjean and focal tensors are independent of voltage. An I-V plot is shown in Figure 1 (See next page). In the high voltage regime (substantially above the onset of turbulent scattering), there is, of course, only one value since the measuring field converts both samples into the same state. At intermediate fields, the measuring field perturbs the textures. Optical observations indicate the onset of the grid deformation[7] at around 8 volts. At approximately 12 volts, localized regions (Grandjean) began to convert to the focal conic texture. Although the degree of conversion increased with time, after long times (minutes), no further conversion was observed, and the resulting film was a mixed texture. Figure 2 (See next page) is a plot of σ_G and σ_F over the extended voltage range. Data were taken 15 seconds

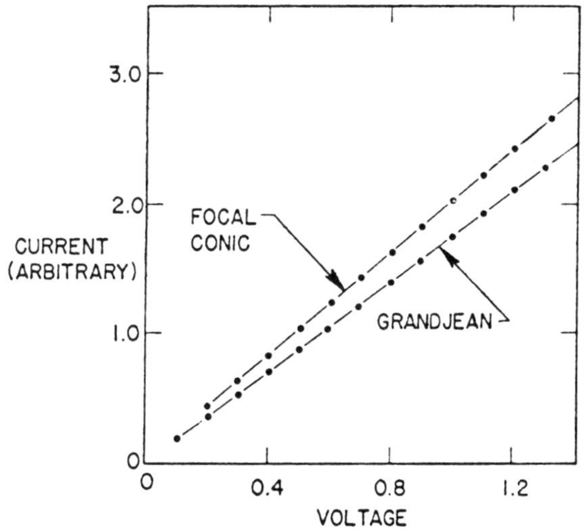

FIGURE 1. Current-voltage characteristics in the low voltage regime.

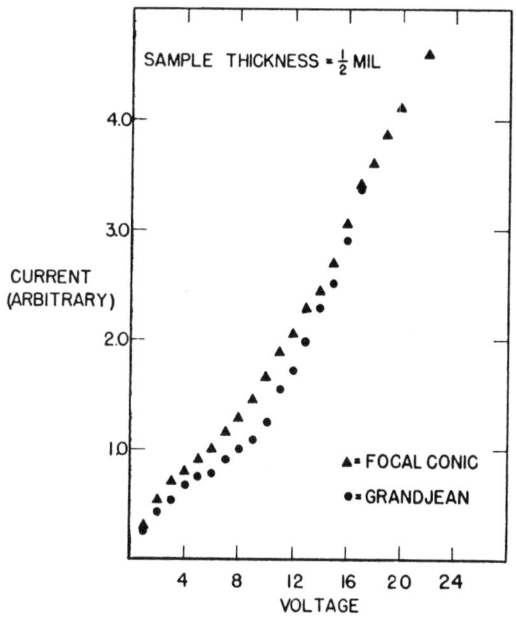

FIGURE 2. Current-voltage characteristics over an extended voltage range.

after application of voltage.

The influence of resistivity on σ_F/σ_G is shown in Figure 3. Resistivity was measured in the low voltage regime in the focal conic texture. The ratio decreases slowly with decreasing resistivity. The effect of temperature on the conductivity ratio is shown in Figure 4 (See next page). The observed behavior is similar to that reported by Diguet et al. in MBBA.[8] Far from the isotropic point, σ_F/σ_G is substantially independent of temperature and has a value of 1.2.

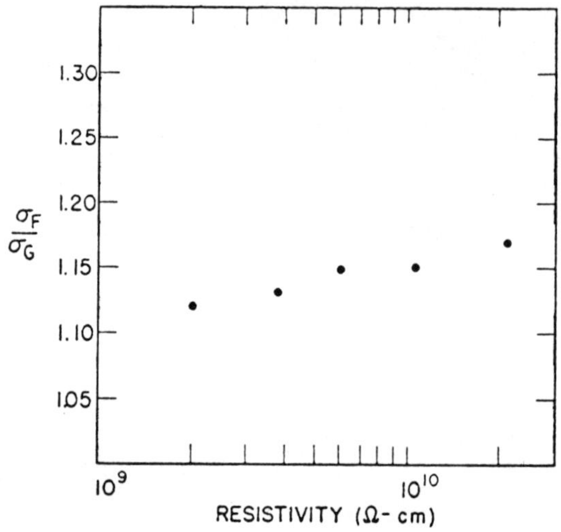

FIGURE 3. Conductivity ratio dependence on resistivity.

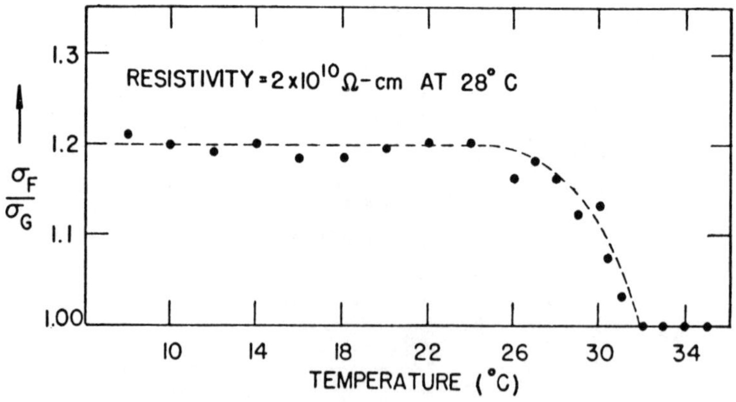

FIGURE 4. Temperature dependence of conductivity ratio.

The calculated value depends both on $\sigma_{\parallel}$ and $\sigma_{\perp}$ for MBBA and on $\sigma_{\parallel}$ and $\sigma_{\perp}$ for COC. The second data are not available, however, since only 20% of the cholesteric is present, presumably the nematic dominates the behavior. Assuming the nematic ratios are appropriate, σ_F/σ_G of 1.2 compares favorably with the data of Diguet[8] et al. and even better with recent data from Eastman Kodak.[9]

CONCLUSIONS

We have determined that substantial conductivity differences exist between the focal conic and Grandjean texture. Furthermore, it has been shown that these differences are consistent with present models of the textures.

BIBLIOGRAPHY

1) T. Svedberg, Ann. Physik 44, 1121 (1914).
2) O. Lehmann, Ann. d. Phys. 2, 676 (1900).
3) P. Chatelain, Bull. Soc. Fr. Miner. Crist. 66, 105 (1943).
4) J. E. Adams, W. Haas and J. Wysocki, JCP 50, 2458 (1969).
5) G. Heilmeier, J. E. Goldmacher, Appl. Phys. Letters 13, 132 (1968).
6) W. Haas, J. E. Adams and J. B. Flannery, Phys. Rev. Letters 24, 577 (1970).
7) W. Helfrich, Appl. Phys. Letters, 531 (1970).
8) D. Diguet, F. Rondelez and G. Durand, C. R. Acad. Sc. Paris, L71B, 954 (1970).
9) Eastman Liquid Crystal Products, JJ-14 (1973).

ELECTRO-OPTICAL PROPERTIES OF IMPERFECTLY ORDERED PLANAR CHOLESTERIC LAYERS

C.J. Gerritsma and P. van Zanten

Philips Research Laboratories

Eindhoven-Netherlands

ABSTRACT

The electro-optical behaviour of imperfectly ordered planar layers of a mixture of 20% cholesteryl chloride and 80% cholesteryl oleyl carbonate (natural pitch p_0 = 0.37 μm) is discussed. Microscopic observations reveal that these layers consist of a number of uniform planar regions with different pitch ($p \neq p_0$), separated by disclinations. With increasing layer thickness these regions become smaller while the differences in pitch decrease. The electric-field-induced blue shift of the selective reflections is suggested as resulting from a successive distortion of the planar regions. The observed decrease in blue shift with increasing layer thickness, as well as the absence of this shift in regions with uniform pitch, demonstrate the correlation between the blue shift and the field-induced periodic distortion.

I INTRODUCTION

Cholesteric layers with the helix axes perpendicular to the boundaries (planar texture) show a characteristic selective reflection of incident light. The wavelength of maximum light reflection is given by $\lambda_{max} = n.p$, while the width of the reflection band is of the order $p\Delta n$ [1,2]. In these relations p, n and Δn denote the pitch of the helix, the mean refractive index, the difference between the extraordinary and ordinary refractive index, respectively. Harper

[3] observed a shift of the reflection peak to shorter wavelengths when an electric field was applied parallel to the helical axes. This blue shift in λ_{max}, which has been ascribed to a field-induced pitch contraction was studied experimentally in more detail by several investigators [4-7]. This situation of a field parallel to the helices has been analyzed theoretically by Meyer [8] and Leslie [9]. For a cholesteric liquid crystal with a positive anisotropy of the dielectric constant ($\varepsilon_{\parallel} - \varepsilon_{\perp} > 0$), Meyer predicts that above a threshold field a conical deformation of the planar texture will occur. This uniform conical deformation is accompanied by a contraction of the pitch resulting in a shift of λ_{max} to shorter wavelengths. The deformation is produced only when $k_{33}/k_{22} < 1$, where k_{33} and k_{22} are the Frank elastic constants for bend and twist, respectively. Leslie's more exact treatment leads to practically the same results, provided that there are many turns of the helix in the layer, i.e. $N = d/p \gg 1$, d being the thickness of the layer.

We have previous [5] measured the blue shift of a mixture of cholesteric esters as a function of the applied electric field. The observed shift was found to be an order of magnitude smaller than predicted by Meyer's theory. Furthermore, the blue shift was preceded by the formation of a periodic grid-like pattern, which indicates that the deformation is not uniform as required by the model. These observations suggest that the observed blue shift is not due to a conical deformation of the texture. Recently [10] we proposed a new explanation in which the shift is not caused by a pitch contraction but results from the field-induced periodic deformation of the planar texture.

In section II a more detailed discussion of the correlation between the periodic distortion and the blue shift is presented, leading to further predictions concerning the blue shift. The material and sample preparation, the measurement of the voltage thresholds for periodic perturbation and the testing of the predictions are discussed in sections III, IV and V, respectively.

II PERIODIC PERTURBATION AND BLUE SHIFT

By now it is well known that an electric field can induce periodic perturbations in the planar cholesteric texture [5,11,12]. These instabilities, predicted by Helfrich [13], occur above a threshold voltage V_c. For the case $\Delta\varepsilon > 0$, the theory [14,15] predicts for this threshold:

$$V_c^2 = f\cdot(6\ k_{22}k_{33})^{\frac{1}{2}}\cdot d/p_0\ , \qquad (1)$$

where p_0 is the value of the natural- or undeformed pitch. The factor f can be different for static and alternating fields [15]. In the case of a purely dielectric deformation $f = 8\pi^3/\Delta\varepsilon$, leading to a frequency independent threshold voltage (until $\Delta\varepsilon$ decreases at higher frequencies due to dielectric relaxation). When space charge and hydrodynamic effects play a role, the factor f is more complicated and directly dependent on the frequency [15]. Equation (1) was verified experimentally for long-pitch cholesterics obtained by doping appropriate nematics [16,17]. However, it was found that equation (1) is no longer quantitatively valid when the helix pitch deviates somewhat from its equilibrium value p_0. Such a situation occurs e.g. in Grandjean textures [18] and in other non-uniform planar layers [10] where the helix is unwound or tightened somewhat.

A wedge-shaped planar layer shows a series of equally spaced disclinations, which follow lines of equal specimen thickness (Grandjean texture). At a disclination there is a discontinuity Δp in the pitch value. Within a strip between two successive disclinations the pitch increases linearly with distance from $p_0 - \Delta p/2$ to $p_0 + \Delta p/2$. As the number of helix turns $N = d/p$ is constant, a linear change in λ_{max} across each strip is observed [19,20]. The difference in N for two successive strips is $|1/2|$ for a single disclination line and $|1|$ for a double disclination. The corresponding jumps in pitch are $\Delta p = p_0^2/2d = p_0/2N$ and p_0/N, respectively. Single disclinations are generally found in samples with a low value of N while also for higher N-values double disclinations arise [21]. If eq. (1) was valid for non-equilibrium pitches too, one would expect the Grandjean pattern to be deformed strip by strip on increasing the voltage. Actually this is not observed.

In a small-angle wedge the periodic deformation starts in each strip at the long-pitch side ($p > p_0$) of each disclination and at almost the same voltage [18]. At increasing voltage, the frontier of the grid deformation spreads smoothly across the strips to the low-pitch sides. Since the deformed parts have different

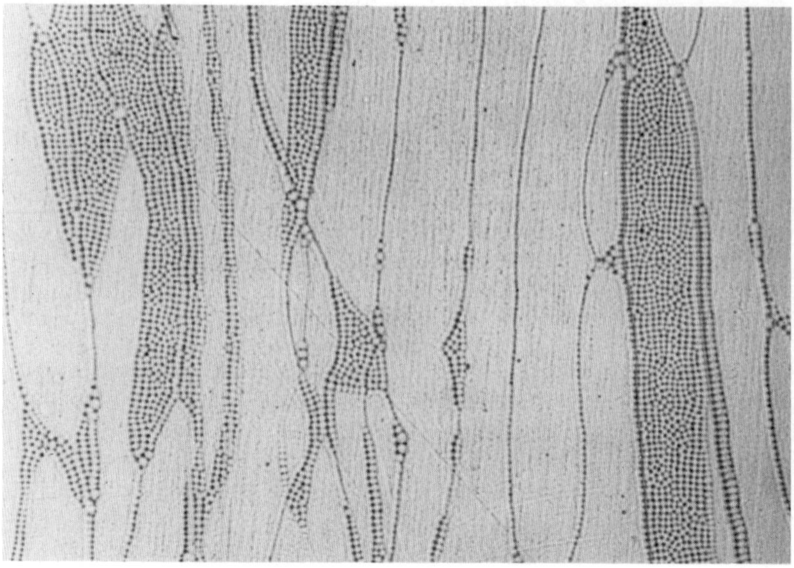

Fig. 1. Microscopic picture of an imperfectly ordered planar texture with periodically perturbed regions ($p > p_0$). Thickness of the layer: 25 μm, applied voltage: 50 V, magnification: 160x.

scattering angles, these parts of the strip will contribute less to the original selective reflection band. This results in a blue shift of this reflection band at increasing voltage. For the cholesteric esters used in previous experiments [5,10] we found differences of 45% in V_c on both sides of a single disclination in a layer with $N \approx 50$. For double disclinations even differences of 100% are reported [18].

In the case of imperfectly ordered layers a similar explanation for the blue shift can be given. Such a layer is built up of a large number of small planar regions with constant number of helix turns [10], separated by disclinations. Now N, which is not necessarily an integer, can vary because of differences in d and random boundary conditions. This leads to a distribution of pitches around p_0. For regions with $p < p_0$, λ_{max} is lower than for regions with $p = p_0$. The opposite is true for regions with $p > p_0$. Again we observed that the periodic deformation begins in the regions with longest pitch and extends to shorter pitch regions at increasing voltage. This effect is shown in fig. 1, where the periodic pattern is present in a number of regions with $p > p_0$, while regions with shorter pitch are still unperturbed. Just as in the case of a Grandjean texture this results in a decrease of the intensity and in a shift of λ_{max} to shorter wavelengths. Again the variations in p are only small (~ 2%), while the corresponding differences in V_c are much greater (~ 50%), leading to a blue shift in a broad voltage range. At higher voltages, when the smallest-pitch regions are also perturbed, the effect becomes less reversible because the long-pitch regions are transformed to the rather stable "focal conic" texture.

If our explanation is correct the blue shift can only occur in non-uniform planar textures. Therefore, the shift should be completely absent within a region of uniform pitch. Furthermore, the shift should decrease with increasing thickness because Δp is inversely proportional to d. These predictions are tested in Section V.

III MATERIAL AND SAMPLE PREPARATION

The material used was a mixture of 20 percent (by weight) cholesteryl chloride (CC) and 80 percent cholesteryl oleyl carbonate (COC). As before, this mixture was chosen because its temperature coefficient $d\lambda_{max}/dT$ is as low as 2 nm/K. At 25°C the line width at half intensity was about 20 nm, while λ_{max} was about 550 nm. In the planar texture at 25°C the dielectric constant parallel to the helices was 2.50, while in the "focal conic" texture (helix axes are parallel to the surface [22]), the dielectric constant perpendicular to the helices was 2.65. From these results we derive $\varepsilon_{\parallel} = 2.8$ and $\varepsilon_{\perp} = 2.5$. As in both textures the helical ordering was not completely uniform, these values of $\varepsilon_{\parallel}$ and $\varepsilon_{\perp}$ are not very accurate. The specific resistance of the material ρ was 10^{13}-10^{14} Ωcm.

Samples of various thickness in the 6-200 μm range were prepared by confining the isotropic material to a sandwich cell of known thickness. The glass plates constituting the cell were polished and coated with a thin layer of transparent conducting tin oxide. After cooling to room temperature a slight displacement of the cover slide always produced a more or less uniform planar texture. The degree of ordering of this texture was found to depend on factors such as uniform surface alignment and layer thickness.

Layers with uniform surface orientation could be obtained by the Chatelain technique of uni-directional rubbing. Especially well aligned samples were obtained when the layer thickness was small (d = 6-15 μm). In general these samples displayed two or three Grandjean disclinations at a mutual distance of about 1-2 cm. It is difficult to avoid these disclinations since the cells are always slightly wedge shaped and the pitch of the helix is small. At larger thickness (d > 15 μm), the Grandjean pattern becomes less regular and the strips are broken up into a number of planar regions. Fig. 1 shows such an imperfectly ordered texture; regions with $p > p_0$ are periodically perturbed by an electric field. At increasing thickness the planar regions become gradually smaller, while the differences in pitch decrease. In the region d = 100-200 μm the boundaries between the discrete planar regions become more and more vague, resulting in a texture which resembles that of a layer supported by one glass plate only.

When the surfaces are not rubbed the surface orientation varies randomly. Now the various planar regions with different pitches are already formed in the thinnest samples. In these imperfectly aligned samples the regions are much smaller, while the boundaries are less defined.

The purity of the cholesteric material can also influence the uniformity of the texture. CC-COC samples contaminated with a few percent of lecithin or oleylalcohol always showed a less perfect ordering than pure samples. This might be ascribed to the surface-active properties of these substances, thus opposing the rubbing-induced surface alignment. Since oleylalcohol is a natural contamination of COC, the effect is also expected in unpurified mixtures.

IV THE VOLTAGE THRESHOLD FOR PERIODIC DEFORMATION

Using cholesteric layers between rubbed surfaces, the threshold for the periodic perturbation on both sides of a single disclination was measured as a function of thickness. The onset of the periodic perturbation is determined with a microscope as the optical appearance of the grid-like pattern. Figure 2 shows the results for the long-pitch side (curve I) and the small-pitch side (curve II) for thicknesses between 7 and 50 μm. As curves I and II have the same slope (0.57 ± 0.05),

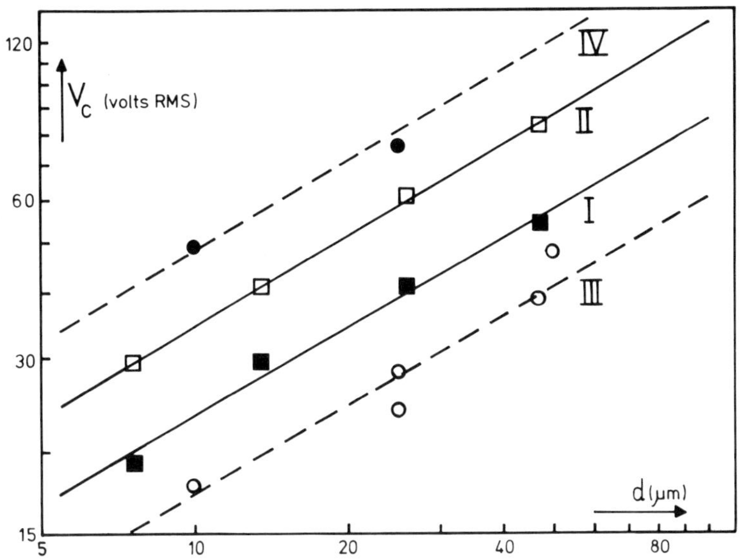

Fig. 2. Thickness dependence of the threshold voltage V_c for periodic deformation of an imperfectly ordered planar texture.
curve I and II: V_c on both sides of a single disclination,
curve III and IV: V_c in regions with longest and smallest pitch, resp. (probably on both sides of a double disclination).

the ratio of V_c on both sides of a single disclination is constant. Its value is 1.5 for N between 15 and 150. The threshold voltage for equilibrium pitch $V_c(p_0)$ is the average of curves I and II because p_0 is located in the middle between two disclinations [20]. Since both curves are parallel, the slope of the log $V_c(p_0)$ vs log d curve is also 0.57 ± 0.05, in fairly good agreement with the value of 0.5 predicted by eq. (1).

In imperfectly ordered samples most of the planar regions have a threshold voltage near curves I and II. This indicates that the width of the pitch distribution is practically equal to the difference in pitch on both sides of a single disclination in a sample of the same thickness. However, in a small number of regions deviations were noted. In some of them λ_{max} is larger than at the long-pitch side of a single disclination, while in the other ones λ_{max} is smaller than at the small-pitch side of this disclination. In the first type of regions V_c was considerably lower than corresponds to curve I, while in the other regions the values of V_c were greater than curve II. These thresholds are also plotted in fig. 2 as a function of thickness. Two curves III and IV drawn parallel to I and II fit these measurements fairly well. The analogy of fig. 2 with Scheffer's observations [18] suggests that curves III and IV correspond to the thresholds on both sides of a double disclination. This is in agreement with the fact that neither type of region was found in the thinnest sample ($d = 7$ μm). The ratio of threshold voltages is 2.6 for the case of curves III and IV. Furthermore, we noted that in samples with $d > 50$ μm the regions correspondin to the small-pitch side of a double disclination (curve IV) were absent, suggesting that they are less stable than the largest-pitch regions. This seems to be supported by the observation that in thinner samples, regions with $p > p_0$ grow at the expense of regions with $p < p_0$ on application of a voltage just below the threshold.

Finally, a comment on the nature of the instability leading to the periodic deformation of the planar texture. For the samples with $\rho \sim 10^{14}$ Ωcm, V_c is frequency-independent of 0 to at least 10^4 Hz. In samples for which ρ was 10^{12} Ωcm, V_c had practically the same value and was also found to be independent of frequency in the same frequency range. This means that the observed deformation is purely static. Therefore the instability is suggested as resulting from the dielectric anisotropy at voltages where space charge effects are still unimportant. This is not surprising in view of the relatively high resistivity of the samples.

V SELECTIVE REFLECTIONS FROM UNIFORM AND NON-UNIFORM PLANAR REGIONS

Based on the correlation between the blue shift and the periodic perturbations (section II) two predictions were made. First, the shift decreases with increasing thickness since Δp is inversely proportional to d. Furthermore, within a uniform pitch region the shift will be completely absent. In the following paragraphs we describe the experimental verification of these predictions. The selective reflections of the various samples were measured accurately at 25° C using an optical arrangement previously described in detail [5].

a. Uniform Pitch Regions.

For the experiment on absence of the blue shift in a uniform pitch region, a well aligned 6 μm sample was prepared using the rubbing technique. This sample displayed only three single Grandjean disclinations at a mutual distance of about 15 mm. The selective reflection was measured at normal incidence; reflected circularly polarized light was detected in the backward direction by means of a half-mirror beam splitter. To avoid interfering specular reflections, the sample cell was provided with a circular polarizer which is only transparant to the light reflected from the cholesteric layer. The selective reflection spectrum of a "uniform" pitch region was obtained by using a light beam with a small diameter (1.3 mm) located in the middle of a strip of the Grandjean texture. In such a small region the variation in pitch was less than 0.25%.

Microscopic observations showed that for this part of the strip the periodic grid-like pattern becomes visible at 21 V. Now λ_{max} and the intensity were measured for voltages below and above V_c. The results of these measurements are given in table I; the accuracy of λ_{max} being about 0.5 nm and the intensity reproducable to about 1%. From table I it is evident that the intensity is practically constant below 21 V, while the intensity decreases considerably above V_c. However, λ_{max} does not shift to shorter wavelengths, in agreement with the prediction.

TABLE I

Voltage	λ_{max} (nm)	relative intensity in percent
0	557.5	100
10	558.0	100
16	558.0	100
18	556.5	97
20	557.0	96
22	557.5	94
24	557.5	89
26	557.5	72
27	557.5	53

b. Imperfectly Ordered Samples.

For the thickness-dependent shift measurements, samples between rubbed surfaces and thicknesses of 25, 100 and 200 μm were prepared. Contrary to the measurements reported above, the cross-section of the light beam was large compared with the diameter of the planar regions. The selective reflection of the 25 μm sample was measured at normal incidence. In the thicker samples the selective reflection can be measured without a circular polarizer, since these samples showed additional reflections at $I \neq R$ (I and R were the angle of incidence of the light beam and angle of observation, respectively). In this situation λ_{max} can be calculated from the centre of the reflection band, λ_{peak}, if I, R and the refractive index ($n = 1.51$) are known.

On increasing the voltage across an imperfectly ordered layer, the intensity of the reflection peak begins to decrease as soon as the periodic perturbation starts in one or more regions with $p > p_0$. At the same time a shift in λ_{max} to shorter wavelengths is observed. The 25 μm and 100 μm sample showed a shift increasing with voltage. For the 200 μm sample, neither a shift nor a periodic perturbation was observed for voltages up to 700 V, where electrical breakdown occurred. Obviously the texture of this sample, which looked very turbid, was not planar.

The maximum blue shift of an imperfectly ordered sample is in principle determined by the maximum difference in pitch between the planar regions. In general the maximum shift is greater for an imperfectly ordered planar texture than for a Grandjean texture. The maximum observable blue shift is determined by the sensitivity of the spectrometer. When comparing the results of imperfectly ordered samples with those of Grandjean textures, the blue shift $\Delta\lambda$ is determined at the voltage V_c of curve II in fig. 2 where the last part of a Grandjean strip is just deformed. The value of $\Delta\lambda$ is expected to be of the same order of magnitude as the difference in λ_{max} on both sides of a single disclination. The latter difference is equal to $n \cdot \Delta p = n \cdot p_0^2/2d$.

The results of our $\Delta\lambda$-measurements are summarized in table II. The data for the 12 μm sample are collected from previous work [5,10], while the threshold voltage for the 100 μm sample, needed to determine $\Delta\lambda$, was obtained by extrapolation of curve II in fig. 2. Table II clearly demonstrates that the shift is of the order

TABLE II

sample thickness d (μm)	blue shift Δλ (nm)	n·Δp (nm)	d·Δλ (μm. nm)
12	12	9.5	144
25	5.5	4.6	138
100	1.4	1.15	140

of the differences in λ_{max} at both sides of a single disclination and is inversely proportional to the specimen thickness.

We have previously reported [10] that the shift of a 12 μm, imperfectly ordered sample was of the order of the line width. This can be explained in the following way. For a uniform pitch region the line width is of the order $p \cdot \Delta n$; for imperfectly ordered samples the spread Δp in pitch values will lead to further broadening of the reflection peak. Because Δp decreases with d, the line width will also have the tendency to decrease with d. This is indeed observed: the width of the reflection peak at half intensity, $\Delta\lambda_{\frac{1}{2}}$, is equal to 23.5, 18.0 and 10.0 nm for the 6, 25 and 100 μm samples, respectively. Now, the blue shift is of the order of $\Delta\lambda_{\frac{1}{2}}$ in layers where the contribution of Δp to the line width is at least of the same order of magnitude as the "natural" line width $p \cdot \Delta n$. This situation is more or less realized in the 12 μm sample; for thicker samples the ratio $n \cdot \Delta p / \Delta\lambda_{\frac{1}{2}}$ is continuously decreasing with increasing d.

VI CONCLUSIONS

In this article we have shown that the blue shift of the selective reflection results from a periodic distortion of the texture, rather than from a pitch contraction. The correlation between the blue shift and the periodic distortion is supported by the following facts:

- the shift is observed to occur in imperfectly ordered samples,
- the onset of the blue shift coincides with the threshold for the largest-pitch regions ($p > p_0$),
- the shift is inversely proportional to the layer thickness, and of the order of magnitude of the difference in λ_{max} on both sides of a single disclination,
- the shift does not exceed the line width of the selective reflection peak.

REFERENCES

1. Hl. de Vries, Acta Cryst. 4, 219 (1951).
2. J. Fergason, Liquid Crystals, Gordon and Breach, New York (1967), p. 89. This work was presented at the 1st Int. Conference on Liquid Crystals, Kent, Ohio, August 1965.
3. W.J. Harper, Liquid Crystals, Gordon and Breach, New York (1967), p. 121.
4. J.R. Hansen and R.J. Schneeberger, IEEE Trans. Electron Devices, Vol. ED-15, 896 (1968).
5. C.J. Gerritsma and P. van Zanten, Mol. Cryst. 15, 257 (1971).
6. G.A. Dir et al., Proc. Intern. Symposium of the Society for information display, Philadelphia (1971) p. 132.
7. N. Oron and M.M. Labes, Appl. Phys. Lett. 21, 243 (1972).
8. R.B. Meyer, Appl. Phys. Lett. 12, 281 (1968).
9. F.M. Leslie, Mol. Cryst. 12, 57 (1970).
10. C.J. Gerritsma and P. van Zanten, Phys. Lett. 42A, 329 (1972).
11. C.J. Gerritsma and P. van Zanten, Phys. Lett. 37A, 47 (1971).
12. F. Rondelez and H. Arnould, C.R. Acad. Sci., Ser. B237, 549 (1971).
13. W. Helfrich, Appl. Phys. Lett., 17, 531 (1970).
14. W. Helfrich, J. Chem. Phys., 55, 839 (1971).
15. J.P. Hurault, J. Chem. Phys., to be published August 1973.
16. F. Rondelez, H. Arnould and C.J. Gerritsma, Phys. Rev. Lett., 28, 735 (1972).
17. T.J. Scheffer, unpublished results.
18. T.J. Scheffer, Phys. Rev. Lett., 28, 593 (1972).
19. P. Kassubek and G. Meier, Mol. Cryst., 8, 305 (1969).
20. C.J. Gerritsma, W.J.A. Goossens and A.K. Niessen, Phys. Lett., 34A, 354 (1971).
21. Orsay Liquid Crystal Group, Phys. Lett., 28A, 687 (1969), J. Phys. C 30, 38 (1969).
22. J.E. Adams, W. Haas and J. Wysocki, J. Chem. Phys. 50, 2458 (1969).

TEMPERATURE DEPENDENCE AND RHEOLOGICAL BEHAVIOR OF THE SHEAR-INDUCED GRANDJEAN TO FOCAL CONIC TRANSITION IN THE CHOLESTERIC MESOPHASE

JOHN POCHAN, PETER ERHARDT AND W. CONRAD RICHARDS

Xerox Corporation, Rochester, New York

This report is a study of the shear induced Grandjean to dynamic focal conic transition by rheological and thermal measurements. Previous structural postulations of the sheared cholesteric mesophases are consistent with the observed data. For a room temperature cholesteric mixture of Cholesteryl Oleyl Carbonate (COC) and Cholesteryl Chloride (CCl) (23% by wt. CCl) activation energies for viscous flow are identical for the Grandjean and dynamic focal conic textures and equal to 15 Kcal/mole. Extrapolation of the dynamic focal conic viscosity region into the isotropic region is continuous. Transient rheological phenomena in the transition region are temperature dependent and are shown to be associated with non-equilibrium tilting of the cholesteric helices and breakdown of the normal Grandjean structure. The activation energy for the shear induced conversion of Grandjean to dynamic focal conic texture is 25 Kcal/mole.

Cholesteric liquid crystals exhibit unusual optical and rheological responses when shear is applied.[1,2] Recent investigations have established shear induced transitions between the Grandjean, dynamic focal conic and homeotropic textures of the cholesteric mesophase.[3,4] The dynamic focal conic state has been so designated because of its relaxation to the true focal conic texture on cessation of shear. The textures have been associated with non-Newtonian, Newtonian, and non-Newtonian behaviors respectively. Normal force measurements on the homeotropic state have demonstrated decreasing molecular aggregation with increasing shear rate.[5] In the limit of very high shear, the aggregate size approaches the molecular lengths associated with cholesteric molecules. These data indicate strong molecular interactions in the mesophase at high shear rates and predict non-aggregated liquid-like character in the limit of very high shear.

The rheological regions defined above can be described in terms of the orientation of the cholesteric helix.[4,6,7] The Grandjean texture has an average helical alignment perpendicular to the shear field. The dynamic focal conic texture has the alignment parallel to the shear field. The dynamic homeotropic texture on the other hand is pictured as tumbling helical fragments which have no preferred macroscopic orientation. These previous studies dealt with the individual textures generated by shear, and their rheo-optical response. Pochan and Marsh noted from spectroscopic measurements that the helical correlations of the Grandjean texture responded to applied shear by tilting with little helical distortion.[4] They postulated that at some critical angle of tilt, this correlated structure began breaking down, producing a mixed state in which material near the container walls maintained the perturbed Grandjean texture while the regions distant from the moving boundaries assumed a structure in which helical correlations were predominantly parallel to the applied shear. This state was called the dynamic focal conic textured state and the relative amounts of textures comprising the state were found to be shear dependent.

The effect of an applied electric field on the shear induced dynamic focal conic to homeotropic texture has been studied,[8] indicating that a constant dipolar interaction with the applied electric field facilitates the transition. The effect of the electric field was interpreted in terms of distortion of the helices in the dynamic focal conic structure. The resulting strain induced in the sheared system, facilitated breakdown of the correlated molecular structure. In this paper we discuss our investigation of the thermal and rheological responses of the Grandjean to dynamic focal conic transition.

Experimental

The liquid crystals used in these experiments are mixtures of cholesteryl oleyl carbonate (COC) (Eastman Kodak) and cholesteryl chloride (CCL), (Sigma Chemical Corporation) produced by melt mixing without further purification. Each material exhibited a melting point within 2°C of literature values. Similar mixtures have been studied optically[9,10] and rheo-optically[4] in this laboratory.

The Weissenberg Rheogoniometer, Model R18, was used for this study. In some of the experiments a pair of transparent glass platens were used in order that the rheological transitions could be visually followed.

Theoretical calculations[11] on the system indicate that in the cone and plate assembly viscous heating is negligible below 100 sec^{-1} shear rate and should only result in temperature changes of ∿ 5°C with prolonged shearing. Thus the changes described here are attributed to structural changes in the material with applied shear.

Results and Discussion

Viscosity-shear data typical of that observed in pure and

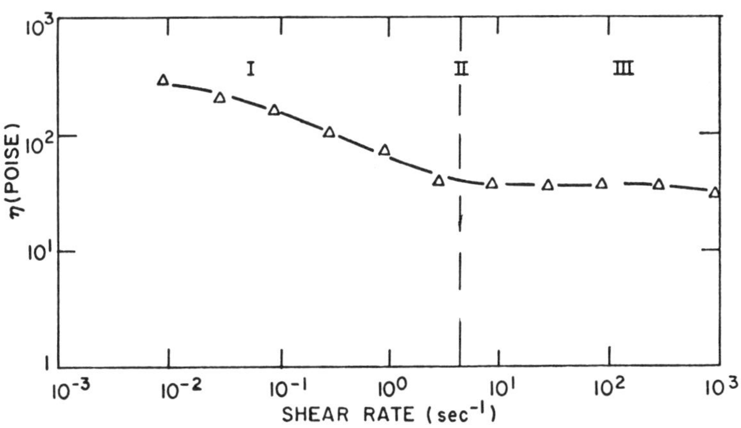

Figure 1. Steady state viscosity versus shear rate for a 23/77 weight percent mixture of CCL/COC, at 17°C. Region I is Grandjean; region III is dynamic focal conic. These regions of dominant texture are separated by a transition zone, II.

mixed cholesteric systems[3,5] is given in Fig. 1 for a 23 weight percent mixture of CCL in COC. Only the low shear rate transition region II, established previously to be the Grandjean to dynamic focal conic transition, is shown. This transition is characterized by the change from non-Newtonian to Newtonian flow due to shear realignment of the correlated helical axes.[4] When the shear rate is lowered through the transition region Newtonian flow is observed down to very low shear rates, unless the system is perturbed by removal of the surface constraints or the sample is purposely melted.

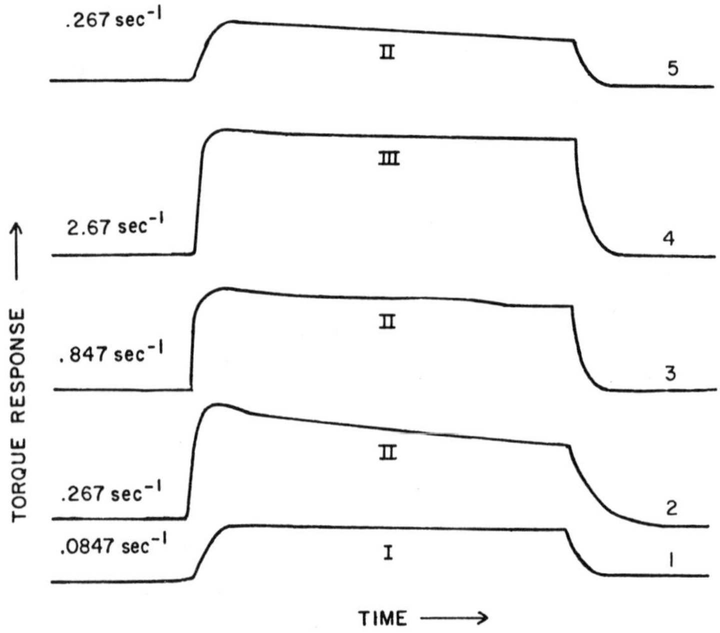

Figure 2. Rheogoniometer torque traces as a function of time. Roman numerals designate regions identified in Figure 1. Curve 5 is a repeat of condition for curve 2 (region II) after establishment of dynamic focal conic texture in region III.

In the transition region the system exhibits yield behavior. This effect is illustrated in Figure 2, where the torque response to continuous shear is shown as a function of time for several shear rates. At low shear rate (curve 1) the torque response initially increases and then remains constant with time indicating no structural changes in the sheared material once steady state shear has been established. The shear rate-torque curve [(2) of Figure 2] shows yield response exhibited by the sample at 0.267 sec^{-1}. Initially a large torque is observed which decays in time to a lower level. The data reported in Fig. 1 for the transition region are for the equilibrium torque values attained at long times. This transient effect is noted also at .847 sec^{-1}, although the changes are not as obvious as at .267 sec^{-1}. This yield response in the transition region appears to be a general phenomenon, and was also observed for a 95.2/4.8 (COC/CCL) mixture, as well as for the pure cholesteric COC. There is a compositional dependence to the transition, and the data indicates that higher values of shear rate are needed to produce breakdown in the Grandjean structure as the amount of CCL is decreased.

The helical pitch of the cholesteric mixtures also changes systematically with composition but a more precise measure of the onset of yielding is needed to establish a correlation with pitch. Curve (4) is a torque time response of the system typical of those observed at shear rates above the transition region. The structural breakdown is not reversible (as long as surface constraints are maintained) and no further yielding is observed.

Curve 5 of Figure 2 represents torque response of the system at .267 sec^{-1} after the system has been exposed to much higher shear. The level of response is identical to the equilibrium response observed in curve 2, indicating that the transient behavior associated with a structural change has already occurred and is no longer observable even at low shear rates. These results verify earlier data[3] which demonstrated the irreversibility of the shear induced transition.

These results when taken with rheo-optical studies present the following picture of the Grandjean to focal conic transition region. Below the critical shear rate the tilted helical structure of the sheared Grandjean texture persists throughout the entire sample thickness. In the transition region, with the application of shear, the helical structures tilt. The angle of tilt in this region is greater than a stable equilibrium value θ_e, and with increasing time the population of tilted helicies that is above θ_e rearrange with their helical axis parallel to the applied shear. As shear rate is further increased, the distribution again rearranges producing more tilted helices with angles greater than θ_e and more of the material rearranges to the layer of dynamic focal conic texture. Once this rearrangement has taken place, only disruption of the surface constraints or melting to the isotropic liquid state and cooling, will reproduce the Grandjean texture at lower shear rates. A limiting thickness of approximately 7μ has been set on this central dynamic focal conic region.[7]

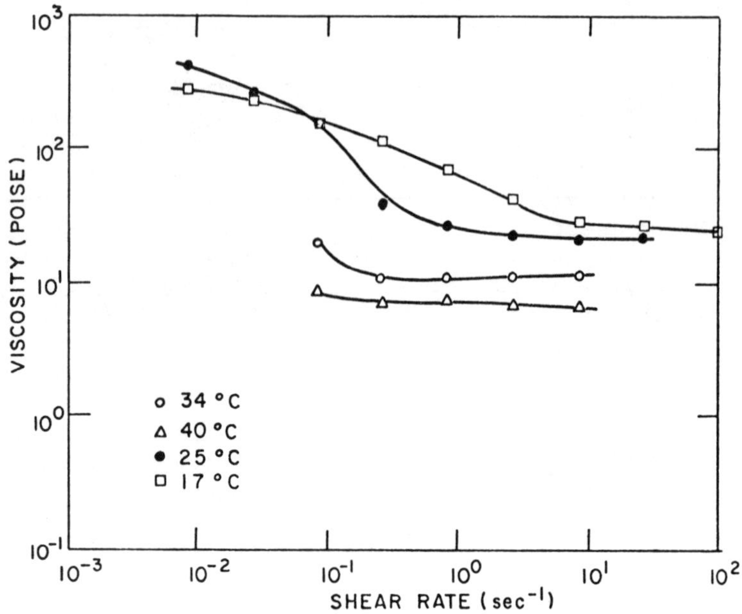

Figure 3. Steady state viscosity versus shear rate for a 23/77 CCL/COC mixtures at temperatures indicated.

The temperature dependence of the Grandjean to focal conic transition has been studied. An example of the viscosity response with shear is shown in Fig. 3 for the 23% CCL in COC mixture. The shear rate at which the transition is observed decreases with increasing temperature. In order to characterize the effect of temperature on the transition, a plot of log (shear rate at inflection) versus reciprocal temperature for data at 17, 25 and 34°C shown in Fig. 4. The activation energy of 25 Kcal/mole is sufficiently different from the viscosity activation energies (15 Kcal/mole) to indicate that the thermally activated flow behavior is different from that associated with the observed shear induced transitions.

Since yield phenomenon is associated with the temperature dependent transition zone, a plot of shear stress versus shear rate was made by matching the curves in the Newtonian region. This is shown in Figure 5.

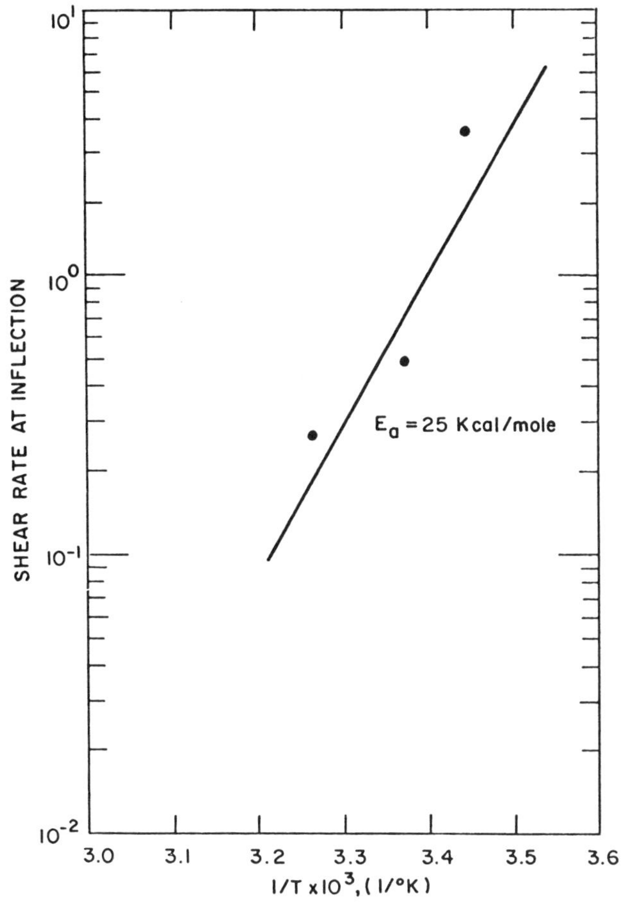

Figure 4. Arrhenius plot for the critical shear rate for the Grandjean to dynamic to focal conic transition, data at 17, 25 and 34°C.

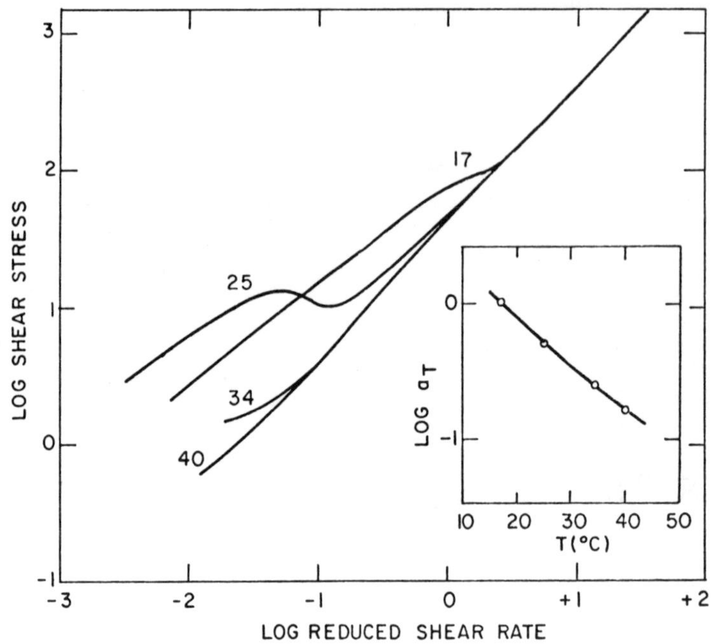

Figure 5. Log shear stress versus log reduced shear rate: Data reduced to 17° data. Inset: log a_T (shift factor) versus temperature.

The shift factor, a_T, shown in the inset, is a smoothly varying function of the temperature. A critical shear stress appears to be necessary to completely establish a layer of dynamic focal conic texture, as evidenced by the Newtonian behavior. This critical shear stress decreases with increasing temperature indicating that thermal energy facilitates completion of the transition from Grandjean to dynamic focal conic texture. In the Grandjean and transition regions the temperature dependence is not clear cut. The shapes of the curves in the low shear rate region vary. The tilted Grandjean texture at 25°C appears to support a higher shear stress than it does at 17°C. It is felt that the 17°C transition is for some reason occurring over an unusually broad range of shear rates, perhaps due to sample history, and that this accounts for the observed crossover in the shear rate dependence of stress. It would be of interest to determine the temperature dependence of the distribution of helix tilt angles in such a case, as this would have some bearing on the degree of cooperativity involved. Thus far only isothermal studies on the helix tilt angle have been done.[4]

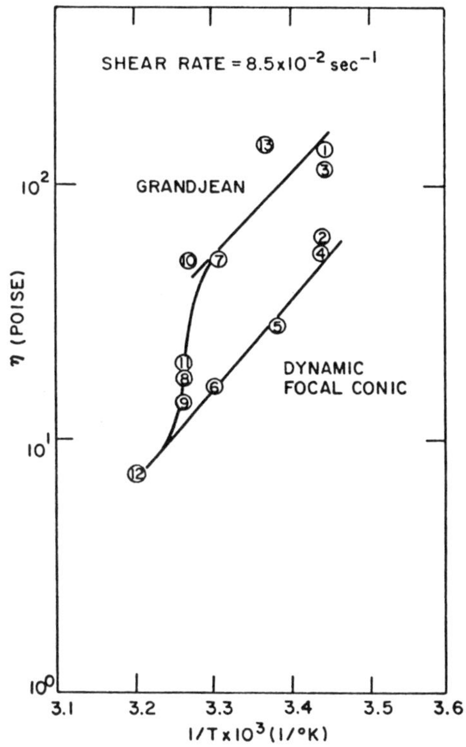

Figure 6. Arrhenius plot of viscosity versus $^1/T$ for various experimental conditions of the Weissenberg.

1. initial reading
2. after shearing at 85 sec^{-1} (above the transition shear).
3. gap opened
4. after shearing at 85 sec^{-1}
5. increase temperature
6. increase temperature
7. gap opened
8. increase temperature
9. after shearing at 75 sec^{-1}
10. gap opened
11. continued shear at .085 sec^{-1}
12. isotropic state
13. cooled to $24^{\circ}C$

The thermal-viscosity history behavior of the system is described in the combined temperature-shear profile shown in Fig. 6. The experiment was conducted under a variety of conditions which are sequentially numbered in the figure and these conditions are listed in the figure caption. Of interest is the irreversibility of the Grandjean to focal conic transition unless the system constraints are removed (anisotropic shear) or it is melted to the isotropic and returned to low shear. Both textures exhibit Arrhenius type flow activation energies, which are the same, within the limits of measurement error, and are 15 Kcal/mole. These values are similar to those measured for other cholesteric materials.[1] The Grandjean viscosity is approximately an order of magnitude higher than the dynamic focal conic viscosity, until the isotropic region is approached. Extrapolation of the dynamic focal conic viscosity coincides with the isotropic point and indicates that at least part of the layered structure of the sheared mesophase has rheological properties similar to the isotropic fluid. This structure could not contain isotropic fluid as the transition from focal conic to homeotropic texture occurs at higher shear rates and produces a homogeneous texture.[3]

Conclusions

In conclusion, this study of the cholesteric to dynamic focal texture of the cholesteric mesophase has revealed rheological behavior that is consistent with the structural response postulated on the basis of rheo-optical studies.[4] Yield behavior is observed in the shear rate region in which the transition takes place, indicating a change in structure and flow properties of the system. Thermal data indicate Arrhenius activation curves for both dynamic focal conic and Grandjean textures with the Grandjean phase being an order of magnitude higher in viscosity. Extrapolation of the dynamic focal conic viscosity into the isotropic melt region coincides with the isotropic viscosity.

Acknowledgement

The authors wish to acknowledge helpful discussions with W.M. Prest.

Literature Cited

(1) R.S. Porter, J.F. Johnson "The Rheology of Liquid Crystals" in Rheology Edited by F.R. Eirich, Academic Press, New York, N.Y. Chap. 5. (1967).

(2) "Liquid Crystals", Proceedings of the International Conference on Liquid Crystals, at Kent State (1965), Gordon & Breach, New York, N.Y. 1967.

(3) J.M. Pochan and P.F. Erhardt, Phys. Rev. Lett., 27, 790 (1971)
(4) J.M. Pochan and D.G. Marsh, J. Chem. Phys., 57, 1193 (1972).
(5) P.F. Erhardt, J.M. Pochan, and W.C. Richards, J. Chem. Phys., 57, 3596 (1972).
(6) D. Marsh and J.M. Pochan, J. Chem. Phys., 57 2835 (1973).
(7) J.M. Pochan and D.G. Marsh, J. Chem. Phys., 57 5154 (1973).
(8) J.M. Pochan, P.F. Erhardt, and W.C. Richards, Mol. & Liq. Crys. (in press).
(9) J. Adams, W. Haas, and J. Wysocki, Phys. Rev. Letters 22, 92 (1969).
(10) J. Adams, W. Haas, and J. Wysocki, J. Chem. Phys. 50, 2458 (1969).
(11) Byrd, Stewart and Lightfoot, "Transport Phenomena", J. Wiley & Sons., Inc. N.Y., N.Y., 1960.

MESOMORPHIC BEHAVIOUR OF OPTICALLY ACTIVE ANILS: 4-n-ALKOXYBENZYLIDENE-4'-METHYLALKYLANILINES

Y. Y. Hsu and D. Dolphin

Department of Chemistry, Harvard University

Cambridge, Massachusetts 02138

INTRODUCTION

The effect of bulky substituents, at the ortho position of a benzene ring, the α-position of a central linkage, as well as at a terminal alkyl chain, upon mesomorphic properties has recently been studied in several classes of liquid crystals.[1-8] The results have shown that in most cases both the solid-mesomorphic and the mesomorphic-isotropic transition temperatures were lowered markedly as compared with the unsubstituted parent compounds. More recently three members of both the active and racemic forms of the 4-alkoxybenzylidene-4'-β-methylbutylaniline series (I)[9] have been prepared to investigate their optical properties. It was found that both the melting and clearing points were about 10-20° lower than that for the corresponding unsubstituted benzylideneanilines.

We report here the synthesis of the other members of this series and two higher homologous series, i.e., γ-methylpentylaniline (II) and δ-methylhexylaniline (III) series, and the effect of the position of the methyl group and chain length on their mesomorphic and optical properties. In addition, the thermodynamic data for the phase transitions in these three series are reported.

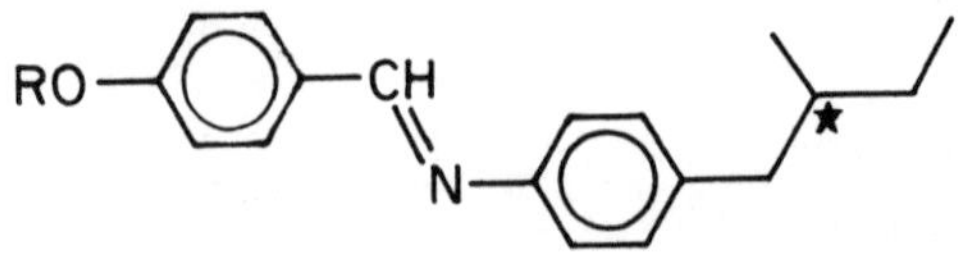

I

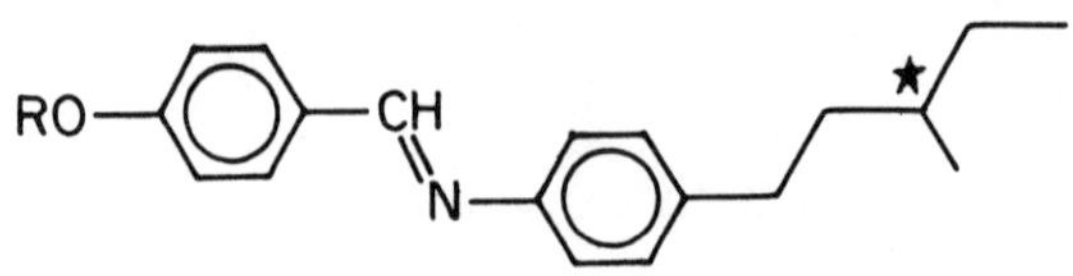

II

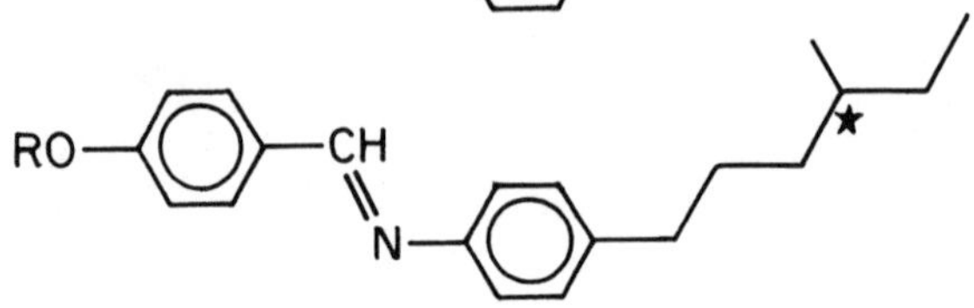

III

RESULTS AND DISCUSSION

Thermal Behaviour

The phase transition temperatures and the calorimetric data are summarized in Table 1. The effects of the introduction of a methyl substituent to the alkyl chain of the aniline moiety upon the phase transition temperature were examined by comparing the solid-mesomorphic and mesomorphic-isotropic transition temperature between the present compounds and the corresponding unsubstituted homologues.[10] The results, shown in Table 2, revealed that the methyl substituent caused both increases and decreases in the solid-mesomorphic transition temperatures for individual compounds, and there was no clear-cut trend in the melting point as molecular structure was regularly varied. However, the effect on the mesomorphic-isotropic transition temperature was striking and only the lowering tendency was observed.

The variation of transition temperature with increasing length of the alkoxy chain for the β-methylbutylaniline series (compounds 1 through 10) is shown in Figure 1. An odd-even alternation of chiral nematic-isotropic transition temperature is apparent. A decreasing then rising curve for even-carbon members is above a rising curve for odd-carbon members. This situation is similar to that for the corresponding unsubstituted homologues.[10] It was found that this series showed one to three monotropic phases for members above the C_4 ether. Compound 6 shows a monotropic smectic phase between the two enantiotropic smectic phases, and compound 9 shows a monotropic chiral nematic as well as two monotropic smectic phases. Compound 2 was previously reported to have a chiral nematic range of 15-60°,[9] however, the present study indicated a transition temperature of 27.8-60° for the same compound. This discrepancy suggests that the previous data was obtained under supercooling conditions.

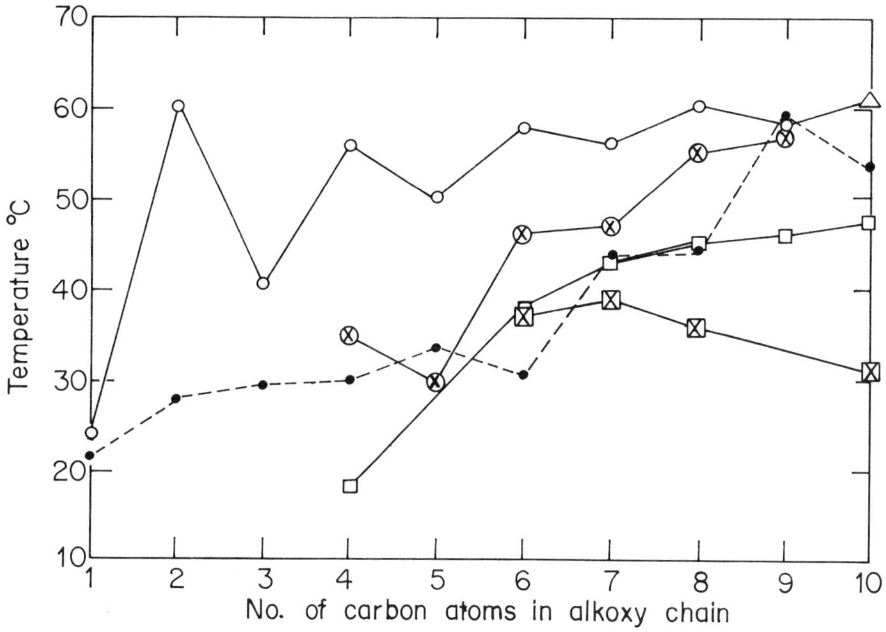

Figure 1. Phase transition diagram for the (R)-4-n-alkoxybenzylidene-4'-β-methylbutylanaline series. • solid-mesomorphic or isotropic; ○ chiral nematic-isotropic; △ smectic-isotropic; ⊗ smectic-chiral nematic; □ smectic$_1$-smectic$_2$; ⊠ smectic$_2$-smectic$_3$.

Figure 2 shows the phase transition diagram for the γ-methylpentylaniline series. The branching methyl group also caused some compounds in this series to exhibit monotropic phases. Compound 14 is an example which shows three monotropic smectic phases. Compounds 19 and 20 show two monotropic phases and the transition temperature for the smectic$_3$ is above the melting point. Their transition temperatures can only be observed optically and by DSC on cooling the melting samples. The smectic tetramorphism was observed for all members above the C_4 ether.

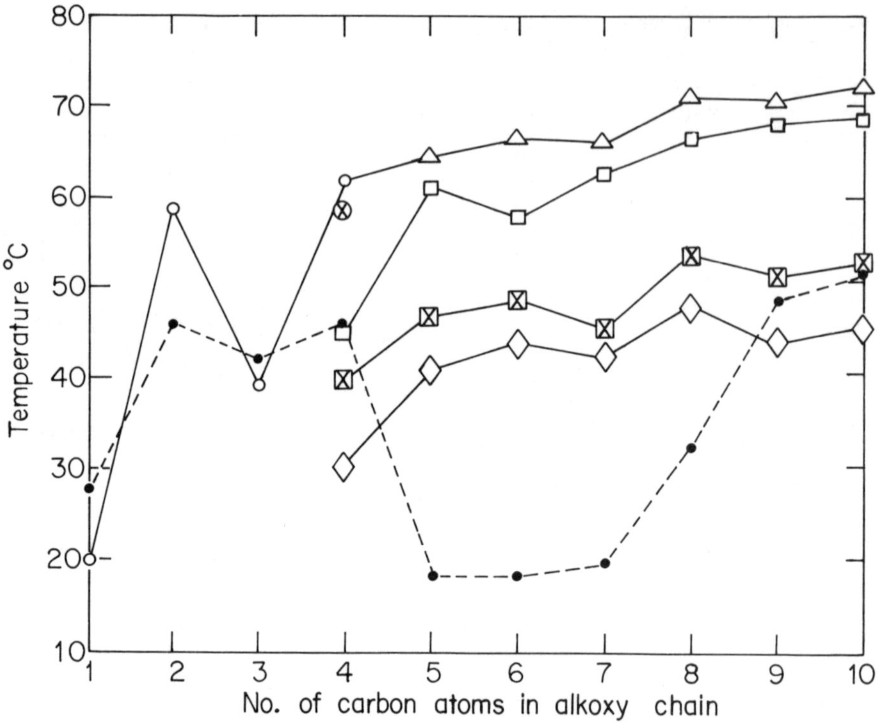

Figure 2. Phase transition diagram for the (R)-4-n-alkoxybenzylidene-4'-γ-methylpentylanaline series. • solid-mesomorphic or isotropic; ○ chiral nematic-isotropic; △ smectic-isotropic; □ smectic$_1$-smectic$_2$; ⊠ smectic$_2$-smectic$_3$; ◇ smectic$_3$-smectic$_4$; ⊗ smectic-chiral nematic.

The phase transition diagram for the δ-methylhexylaniline series is given in Figure 3. In this series compounds 22 and 23 exhibit a wide chiral nematic range at subambient and room temperature (8.5-51 and 14-30.5° respectively). The racemic form of compound 22 (i.e., compound 31) with transition temperatures of 8-49° exhibits a wider nematic range than does the well-known room temperature nematic MBBA (21-47°). All members of this series except for compounds 21, 22 and 23 show enantiotropic phases. The solid-mesomorphic transition temperatures for compounds 22, 24, 25 and 26 could not be determined by DSC even upon cooling the samples down to -120°. However, the same transition temperature for compound 23 was obtained by microscopic observation from the partially crystallized sample previously frozen at -20° for three days.

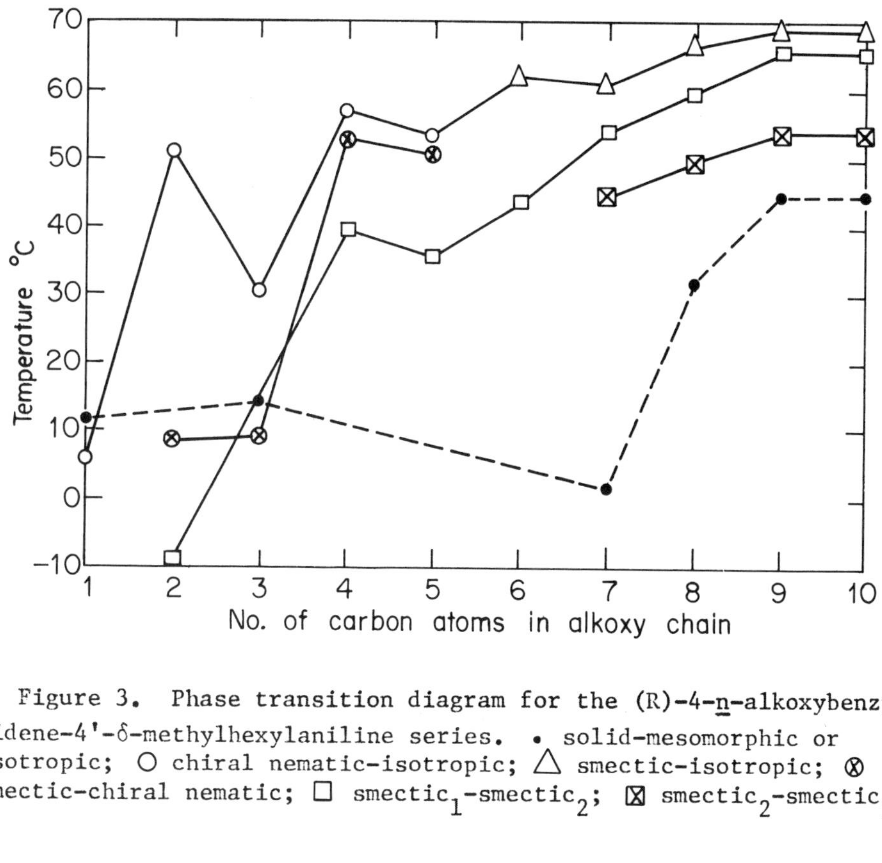

Figure 3. Phase transition diagram for the (R)-4-n-alkoxybenzylidene-4'-δ-methylhexylaniline series. • solid-mesomorphic or isotropic; ○ chiral nematic-isotropic; △ smectic-isotropic; ⊗ smectic-chiral nematic; □ $smectic_1$-$smectic_2$; ⊠ $smectic_2$-$smectic_3$.

Table 1. Physical Constants of (R)-4-n-Alkoxybenzylidene-4'-methylalkylanilines

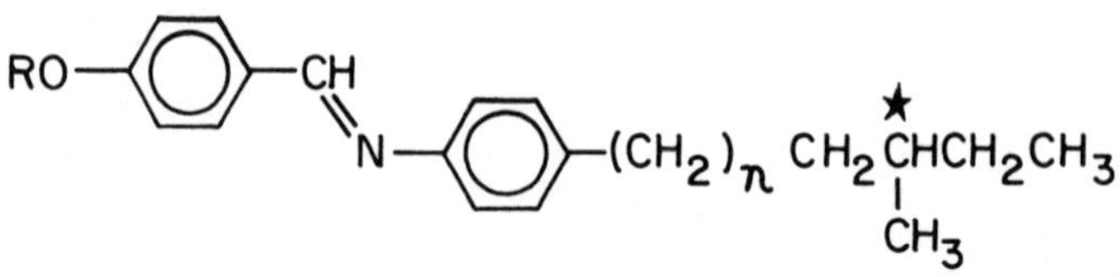

No.	R	n	Transition	Temp. °C	ΔH Kcal/mol	ΔS Cal/mol/°K
1	CH_3	0	K-Ch	21.5	4.44[h]	15.05
			Ch-I	24		
2	C_2H_5	0	K-Ch	27.8[b]	3.55	11.8
			Ch-L	60	0.11[c]	0.33
3	C_3H_7	0	K-Ch	29.7	4.27	14.09
			Ch-I	40.5	0.16	0.5
4	C_4H_9	0	K-S_1	30	4.16	13.72
			S_1-Ch	35	0.18	0.58
			Ch-I	56	0.17	0.52
			S_1-S_2[a]	18		
5	C_5H_{11}	0	K-Ch	33.4	3.75	12.24
			Ch-I	50.1	0.26	0.81
			Ch-S_1[a]	30.1		
6	C_6H_{13}	0	K-S_3	30.4[d]	3.44	11.33
			S_3-S_1	37.9	1.33	4.27
			S_1-Ch	46.2	0.5	1.56
			Ch-I	58.1	0.32	0.96
			S_1-S_2[a]	37.5		
			S_2-S_3[a]	36.9		

Table 1 Continued

No.	R	n	Transition	Temp. °C	ΔH Kcal/mol	ΔS Cal/mol/°K
7	C_7H_{15}	0	$K-S_1$	43.6[d]	6.57[h]	20.74
			S_1-Ch	46.9		
			Ch-I	56.1	0.36	1.09
			S_1-S_2[a]	42.9		
			S_2-S_3[a]	39		
8	C_8H_{17}	0	$K-S_1$	44.1[d]	5.59	17.61
			S_1-Ch	55.5	0.48	1.46
			Ch-I	60.3	0.41	1.23
			S_1-S_2[a]	44.8		
			S_2-S_3[a]	36		
9	C_9H_{19}	0	K-I	59.3	7.58	22.8
			I-Ch[a]	58.3		
			$Ch-S_1$[a]	57		
			S_1-S_2[a]	46.2		
10	$C_{10}H_{21}$	0	$K-S_1$	53.8	6.16	18.84
			S_1-I	61.7	1.03	3.07
			S_1-S_2[a]	47.4		
			S_2-S_3[a]	31		
11	CH_3	1	K-I	27.6	4.7	15.63
			I-Ch[a]	19.9		
12	C_2H_5	1	K-Ch	46	5.03	15.7
			Ch-I	58.7	0.09	0.28
13	C_3H_7	1	K-I	42	4.27	13.5
			I-Ch[a]	39.1		
14	C_4H_9	1	$K-S_1$	46.1[d]	3.1	9.72
			S_1-Ch	58.5	0.66	1.98
			Ch-I	61.5	0.4	1.19
			S_1-S_2[a]	45		

Table 1 Continued

No.	R	n	Transition	Temp. °C	ΔH Kcal/mol	ΔS Cal/mol/°K
			S_2-S_3 [a]	39.9		
			S_3-S_4 [a]	30.1		
15	C_5H_{11}	1	K-S_4	18.3[d]	2.9	9.95
			S_4-S_3	41.1	0.08	0.24
			S_3-S_2	47.1	0.46	1.44
			S_2-S_1	60.9	0.25	0.75
			S_1-I	64.3	1.11	3.3
16	C_6H_{13}	1	K-S_4	18[d]		
			S_4-S_3	44.1	0.1	0.31
			S_3-S_2	48.8	0.6	1.86
			S_2-S_1	57.5	0.19	0.57
			S_1-I	66.1	1.23	3.61
17	C_7H_{15}	1	K-S_4	19.4[d]	3.19	10.91
			S_4-S_3	42.5	0.13	0.41
			S_3-S_2	45.5	0.59	1.85
			S_2-S_1	62.5	0.32	0.95
			S_1-I	65.8	1.31	3.86
18	C_8H_{17}	1	K-S_4	32.4[d]	4.37	14.35
			S_4-S_3	47.7	0.15	0.46
			S_3-S_2	53.3	0.61	1.87
			S_2-S_1	66.4	0.27	0.8
			S_1-I	71.1	1.35	3.92
19	C_9H_{19}	1	K-S_2	48.8[d]	6.44	20
			S_2-S_1	67.6	0.35	1.01
			S_1-I	70.4	1.59	4.64
			S_2-S_3 [a]	51.6		
			S_3-S_4 [a]	44.1		

Table 1 Continued

No.	R	n	Transition	Temp. °C	ΔH Kcal/mol	ΔS Cal/mol/°K
20	$C_{10}H_{21}$	1	$K-S_2$	51.2[d]	5.98	18.45
			S_2-S_1	68.1	0.12	0.36
			S_1-I	72.1	1.5	4.35
			S_2-S_3[a]	52.9		
			S_3-S_4[a]	45.4		
21	CH_3	2	K-I	11.9	2.88	10.12
			I-Ch[a]	5.8		
22[e]	C_2H_5	2	K-Ch	f		
			Ch-I	51	0.15	0.46
			$Ch-S_1$[a]	8.5		
			S_1-S_2[a]	-9.5		
23[e]	C_3H_7	2	K-Ch	14[g]		
			Ch-I	30.5	0.21	0.7
			$Ch-S_1$[a]	9.0		
24	C_4H_9	2	$K-S_2$	f		
			S_2-S_1	39.4	0.6	1.92
			S_1-Ch	52.9	0.75	2.29
			Ch-I	57	0.37	1.11
25	C_5H_{11}	2	$K-S_2$	f		
			S_2-S_1	35.5	0.68	2.2
			S_1-Ch	50.9	0.58	1.8
			Ch-I	53.1	0.63	1.94
26	C_6H_{13}	2	$K-S_2$	f		
			S_2-S_1	43.3	0.58	1.84
			S_1-I	61.9	1.21	3.62
27	C_7H_{15}	2	$K-S_3$	1.5[d]		
			S_3-S_2	44.1	0.53	1.68
			S_2-S_1	53.9	0.17	0.51
			S_1-I	60.4	1.18	3.53

Table 1 Continued

No.	R	n	Transition	Temp. °C	ΔH Kcal/mol	ΔS Cal/mol/°K
28	C_8H_{17}	2	$K-S_3$	31.1[d]	4.72	15.51
			S_3-S_2	49.7	0.82	2.54
			S_2-S_1	59.6	0.19	0.58
			S_1-I	66.1	1.37	4.03
29	C_9H_{19}	2	$K-S_3$	44.1	6.13	19.32
			S_3-S_2	53.7	0.65	1.98
			S_2-S_1	65.9	0.18	0.52
			S_1-I	68.9	1.45	4.25
30	$C_{10}H_{21}$	2	$K-S_3$	44.4[d]	5.28	16.63
			S_3-S_2	53.8	1.03	3.14
			S_2-S_1	65.1	0.31	0.91
			S_1-I	68.3	1.63	4.77
31[e]	C_2H_5	2	K-N	f		
			N-I	49	0.15	0.47
			$N-S_1$[a]	8		
			S_1-S_2[a]	-8.5		

a. Monotropic phase transition; b. 15° from reference 9; c. 0.12 Kcal/mol from reference 9; d. Values obtained from DSC; e. Temperatures were determined using a Köfler heating-cooling stage. Sandwiched samples, previously frozen at -20°C for three days, were used; f. Temperatures could not be measured due to supercooling of the samples; g. Temperature was obtained from the partially crystallized sample; h. Values including the neighbouring phase transition; K= Crystalline; S= Smectic; Ch=Chiral nematic; N= Nematic; I= Isotropic.

Table 2. The Melting Points and Clearing Points of 4-Alkoxybenzylidene-4'-alkylanilines and 4'-methylalkylanilines

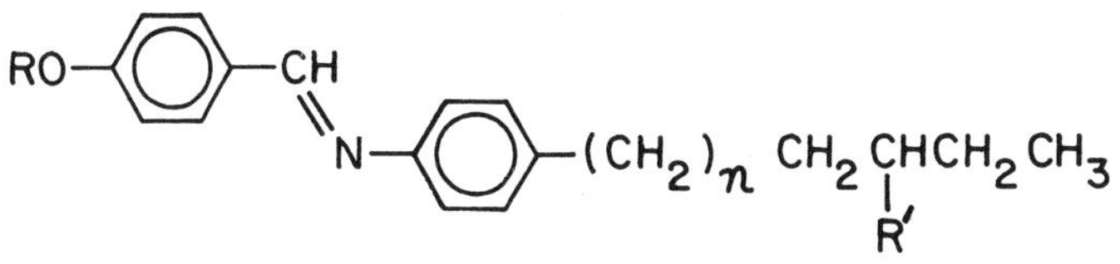

R	n	Transition	Temperature °C R' =H[a]	R' =CH_3	Difference
1	0	K-N,Ch[b]	20	21.5	-1.5
		N,Ch-I	47.3	24	23.3
2	0	K-N,Ch	35.5	27.8	7.7
		N,Ch-I	79	60	19
3	0	K-N,Ch	39.5	29.7	9.8
		N,Ch-I	56	40.5	15.5
4	0	K-S_1	36.5	30	6.5
		N,Ch-I	73.5	56	17.5
5	0	K-S_2,Ch	24	33.4	-9.4
		N,Ch-I	68.5	50.1	18.4
6	0	K-S_3	34	30.4	3.6
		N,Ch-I	77.5	58.1	19.4
7	0	K-S_3,S_1	36	43.6	-7.6
		N,Ch-I	76.5	56.1	20.4
8	0	K-S_3,S_1	40	44.1	-4.1
		S_1,Ch-I	82	60.3	21.7
9	0	K,S_3,I	50.5	59.3	-8.8
		S_1,K-I	82.5	59.3	23.2
10	0	K-S_3,S_1	40	53.8	-13.8
		S_1-I	85	61.7	23.3
1	1	K-N,I	39.3	27.6	11.7
		N,K-I	64	27.6	36.4

Table 2 Continued

R	n	Transition	Temperature °C $R'=H^a$	$R'=CH_3$	Difference
2	1	K-N,Ch	63.2	46	17.2
		N,Ch-I	91	58.7	32.3
3	1	K-N,I	32.6	42	-9.4
		N,K-I	71.6	42	29
4	1	K-S_1	40.3	46.1	-5.8
		N,Ch-I	86.3	61.6	24.7
5	1	K-S_3,S_4	24.2	18.3	5.9
		N,S_1-I	78	64.3	13.7
6	1	K-S_2,S_4	36	18	18
		N,S_1-I	85.4	66.1	19.3
7	1	K-S_2,S_4	18.6	19.4	-0.8
		N,S_1-I	83.4	65.8	17.6
8	1	K-S_2,S_4	45	32.4	12.6
		S_1-I	88.5	71.1	17.4
9	1	K-S_2	35.4	48.8	-13.4
		S_1-I	88	70.4	17.6
10	1	K-S_2	39.4	51.2	-11.8
		S_1-I	91.1	72.1	19
1	2	K-N,I	32.7	11.9	20.8
		N,K-I	53.5	11.9	41.6
2	2	K-N,Ch	37	8.5	28.5
		N,Ch-I	80.3	51	29.3
3	2	K-N,Ch	38.6	9	29.6
		N,Ch-I	62.5	30.5	32
4	2	K-S_4,S_2	1.5	?	
		N,Ch-I	76	57	19
5	2	K-S_4,S_2	32.4	?	
		N,Ch-I	73.5	53.1	20.4

Table 2 Continued

R	n	Transition	Temperature °C $R'=H$[a]	$R'=CH_3$	Difference
6	2	$K-S_2$	9	?	
		N,S_1-I	79.5	61.9	17.6
7	2	$K-S_2,S_3$	37.8	1.5	36.3
		S_1-I	80.5	60.4	20.1
8	2	$K-S_2,S_3$	43.3	31.1	12.2
		S_1-I	86	66.1	19.9
9	2	$K-S_2,S_3$	44	44.1	-0.1
		S_1-I	87	68.9	18.1
10	2	$K-S_2,S_3$	29.8	44.4	-14.6
		S_1-I	88	68.3	19.7

a. Values from reference 10; b. K-N,Ch=Crystal-nematic transition for unsubstituted compounds ($R'=H$) and crystal-chiral nematic transition for methylsubstituted compounds ($R'=CH_3$); K=Crystalline; S=Smectic; Ch=Chiral nematic; N=Nematic; I=Isotropic.

Transition Heats

Figures 4, 5 and 6 show the relationship between the entropy change and the alkoxy chain length. As in the homologous series of cholesteric mesophases,[11] the entropy change for the chiral nematic-isotropic transition of the present compounds shows a smooth increase with increasing length of alkoxy chain. It was found that the smectic-isotropic transition heats are several times higher than the chiral nematic-isotropic transition and also show an increasing tendency with increase in molecular weight. For the γ-methylpentylaniline series the smectic$_3$-smectic$_2$ and smectic$_4$-smectic$_3$ transition heats show a smooth curve, while the smectic$_2$-smectic$_1$ transition heats indicate a small odd-even alternation. For the δ-methylhexylaniline series the smectic$_3$-smectic$_2$ transition heats shows a big odd-even effect.

The transition entropy for smectic$_2$-smectic$_1$ phase transition of ethers below C_6 is higher than that of ethers above C_7. This suggested that they have different order in the $smectic_2$ phase and that this effect results from their different preceding phases, i.e., crystalline phase for ethers below C_6 and $smectic_3$ phase for ethers above C_7. The solid-smectic transition temperatures for compounds 16 and 27 were determined using DSC by reheating the sample from -120°. The peak area was only about one tenth of that for general solid-smectic transition. This indicated that the smectic phase was only partially crystallized on cooling to -120°. Therefore, such peak areas were not used to measure the transition heats.

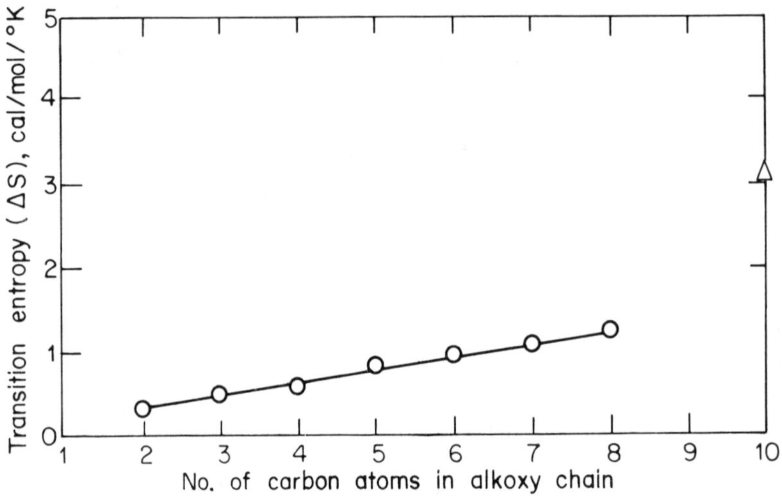

Figure 4. Transition entropy for the (R)-4-n-alkoxybenzylidene-4'-β-methylbutylaniline series. ○ chiral nematic-isotropic; △ smectic-isotropic.

Mesophase Textures

The mesophase textures were observed by sandwiching samples between the slide and cover glasses under a Reichert polarizing microscope equipped with a Mettler FP 52 heating-cooling stage. In most cases definite texture changes corresponding to the phase changes measured by DSC were observed optically. In some cases observation of a texture change during transition from solid to smectic phase was difficult for the compounds which exhibited supercooling behaviour. The texture identifications are based on the classification systems reported by Sackman and Demus.[12]

For the β-methylbutylaniline series all $smectic_1$ phases, except for compound 5, show a simple fan-shaped texture or a spherulitic fan texture and are classified as smectic A. All $smectic_2$

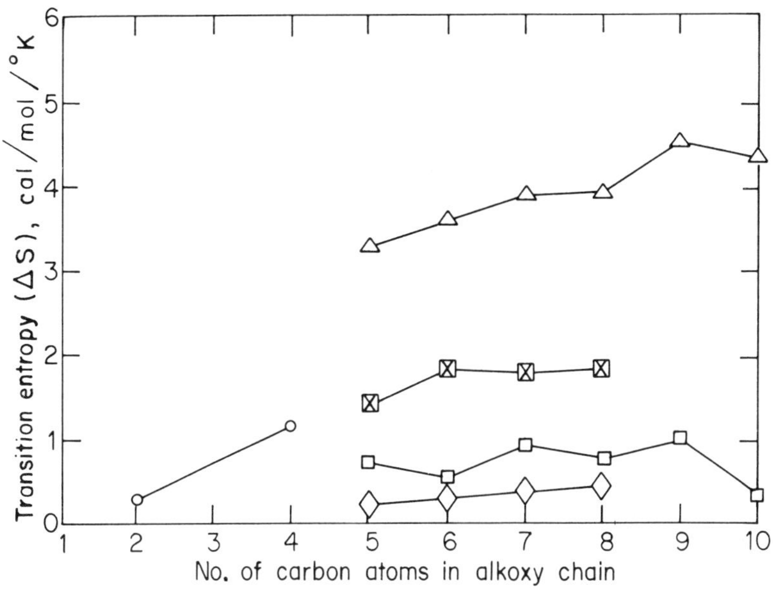

Figure 5. Transition entropy for the (R)-4-n-alkoxybenzylidene-4'-γ-methylpentylaniline series. ○ chiral nematic-isotropic; △ smectic-isotropic; □ $smectic_1$-$smectic_2$; ⊠ $smectic_2$-$smectic_3$; ◇ $smectic_3$-$smectic_4$.

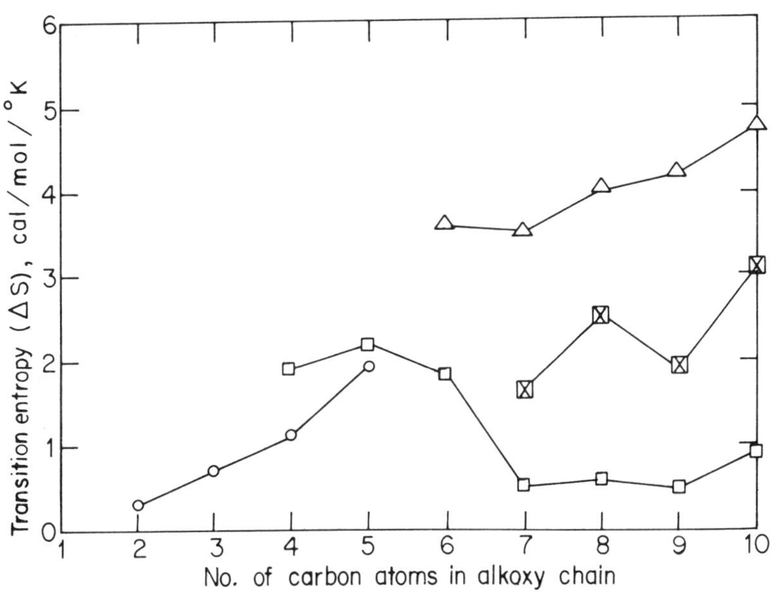

Figure 6. Transition entropy for the (R)-4-n-alkoxybenzylidene-4'-δ-methylhexylaniline series. ○ chiral nematic-isotropic; △ smectic-isotropic; □ $smectic_1$-$smectic_2$; ⊠ $smectic_2$-$smectic_3$.

phases, except for compound 4, show a broken fan-shaped texture in which many stripes grow along the blades and are designated as smectic C. All $smectic_3$ phases, the $smectic_2$ phase for compound 4 and the $smectic_1$ phase for compound 5 exhibit a mosaic texture and are tentatively identified as smectic B (or H?). It was interesting to observe that compound 6 showed only a mosaic texture and a simple fan-shaped texture on heating the sample, however on cooling it gives an additional short range of broken fan-shaped texture (smectic C) between these two textures. Figure 7 illustrates the typical texture changes for this series. Figure 7a shows the thread-like chiral nematic phase. Figure 7b shows the simple fan-shaped texture. Figure 7c shows the broken fan-shaped texture and Figure 7d shows a mosaic texture.

For the γ-methylpentylaniline series all $smectic_1$ phases show a simple fan-shaped texture (smectic A). All $smectic_2$ phases show a broken fan-shaped texture (smectic C) which, however, is different from the $smectic_2$ phase shown in the β-methylbutylaniline series. In this $smectic_2$ phase the simple fan-shaped texture of the $smectic_1$ phase became broken by many stripes across the blades. All $smectic_3$ and $smectic_4$ phases are unidentified textures. The typical texture changes for this series are illustrated in Figure 8. Figure 8a shows a simple fan-shaped texture (smectic A). Figure 8b shows the $smectic_2$ phase with a broken fan-shaped texture. Figures 8c and 8d show the unidentified $smectic_3$ phases on cooling from the $smectic_2$ phase and on heating the $smectic_4$ phase respectively. At the S_2S_3 and S_4S_3 transition temperatures a discontinuous change was observed but rather a gradual change was shown in their appearance depending on their preceding texture. Figure 8e shows the unidentified $smectic_4$ phase. The smectic mesophase textures for the δ-methylhexylaniline series, except for compound 23 and 24, shown in Figure 9 are similar to that for the γ-methylpentylaniline series in the corresponding phases. Compound 23 shows a mosaic like texture for the $smectic_1$ phase. Compound 24 exhibits a simple fan-shaped texture (smectic A) for $smectic_1$ phase, while the $smectic_2$ phase has a fan-shaped texture with diminished discontinuities (smectic B).

Optical Properties

The refractive indices of the racemic form of compound 22 (i.e., compound 31) were measured at 6328 Å with a Pulfrich Refractometer (Bellingham & Stanley Ltd.) using the same method described previously.[9] The ordinary and extraordinary indices, as a function of temperatures, are shown in Figure 10. It was observed that compound 31 had smaller refractive indices than the racemic form of compound 2.[9]

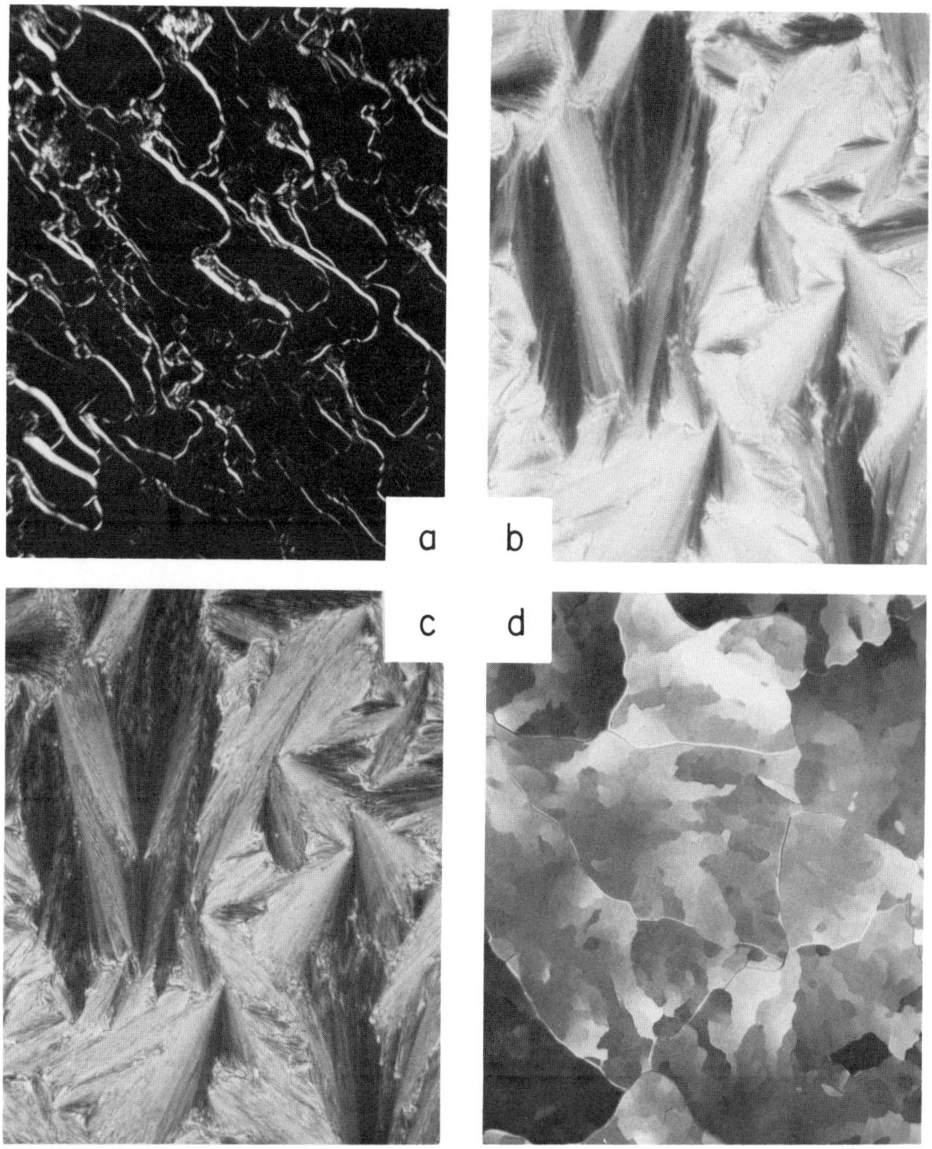

Figure 7. Mesomorphic textures of compound 8,
(100 X, crossed polarizers)
a. Chiral nematic at 59.5°C; b. S_1=S_A at 50.5°C;
c. S_2=S_C at 43.9°C; d. S_3, unidentified mosaic texture at 33.4°C

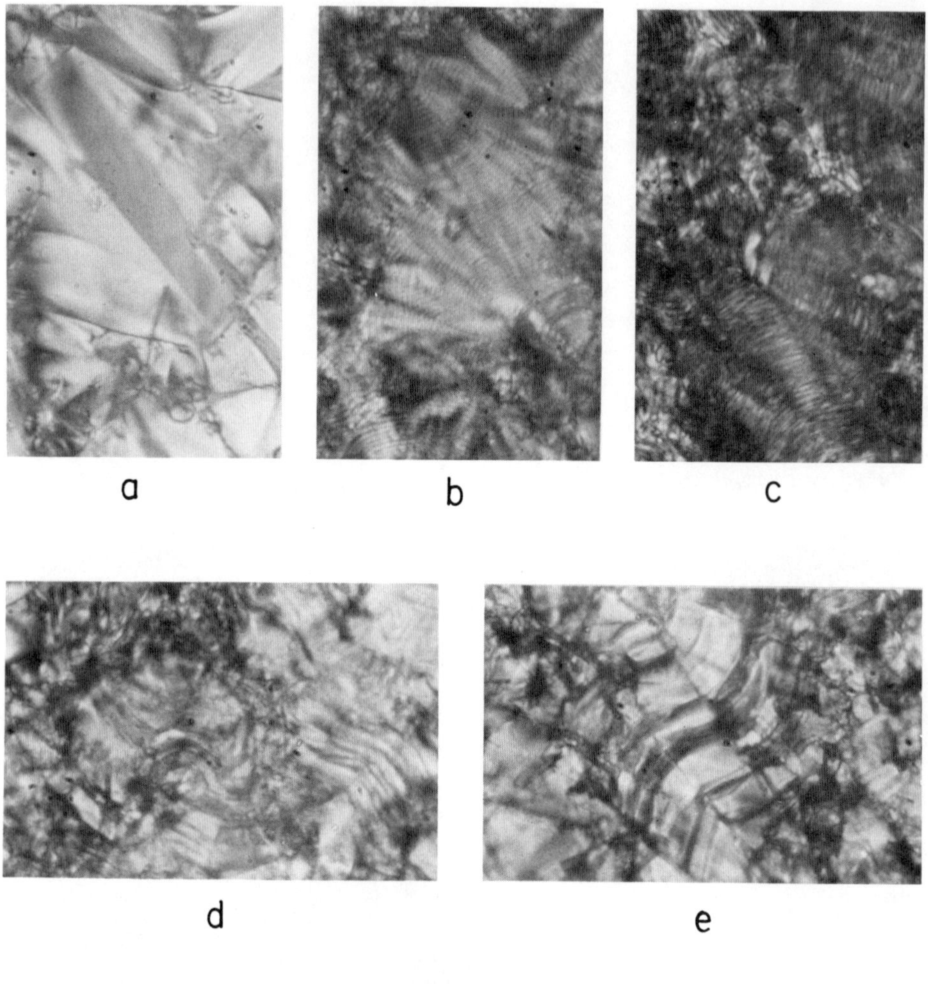

Figure 8. Smectic textures of compound 18,
(250X, crossed polarizers)
a. $S_1=S_A$ at 68°C; b. $S_2=S_C$ at 65.6°C;
c. and d., S_3, unidentified textures cooling from S_2 at 51°C and heating from S_4 at 49.2°C respectively; e. S_4, unidentified texture at 44.6°C.

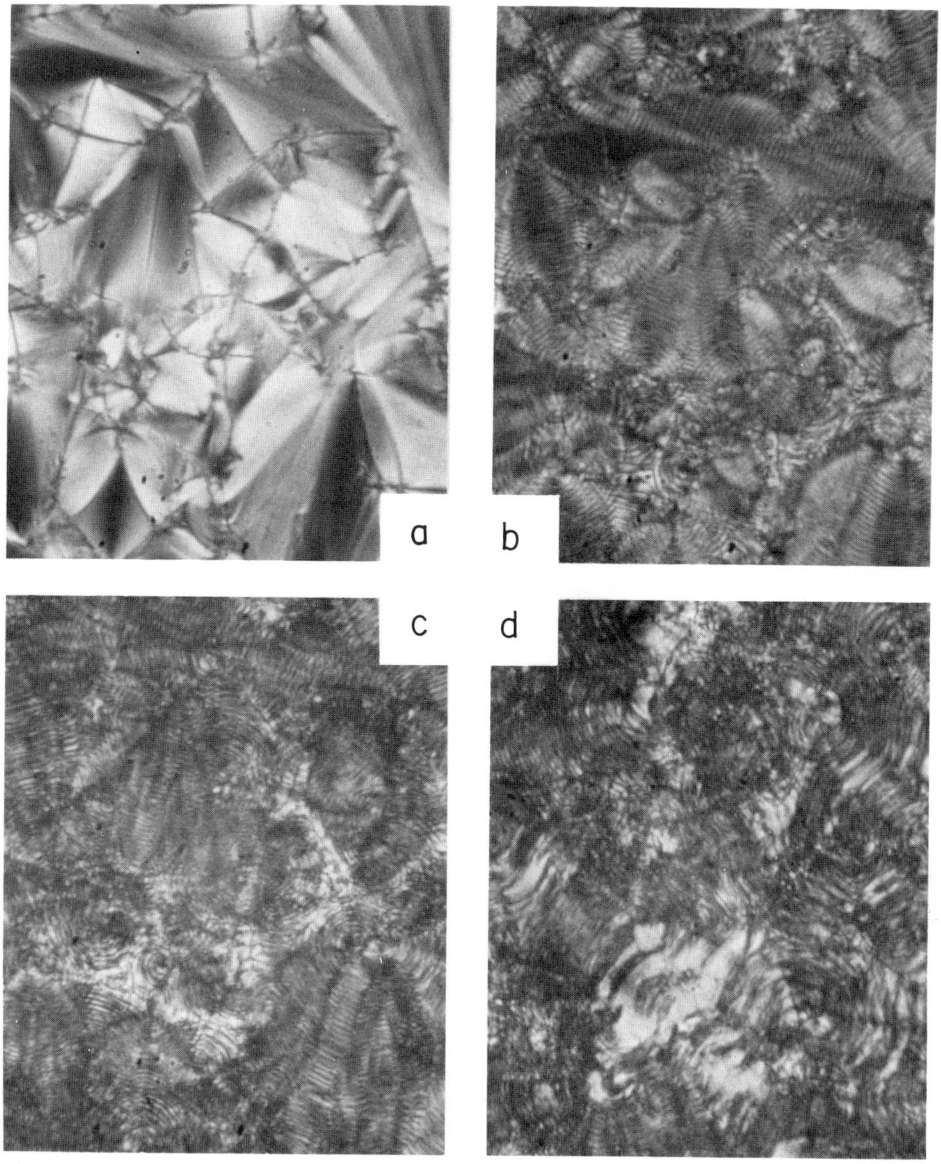

Figure 9. Smectic textures of compound 27, (250X, crossed polarizers)

a. $S_1 = S_A$ at 55°C. b. $S_2 = S_C$ at 51°C.

c. and d. S_3 (unidentified texture) at 42.5° and 27°C respectively.

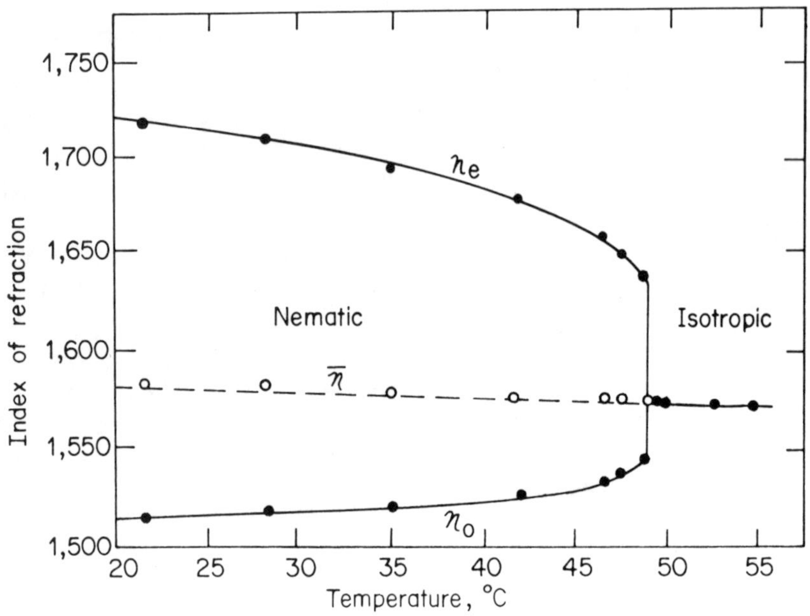

Figure 10. Refractive indices of 4-ethoxybenzylidene-4'-δ-methylhexylaniline.

As previously reported the optical pitch of compound 2 lies within the absorption region of the molecule itself. In order to measure the optical pitch of this compound various mixtures, composed of the active and racemic forms, were prepared and their optical pitch determined. An extrapolation of the reflection center wavelength to 100 percent of compound 2 gives 3570 Å for its optical pitch such that its actual pitch is then 3570/1.60=2230 Å.

In order to study the effect of the chain length, and the position of the methyl substituent, on the helical pitch, compounds 22 and 23 with subambient chiral nematic phase were chosen to measure their optical pitch at room temperature.[9] It was found that the optical pitch of these two compounds lies in the far infra-red region and could be measured directly with the pure sample. The results show that compound 22 has an optical pitch of 15,500 Å (reflection width 16240-14760 Å) and compound 23 has an optical pitch of 14870 Å (15120-14620 Å). The actual pitch for compound 22 is thus 15500/1.58=9810 Å. This shows that the introduction of an ethylene group to compound 2, between the ring and the asymmetric center, caused the actual pitch to increase 7580 Å.

CONCLUSION

The introduction of a methyl group to the alkyl chain, of the aniline moeity, of alkoxybenzylidene alkylanilines has shown a dramatic effect upon their mesomorphic properties. A large depression in clearing point for all compounds, and the melting points for most of the compounds, was observed. Two new compounds with a wide chiral nematic or nematic range at subambient temperature were demonstrated. The monotropic polymorphism was mostly found in the β-methylbutylaniline series. The effect of the longer chain of the alkylaniline and the further methyl position from the ring caused a decrease in their refractive indices and an increase in the optical pitch. It was also observed that most of the compounds show supercooling behaviour.

EXPERIMENTAL

The transition temperatures and the thermodynamic properties were determined with a differential scanning calorimeter (Perkin-Elmer DSC-2) using a 10-15 mg sample size and a heating rate of 5 or 10° per minute. The transition temperatures were within 1° when compared to those obtained from microscopic observation. The mesomorphic textures were examined with a Reichert polarizing microscope. The indices of refraction were determined at 6328 Å using a Pulfrich Refractometer, and the helical pitch was measured using a Cary 17 Spectrometer.

ACKNOWLEDGEMENTS

This work was supported in part by the National Science Foundation under grants GH-33576 and GH-34401, and by the Advanced Research Projects Agency under grant DAHC-15-67-C-0219.

REFERENCES

1. G. W. Gray, Mol. Cryst. and Liq. Cryst., 7, 127 (1969).

2. Z. G. Gardlund, R. J. Curtis and G. W. Smith, Chem. Comm., 202 (1973).

3. W. R. Young, I. Haller and D. C. Green, J. Org. Chem., 37, 3707 (1972).

4. R. J. Cox, Mol. Cryst. and Liq. Cryst., 19, 111 (1972).

5. W. R. Young, A. Aviram and R. J. Cox, J. Am. Chem. Soc., 94, 3976 (1972).

6. M. M. Leclercq, J. Billard and J. Jacques, Mol. Cryst. and Liq. Cryst., 8, 367 (1969). 10, 429 (1970).

7. G. W. Gray and K. J. Harrison, Mol. Cryst. and Liq. Cryst., 13, 37 (1971).

8. J. van der Veen and A. H. Grobben, Mol. Cryst. and Liq. Cryst., 15, 239 (1971).

9. D. Dolphin, Z. Muljiani, J. Cheng and R. B. Meyer, J. Chem. Phys., 58, 413 (1973).

10. Y. Y. Hsu, Ph.D. Dissertation, Kent State University,(August, 1972).

11. W. Elser, J. L. W. Pohlmann and P. R. Boyd, Mol. Cryst. and Liq. Cryst., 20, 77 (1973).

12. H. Sackmann and D. Demus, Mol. Cryst. and Liq. Cryst., 2, 81 (1966).

COMMENTS ON THE RELAXATION PROCESS IN THE CHOLESTERIC-NEMATIC TRANSITION

R.A. Kashnow, J.E. Bigelow, H.S. Cole, and C.R. Stein

General Electric Corporate Research and Development

ABSTRACT

The relaxation process by which a field-aligned nematic state returns to a helicoidal structure has been the subject of several studies. In this paper, we report some new transient optical and capacitive measurements which emphasize the dependence of the relaxation on sample boundary conditions. Homeotropic samples exhibit a delay time after field removal, followed by nucleation of the decay sequence about isolated points. During the decay, the capacitance decreases monotonically, but the light-scattering transient is structured even in the absence of polarizing optics. A spiral domain structure is observed for these samples. For samples with parallel boundary conditions, the transient capacitance data exhibit a minimum which supports earlier inferences of a tilted-Grandjean-planar structure.

INTRODUCTION

The field-induced cholesteric-nematic transition[1-4] involves transformations between a field-aligned nematic state and various helicoidal structures, the details of which depend upon sample boundary conditions. For the planar liquid crystal layers frequently employed in electro-optic applications, the sample boundaries are fixed with the director either parallel or perpendicular to the bounding substrates.[5] In this paper, we report some experimental observations which indicate the dependence on these boundary conditions of the relaxation processes by which the field-aligned state decays to a helicoidal configuration upon removal of the field.

Wysocki et al.,[6,7] and more recently Ohtsuka and Tsukamoto,[8] have observed a transient optical rotation during the relaxation process, from which they have inferred that a (tilted) Grandjean-planar-like structure occurs as a transient. Some transient capacitance measurements reported below are consistent with such an inference for samples with parallel boundary conditions, but our observations on samples with perpendicular boundary conditions suggest a different sequence.

Following a brief section on experimental details, we discuss steady-state measurements of the voltage-dependence of sample capacitance, which complement direct microscopic observations of the director configurations for chiral nematic samples with each type of boundary condition. Following that, we present transient optical and capacitive data which can be related to the steady-state data. In the final section, we discuss some director distributions which might account for the observations.

EXPERIMENTAL DETAILS

The present studies were made at room-temperature on samples of positive dielectric anisotropy which comprised binary mixtures[9] of p-methoxybenzylidene-p-n-butylaniline (MBBA) and p-ethoxybenzylidene-aminobenzonitrile (PEBAB) to which small amounts of cholesteryl oleyl carbonate (COC) were added for optical activity. From measurements of stripe spacing in the domain texture,[10] the pitch (p_o in μm) was determined to vary with COC concentration (c in weight fraction) approximately as $(p_o/2)^{-1} = 4.5$ c. Parallel boundary conditions were obtained by unidirectional rubbing of tin-oxide-coated glass plates and doping with small amounts of bifunctional agents such as 1,5 dicyanopentane.[9] Perpendicular conditions were achieved with monopolar dopants, sometimes with silane coupling agents;[11] spontaneous homeotropy frequently occurred but we did not depend on it.

Capacitance measurements were made with a laboratory-constructed admittance meter which incorporated a lock-in-amplifier and which was undisturbed by the application of large ac bias voltages to the sample. The ac bias frequencies were chosen to avoid hydrodynamic effects and the capacitance-measuring frequencies were chosen to avoid dielectric dispersion; typically, values of 350 Hz and 3 kHz, respectively, were suitable.

STEADY-STATE OBSERVATIONS

Figure 1 shows the voltage-dependence of the effective dielectric constants (inferred from capacitance measurements in calibrated sample cells) for two samples, one with parallel (a) and one with perpendicular (b) boundary conditions. The sample composition comprised COC, MBBA, and PEBAB with a ratio of parts by weight of 0.02: 0.93: 0.05, respectively. The quiescent half-pitch was $p_o = 11.0\ \mu m$ and the sample thickness was $d = 20.0\ \mu m$. The ordinate in Fig. 1 is $\Delta\varepsilon(V)$, where $\varepsilon_{eff}(V) = \varepsilon_t + \Delta\varepsilon(V)$; $\Delta\varepsilon(V > V_c) = \varepsilon_p - \varepsilon_t$. The subscripts p and t denote parallel and transverse components of the (low-frequency) dielectric constants; for this sample $\varepsilon_p = 6.36$ and $\varepsilon_t = 5.65$. The critical voltage $V_c = E_c d$ for this sample was about 18 vrms.

For samples with parallel boundary conditions, the sequence of states observed for a monotonically increasing field is well-known.[12] The quiescent state ($\Delta\varepsilon = 0$) is the Grandjean planar one which is perturbed first to a grid pattern instability[13] (at about 7 vrms in Fig. 1) and subsequently (at 10 vrms) to the strong light-scattering state which has been variously referred to as the "undisturbed" state,[14] the "fingerprint" texture,[12] or the domain texture.[15] In this domain state, the helical axis is approximately parallel to the bounding planes, but an array of disclinations is required to accommodate the helicoidal structure to the boundary conditions.[15] Neglecting the latter, the effective dielectric constant for this condition is simply $\varepsilon_{eff} = (\varepsilon_p + \varepsilon_t)/2$. Microscopic observations confirmed the occurrence of that state for the point at $\Delta\varepsilon(V) = \Delta\varepsilon(V > V_c)/2$ in Fig. 1(a). For a further increase in voltage, the pitch dilates and ultimately diverges as nematic ordering obtains for $V = V_c$ (for which $\varepsilon_{eff} = \varepsilon_p$). Actually, the parallel boundary condition persists throughout a layer near the boundaries,[6] the thickness of which diminishes with increasing field[16] above the threshold. This is barely perceptible in the capacitance measurements, but optical observations (e.g., of the conoscopic figure) evidence its existence. This layer is important in the transient relaxation, as discussed below.

For perpendicular boundary conditions, the sample in its quiescent state assumes not the Grandjean structure, but rather a more complicated domain pattern, a typical microscopic view of which is shown in Fig. 2(a). The corresponding capacitance measurement (Fig. 1b) shows that $\Delta\varepsilon(0) \simeq 0.3\ \Delta\varepsilon(V > V_c)$ for this condition. At about 5.3 vrms, a light-scattering domain structure forms (Fig. 2b) which exhibits a spiral configuration which appears to correspond to that described by Bouligand.[17]

TRANSIENT RELAXATION OBSERVATIONS

The nematic-cholesteric relaxation exhibits considerable structure in its light-scattering transients, as shown in some typical traces in Figure 3, where the dependence on sample thickness as well as that on boundary condition is emphasized. These traces were

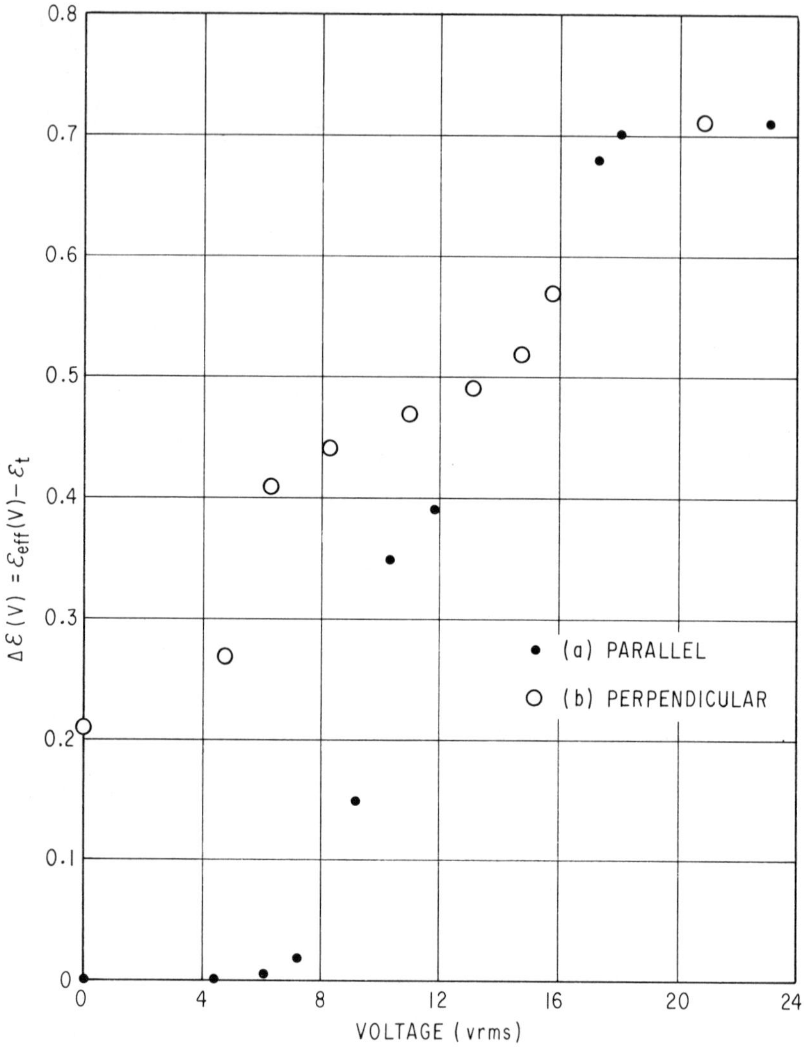

Fig. 1. Voltage dependence of the effective dielectric constant for a chiral nematic (2 percent COC) with (a) parallel and (b) perpendicular boundary conditions. The capacitance was measured at 3 kHz; bias voltage applied at 350 Hz.

measured by monitoring the time-dependence of zero-order-transmission of a normally-incident, unpolarized He-Ne laser beam (632.8 nm) following the sudden removal (at t = 0) of a 45 vrms, 1 kHz signal. The samples comprised 0.05, .76, and .19 parts by weight of COC, MBBA, and PEBAB, respectively, in wedge-shaped samples with nominal thicknesses of 15 to 50 μm. We note two important aspects of these

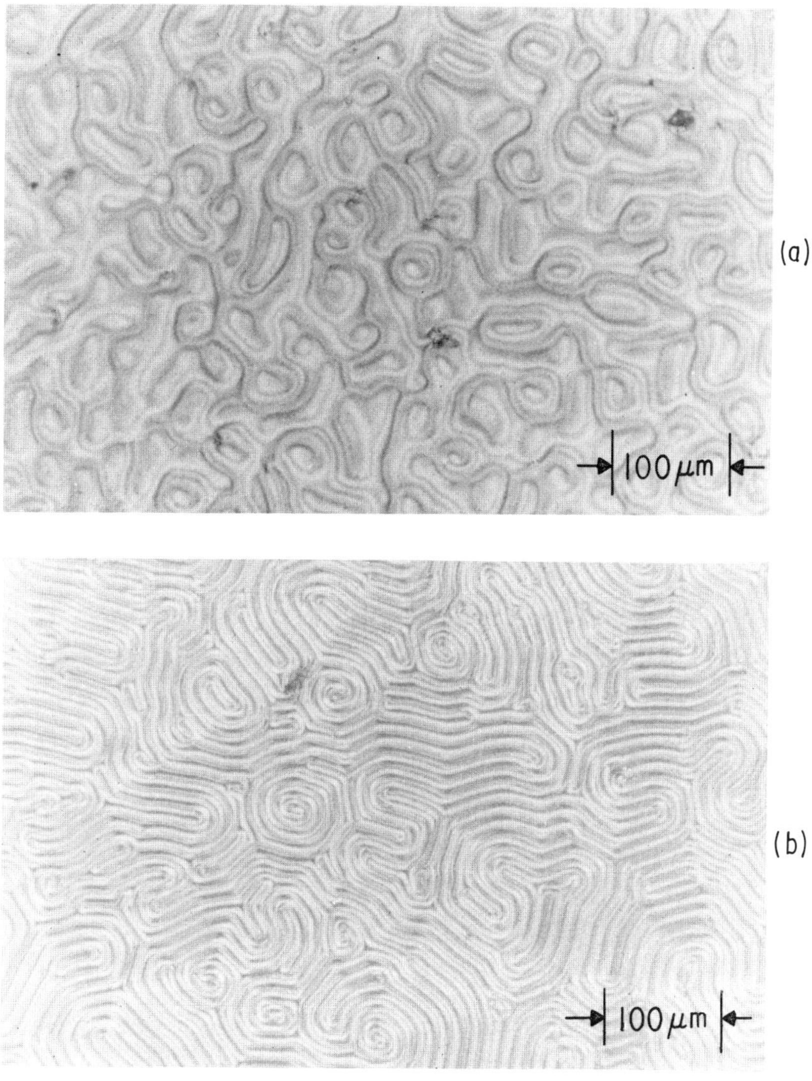

Fig. 2. Microphotographs (with parallel polarizers) of the domain structures in a chiral nematic (2 percent COC) with perpendicular boundary conditions: (a) quiescent state; (b) light-scattering state (5.3 vrms, 350 Hz).

traces. First, the sample with perpendicular boundary conditions (3a) exhibits a real delay time following field removal. Second, since no polarizing optics were used in the measurements, the "hump" in the transient scattering curve evidences a diminution in the light-scattering due to birefringent effects, rather than, for example the effects of transient optical rotation on transmission through external polarizers.[6-8] These qualitative features were reproduced for a variety of experimental conditions, including non-normal incidence and incoherent light of various visible wavelengths, as well as for samples of widely differing compositions. Samples of shorter pitch, of course, exhibit these features on a much shorter time scale.[18] The ratio of pitch to sample thickness can be important, however: For $p_o/d > 1$, the decay time for homeotropic samples increases sharply with decreasing thickness;[19] for $p_o/d \ll 1$, the influence of the boundaries is less prominent.

Transient capacitance measurements were made more conveniently on slower samples. Figure 4 shows the transient capacitance (a) and the zero-order-transmission (b) for the parallel sample for which steady-state data were given in Fig. 1a. The ordinate scale on the capacitance trace was calibrated so that the zero corresponds to the quiescent (Grandjean-planar) sample capacitance. With reference to Fig. 1a, then, values of $\Delta\varepsilon$ can be assigned to this trace, as follows. $\Delta\varepsilon$ decreases from its maximum value of 0.71 at t=0 to a relative minimum of $\Delta\varepsilon$ = 0.04 at t = 1.3 s, and then increases to a relative maximum at $\Delta\varepsilon$ = 0.15 at t $\simeq$ 5 s; ultimately (on a longer time scale than shown in this figure), $\Delta\varepsilon$ approaches zero. The optical trace (4b) was taken with a fiber-optic microscope eyepiece probe. Taken together, these data are consistent with an initial transient relaxation to a slightly-tilted Grandjean structure, followed by the formation of disclinations which underlie the strong light-scattering domain state. The ultimate relaxation of this defect state to the quiescent one has recently been discussed by Hulin.[20]

Figure 5 shows comparable traces for the perpendicular sample of Fig. 1b. Here it is clear that the capacitance decreases monotonically, nearly to its quiescent value, in the initial part of the relaxation, while the optical trace exhibits the structure which has been shown more clearly in Fig. 3a. Cinemicrophotographic studies of this relaxation have shown that the decay is initiated by nucleation about isolated points which often appear to be defects pinned to a sample surface. This corresponds to the initial delay period in the relaxation process. In polarized light, the initial relaxation appears as an array of sets of concentric contour lines which spread outward from the nucleation site. As these isolated regions grow together, the light-scattering is temporarily diminished, corresponding to the "hump" in the optical trace. An array of disclinations subsequently forms which causes a concomitant increase

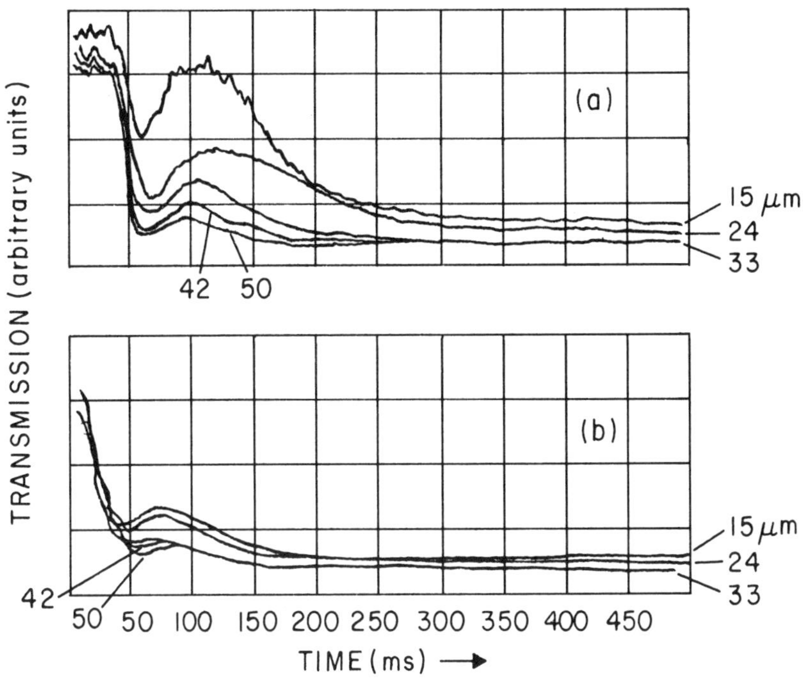

Fig. 3. Zero-order-transmission (no polarizing optics) vs. time after field removal, showing transient relaxation of 5 percent COC samples with (a) perpendicular and (b) parallel boundary conditions, for sample thicknesses between 15 and 50 μm.

in light-scattering. As in the parallel case, an ultimate relaxation to a quiescent state occurs on a longer time scale.

DISCUSSION

In this section, we speculate on a qualitative model for the director configuration during the relaxation process. Figure 6 shows a cross-sectional sketch of the evolution of a possible director distribution in the vicinity of a nucleation site, which we suppose to be an extrinsic surface defect at which a local parallel boundary condition is favored in distinction to the perpendicular condition elsewhere. The solid lines represent the director in the plane of the drawing; the notation of Kléman and Friedel[21] is used to indicate the tilt of the director with respect to the drawing. This sequence can evolve to a spiral configuration such as that described by Bouligand,[17] as disclinations (marked X in Fig. 6)

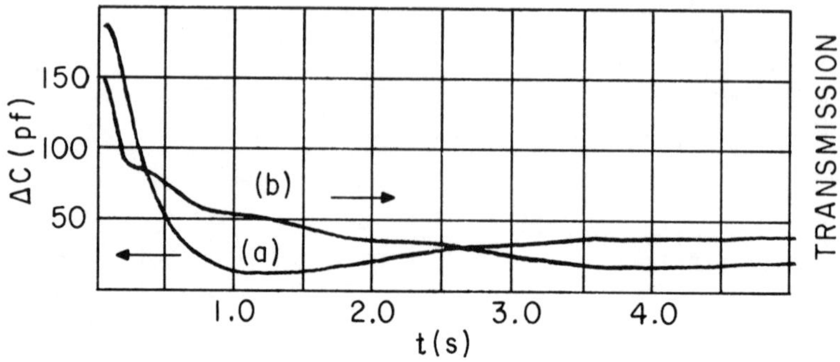

Fig. 4. Transient capacitance (a) and light-transmission (b) for a 2 percent COC sample with parallel boundary conditions (see Fig. 1a).

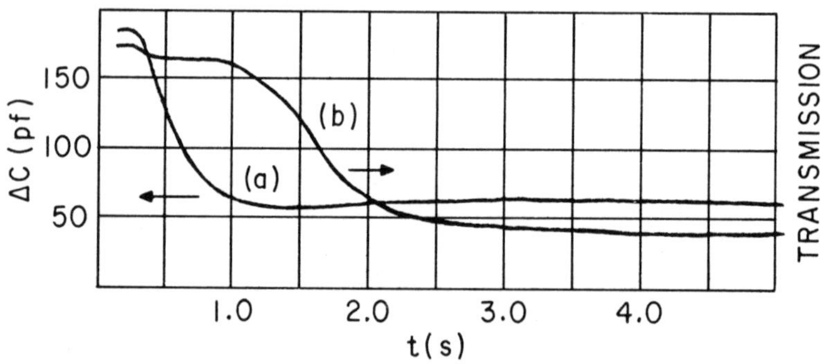

Fig. 5. Transient capacitance (a) and light-transmission (b) for a 2 percent COC sample with perpendicular boundary conditions (see Fig. 1b).

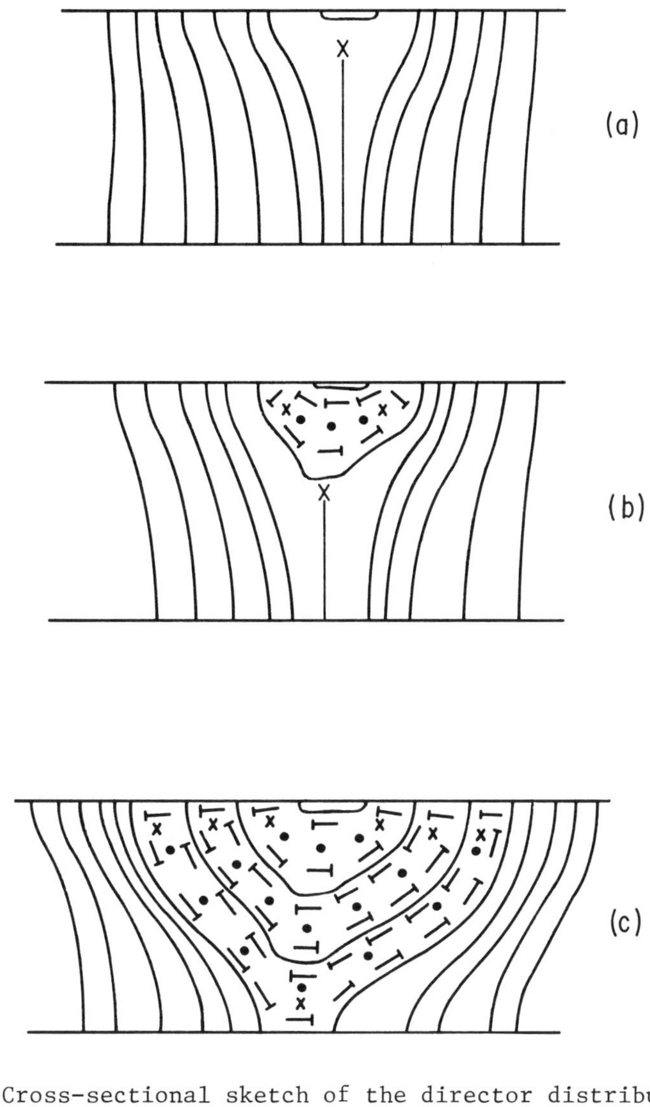

Fig. 6. Cross-sectional sketch of the director distribution near a single nucleation site during the beginning of relaxation of the aligned nematic state for a sample with perpendicular boundary conditions. The closely-spaced solid lines outside a twisted area denote nematic ordering. "X" denotes the intersection of a disclination line with the plane of the drawing.

split to form a defect structure near the surfaces, in analogy to the domain structure described by Cladis and Kleman.[15] Figure 7 is a cinemicrophotographic frame which shows a long-pitch sample in the process of such a relaxation. The inner region displays the beginnings of the spiral configuration whereas the surrounding contour lines are due to birefringence (in crossed polarizers) in the surrounding tilted director distribution (e.g., as in Fig. 6b).

For the case of parallel boundary conditions, large areas of the surface may act as defect sites about which the decay can nucleate. This would account for the relative absence, in that case, of a delay time following field removal. The initial transient state might then correspond to the conical deformation described by Meyer,[4] for which the director distribution is $\hat{n} = (\cos\rho \cos 2\pi z/p_o, \cos\rho \sin 2\pi z/p_o, \sin\rho)$, assuming that the field was applied in the z-direction and $\pi/2 - \rho$ is the tilt angle relative to the aligned direction. We may then estimate, very roughly, the minimum value of ρ attained during relaxation, using transient capacitance measurements such as shown in Fig. 4a. For that sample, if $\epsilon_{eff} = \epsilon_t \cos^2\rho + \epsilon_p \sin^2\rho$ and since $\Delta\epsilon = 0.04$ for the first relative minimum, we find $\rho \simeq 14°$. The relaxation to the Grandjean state does not progress to completion but rather is disturbed by the formation of disclinations which characterize the light-scattering domain state; only after these relax does the Grandjean planar structure obtain in quiescence.

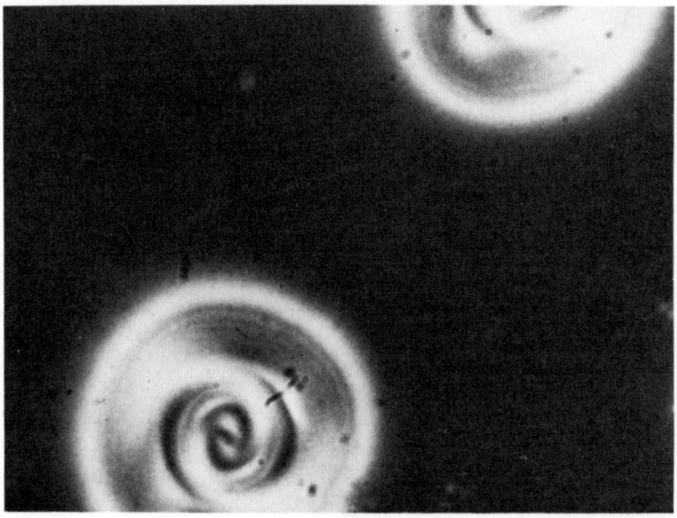

Fig. 7. A frame from a microphotographic movie (in crossed polarizers) showing one aspect of the relaxation of a long-pitch (1.25 percent COC) sample with perpendicular boundary conditions.

Acknowledgment. We thank G.H. Glover for his valuable advice in connection with the capacitance measurements.

REFERENCES

1. E. Sackmann, S. Meiboom, and L.C. Snyder, J. Am. Chem. Soc. 89, 5981 (1967).

2. J.J. Wysocki, J. Adams, and W. Haas, Phys. Rev. Lett. 20, 1024 (1968).

3. P.G. de Gennes, Sol. State Comm. 6, 163 (1968).

4. R.B. Meyer, Appl. Phys. Lett. 12, 281 (1968).

5. The dynamics of the transition in samples with one free boundary have been studied by J. Prost and H. Gasporoux, Mol. Cryst. Liq. Cryst., to be published.

6. J.J. Wysocki, J. Adams, and D.J. Olechna, "Liquid Crystals and Ordered Fluids," ed. J.F. Johnson and R.S. Porter, (Plenum Press, Inc., New York, 1970), p. 419.

7. J.J. Wysocki, Mol. Cryst. Liq. Cryst. 14, 71 (1971).

8. T. Ohtsuka and M. Tsukamoto, Japanese J. Appl. Phys. 12, 22 (1973).

9. R.A. Kashnow and H.S. Cole, Fourth International Liquid Crystal Conference, Kent, Ohio, 1972; Mol. Cryst. Liq. Cryst., to be published.

10. R.B. Meyer, Appl. Phys. Lett. 14, 208 (1969).

11. F.J. Kahn, Appl. Phys. Lett. 22, 386 (1973).

12. F. Rondelez and J.P. Hulin, Sol. State Comm. 10, 1009 (1972), and references therein.

13. C.J. Gerritsma and P. Van Zanten, Phys. Lett. 37A, 47 (1971).

14. J. Adams, W. Haas, and J. Wysocki, Mol. Cryst. Liq. Cryst. 8, 9 (1969).

15. P.E. Cladis and M. Kleman, Mol. Cryst. Liq. Cryst. 16, 1 (1972).

16. P.G. de Gennes, Mol. Cryst. Liq. Cryst. 7, 325 (1969).

17. Y. Bouligand, J. de Phys. 33, 525 (1972).

18. E. Jakeman and E.P. Raynes, Phys. Lett. 39A, 69 (1972) report very short decay times which may characterize the slope of the initial change in light transmission.

19. This probably corresponds to the "self-orientation" effects described by S. Sato and M. Wada, Japanese J. Appl. Phys. 11, 1566 (1972).

20. J.P. Hulin, Appl. Phys. Lett. 21, 455 (1972).

21. M. Kléman and J. Friedel, J. de Phys. 30C, 4 (1969).

EFFECTS OF DETERGENTS ON ISOLATED RAT LYMPHOCYTE PLASMA MEMBRANES

D. N. Misra, C. T. Ladoulis, L. W. Estes, T. J. Gill III

Department of Pathology, University of Pittsburgh School of Medicine, Pittsburgh, Pennsylvania 15261

For several years we have been studying the genetic control of the immune response in inbred rats by immunochemical procedures for assaying antibody and by kinetic analysis of the production of antibody forming cells. These studies have been carried out using a chemically defined synthetic polypeptide antigen poly($Glu^{52}Lys^{33}Tyr^{15}$) in high and low responder strains of rats (1). We are now purifying lymphoid cells in order to explore the chemical basis of the genetic control mechanisms by isolating and characterizing the membrane components which may be involved in the interaction with antigen. Since the membrane proteins and glycoproteins are integral components of the liquid crystalline structure of the plasma membrane, their isolation depends upon dissociation from their native semicrystalline environment into a liquid phase suitable for studies of their binding activity, molecular structure and biological function. The methods of dissociation employ certain assumptions about the effects of solubilization on the structure and function of membrane components. Although the various detergents used for isolation of membrane components are classified as anionic, cationic or nonionic, there is no reliable means of predicting the results of solubilization either in terms of the degree of solubilization or with respect to their selectivity for different macromolecules. For this reason, the selection of these agents is based upon empirical results. In order to carry out the detailed structural analysis of the proteins and glycoproteins from the lymphocyte plasma membrane, we have studied the effects of three different detergents on the degree of solubilization and on the state of the solubilized products.

Lymphocyte plasma membranes are isolated in the following manner. The thymus or spleen organs are surgically removed, finely minced and dissociated cells are suspended in cold tris-buffered 0.15 M NaCl at pH 7.4. The lymphocytes are purified by flotation on a 39% hypaque-9% ficoll (1:1) gradient after centrifugation at 400 x g for 30 minutes at room temperature. The cells are washed in tris-buffered 0.15 M NaCl and homogenized in a Potter-Elvejhem homogenizer at 4° C until more than 85% of the cells are disrupted, as measured by trypan blue dye exclusion or by phase microscopy. The plasma membranes are isolated after a series of successive sedimentation steps to remove intact nuclei, unbroken cells, mitochondria and rough endoplasmic reticulum, and then the final step is isopycnic density gradient centrifugation in a 20-50% sucrose gradient at 90,000 x g for 18 hours at 4° C. The isolated plasma membranes recovered from the 30-40% interface of the sucrose gradient are smooth membrane vesicles free of contamination by ribosomes or debris (Figure 1). The purity of these plasma membranes was assayed by 5'-nucleotidase activity, which is a reliable marker for plasma membrane although it is not exclusively localized to these membranes (2,3). The specific activity in this fraction is 12 - 13 times greater than in the homogenate, and this result is comparable to that obtained by others with isolated plasma membranes of pig or human lymphocytes (4,5).

In order to characterize those membrane components which are involved in control of immune responsiveness, we have analyzed the solubilized lymphocyte plasma membranes by electrophoresis in polyacrylamide gels containing

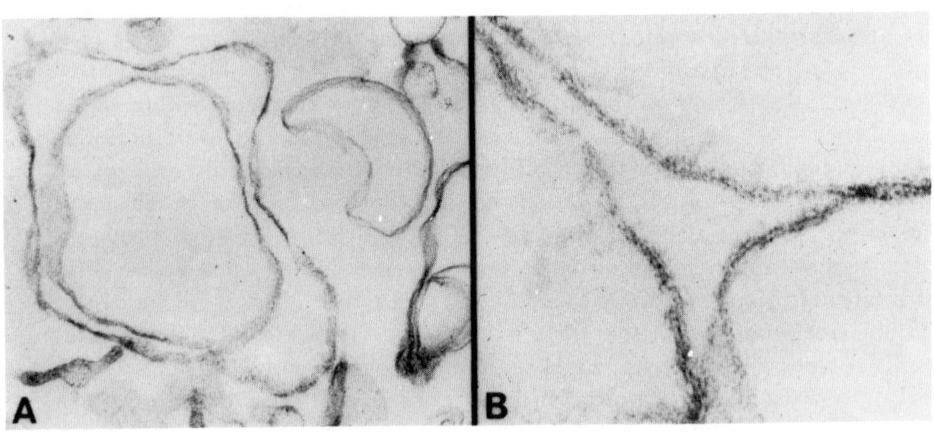

Figure 1. Electron micrographs of purified plasma membrane containing smooth vesicles (A) and displaying the typical trilaminar structure (B).
A - 55,000X; B - 181,000X

0.5% sodium dodecyl sulphate (SDS). The initial experiments involved solubilization of membrane vesicles by heating in 2% SDS at 100° C for 5 minutes and then electrophoresis of the solubilized sample in 7.5% polyacrylamide gels containing 0.5% SDS. This method was standardized and calibrated with 8-10 proteins of known molecular weight in order to estimate the molecular weight and size distribution of the membrane proteins and glycoproteins. Membrane protein analysis was made by staining the gels with Coomasie brilliant blue for protein and with periodic acid-Schiff for carbohydrate. Glycolipids were detected by staining with Sudan Black. Electrophoretic analyses were also carried out after labelling intact cells with radioactive iodine using the lactoperoxidase method since the interaction between antigen and the immunocompetent cell apparently involves the external protein or glycoprotein components (6). The analysis of the external membrane protein distribution was made by slicing the gels after electrophoresis into 64-65 equal fractions and counting them with an automated gamma counter.

When the plasma membranes of thymic and splenic lymphocytes are compared in stained gels, about 25 bands are observed, and there are small but significant quantitative as well as qualitative differences between thymic and splenic membranes. Specifically, the splenic lymphocyte membranes contain small amounts of high molecular weight components in the region above 200,000 daltons which are absent from thymic lymphocytes. The 160,000 dalton component is much more prominent in spleen than in thymus. Furthermore, the glycoprotein patterns for the two membranes are different. The most striking and consistent difference is that the thymic lymphocyte plasma membrane contains an unique 25,000 dalton glycoprotein. In splenic lymphocyte membranes, there is little or no corresponding carbohydrate-containing component in this region. The results of these electrophoretic analyses are illustrated in Figure 2.

Since we wish to isolate membrane components for further physical chemical and immunochemical analysis, sodium dodecyl sufphate could not be employed as the solubilizing agent. This detergent binds to proteins in considerable amounts and interferes with their reactivity due either to steric hindrance by SDS or to denaturation of the protein (7). For this reason Triton X-100, 1% at room temperature, was used for the solubilization of isolated plasma membranes, since this kind of nonionic detergent is reported not to interfere with the specific interactions of antibodies directed against membrane proteins (8).

The membrane components of interest in immunological reactions are the external glycoproteins and proteins. When thymic or splenic lymphocytes were externally labeled with radioactive iodine using lactoperoxidase,

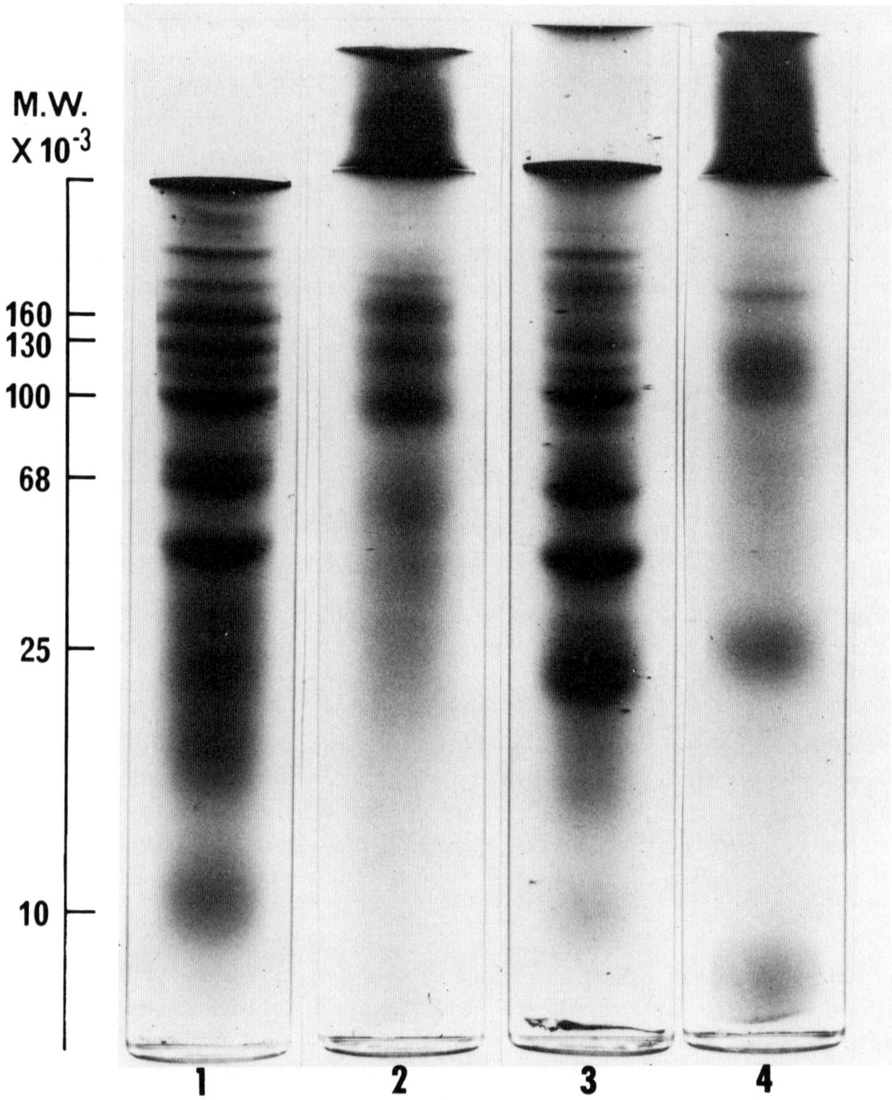

Figure 2. - Polyacrylamide gel electrophoretic patterns of SDS-solubilized membranes. Gels 1 and 2 are from splenic lymphocyte membranes and gels 3 and 4 are from thymic lymphocyte membranes. Gels 1 and 3 are stained with coomassie blue for protein, and the other two are stained with PAS for carbohydrate. The protein gels show that some differences are significant although there is an overall similarity in the band patterns. The thymic membrane has a prominent 25,000 daltons component which is a glycoprotein (Gel 4). The glycoprotein patterns of the two membranes are quite different.

solubilized in SDS and then analysed by polyacrylamide gel electrophoresis, typical patterns were obtained (Figure 3, curves A and B). These profiles demonstrate that there is a prominent, common external membrane protein of 120,000-125,000 daltons in both thymic and splenic lymphocytes. The striking observation, however, is that the prominent thymic 25,000 dalton glycoprotein seen in stained gel patterns is externally labeled. Thus, when membranes are solubilized in SDS the characteristic difference between thymic and splenic lymphocytes is the presence in thymus of an unique external membrane glycoprotein. However, when similar analyses were performed with plasma membranes solubilized in Triton X-100 (1%) at room temperature, the 25,000 daltons thymic glycoprotein was not recovered (Figure 3, Curve C). Since the conditions of external labeling and electrophoretic analysis were identical, the difference is due to the effects of the detergents used for initial solubilization. Either Triton X-100 does not completely solubilize this glycoprotein or the combined effect of Triton X-100 and SDS used for electrophoresis leads to precipitation and loss of this glycoprotein from the sample.

Because of the differences in the effects of these two detergents, experiments were carried out to compare their solubilization effects under different conditions. In addition, sodium deoxycholate (DOC) was also used, since this detergent does not bind to proteins in significant quantities (9) and may be an alternative agent for studying the interaction between solubilized membrane proteins and antibodies or antigens. We chose several parameters to analyze: (1) the final concentration of detergents was varied over the range 0.1-2.0%; (2) studies were made at ratios of 5, 10 and 20 μg detergent/μg membrane protein, since we observed that the effects of these detergents were apparently related in some manner to the weight ratio of detergent to sample protein; and (3) the solubilization effects were studied with heating at 100° C for 5 minutes and at room temperature for 30 minutes. Since all the membrane isolations were carried out at pH 7.4, this pH was maintained constant throughout these experiments. The effects of the different detergents under these conditions were analyzed by measuring the extent of solubilization of the externally labeled membrane components and by electron microscopy of the solubilized components using the negative contrast technique. Previous experiments with labeled membranes showed that more than 97% of the membrane solubilized in 2% SDS by heating at 100° for 5 minutes entered a 7.5% polyacrylamide gel for which the exclusion limit was determined to be about 300,000 daltons. The definition of complete solubilization was the recovery of 100% of the radioactivity in the supernatant after centrifugation of the treated sample at 90,000 x g for 60 minutes at 20° C.

The results of these experiments at final concentrations of 0.1-0.2% and 1-2.0% are summarized in Table 1. SDS at room temperature completely

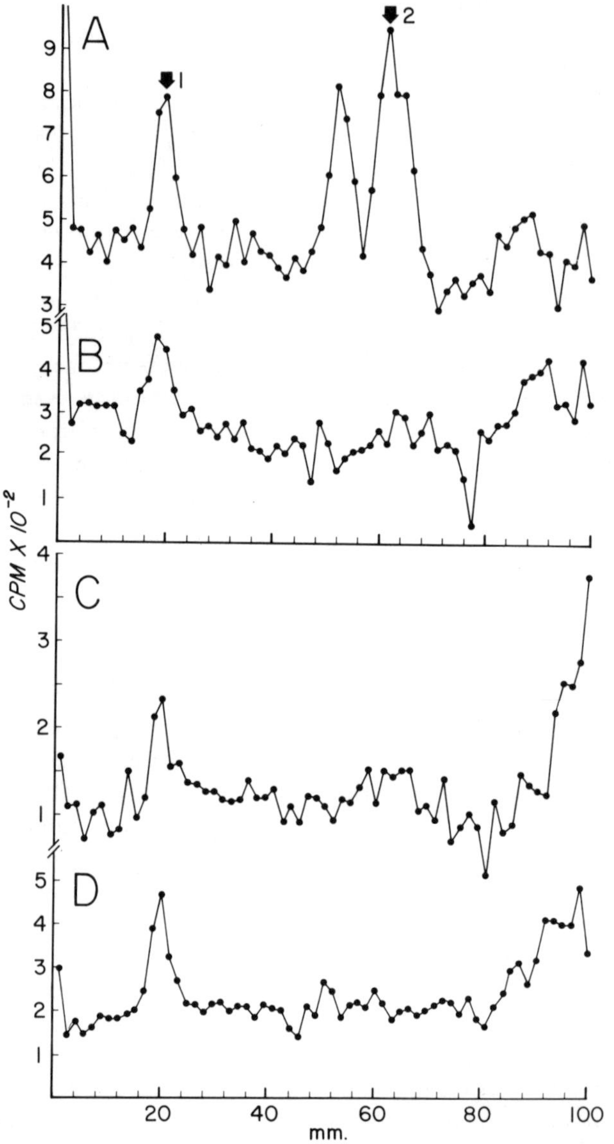

Figure 3. Radioactivity profiles of thymic and splenic lymphocyte membranes after polyacrylamide gel electrophoresis. Curves A and B are for SDS dissolved samples and C and D are for Triton X-100 dissolved samples. Curves A and C are for thymic and B and D for splenic membranes. All curves show a prominent surface protein of about 125,000 daltons molecular weight **(1)**. Another surface protein of about 25,000 daltons molecular weight is found in thymic membrane when dissolved in SDS **(2)**. This protein corresponds to the glycoprotein band found on the stained gels of thymic membrane (Fig. 2).

Table 1
Effects of Detergent Concentration, Detergent/Protein Ratio and Experimental Conditions on Solubilization of Isolated Lymphocyte Plasma Membranes

Detergent	Treatment	μg detergent per μg protein	Per cent of solubilization at final detergent concentration of 0.1-0.2%	1.0-2.0%
SDS[a]	100° for 5 min.	5	94[c]	98[c]
		10	100	--
		20	99	99
SDS	Room Temp. for 30 minutes	5	98	100
		10	99	--
		20	99	100
Triton X-100	100° for 5 min.	5	99	98
		10	98	--
		15	98	98
Triton X-100	Room Temp. for 30 minutes	5	94	98
		10	96	--
		20	97	98
DOC[b]	100° for 5 min.	5	80	98
		10	85	--
		20	97	98
DOC	Room Temp. for 30 minutes	5	73	96
		10	84	--
		20	88	97

[a] - Sodium dodecyl sulphate

[b] - Sodium deoxycholate

[c] - Each figure represents an average of 2-3 different experiments

solubilized the membrane at both concentrations of detergent and at all weight ratios of detergent to membrane protein. When the reaction was carried out at 100° for 5 minutes, there was complete solubilization at 10 or 20 μg per μg of protein regardless of detergent concentration. At a weight ratio of 5 μg of SDS per μg of protein, however, there was slightly more solubilization at the higher detergent concentration. When membranes were solubilized in Triton X-100 at room temperature, less solubilization was achieved at 0.2% concentration, and it was proportional to the weight ratio of Triton X-100 to protein (94% at 5 μg/μg and 97% at 20 μg/μg). At 2% Triton X-100, there was about 98% solubilization. The results with heating the samples in Triton X-100 were the same at both concentrations and in all ratios of detergent to protein. When samples were solubilized in DOC at room temperature, there was less solubilization than with either SDS or Triton X-100. At room temperature, there was more solubilization at both concentrations with a higher ratio of DOC to protein, and at the same ratio of detergent to protein there was more solubilization at 2% than at 0.2% final detergent concentration. When samples in DOC were heated, there was generally greater solubilization at the higher concentration of DOC. At 0.2% DOC there was a marked increase in solubilization with increasing ratio of detergent to protein (from 80 to 96%), but at 2.0% DOC concentration virtually complete solubilization (98-99%) was obtained at all three ratios. In summary, there is complete solubilization by SDS at room temperature or above, whereas solubilization with Triton X-100 or DOC is dependent to some extent upon the temperature of solubilization, the concentration of detergent, and the weight ratio of detergent to membrane protein. In view of the fact that solubilization with SDS is usually carried out at an elevated temperature, it is interesting to note that at room temperature complete solubilization was achieved regardless of concentration or weight ratio of detergent to protein, and at elevated temperature the results suggest that there is slightly less solubilization at low SDS concentration with the lowest ratio of detergent to protein. In contrast, the results of solubilization in Triton X-100 and DOC showed increased solubilization with elevation of temperature, increased concentration of detergent, or increased ratio of detergent/membrane protein. This conclusion is applicable only to the solubilization of the radioactively labeled external membrane proteins, but they are the proteins of primary interest in the study of interaction of the cells surface with macromolecules.

We observed the solubilization effects of these detergents by negative contrast electron microscopy in order to gain some independent information about the conformation of the solubilized components of the lymphocyte plasma membranes. The isolated membrane vesicles were solubilized at concentrations of 0.2% or 2% detergent and at a constant ratio of 10 μg of detergent per μg of protein. The effects of solubilization were determined both before and

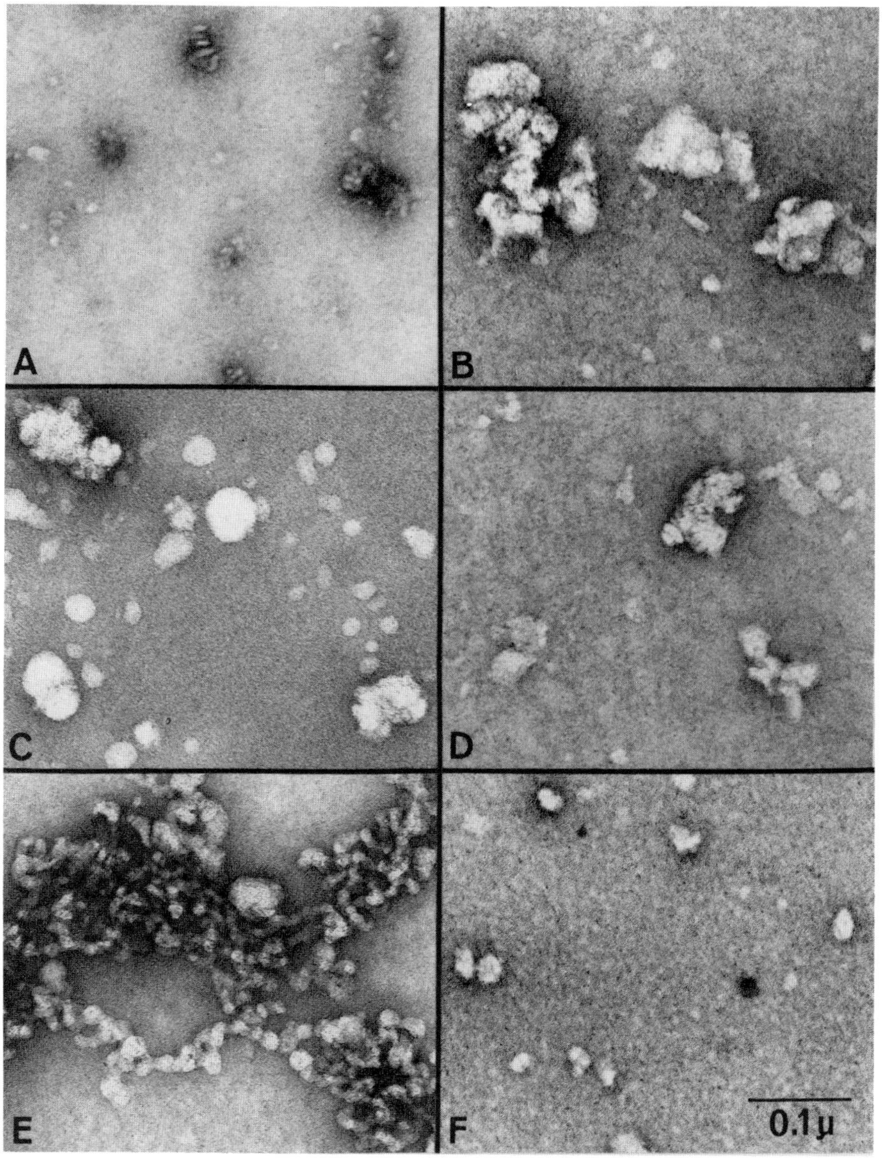

Figure 4. Electron micrographs of rat lymphocyte membranes dissolved in SDS (A and B), Triton X-100 (C and D) and DOC (E and F). Detergent concentration is 0.2% and detergent to protein weight ratio is 10:1. Samples A, C and E are undialysed and the rest are dialysed. All three detergents give different membrane products, and there is a remarkable difference between the dialysed and undialysed samples in the case of SDS and DOC treatment.

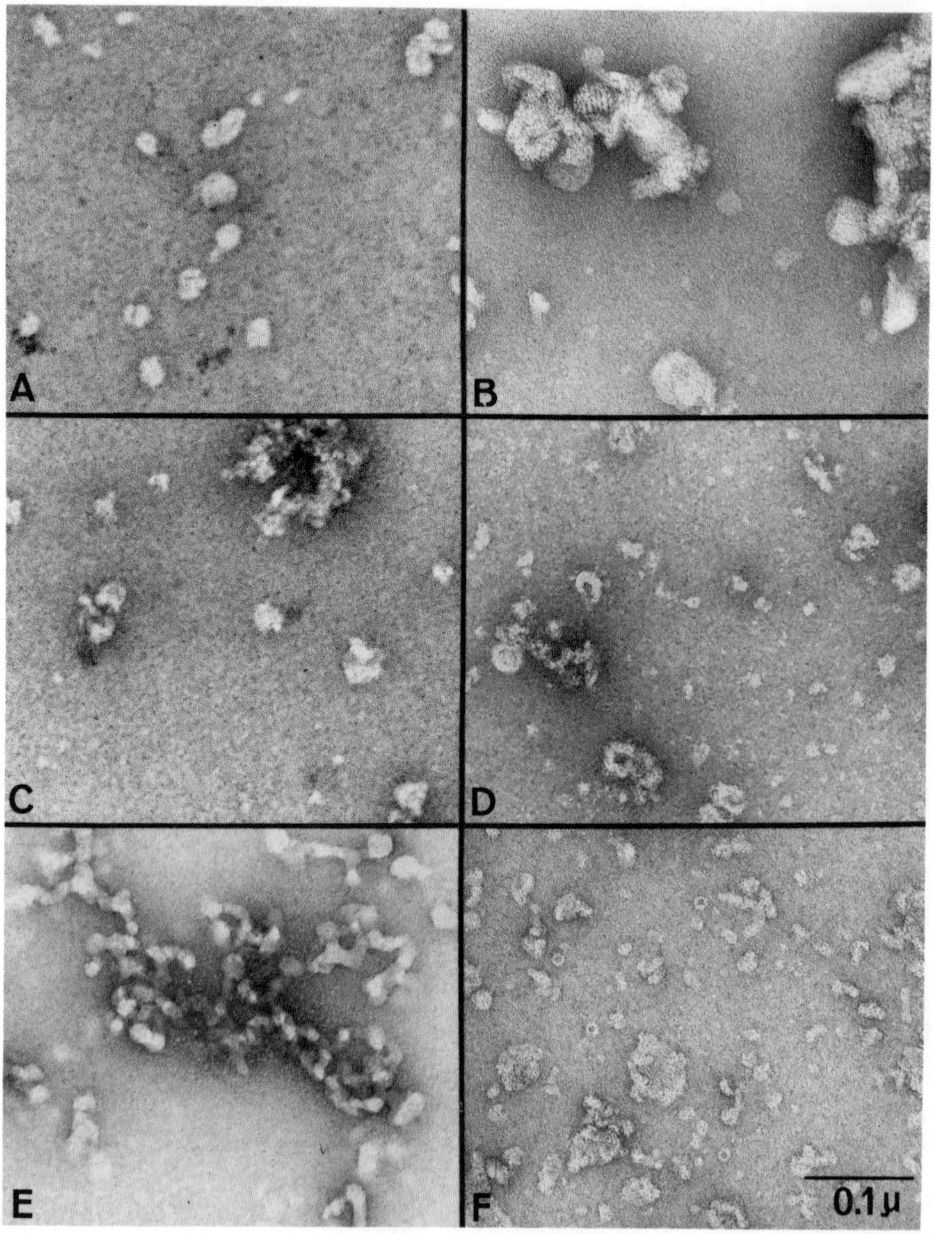

Figure 5. Electron micrographs of rat lymphocyte membranes dissolved in SDS (A and B), Triton X-100 (C and D) and DOC (E and F) all at 2% concentration and at a detergent to protein weight ratio of 10:1. The samples A, C and E are micrographs of undialysed and the rest are dialysed. The products and the differences among them are the same as in Fig. 4.

after 48 hours of dialysis which was used to reduce the concentration of detergent. The results of these experiments at 0.2% and 2.0% detergent concentration are illustrated in Figures 4 and 5, respectively. At the lower concentration, the undialyzed SDS-solubilized membrane preparation showed finely dispersed particles which were occasionally stacked in rouleaux formation. When the samples were dialyzed, large irregular aggregates were formed. At low concentration, undialyzed DOC-solubilized membranes formed large interlacing irregular filaments, particles or strands. After dialysis, the preparation was finely dispersed into discrete amorphous particles of irregular size. Finally, undialyzed samples in low concentration of Triton X-100 revealed a heterogenous population of globular particles, some being larger than those obtained with SDS. After dialysis, there was little, if any, change in the appearance of these particles. Similar results were obtained with all three detergents at concentrations of 2% (Figure 5). Thus, at these two concentrations the three detergents have different effects in solubilizing the plasma membranes of lymphoid cells: (1) the removal of SDS by dialysis resulted in large aggregates with occasional substructure; (2) the dialysis of DOC solubilized membranes resulted small particles; and (3) the dialysis of Triton X-100 solubilized samples had little or no effect on the particle size or shape.

The method of solubilization by these three detergents may differ significantly enough to effect qualitatively, as well as quantitatively, the membrane components. The results illustrated in Figure 3 suggest that Triton X-100 solubilization differs from that of SDS and, despite complete solubilization by other criteria, the effect of Triton X-100 may differ significantly from the effects of SDS or DOC. The different microscopic appearances of the solubilized membranes suggest that the divergent results obtained with thymic and splenic lymphocyte membranes were due to the reduced solubility of the thymic lymphocyte glycoprotein in Triton X-100.

Summary

There is a unique 25,000 dalton membrane glycoprotein on thymic but not splenic lymphocyte plasma membranes of the rat. This component is recovered from isolated plasma membranes by solubilization in sodium dodecyl sulphate but not in Triton X-100. Studies on the effects of detergents showed that solubilization is quantitatively more complete with SDS than with Triton X-100 or deoxycholate. Less quantitative solubilization is associated with the formation of larger micellar particles. The effects of solubilization by different detergents are both quantitative and selective. Thus, the use of a specific detergent may obscure important biochemical differences between the membranes of functionally different cells, and several detergents should be tried in each new type of experiment.

Acknowledgments

This work was supported by grants from the National Science Foundation (GB 30826X), the U. S. Army Medical R & D Command (DADA 17-73-C-3020) and the National Institutes of Health (5 TOI GM 00135).

References

1. T. J. Gill III, H. W. Kunz, D. J. Stechschulte and K. F. Austen, J. Immunol., 105, 14 (1970).

2. R. Coleman and J. B. Finean, Protoplasma, 63, 172 (1967).

3. C. C. Widnell, J. Cell Biol., 52, 542 (1972).

4. D. Allan and M. J. Crumpton, Biochem. J., 120, 133 (1970).

5. D. Allan and M. J. Crumpton, Biochim. Biophys. Acta, 274, 22 (1972).

6. D. R. Phillips and M. Morrison, Biochem. Biophys. Res. Comm. 40, 284 (1971).

7. C. A. Nelson, J. Biol. Chem., 246, 3895 (1971).

8. J. W. Uhr and E. S. Vitetta, Fed. Proc., 32, 35 (1973).

9. J. Philippot, Biochim. Biophys. Acta, 225, 201 (1971).

HETEROCYCLIC LIQUID CRYSTALS AND SOME AIR FORCE APPLICATIONS OF MESOMORPHIC COMPOUNDS

Rudolph A. Champa
SAMSO/RSSE, P.O. Box 92960
Worldway Postal Center
Los Angeles, California 90009

INTRODUCTION

Applications of liquid crystals continues to provide the impetus for research of new materials with improved physical properties. Performance requirements vary to a considerable extent depending on application; more specifically those technical applications of interest to the Department of Defense (DOD) are unique because of the more dynamic environments to which they would necessarily be exposed. Such things might include useful temperature ranges of operation (typically -50C to +80C), radiation (EM and nuclear), vibration, g-forces, response times, stability etc.

The DOD organizations have been interested in display and computer applications of liquid crystals for several years. More recently interest in their use as diagnostic probes and in NDT has been revived with the continued improvements in material properties and understanding of their behavior and response. Both of these areas, however, still require further work toward a better understanding of the mechanisms involved in liquid crystalline devices, time constants of materials and the sensitivity of liquid crystals (LCs) to respond to energy inputs.

Work at the Air Force Materials Lab (AFML) has centered on a search for new nematic liquid crystals and an improvement in the operations of potential devices to aid in data computation, display and storage. Additionally, the lab has recently become interested in the extent to which a liquid crystal could practically be used in laser diagnosis.

Thus, this paper will address the progress made in heteroatomic

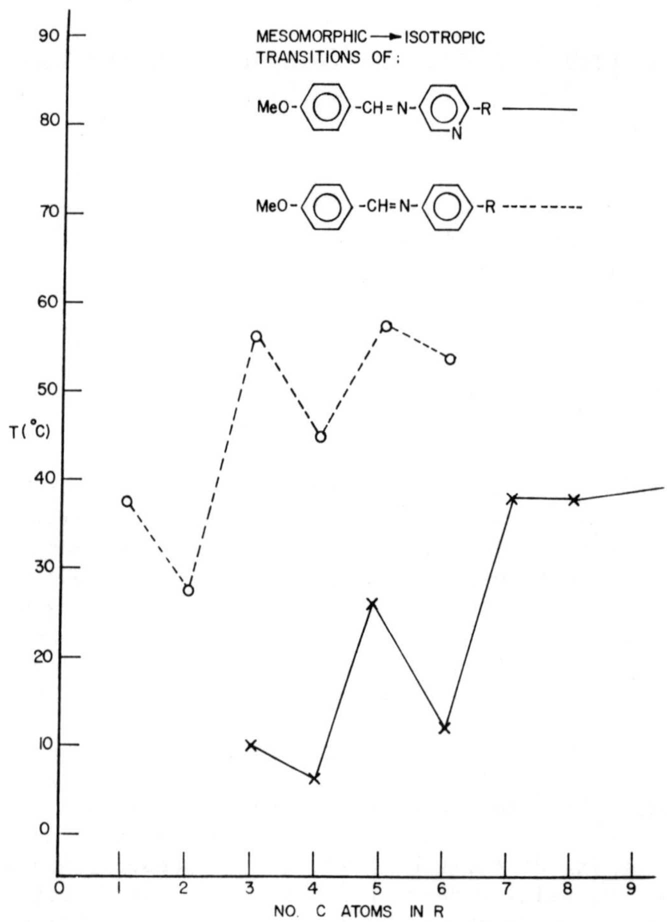

MESOMORPHIC → ISOTROPIC
TRANSITIONS OF:
MeO-⟨⟩-CH=N-⟨N⟩-R
MeO-⟨⟩-CH=N-⟨⟩-R
T(°C)
90
80
70
60
50
40
30
20
10
0
1
2
3
4
5
6
7
8
9
NO. C ATOMS IN R

Figure 1

variation of molecular structure to improve the material properties of liquid crystals. Additionally, current known applications will be identified and some comment will be made about their future promise and problem areas yet to be worked.

DISCUSSION

Synthetic Studies

To date there have been only a limited number of heteroatomic species that have exhibited a liquid cyrstalline phase. One reason for this is the relative difficulty involved in synthesis of the heteroatomic precursors to the LC. The most common heteroatom has been heterocyclic nitrogen although a few oxygen and sulfur derivatives have been reported[1]. Mesomorphic transitions at or below room temperature have only recently been reported with pyridine anils of the class below[2] and have generally been monotropic with respect to

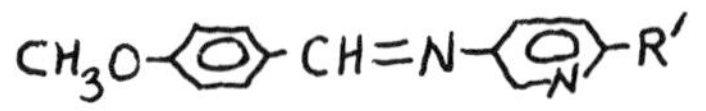

the nematic phase; this transition has been as low as 6° C for R'= n-C_4H_9 (see figure 1).

It is known that a variety of parameters govern the occurence of a mesomorphic phase: geometry, conformation, lateral to terminal attractive force ratio, length to breadth ratio and the presence and position of dipoles. The attractive feature of utilizing a heteroatom is the introduction of a dipole without increasing the molecular breadth. It was found in an AFML study of some 3-N-(4'-alkoxybenzilideneamino)-6-n-alkylpyridines that the pyridine anils generally melted lower than the carbocyclic analogs and all the corresponding mesomorphic transitions were lowered by an average 40°C. The lower melting points resulted most probably from the lower symmetry of the pyridine anils; the lower clearing points from a combination of symmetry and heteroatomic dipolar alteration of the molecular polarizability.

Although no purely enantiotropic nematic resulted from the above study mixtures were prepared which exhibited nematic phases and existed as such at temperatures below -30C for several days. A typical binary mixture consisted of 50 wt % heterocyclic anil and 50 wt % of the corresponding carbocyclic moiety. Lack of accurate equipment for very low transition temperature determination (in bulk) coupled with degradation of the material as a function of time limited the usefulness of any extensive investigation. The main mode of degradation appeared to be hydrolysis of the anil linkage. In cases where the pyridine ring was unsubstituted, hydrolysis appeared to be complete (i.e. only starting materials were identified) after several months. The materials had been stored in closed containers and kept essentially in the dark, although no special precautions were taken during initial specimen collection or subsequent retrieval of samples (e.g. dry box, gas purging etc.).

Dynamic Scattering Studies

Although the factors determining which chemical materials will be nematic are not completely understood, they have been studied for some fifty years. Those factors influencing dynamic scattering, however, have only been pursued during the last few years.

The usual procedure for assessing liquid crystal response (i.e. time constants and contrast) has been to utilize an experimental configuration similar to that shown in figure 2[3]. In this particular study the light source was modified from the diagram by placing it in a box with a small hole in it to approximate more closely a collimated source thus minimizing variations due to light source fluctuations. Utilizing this set-up, typical rise times (t_r) and decay times (t_d) for standard MBBA are shown in figure 3; figure 4 shows a typical scope trace. In this discussion t_r and t_d are defined as points B and D respectively.

By comparison to a pure component, a 50/50 mixture (by wt) of MBBA/EBBA showed essentially the same t_r within the limits of error. Replacing MBBA with the heterocyclic liquid crystal 3-N-(4'-methoxybenzilideneamino)-6-n-butylpyridine (MBBP) resulted in little change in t_r but extended t_d by a factor of 3 (see figure 5). The reason for the increased decay times may be speculatively attributed to at least three factors: 1) greater interaction of the heterocyclic moiety with the electric field, 2) a greater interaction with the surface of the electrodes, and 3) a greater propensity for ionic saturation.

APPLICATIONS

Electronic devices perform many of the vital tasks in today's sophisticated systems. Perhaps a major application of electronis is in extending and complementing man's senses and actions, particularly in the US Air Force regime of high speed flight and space exploration where man's reaction time and sensitivities limit the behavior of the man-machine system. It has thus become necessary to extend and complement man's ability with electro-optical systems, high speed computers and sophisticated communications. To achieve this requires an extensive and complex array of electronic and electro-optic components; often these are required on-board where weight, reliability, and power efficiency are of prime importance. The concern for weight and power efficiency is evident from the large number of government and industrial efforts in miniaturization of devices and components. Liquid crystals serve these goals well: they require only microwatts of power at the readout stage, have excellent contrast in a dynamic light environment, are relatively stable, very cheap, offer several shades of gray and have a no-power, long term

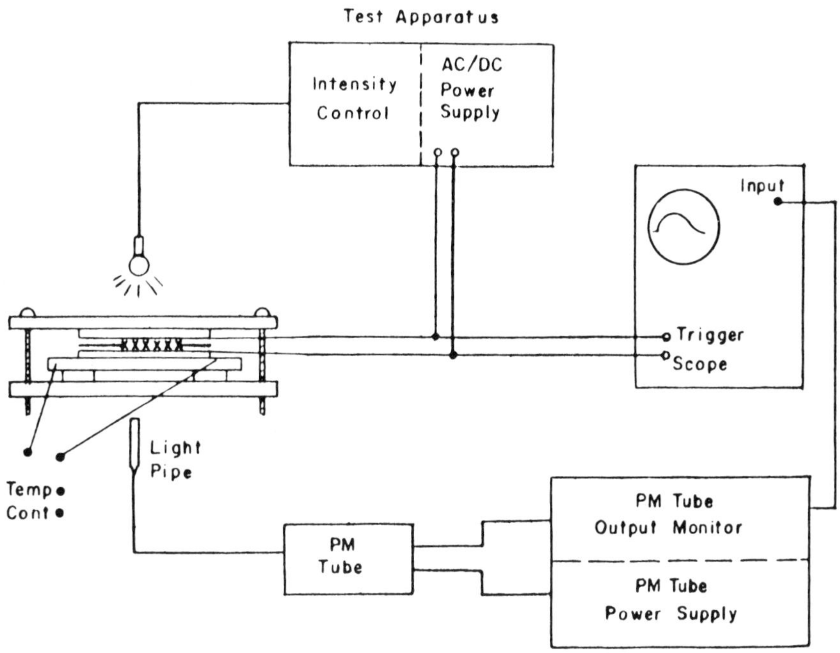

Figure 2. Experimental Setup

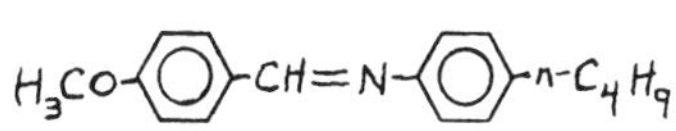

For a 25μ Thickness

D. C. Voltage	T_{Rise} (Millesec)	T_{Decay} (Millesec)
44	40 ± 5	65 ± 5
67	16 ± 4	55 ± 5
90	9 ± 3	50 ± 5
136	4 ± 2	40 ± 5

Figure 3. Effect of Applied Field

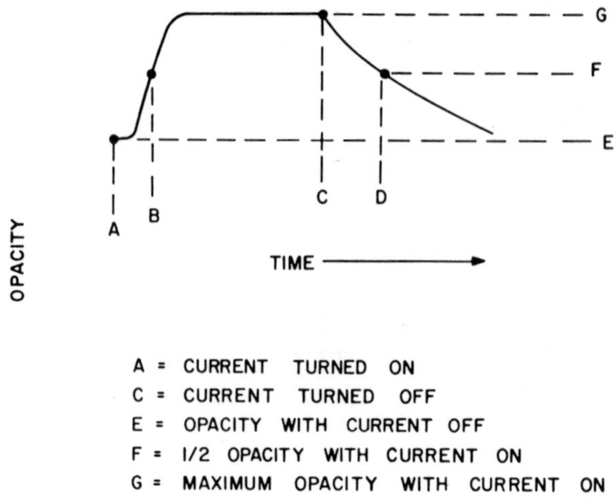

Figure 4. Typical Scope Trace

50/50 wt % MBBA/EBBA

Voltage (dc)	t_r (msec)	t_d (msec)
15	~90	~215
30	~28	---
45	~14	~185

50/50 wt % MBBP/EBBA

Voltage (dc)	t_r (msec)	t_d (msec)
15	~80	~160
30	~30	~325
45	~17	~700

Figure 5. Comparative Response Times of Heterocyclic and Carbocyclic Mixtures

storage capability with a natural short storage compatible with TV frame rates and computer requirements.

The second area of LC application is diagnostic and, in fact, was the first area to be exploited. The cholesteric mesophase is the one primarily used for thermographic diagnostics and has been used to check structural bonding, uniformity of system operation, solid state electronic boards, laser beam profiles, real time interferometry and holography, acoustical and RF magnetic visualization, and medical diagnostics such as tumor detection.

The need for nematic liquid crystals for practical applications (especially display) has provided the impetus to search for materials that will function over a wide temperature range and well below room temperature. This work has now been in progress for the last three years and until this year only one confirmed material existed (MBBA) as a liquid crystal at (and below) ambient temperatures. Recently, two reports by Dolphin and Meyer[4] on p-cyano-p'-alkoxy or acyloxybiphenyls and Gray[5] on β-methoxybutylanilines have appeared to add to this list. However, several researchers in government, industry and universities have claimed success in achieving mil spec temperature requirements with mixtures of materials and with improved lifetimes and time constants.

Electro-optic devices. An Air Force technical report[6] of a year ago identified several applications such as digital displays, light valves, UV/visible imaging and computer storage, high resolution matrix displays and large scale displays for command and control information. It appeared at that time that the industrial researchers were very close to the marriage of large scale integrated circuitry (LSI) with LC technology. Unfortunately, that marriage hasn't occurred yet but it is clear that the electronic package is the main problem. The key LC material problem appears to be lifetime.

A further information aspect of the above technology which meets another important USAF need, information storage, is the real time inputs for optical computing offered by LCs on CdS[7]. This optically addressed light valve display offers impressive performance figures for cells that transfer spatially distributed information from incoherent to coherent light beams with coherent light efficiencies as high as 92%. Both Navy and NASA have been keenly interested in this application. Part of this interest is due to the fact that LCs are amongst precious few materials with high polarizabilities and relatively large polarizability anisotropies capable of operation as thin films or at the molecular level for use in computer memory, information storage and electro-optical devices.

Diagnostic applications. This area is still very much in need of increased application engineering. Such areas as on-board optical sensors, LCs used in conjunction with fiber optics for remote temperature sensing, NDT of composites, laser diagnostics, aerodyanamic diagnostics, LC use as a "strain gage" in dynamic mechanical tests and holographic inspection are of critical importance to systems development and design engineers. Innovative research in applying liquid crystals to these problems is lacking. Although researchers in the area may place this technology deficiency on lack of adequate support funding, the onus is really on the scientists to provide the systems people with data to convince them of the utility and cost effectiveness of liquid crystals.

More work is required in exploiting encapsulated liquid crystals from the standpoint of sensitivity, resolution and temperature range of operation and simplicity of application. For example, in real time laser beam profile analysis, in flaw detection of laser window materials and in real time aerodynamic visualization of boundary layer phenomenon over an entire surface where extrapolation from finite point thermocouple data can be alleviated, the funding agency program managers need to be convinced that there is more than just "potential real life" application of liquid crystals. If this doesn't occur, the government funded liquid crystal programs will continue to decline.

REFERENCES

1. W.R. Young, I. Haller, and L. Williams, "Liquid Crystals and Ordered Fluids", edited by J.F. Johnson and R.S. Porter, Plenum Press, New York, 1970, p.383.

2. R.A. Champa, Mol. Cryst. Liq. Cryst., 19, 233(1973), and references therein.

3. M.A. Berwick, R.A. Champa et. al., AFML-TR-71-72, July 1971.

4. J. Dolphin, R. Meyer, J. Chem. Phys., 58, 413(1973).

5. G. Gray, 9, 130(1973).

6. R.A. Champa, AFML TR-72-77, May 1972.

7. T.D. Beard, private communication.

CATALYSIS IN MICELLAR AND LIQUID CRYSTALLINE PHASES

S. Friberg and S.I. Ahmad [+]

The Swedish Institute for Surface Chemistry

S-114 28 Stockholm, Sweden

ABSTRACT

The reaction rates were determined of p-nitrophenyl laurate hydrolysis in a lameller liquid crystalline phase containing water, cetyl trimethyl ammonium bromide and hexanol. The pronounced increase in the rate of reaction taking place in the water content range 50 - 70 percent (W/W) was related to a reduction in activation energy and with changes in the structure of the phase.

INTRODUCTION

The catalytic affect of the association of surfactant molecules into micelles has been investigated for more than a decade following the pioneering work by Duynstee and Grunwald[1] in 1959, and extensive reviews of the progress of the catalysis by normal micelles in aqueous solutions have been published,[2 - 4].

Surfactants may, however, also form reversed micelles with water in nonaqueous solvents[5] and also liquid crystalline phases[6]. Earlier reports demonstrated the possibilities of catalytic effects by reversed micelles[7] and in liquid crystalline phases[8] on the hydrolysis of p-nitrophenol laurate.

With regard to the fact that our results showed large variations in the hydrolysis rate, with the composition of the

+ Present address: School of Pharmacy, Univ. of Wisconsin, MADISON, Wisconsin 53706, U.S.A.

liquid crystalline phase[8] and to the fact that the rate of ester hydrolysis in a liquid crystal composed of water and nonionics has been shown, to be lower[9] compared with that in solution; we found an investigation to be worth while on structural changes of the liquid crystalline phase, which could be related to the reaction rate.

EXPERIMENTAL

Materials

Analytical reagents p-nitrophenyl laurate (Schuchardt, München), 1-hexanol (Fluka, Switzerland) and cetyltrimethylammonium bromide (Merck, Germany) were used. The water was distilled twice and buffered to pH 11.05 with 0.1 M sodium phosphate buffer.

Reaction Rates

The reaction rate was followed from the formation of phenolate ion which was determined by the increase in extinction at 400 mμ by means of a Zeiss PMQ spectrophotometer. The thermostated cells were 0.2 mm thick. Initial concentration of p-nitrophenyl laurate was 5×10^{-4}M in all cases. The solutions were buffered to pH 11.05 with phosphate buffer.

The reaction rates were determined in the liquid and liquid crystalline phases for a number of series with a constant alcohol-CTAB ratio. The phases were highly viscous, and a special mixing procedure had to be applied in order to achieve reproducible results. 1 g of the differens components according to the actual composition minus equivalent weight of 0.2 ml hexanol were mixed together and allowed to equilibrate for 24 hrs at 25°C. Hexanol (0.2 ml) containing an amount of p-nitrophenyl laurate so as to achieve a total concentration of 5×10^{-4}M was then added under gentle mixing. The initial turbidity of the mixing disappeared after 2 - 3 min. and the determinations of extinction commenced.

The pseudo-first-order reaction rate constant was then determined in the usual manner from the slope of extinction vs. time. Since the water content varied considerably within each phase, second-order "rate constants" were calculated by division with the molar fraction of water. For the determination of activation energies, reaction rates at different temperatures were determined using thermostated cells with a thickness of 0.2 mm.

X-Ray Diffraction

A Kiessig's camera was used with nickel-filtered copper radiation ($\lambda = 1.542$ A) at 16 mA and 40 kV, The distance between specimens in a thin-walled ($\emptyset = 0.1$ mm) glass capillary tube and the film was 500 mm, occasionally changed to 200 mm.

RESULTS

Two series of samples were used for the determination. In one series, the hexanol: CTAB molecular ratio was kept constant and equal to 1.8, and in the other the hexanol content was varied while the water: CTAB ratio was constant at 20:1. The composition of the samples and the region of the liquid crystalline phase are observed in figure 1.

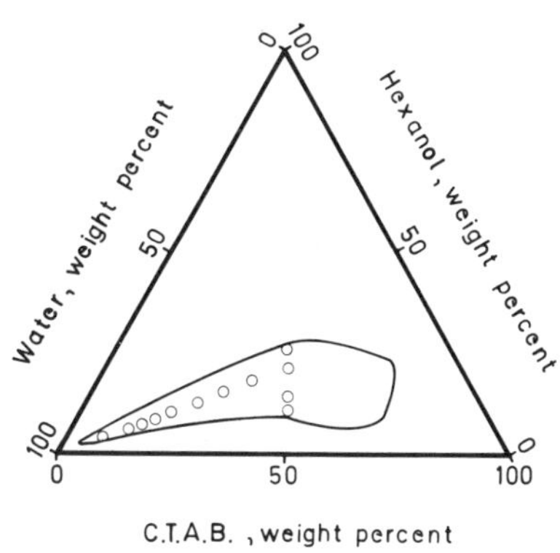

Figure 1. The region of the lamellar liquid crystalline phase and the composition of the samples.

Reaction Rates and Activation Energy

The reaction rate at 25°C and the and the activation energy for the reaction are shown in figure 2 as functions of water content for a constant hexanol: CTAB ratio. The pronounced increase in reaction rate with water contents between 50 and 70 percent is notable. The second order rate

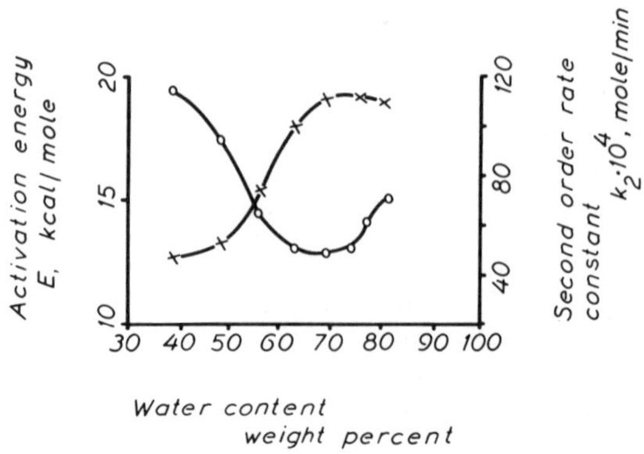

Figure 2. Activation energy (⊖) and second order rate constant (X) versus water content at a hexanol:CTAB ratio of 1:2 (W/W)

constant increased more than twice in the mentioned range. The increase in reaction rate was accompanied by a reduction in the activation energy, from the highest value of about 19.1 kcal/mole with 40 percent water to a minimum of 13 kcal/mole with 70 percent water.

Changes with alcohol content were uniform. An increase in alcohol content from 10.5 to 27 percent by weight more than doubled the reaction rate but changed the activation energy by only a few percent according to figure 3.

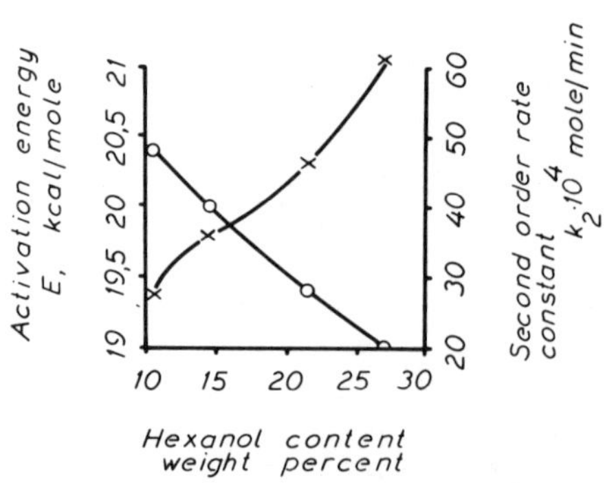

Figure 3. Second order rate constant (X) and activation energy (⊖) versus hexanol content at a water:CTAB (W/W) ratio of 20:1.

X-Ray Diffraction

The typical distribution of reflections of a lamellar phase (d ratios 1: 1/2: 1/3: ....) were obtained at all three temperature levels. The results were used to calculate the various parameters: viz. the thickness of the polar part of the amphiphilic layer, d_a: the thickness of the water layer, d_w; and the area per hydrophilic group in the layer, S. For calculation, the following formulas were used:

$$d_a = v_a d,$$

$$d_w = v_w d \text{ and}$$

$$S = 2v_a \cdot \bar{v} \cdot 10^{24} / d_a N n_a$$

where v_a and v_w are the volume fractions of water and amphiphile. d is the lamellar layer thickness, $\bar{v}$ is the specific volume of the specimen, N Avogadro's number and n_a the number of amphiphiles per gram of the phase.

The distance between layers, d, versus the reciprocal of the volume fraction of the amphiphilic was linear up to a water percentage of 74, figure 4.

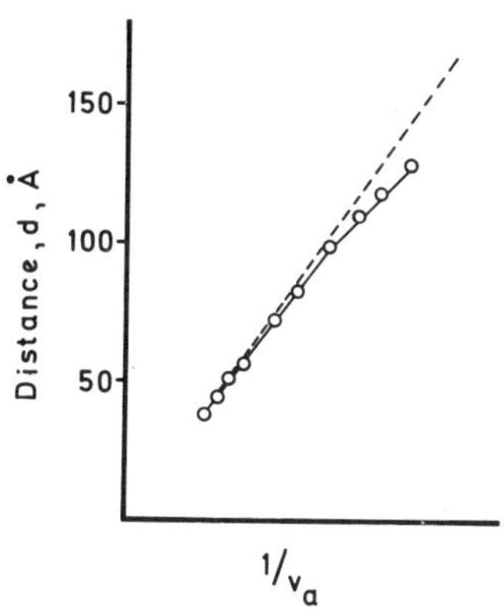

Figure 4. The distance between layers for the lamellar phase with buffered water (⊙) compared to unbuffered water (- - -).

At this composition, a change of gradient was observed, showing a lower increase in d with water contents at values in excess of 74 percent (W/W). The higher order reflections could not be obtained at higher water contents, and the lamellar structure is not confirmed for that range. The following considerations assume a lamellar structure.

The thickness of the water layer d_a increased linearly from 9 Å to 106 Å with a slight reduction of increase at 74% water. The thickness of the amphiphilic layer increased slightly to the corresponding value and showed a steep decrease at higher water contents, whilst the area per molecule showed a reciprocal behaviour, figure 5.

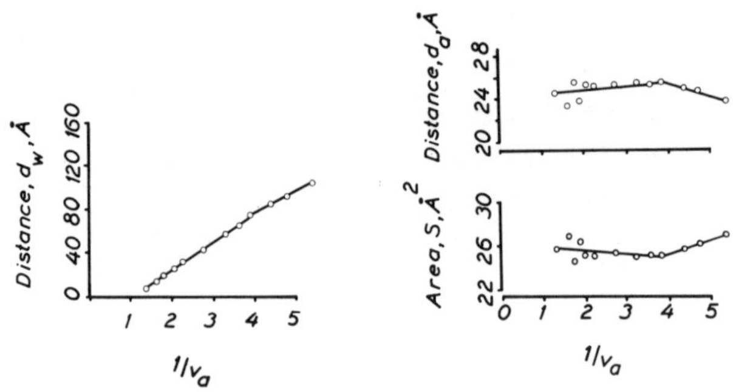

Figure 5. Thickness of the water layer (d_w); the amphiphile layer (d_a) and the area per amphiphilic molecule (S).

The parameter changes with hexanol content at compositions with a constant water: CTAB ratio were linear and small; they are not presented.

DISCUSSION

The result have demonstrated the sensitivity of reaction rates to changes in the structure of the liquid crystalline phase. Of the possible factors which may influence the reaction rate [2,10] for this case, the increase in rate observed when the water content is raised from 40 to 70 percent is accompained by a reduction in the activation energy of the reaction. The two curves in figure 2 display a striking inverse relationship.

The main question raised at the initiation of the investigation was concerned with concentration-dependent structure changes of the liquid crystalline phase and their specific influence on the reaction rate. The liquid crystalline region in the system water, hexanol and CTAB has earlier been described as having a lamellar stucture [11]. A comparison of our results with those obtained earlier [11] (fig. 4) exhibit differences. The linear relationship of d versus $1/v_a$ over the whole concentration range was now replaced by two straight lines intersecting at 75 percent water content and with a slightly lower value of the spacing.

The reduced spacing between the layers was an expected result of the addition of the buffer, which should have an influence on the distribution of the electric double layer potential in the aqueous part of the lamella. The deflection of the curve d versus $1/v_a$ could be interpreted as a sign of a structure change; our present experimental basis does not allow an unambiguous determination. The structure persisting with water contents in excess of 74 percent exhibits lower swelling than the structure with lower water contents. The relationship d versus log $1/v_a$ had a gradient equal to 0.8, which is close to "two-dimensional swelling".

The results permit the conclusion that a structure change in the liquid crystallinge phase takes place at a water content of 74 percent, and that further addition of water in excess of that value penetrates the lamella to a higher degree than is the case with lower percentages of water.

The change of reaction rate is, however, not specifically associated with the structure change; the enhancement took place in the range 40 to 70 percent water. The rate increase is not dependent on the structure change, but is observed when the structure is stable, prior to its change. Further investigations on the specific distribution of energy and the mobility of various parts of the molecules will show the relation to the changes of the reaction rate in more detail.

On varying the alcohol range (from 10 to 27 percent) the reaction rate was increased so that the value at the upper limit was 2.1 times that at the lower. This increase is identical in magnitude to that where the water percentage is increased from 40 to 70 percent. The present change due to alcohol addition is, however, not accompanied by a correspondingly large decrease in the activation energy, a reduction of 7% compared to 35%. In order to understand the difference it is necessary to consider the changes taking place when water and hexanol are added to the structure.

High alcohol content in the lamellar phase gives rise to[11] one-dimensional swelling with the addition of water, while a reduction in alcohol content causes the swelling with added water to disappear [6]. A reasonable interpretation is that the added alcohol prevents the penetration of water in between the amphiphilic molecules in the layer structure.

At first it might appear that the exclusion of water from in between the amphiphilic molecules should have a reducing effect on the reaction rate diminishing the proximity effect[10]. This is a reasonable assumption, but other factors than the proximity of the two reactants are operative in the reaction. The reactants and the transition states interact with each other and with the molecules forming the liquid crystalline matrix in which the reaction takes place.

The presence of the matrix essentially constitutes a micro-separation into hydrophobic and hydrophilic layers, giving rise to a gradient of hydrophilicity in the system. This gradient is reasonably expected to have an orientation effect on the substrate and on the various transient states. The specific influence of the orientation has, however, been shown not to have a positive effect on the reaction rates. On the contrary, liquid crystals composed of nonionics and water[10] gave lower reaction rates than those in a solution.

In view of these facts, it appears reasonable to focus the attention on the properties of the electric double layer and its influence on the reaction. The cationic amphiphile will give rise to an increased concentration of negative ions at the region of the polar groups, changing the apparant pH value. The usual assumption of a Boltzmann distribution gives

$$n_{i,OH} = n_{bulk,OH} \exp\left(e\psi_1 / kT\right)$$

On comparison with the conditions at an interface, a calculation of the pH increase is possible.

$$pH_{surf.} = pH_{bulk} + \frac{\psi_o}{2.303\ kT}$$

A difference in potential of magnitude 50 mV should give a pH difference of the magnitude one unit.

The influence of the added hexanol may in view of these considerations be attributed to changes in the electric double layer. The hexanol molecules were attached to the amphiphilic layer by means of hydrophobic interaction. Their presence reduced the water penetration between the cationic molecules. The reduced water concentration was reasonably followed by a reduced concentration of negative ions in the Stern layer. A higher concentration of negative ions was obtained in the aqueous part giving rise to increased potential difference and increased differences in the pH value. An increase of the pH value will give a corresponding increase of the reaction rate. In the present investigation the reaction rate was doubled; equal to an increase of pH by 0.3 units.

The investigations will continue with determinations of the substrate localization in the layers under varied experimental conditions.

REFERENCES

1. E. F. J. Duynstee and E. Grunwald, J. Am. Chem. Soc., 81, 450 (1959).
2. E. H. Cordes and R. B. Dunlop, Accounts Chem. Res., 2, 329 (1969).
3. H. Moravetz, Advan. Catal., 20, 341 (1969)
4. E. J. Fendler and J. H. Fendler, Advan. Phys. Org. Chem., 8, 271 (1969).
5. P. Ekwall, L. Mandell and P. Solyom, J. Colloid Interface Sci., 35, 266 (1971).
6. F. Husson, H. Musstachi and V. Luzzati, Acta Cryst., 13, 668 (1960).
7. S. Friberg and S. I. Ahmad, J. Phys. Chem., 75, 2001, (1971).
8. S. I. Ahmad and S. Friberg, J. Am. Chem. Soc., 94, 5196 (1972).
9. K. S. Murthy and E. G. Rippie, J. Pharm. Sci., 59, 459 (1970).
10. C. A. Bennton and B. Wolfe, J. Am. Chem. Soc. In press.
11. P. Ekwall, L. Mandell and K. Fontell, J. Colloid Interface Sci., 29, 639 (1969).

LIQUID CRYSTAL DYNAMICS AS STUDIED BY EPR AND NMR

I.Zupančič, M.Vilfan, M.Šentjurc, M.Schara,
F.Pušnik, J.Pirš and R.Blinc

J.Stefan Institute, University of Ljubljana
61001 Ljubljana, Jamova 39, P.O.B. 199/IV, Yugoslavia

Magnetic resonances have proved to be a convenient tool to study the problem of local molecular dynamics in liquid crystals. In this paper three different approaches will be presented.

In the first part the influence of order fluctuation and molecular diffusion on the spin relaxation will be discused and the results compared with the experimental data.

In the second study the measurements of translational self-diffusion of liquid crystal molecules is given. The main emphasis is on the technique by which the determination of the diffusion coefficient is possible. The problems are with the short spin-spin relaxation times which prevent the application of the non-modified field gradient methods.

The efforts to gain an insight into the molecular orientational fluctuations and rotational diffusional motions will be presented in the third part, using EPR technique. Here the temperature dependence of the correlation time through isotropic, nematic and smectic phase is given. From the angular dependence of the linewidth, two motions with different dynamics have been determined.

1. NUCLEAR SPIN RELAXATION AND ORDER PARAMETER FLUCTUATIONS

Interest in spin relaxation in the nematic liquid crystalline phase grew after it was recognized that order fluctuations may play a dominant role in the relaxation mechanism in these systems. A measurement of the spin-lattice relaxation time T_1 may yield the spectral density $I(\omega)$ of these fluctuations and thus give some information on the auto-correlation function for the molecular orientation.

The order parameter fluctuations are collective excitations and the spectrum of these excitations has at least one branch with relaxation times which are comparable to the NMR Larmor frequencies. Unique to liquid crystals it was found that the frequency dependence of T_1 should be of the form[1)]

$$T_1^{-1} = A\,\nu^{-1/2} + B \qquad (1.1)$$

where ν is the nuclear Larmor frequency and A and B are constants. In normal isotropic liquids, on the other hand, T_1^{-1} is either not frequency dependent or exhibits a different type of the frequency dependence[2,3)]

$$T_1^{-1} = C - F\,\nu^{1/2} \qquad (1.2)$$

Here the constant F is different from zero only in very viscous liquids where molecular diffusion is slow.

The basic facts about the study of order parameter fluctuations by NMR can be sketched as follows:

The free energy expansion in terms of the nematic order parameter $\eta = \eta(r,t)$ is given by

$$F = F_o + \frac{1}{2}\,A\eta^2 + C\left[\nabla\eta\right]^2 \qquad (1.3)$$

if anharmonic terms can be neglected. Expanding the local value of the order parameter into a Fourier series

$$\eta(\vec{r}) = \sum_{\vec{q}} \eta_{\vec{q}}\; e^{i\vec{q}\vec{r}} \qquad (1.4)$$

we can express the extra free energy due to order fluctuations with different wave numbers $\vec{q}$ as:

$$\Delta F = \frac{1}{2}\sum_{\vec{q}} A\left|\eta_{\vec{q}}\right|^2\left[1 + \vec{q}^{\,2}\xi^2\right] \qquad (1.5)$$

where the correlation length ξ is introduced as

$$\xi^2 = \frac{C}{A}\;\frac{C}{a(T - T_o)} \qquad (1.6)$$

In the harmonic approximation the rate equations for the order fluctuations with different wave numbers are independent of each other:

$$\frac{d\eta_{\vec{q}}}{dt} = -\Gamma \frac{\partial \Delta F}{\partial \eta_{\vec{q}}} = -\Gamma \cdot A \cdot |\eta_{\vec{q}}| \cdot (1 + q^2\xi^2) \tag{1.7}$$

The corresponding wave vector dependent relaxation time is obtained as

$$\tau_{\vec{q}} = \frac{\Gamma^{-1}}{A(1 + q^2\xi^2)} \tag{1.8}$$

and shows a critical slowing down for $\vec{q} = 0$ as $T \rightarrow T_o$.

The spectral density of the order fluctuations is then obtained as:

$$I_{\vec{q}}(\omega) = \int_{-\infty}^{+\infty} \langle\eta_{\vec{q}}(0)\eta_{-\vec{q}}(t)\rangle e^{i\omega t}\, dt = \overline{|\eta_{\vec{q}}|^2}\; \frac{2\tau_{\vec{q}}}{1 + (\omega\tau_{\vec{q}})^2} \tag{1.9}$$

where

$$\overline{|\eta_{\vec{q}}|^2} = \frac{k_B \cdot T}{A|1 + q^2\xi^2|} \tag{1.10}$$

Inserting expressions (1.8) and (1.10) into eq. (1.9) we finally find that:

$$I_{\vec{q}}(\omega) \propto \frac{1}{\omega^2 + \{\frac{C}{\Gamma}(q^2 + \frac{1}{\xi^2})\}^2} \tag{1.11}$$

We may now distinguish two different cases:

(i) $\omega < \omega_o = \dfrac{C}{\Gamma\xi^2}$ ... (small frequencies)

and

(ii) $\omega > \omega_o = \dfrac{C}{\Gamma\xi^2}$ ... (large frequencies).

In the first case of *small nuclear Larmor frequencies* (i), the frequency spectrum $I_{\vec{q}}(\omega)$ is *white*. The nuclear spin-lattice relaxation rate

$$\frac{1}{T_1} \propto \int_0^{q_{max}} I_{\vec{q}}(\omega)\, 4\pi q^2 dq \propto \xi \propto \frac{1}{(T - T_o)^{1/2}} \neq f(\omega) \tag{1.12}$$

is independent of ω but strongly increases with decreasing temperature as the stability limit is approached.

In the *second case of large nuclear Larmor frequencies (ii)*, the *frequency spectrum is Lorentzian*. The spin-lattice relaxation rate

$$\frac{1}{T_1} \propto \int_0^{q_{max}} I_q(\omega)\, 4\pi q^2 dq \propto \frac{1}{\sqrt{\omega}} \neq f(T - T_o) \qquad (1.13)$$

is now frequency dependent but does not show a critical dependence on temperature.

It should be noticed that in the nematic phase where ξ is large we are practically always in this second limit (ii). In case (i) measurements of T_1 allow a determination of the coherence length of the order fluctuations.

As it can be seen from Fig.1, the Zeeman spin-lattice relaxation time T_1 in nematic PAA follows a frequency dependence like that of equation (1.1) - i.e. order fluctuations -, where-as T_1 in nematic MBBA (Fig.2) shows the dependence of eq.(1.2) - i.e. diffusion mechanism. This demonstrates that in some viscous liquid crystals slow intermolecular diffusion effects, predicted by Torrey, dominate over the order parameter relaxation, whereas in liquid crystals of low viscosity order fluctuations represent the dominant relaxation mechanism.

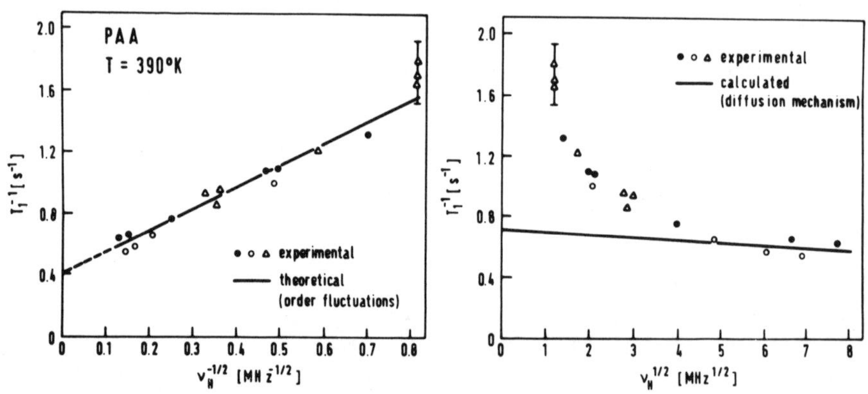

Fig.1. Frequency dependence of the proton spin-lattice relaxation time T_1 in PAA at T = 390°K. Solid circles, our measurements; open circles, from reference 4; triangles, from reference 5. The measured values are plotted vs $\nu^{-1/2}$ to show the predicted linear dependence for the order fluctuation mechanism and vs $\nu^{1/2}$ together with the contrasting linear dependence (solid line) at low frequencies predicted for a diffusion mechanism

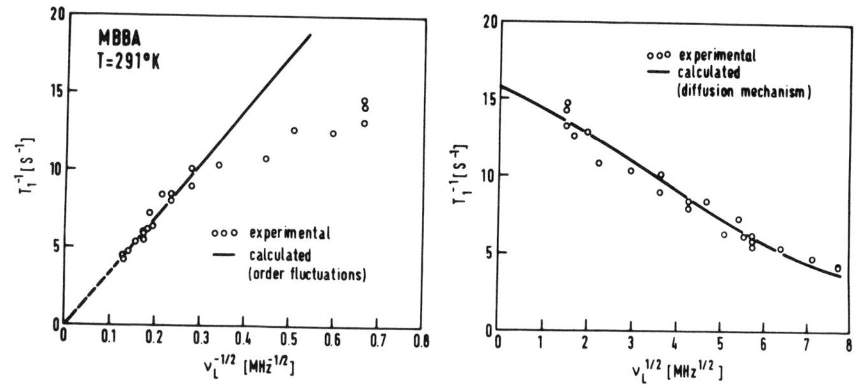

Fig.2. Frequency dependence of the proton spin-lattice relaxation time T_1 in MBBA at T = 291°K. The measured values are plotted vs $\nu^{1/2}$ to show the predicted linear dependence at low frequencies for diffusion controlled relaxation and vs $\nu^{1/2}$ together with the contrasting linear dependence predicted for the order fluctuations mechanism. The solid line in the $\nu^{1/2}$ plot is the dependence predicted from the Torrey theory.

The frequency dependence of the dipolar spin-lattice relaxation times were as well studied. Both in PAA and in MBBA the ratio of the Zeeman and the dipolar spin-lattice relaxation times equals 3 over the whole nematic range: $(T_{1Z}/T_{1D}) = 3$. This demonstrates that not only order fluctuations but molecular diffusion as well is correlated over distances which are large as compared to the molecular dimensions.

2. MEASUREMENT OF SELF-DIFFUSION IN LIQUID CRYSTALS

It has been already mentioned that the most basic property of liquid crystals is that they flow while sustaining an orientationally ordered structure. Very little, however, is known on the nature of this process on a microscopic basis. This is at least partly due to a lack of suitable technique which would allow a fast and accurate determination of the translational self-diffusion coefficients in these systems. The proton magnetic-resonance spin-echo method would be ideally suited for such a purpose if it would work. This method can be successfully applied for the determination of self-diffusion coefficients only if the extra damping of the transverse nuclear magnetization due to the change in the Larmor frequency as a result of migration of the molecules across the inhomogeneous applied magnetic field is larger than the inherent damping due to spin-spin interactions (T_2). In liquid crystalline systems the dipolar broadening is not completely averaged out by molecular motion as in isotropic liquids, and therefore T_2 is so short that the classical NMR method fails.

Doane and Parker[6)] have partially removed dipolar interactions by working with an oriented smectic liquid crystal at the magic angle ($\Theta = 55^\circ$). This method can be applied only in some special cases, and it does not represent a general solution of the problem.

Blinc, Pirš and Zupančič[7)] have proposed the use of a new technique which overcomes the above-mentioned difficulties and which allowed a direct NMR determination of translational self-diffusion coefficients in nematic liquid crystals. The technique used represents a superposition of a Waugh-type multiple 90° rf pulse sequence, averaging out dipolar interactions, a pulsed linear magnetic-field-gradient sequence, and a relatively slow Carr-Purcell train of 180° rf pulses.

The detailed pulse sequence is shown in Fig.3.

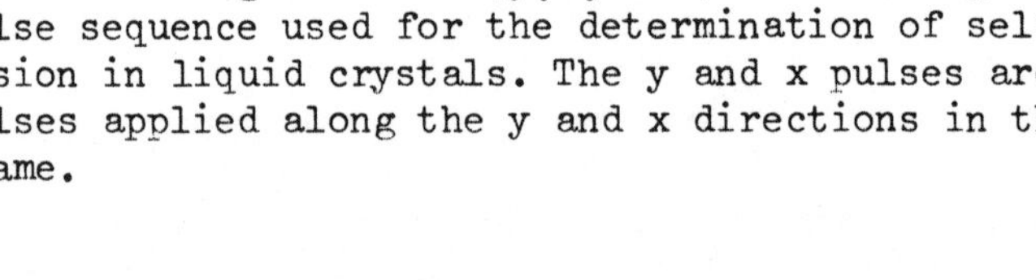

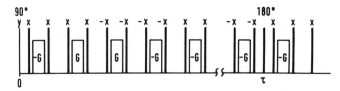

Fig.3. Schematic diagram of rf ($y, x, 180^\circ$) and field-gradient (G) pulse sequence used for the determination of self-diffusion in liquid crystals. The y and x pulses are 90° pulses applied along the y and x directions in the rotating frame.

For experimental convenience we used the simplest Waugh sequence, which averages out dipolar interactions. It consist of a train of equally spaced 90° pulses which are shifted in phase with respect to the first 90° pulse by $\frac{1}{2}\pi$. The spacing between the first 90° pulse, applied along the y direction in the rotating frame, and the second 90° pulse, applied along the x direction, is $\frac{1}{2} t_o$, whereas the spacing between all other x pulses is t_o. Since four 90° pulses represent an identity transformation,

$[\exp(-\frac{1}{2}i\pi I_x)]^4 = 1$, the cycle time t_c is $t_c = 4t_o$. To prevent an accumulation of phase errors and to make the experimental setup less critical, the phases of each second pulse cycle are reversed so that the Waugh sequence we used is y, x, x, x, x, -x, -x, -x, -x, etc.

It has been shown by Haeberlen and Waugh[8,9,10,11] that if the cycle time t_c becomes small as compared with T_2, the spin system behaves over long times Nt_c - where N is the number of cycles - as if under the influence of an effective time-independent average Hamiltonian H which can be expressed in terms of a rapidly converging series:

$$\overline{H} = \sum_{\ell=o} \overline{H}^{(\ell)} , \tag{2.1}$$

where

$$\overline{H}^{(0)} = \sum_{i=1}^{\kappa} A_i , \tag{2.2a}$$

$$\overline{H}^{(1)} = \frac{1}{2} \sum_{i<j} [A_i, A_j] , \tag{2.2b}$$

$$\overline{H}^{(2)} = \frac{1}{12} \{ \sum_{i<j} [(A_i - A_j), [A_i, A_j]] + 2 \sum_{i<j<k} [A_i, [A_j, A_k]] + [[A_i, A_j], A_k] \}, \tag{2.2c}$$

$$A_i = \frac{t_{i+1} - t_i}{t_c} H(t), \quad t_i < t < t_{i+1} \tag{2.3}$$

In deriving the above expressions it was assumed that the Hamiltonian H(t), as transformed by the rf pulses, is constant during the H intervals (t_i, t_{i+1}) of the cycle t_c. The results (2.2a-c) can be most simply obtained from the theorem of Liapunov[11] in the theory of systems of linear differential equations with periodic coefficients.

The first y pulse transforms the spin Hamiltonian into the rotating frame, and the Waugh cycle of x pulses partially averages out the dipole-dipole interaction Hamiltonian H_D. Though $\overline{H}_D(0)$ as given by (2.2a) and (2.3) is nonzero, it commutes with I_x, $[H_D^{(0)}, I_x] = 0$, so that in this order the dipolar contribution to the second moment term in the free induction decay vanishes. Since $\overline{H}_D^{(1)} = 0$, it is only $\overline{H}_D^{(2)}$ which gives a dipolar contribution to the decay of the precessing nuclear magnetization. The decay time is thus significantly lengthened. The above Waugh sequence, however, also averages out the static field-gradient term in the Hamiltonian. During the five intervals of the cycle, $0 < t < \frac{1}{2} t_o$, $\frac{1}{2} t_o < t < \frac{3}{2} t_o$, $\frac{3}{2} t_o < t < \frac{5}{2} t_o$, $\frac{5}{2} t_o < t < \frac{7}{2} t_o$, $\frac{7}{2} t_o < t < 4t_o$, the values of the field-gradient term are ΔI_z,

ΔI_x, $-\Delta I_z$, $-\Delta I_x$,ΔI_z, respectively, and the average over the cycle vanishes. If, however, the field gradient is applied in pulses so that $\Delta = \omega - \omega_o$ is different from zero only in the second and fourth intervals and its sign in the fourth interval is just opposite to the one in the second interval, the effect of the field gradient does not average out.

The shortening of the decay time due to the presence of the pulsed field-gradient sequence is illustrated in Fig. 4.

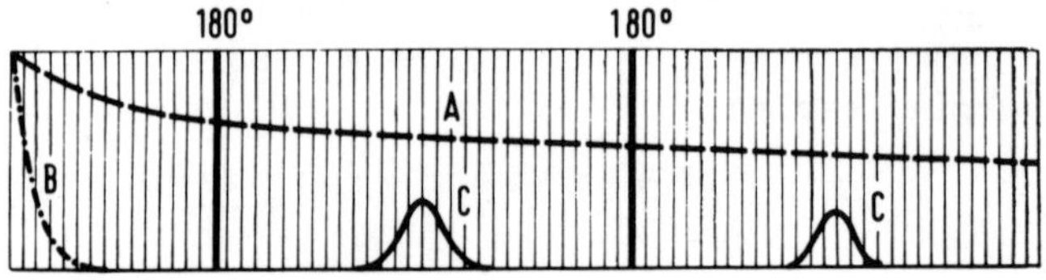

Fig.4. "Waugh" signal envelope without (A) and with (B) applied field-gradient pulses. Echoes appear (C) when a slow Carr-Purcell train of 180° pulses is superimposed on the Waugh sequence in the presence of gradient pulses.

Here, the broken lines show the Waugh echo envelope without an applied field gradient; the dotted line shows the same envelope in the presence of the field gradient pulses. The loss of magnetization caused by the field gradient is, however, not irreversible. The magnetization can be recovered in the form of echoes which appear at times $(2n)\tau$ if a slow Carr-Purcell train of 180° pulses is applied at times $n\tau$, where τ is an integral multiple of the cycle time $t_c = Nt_o$.

The echo maxima are given by

$$m(2n\tau) = m(0)\, \exp\left[-\, 2n\tau/T_2^x - \frac{1}{12}\, \gamma^2\delta^2\vec{G}\, \overleftrightarrow{D}\, \vec{G}(2n\tau)\right], \qquad (2.4)$$

where δ is the total time the field gradient $\vec{G}$ = grad H_z is on during the interval $(2n\tau)$ and $\overleftrightarrow{D}$ is the self-diffusion-constant tensor. The effective decay time T_2^x is limited by the second-order dipolar contribution $[\bar{H}_D{}^{(2)}]^2$, the mixed field-gradient-dipolar terms which, as well, occur in second order, and the spin-lattice relaxation time in the rotating frame, $T_{1\rho}$. By measuring the spin-echo amplitudes, as a function of the width of the field-gradient pulses, the various components of $\overleftrightarrow{D}$ can be obtained.

As an example we measured partially deuterated MBBA (para-methoxybenzilidenebutylaniline)which is nematic at room temperature and where T is of the order of 100 μsec, Fig.5. The value of t_o we used is 25 μsec. The amplitude of the field-gradient pulses was about 50 G/cm and the gradient pulse width was varied between 5 and 14 μsec. Echoes could be still seen 60 - 100 msec after the first y pulse, demonstrating that T_2^x is of the order of $T_{1\rho}$ and hence much longer than T_2. From the dependence of the echo amplitudes on the widths of the gradient pulses, the components of the self-diffusion tensor parallel and perpendicular to the magnetic field, are obtained. The measured anisotropy at room temperature is $D_\parallel / D_\perp \approx 1.3$.

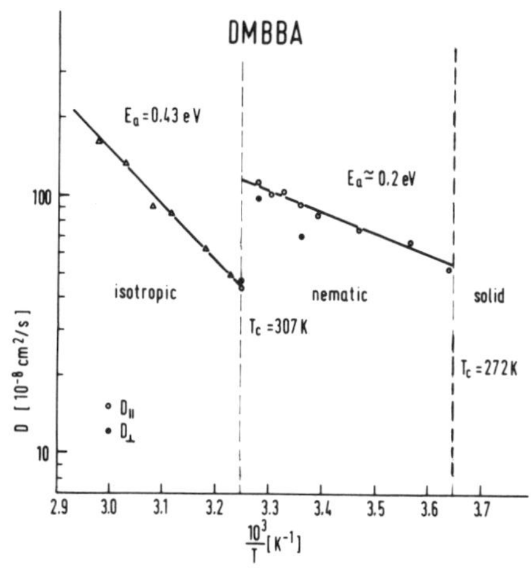

Fig.5. Temperature dependence of the self-diffusion coefficient in partially deuterated MBBA.

3. THE EPR STUDIES OF THE MOLECULAR FLUCTUATIONS IN THE NEMATIC AND SMECTIC MESOPHASE

In the EPR studies the paramagnetic centers have to be dissolved in the liquid crystal matrix, and the anisotropic magnetic interactions, usually electron - nucleus, are used to trace the static as well as dynamic properties of the liquid crystal.

The molecular dynamics has been already analysed in some nematic and smectic liquid crystal mesophases[12,14]. The well known expression for transverse spin-lattice relaxation time, which is proportional to the linewidth is:

$$\frac{1}{T_2} = \frac{1}{\hbar^2} \int_0^\infty \overline{\langle i | [H_1^*(t), [H_1^*(t-\tau), S_+]] | j \rangle dt} \qquad (3.1)$$

where $H(t) = H_o + H_1(t)$, $H_1^{*} = e^{iH_o t/\hbar} H_1 e^{-iH_o t/\hbar}$

and $H_1(t) = \sum_{\lambda} F_{\lambda}^{LM} D_{MM'}^{L}(\alpha,\beta,\gamma) K_{\lambda}^{LM'}$. F_{λ}^{LM} are the operators of the lattice and K_{λ}^{LM} that of the spins, while $D_{MM'}^{L}(\alpha,\beta,\gamma)$ is the rotational transformation matrix relating the molecular and laboratory coordinate system.

When the averaging is performed for the rapid motion we get the expression

$$\frac{1}{T_2} = a_o + a_1 m_I + a_2 m_I^2 \tag{3.2}$$

where m_I is nuclear spin projection on the magnetic field direction and a_o, a_1, and a_2 are coefficients including the spectral densities: the dynamical as well as structural characteristics which define linewidth.

From the difference between the linewidth of the lines which correspond to the transitions with $m_I = 1/2$ and $m_I = -1/2$ the coefficient a_1 can be determined

$$a_1 = 2\Delta g b \beta B_o/\hbar\ (\overline{\cos^4\theta} - \overline{\cos^2\theta}^{\,2})\tau_c = 2\Delta g b \beta B_o/h\ (\overline{\cos^4\theta} - \overline{\cos^2\theta}^{\,2})_{isotr}\Phi_o(\gamma)\tau_c \tag{3.3}$$

The signs used means the following: $\gamma = \frac{1}{2}\overline{(3\cos^2\theta - 1)}$ is ordering parameter, Δg and b are anisotropic parts of tensor g and hyperfine tensor respectively, β is Bohr's magneton, B_o is the magnitude of the applied magnetic field, θ is the angle between the long molecular axis and magnetic field direction and $\Phi_o(\gamma)$ is the function given by S.Glarum and J.Marshal[12] and is calculated with a distribution function $e^{\lambda\cos^2\theta}$. To get (3.3) we calculated integral in (3.1) by introducing a simple correlation function $e^{-|\tau|/\tau_c}$ with the same correlation time τ_c for all averages. One can get a_1 directly from the measurements and using the expression (3.3) the correlation time can be calculated.

A temperature dependence of the correlation time for the vanadylacetylacetonate (VAA) paramagnetic complex dissolved in 4-4'-di-n-heptyloxyazoxybenzene (PAH), and two from the homologue serie of 4-n-alkoxybenzilidene-4'-phenylazoaniline with n = 1 and n = 10, (C-1 and C-10) is given on Fig. 6,7, and 8. The observed slowing down of the rotational motions is proportional to the viscosity changes with temperature. In all cases a singular increase of τ_c at the transition points was observed. It resembles the viscosity peak at the nematic-isotropic transition[13] and could correspond to increased molecular order fluctuations at the transition points, discussed in the first part of this paper.

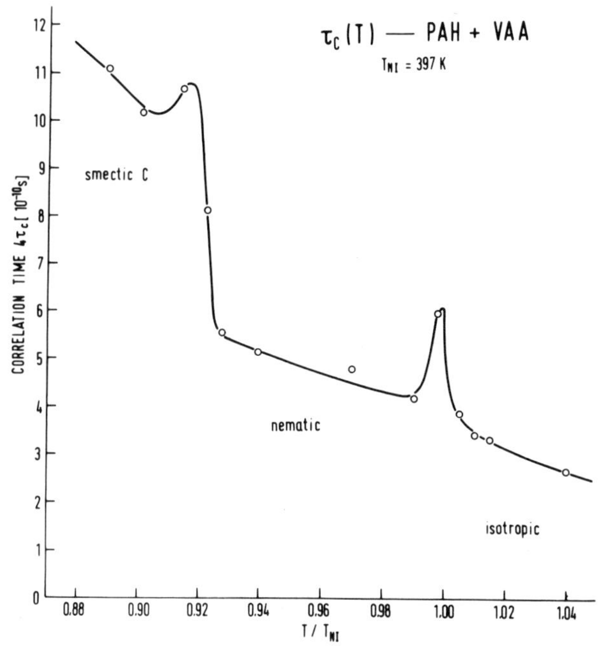

Fig.6. Temperature dependence of the correlation time for VAA molecules dissolved in liquid crystal PAH

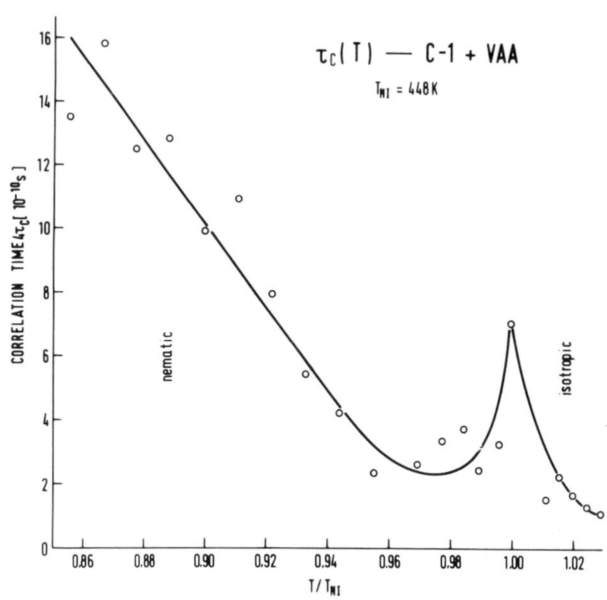

Fig.7. Temperature dependence of the correlation time for VAA molecules dissolved in liquid crystal C-1

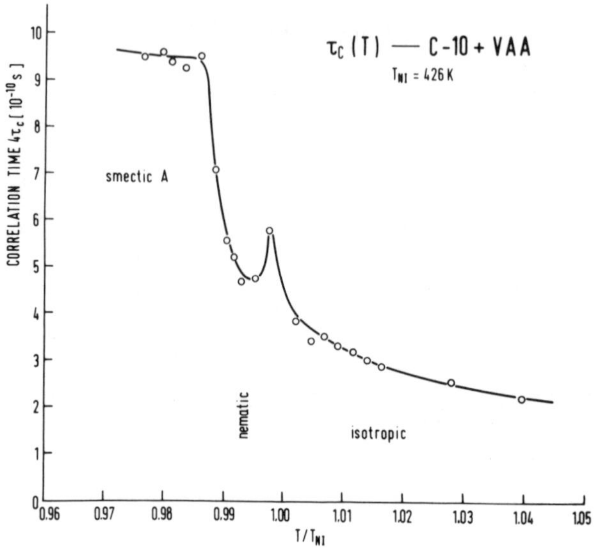

Fig.8. Temperature dependence of the correlation time for VAA molecules dissolved in liquid crystal C-10

At the nematic - smectic transition point the correlation time increases for a factor two due to the obviously reduced rotations in this phase. These results are less reliable especially in C-10 smectic phase, where in the applied magnetic field of approximately 3000 G the orientational order of nematic phase is not extended in the smectic phase.

In order to trace the effect of orientation on τ_c determination, the angular dependence of linewidth in the smectic phase of PAH and C-5 was measured (Fig.10). In this case cosθ in eq.3.3 is given by the expression:

$$\cos\theta = \sin\Omega \sin\Theta \cos\Psi + \cos\Omega \cos\Theta \quad (3.4)$$

The angles θ,Θ,Ψ and Ω are presented on Fig.9.

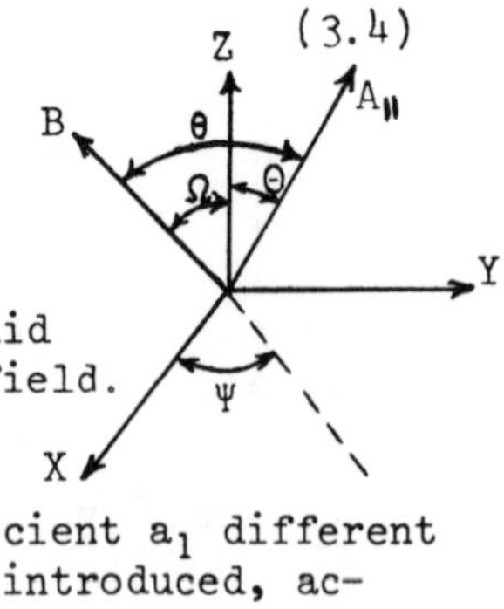

Fig.9. Orientation of the VAA molecule in the laboratory coordinate system x,y,z - where z is the optical axis of the liquid crystal and $\vec{B}$ is the applied magnetic field.

In deriving the expression for the coefficient a_1 different correlation times for different averages were introduced, according to G.Luckhurst[14].

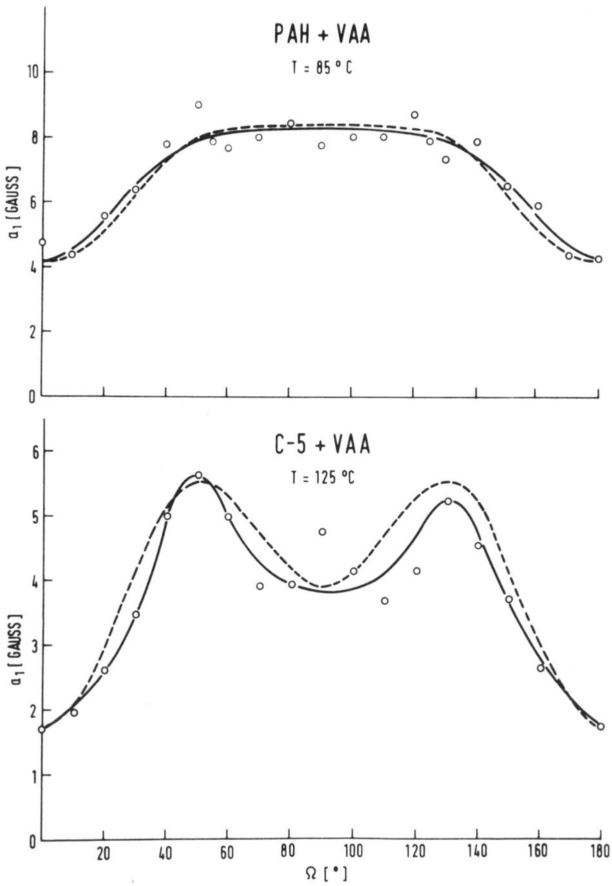

Fig. 10. Angular dependence of the coefficient a_1 for VAA molecule dissolved in liquid crystals PAH and C-5 (—— experimental curve, ---- calculated curve)

From eq. 3.1 one can see that before integration over τ and averaging, the expression for a_1 has a form

$$a_1 = 2\Delta g b \beta B_o/\hbar \; \frac{1}{2} \int_{-\infty}^{\infty} d\tau \; e^{i\omega\tau} \overline{(\cos^2\theta(t) - \overline{\cos^2\theta})(\cos^2\theta(t-\tau) - \overline{\cos^2\theta})} \quad (3.5)$$

Here eq. 3.4 is used for the transformation of the angle θ. Correlation functions of the same form but with different correlation times have been used for different angular terms.

$$\cos^2\theta(t) - \overline{\cos^2\theta} = \sin^2\Omega[\Theta_1(t)\Psi_1(t) + \overline{\sin^2\Theta}\;\Psi_1(t) + \frac{1}{2}\,\Theta_1(t)] + \sin^2\Omega\;\Theta_2(t)\cos\Psi(t) - \cos^2\Omega\;\Theta_1(t),$$

where $\Theta_1 = \sin^2\Theta - \overline{\sin^2\Theta}$, $\Theta_2 = \sin\Theta\cos\Theta$, $\Psi_1 = \cos^2\Psi - \overline{\cos^2\Psi}$

and finally:

$$a_1 = 2\Delta g b \beta B_o/\hbar \; \{[\frac{1}{4}\sin^4\Omega + \cos^4\Omega - \frac{1}{8}\sin^2 2\Omega]\;\overline{\Theta_1^2}\;\tau_\Theta + \frac{1}{8}\sin^4\Omega\;\overline{\sin^2\Theta}^2\;\tau_\Psi + \frac{1}{8}\sin^4\Omega\;\overline{\Theta_1^2}\;\tau_2 + \frac{1}{2}\sin^2 2\Omega\;\overline{\Theta_2^2}\;\tau_3\} \quad (3.7)$$

where τ_Θ and τ_Ψ represents correlation times for the fluctuations in the angle Θ and Ψ respectively (Fig. 11) and τ_2 and τ_3 are expected to be the functions of τ_Θ and τ_Ψ.

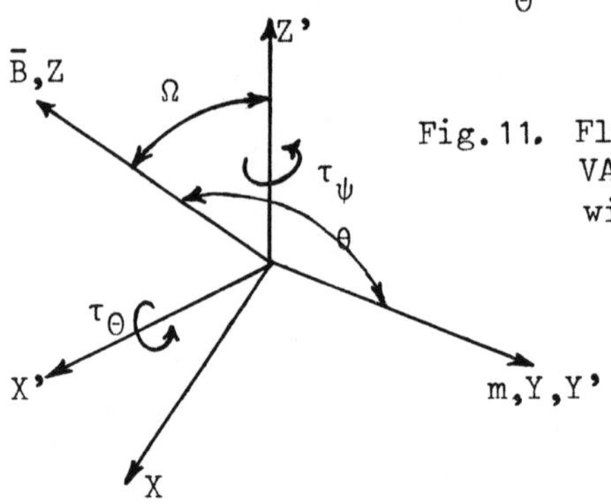

Fig. 11. Fluctuations of the molecule VAA in θ and Ψ in relation with the coordinate system

If the ordering parameter for liquid crystal is known, the values for different correlation times can be chosen in such a way that the experimental curve for a_1 can be reproduced. For PAH the ordering parameter derived from the hyperfine splitting constant at T = 85°C is γ = 0.52; for C-5, γ = 0.75 at T = 125°C was obtained from NMR measurements[15]. Using this values for ordering parameters and with correlation times:

PAH: $\tau_\Theta = 2{,}0 \cdot 10^{-10}$s C-5: $\tau_\Theta = 2{,}4 \cdot 10^{-10}$s

$\tau_\Psi = 1{,}5 \cdot 10^{-10}$s $\tau_\Psi = 0{,}6 \cdot 10^{-10}$s the dotted lines on Fig. 10 were derived. In our calculations the assumptions that the molecular order does not change with the angle between magnetic field and optical axis could be incorrect. It could be also that the distribution lose his isotropy in Ψ. This problem was discussed previously[16,17],but it seems that the corrections would be small.

One can see that the angular variations of the linewidth can be explained with two correlation times which correspond approximately to the fluctuations of the rotational motions perpendicular to the long molecular axis and around the long molecular axis.

When the angle between the magnetic field and the optical axis is zero the results (Fig. 6,7,8) can be explained by a single correlation time τ_Θ, similar as in the nematic phase.

REFERENCES

1) J. Doane and O.L. Johnson, Chem.Phys.Lett. 6, 291 (1970)
2) J.F.Harmon and B.H.Müller, Phys.Rev. 182, 400 (1969)
3) H.C.Torrey, Phys.Rev. 92, 962 (1953)
4) J. Doane, J.J.Visintainer, Phys.Rev.Letters 23, 1421 (1969)
5) M.Weger and B.Cabane, J.Physique 30, C4-72 (1969)
6) A.Abragam, The principles of Nuclear Magnetism (Oxford Univ. Press, 1961)
7) R.Blinc, V.Dimic, J.Pirš, M.Vilfan and I.Zupančič, Mol.Cryst. and Liq.Cryst. 14, 97 (1971)
8) J. Doane and R.S.Parker, Presented at the 17th Colloque Ampère, Turku, Finland, 1972 (to be published)
9) E.D.Ostroff and J.S.Waugh, Phys.Rev.Lett. 16, 1097 (1966)
10) V.Haeberlen and J.S.Waugh, Phys.Rev. 175, 453 (1968)
11) A.M.Liapunov, Collected Papers (Academy of Sciences SSSR, Moscowq 1956)
12) S.Glarum, and J.Marshal, J.Chem.Phys. 46, 55 (1967)
13) G.W.Gray, Molecular Structure and the Properties of Liquid Crystals (Academic Press, 1962)
14) G.R.Luckhurst and A.Sanson, Mol.Phys. 24, 1297 (1972)
15) J.Doane, R.Parker, B.Cvikl, D.Johnson and D.Fishel, Phys.Rev. Letters 28, 1694 (1972)
16) M.Schara and M.Šentjurc, Solid State Commun. 8, 593 (1970)
17) M.Šentjurc and M.Schara, Mol.Cryst. and Liq.Cryst. 12, 133 (1971)

INFLUENCE OF MOLECULAR STRUCTURAL CHANGES ON THE MESOMORPHIC BEHAVIOR OF BENZYLIDENEANILINES

Zack G. Gardlund, Ralph J. Curtis and George W. Smith

General Motors Research Laboratories

Warren, Michigan 48090

INTRODUCTION

Since the discovery by Heilmeier, Zanoni and Barton[1] that liquid crystals with negative dielectric anisotropy exhibit dynamic scattering, there has been considerable interest in this class of materials. The Schiff Bases resulting from the reaction between para-substituted anilines and para-substituted benzaldehydes have been of particular interest, especially since the report[2] that N-(4-methoxybenzylidene)-4'-n-butylaniline is nematogenic at room temperature. The research reported here is the result of a detailed study of the effects of small changes in structure on the transition temperatures of benzylideneanilines. Specifically, two series of compounds (I) and (II) are described. In both series the alkoxy group is systematically varied from methoxy to heptyloxy. In series I, the alkyl groups used were n-C_4H_9, n-C_5H_{11}, n-C_6H_{13}, n-C_7H_{15} and n-C_8H_{17} while in series II, the alkyl groups were C_2H_5, n-C_4H_9, n-C_6H_{13} and n-C_8H_{17}.

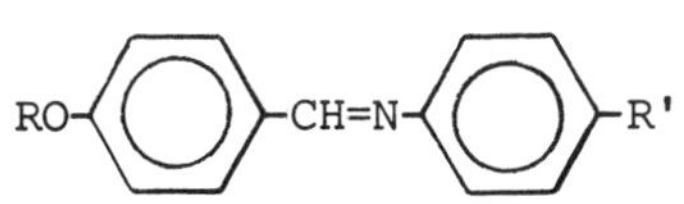

I

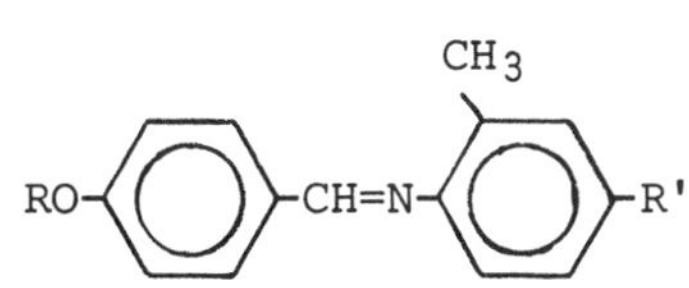

II

R = CH_3, C_2H_5, n-C_3H_7 → n-C_7H_{15}

R' = n-C_4H_9 → n-C_8H_{17}

R = CH_3, C_2H_5, n-C_3H_7 → n-C_7H_{15}

R' = C_2H_5, n-C_4H_9, n-C_6H_{13}, n-C_8H_{17}

Whereas series I shows the effects of changes in molecular length and polarizability, series II shows the very dramatic effect of molecular broadening as well. The results reported previously[2-8] for compounds homologous to series I have been included and comparisons made where possible. Some of the results reported here have appeared in preliminary communications.[9,10] A detailed examination of the textures observed as well as an interesting textural transition phenomenon are discussed elsewhere.[11]

EXPERIMENTAL

The transition temperatures were determined with a calibrated Perkin-Elmer Differential Scanning Calorimeter (DSC) 1B as well as a DuPont 900 Differential Thermal Analyzer (DTA) in the DSC mode. In general the DTA and DSC results were in excellent agreement. The liquid crystal textures were observed with a Leitz Wetzlar Dialux-Pol polarizing microscope and a Mettler FP52 microfurnace for sample temperature control. The p-alkoxybenzaldehydes were purchased from Eastman Organic Chemicals Company and the p-n-butylaniline from Aldrich Chemical Company. The other p-n-alkylanilines and -toluidines were synthesized by rearrangement of the requisite N-n-alkylanilines and -toluidines as described below. Purity of the products was determined by vapor phase chromatography in a F&M Gas Chromatograph with flame detection and a 10% silicone rubber (SE-30) packed column. Nuclear magnetic resonance spectra were determined by means of a Varian HS-100D Spectrometer and a Perkin Elmer 337 Infrared Spectrometer was used for infrared analysis.

Synthesis of N-n-Alkylanilines and -2-Methylanilines

A procedure similar to that of Hickinbottom[12] was used. A solution (3:1 molar ratio) of distilled aniline (toluidine) and n-alkylbromide was refluxed overnight. After cooling to room temperature the reaction mixture was made basic with dilute aqueous ammonia. The aqueous layer was repeatedly washed with pentane. The pentane washings were combined with the organic layer and treated with excess 50% aqueous zinc chloride. Extraction of the resulting solid with pentane, followed by drying with anhydrous potassium carbonate and flash evaporation of the pentane yielded crude N-n-alkylaniline (toluidine). A pure compound was obtained by fractional distillation on a spinning band column. The yields and boiling ranges were as follows (compound, yield, b.p. °C, literature[12] b.p. °C): N-n-amylaniline, 93-97%, 125°/16 mm, 127-128°/16 mm; N-n-hexylaniline, 87%, 145°/16 mm, 158°/28 mm; N-n-heptylaniline, 90%, 160°/16 mm, 160°/21 mm; N-n-octylaniline, 98%, 99°/.01 mm, 177-178°/25 mm; N-ethyl-2-methylaniline, 67%,

75°/3.5 mm, none; N-n-butyl-2-methylaniline, 80%, 102°/3.4 mm, none; N-n-hexyl-2-methylaniline, 81%, 110°/1.4 mm, none; N-n-octyl-2-methylaniline, 83%, 155°/4.3 mm, none.

Synthesis of 4-n-Alkyl and 4-n-Alkyl-2-Methylanilines

The reaction conditions were similar to those of Hickinbottom[12] but the work-up procedure has been improved. A mixture (2:1 molar ratio) of N-n-alkylaniline or -2-methylaniline and anhydrous cobalt chloride was heated with stirring at 210°C for 16 hours. Care must be used to obtain anhydrous cobalt chloride which has a characteristic powder blue appearance. After cooling to room temperature the dark blue viscous mixture was dissolved in dilute aqueous hydrochloric acid. After being made basic with dilute aqueous ammonia the organic layer was removed and combined organic-ether layers were washed with water and treated with excess 50% aqueous zinc chloride. Unrearranged N-n-alkylaniline or -2-methylaniline was removed from the salt by extraction with pentane. The zinc chloride salt was then decomposed with aqueous ammonia. The organic layer was combined with pentane washings of the aqueous layer. After drying the compounds were obtained in pure form by fractional distillation on a spinning band column. Nuclear magnetic resonance (NMR) spectroscopy proved the 4-n-alkyl-substitution in these compounds. Proton resonance assignments for the 4-n-alkylanilines were made as follows: 6.7 τ, (singlet, $-NH_2$); 7.5 τ, (triplet, $-CH_2-$ adjacent to ring); 8.6 τ, [multiplet, $(CH_2)_n$]; 9.1 τ, (singlet, $-CH_3$); 3.6 τ, (A-B quartet, proton ortho to $-NH_2$); 3.2 τ, (A-B quartet, proton meta to $-NH_2$). The NMR spectrum of the 4-n-alkyl-2-methylanilines differed in the aromatic region as shown in Figure 1. In Figure 1, a is the aromatic portion of the spectrum of 4-n-butylaniline, whereas b is the same portion of the spectrum for 4-n-butyl-2-methylaniline. The 2,6 protons in the unmethylated compound are magnetically equivalent, resonate at 3.6 τ, and appear as superimposed doublets. Similarly the 3,5 protons resonate at 3.2 τ and also appear as superimposed doublets. In the methyl-substituted compound, spectrum 1b, proton 6 still appears as a doublet at 3.6 τ and proton 5 is a doublet at 3.2 τ with proton 3 superimposed on it. This doublet nature of proton 5 was proven by spin decoupling. Each peak resulting from the resonance of proton 5 is split into a doublet by meta spin-spin coupling with proton 3. If proton 3 was in the 2 position, then the additional splitting would be observed on the doublet from resonance of the 6 proton instead of on the doublet from the resonance of the 5 proton.

The yields and boiling points for these compounds were as follows (compound, yield, b.p. °C, literature,[12] b.p. °C): 4-n-amylaniline, 22%, 130°/16 mm, 130°/16 mm; 4-n-hexylaniline, 25%, 125°/3 mm, 146-148°/17 mm; 4-n-heptylaniline, 28%, 140°/1 mm, 159°/18 mm;

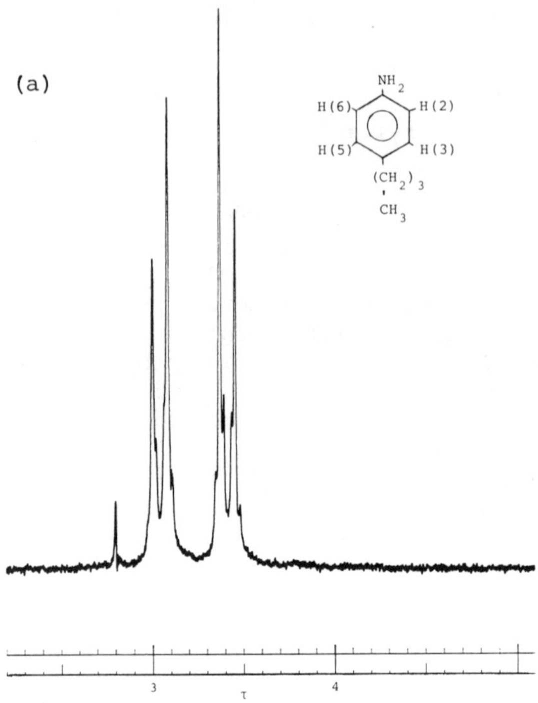

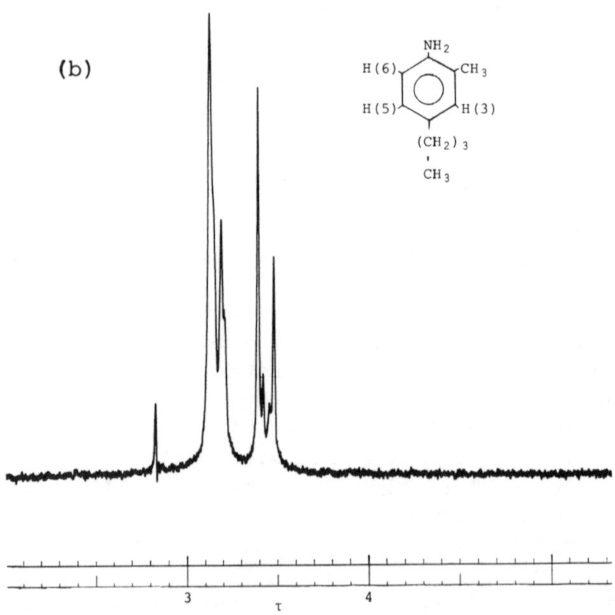

Fig. 1. Aromatic portion of NMR spectra: (a) 4-n-butylaniline, (b) 2-methyl-4-n-butyaniline.

4-n-octylaniline, 38%, 105°/.01 mm, 170°/17 mm; 4-ethyl-2-methylaniline, 48%, 88°/3.8 mm, none; 4-n-butyl-2-methylaniline, 38% 108°/3 mm, none; 4-n-hexyl-2-methylaniline, 23%, 92°/.02 mm, none; and 4-n-octyl-2-methylaniline, 24%, 110°/0.2 mm, none.

Synthesis of N-(4-Alkoxybenzylidene)-4-n-Alkyl- and -4-n-Alkyl-2-Methylanilines

These compounds were prepared by dissolving equal molar amounts of the 4-alkoxybenzaldehyde and 4-n-alkyl- or 4-n-alkyl-2-methylaniline in absolute ethanol and refluxing for four hours. The products were repeatedly crystallized from absolute ethanol to a constant melting point. Elemental analyses (C, H, N) agreed with theoretical calculations (±0.2%) and the molecular weights agreed with empirical molecular weights (±2%).

EFFECT OF STRUCTURE ON TRANSITION TEMPERATURES OF N-(4-ALKOXYBENZYLIDENE)-4-ALKYLANILINES

Although the research reported here has been concerned with the compounds having four to eight carbon atoms in the n-alkyl chain, data is available for some of the lower homologs, i.e., R' = H, CH_3, C_2H_5, n-C_3H_7. Two of the parent compounds, R = CH_3, R' = H[2] and R = n-C_4H_9, R' = H[3], have been studied. The butoxy compound reportedly showed a smectic phase from 22 to 44°C and a monotropic nematic phase at 20°C. The methoxy compound was not mesomorphic. More data is available for the methyl homolog but again it is incomplete. The methoxy compound has a crystalline melting point of 92-93°C[2] or 91°C[3] and a monotropic nematic transition at 38°C[2] or 39°C.[3] The ethoxy compound was also non-enantiotropic[2] with a crystalline melting point of 94°C and a monotropic nematic transition at 80°C. The other known member of the methyl series,[3] R = n-C_4H_9, was enantiotropic nematic over the range 66-68°C and showed a monotropic smectic phase at 49°C. The N-(4-alkoxybenzylidene)-4-ethylanilines have received more attention. Three sets of authors agree as to the crystalline melting point of the methoxy compound, 57°C,[2] 56°C,[3] 56.5°C,[4] and the appearance of a monotropic nematic phase, 28°C,[2] 32°C,[3] 32.5°C.[4] The ethoxy compound has a monotropic nematic phase at 67°C[2] or 63°C[3] and a crystalline to isotropic transition at 70°C[2] or 69°C.[3] In the n-propyl series the methoxy compound showed enantiotropic nematic behavior from 42-57°C,[2] 44.6-61.2°C,[4] and 45.5-60.5°C.[7] Kelker and Scheurle[2] reported similar behavior, 76-97°C, when R = n-C_3H_7, R' = n-C_3H_7. Murase[7] disagreed on the clearing point, 87.5°C, but agreed on the nematic transition temperature. Recently Murase has reported[7] several additional members of the ethyl and n-propyl series. With the exception of one compound, R = n-C_3H_7,

R' = C_2H_5, all were mesomorphic. The ethyl compounds where R = n-C_4H_9, n-C_5H_{11} and n = C_6H_{13} as well as the n-propyl compounds where R = CH_3 through n-C_5H_{11} were all nematogenic according to Murase.[7] This is in contrast to the work of Flannery and Haas[3] as well as that of DeVries and Fishel,[6] who reported a smectic phase when R = n-C_4H_9, R' = C_2H_5. Murase reported the appearance of enantiotropic smectic phases as well as nematic for the higher members of each series, i.e., R>n-C_6H_{13}, R' = C_2H_5 and R>n-C_5H_{11}, R' = n-C_3H_7.

The transitions for the benzylideneanilines with R = $CH_3 \longrightarrow$ n-C_7H_{15} and R' = n-$C_4H_9 \longrightarrow$ n-C_8H_{17} are listed in Table I. The symbols K, S, N and I represent crystalline, smectic, nematic and isotropic phases, respectively. Subscripts refer to the various polymorphic smectic states. Monotropic phases are indicated by parentheses. Comparisons with the existing literature are made where possible. The results of other workers are enclosed in brackets. Agreement is good with some exceptions in the N-(4-alkoxybenzylidene)-4-n-butylaniline series, notably when R = n-C_3H_7. A monotropic smectic transition is observed at 23.3°C. When R = n-C_4H_9 Flannery and Haas[3] observed a K→S transition at 7°C which agrees with the transition reported here. Dietrich and Steiger[8] did not report a low temperature smectic phase. The appearance of three smectic phases for the butoxy compound is reported here for the first time; one of which has a range of 0.5°C. Disagreement with previous workers is also found when R = n-C_5H_{11}. The first smectic phase appears at 12°C but has been previously reported at 24°C[3] and 49.6°C.[8] In the case where R = n-C_6H_{13} three smectic phases have not been reported before. The K→S transition is reported here to occur at 10.0°C rather than 58°C[8] or 59.5°C.[3]

In general, regardless of the length of the alkyl chain R', compounds with short chain alkoxy groups are nematogenic. The higher homologs all show smectic in addition to nematic phases with the exception of the n-C_6H_{13} compound when R = n-C_7H_{15} and the n-C_8H_{17} compound when R = n-C_7H_{15}. These compounds show three and two smectic phases, respectively, but no nematic phases. The results for the compounds with long alkyl groups are different from those reported[2,3,4,7] for compounds with R' = H, CH_3, etc. Although there are inconsistencies the shorter chain compounds are not mesomorphic or are only monotropic nematic when the alkoxy group is small. In general nematogenic behavior was observed up to and including R = n-C_5H_{11} or n-C_6H_{13} depending on the length of R'. In only one instance was there more than one smectic phase for these compounds. Kelker and Scheurle[2] noted two smectic phases when R = n-C_9H_{19}, and R' = n-C_3H_7.

TABLE I

MESOMORPHIC TRANSITION TEMPERATURES OF N-(4-ALKOXYBENZYLIDENE)-4-n-ALKYLANILINES

Substituent		Transition Temperature (°C)							
R	R'	K→N	K→S	S_5-S_4	S_4-S_3	S_3-S_2	S_2-S_1	S-N	N-I
CH_3	n-C_4H_9	22.2[22[2],17[3],20[4]]							45.9[47.5[2],39.5[3],47.3[4]]
C_2H_5	n-C_4H_9	35.3[37[2,3],35.5[4]]							79.0[80[2],83[3],79[4]]
n-C_3H_7	n-C_4H_9	41.1[42[7],41[8]]						(23.3[a])	55.7[55[7],58.9[8]]
n-C_4H_9	n-C_4H_9		8.0[7[3],46[8]]			41.0[41[3]]	45.2	45.7[46[3],59[8]]	74.7[75[3,8]]
n-C_5H_{11}	n-C_4H_9		12.0[24[3],49.6[8]]				52.1[52[3],52.5[8]]	52.4[54[3],57.8[8]]	69.0[71[3],70.2[8]]
n-C_6H_{13}	n-C_4H_9		10.0[?[3],58[8]]			55.4[59.5[3]]	58.2[60.5[3]]	69.2[70[3],70.4[8]]	77.0[78[3,8]]
n-C_7H_{15}	n-C_4H_9		32.2[29[3]]			63.3[64[3],63[8]]	64.8[66.5[3],64.6[8]]	74.1[76[3],74.1[8]]	76.2[77.5[3],76.4[8]]
n-C_8H_{17}	n-C_4H_9		39.5[b][39.5[3],65.4[8]]			62.2[b][66[3]]	66.7[b][69.5[3]]		80.7[c][83.5[3],80.5[8]]
CH_3	n-C_5H_{11}	39.7[38[1],39.3[4]]							62.8[58[1],64[4]]
C_2H_5	n-C_5H_{11}	63.3							90.4
n-C_3H_7	n-C_5H_{11}	32.7[32.6[4]]						(23.6[a])[21.5[4]]	71.1[71.6[4]]
n-C_4H_9	n-C_5H_{11}		28.0				41.5	44.4	84.6
n-C_5H_{11}	n-C_5H_{11}		29.0		47.5	48.0	52.7	53.6	77.5
n-C_6H_{13}	n-C_5H_{11}		40.0			45.2	61.6	75.2	85.2
n-C_7H_{15}	n-C_5H_{11}		23.0		58.0	64.4	68.3	79.6	83.2
CH_3	n-C_6H_{13}	35.3							53.7
C_2H_5	n-C_6H_{13}	39.6							80.2
n-C_3H_7	n-C_6H_{13}	40.7						(19.8[a])	62.8
n-C_4H_9	n-C_6H_{13}		26.0				47.3	54.7	76.9
n-C_5H_{11}	n-C_6H_{13}		36.0	40.8	43.4	51.0	52.8	61.4	73.0
n-C_6H_{13}	n-C_6H_{13}		15.0			35.0	62.9	77.0	80.7
n-C_7H_{15}	n-C_6H_{13}		40.4			66.9	69.7		80.2[c]
CH_3	n-C_7H_{15}	27.3[25.5[4]]							62.8[61.4[4]]
C_2H_5	n-C_7H_{15}	52.5							86.2
n-C_3H_7	n-C_7H_{15}	30.6[29.2[4]]						(20.0[a])	70.0[69.0[4]]
n-C_4H_9	n-C_7H_{15}		20.0			29.0	48.8	56.5	83.3
n-C_5H_{11}	n-C_7H_{15}		29.5		37.3	52.1	55.4	64.0	78.0
n-C_6H_{13}	n-C_7H_{15}		27.0				66.0	81.3	85.3
n-C_7H_{15}	n-C_7H_{15}		33.0		55.0	69.0	72.0	83.7	84.0
CH_3	n-C_8H_{17}	49.3							58.6
C_2H_5	n-C_8H_{17}	47.6							80.6
n-C_3H_7	n-C_8H_{17}	38.9[36.0[4]]						(19.4[a])	65.1[64.4[4]]
n-C_4H_9	n-C_8H_{17}		33.0				49.5	63.7	79.0
n-C_5H_{11}	n-C_8H_{17}		43.2				53.6	67.2	74.6
n-C_6H_{13}	n-C_8H_{17}		29.0				66.3	81.7	82.5
n-C_7H_{15}	n-C_8H_{17}		48.1				69.7		82.8[c]

[a] Monotropic Smectic Phase

[b] Sample provided by J. B. Flannery and purified by us.

[c] Smectic - Isotropic transition

The effect of molecular structure on transition temperatures and stability of the various phases is shown in Figure 2 through 10. When R is methyl (Figure 2) or ethyl (Figure 3), the compounds are nematogenic regardless of alkyl chain length. Increased alkyl group polarizability has little effect on the temperature range of the nematic phase with the exception of the last member in the series where R = CH_3. The expected even-odd effect for the nematic-isotropic transition is observed. This behavior has been discussed in detail elsewhere for analogous compounds.[7,13] For the n-propoxy compound (Figure 4) the nematic phase still predominates but a monotropic smectic phase is observed. The alkoxy group is now sufficiently long and thus polarizable to overcome residual terminal cohesive forces. The interactions, however, are not sufficiently strong to be effective when the compound melts. They are only effective during cooling from the nematic phase. The insertion of one more methylene into the alkoxy group has a large effect as shown for the butoxy compound (Figure 5). End group polarizability is now large enough to give rise to enantiotropic smectic phases. The crystal lattice energetics have been decreased to such an extent that a crystalline to smectic transition is observed at 8.0°C. The effect of increased alkyl group chain length on nematic phase stability becomes very evident in Figure 6. As is shown here and in the other figures, the crystalline to mesomorph transition temperature is unpredictable as is the effect of structure on smectogenic behavior. For example, N-(4-n-amyloxy-benzylidene)-4-n-hexylaniline is the only compound to have five smectic phases. Further increasing the length of the alkoxy group to six carbon atoms (Figure 7) continues to decrease nematic stability. It is interesting to note that the number of smectic phases increase with increasing alkoxy group chain length up to and including amyloxy. With increased alkoxy chain length the number of smectic phases decreases. The transition temperatures for the heptyloxy compounds are shown in Figure 8. The nematic phase exists only over a very small temperature range. In fact no nematic behavior is observed for two members of the series. Overall smectic phase stability remains large.

In Figures 9 and 10 the effects of small changes in the alkoxy group while the alkyl group remains constant are shown for two series of compounds. The even-odd effect in the N→I transition as well as the decreased stability of the nematic phase with increased polarizability is shown. The effect of increased chain length on the crystalline melting point, appearance of the monotropic smectic phase and stability of the smectic phases is also clear.

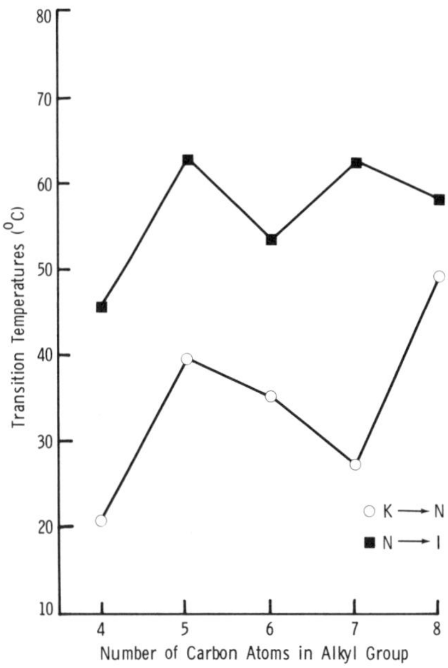

Fig. 2. Transition temperatures for N-(4-methoxybenzylidene)-4-alkylanilines.

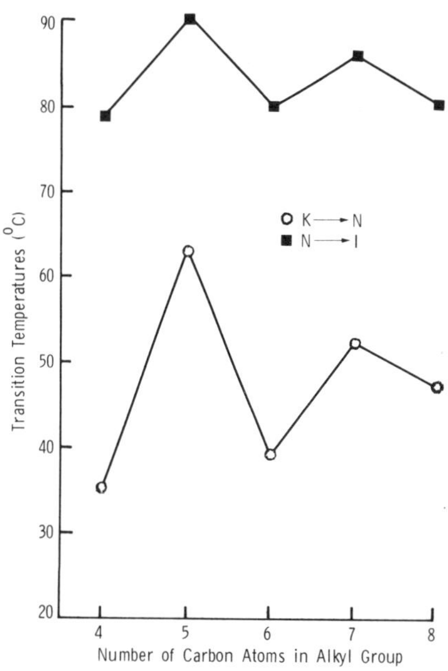

Fig. 3. Transition temperatures for N-(4-ethoxybenzylidene)-4-alkylanilines.

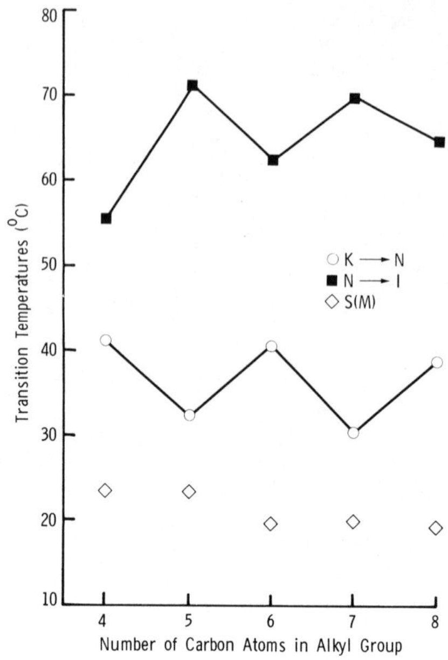

Fig. 4. Transition temperatures for N-(propoxybenzylidene)-4-alkylanilines.

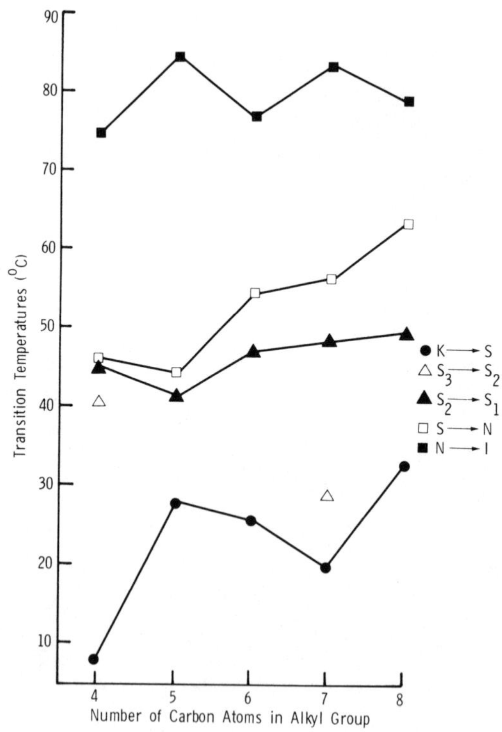

Fig. 5. Transition temperatures for N-(4-butoxybenzylidene)-4-alkylanilines.

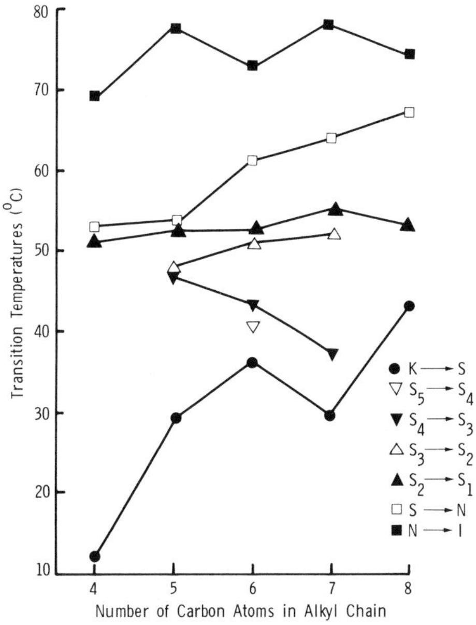

Fig. 6. Transition temperatures for N-(4-amyloxybenzylidene)-4-alkylanilines.

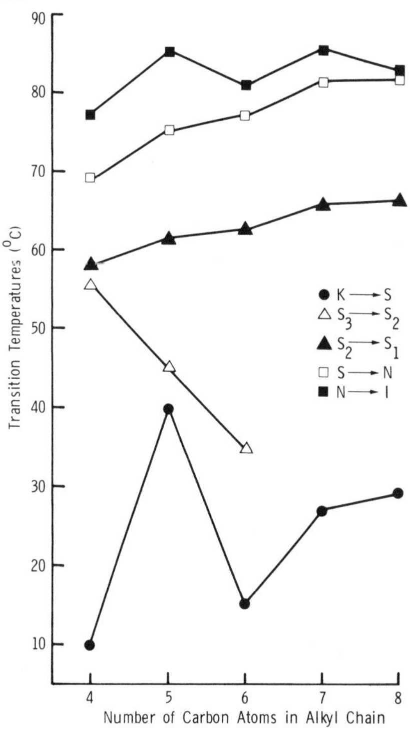

Fig. 7. Transition temperatures for N-(4-hexyloxybenzylidene)-4-alkylanilines.

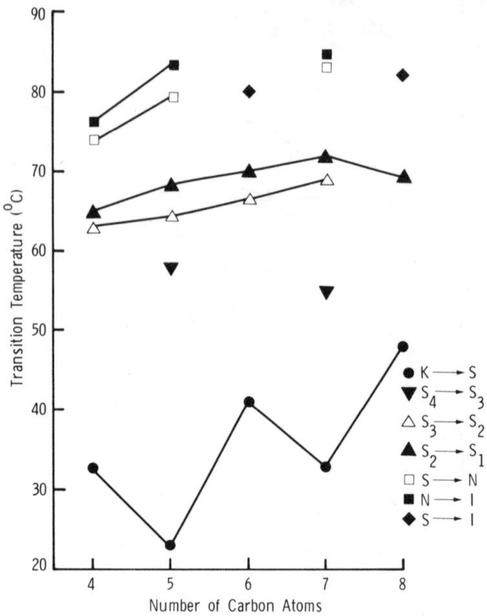

Fig. 8. Transition temperatures for N-(4-heptyloxybenzylidene)-4-alkylanilines.

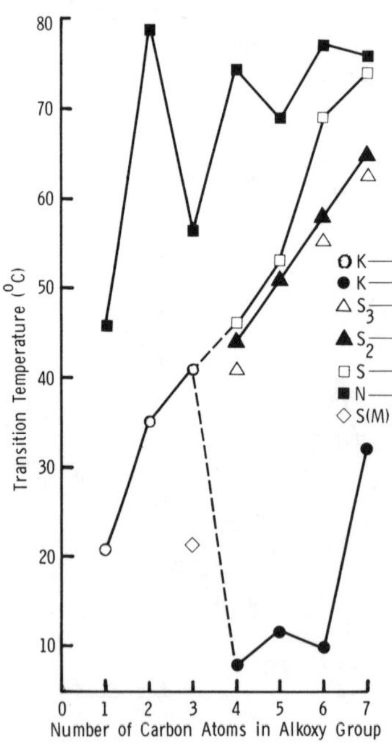

Fig. 9. Transition temperatures for N-(4-alkoxybenzylidene)-4-butylanilines.

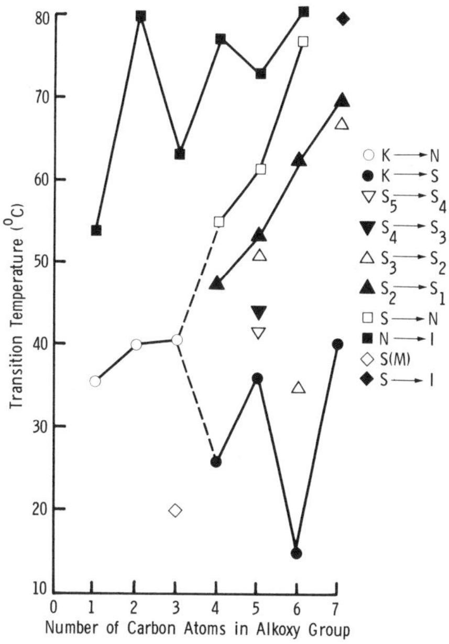

Fig. 10. Transition temperatures for N-(4-alkoxybenzylidene)-4-hexylanilines.

EFFECT OF STRUCTURE ON TRANSITION TEMPERATURES OF N-(4-ALKOXYBENZYLIDENE)-4-n-ALKYL-2-METHYLANILINES

The research of Gray[13] and others[14-17] has shown that at least in high melting compounds molecular broadening can lower crystalline to nematic transition temperatures. Also, van der Veen and Grobben[18] have shown that molecular broadening of 4,4'-dialkoxy- and 4-alkoxy-4'-acryloxybenzylideneanilines will yield in most cases compounds that are monotropic nematic liquid crystals at room temperature. Two compounds were nematogenic over temperature ranges of 51-68°C and 61-64°C.

Insertion of a methyl group into the aniline portion of the N-(4-alkoxybenzylidene)-4-n-alkylanilines described earlier will also increase the molecular breadth of these compounds. If, as shown by van der Veen and Grobben,[18] the benzene rings in the unsubstituted compound have a twist angle of 40-60°, then the introduction of a bulky methyl group will not only make the molecule more rigid but also increase the twist angle. Based on this assumption of an increased twist angle the ortho-methyl group will increase the molecular breadth of the molecule by about 0.7 to 0.9 Å. The molecular broadening has produced three effects as shown in Table II. Here again R has been systematically varied from CH_3 through n-C_7H_{15}. The 4-alkyl group, R', however, has been changed in increments of two carbon atoms, i.e., C_2H_5, n-C_4H_9, n-C_6H_{13} and n-C_8H_{17}. The temperature at which a nematic phase is observed has been lowered. Most of the compounds are monotropic nematic with two of them, R = n-C_3H_7, R' = C_2H_5, and R = CH_3, R' = n-C_4H_9, having very low transition temperatures of -12.8°C and -11.9°C, respectively. For these monotropic compounds the intermolecular forces have been reduced to such an extent that ordering may take place only during cooling from the melt. This is similar to the results observed with the n-propoxy compounds of the unsubstituted benzylideneanilines. Several compounds are nematogenic at and below room temperature with new low crystalline to nematic transition temperatures. The crystal lattice packing in these compounds is less favorable than that of the unsubstituted compounds and a lower melting point results. End group polarizability is sufficient, however, for ordering into a nematic phase. Note that when R' is C_2H_5 none of the compounds studied were nematogenic.

TABLE II

TRANSITION TEMPERATURES (°C) OF THE N-(4-n-ALKOXYBENZYLIDENE)-4-n-ALKYL-2-METHYLANILINES

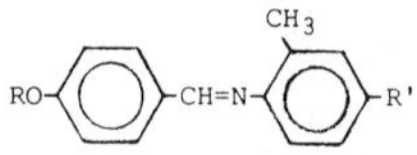

Compound		Temperature of Transition to	
R	R'	Nematic	Isotropic
CH_3	C_2H_5	---	51.3
C_2H_5	C_2H_5	12.5 (M)	49.8
n-C_3H_7	C_2H_5	-12.8 (M)	42.7
n-C_4H_9	C_2H_5	12.4 (M)	38.8
n-C_5H_{11}	C_2H_5	5.4 (M)	31.9
n-C_6H_{13}	C_2H_5	18.8 (M)	40.1
n-C_7H_{15}	C_2H_5	17.3 (M)	30.1
CH_3	n-C_4H_9	-11.9 (M)	16.0
C_2H_5	n-C_4H_9	21.0 (M)	47.7
n-C_3H_7	n-C_4H_9	-4.1 (M)	29.7
n-C_4H_9	n-C_4H_9	15.7 (M)	20.7
n-C_5H_{11}	n-C_4H_9	9.7 (M)	28.7
n-C_6H_{13}	n-C_4H_9	15.2	20.2
n-C_7H_{15}	n-C_4H_9	16.3	16.7
CH_3	n-C_6H_{13}	---	29.7
C_2H_5	n-C_6H_{13}	24.2	24.8
n-C_3H_7	n-C_6H_{13}	2.3 (M)	14.2
n-C_4H_9	n-C_6H_{13}	21.4 (M)	25.6
n-C_5H_{11}	n-C_6H_{13}	---	32.0
n-C_6H_{13}	n-C_6H_{13}	17.5	24.1
n-C_7H_{15}	n-C_6H_{13}	---	32.8
CH_3	n-C_8H_{17}	---	43.9
C_2H_5	n-C_8H_{17}	27.2 (M)	30.3
n-C_3H_7	n-C_8H_{17}	9.2 (M)	24.5
n-C_4H_9	n-C_8H_{17}	17.8	24.2
n-C_5H_{11}	n-C_8H_{17}	---	39.4
n-C_6H_{13}	n-C_8H_{17}	28.1	28.8
n-C_7H_{15}	n-C_8H_{17}	27.2	27.6

(M) - Monotropic nematic

However, when R' was n-C_4H_9 or larger nematogenic behavior was observed if the alkoxy group was sufficiently large. The increase in molecular broadening also decreases nematic phase stability and inhibits the smectic phase formation observed in the unsubstituted benzylideneanilines.

SUMMARY

A systematic study has been made of the effects of small changes in molecular structure on the mesomorphic behavior of 35 N-(4-n-alkoxybenzylidene)-4-n-alkylanilines and 28 N-(4-n-alkoxybenzylidene)-4-n-alkyl-2-methylanilines. It has been shown that the polarizabilities of the alkoxy and alkyl groups are of primary importance in determining mesomorphic character. Small increases in molecular breadth have three pronounced effects: 1) lowering of the temperature at which the nematic phase appears, 2) decreasing the stability of the nematic phase, and 3) destroying any smectogenic behavior regardless of alkoxy or alkyl group chain length.

ACKNOWLEDGEMENTS

The authors wish to thank Messrs. W. Florance, M. Myers, W. Lee, D. Dungan and L. Melkvik for their technical assistance in this research.

REFERENCES

1. G. H. Heilmeier, L. A. Zanoni and L. A. Barton, Proc. IEEE, 56 (7), 1162 (1968).
2. H. Kelker and B. Scheurle, Angew. Chem. Intern. Edit., 8, 884 (1969); 9, 962 (1970).
3. J. B. Flannery and W. Haas, J. Phys. Chem., 74, 3611 (1970).
4. D. L. Fishel and Y. Y. Hsu, J. Chem. Soc. (D), 1971, 1557.
5. G. C. Fryburg, E. Gelerinter and D. Fishel, Mol. Cryst. Liquid Cryst., 16, 39 (1972).
6. A. DeVries and D. Fishel, ibid, 16, 311 (1972).
7. K. Murase, Bull. Chem. Soc. Japan, 45, 1772 (1972).
8. H. J. Dietrich and E. Steiger, Mol. Cryst. Liquid Cryst., 16, 263 (1972).
9. G. W. Smith, Z. G. Gardlund and R. J. Curtis, ibid, 19, 327 (1973).

10. Z. G. Gardlund, R. J. Curtis and G. W. Smith, J.C.S. Chem. Commun., 1973, 202.

11. G. W. Smith and Z. G. Gardlund, J. Chem. Phys., In Press.

12. W. J. Hickinbottom, J. Chem. Soc., 1937, 1119.

13. G. W. Gray, "Molecular Structure and the Properties of Liquid Crystals," Academic Press, London and New York (1962).

14. S. L. Arora, L. Fergason and T. R. Taylor, J. Org. Chem., 35, 4055 (1970).

15. R. J. Cox, Mol. Cryst. Liquid Cryst., 19, 111 (1972).

16. W. R. Young, A. Aviram and R. J. Cox, Angew. Chem. Intern. Edit., 10, 410 (1971).

17. W. R. Young, I. Haller and D. C. Green, J. Org. Chem., 37, 3707 (1972).

18. J. van der Veen and A. H. Grobben, Mol. Cryst. Liquid Cryst., 15, 239 (1971).

DOMAIN FORMATION IN HOMOGENEOUS NEMATIC LIQUID CRYSTALS

J. M. Pollack and J. B. Flannery

Rochester Corporate Research Center
Xerox Corporation
Rochester, New York 14644

A new electrohydrodynamic mode in nematic liquid crystals (NLCs) was reported recently by Vistin[1], and Greubel and Wolff.[2] The boundary value problem describing the effect was solved by Penz and Ford[3], based on the conduction-induced alignment mechanism of Helfrich.[4] Known as the variable grating mode (VGM), this dc effect in planar electroded cells of NLCs with negative dielectric anisotropy is characterized as follows: In thin cells (<10μm) containing high resistivity (>10^{10}Ωcm) material, parallel domains which resemble the well-known Williams striations[5] are observed in the plane of the sample. The onset of domain structure occurs at a characteristic threshold voltage of 5-10 V, regardless of cell thickness L, but the linear domain density or spatial frequency, ν, is inversely proportional to L near threshold. The domains consist of regions of vortex-type fluid motion, and are visible by virtue of the spatially periodic variation of the anisotropic refractive index of the NLC and its director. These domains act as a volume phase grating whose relative phase factor $\phi = \pi L \nu/2$ varies linearly with voltage above threshold.[3]

The present work was undertaken to determine systematically the voltage and cell thickness dependences of diffraction efficiencies, grating spatial frequencies, and transient responses of typical NLCs in the VGM. Of particular interest was the nature of the volume phase grating, the response times for thin NLC cells, and the relationship of homogeneous zero-field alignment of the NLC to electrooptic behavior. Such information is useful in assessing the potential of the VGM for light deflection, and for display and image processing applications requiring low-noise spatial filtering.

EXPERIMENTAL

The room-temperature NLCs were mixtures of terminally-substituted aromatic azoxy compounds from E. Merck. The first, N4, is the eutectic mixture of

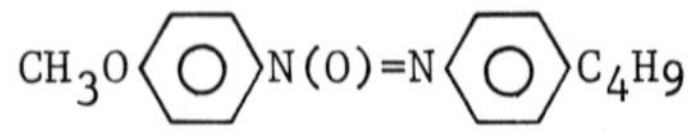

and

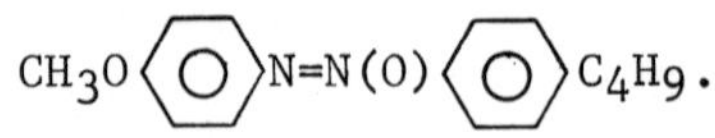

The second, N5, is a mixture of N4 and the eutectic of

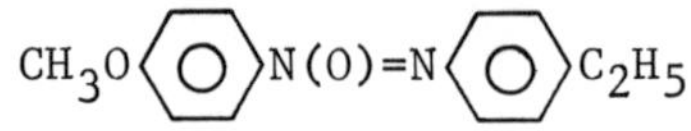

and

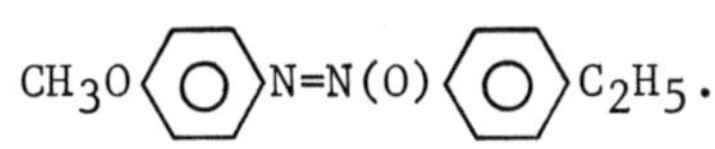

These materials, used as received, are characterized by high resistivity ($\rho \geq 10^{10}\Omega$cm),negative dielectric anisotropy ($\varepsilon_{||}-\varepsilon_{\perp} = -0.2$),and dielectric constants of 5.3 (N4) and 5.5 (N5).[6]

Cells for observation and measurement were constructed in the conventional parallel plate capacitor configuration using plate glass of 0.25 in thickness coated with a transparent conductive layer of indium oxide. Both electrodes were selectively etched to provide a circular active area of 1cm^2. Prior to assembly, plates were rigorously cleaned, and orientation rubbed without the use of surfactants, as described by Chatelain[7], to promote homogeneous alignment of the NLC. Cells of thickness 6-15μm employed Mylar strips of appropriate thickness for spacers. For cells thinner than 6μm, polystyrene latex microspheres (Particle Information Service) placed outside the active area were used. Filling of cells was by capillary action. Coplanarity of the confining electrodes over the active area was determined by observation of optical fringes using a Van Keuren helium lamp illuminator. Measured variance was < 0.5 optical fringe (0.15μm). Measurements of cell thickness were made by capacitive techniques using a General Radio Impedance Bridge Model 1615A both before (air) and after filling with NLC. Computed thicknesses for filled and unfilled cells generally agreed within ten percent. After filling, cells were sealed with an epoxy resin ("Epoxypatch," Dexter Corporation) and electrical connections made in the conventional manner. All subsequent measurements were made at ambient temperature, without any additional temperature control of the NLC cells.

Microscopic observations were made using a Leitz Ortholux

Polarizing Microscope. The diffraction efficiency of voltage-induced phase gratings was measured as the relative decrease in intensity of the zero order transmitted beam of a Spectra Physics 133M 5mw He-Ne laser used for normal illumination of the NLC cell. Light intensities were monitored using a Spectra Physics 401 Power Meter, or with a Fairchild K1719 PMT operated at 1200 V with appropriate optical filtering to ensure a linear response characteristic. Measurements of first and second order diffracted intensities required appropriate spatial filtering in the input plane of the detector.

Spatial frequencies of stationary domains were determined from the Bragg relationship using measured angular deflections of the first order diffracted beam as voltage to the NLC cell was varied incrementally.

Rise and decay times were measured as the change in intensity of the zero order beam, displayed in reciprocal fashion on a Tekronix 564 oscilloscope. A decrease in zero order intensity corresponded to a rise of intensity in higher spectral orders, and conversely. Triggering the NLC cell was accomplished with a high slewing speed switch using TTL logic with a switching time of less than 120μsec over the range of 0-60 V dc.

RESULTS AND DISCUSSION

For both N4 and N5, polarized microscopy confirmed that uniform homogeneous alignment was obtained in the zero-field state without the use of surfactants, over the entire cell area. The degree of alignment was superior for N5 compared with N4, and generally better in the thinnest cells, as judged by the level of extinction between crossed polarizers. In the voltage-activated state parallel domains were observed, the long axes of which were perpendicular to the direction of rubbing. An example of the texture observed near threshold in N5 for light polarized parallel to the domain axis is shown in Figure 1. For both N4 and N5, the visibility of domains was greatest when incoming light was polarized parallel to the domain axis. This is consistent with the finding of Vistin[1] for another material, but at odds with Greubel and Wolff[2] for N4. The observed domain pattern was stationary at a fixed voltage, but the domain width decreased with increasing voltage. In all cases, the parallelism and the length of domains were greater as the uniformity of zero-field homogeneous alignment was improved. Between crossed polarizers the domain birefringence was voltage-variable as well. Turbulence was not seen in any cell for applied voltages up to 60 V, nor for N5 in the thinnest (2.8μm) cells to biases as high as 120 V. In the latter instance, however, the domain pattern was superseded by other stationary

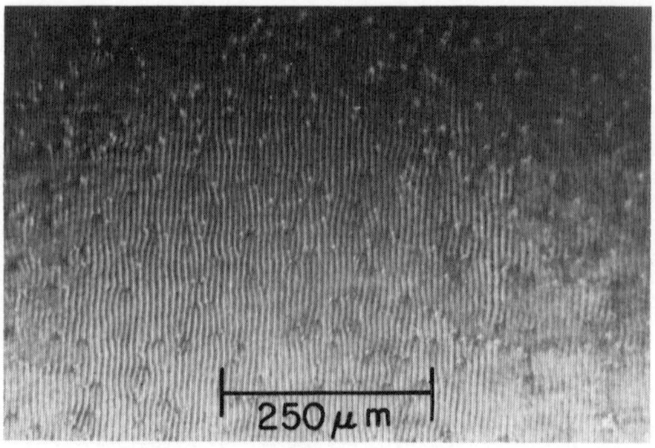

Figure 1. Photomicrograph of parallel domains observed in N5 at 10 V dc in a cell of thickness 8μm for light polarized parallel to the domain axis.

textures as the domain spatial frequency exceeded optically resolvable limits.

Nonpolarized laser illumination (6328Å) of the NLC cells confirmed the microscopic observations for N4 and N5 in the range 0-60 V. Above the voltage threshold in a transmission projection plane, partial diffraction rings were observed for cells with L > 4μm, while in those thinner, only discrete diffraction spots resulted. The latter are indicative of superior alignment in thin cells. Turbulence and, hence, random scattering was not in evidence in this voltage range (increased spatial noise was seen at higher voltages), and thus, it was inferred that measured decreases in zero order transmitted light intensity corresponded only to increased intensity in the higher spectral orders. The measured diffraction efficiency for both N4 and N5 in 8μm cells is shown as a function of applied bias in Figure 2. The pertinent observations are i) both materials exhibit a voltage threshold (5-10 V) and ii) oscillations in total diffraction efficiency accompany voltage variations.

The threshold, and the diffraction extrema were reproduced for either increasing or decreasing applied voltage over the range investigated. The onset of diffraction occured at lower voltage for N4, while the maximum total diffraction efficiency was about 50 percent for both materials in 8μm cells. A similar efficiency was reported[2] elsewhere for unaligned samples of N4, although the applied voltage and cell dimension were unspecified.

The dependence of diffraction efficiency on applied voltage

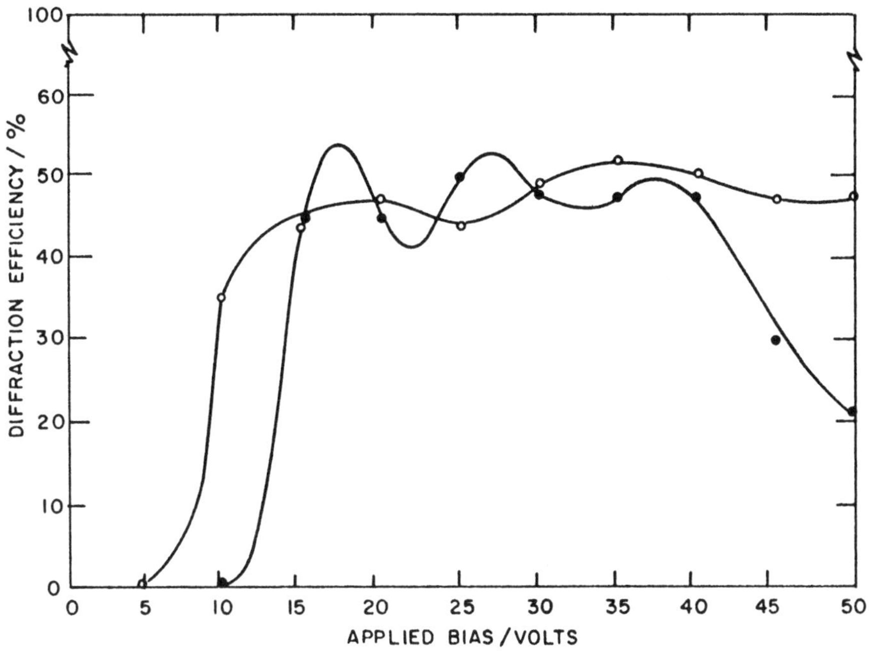

Figure 2. Plot of diffraction efficiency bersus applied voltage to 8μm cells of N4 (O) and N5 (●).

was measured for N5 cells with thicknesses of 2.8-14.6μm. The results are compared in Figure 3. The voltage thresholds for diffraction appear to vary with thickness, but experience indicates this may be a random, rather than systematic effect. The peak diffraction efficiencies increased regularly with L, and were generally maximal between 11-17.5 V. For the thinnest (2.8μm) cell, 86 percent of the incident light was deflected into nonzero spectral orders. The peak diffraction efficiency was highest for those cells with best homogeneous alignment in the quiescent state, and domain parallelism and length, when activated. These latter conditions, as noted, are satisfied in the thinnest cells.

The domain density or spatial frequency, ν, was determined for various cell thicknesses using N5, and for N4 at 8μm, as a function of applied voltage. The results are shown in Figure 4. Additional comparison data for N4 are from Greubel and Wolff.[2] The voltage for precise measurement of deflection was approximately 10 V for both N4 and N5. Inspection of Figure 4 shows that domain density

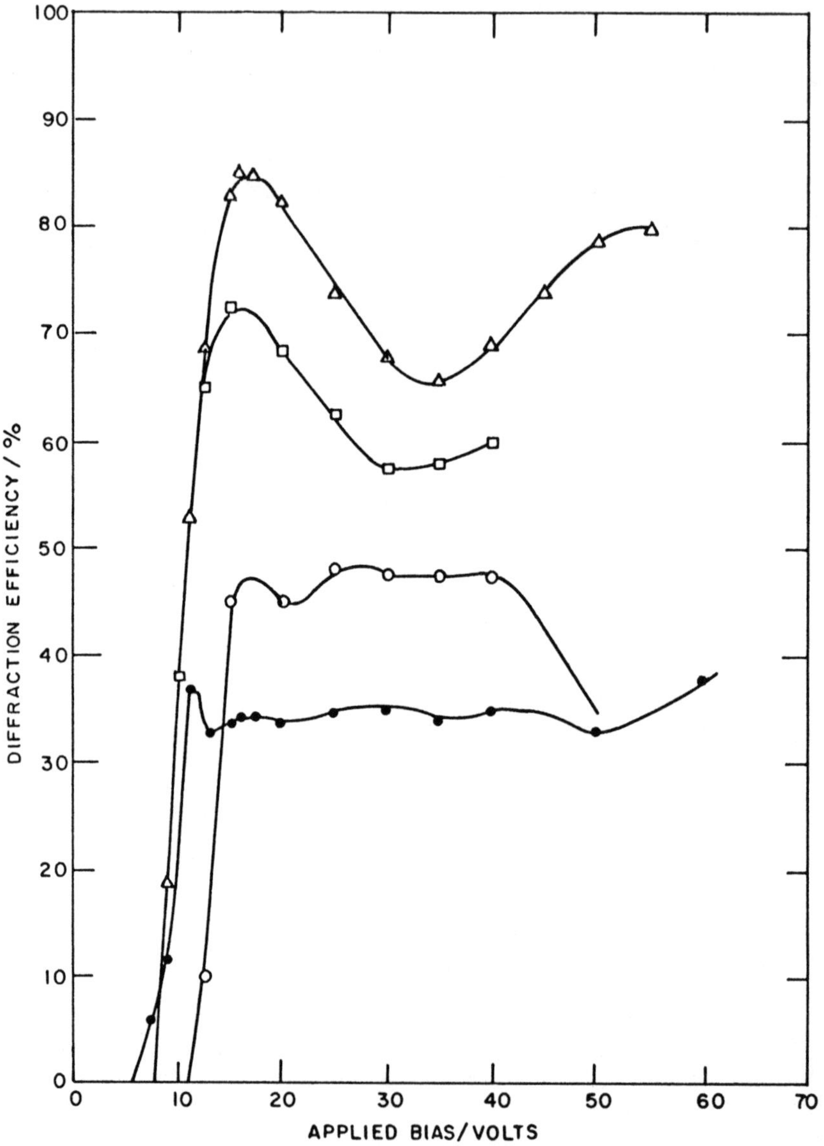

Figure 3. Plot of diffraction efficiency versus applied voltage for N5: Δ, 2.8μm; □, 3.7μm; O, 8μm; ●, 14.6μm.

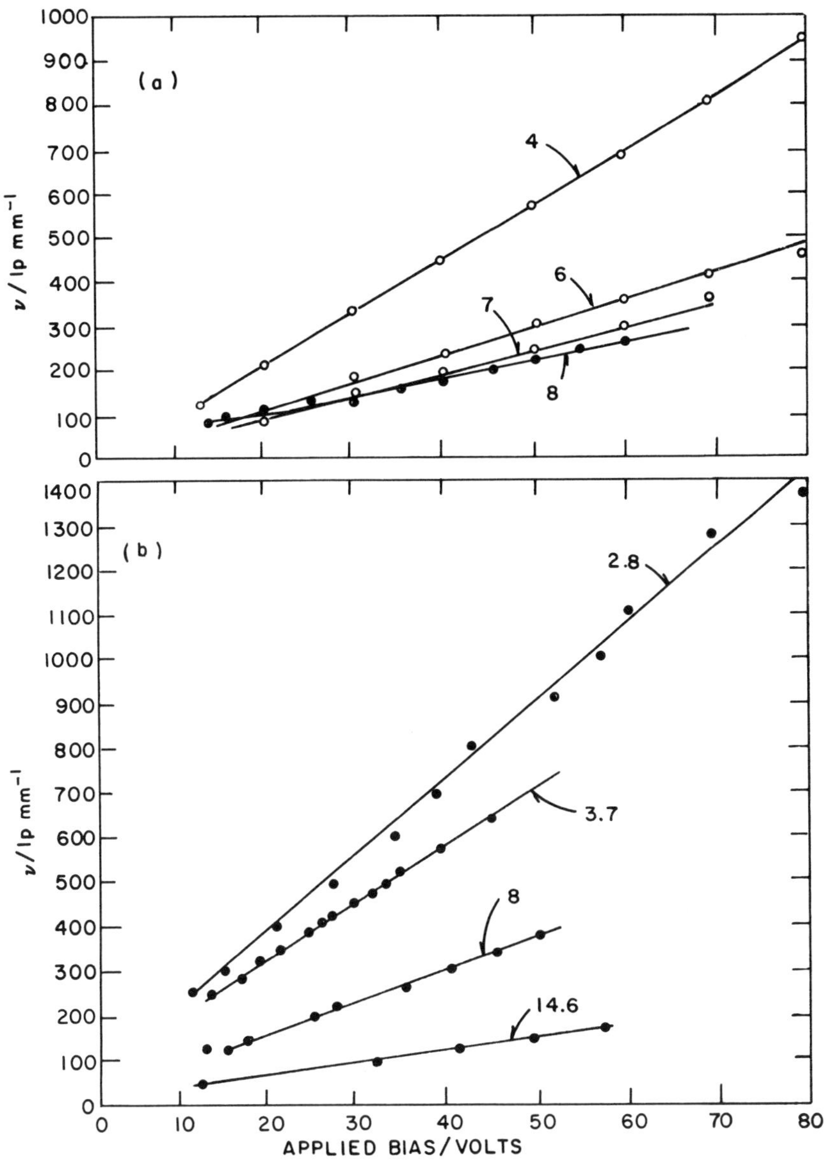

Figure 4. (a) Plot of domain spatial frequency (lp mm^{-1}) versus applied voltage for N4. Numerals indicate cell thickness in μm. Sources: O, reference 2; ●, this work. (b) Plot of domain spatial frequency versus applied voltage. Numerals indicate cell thicknesses. Source: ●, this work.

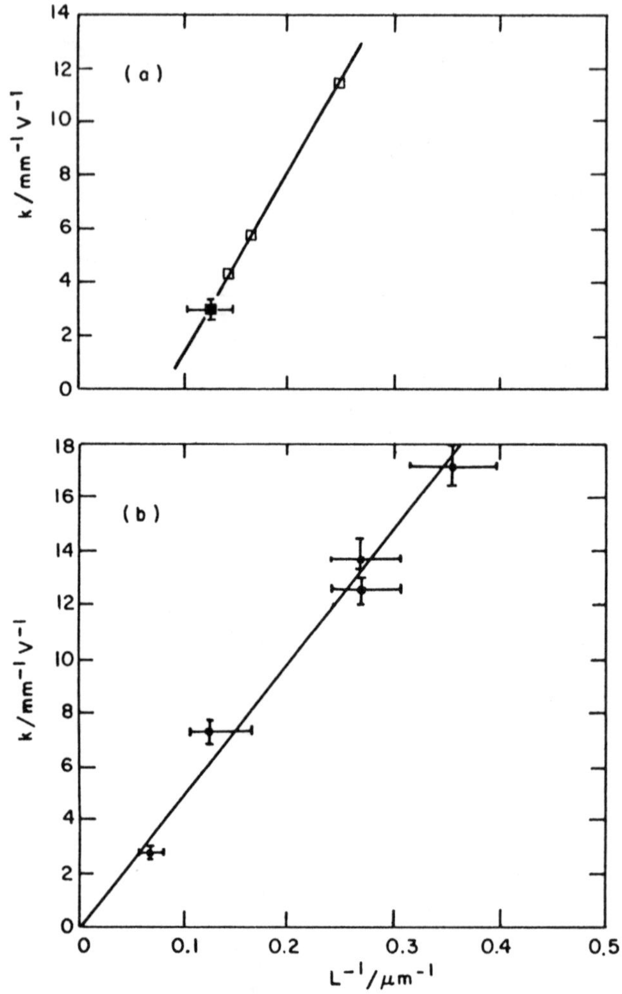

Figure 5. (a) Plot of k versus L^{-1} for N4: □, reference 2; ■, this work. (b) Plot of k versus L^{-1} for N5: ●, this work. Error bars represent ± 10% uncertainty in L, and ± 5% in k.

increased linearly with applied voltage above threshold, but that the rate of increase depended inversely on cell thickness. In fact, the domain density appears to be simply field dependent. For all N4 and N5 cells in the present investigation, the domain density at threshold and for ∿ 5 V above threshold was closely approximated by the inverse cell thickness, L^{-1}. Noteworthy in Figure 4(b) is the high spatial frequency (∿ 1300 lp mm^{-1}) characteristic of a 2.8μm cell at 70 V. As a consequence, measurement of deflection angle for this cell beyond ∿ 80 V was frustrated by a limit on the available angular window imposed by the refractive index of the exit plate glass.

The relationship of ν with V, and the threshold values of both parameters, ν_o and V_o, are adequately described by the linear equation:

$$\nu = k(V-V_o) + \nu_o \quad (1)$$

for $V \geq V_o$, where k is the first order derivative dependent upon L^{-1}, but otherwise invariant. Evaluation of k follows simply from slopes of the linear plots in Figure 4.

The relative phase factor, Φ, for a volume phase grating[3] is

$$\Phi = \frac{\pi}{2} L \nu \quad (2)$$

Substitution from (1) gives

$$\Phi = k' (V-V_o) + \Phi_o \quad (3)$$

where $k' = \frac{\pi}{2}Lk$ and $\Phi_o = \frac{\pi}{2}L\nu_o$. It follows from eq (3) that

$$k = \frac{\nu - \nu_o}{V - V_o} = \frac{2k'}{\pi} L^{-1} \quad (4)$$

Thus, k should depend linearly on L^{-1} with a first derivative of $(2k'/\pi)$. The result of a graphical analysis of k versus L^{-1} are shown in Figure 5, for N4 and N5. Inspection shows that within experimental precision, the predicted linear variation of k with L^{-1} is obtained. A k = 0 limit for large L is indicated by eq (4) and observed in the present work for N5 (Figure 5(b)). However, the available data for N4 (Figure 5(a)) do not conform to the constraint of a (0,0) intercept. The origin of this discrepency is not presently known, but may represent systematic errors in L for the data from reference 2.*

---

* It is the experience of these authors that reliance on nominal

The distribution of intensities, I_n, in the far-field diffraction pattern for light transmitted through a pure sinusoidal phase grating is given by the Bessel functions of integral order[8]:

$$I_n \propto [J_n(\phi)]^2 \qquad (5)$$

where the argument or phase factor ϕ is proportional to Φ and depends on the domain density and cell dimension according to eq (2).

Using the theory of phase modulation, the diffraction efficiency DE in any order n can be expressed directly in terms of the phase difference ϕ:

$$DE_n = J_n^2(\phi) \qquad (6)$$

Also, it follows from the properties of the Bessel function that the total efficiency on all non zero orders is given by

$$DE_o = 1 - J_o^2(\phi) = \sum_{n=1}^{\infty} J_n^2(\phi) \qquad (7)$$

A plot of the normalized diffraction efficiency (DE_o/DE_{omax}) versus ϕ (eq (6)) for N5 is shown in Figure 6(a). For comparison, the computed variation of DE_o from eq (7) is also shown. Measured first and second order diffraction efficiencies are compared with those computed from eq (6) in Figures 6(b) and 6(c).

In all cases, maxima in the experimental diffraction efficiency versus voltage curves were made to coincide with the corresponding maxima in the computed DE curves from eqs (6) and (7). This "best-fit" procedure allowed empirical determination of a ϕ - V relationship which has the form:

$$\phi = \alpha(V-V_o) + \phi_o \qquad (8)$$

where $\alpha = 0.370\ V^{-1}$ and $\phi_o = 1.80$. These numerical constants

---

spacer dimensions for a determination of cell thickness can be misleading. Thus, the anomaly shown in Figure 5(a) is presently being investigated. Another concern with the data of ref. 2 is (Figure 4) the experimental designation of a domain density of $\sim$ 120 mm^{-1} for a 4μm cell at $\sim$ 12.5 V. Present experience with similarly dimensioned cells of N4 indicates a minimum $\nu = 250 \pm 50\ mm^{-1}$.

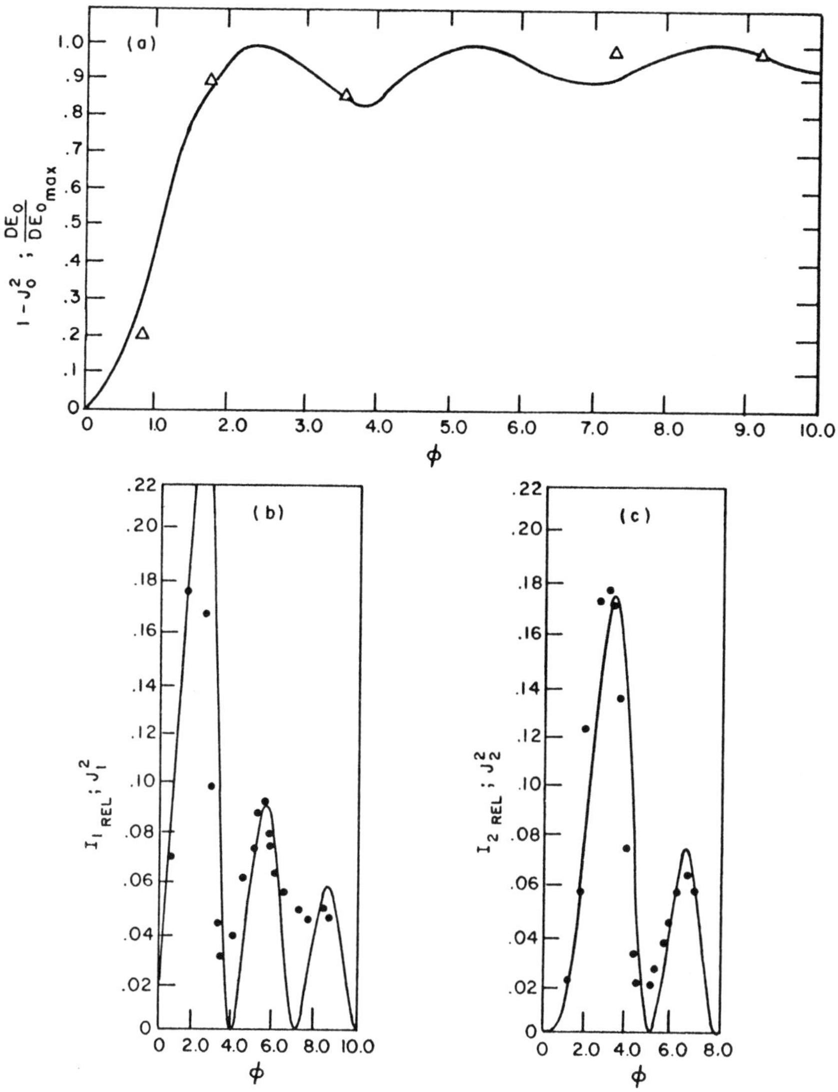

Figure 6. (a) Plots of normalized all-orders diffraction efficiency (Δ) and $1 - J_o^2$ versus phase factor ϕ. (b) Plots of measured first order diffracted intensity and J_1^2 versus ϕ. The computed curve is shown as a solid line, and experimental data are shown as points. (c) Plots of measured second order diffracted intensity (points) and computed J_2^2 (solid line) versus ϕ. Experimental data are for N5, L = 8μm.

are to be compared with those predicted by eq (3):

$$k' = \frac{\pi}{2}Lk = 0.092\ V^{-1}$$

and

$$\Phi_o = \frac{\pi}{2}L\,\nu_o = 1.57$$

The correlation is regarded as favorable in view of the assumption of a pure phase grating (eq (2)). That this assumption can be only approximate is clear on recalling (Figure 3) that measured peak diffraction efficiencies depend strongly on the thickness of the NLC cell. Attempts to better quantify this correlation using thin NLC cells are in progress.

The dynamic behavior of N5 was also investigated for cells with L = 2.8 - 14.6μm in the range 0-60 V. Typical traces of rise and decay curves are shown in Figure 7. Certain general features are notable. Rise curves for all cells at lower voltages displayed the "shoulder" evident in Figure 7(a). This generally disappeared at higher voltages. An indication period, t_i, which varied inversely with voltage, always preceded the period, t_r, for rise of diffraction. In the high voltage range, DE peaked at values as much as 30% greater than the steady-state condition. This DE maximum and subsequent relaxation to the steady-state were attained in times ranging from 3-100 msec, depending on voltage and cell thickness.

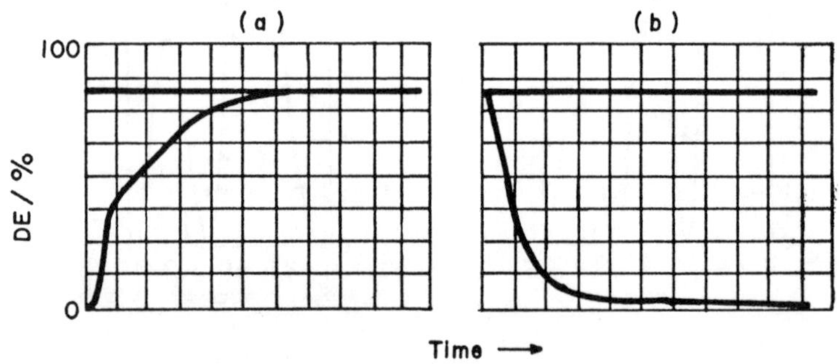

Figure 7. (a) Oscilloscope trace of temporal decrease of relative zero order intensity after application of bias voltage. Time scale is 2 msec/div. (b) Oscilloscope trace of temporal increase of relative zero order intensity after removal of applied bias and grounding both cell electrodes. Time scale is 1 msec/div. Data are for N5, L = 2.8μm, at 15 V. The horizontal scan in both traces indicates the steady-state DE.

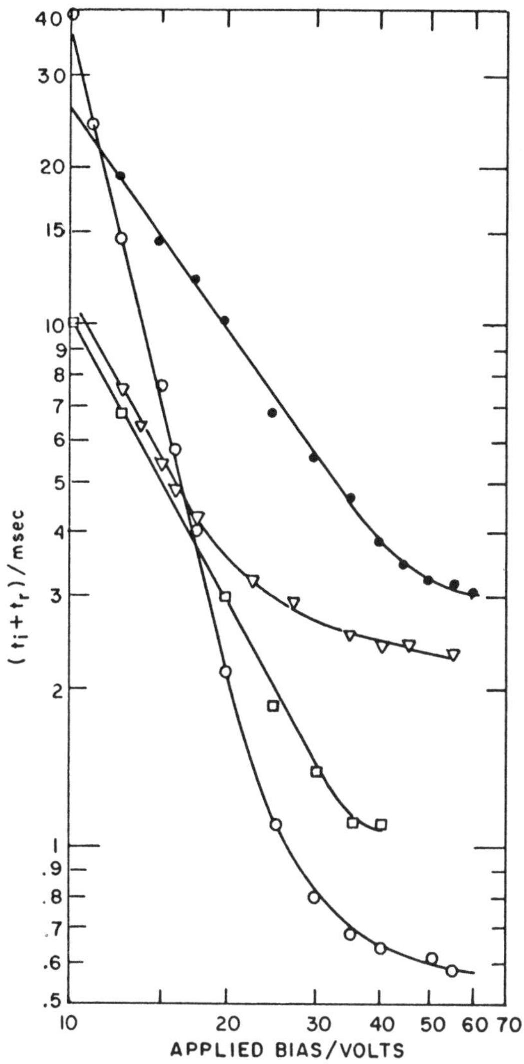

Figure 8. Plots of $(t_i + t_r)$ versus applied bias for N5: L: O, 2.8μm; □, 3.7μm; ∇, 7.8μm; ●, 14.6μm. Slopes of linear regions: 2.8μm, 4.0; 3.7μm, 1.8; 7.8μm, 1.8; 14.6μm, 1.4.

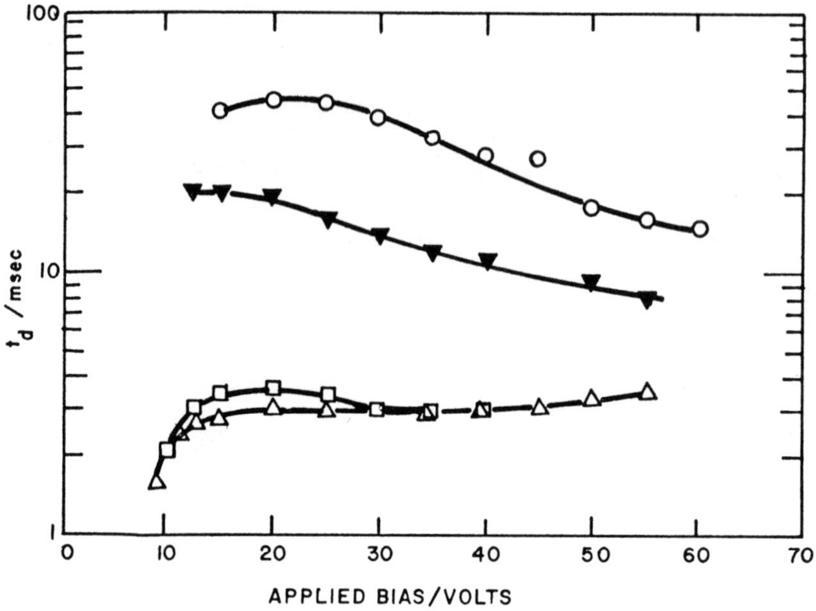

Figure 9. Plots of t_d versus applied bias for N5. Δ, 2.8μm; □, 3.7μm; ▼, 7.8μm; O, 14.6μm.

Defining t_r as the period encompassing the onset of diffraction to 90 percent of the steady-state efficiency, the relationship between $(t_i + t_r)$ and applied bias was determined. The results are shown in Figure 8. Below about 30 V the rise times varied with voltage according to

$$(t_i + t_r) \propto V^{-n} \qquad (9)$$

where n depended strongly on thickness, and ranged from 1.4 - 4.0. These results are to be contrasted to the rise time dependence on E^{-2} for diffraction in N4,[2] to the similar field dependence for rise of turbulence in MBBA,[9] and the same result for rise of turbulent electrohydrodynamic instability in isotropic liquids.[10] The origin of this difference, particularly in the thinnest cell (2.8μm) is not presently known, but is under investigation. Finally, it is important to note that for thin cells of N5, as with N4,[2] that rise times can be as low as fractional msec.

Decay of diffraction was spontaneous on removal of electrical bias, and return to the homogeneous state occurred at a rate depending on voltage and cell thickness. Since the decay curves did not conform to a simple functional relationship, t_d was defined as the period from removal of voltage to ten percent of total diffraction efficiency. The results are shown in Figure 9. The dependence of t_d on voltage is not strong, but the effect of cell thickness is significant. For a bias of 20 V, it was found that t_d varied approximately with L^2. The same thickness dependence was found for decay of the turbulent mode in other materials, and is qualitatively understood in terms of diffusion-limited molecular reorientation.[11]

ACKNOWLEDGMENTS

The authors acknowledge gratefully assistance and helpful discussions with G. Dir, D. Kermisch, and J. Wysocki.

REFERENCES

1. L. K. Vistin, Kristallografiya 15, 594 (1970) [Sov. Phys. Crystallogr. 15, 514 (1970)].

2. W. Greubel and U. Wolff, Appl. Phys. Lett., 19, 213 (1971).

3. P. A. Penz and G. W. Ford, Phys. Rev. A, 6, 414, (1972).

4. W. Helfrich, J. Chem. Phys., 51, 4092 (1969).

5. R. Williams, J. Chem. Phys., 39, 384 (1963).

6. L. Pohl, R. Steinstrasser and B. Hample, 4th Int. Liquid Crystal Conf., Kent, Ohio (1972).

7. D. Chatelain, Bull, Soc. Fr. Mineral Cristallogr., 60, 1580 (1937).

8. J. A. Ratcliffe, Rep. Prog. Phys., 19, 188 (1956).

9. H. Koelmans and A. M. Boxtel, Molec. Cryst. Liquid Cryst., 12, 185 (1971).

10. J. Filippini, J. Gosse, J. Lacroise, and R. Tobazeon, C. R. Acad. Sci. Paris, 269B, 736 (1969).

11. G. H. Heilmeier, L. A. Zanoni and L. A. Barton, Proc. IEEE, 56, 1162 (1968).

PHASE DIAGRAM OF MIXED MESOMORPHIC BENZYLIDENEANILINES - MBBA/EBBA

George W. Smith, Zack G. Gardlund, and Ralph J. Curtis
General Motors Research Laboratories, Warren, Michigan 48090

ABSTRACT

The crystal-nematic transition of binary mixtures of MBBA and EBBA is strongly sensitive to thermal history. A metastable phase (β-phase) has a eutectic-like minimum melting point of about -14°. However, a sample maintained at -18°C for about one day will convert to a stable phase (α-phase) with a minimum melting temperature of 13°C. The stable phase seems to be a solid solution. The crystal-nematic latent heat of transformation L_{KN} for the α-phase is linearly dependent on concentration, but that for the β-phase has a minimum. The concentration-dependence of the nematic-isotropic transition temperature is similar to that reported elsewhere for other systems. The nematic-isotropic latent heat of transition is essentially linearly dependent on concentration.

INTRODUCTION

Although the binary system N-(p-methoxybenzylidene)-p-n-butylaniline/N-(p-ethoxybenzylidene)-p-n-butylaniline (MBBA/EBBA) is well-known as an extended-range room-temperature nematic liquid crystal, relatively little has been published on its phase behavior.[1-4] Previous publications gave the minimum melting composition and transition temperatures,[1] transition temperatures for a 50 mole percent mixture,[2] latent heats[3] and transition temperatures[1,3,5-8] of the pure components, and a phase diagram of the binary mixtures.[4]

In this paper we report crystal-nematic (KN) and nematic-isotropic (NI) transition temperatures for MBBA/EBBA and latent heats determined by differential scanning calorimetry (DSC) using a Perkin-Elmer DSC 1B instrument. We are aware that discrepancies can exist between thermal data obtained by DSC and the more accurate values from adiabatic calorimetry.[9] However, we feel that the use of DSC is suitable for comparisons of similar systems. Furthermore, much of the published thermal data on liquid crystals has been obtained by DSC.[10] As we shall see, considerable discrepancies exist between our work and previous studies. In fact it will become evident that the extreme melting point minimum reported previously[1,2,4] is characteristic of a metastable phase and that the stable high-temperature phase (α-phase) of the binary mixtures is apparently a solid solution.

SAMPLE PREPARATION

We studied samples prepared in several different ways using two DSC instruments. Samples were prepared from as-received or recrystallized MBBA and EBBA. Some samples were mixed on a low frequency mechanical oscillator for 3 hours. Others were mixed by various combinations of shaking on a "Wiglbug" and stirring (with heating). The thermal behavior of the various samples was insensitive to preparation technique.

CRYSTAL-NEMATIC TRANSITION TEMPERATURE

The melting behavior depends sensitively on thermal history of the sample. This is illustrated by the phase diagram of Fig. 1. The dependence of T_{KN}, (the KN transition temperature) upon x (the mole fraction of MBBA) is given by the lower curves of the figure. A sample crystallized from the nematic phase and shortly thereafter remelted gives T_{KN} values indicated by the solid circles. For MBBA concentrations in the range 0.37 to ~0.55 a lower temperature transition (presumably a crystal-crystal transition) is also observed (closed triangles). However, at times, in the range $x \approx 0.5$ to 0.7 the low temperature transition is the only one seen and is then interpretable as a KN transition (lowest group of closed circles). If the sample is held below T_{KN} for several hours before remelting, a second, smaller DSC peak is observed at higher temperatures (open triangles in Fig. 1). The cause of this peak is unclear,

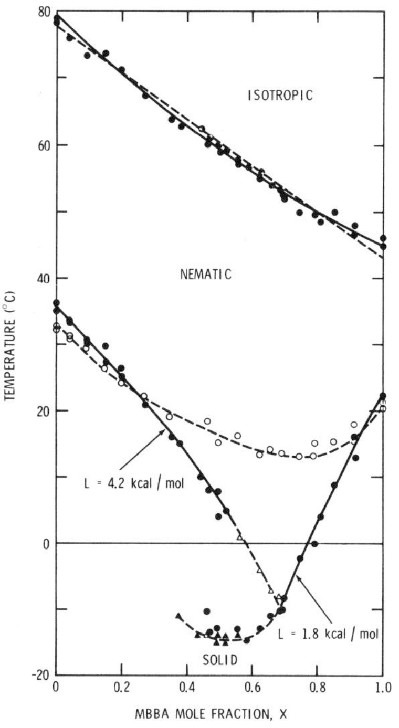

FIG. 1. Crystal-nematic (KN) and nematic-isotropic (NI) transition temperatures as a function of MBBA mole fraction, x. The lower curves indicate KN temperatures for a metastable (●) and stable (o) solid phase. A possible low temperature solid-solid transition (▲) and a low energy transition (Δ) are also indicated. Latent heats for EBBA and MBBA derived from Schröder-van Laar equations are shown.

but it may be due to melting of small regions of a second solid phase coexisting with the more abundant, lower-melting solid. Apparently, however, both of these phases are metastable with respect to a still higher-melting solid phase (see next paragraph). The melting behavior indicated by the closed circles in Fig. 1 is qualitatively similar to that described in Refs. 1 and 4 (although Ref. 4 reports much lower transition temperatures than do the present work or Ref. 1).

A second type of melting behavior is obtained if the samples are maintained at a temperature of about -18°C for about a day or longer. T_{KN} values for mixtures subjected to this "cold treatment" are given by the open circles in Fig. 1. This melting behavior may indicate a solid solution. The dashed curve of Fig. 1 is a best fit of a quartic to the data. A cold-treated sample which has been melted and then supercooled to recrystallize it will exhibit the melting behavior described in the previous paragraph (unless subjected to a second cold treatment). We conclude that the high-melting phase is the thermally stable high temperature solid-phase (α-phase). This view is supported by the fact that a cold-treated sample (even one with $x \approx 0.7$) will remain solid indefinitely at 5°C, well above the minimum melting temperature of the metastable phases.

CRYSTAL-NEMATIC LATENT HEATS

Using the Schröder van-Laar equations[11] we have calculated latent heats of melting of pure EBBA and MBBA from the melting curves of the metastable phases of Fig. 1. The best fit Schröder-van Laar curves are shown in Fig. 1, along with the derived latent heats L; these values of L (4.2 Kcal/mol for EBBA and 1.8 Kcal/mol for MBBA) must be taken with reservation since they correspond to metastable phases. They do not agree well with the values measured by Barrall, et al,[3] (6.4 and 3.25 Kcal/mol). (The difference likely illustrates the limitations of theory - R.S.P.)

From the areas of our DSC melting peaks we have determined rough[12] crystal-nematic latent heats L_{KN}. These are plotted in Fig. 2 as a function of x for both the low-melting metastable phase (closed circles) and high-melting cold-treated stable phase (open circles). It is apparent that the cold treatment does not influence the latent heat (or melting point) of pure MBBA whereas both L_{KN} and T_{KN} of pure EBBA are affected. This behavior is in accord with the fact that pure EBBA is known to possess solid polymorphism whereas MBBA does not.[3] Our measured value of L_{KN} for stable EBBA is 4.1 Kcal/mole - in agreement with that found above from Schröder-van Laar but in disagreement with the value of Barrall, et al. On the other hand our measured value of L_{KN} for pure MBBA (3.3 Kcal/mol) is in excellent agreement with result of Barrall et al but in poor agreement with the Schröder-van Laar calculation.

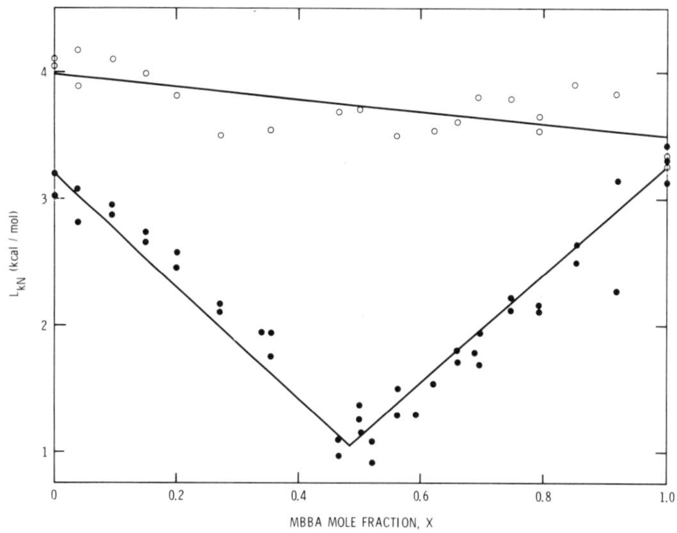

FIG. 2. Crystal-nematic latent heats of transition for stable (o) and metastable (●) MBBA/EBBA phases.

For the metastable solid phase, a minimum in L_{KN} is seen at about $x \approx 0.5$; this behavior is somewhat similar to that of eutectic-forming binaries. However, the minimum value of L_{KN} is about 1 Kcal/mol rather than 0. Although the experimental scatter is large, a straight line fits the concentration-dependence of L_{KN} for the stable phase of the mixtures. The concentration dependence of T_{KN} for the stable phase is not incompatible with that for certain solid solutions,[13] and certainly the linear behavior of L_{KN} is reasonable for such a system.

NEMATIC-ISOTROPIC TRANSITION

As is seen in Fig. 1, the concentration dependence of T_{NI} is rather typical.[1,11,14] Humphries, et al,[14] have described a theory which can give a concentration dependence of T_{NI} which is either linear or has a positively or negatively curved departure from linearity.

As we see in Fig. 1, T_{NI} shows curved rather than linear (dashed line) behavior. The solid line is a best-fit quadratic through the data.

In Fig. 3 is plotted the x-dependence of L_{NI}. A slightly better fit to the data is given by a quadratic (solid) rather than linear (dashed) curve. The values of L_{NI} for pure EBBA and MBBA from this fit are 125 and 74 cal/mol respectively in reasonable agreement with the 148 and 79 cal/mol reported by Barrall, et al.[3] The close approximation to linearity is reasonable for a transition between two miscible systems.

CONCLUSIONS

The fact that the stable high temperature solid phase (α-phase) of MBBA/EBBA mixtures has a minimum melting temperature of about 13°C - about 25° higher than the minimum melting point previously reported for the metastable phase of this system - indicates that this binary may be unsuitable for low temperature applications. The long time (~1 day) required for conversion

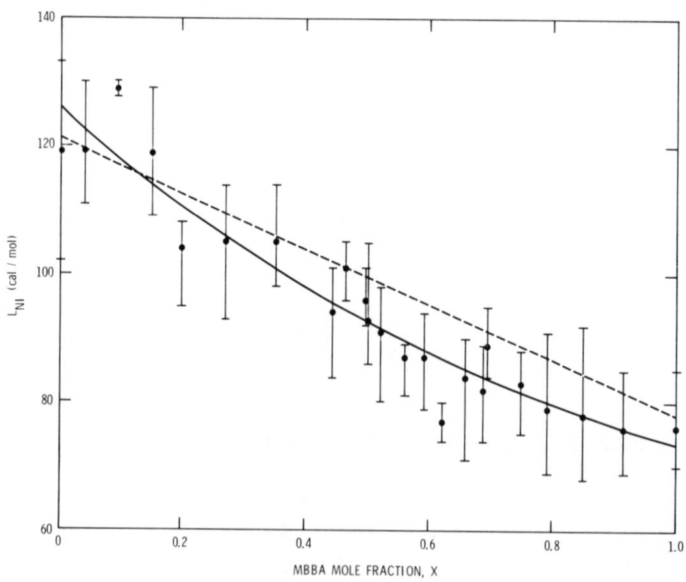

FIG. 3. Nematic-isotropic latent heats of transition for MBBA/EBBA binary mixtures.

in the solid to the stable phase probably accounts for the fact that previous papers report only T_{KN} values characteristic of the metastable phase and not those of the stable phase. The long time constant for solid-solid conversion also allows short-term operation of devices below 13°C. The nematic-isotropic transition behavior for MBBA/EBBA is similar to that reported for other systems.

REFERENCES

1. D. L. Fishel and Y. Y. Hsu, J. Chem. Soc. D 1971 1557.
2. Riedel-deHaën Organic Chemical Catalog, 1970.
3. E. M. Barrall II, K. E. Bredfeldt, and M. J. Vogel, Fourth International Liquid Crystal Conference, Kent State University, Kent, Ohio, 21-25 August 1972.
4. S. Kobayashi, T. Shimojo, K. Kasano, and I. Tsunda, Digest of Technical Papers, 1972 Society for Information Displays International Symposium, 6-8 June 1972, San Francisco; p. 68.
5. H. Kelker and B. Scheurle, Angew. Chem. 81, 903 (1969) [Internat. Ed. 8, 276 (1969)].
6. H. Kelker, B. Scheurle, R. Hartz, and W. Bartsch, Angew. Chem. 82, 984 (1970). [Internat. Ed. 9, 962 (1970)].
7. J. B. Flannery and W. Haas,J.Phys.Chem. 74,3611(1970).
8. G. W. Smith, Z. G. Gardlund, and R. J. Curtis, Mole Cryst. and Liq. Cryst. 19, 327 (1973).
9. J.T.S. Andrews (private communication).
10. See, for example: R. S. Porter, E. M. Barrall II, and J. F. Johnson, Accounts of Chem. Res. 2, 53 (1969); R. D. Ennulat in "Analytical Calorimetry, Vol. I", R. S. Porter and J. F. Johnson, eds., Plenum Press (N.Y.), 1970, p.219 ff.
11. E. C. Hsu and J. F. Johnson, Mole Cryst. and Liq. Cryst. 20, 177 (1973).
12. Values in certain concentration ranges (x~0.4 to 0.5, x~0.7 to 0.95) are particularly rough because of structure in the DSC peaks.
13. J. H. Hildebrand and R. L. Scott, "The Solubility of Nonelectrolytes", (Reinhold, New York, 1950), p.300 ff.
14. R. L. Humphries, P. G. James, and G. R. Luckhurst, Far. Soc. Symposium No. 5, 107 (1971).

SOME MECHANISTIC ASPECTS OF THE CHOLESTERIC LIQUID CRYSTAL INDUCED CIRCULAR DICHROISM (LCICD) PHENOMENON

F. D. Saeva

Xerox Corporation, Rochester Research Center

Webster, New York 14580

Abstract

Mechanistic aspects of the LCICD phenomenon have been investigated as an extension of our initial observation that achiral molecules dissolved in cholesteric liquid crystals display circular dichroism in the region of their electronic transitions.

LCICD has been previously found to be dependent on the chirality of the cholesteric helix, orientations of electric transition dipoles and λ_o of the cholesteric pitch band. More recently the LCICD intensity has been observed to be a function of the pitch of the cholesteric mesophase, temperature and texture. The sign of the LCICD has been found to be dependent on the cholesteric matrix as well as on its chirality. The results of solute concentration studies will also be presented along with a discussion of the origin of LCICD in terms of the following two proposed mechanisms:

1. exposure of the dissolved solute to a helical ordering of liquid crystal molecules, and

2. helical ordering of solute within the cholesteric structure.

INTRODUCTION

The ordering ability of liquid crystals has been utilized to obtain polarization data, i.e., direction of transition moments, for oriented solute molecules in aligned nematic[1] and cholesteric mesophases.[2] The preferred alignment of the solute in liquid crystals can be determined both from nmr[3-5] and electronic absorptions studies.[1,2] It is concluded that solute molecules align in the best packing arrangement from steric considerations independent of the chemical structure of the liquid crystal. Planar molecules will orient their long axis parallel to the long axis of the liquid crystal molecules. These studies, however, provide more information about the solute than the internal molecular structure of the liquid crystal matrix itself.

The purpose of this contribution is to describe some mechanistic aspects of the cholesteric liquid crystal induced circular dichroism (LCICD) phenomenon and indicate the information LCICD provides both about the solute as well as the internal structure of cholesteric mesophases.

Cholesteric Liquid Crystal Induced Circular Dichroism (LCICD)

We have shown recently that achiral molecules dissolved in cholesteric mesophases display intense circular dichroism (CD) in the region of their absorption bands.[6] The existence and sign of this extrinsic CD is dependent on the presence of a helical structure and its chirality,[6] respectively.

Subsequent investigations have shown the LCICD sign to be dependent on the polarization of the electronic transitions within the solutes.[7,8] Figure 1 presents the LCICD and absorption spectrum of pyrene (1) in a cholesteric mesophase composed of a cholesteryl nonanoate (2)-cholesteryl chloride (3) mixture.

The 0-0 band for the ^1L_b, ^1L_a and ^1B_b transitions in pyrene occur at 372, 339 and 277 nm respectively. The 0-0 band of the ^1L_b and ^1B_b transitions, which are known to be transversely polarized (short axis in-plane),[9,10] show positive circular dichroism (i.e., $\varepsilon_L > \varepsilon_R$) in a right-handed helicoidal cholesteric mesophase consisting of 70/30 wt %, 2/3. All the vibrational bands with the ^1L_b transition appear to be of the same polarization in contrast to the ^1B_b transition which appears to be of mixed polarization[11,12] or may contain overlapping transitions, which are of opposite CD sign, as indicated by the lack of match between the absorption and LCICD spectra. The ^1L_a transition, on the other hand, is longitudinally polarized (long axis in-plane) and shows negative circular dichroism.

LCICD can also be used to detect hidden transitions provided the two overlapping transitions are of different polarization. The LCICD of anthracene indicates the presence of the normally unobserved ^1L_b transition. Figure 2 presents the circular dichroism

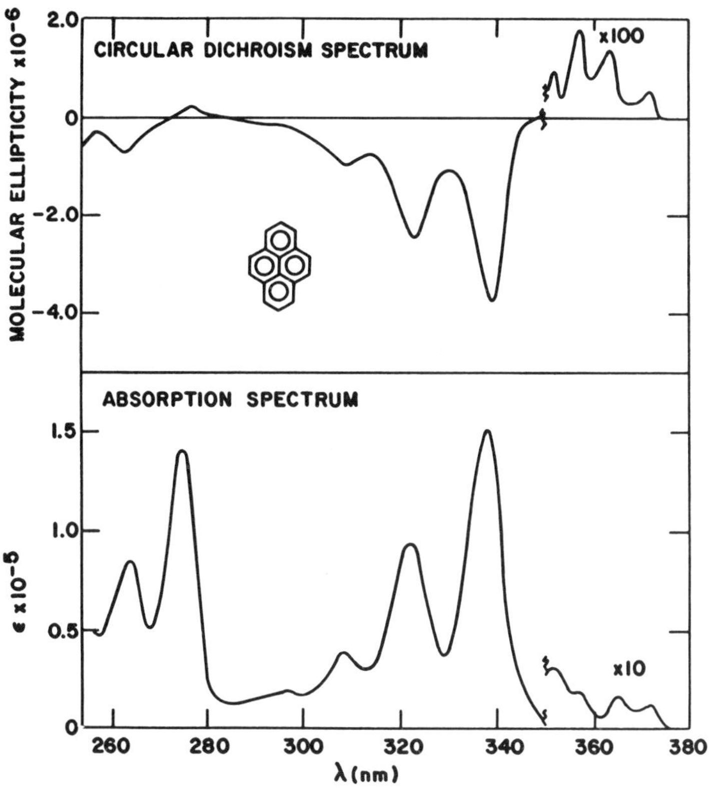

Figure 1. Circular dichroism and absorption spectra for pyrene in 70/30 wt % cholesteryl nonanoate(2)/cholesteryl chloride(3) (λ_o of the cholesteric pitch band = 6300 Å).

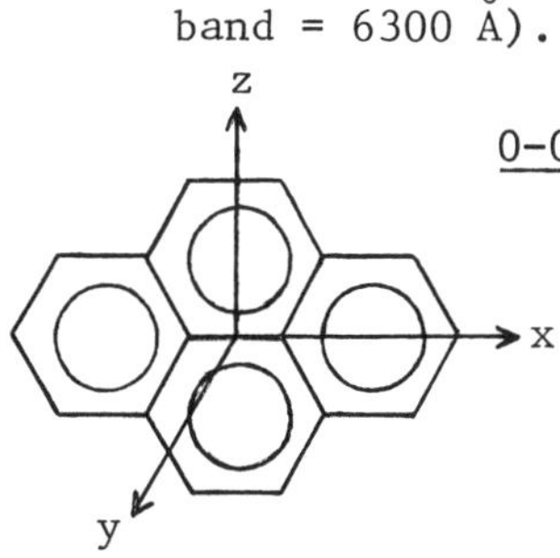

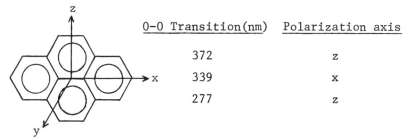

0-0 Transition(nm)	Polarization axis
372	z
339	x
277	z

and absorption spectrum for anthracene in 70/30 wt % 2/3. The ^1L_a transition in anthracene (0-0 at 379 nm) is transversely polarized[13] (short axis in-plane), and gives rise to positive circular dichroism. The relative intensities of the LCICD bands in anthracene (between 300 - 400 nm) do not follow its absorption spectrum. This may be a result of mixed polarization within the ^1L_a transition or an overlapping with the normally unobserved ^1L_b transition which is predicted to be longitudinally (long axis in-plane) polarized[13] and show negative circular dichroism following the experimental observations with pyrene.

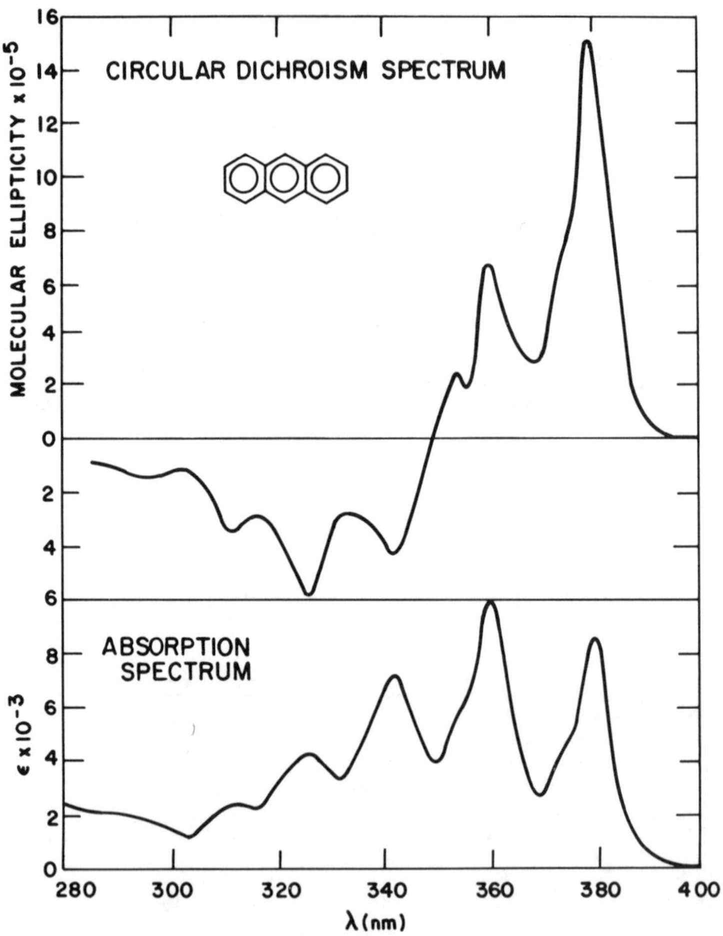

Figure 2. Circular dichroism and absorption spectra for anthracene in 60/40 wt % cholesteryl chloride (3)/ cholesteryl nonanoate (2) (p = 9μ).

Cholesteric mesophases may exist in either the Grandjean (helix axes parallel to the optic axis) or focal conic texture (helix axes perpendicular to the optic axis) or some combination of the two textures depending on the conditions used for sample preparation. Since there is a 90° change in molecular orientation between the two textures, we decided to investigate the intensity of LCICD of pyrene in the 60/40 3/2 as a function of the texture of the cholesteric mesophase. As the Grandjean is transformed into the focal conic texture, by means of an electric field, the LCICD intensity decreased for a longitudinally polarized electronic transition. This decrease in intensity is in line with what one would expect from a decrease in molecular absorptivity as a result of the change in molecular orientation. The sign of LCICD did not change with the texture change. Since the intensity of LCICD was found to be dependent on texture of the cholesteric mesophase we have attempted to work with the Grandjean (planar) texture exclusively.

More recently the LCICD sign has been observed to be dependent on the position of λ_o of the cholesteric pitch band relative to the wavelength of the absorption band.[14] For a right-handed helical cholesteric mesophase composed of cholesteryl chloride-cholesteryl nonanoate mixtures the LCICD sign for a transition moment with a preferred orientation parallel with the long axis of the liquid crystal molecules is summarized in Table 1.

TABLE 1
Sign of LCICD as a Function of λ_o

Helix sense	λ_o/λ_{ab} [a]	LCICD sign [b]
Right-handed	> 1	-
	< 1	+
Left-handed	> 1	+
	< 1	-

a. λ_{ab} = wavelength of absorption; λ_o = reflective wavelength of the cholesteric pitch band.
b. For cholesteric mesophases composed of cholesteryl derivatives only since LCICD sign has been found to be a function of the liquid crystal matrix.

Recent experiments designed to provide further mechanistic information concerning the LCICD phenomenon have led to a study of LCICD intensity as a function of pitch of the cholesteric helical structure. In this investigation the pitch of the cholesteric mesophase was altered by means of a dc electric field while mesophase texture, composition, temperature and solute concentration were held constant.

The LCICD intensity of pyrene, in a 70/30 wt% 2/3 supercooled cholesteric mesophase, was monitored as the λ_0 of the cholesteric pitch band was varied between 5979 and 6378Å. The experimental results are presented in Table 2. The molecular ellipticity, [Θ],* of pyrene was found to be a hyperbolic function of pitch of the cholesteric mesophase over the range of pitches studied and the intensity of the LCICD decreased as the helix pitch decreased. A decrease in pitch results from an increase in the angle between the 'nematic like' layers in the cholesteric mesophase. The dependence of LCICD intensity on pitch is in agreement with previous suggestions[6] and theoretical calculations.[8] The variation of the LCICD intensity with pitch was studied over a limited range of pitches and as the pitch increases to infinity the circular dichroism must decrease and eventually reach zero intensity.[6,8]

The combined influence of temperature and pitch on the LCICD intensity of pyrene in 70/30 wt % 2/3 was investigated and, the LCICD of pyrene found to decrease in non-linear fashion as the temperature was increased from 29 to 55°C. A movement of λ_0 of the cholesteric pitch band to longer wavelength also accompanied the change in temperature. The influence of temperature on LCICD intensity dominates over the pitch dependence. This conclusion is based on the previous discussion of the pitch dependence where the LCICD intensity was found to increase with increasing pitch. The observed decrease of LCICD is presumably a result of weaker electronic and/or magnetic interaction between the solute and liquid crystal molecules at higher temperatures due to a random ization of the solute within the cholesteric mesophase.

The intensity of solute LCICD was found to be a linear function of solute concentration, between 8.38×10^{-4} to 9.6×10^{-6} gm/cc for pyrene[14] in 70/30 wt % 2/3, and sample thickness. The solute concentration independence of the molecular ellipticity suggests the unimportance of solute-solute interactions that do not involve a coupling through the matrix.

* Molecular Ellipticity $[\Theta] = \frac{\Theta \, MW}{10 \, \ell \, c}$ (degree -cm^2 per decimole)

Θ = observed ellipticity in degrees

MW = molecular weight

ℓ = pathlength in cm

c = concentration in gm/cc

TABLE 2

LIQUID CRYSTAL INDUCED CIRCULAR DICHROISM (LCICD) OF PYRENE AS A FUNCTION OF PITCH OF THE CHOLESTERIC MESOPHASE

$[\Theta]^a \times 10^{-5}$	$\lambda_o{}^b$ (A°)	Applied Voltage[c]
2.93	6378	0
2.90	6360	10
2.88	6339	20
2.82	6279	30
2.66	6168	40
2.07	5979	50

a. molecular ellipticity in units of degree $-cm^2$ per decimole
b. pitch (p) of the helicoidal cholesteric helix can be obtained by using the following expression: $\lambda_o = np$, where n is the refractive index and λ_o the reflective wavelength of the cholesteric pitch band.
c. voltage applied to a 1 mil film of 2.85×10^{-3}M pyrene in 70/30 wt % 2/3.

Spectroscopic and Conformational Effects in Some Substituted Benzenes

The LCICD spectra of benzene and a series of six mono-substituted benzenes were measured in a single, right-handed helical, cholesteric mesophase along with their electronic spectra with the object of comparing LCICD with magnetic circular dichroism studies on the same compounds.

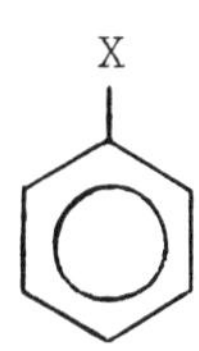

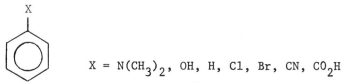

X = $N(CH_3)_2$, OH, H, Cl, Br, CN, CO_2H

LCICD, within the 0-0 band of the ^1L_b transition, for the mono-substituted benzenes was found to be of negative sign when the substituent is ortho, para directing, except for dimethylamino, and of positive sign when the substituent is meta directing. The correlation between LCICD sign and phenyl substituent electronic behavior is similar to that found in MCD studies,[15] where dimethylamino was also found to behave anomalously.

The sign of the LCICD of all the vibrational bands within the ^1L_b transition for hydroxyl and methoxy substituted compounds were of negative sign. As the ortho, para directing ability of the substituent decreases both negative (0-0) and positive CD bands are observed indicating mixed polarization,[11,12] or a result of overlapping with higher electronic states. Strongly meta directing substituents on benzene such as cyano and carboxyl, on the other hand, produce positive CD throughout the entire ^1L_b transition.

The change in LCICD sign as a function of aromatic ring substituent is suggested to be a result of a spectroscopic variation and not a conformational effect of the solute in the liquid crystal. We have assumed that all the mono-substituted benzenes have the same preferred conformation, i.e., align their long molecular axis with the long axis of the liquid crystal molecules.

The LCICD intensity in the mono-substituted benzenes appears to be proportional to the molecule absorptivity, i.e., high values of $[\theta]$ are associated with high ϵ values. Substituents such as chloro, bromo and cyano on benzene give LCICD bands within the ^1L_b transition of lower CD intensity than hydroxyl, methoxy and carboxyl.

Subsequent LCICD investigations of a variety ortho, meta and para-disubstituted benzenes, in a cholesteric mesophase of a known helix sense (right-handed) showed the interesting para effect shown in Figure 3. Figure 3 presents the LCICD and absorption spectrum for ortho, meta and para substituted methoxy chlorobenzene. Within the ^1L_b electronic transition the para-isomer exhibits oppositely signed CD from its ortho and meta isomers. Phenyl ring substituent combinations of chloro-methoxy, chloromethyl, methoxy-methoxy,[16] methoxy-methyl, methyl-methyl,[16] and chloro-chloro[16] show a correlation between the sign of the CD of the ^1L_b transition and the position of substitution. The ortho and meta isomers show negative CD bands while the para isomer shows positively signed bands. However, when the substituent combinations are hydroxy-methyl, and chloro-cyano this correlation breaks down. Since MCD measurements on the meta and para isomers of xylene and dichlorobenzene[17] do not show a difference in CD sign between these isomers we believe the para-effect is the result of a conformational variation between the para and ortho-meta isomers and not a spectroscopic one. From the previously proposed quandrant rule[7] the orientation of the ^1L_b transition moment in the para-isomers, that exhibit the above effect, is ex-

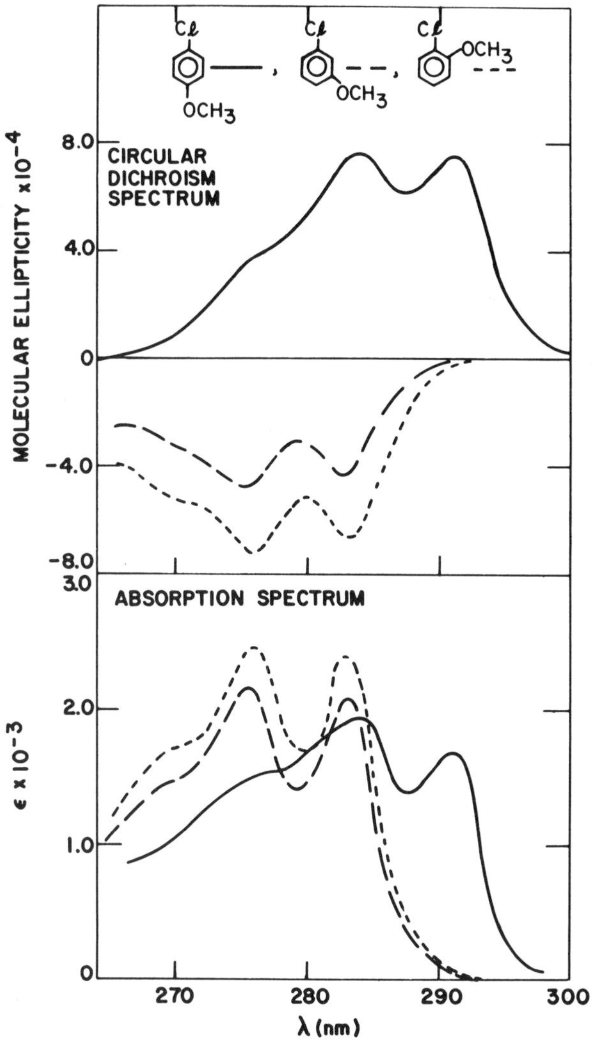

Figure 3. Circular dichroism and absorption spectra for o, m, p-methoxy chlorobenzene in 70/30 wt % cholesteryl nonanoate (2)/cholesteryl chloride (3). Pitch band λ_0 for the o, m, p-isomers is 6360, 6370 and 6430 Å, respectively.

pected to lie perpendicular to the long molecular axis of the cholesteryl molecules, while the average direction of the ^1L_b transition in the ortho and meta isomers, on the other hand, is expected to lie parallel with long molecular axis of solvent molecules.

In general, large rotational strengths are again associated with the intensity of the transition and unusually high intensity LCICD is associated with presence of methoxy groups.

CONCLUSION

Experimentally the sign of the LCICD for a solute is dependent on the following:

1. the chirality of the cholesteric helix
2. the polarization of absorption bands
3. preferred conformation of the solute with respect to the long molecular axis of the liquid crystal molecules
4. the position of λ_{ab} absorption relative to λ_o of the cholesteric pitch band.

The LCICD intensity, on the other hand, is a function of pitch of the cholesteric helix and temperature.

The concentration independence of the molecular ellipticity of the solute over the concentration range studied indicates that solute-solute interactions are not essential for LCICD. The origin of LCICD for solutes in a cholesteric mesophase at low concentrations is attributed to electronic and/or magnetic interactions with the cholesteric matrix molecules and not from simply the helical ordering of the solute although the sign of the LCICD is dependent on the anisotropic ordering of the solute.

The LCICD behavior of solutes in thermotropic cholesteric mesophases at low concentration agree with the results of theoretical studies of CD behavior of cholesteric mesophases, within the region of the absorption of the liquid crystal molecules themselves.[18] The theoretical study extends the theory of electromagnetic radiation in non-absorbing cholesteric liquid crystals to the absorbing case by adding a frequency dependent complex distribution to the spiraling dielectric tensor of the liquid crystal. It is believed that LCICD arises from a coupling of solute and solvent chromophores, which is dependent on electronic and magnetic properties of each chromophore and their relative orientations. It is expected that randomly oriented solute molecules in cholesteric mesophase would exhibit LCICD.

LCICD studies on benzene and some of its substituted derivatives show both spectroscopic and conformational effects as a function of aromatic ring substitution. The sign of the CD within the ^1L_b electronic transition in monosubstituted benzenes shows variations which depend on the electronic behavior (<u>ortho</u>, <u>para</u> or <u>meta</u> directing properties) of the substituent, similar to that found in MCD studies. We conclude the change in LCICD sign as a function of aromatic ring substituent is a result of a spectroscopic change within the benzene electronic transitions and not a result of a conformational variation, as a function of substituent, in the liquid crystal.

The LCICD behavior of some _ortho_, _meta_ and _para_ disubstituted benzene derivatives show an interesting trend where the _para_-isomer possesses oppositely signed CD bands from the less symmetrical _ortho_ and _meta_ derivatives. We believe the latter phenomena to be as a result of conformational variation between the isomers in a cholesteric mesophase of a single chirality.

ACKNOWLEDGEMENT

Stimulating discussions with Drs. W.H.H. Gunther, G. Johnson, H. Gibson, H.A. Scheraga, J.E. Kuder and technical assistance by P.E. Sharpe and G.R. Olin are gratefully acknowledged.

REFERENCES

1. G.P. Ceasar and H.B. Gray, _J. Am. Chem. Soc._, _91_, 191 (1969).
2. E. Sackmann, _ibid_, _90_, 3569 (1968).
3. A.D. Buckingham and K.A. McLauchlan, '_Progress in Nuclear Magnetic Resonance Spectroscopy_', Vol. 2, Pergamon Press, Oxford, 1967.
4. A. Saupe, _Proc. Intern. Conf. Liq. Cryst._, _207_, (1965).
5. G. Englert and A. Saupe, _ibid_, 183 (1965).
6. F.D. Saeva and J.J. Wysocki, _J. Am. Chem. Soc._, _93_, 5928 (1971).
7. F.D. Saeva, _ibid_, _94_, 5135 (1972).
8. E. Sackmann and J. Voss, _Chem. Phys. Lett._, _14_, 528 (1972).
9. R.M. Hochstrasser, _J. Chem. Phys._, _33_, 459 (1960).
10. V.H. Zimmermann and N. Joop, _Zeitschrift für Electrochemie_, 138 (1960).
11. O.E. Weigang, Jr., _J. Chem. Phys._, _43_, 3609 (1965).
12. J. Horwitz, E.H. Strickland and C. Billups, _J. Am. Chem. Soc._, _91_, 184 (1969).

13. J.B. Birks, '*Photophysics of Aromatic Molecules*,' Wiley-Interscience, London 1969, p. 70, 526.

14. F.D. Saeva, P.E. Sharpe and G.R. Olin, submitted for publication.

15. J.G. Foss and M.E. McCarville, *J. Am. Chem. Soc.*, *89*, 30 (1973).

16. Only the meta and para isomers were studied.

17. D.J. Shieh, S.H. Lin and H. Eyring, *J. Phys. Chem.*, *77*, No. 8, 1031 (1973).

18. G. Holzworth and N.A. Holzworth, *J. Opt. Soc. Am.*, *63*, 324 (1973).

Effect of Cholesteryl Alkanoate Structure on the Pitch of the Cholesteric Mesophase

Harry W. Gibson, John M. Pochan and DarLyn Hinman

Rochester Research Center

Xerox Corporation   Webster, New York 14580

Abstract

The pitches of a series of new cholesteryl alkanoates derived from branched alkanoic acids, some of them optically active, have been determined from extrapolation of pitches of binary mixtures of cholesteryl oleyl carbonate (COC). The pitches of 2-branched and linear cholesteryl alkanoates show a maximum as a function of the heats of fusion (ΔH_f) of the pure crystalline materials. The pitches of the 3- and 4- branched alkanoates increase with ΔH_f. In linear alkanoates the pitch is directly proportional to the interfacial energy between the cholesteric mesophase and the crystal. These results relate crystal and mesophase structure.

Interpreted in terms of the volume requirements of the alkanoate chain, the dominant effect of branching with a single carbon is to make the alkanoate moiety more rigid through restriction of rotation about C-C single bonds. This decreases the volume requirement, allowing better crystal packing and lower angular displacements between adjacent molecules in the cholesteric helix, hence large pitches. A steric bulk increase is seen via decreased pitch when branches are lengthened or occur near the end of the alkanoate chain. These factors effect crystal and mesophase stability similarly up to a certain point, where chain folding may occur in the crystal, but not the mesophase. The structural generalizations are, however, subject to dramatic changes by alteration of configuration in chiral alkanoates. The dramatic effects of conformational and configurational factors on pitch are indicative of the potential of these systems for detailed study of these factors once the nature of the cholesteric mesophase is more fully understood.

INTRODUCTION

For some time now we have been working with cholesteryl alkanoates. Our initial studies involved nucleation of crystals from supercooled liquid crystalline phases.[1,2] To obtain nearly compensated, i.e., high pitch, materials, mixtures of right and left-handed liquid crystals are used. These mixtures can be supercooled for longer periods of time than pure high temperature systems but not indefinitely. From our work, parameters derived from nucleation rate data correlated with chain length, but overall nucleation rates did not. The second stage of our work, therefore, namely assessing structural affects on transition temperature and behavior in binary mixtures,involved the synthesis and investigation of a series of branched cholesteryl alkanoates.[3] This work indicated that the transition temperatures were influenced by steric effects near the carbonyl group. This prompted us to investigate the effect of placing a chiral (optically active) center at various positions of the alkanoate group. The synthesis of the necessary precursors and the resultant esters will be separately reported;[4] the behavior of binary mixtures of these with cholesteryl nonanoate will be reported shortly.

Right-handed liquid crystals, i.e., those transmitting right-handed circularly polarized light at λ_o, are not readily available. Until recently,[5,6] the most well known were the cholesteryl halides. The chloride is the most stable, but it undergoes thermal dehydrochlorination to cholestadiene.[7] This is, of course, not desirable because it effects optical and electrical properties. Other such materials were, therefore, sought. The availability of a series of cholesteryl alkanoates, some derived from chiral acids, afforded the opportunity to study the consequences of structural and chiral alterations on the pitch of these liquid crystals in the belief that this information would be helpful in understanding the bonding forces of the cholesteric mesophase. This report is an account of that investigation.

EXPERIMENTAL

The alkanoates used in this study were those previously synthesized and purified.[3,4] Heats of fusion were determined as previously described[2,3] on a Perkin Elmer DSC-1B. The pitch measurements were performed by preparing binary mixtures of each of the alkanoates with cholesteryl oleyl carbonate (COC) which had a cholesteric mesophase range of 16° - 38°C and a λ_o of 338 nm. The reflection bands were determined on a Cary 15 spectrometer. All measurements were at ambient temperature. (Cholesteryl pivalate[3] and 1-adamantanecarboxylate[3] were insoluble in COC.) The results for several concentrations were then plotted in the form $1/\lambda$ versus weight percent alkanoate, a technique previously utilized for

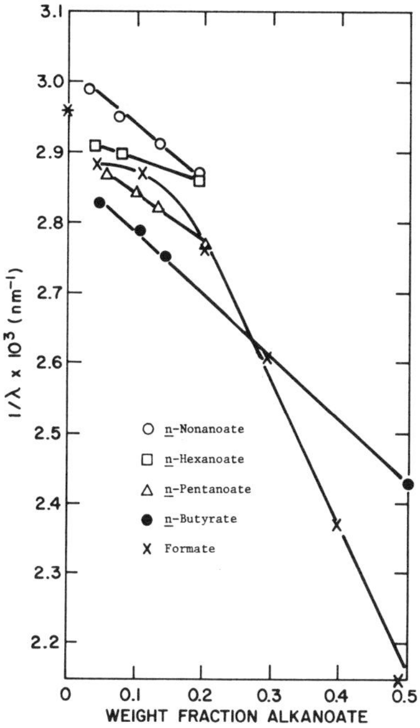

Figure 1. Inverse of Reflective Wavelength versus Weight Fraction Cholesteryl Alkanoate in Cholesteryl Oleyl Carbonate

determination of pitches of materials which cannot be determined either due to inaccessible wavelengths or inaccessible mesomorphic ranges.[8] This technique assumes that the inverse pitches are additive as shown in Equation 1.

$$1/\lambda = \Sigma(w_i/\lambda_o^i) \qquad (1).$$

> w_i = weight fraction of component i, λ_o^i = reflective wavelength of component i, and equals $2\eta P$, η being the refractive index and P the pitch

This in turn essentially means that the two components do not influence each other but act merely as "inert" diluents.

RESULTS

The data obtained in the present work are presented graphically in Figures 1-3. The plots are linear within experimental error if the value for pure COC is not included. Examination of this point reveals that close to 0% of alkanoate the curve is discontinuous. The curvature and discontinuity near 100% COC suggest that use of Eq. 1 to estimate pitches is in general of questionable accuracy since curvature and/or discontinuities probably occur(s) near pure second component. Such curvature has been previously reported.[5,8] Nonetheless, the method does provide a relative estimate if all pitches are determined in a given matrix as we have done. The extrapolated λo's of COC and the alkanoates determined from the straight lines are recorded in Table 1 along with other pertinent data.

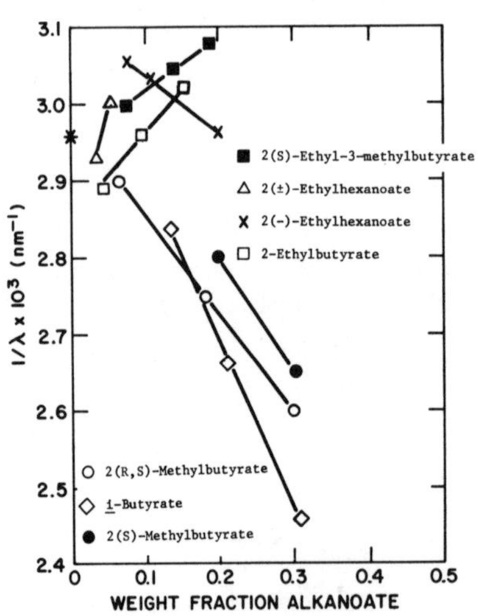

Fig. 2

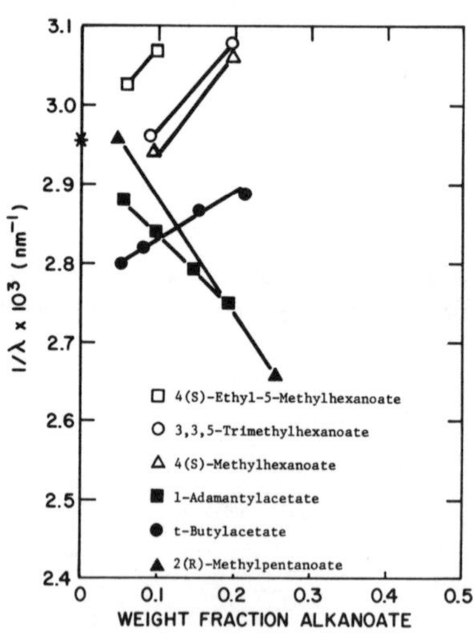

Fig. 3

Inverse of Reflective Wavelength versus Weight Fraction Cholesteryl Alkanoate in Cholesteryl Oleyl Carbonate

TABLE 1

Pitch Determination of Cholesteryl Alkanoates from Binary Mixtures with COC[a]

Alkanoate	SLOPE x 10^4 (nm)	r[b]	n[c]	Fig.	λ_o^{COC} (nm)	λ_o^{ALK} (nm)	ΔH_f (Kcal/mole)	Code Figs. 4-6
Formate	-20.9	0.9973	4[d]	1	314	910	5.19	1
n-Butyrate	- 8.94	0.9985	5	1	359	529	5.07	2
n-Pentanoate	- 6.92	0.9998	4	1	344	451	4.50	3
n-Hexanoate	- 3.30	0.9961	3	1	342	385	6.45	4
n-Nonanoate	- 7.60	0.9950	4	1	332	444	5.57	5
n-Myristate	-	-	-	-	-	201[e]	11.2[f]	6
i-Butyrate	-22.1	0.9994	3	2	319	1081	6.36	7
2(S)-Methylbutyrate	-14.7	-	2	2	324	617	5.45	8
2(R,S)-Methylbutyrate	-12.7	1.000	3	2	335	585	4.98	9
2(R)-Methylpentanoate	-14.6	-	2	3	330	637	7.17	10
2-Ethylbutyrate	11.7	0.9973	3	2	352	249	4.26	11
2(S)-Ethyl-3-Methylbutyrate	2.50	0.9950	3	2	335	309	5.79	12
2(-)-Ethylhexanoate	- 7.17	0.9992	3	2	323	420	3.03	13
2(±)-Ethylhexanoate	60	-	2	2	366	115	3.13	14
4(S)-Ethyl-5-methylhexanoate	9.76	-	2	3	336	252	3.86	15
4(S)-Methylhexanoate	11.6	-	2	3	354	250	5.87	16
t-Butylacetate	5.73	0.9889	4	3	361	299	6.73	17
1-Adamantylacetate	- 9.27	0.9994	4	3	342	500	7.62	18
3,3,5-Trimethylhexanoate	11.2	-	2	3	350	251	5.15	19

[a] By least squares linear fit of $1/\lambda_o$ versus weight fraction alkanoate excluding point for pure COC; λ_o's calculated from extrapolated $1/\lambda$ at 0 and 1.00 weight fraction alkanoate. ΔH_f from Ref. 3 or present work.

[b] r = correlation coefficient. [c] n = number of points. [d] Using data for 0.198 to 0.493 weight fraction formate.

[e] From Ref. 10. [f] From Ref. 11.

The extrapolated λ_o of COC and the slope of the $1/\lambda$ versus composition plot are related, the former generally increasing with the latter as shown in Figure 4. The extrapolated λ_o of COC, however, generally decreases as the λ_o of the alkanoate increases as shown in Figure 5. These relationships seem to hold regardless of the structure or chirality of the alkanoate and describe the interactive behavior of the two components. It can be seen that only at zero slope (Fig. 4) or λ_o (alkanoate) = λ_o (COC) (Fig. 5) is the extrapolated value for COC identical to that observed; in other words the more different the two pitches the more non-linear the behavior near 100% COC.

The pitch is also a function of ΔH_f, the heat of fusion, as shown in Fig. 6. The data breaks down into three categories; 1)alkanoates branched at the 2-position (adjacent to the carbonyl), 2)alkanoates branched at position other than the 2-position and 3) straight chain alkanoates. λ_o in the 3- and 4- branched series increases continuously with ΔH_f, there are maxima in λ_o in the straight chain compounds at ∿ 5.2 kcal/mole and 6.5 kcal/mole in the 2-branched series.

In line with the above results relating crystal and mesophase structures is the direct correlation of pitch of the linear alkanoates and the average molecular interfacial energy (σ) (Fig. 7). These interfacial energies were obtained independently.[1,2,9] Pitch data from other investigators[10] have also been included.

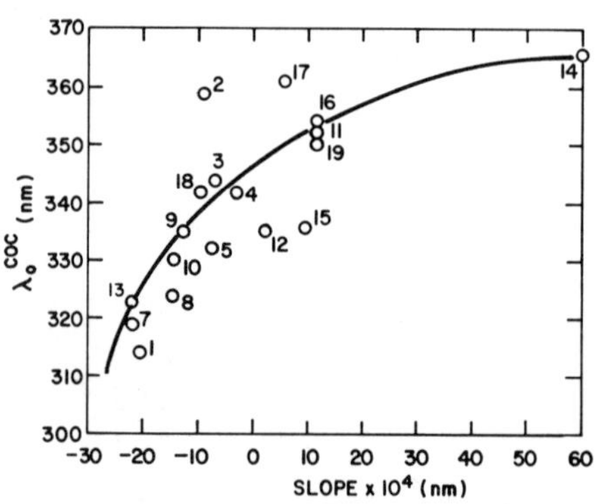

Fig. 4. Extrapolated Reflective Wavelength for Cholesteryl Oleyl Carbonate versus Slopes of Figures 1-3. See Table 1 for code.

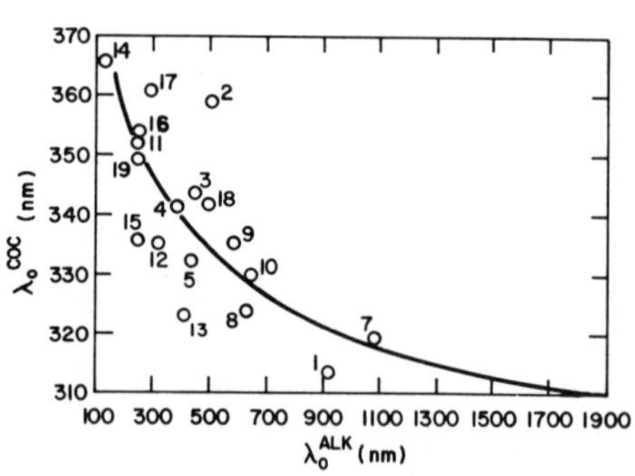

Fig. 5. Extrapolated Relfective Wavelength of Cholesteryl Oleyl Carbonate versus Extrapolated Reflective Wavelength of Cholesteryl Alkanoates. See Table 1 for code.

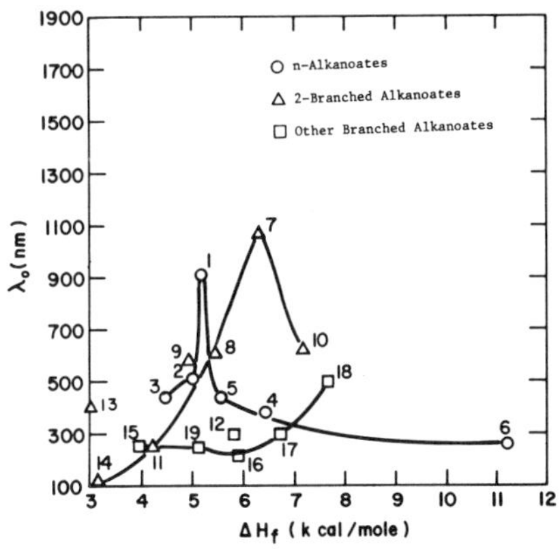

Fig. 6. Extrapolated Reflective Wavelength Versus Heat of Fusion for Cholesteryl Alkanoates. See Table 1 for code.

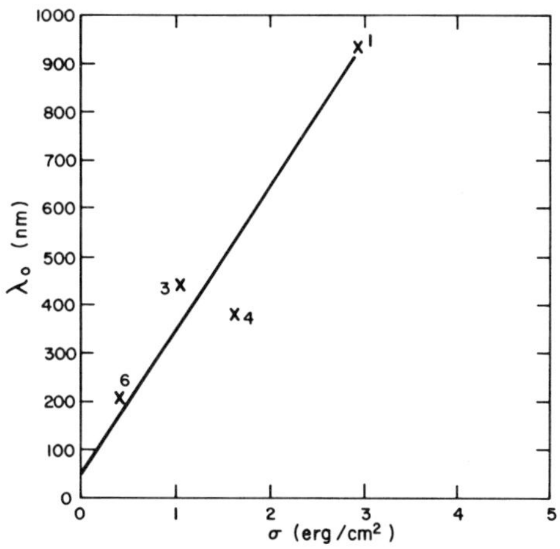

Fig. 7. Reflective Wavelength versus Interfacial (Cholesteric-Solid) Energy for Cholesteryl n-Alkanoates. See Table 1 for code.

DISCUSSION

Non-linear Behavior of Pitch Near Extremes of Composition

The fact that some of the plots of $1/\lambda$ versus composition of Figs. 1-3 are non-linear near 100% COC is not too surprising if one considers that steric effects play a role in determining how the molecules interact with one another in the cholesteric mesophase of the pure alkanoates and that this should also be true in mixtures. If it is assumed that intermolecular distance along the helical axis in the mesophase is constant and only the dihedral angle between adjacent molecules varies, then it is easy to see why materials with non-identical pitches interact in a non-averaging manner. They have different steric packing requirements, hence different angular constraints. Interaction of a cholesteryl alkanoate with itself (A-A) will be different from the interaction with another cholesteryl ester (A-B) and the A-B interaction can differ from the average of A-A and B-B. $1/\lambda$ is a linear function of composition at intermediate compositions (5-95% except for formate) due to statistical changes in the number of A-A, A-B and B-B interactions. This treatment is oversimplified in that it does not consider ternary interactions such as ABA or longer range influences.

The results of Figs. 4 and 5 show that the extrapolated λo for COC is high if λo of the alkanoate is less than that of pure COC, and low if it is greater than λo COC, i.e., the extrapolated λo COC is correct only if the slope of Fig. 4 is about zero. This is interpreted to mean that for alkanoates with large pitches, i.e., small angular displacements in the helix, interaction with COC (A-B) gives an angle larger than in pure COC, causing the extrapolated λo(COC) to be lower than observed for pure COC. Similarly in the case where the alkanoate pitch is less than that of COC the extrapolated λo(COC) is high. These results are difficult to rationalize in further detail since the steric and conformational requirements of the alkanoates and COC are not well enough understood (see below).

These results are generally supportive of the conclusions drawn in our previous work, namely a) that steric effects in the neighborhood of the carbonyl moiety play a major role in the cohesive interaction of the cholesteric and b) that because of this, steric alterations in this region can greatly alter the liquid crystalline properties.[3]

Relationship of Pitch and Heat of Fusion

In our previous work[3] we showed that the heat of fusion (ΔH_f) of the cholesteryl alkanoates was directly related to their structure; ΔH_f increases as branching is moved closer to the carbonyl and as the extent of branching near the carbonyl increases. We interpreted these findings in terms of decreased conformational freedom about C-C single bonds as branching increased, leading to a more organized crystal structure. These same two structural variations led to decreased thermal stability of the cholesteric relative to the smectic mseophase and this was interpreted as being due to steric alteration of the carbonyl-carbonyl intermolecular cohesive forces.

The results of Fig. 6 show that the most striking dependence of pitch on ΔH_f is due to branching at the 2-position. This series shows the greatest change in λo of the present group of compounds. The 3-branched and 4-branched alkanoates fall on another line. This, we believe, indicates that the cholesteric behavior as manifested in λo is altered to about the same extent by branching at the 3 and 4-positions of the alkanoate. Our interpretation is that branching at the 2-position has a greater influence on the stacking ability in the cholesteric mesophase than does branching further out on the chain. Conversely, in the crystal, branching further out on the chain has a larger stabilizing effect as manifested by the larger ΔH_f's observed in the 3 and 4-branched series relative to the 2-branched esters. The similarity of the shapes of the curves suggests that the same mechanisms are operative in two branched series but that in the 3- and 4-branched esters these are damped relative to the 2-series. The 2(-)-ethylhexanoate is somewhat anomalous. Again these results corroborate our earlier work[3] and suggest that the carbonyl group plays a major role in the cholesteric mesophase, more so than in the crystalline phase since branches at the 2-position effect the cholesteric phase most strongly while branches at the 3- and 4-positions effect the crystal stability more strongly.

If it is assumed that for these compounds the molecules are parallel in the crystal, it follows that ΔH_f is a measure of how much resistance the alkanoate chains offer to stacking one over the other. For high ΔH_f the molecules offer little resistance to such a constraint in the crystal: likewise in the cholesteric mesophase they are able to adopt a helical structure in which the angular displacement from one molecule to the next is small, giving a large pitch. Thus, λo should increase with ΔH_f.

This is observed for the 3- and 4-branched alkanoates and up to a point for the 2-branched and unbranched esters. The decrease in λo above a certain ΔH_f in the 2-branched and straight chain alkanoates is related to the length of the chains. This is apparently due to an increase in the steric requirements of the longer alkanoates which increases the angular displacement between adjacent molecules in the helix and decreases the pitch.[8,10] In the crystal apparently the organization of the longer chains increases the order and the stability perhaps by chain folding. This would mean that the chains in the longer alkanoates are extended in the more mobile cholesteric phase, but are folded in the crystal. This may be why addition of shorter alkanoates to nonanoate causes little disruption of the crystal, but in the reverse case substantial disruption occurs as manifest in lower ΔH_f and melting point.[3] A discontinuity in ΔH_f as a function of chain length occurs, in the straight chain alkanoates.[11]

Pitch-Interfacial Energy Relationship

If it is assumed that the molecules lie parallel to one another in the crystal, the smaller the interfacial energy difference between the cholesteric mesophase and the solid, the smaller the angular displacement between molecules in the helix and the larger the helix. This is the opposite of the results for the unbranched alkanoates shown in Fig. 7, but it is what would be expected for the branched alkanoates. As noted above the relationship between λo and ΔH_f also appears to be anomalous, perhaps due to chain folding in the crystal and chain extension in the cholesteric phase of the longer alkanoates.

The value of σ used is a mathematical average of lateral free molecular surface energy and end-to-end interaction energy. In the cholesteric mesophase, one would expect lateral energy to dominate the helical structure. This is apparent from Fig. 7. Increasing the chain length linearly affects the pitch of the pure alkanoates; the longer the chain, the smaller the pitch. If end-to-end interactions dominated; it would be expected that the pitches of the materials would be independent of chain length.

We speculate on the basis of the correlation of Fig. 7 that this may be a way of obtaining molecular surface free energies without doing the tedious experiments involved in their determination. By measuring the pitches of a series of materials, and the surface free energies of two or three of them, the surface energies of the others might then be estimated.

Structural Effects-Conformational and Configurational

Our initial aim in this work was to try to correlate pitch with variations in alkanoate structure. A priori it was anticipated that two interrelated effects would manifest themselves: 1) conformational, i.e., stereochemical changes based on rotations about C-C single bonds and 2) steric bulk effects. The conformational effect actually is comprised of two contributors; 1a) an increase in the barrier to rotation about the two C-C single bonds adjacent to the branch and 1b) the emergence of a preferred conformer, i.e., one of the three possible non-eclipsed rotamers will be of lowest energy. By steric bulk effect we mean an increase in the total volume occupied by the alkanoate chain as it rotates and vibrates.

The conformational effect of branching would serve to make the alkanoate chain less flexible than its unbranched homologs. This increased rigidity would lessen the total volume occupied by the branched chain during its thermal motions (rotation and vibration). The decreased volume requirement due to branching thus is analogous to that achieved by a decrease in temperature. The anticipated result is also the same- an increase in the stability of the crystal as measured by ΔH_f and an increase in the pitch due to a decrease in the angle between adjacent molecules in the cholesteric helix.

Conversely, however, the stabilization of a conformation not conducive to intermolecular bonding in the crystal and mesophase could also result. This possibility is closely related to the steric bulk effect. In the case of a preferred conformation, a projecting alkyl group could interfere with bonding.

The interrelationship of these two effects depends on the number, position, length of the branches and the chirality of the alkanoate moiety. For example, introduction of a branch next to the carbonyl group is expected to decrease the volume occupied by the hexanoate. A branch on the 5-carbon will lessen the conformational mobility, but because it is near the end of the chain this effect is less and the increase in steric bulk may have the overall effect of increasing the total volume. Increasing branch lengths will increase conformational flexibility, hence occupied volume. Increasing the number of branches will decrease the flexibility but will also tend to add steric bulk. Changes in configuration in chiral alkanoates will, of course, alter the spatial relationships of the groups in the alkanoate portion and with that their spatial relationship to the cholesteryl moiety.

The pitch of the cholesteric mesophase and ΔH_f of the crystal phase should be directly related to the volume requirements of the alkanoate. A large volume requirement should lead to a large angular displacement between adjacent molecules in the helix, hence lower pitch, and lower stability in the crystal, hence lower ΔH_f.

The results in Table 1 indicate that for the compounds included in this study the major effect of branching on pitch is that arising from an increase in conformational rigidity. An increase in pitch occurs. A decrease in pitch would occur if the major effect of branching were an increase in total occupied volume due to the added steric bulk. The introduction of branching next to the carbonyl group, as exemplified by comparison of isobutyrate to _n_-butyrate, causes a dramatic increase in λo. Likewise, both 2(S)-methylbutyrate and 2(R,S)-methylbutyrate have larger λo's than _n_-butyrate or _n_-pentanoate and the trend continues with 2(R)-methylpentanoate relative to _n_-pentanoate or _n_-hexanoate. As argued above, branching further out on the alkanoate chain seems to influence the pitch more from a steric bulk increase, at least as deduced from comparison of 4(S)-methylhexanoate to _n_-hexanoate and 2(R)-methylpentanoate.

Lengthening a branch, as expected on the grounds of both increased flexibility and steric bulk, does cause a decrease in λo as can be seen by comparing the series isobutyrate, 2(R or R,S)-methylbutyrate, 2-ethylbutyrate. Increasing the number of branches decreases λo as indicated by the results for 2(S)-methyl-3-ethylbutyrate vs. 2(S or R,S)-methylbutyrate and 3,3,5-trimethylhexanoate vs. t-butylacetate. This, of course, is analogous to the decrease in pitch with increasing length of the straight chain alkanoates.

The relative λo's for the t-butylacetate (tBA), 1-adamantylacetate (AA) and 3,3,5-trimethylhexanoate (TMH), all having two branches at the 3-carbon of the alkanoate, provide some more subtle information. The relative λo's of tBA and TMH are as expected on the basis of increased chain length and increased branching in TMH relative to tBA. AA, on the other hand, has a larger λo than both the others even though it has longer branches. Its branches, however, are rigid due to their being part of a tricyclic ring system. Therefore, it would be more rigid than TMH. Nonetheless, it is difficult to understand why λo for AA is so much greater than tBA in view of the greater bulk of AA. The answer may lie in the fact that in tBA the methyl groups are free to rotate while in AA there is no rotation possible in the adamantyl unit. It is known that eclipsing of the carbonyl bond (C = O) with alkyl groups is favored over eclipsing with hydrogen.[12] The adamantyl group may present less steric hindrance to this eclipsing than the t-butyl group and thus have less contribution from other rotamers.

It is interesting to note that these same principles apply to the steryl chlorides - cholesteryl (1), compesteryl(2) and sitosteryl(3)

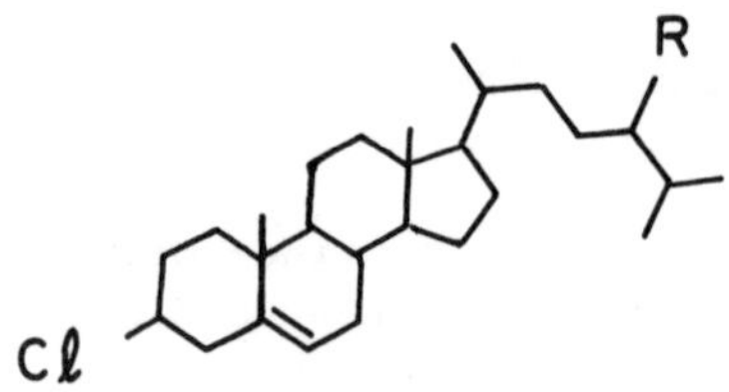

1, R = H
2, R = CH_3
3, R = C_2H_5

The λo's are 488, 256 and 238nm, respectively.[13] Introduction of the methyl group near the end of the 17β side chain leads to a decreased pitch because of the increase in total volume occupied by the side chain. Similarly extension of the side chain to two carbons causes a further decrease in pitch. The pitch of steroidal alkanoates, then, depends upon the total volume occupied by the alkyl chains on each end of the rigid tetracyclic core. These volumes are apparently subject to the same structural effects.

It is also worth mentioning that while these arguments have been made in terms of the angular displacement between molecules in the helix, the possibility of changes in the intermolecular distance changes cannot be overlooked even though such changes seem to be relatively minor.[13] Whether the pitch changes are due solely to changes in intermolecular distance or changes in angular displacement or a combination of these the arguments presented here still hold.

The effect of changes in chirality of the alkanoate portion can be very dramatic as is the case with the 2-ethylhexanoates. The ester from the levorotatory (-) acid has λo = 420nm, while that from the racemic (±) acid has λo = 115nm. If the λo of the ester from the racemic acid obeys Eq. 1, then the ester from the dextrarotary acid has λo = 60.2nm. Similarly, assuming λo = 2ηP, the 2(S)-methylpentanoate reportedly has λo = 268nm, the 2(R)-methylpentanoate λo = 578nm and the ester from the d,ℓ-acid λo = 455nm.[14] We obtain λo = 637nm for the ester from 2(R)-methylpentanoic acid by extrapolation. We did not examine the esters from the R or R,S forms. The 2(S)methylbutyrate has λo = 617nm, while the ester from the racemic acid has λo = 585nm; based on Eq. 1 2(R)-methylbutyrate has λo = 555nm. From these results it is obvious that large variations in pitch can result from changes in configuration at the carbon adjacent to the carbonyl moiety. These results prescribe caution in interpreting structural effects in chiral molecules, especially in comparison to achiral analogs.

It would be advantageous to be able to correlate configuration and pitch, i.e., to be able to identify a given configuration with a high or low pitch. However, in order to do this, much more detailed knowledge of the conformation of the ester moiety, especially at the carbon-alkyl bond, would be required. This same conclusion has been reached in regard to circular dichroism of liquid crystalline systems.[15]

Perhaps through systematic study of the results of configurational changes on pitch and circular dichroism effects[16] more detailed knowledge can be acquired. Such knowledge would be of great practical importance for it appears that these mesomorphic properties are sensitive to subtle conformational changes. This subtlety is demonstrated by the twentyfold change in angular displacement in the helix between the esters from the (-) and (+) forms of 2-ethylhexanoic acid; if an intermolecular distance of 5 A (0.5nm) is assumed, the angle for the ester from the (-) acid is 1.38×10^{-1} degrees or 8.3 minutes, while that of the ester from the (+) acid is 3.0 degrees.

An attractive possibility suggested by the present results is that left and right-handed cholesterics could be prepared from the two pure enantiomers of a chiral acid.

CONCLUSIONS

In summary, this study indicates that pitch determination by extrapolation of data from binary mixtures is not rigorously valid in view of nonlinearity near the extremes. These deviations arise from intermolecular interactions, and the extent of deviation is a function of the dissimilarity of the pitches of the two components. The pitches of homologous series of the alkanoates are continuous functions of the heat of fusion. In the straight chain series they are also funcitons of the molecular interfacial (mesophase-crystal) free energy.

In structural terms the pitch generally increases as branching is introduced due to decreased chain flexibility by means of decreased rotation about C-C single bonds. This usually offsets the increase in steric bulk, although the effects of steric bulk are observed as decreased pitches when branches are lengthened or occur near the end of the chain. These general trends, however, are subject to large changes in chiral alkanoates. It is not possible to assess these configurational effects at this time due to lack of detailed knowledge of the conformation of the alkanoate portion of the molecule. The results of this study provide an insight into the importance of conformational effects in mesophases; in fact, mesomorphic properties are probably one of the most sensitive probes of conformation. Very little is known about these effects which govern mesomorphic properties. In view of the potential for understanding such properties, continued effort in this direction seems desirable.

REFERENCES

1. J.M.Pochan and H.W.Gibson, J. Amer. Chem. Soc., 93, 1279 (1971).
2. J.M.Pochan and H.W.Gibson, J. Amer. Chem. Soc., 94, 5573 (1972).
3. H.W.Gibson and J.M.Pochan, J. Phys. Chem., 77, 837 (1973).
4. H.W.Gibson and F.C.Bailey, to be published.
5. L.B.Leder, J. Chem. Phys., 58, 1118 (1973).
6. J.Y.C.Chu, Xerox Corp., private communication, to be submitted to Mol. Cryst. Liq. Cryst.
7. L.C.Scala and G.D.Dixon, Liq. Cryst. Mol. Cryst., 10, 411 (1970).
8. J.E.Adams, W.Haas and J.J.Wysocki, Proced. Amer. Chem. Soc. Symp. Ordered Fluids Liq. Cryst. 1969, Plenum Press, New York (1970) p. 463.
9. S.A.Jabarin and R.S.Stein, J. Phys. Chem., 77, 409 (1973).
10. H.Baessler and M.M.Labes, J. Chem. Phys., 52, 631 (1970).
11. E.M.Barrall, II, J.F.Johnson and R.S.Porter, Mol. Cryst. Liq. Cryst., 8, 27 (1969).
12. G.J.Karabatsos and N.Hsi, J. Amer. Chem. Soc., 87, 2864 (1965).
13. L.B.Leder, J. Chem. Phys., 54, 4671 (1971).
14. H.Hakemi and M.M.Labes, J. Chem. Phys., 58, 1318 (1973).
15. F.D.Saeva, Mol. Cryst. Liq. Cryst., in press.
16. F.D.Saeva, J. Amer. Chem. Soc., 94, 5135 (1972).

ON THE THEORIES OF OPTICAL REFLECTION FROM CHOLESTERIC LIQUID CRYSTAL FILMS

J.Shashidhara Prasad
Department of Physics, Manasagangotri
University of Mysore, Mysore 570006, India

INTRODUCTION

One of the unique optical properties of cholesteric liquid crystals, viz., the optical reflection has been theoretically dealt by a number of workers, using two different models: one based on the Oseen's dielectric continuum model; and the other by using the difference equations for wave propagation in a periodic structure (as in Darwin's dynamical theory of X-ray diffraction). de Vries(1) and Aihara and Inaba(2) have used the first model to explain the optical reflection at normal incidence. Taupin(3); Berreman and Scheffer(4) have treated the case of oblique incidence and have calculated by numerical methods the reflected and transmitted intensities. Dreher, Meier and Saupe(5) have solved the case of reflection by cholesteric liquid crystals for arbitrary angles of incidence. Chandrasekhar and Srinivasa Rao(6) and Chandrasekhar and Shashidhara-Prasad(7) have discussed the reflection for normal incidence using the second model. In fact, both the approaches are shown to be fully equivalent by Nityananda Rajaram(8).

Essentially, the various theoretical discussions predict a flat topped maximum for reflection from a specimen of infinite thickness, signifying a total reflection; But for a film of finite thickness, there exists secondary maxima on either side of the principal maximum in the region of selective reflection; the wavelength seperations from the central peak to the secondary peaks become larger when the thickness is reduced. A much thinner specimen shows a broadened out maximum, having a very small value for reflection with no secondaries.

THEORETICAL FIT FOR THE DATA ON OPTICAL REFLECTION FROM MIXED CHOLESTERIC LIQUID CRYSTALS

CASE 1: Fergason(9) has reported circular dichroism measurements for a mixture of cholesteryl benzoate, cholesteryl acetate and cholesteryl palmitate. The data for the 10 microns thick specimen gives a reflection maximum at 5230 Å with the total width of reflection for an infinitely thick specimen to be 150 Å. An attempt has been made to fit the reflection data obtained by using the relation between reflection and circular dichroism, D = R/2-R. The observed values of λ_o, the wavelength of maximum reflection and the width of total reflection $\Delta\lambda$ give a value of 3487 Å for the pitch.

The expression given by Chandrasekhar and Shashidhara Prasad(7) for reflection from a film of finite thickness has been used;

$$R = Q^2 \left(\varepsilon^2 + \xi^2 \coth^2 m\xi \right)^{-1}$$

where $\xi \simeq \pm (Q^2 - \varepsilon^2)^{\frac{1}{2}}$

$$Q^2 = (Q_o^2/\varepsilon^2)\sin^2\varepsilon$$

$$\varepsilon = -(2\pi/\lambda)(\lambda - \lambda_o)$$

and m the number of planes, is given by thickness/pitch as used in deriving the theory. The width of total reflection $\Delta\lambda$ is given by, P n = $Q_o\lambda_o/\pi$. Also $n_d P = \lambda_o$, where n_d is the refractive index for right circularly polarized light, λ_o, the wavelength of maximum reflection and P = pitch.

As we see in figure 1, the theory gives a principal maximum with the secondaries on either side for a specimen of thickness 10 microns, and with n_d value assumed to be 1.5, the order of refractive index of glass, which in turn gives the value of Δ n. m, the

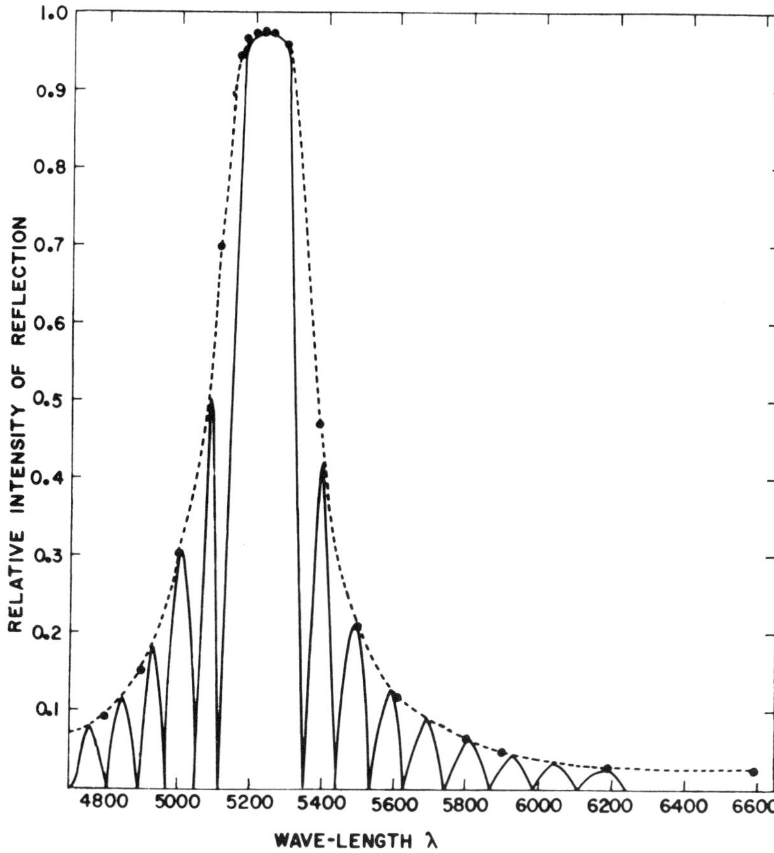

FIGURE 1: Computed and experimental data of reflection for 10 microns thick specimen of the mixture of cholesteryl benzoate, cholesteryl acetate and cholesteryl palmitate (from Fergason(9)). ——— theoretical; ● experimental points; - - - envelope of the primary and secondary maxima.

number of layers turns out to be 29. The experimental points lie on the envelope of primary and the secondary maxima, which could not have been seen clearly by the above author (see Berreman and Scheffer(4)), but which are expected both theoretically and experimentally (4,5,7,10,11) in the case of single domain cholesteric liquid crystal films of finite thickness.

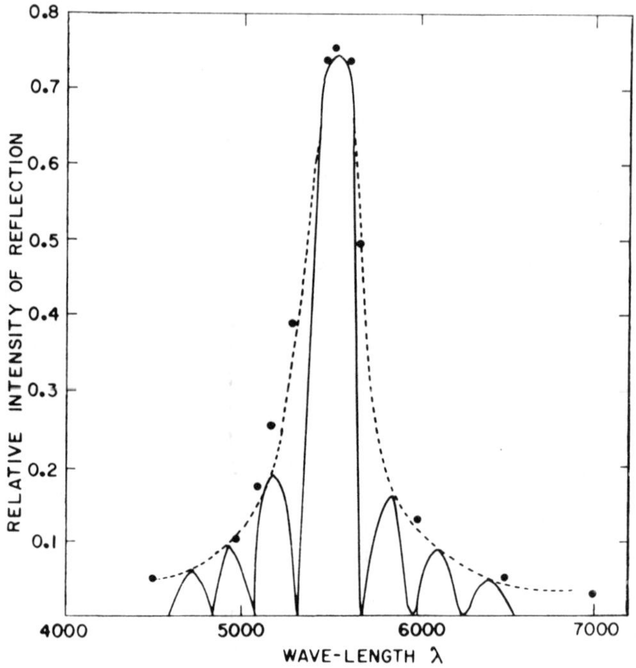

FIGURE 2: Computed and experimental data of reflection for 5 microns thick specimen of the mixture of cholesteryl nananoate and cholesteryl chloride(from Adams et al.(12)). ——— theoretical; ● experimental points; - - - envelope of the primary and secondary maxima.

CASE 2: Adams et al.(12) have reported the reflection measurements for a mixture of 70% of cholesteryl nananoate and 30% of cholesteryl chloride. The data for 5 microns thick specimen gives a reflection maximum at 5570 Å, with the total width of reflection for an infinitely thick specimen to be 250 Å. The angle of incidence is equal to the angle of observation and is equal to 45°. Modifying the theory for oblique incidence with the Bragg angle θ_B in the Bragg condition $2n_dP \sin\theta_B = n\lambda_o$, for the best reflection, to be given by

$$\sin\theta_B = \cos\frac{1}{2}\left[\sin^{-1}(\sin\phi_i/n_d) + \sin^{-1}(\sin\phi_s/n_d)\right]$$

(ref. 12).

The theoretical curve for a specimen of 5 microns thickness (Fig. 2) gives a principal maximum with the secondaries on either side. As in the previous case, the experimental points fall on the envelope of the secondary and the primary maxima. The expected secondary maxima have not been observed experimentally.

DISCUSSION

The experimentally observed curves of reflection for films of finite thickness are shown by the above theoretical fits to be the envelopes of the primary and secondary maxima which are to be expected for films of finite thickness. In fact the experimentally observed reflection peak and the width of reflection agree very well with those given by the envelopes, thereby showing that the experimental measurements are not made on perfectly single domain specimens. The observed experimental curves seem to be the averaged out effect from different domains of the polydomain specimens. The height of the reflection band and the shape are essentially due to the finiteness of the thickness of the specimen and the specimen not being single domain. This clearly refutes the theoretical interpretation of Subramanyam(13), who envisages cholesteric absorption to be the reason for the shape and height of the reflection band, without considering the finiteness of the thickness of the specimen.

The author wishes to thank Professor D.Krishnamurti for valuable discussions and Professor B.Sanjeevaiah, Head of the Department of Physics, University of Mysore, Mysore, for encouragement.

REFERENCES

(1) de Vries, Hl., Acta Cryst., 4, 219 (1951).
(2) Aihara, M. and Inaba, H., Optics Commun., 3, 77(1971)
(3) Taupin, D., Journal de Physique., 30, c4-32 (1969).
(4) Berreman, D.W. and Scheffer, T.J., Phys. Rev. Letters 25, 577 (1970).
(5) Dreher, R., Meier, G. and Saupe, A., Mol. Cryst. and Liq. Cryst., 13, 17 (1971).
(6) Chandrasekhar, S. and Srinivasa Rao, K.N., Acta Cryst., A24, 445 (1968).
(7) Chandrasekhar, S. and Shashidhara Prasad, J., Mol. Cryst. and Liq. Cryst., 14, 115 (1971).

(8) Nityananda Rajaram: 'On the theory of propagation of light by cholesteric liquid crystals' Mol. Cryst. and Liq. Cryst. (in press).
(9) Fergason, J.L., 'Liquid Crystals', Gordon and Breach, New York-London-Paris, edited by Brown, G.H., Diennes, G.J. and Labes, M.M., p 89 (1967).
(10) Shashidhara Prasad, J. and Madhava, M.S.,: 'Optical rotatory dispersion of cholesteric liquid crystals' Mol. Cryst. and Liq. Cryst. (in press).
(11) Shashidhara Prasad, J (Mysore University Journal 1973).
(12) Adams, J.E., Haas, W. and Wysocki, J., J. Chem. Phys. 50, 2458(1969).
(13) Subramanyam, S.V., Applied Optics, 10, 317 (1971).

SOME MESOMORPHIC PROPERTIES OF N-(p-AZIDOBENZYLIDENE) - ANILINES

Craig Maze, Henry G. Hughes

Motorola Semiconductor Prod.

Phoenix, Arizona 85008

INTRODUCTION

Synthesis of several p-substituted alkyl and alkoxy azidobenzylidene anilines was carried out to study their mesomorphic transition temperatures and the affect made upon these transitions by the linear azide moiety.

The synthetic work involved preparation of azidobenzaldehyde followed by condensation azidobenzaldehyde with commericially available alkyl and alkoxy anilines to form the schiff bases used in this study.

SYNTHESIS

The first step in the reaction scheme is the simultaneous oxidation of $-CH_3$ and reduction of $-NO_2$ on I using sodium polysulfide and base. Oxidation of the methyl group is a two step reaction[1,2] which involves the evolution of H_2S followed by a dehydration step leaving p-nitrobenzaldehyde. After reducing $-NO_2$, forming II, the first step is complete. p-Animobenzaldehyde II was extraced with chloroform and stored at 0^oC for no longer than twenty-four hours since it tends to condense with itself and form polymeric impuritites.

Diazotization was carried out in cold chloroform followed by the addition of urea to destroy any excess reagent before proceeding with azidozation. The final step consisted of adding sodium azide to form IV followed by extraction of III from the water layer using ether. The product, azidobenzaldehyde, was kept refrigerated until used.

Condensation of IV with commerically available anilines was carried out in the usual straightforward manner to form the imines VII used in this study. Benzene was used as the solvent and the water formed was collected in a Dean-Stark trap. Purification of the imines was by recrystallization from acetonitrile or petroleum ether followed by a cold methanol wash. Vacuum distillation was tried, but discarded as it appeared that some thermal decomposition was taking place in the still.

Eleven compounds were synthesized, six alkoxy, $Z = C_nH_{2n+1}O-$, n = 1,6, and five alkyl, $Z = C_nH_{2n+1}$ n = 1,5.

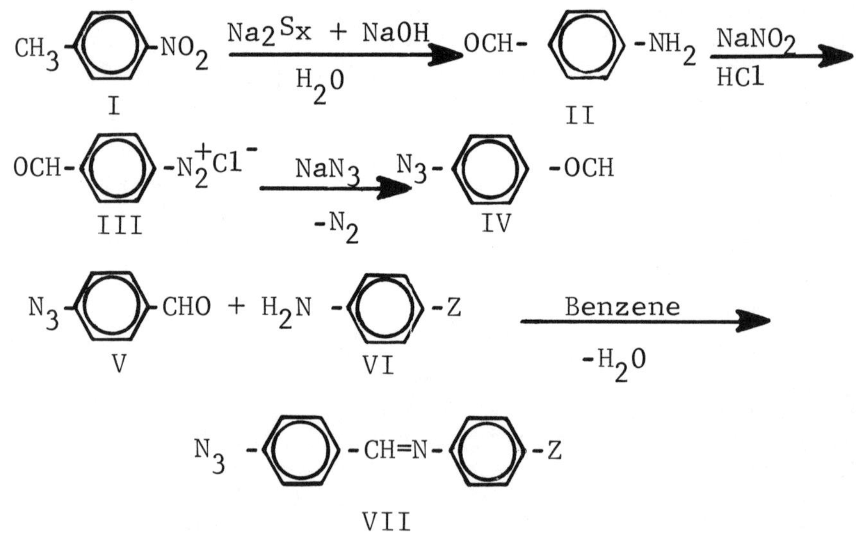

RESULTS AND DISCUSSION

Mesophase transitions temperatures for the aforementioned compounds appear in Table I. These were determined using either a Mettler FP-52 hot stage or a Perkin-Elmer DSC-2 scanning calorimeter at heating rates of no more than five degrees/min.

Melting points for the alkoxy series varied from 122 to 98°C, and the first compound in the series showed no mesophease. Compounds two and three were monotropic nematic, four a questionable monotropic smectic, and the others smectic and nematic on both heating and cooling. Nematic ranges are narrow on heating, being no more than two degrees, and would probably wash out if the alkoxy chain were extended by one or two carbon atoms.

I. TRANSITION TEMPERATURES

$$Z = -OC_nH_{2n+1}$$

NO.	n	HEATING	COOLING
1	1	122 C	104
2	2	114	108 N 95
3	3	115	97 N 92
4	4	107 N 109	108 N 98 S(?) 97
5	5	98 S 101 N 105	105 N 103 S 80
6	6	99 S 108 N 109	108 N 106 S 83

$$Z = -C_nH_{2n+1}$$

NO.	n	HEATING	COOLING
7	1	68	63
8	2	51	52
9	3	33 S 41 N 57	54 N 39 S 3
10	4	25 S 54 N 55	53 N 51 S 5
11	5	32 S 60 N 67	64 N 57 S 10

S = Smectic

N = Nematic

None of the alkyl series exhibited any monotropic behavior and all were smectic which was unexpected. Compound 10 is shown in the table as having a one degree nematic range; however, the nematic schlieren texture could only be observed under crossed polarizers as a transient appearing during isothermal melting. Moreover, no peak was seen on the DSC for this transition.

Melting points for all of these alkyl compounds were low running from 25 to 68C on heating, and supercooling to between three and ten degrees before freezing. As a consequence, compound 10 is a room temperature smectic, and based on an extrapolation of these results, the hexyl homolog of this series should also melt near room temperature.

The azide group gives rise to the smectic A phase, as observed under the microscope, which can be attributed to its linear structure and attendent closer packing in the mesophase. However, melting temperatures for the alkyl series are low indicating that lateral attractive forces are not too large probably due to the lack of intermolecular oxygen-hydrogen bonding. Compounds of the alkoxy series on the other hand all have much higher melting points, and show a greater tendency to form nematic mesophases relative to their alkyl counterparts. Moreover, behavior of the alkoxy series more nearly follows that of other liquid crystalline substances of similar structure indicating that the alkoxy oxygen atom plays a dominant role in determining both melting temperatures and mesophase transition temperature patterns.

REFERENCES

1. E. Campaigne, W. M. Budde, and F. Schaefer, Org. Syn. Coll. 4, 31 (1963).

2. H. G. Beard and H. H. Hodgson, J. Chem. Soc., 1944, 4.

STABLE, LOW MELTING NEMATOGENS OF POSITIVE DIELECTRIC ANISOTROPY FOR DISPLAY DEVICES

G.W. Gray, K.J. Harrison and J.A. Nash*
J. Constant, D.S. Hulme, J. Kirton and E.P. Raynes+
*Department of Chemistry, University of Hull, Hull, England
+Royal Radar Establishment, Malvern, England

INTRODUCTION

During the last few years, the application of liquid crystalline compounds in display devices based on a number of different electro-optical effects[1] has made the availability of a range of new mesogens desirable. Consequently, much effort has been devoted to the synthesis of mesogens with suitably low melting points, for display device application. Such compounds must, however, satisfy several other quite stringent requirements if display devices based on the dynamic scattering[2], twisted nematic [3] or cholesteric-nematic phase change[4] effects are to compete with the more conventional modes of display.

At present, nematogens and cholesterogens of both positive and negative dielectric anisotropy are required, and in addition, a mesomorphic range of -20°C to +65°C is desirable if the display devices are to realise their full potential. Of great importance too is the fact that the materials used should be chemically, photochemically and electrochemically stable; these requirements should be linked closely with the compounds being safe to handle.

It is evident from surveying[5] the majority of known nematogens with reasonably low C-N temperatures, that the systems have the general formula

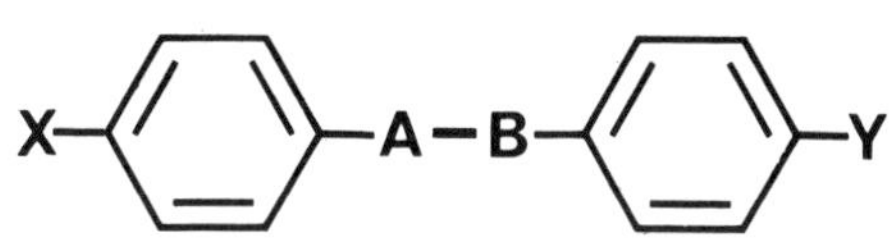

where two p-phenylene rings are linked through a central group A — B, usually containing a multiple bond to preserve the rigidity and linearity of the molecule. The p-substituents X and Y may vary widely, but the lowest C-N temperatures have been observed with X = n-alkyl and Y = n-alkoxy or vice versa, or X and Y = n-alkyl groups. The terminal group -O.CO.Alkyl has also been used with some success. Furthermore, alkyl chain lengths of 4,5 and 6 carbon atoms appear to give the lowest melting materials. The systems are therefore of reasonably low molecular weights and examples of the central groups represented as A — B in the general formula are:

-N=N-	Azo compounds[6]	$-N(\rightarrow O)=N-$	Azoxy compounds[6,7]
-CH=N-	Schiff's bases[8]		
-C≡C-	Tolanes[10]	$-CH=C(Cl)-$	α-Chlorostilbenes[9]

An exception to this general type of central group is $-\overset{\overset{O}{\|}}{C}-O-$ [11] as used in the 4,4'-disubstituted phenylbenzoates and related esters.

Although the range of systems at present available affords materials of both positive and negative dielectric anisotropy, most of the compounds have one or more disadvantage which, in the long term, makes their display device application dubious[5]. Schiff's bases such as 4-methoxybenzylidene-4'-n-butylaniline (MBBA) and its mixtures with homologues such as the 4-ethoxy analogue (EBBA) have attracted much attention for their application in dynamic scattering devices, but Schiff's bases in general are yellow, prone to polymerisation and oxidation, and extremely labile to hydrolysis. Cyano-substituted Schiff's bases[12] (A — B = -CH=N-, X = n-alkyl or n-alkoxy, Y = -C≡N) have found application in twisted nematic devices[3,13], but they too suffer from the disadvantages mentioned above. Furthermore, where two differently disubstituted Schiff's bases are used to obtain a lower melting mixture, cleavage of the Schiff's base linkage and recombination of the fragments eventually gives an equilibrium mixture of the four possible Schiff's bases with properties different to that originally formulated.

Azoxy compounds are certainly more stable than Schiff's bases, but are quite highly coloured[14]. Azo compounds are most unsatisfactory, being not only highly coloured but also susceptible to oxidation and isomerisation.

The tolanes and α-chlorostilbenes are colourless, but the latter materials are highly photochemically unstable, particularly to UV light. The stability of the tolane system has not been reported on in definite terms.

The 4,4'-disubstituted phenylbenzoate esters are also colourless and in admixture give low C-N temperatures and wide nematic ranges[11]. However, an ester function is always susceptible to hydrolysis and other forms of nucleophilic attack.

The problems of colour, and of chemical and photochemical instability in these systems are due almost entirely to the central A — B linkage. We reasoned therefore, that elimination of this central linkage would make the system more stable, both chemically and photochemically, and the reduction in the extent of conjugation in the molecule would eliminate or at least greatly reduce the colour.

The resulting system, without the central A — B group, is a 4,4'-disubstituted biphenyl.

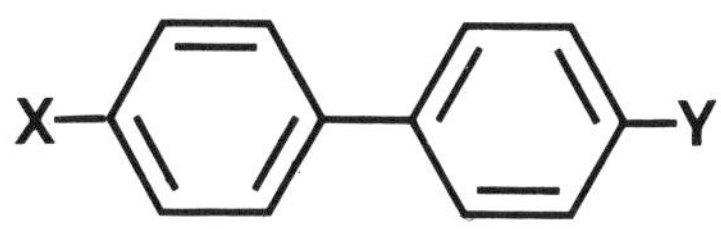

As a result of earlier work by Gray et al, it has been established for many years that such biphenyl systems are highly mesogenic, examples of liquid crystalline compounds being provided by the 4'-n-alkoxybiphenyl-4-carboxylic acids[15], 4-p-n-alkoxybenzylideneaminobiphenyls[16], 4,4'-di-(p-n-alkoxybenzylideneamino) biphenyls[16], n-alkyl 4'-n-alkoxy-biphenyl-4-carboxylates[15] etc. The melting points of these compounds were often high and in some cases the compounds were purely smectic. Simpler terminal groups were therefore necessary if low melting mesogens were to be obtained. Moreover, since the primary requirement was for low melting nematogens, these groups would have to be of a type such that nematic properties were strongly favoured. We therefore chose to investigate a number of 4'-n-alkyl-4-cyanobiphenyls and 4'-n-alkoxy-4-cyanobiphenyls[17]. In this way, a high nematogenic character was combined with a simple, low molecular weight system having no central group A — B, so affording the best chance of producing stable, low melting nematogens. In addition, these systems would be of positive dielectric anisotropy and therefore any nematogens produced would be of value for devices based on the twisted nematic or phase change effects, the two electro-optical effects of most interest to our group. For the same reasons a number of 4'-n-alkoxy-4-nitrobiphenyls were also investigated.

RESULTS

The melting points and mesomorphic transition temperatures for the various 4,4'-disubstituted biphenyls are given in Table 1.

Table 1 Mesomorphic Transition Temperatures for the Compounds:

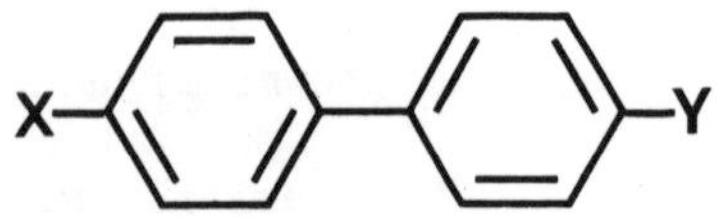

X	Y	C-S, N or I Temp. (°C)	S-N Temp. (°C)	N-I Temp. (°C)
n-C_4H_9	CN	46.5	-	(16.5)
n-C_5H_{11}	CN	22.5	-	35
n-C_6H_{13}	CN	13.5	-	27
n-C_7H_{15}	CN	28.5	-	42
n-C_8H_{17}	CN	21	32.5	40
n-C_9H_{19}	CN	40.5	44.5	47.5
n-C_3H_7O	CN	71.5	-	(64)
n-C_4H_9O	CN	78	-	(75.5)
n-$C_5H_{11}O$	CN	48	-	67.5
n-$C_6H_{13}O$	CN	58	-	76.5
n-$C_7H_{15}O$	CN	53.5	-	75
n-$C_8H_{17}O$	CN	54.5	67	80
n-$C_5H_{11}O$	NO_2	54.5	-	(<42)
n-$C_6H_{13}O$	NO_2	67	-	(32.5)
n-$C_7H_{15}O$	NO_2	36.5	(30.5)	38.5
n-$C_8H_{17}O$	NO_2	49 and 51.5*	49.5	51.5

C = crystal; N = nematic; S = smectic; I = amorphous, isotropic liquid.

Temperatures in parentheses are for monotropic transitions.

* This compound exists in two dimorphic forms with distinct melting points.

The transition temperatures were determined by optical microscopy, using a polarising microscope in conjunction with a heated stage (C Reichert, Optische Werk A G, Wien, Austria). Mesophase-mesophase and mesophase-amorphous, isotropic liquid transitions were precisely reversible. Temperatures were measured with an accuracy of ±0.25°C. A number of polymorphic transitions were observed and are summarised below.

1. When the nematic melt of 4'-n-heptyl-4-cyanobiphenyl is solidified and the solid reheated, melting occurs at 15°C, but the higher melting solid (mp 28.5°C) is seen to grow in the nematic phase.

2. When the nematic melt of 4'-n-hexyloxy-4-cyanobiphenyl is solidified and the solid reheated, a melting point of 44°C is obtained. However, on leaving this solid to stand for several minutes, the more stable solid, mp 58°C is obtained.

3. When the nematic melt of 4'-n-heptyloxy-4-cyanobiphenyl is solidified, the solid obtained has a melting point of 47.5°C but gradually reverts on standing or on slow heating to the more stable solid with mp 53.5°C.

4. When the smectic phase of 4'-n-nonyl-4-cyanobiphenyl is solidified, a metastable solid (mp 29.5°C) is sometimes obtained, but on leaving this to stand it reverts to the more stable form (mp 40.5°C).

All the above compounds were examined by differential thermal analysis (DTA) using a low temperature thermal analyser (Stanton Redcroft Limited, Copper Mill Lane, London SW17), to check on the number of transitions and the transition temperatures obtained by optical microscopy. Where polymorphic solid changes had been detected by microscopy, these were confirmed by DTA. Very good agreement was observed between the temperatures measured by optical microscopy and DTA.

The enthalpies of transition ($\Delta\underline{H}$) for the 4'-substituted 4-cyanobiphenyls were determined by differential scanning calorimetry (DSC) (Du Pont Co. (UK) Limited, Hitchin, Herts). Results for the enthalpies of melting are given in Table 2; for these C-S, N or I transitions, the $\Delta\underline{H}$ values relate to the most stable (highest melting) solids when polymorphism occurs.

The enthalpies for the N-I transitions were typically in the range 0.1-0.3 kcal mol^{-1}. The enthalpies for the S-N transitions were very small and either equal to or less than the enthalpy for the N-I transition for the particular compound. The smectic phase exhibited by a number of the 4,4'-disubstituted biphenyls now under discussion has not yet been fully characterised. The small enthalpies of the S-N transitions might suggest that smectic C to nematic transitions[18] may be occurring. However, the apparent positive uniaxial interference figures shown by the smectic phases

Table 2 Enthalpies of Melting for the Compounds:

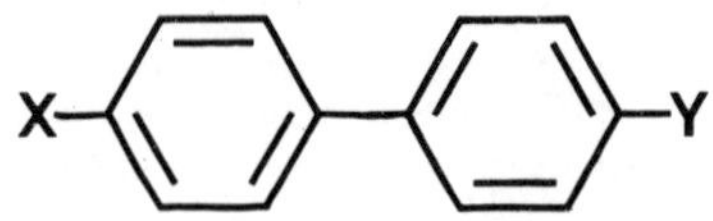

X	Y	C–S, N or I $\Delta\underline{H}$ (kcal mol^{-1})
n-C_4H_9	CN	5.5
n-C_5H_{11}	CN	4.1
n-C_6H_{13}	CN	5.8
n-C_7H_{15}	CN	6.2
n-C_8H_{17}	CN	5.3
n-C_9H_{19}	CN	8.0
n-C_3H_7O	CN	4.6
n-C_4H_9O	CN	5.6
n-$C_5H_{11}O$	CN	6.9
n-$C_6H_{13}O$	CN	7.1
n-$C_7H_{15}O$	CN	6.9
n-$C_8H_{17}O$	CN	5.9

of these biphenyl compounds do not support this phase being classified as a smectic C (unless the tilt angle is very small), since the latter phase is biaxial, though sometimes only weakly. These uncharacterised smectic phases exhibit both homogeneous and homeotropic textures and in some cases a fan-type texture. Miscibility studies with available compounds exhibiting known S_A, S_B and S_C phases have so far failed to produce a positive identification of the uncharacterised smectic states, although they do not appear to be either smectic A or smectic B phases. The somewhat inconclusive nature of the miscibility studies is largely due to the fact that no standard compounds are available which exhibit smectic phases in the required temperature ranges. Further work on the characterisation of the smectic phase is in progress and includes X-ray studies.

Fig 1 shows the plot of the transition temperatures against the number of carbon atoms in the alkyl chain for the series of 4'-n-alkyl-4-cyanobiphenyls.

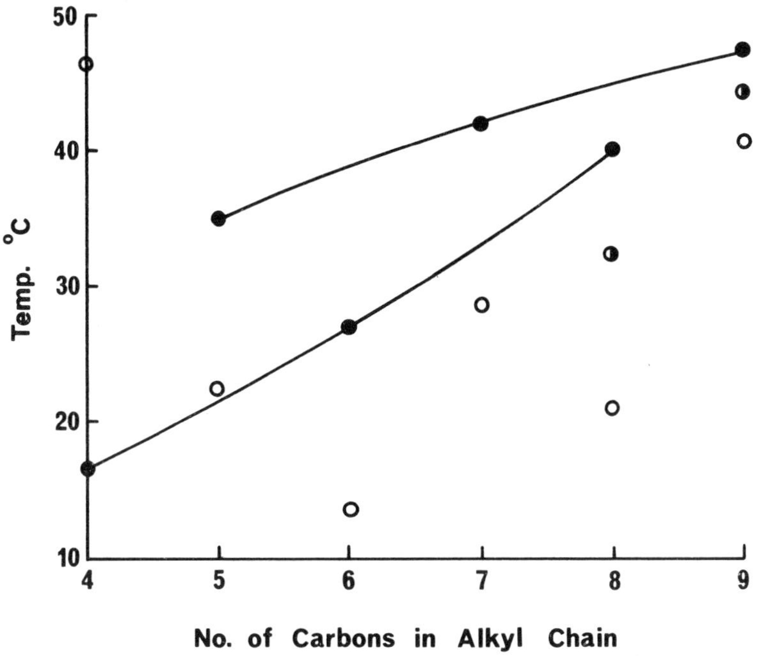

Fig 1 Plot of transition temperatures against number of C atoms in the alkyl chain for the 4'-n-alkyl-4-cyanobiphenyls (key shown in Fig. 2).

The N-I temperatures for the alkyl compounds (Fig 1) lie on two smooth, rising curves, that for the odd homologues lying above the curve for the even homologues. The two curves, one slightly concave and the other slightly convex, are converging. The apparent curve for the S-N temperatures must rise very steeply since no smectic phase occurs for the n-heptyl compound above -8°C. For higher homologues this "curve" will presumably merge with the converging N-I curves, producing purely smectic members of the series.

Fig 2 shows the plot of the transition temperatures against the number of carbon atoms in the alkoxy group for the series of 4'-n-alkoxy-4-cyanobiphenyls.

The N-I temperatures for the alkoxy compounds (Fig 2) again lie on two smooth, rising curves which are converging. The curve for the even homologues lies above the curve for the odd homologues.

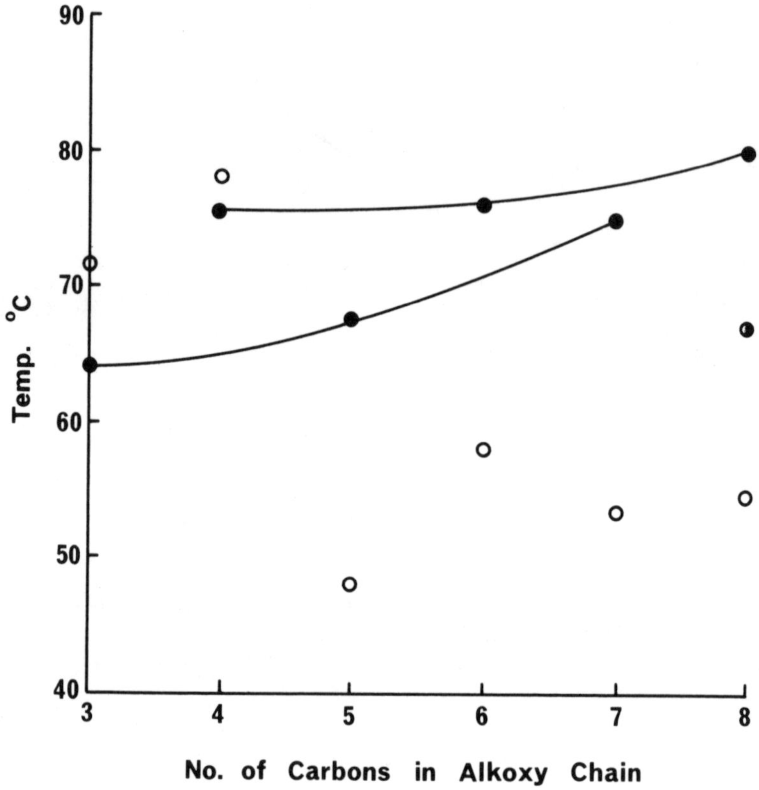

Fig 2 Plot of transition temperatures against number of C atoms in the alkoxy group for the 4'-n-alkoxy-4-cyanobiphenyls.

Key for Fig 1 and Fig 2

○ C–S, N or I

◐ S–N

● N–I

These curves are both concave and presumably will reach a maximum and fall again for higher members of the series which are presently under preparation.

The general increase in nematic thermal stability as the homologous series are ascended may be attributed to the significant overall increase in axial polarisability of the molecule as each successive methylene unit is added. This is a general trend[19] for series of compounds which give N-I temperatures which are fairly low.

The alternation of the N-I temperatures may be explained in terms of the anisotropy of the polarisability of the molecules. Each successive methyl unit makes a different contribution to the axial polarisability of the molecule dependent upon whether the extension in chain length is to an odd or an even carbon chain. The axial polarisability on changing from an even to an odd carbon chain in the n-alkyl series (or from an odd to an even carbon chain in the n-alkoxy series) should be increased more than on changing from an odd to an even carbon chain in the n-alkyl series (or from an even to an odd carbon chain in the n-alkoxy series). This is supported by the fact that the upper N-I curves in the two plots are constituted by the temperatures for the even carbon chain ethers and the odd carbon chain alkyl compounds.

The average nematic thermal stability for the series of alkoxy compounds is considerably higher than the average nematic thermal stability for the corresponding series of alkyl compounds. This is in line with previous observations[20] and is to be expected since the additional ether linkage will lead to a significant change in the polarisability of the molecule.

The melting points of the 4'-n-alkyl-4-cyanobiphenyls are in every case lower than those of the corresponding 4'-n-alkoxy-4-cyanobiphenyls. In both series, there appear to be no regular trends in the melting points except that of a general increase as the homologous series are ascended.

When the cyano-group is replaced by a nitro-group in the n-alkoxy compounds there is an overall decrease in both nematic and smectic thermal stabilities. In the former case this is attributable to the decrease in the polarisability of the molecule and, in the case of the smectic phases, also to the smaller dipole moment of the nitro compound.

For the nitro-n-heptyloxy and nitro-n-octyloxy compounds, the melting points are lower than for the corresponding cyano compounds. The situation is, however, reversed for the n-pentyloxy and n-hexyloxy compounds. However, the decreases in nematic thermal stabilities for the nitro compounds compared with the corresponding cyano compounds are so great that the former compounds show either monotropic or marginally enantiotropic nematic behaviour. There seemed little merit, therefore, in studying other nitro-n-alkoxy compounds or the series of nitro-n-alkyl compounds. Furthermore, the four nitro compounds prepared are distinctly yellow in colour.

From Table 1, it is clear that there are two compounds, namely the cyano-n-hexyl- and the cyano-n-pentyl-biphenyls, which are room temperature nematogens under normal conditions; moreover, the cyano-n-heptylbiphenyl, once melted, remains nematic for many months. In addition, the cyano-n-octylbiphenyl is a room temperature smectogen. These colourless compounds therefore provide very valuable materials for the room temperature study of both smectic and nematic phases consisting of pure, single compounds. However, the nematic ranges of the lower melting compounds are obviously not great, and in addition, a really low melting ($<0^{o}C$) material has not been obtained. To this end we have investigated a number of binary, ternary and quaternary mixtures incorporating the low melting characteristics of the alkyl compounds and the reasonably high nematic thermal stabilities of the alkoxy compounds. We have been very selective in the choice of mixtures used, compatible with maintaining as high a nematic thermal stability as possible. The results for some selected mixtures of 4'-n-alkyl- and/or 4'-n-alkoxy-4-cyanobiphenyls are given below (reference is made only to the 4'-substituent); none of the mixtures is claimed to correspond to a eutectic composition. The temperatures of transition were measured by optical microscopy, using a polarising microscope in conjunction with a cold stage (C Reichert, Optische Werk A G, Wien, Austria) capable of being cooled to $-55^{o}C$.

1. n-Pentyl (56 mole %): n-Heptyl (44 mole %)

C-N, $0.5^{o}C$; N-I, $37^{o}C$. Crystallisation occurs only slowly at $-30^{o}C$, although the nematic phase may be rapidly cooled to $-55^{o}C$ without crystallisation.

2. n-Heptyl (48.5 mole %): n-Octyl (51.5 mole %)

C-N, $0.5^{o}C$; N-I, $40^{o}C$. Crystallisation occurs on rapid cooling to $-20^{o}C$.

3. n-Pentyloxy (55 mole %): n-Heptyloxy (45 mole %)

C-N, $21^{o}C$; N-I, $70.5^{o}C$. Crystallisation occurs slowly at $4^{o}C$.

4. n-Pentyl (54 mole %): n-Heptyl (46 mole %). To this binary mixture (76%) was added n-Heptyloxy (24%).

C-N, ? ; N-I, $47^{o}C$. To date this material has not been obtained crystalline, either by slow or rapid cooling to -55°C, or on storage at -15°C for in excess of 3 months.

5. n-Pentyl (55 mole %): n-Heptyl (45 mole %). To this binary mixture (58%) was added a binary mixture (42%) of n-Pentyloxy (55 mole %): n-Heptyloxy (45 mole %).

C-N, ? ; N-I, 51°C. This material too has not been obtained crystalline either by slow or rapid cooling to -55°C or on storage at -15°C for in excess of 3 months.

6. n-Heptyl (48.5 mole %): n-Octyl (51.5 mole %). To this binary mixture (60%) was added a binary mixture (40%) of n-Pentyloxy (55 mole %): n-Heptyloxy (45 mole %).

C-N, -2°C; N-I, 52.5°C. Crystallisation is initiated at -40°C by disturbing the nematic phase.

Where the n-octyl compound has been used in mixtures, no smectic properties were observed at the compositions studied.

Further work (to be published) on even better mixtures is in progress and the enthalpies and temperatures of melting for the individual compounds are being used for the calculation of eutectic compositions. Where calculations have been made for the compositions quoted above, good agreement between the predicted transition temperatures (C-N) and the measured temperatures (C-N) has been obtained (calculated eutectic composition n-pentyl:n-heptyl is 59:41 mole %; observed C-N, -2°C; N-I, 37°C).

The 4'-substituted 4-cyanobiphenyls have therefore yielded a number of colourless low melting nematogens. These materials are now being prepared by BDH Chemicals, Poole, Dorset, England.

In the quest for even lower melting and more thermally stable nematogens, the fluorene analogues have also been investigated, albeit to a limited extent. This was work carried out because Gray et al[16] observed that in some cases, though not all, substituted fluorenes had lower melting points and nematic thermal stabilities than the corresponding substituted biphenyls. Fluorenes, like biphenyls, would also be chemically/photochemically more stable and less coloured than systems containing central A — B linkages. Two 2,7-disubstituted fluorenes were prepared, by methods analogous to those described later for the biphenyl analogues, and examined by hot-stage microscopy. The constants for the compounds are given in Table 3.

In both cases, the melting points of the fluorene compounds are higher than those of the corresponding biphenyl compounds; for the n-butyl compound the increase in melting point is 17°C, and for the

Table 3 Transition Temperatures for the Compounds:

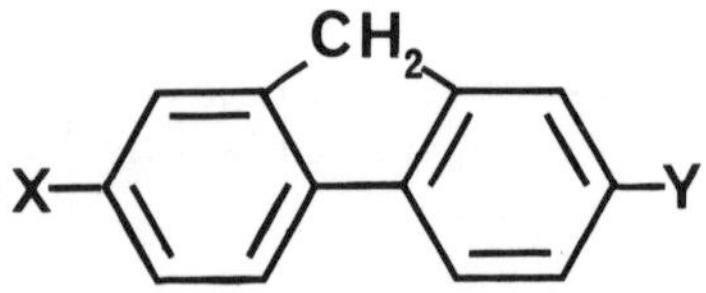

X	Y	C-I Temp. (°C)	N-I Temp. (°C)
$n\text{-}C_4H_9$	CN	63.5	[-23]
$n\text{-}C_6H_{13}O$	CN	106.5	[<82]*

Temperatures in parentheses are for monotropic transitions

* No nematic was, in fact, observed even on rapid chilling, crystallisation occurring at 82°C.

n-hexyloxy compound the increase in melting point is 48.5°C. The nematic thermal stability of the n-butyl-fluorene is considerably reduced compared with that of the n-butyl-biphenyl, the decrease in the N-I temperature being 39.5°C. It is not possible to comment on whether the "nematic" thermal stability is increased or decreased on going from the n-hexyloxy-biphenyl to the n-hexyloxy-fluorene, since the isotropic melt of the latter compound does not supercool sufficiently. However, it seems that the fluorene system is much inferior to the biphenyl system as far as producing low melting nematogens is concerned.

DISCUSSION

All the cyano-n-alkyl and cyano-n-alkoxy compounds and their mixtures are colourless and show no signs of becoming coloured on storage under normal conditions.

A number of these compounds are nematic at room temperature and therefore allow, for the first time, stable room temperature nematogens to be tested as single component systems for twisted nematic display device application. In this context, 4'-n-pentyl-4-cyanobiphenyl (PCB) has been used for exhaustive testing. The material appears to be chemically and thermally very stable (purification is readily achieved by reduced pressure distillation), the N-I temperature of the bulk material remaining constant at 35°C after storage for many months. Even more significant is the fact that a completely exposed thin film of PCB, on a microscope slide, showed a decrease of <1°C of the N-I temperature when left in the laboratory atmosphere for one week. This contrasts with MBBA for which a fall of as much as 3°C is observed when a sample is exposed to the laboratory atmosphere for a matter of minutes.

To further check the stability of PCB it was deliberately subjected to harsh test conditions. Cells containing PCB, sealed only with ordinary "araldite" or "Torr-seal" (epoxys), were heated for one week in an autoclave at 40°C in an atmosphere saturated with water vapour. This treatment affected the N-I transition temperature of PCB by <1°C, whereas MBBA and cyano Schiff's bases contained in similar cells and treated in the same way were converted to isotropic liquids at room temperature. Finally, the photochemical stability of PCB was demonstrated by exposure to UV radiation (from a mercury vapour lamp) for 4h; this caused no visible deterioration of the film or decrease in the N-I temperature. The α-chlorostilbenes suffer large decreases in their N-I temperatures after only a few seconds' exposure and become red.

The electrical properties of PCB and its homologues, the alkoxy analogues and their mixtures have been measured (17). For PCB, the low frequency dielectric constants, at 25°C, are $\varepsilon_{\parallel} = 17$ and $\varepsilon_{\perp} = 6$; positive dielectric anisotropy, $\varepsilon_a = +11$. The resistivity was typically $10^{10}\,\Omega$ cm at 100 Hz and orientated samples gave a resistivity ratio ($\rho_{\perp}/\rho_{\parallel}$) of 1.42. Controlled alignments of thin nematic films were obtained easily. For the n-heptyl analogue, $\varepsilon_a = +10$ and the resistivity ratio was 1.04. In the case of quaternary mixture 6, at 25°C, $\varepsilon_{\parallel} = 17$ and $\varepsilon_{\perp} = 6$ ($\varepsilon_a = +11$) and the resistivity ratio was 1.52 at 100 Hz. All the materials appear to be suitable for twisted nematic and (with a suitable additive) cholesteric-nematic phase change devices.

Twisted Nematic Device

PCB performed exceptionally well in showing the twisted nematic electro-optical effect. A 12 µm layer of PCB has a low threshold of 1.1 Vrms (50 Hz), independent of layer thickness and frequency up to 100 kHz. This compares well with the lowest reported threshold value of 0.9 Vrms[13] for a twisted nematic material. On AC application, using a gated 10 kHz signal, the delay and rise times at 3 Vrms are 0.15s and 0.1s respectively. These times decrease to 0.02s and 0.01s respectively at 10 Vrms. The decay time is 0.25s independent of applied voltage. The threshold voltage for 4'-n-heptyl-4-cyanobiphenyl (12 µm) was also 1.1 Vrms.

Quaternary mixture 6 also functioned well in a twisted nematic device. On AC application, a threshold voltage of 1.2 Vrms was found. It was possible with this mixture to test the twisted nematic device at a temperature as low as $-10^{o}C$ when, on AC application, the threshold voltage was still 1.2 Vrms, but understandably the response times were slower.

Cholesteric-Nematic Phase Change Device

A mixture of PCB containing 10% by weight of a readily available cholesterogen such as cholesteryl chloride has the required properties to exhibit the phase change effect. For liquid crystal films of thickness 6.2 µm, 12.5 µm and 25 µm, the rms voltages required for completion of the transition , using a 10 kHz sine wave signal, were 2.8, 5.4 and 10.4 Vrms respectively (frequency independent). These voltages represent threshold fields from 4.5 to 4.2×10^{5} V/m and are approximately one third of that required for equivalent Schiff's base mixtures[27]. This provides the basis for scattering devices with a contrast ratio comparable with that shown by dynamic scattering devices, but operating at a few volts, eg 5V for a 12 µm layer. For a 7 µm thick layer, using a gated 10 Vrms, 10 kHz sine wave signal, a rise time of 0.1s and a decay time of 0.02s were observed using glass cleaned and then rinsed in methanol. Further investigations of the effects of surface treatments on the decay time are being made.

EXPERIMENTAL

4'-n-Alkyl-4-cyanobiphenyls

The preparation of 4'-n-butyl-4-cyanobiphenyl is given in full. The results for the cognate preparations of the remaining five n-alkyl homologues are given in the appropriate Tables.

<u>4'-n-Butanoyl-4-bromobiphenyl</u>. n-Butanoyl chloride was prepared from n-butanoic acid and thionyl chloride by the standard technique[21]. The acid chloride was purified by distillation and the fraction collected with bp $102^{o}C$. The yield was almost quantitative.

To commercially available 4-bromobiphenyl (18.65g, 0.08 mole) and freshly crushed anhydrous aluminium trichloride (14g, 0.1 mole) dissolved in dry "Analar" nitrobenzene (85 ml), n-butanoyl chloride (10.6g, 0.1 mole) was added dropwise, the temperature of the stirred reaction mixture being maintained below $20^{o}C$ (approximately 20m). The reaction mixture was stirred for 18h at room temperature and poured onto a mixture of ice (110g), water (25ml) and concentrated hydrochloric acid (50ml). The resultant mixture was stirred for

30m and the nitrobenzene layer was separated off, chloroform being added to effect a more efficient separation. The organic layer was washed with water and the chloroform removed by distillation under reduced pressure. The nitrobenzene was removed by steam distillation. The solid residue was extracted into benzene, and the extract dried over anhydrous sodium sulphate. The benzene was removed by distillation under reduced pressure and the solid residue crystallised from absolute ethanol (250ml) to constant melting point. A small amount of animal charcoal was added to remove coloured impurities. The yield of product, mp 104-104.5°C, was 18.2g (75%).

The product was shown by infra-red spectroscopy and nuclear magnetic resonance spectroscopy to be the required material. The purity of the material was checked by tlc on silicic acid with chloroform as solvent. The product showed a single spot with R_F = 0.53. Finally, the product was submitted for combustion analysis (Found: C, 63.4; H, 4.9; Br, 26.0. Calc. for $C_{16}H_{15}BrO$: C, 63.4; H, 5.0; Br, 26.3%).

The results of the cognate preparations of other 4'-n-alkanoyl-4-bromobiphenyls are given in Table 4.

4'-n-Butyl-4-bromobiphenyl. To 4'-n-butanoyl-4-bromobiphenyl (16.4g, 0.054 mole) in diethylene glycol (56ml), 90% hydrazine hydrate (5.6ml) and potassium hydroxide (7.3g, 0.13 mole) were added, and the reaction mixture was heated at 100°C until the potassium hydroxide had dissolved. After heating under reflux for a further 1h, the temperature was raised to 180°C by distillation of approximately 10ml of solvent, and the temperature maintained there for 3h. On cooling, the organic material was extracted into benzene and the extract washed with water before drying over anhydrous sodium sulphate. The benzene was removed by distillation under reduced pressure and the residue was crystallised from absolute ethanol (400ml). A hot filtration was incorporated in the crystallisation to remove small traces of yellow azine. The yield of product, mp 102-102.5°C, was 12.4g (79%).

The product was shown by infra-red spectroscopy to be the required material. The purity of the product was checked by tlc on silicic acid with chloroform as solvent, a single spot with R_F = 0.57 being obtained. Finally, the product was submitted for combustion analysis (Found: C, 66.5; H, 5.9; Br, 30.3. Calc. for $C_{16}H_{17}Br$: C, 66.5; H, 5.9; Br, 27.6%).

The results of the cognate preparations of other 4'-n-alkyl-4-bromobiphenyls are given in Table 5.

Table 4 Data for 4'-n-Alkanoyl-4-bromobiphenyls

n-Alkanoyl Group	Vol. of absolute ethanol for crystallisation (ml)	Yield (%)	mp (°C)	Analysis Found (%) C	H	Br		Calc. (%) C	H	Br
pentanoyl	350	47	98–98.5	64.6	5.4	24.9	$C_{17}H_{17}BrO$	64.4	5.4	25.2
hexanoyl	350	75	108.5–109.5	65.1	5.8	24.4	$C_{18}H_{19}BrO$	65.3	5.8	24.1
heptanoyl	350	79	99–100	66.1	6.1	22.9	$C_{19}H_{21}BrO$	66.1	6.1	23.1
octanoyl	400	50	104.5–105	66.7	6.5	21.9	$C_{20}H_{23}BrO$	66.9	6.4	22.2
nonanoyl	350	83	107–107.5	67.7	6.9	21.7	$C_{21}H_{25}BrO$	67.6	6.8	21.4

Table 5 Data for 4'-n-Alkyl-4-bromobiphenyls

n-Alkyl Group	Solvent for crystallisation	Yield	mp	Analysis						
				Found (%)				Calc. (%)		
		(%)	(°C)	C	H	Br		C	H	Br
Pentyl	Absolute ethanol	62	95–96	67.2	6.5	26.1	$C_{17}H_{19}Br$	67.3	6.3	26.4
Hexyl	Absolute ethanol	55	92.5–93	68.4	6.8	25.1	$C_{18}H_{21}Br$	68.1	6.7	25.2
Heptyl	Absolute ethanol	62	93–94	68.9	7.0	23.9	$C_{19}H_{23}Br$	68.9	7.0	24.1
Octyl	Absolute ethanol	62	87–89	69.6	7.2	22.7	$C_{20}H_{25}Br$	69.6	7.3	23.1
Nonyl	Light petroleum (bp 40–60°C) and Distilled	34.5	87.5	70.4	7.5	22.0	$C_{21}H_{27}Br$	70.2	7.6	22.2

4'-n-Butyl-4-cyanobiphenyl[22]. A mixture of 4'-n-butyl-4-bromobiphenyl (6.0g, 1 mol), cuprous cyanide (5.57g, 1.5 mol), dried in an oven at 160°C for 4h, and dry dimethylformamide (75ml) was heated under reflux for 12h. On cooling, the reaction mixture was poured into a mixture of hydrated ferric chloride (6.6g), concentrated hydrochloric acid (1.5ml) and water (75ml) and the resultant mixture stirred at 60-70°C for 20m. The organic material was extracted into chloroform, and the extract washed successively with 5N-hydrochloric acid, water, 10% aqueous sodium hydroxide, and water and dried over anhydrous sodium sulphate. The chloroform was removed by distillation under reduced pressure and the solid residue was purified by column chromatography on silicic acid using chloroform to elute the column. Before column chromatography, a small amount of insoluble amide was removed by filtration. The major impurity in the product was unreacted bromo compound and this was eluted from the column in advance of the cyano compound. The fraction containing the cyano compound was shown by tlc on silicic acid (chloroform as solvent) to be single component with an $R_F = 0.54$. The product was crystallised from light petroleum (bp 40-60°C) at -78°C. In order to remove last traces of solvent and possibly impurity, the material was distilled under reduced pressure using a sublimation apparatus with a cup on the end of the cold finger; the fraction was collected at a bath temperature of 115-120°C at 0.3mm. The yield of colourless product was 2.72g (56%).

The product was shown by infra-red spectroscopy and mass spectrometry to be the required material. In addition to tlc, the purity of the final product was checked by glc. The following column was used: 1.5m long x 2mm internal diameter glass column packed with 3% OV225 on Gas Chrom Q (100-120 mesh) at a temperature of 230°C using a nitrogen gas flow of 12 ml/m. The purity was 99.6%. Finally, the product was submitted for combustion analysis (Found: C, 87.2; H, 7.1; N, 6.1. Calc. for $C_{17}H_{17}N$: C, 86.8; H, 7.3; N, 6.0%).

The results of the cognate preparations of the other 4'-n-alkyl-4-cyanobiphenyls are given in Table 6.

All the compounds in Table 6 have been shown by glc to have a purity of >99%.

During the synthetic programme, it was found that the crystallisation step in no way improved the purity of the materials. For the last three compounds prepared, the product from column chromatography was distilled without crystallisation.

Table 6 Data for 4'-n-Alkyl-4-cyanobiphenyls

n-Alkyl Group	Yield (%)	Distillation Temp. (bath) (°C/mm)	Solvent if Product Crystallised	Analysis: Found (%) C	H	N		Calc. (%) C	H	N
Pentyl	53	140–150/0.5	light petroleum (bp 40–60°C) at −78°C	86.5	7.4	5.5	$C_{18}H_{19}N$	86.7	7.7	5.6
Hexyl	30	130–140/0.3	-	86.3	8.0	5.3	$C_{19}H_{21}N$	86.7	8.0	5.3
Heptyl	48	130–140/0.3	-	86.6	8.4	5.0	$C_{20}H_{23}N$	86.6	8.4	5.1
Octyl	35	130–140/0.3	light petroleum (bp 40–60°C) at −78°C	86.2	8.8	4.8	$C_{21}H_{25}N$	86.5	8.7	4.8
Nonyl	32	130–140/0.2	-	86.3	9.1	4.3	$C_{22}H_{27}N$	86.5	8.9	4.6

4'-n-Alkoxy-4-cyanobiphenyls

4'-Nitro-4-bromobiphenyl. This compound was prepared by the method of Le Fèvre and Turner[23]. The yield of product, crystallised from 80% acetic acid was 58%.

4'-Amino-4-bromobiphenyl. This compound was prepared by the method of Gray et al[24]. The yield of product, crystallised from absolute ethanol, was 91%.

4'-Hydroxy-4-bromobiphenyl. To 4'-amino-4-bromobiphenyl (30g, 0.12 mole) at 90°C, 95% acetic acid (180ml), preheated to 90°C, was rapidly added with stirring. To the stirred solution at 90°C, sulphuric acid (180ml of 40% w/w), preheated to 90°C, was rapidly added and the temperature of the mixture quickly lowered by use of an ice-salt bath. In this way, the amine salt was obtained in a finely divided form.

Whilst maintaining the temperature between 0-5°C, a solution of sodium nitrite (20.8g, 0.301 mole) in water (50ml) was added dropwise to the stirred suspension. After the addition, the suspension was stirred for 20m at 0-5°C, after which urea was added to destroy the excess of sodium nitrite. 95% Acetic acid (200ml) was then added to effect complete solution of the diazonium salt.

The cold diazonium salt solution was added dropwise (during 20m), slowly initially and then faster, to boiling sulphuric acid (100ml of 40% w/w). The reaction mixture was heated under reflux for 15m, cooled, and diluted with an equal volume of water. The solid deposited was filtered off and digested with hot N-sodium hydroxide (1500ml). Any insoluble material was filtered off and the filtrate acidified. The product was filtered off and washed with water before drying. The product, mp 166-7°C, was obtained in yields of 70-80% over a number of runs using quantities of starting material ranging from 6g to 30g.

4'-n-Alkoxy-4-bromobiphenyls. To 4'-hydroxy-4-bromobiphenyl (8.3g, 0.033 mole) dissolved in cyclohexanone (70ml), freshly crushed anhydrous potassium carbonate (18.2g, 0.132 mole) and the appropriate n-alkyl bromide (0.053 mole) were added. The reaction mixture was stirred efficiently and heated under reflux for 4h, after which time the colour had changed from red to a straw colour and the potassium carbonate adhered to the walls of the flask. The inorganic material was filtered off, washed with ether, and the washings combined with the filtrate. The solvent was removed by distillation under reduced pressure, the last traces of cyclohexanone being removed under high vacuum at 100°C. The residue was crystallised to constant melting point from a suitable solvent.

The products were shown by infra-red spectroscopy to be the required materials and the purities were checked by tlc on silicic acid using chloroform as solvent. Finally, the products were submitted for combustion analysis.

The results of the preparations of the 4'-n-alkoxy-4-bromobiphenyls are given in Table 7.

4'-n-Alkoxy-4-cyanobiphenyls. The cyanation of the five 4'-n-alkoxy-4-bromobiphenyls was carried out in an analogous way to that of the bromo-butyl compound. However, the reaction time was varied for the individual alkoxy compounds in order to establish the optimum heating time. This was, in fact, 12h and this time was used for the cyanation of the six bromo-alkyl compounds. The work up procedure for obtaining the cyano-alkoxy compounds was analogous to that for the cyano-alkyl compounds. Purification was effected by column chromatography and gave materials each of which showed a single spot on tlc (silicic acid with chloroform as solvent). The products were crystallised from a suitable solvent and finally distilled under reduced pressure to yield colourless products. Each product was shown by infra-red spectroscopy and mass spectrometry to be the required material. The purity was checked by glc and shown to be >99% in every case. Finally, each product was submitted for combustion analysis.

The results of the preparations of the five 4'-n-alkoxy-4-cyanobiphenyls are given in Table 8.

During the course of these preparations it was found that the crystallisation step did not serve a useful purpose and only wasted valuable material. With the n-pentyloxy and n-heptyloxy compounds, the chromatographed material was simply distilled as a final purification step.

In addition to preparing five 4'-n-alkoxy-4-cyanobiphenyls by the synthetic route given above, one other compound 4'-n-propyloxy-4-cyanobiphenyl was prepared by an alternative synthetic route to provide a comparison for the future preparation of these materials.

4'-Hydroxy-4-cyanobiphenyl. The cyanation of 4'-hydroxy-4-bromobiphenyl was carried out in an analogous way to that of the 4'-n-alkoxy- and 4'-n-alkyl-4-bromobiphenyls. In this particular case the reaction time was 6h. The work up procedure was the same except that the chloroform extract was washed with 10% aqueous sodium bicarbonate, and not sodium hydroxide. Purification was effected by crystallisation. The solid was dissolved in the minimum volume of hot acetone and sufficient light petroleum (bp 40-60°C) added to produce a turbidity. The solution was

Table 7 Data for 4'-n-Alkoxy-4-bromobiphenyls

n-Alkoxy Group	Solvent for Crystallisation	Yield (%)		Analysis Found (%) C	H	Br		Calc. (%) C	H	Br
Butyloxy	A	53	138.5–139	62.8	5.5	26.3	$C_{16}H_{17}BrO$	63.0	5.6	26.2
Pentyloxy	B	50	132.5–133	63.7	6.1	24.4	$C_{17}H_{19}BrO$	64.0	6.0	25.0
Hexyloxy	C	72	128–128.5	64.7	6.3	24.9	$C_{18}H_{21}BrO$	64.9	6.3	24.0
Heptyloxy	D	65	126–126.5	65.8	6.8	22.8	$C_{19}H_{23}BrO$	65.7	6.7	23.0
Octyloxy	E	53	124–125.5	66.6	6.9	23.2	$C_{20}H_{25}BrO$	66.5	6.9	22.2

A Ethanol/Light Petroleum (bp 40–60°C)

B Methanol/Ethanol

C Ethanol

D Benzene/Light Petroleum (bp 60–80°C)

E Benzene/Light Petroleum (bp 40–60°C)

Table 8 Data for 4'-n-Alkoxy-4-cyanobiphenyls

n-Alkoxy Group	Reaction Time (h)	Yield (%)	Solvent for Crystallisation	Distillation Temp. (bath) (^oC/mm)	Analysis Found (%) C	H	N		Calc. (%) C	H	N
Butyloxy	8	41	A	115–120/0.3	81.3	6.8	5.7	$C_{17}H_{17}NO$	81.3	6.8	5.6
Pentyloxy	12	35	-	140–150/0.1	81.6	7.1	5.1	$C_{18}H_{19}NO$	81.5	7.2	5.3
Hexyloxy	20	38	B	135–140/0.3	81.6	7.6	5.1	$C_{19}H_{21}NO$	81.7	7.6	5.0
Heptyloxy	12	45	-	170/0.1	82.0	8.0	4.9	$C_{20}H_{23}NO$	81.9	7.9	4.8
Octyloxy	5.5	28	C	125–135/0.9	81.9	8.3	4.6	$C_{21}H_{25}NO$	82.1	8.1	4.6

A Benzene/Light Petroleum (bp 40–60°C)

B Light Petroleum (bp 40–60°C) at –78°C

C Light Petroleum (bp 40–60°C)

Table 9 Data for 4'-n-Alkoxy-4-nitrobiphenyls

n-Alkoxy Group	Solvent for Crystallisation	Yield (%)	Analysis Found (%) C	H	N		Calc. (%) C	H	N
Pentyloxy	toluene-light petroleum (bp 40-60°C)	68	71.8	6.7	4.8	$C_{17}H_{19}NO_3$	71.6	6.7	4.9
Hexyloxy*	light petroleum (bp 40-60°C) (x 2)	50	72.2	7.0	4.6	$C_{18}H_{21}NO_3$	72.2	7.0	4.7
Heptyloxy	toluene-light petroleum (bp 40-60°C) (x 3)	51	72.9	7.5	4.3	$C_{19}H_{23}NO_3$	72.8	7.3	4.5
Octyloxy*	light petroleum (bp 40-60°C) (x 2)	67	73.3	7.6	4.3	$C_{20}H_{25}NO_3$	73.4	7.6	4.3

* Crystallisation alone was not effective in removing traces of impurity and the materials were chromatographed on silicic acid using chloroform to elute the column.

refrigerated, whereon the product crystallised. The yield, mp 188–195°C, was 29%. The material was further purified by sublimation at 160°/0.3mm (bath temp.), whereon a white solid, mp 196–199°C, was obtained (lit. mp 196–199°C)[25]. The material was submitted for combustion analysis (Found: C, 79.8; H, 4.9; N, 7.2. Calc. for $C_{13}H_9NO$: C, 80.0; H, 4.7; N, 7.2%).

4'-n-Propyloxy-4-cyanobiphenyl. The alkylation of 4'-hydroxy-4-cyanobiphenyl with n-propyl bromide was carried out as for the alkylation of 4'-hydroxy-4-bromobiphenyl. After working up the reaction mixture, the solid obtained was purified by column chromatography on silicic acid using chloroform to elute the column. By tlc on silicic acid (chloroform as solvent) the cyano compound showed a single spot. After crystallisation from toluene-light petroleum (bp 40–60°C), the product was distilled and the fraction collected at a bath temperature of 145–150°C at 0.2mm. The yield of 4'-n-propyloxy-4-cyanobiphenyl was 41%.

The product was shown by infra-red spectroscopy and mass spectrometry to be the required material. The purity was shown by glc to be >99%. Finally, the material was submitted for combustion analysis (Found: C, 81.2; H, 6.4; N, 6.0. Calc. for $C_{16}H_{15}NO$: C, 81.0; H, 6.4; N, 5.9%).

4'-n-Alkoxy-4-nitrobiphenyls

Four of these compounds were prepared by the alkylation of 4'hydroxy-4-nitrobiphenyl, this compound having been synthesised by the method of Jones and Chapman[26].

The alkylation procedure was analogous to that described for the 4'-hydroxy-4-bromobiphenyl. After working up the reaction mixture, the product was crystallised from a suitable solvent until the material showed a single spot by tlc on silicic acid (chloroform as solvent). Each compound was shown by infra-red spectroscopy and mass spectrometry to be the required material. Finally, the materials were submitted for combustion analysis.

The results of the preparations of the four 4'-n-alkoxy-4-nitrobiphenyls are given in Table 9.

ACKNOWLEDGEMENT

This work was carried out under contract to the Ministry of Defence.

References

1. Elliott, G. Chem. in Brit. 1973, 9(5), 213.

2. Heilmeier, G.H., Zanoni, L.A. and Barton, L.A. Proc. IEEE 1968, 56, 1162.

3. Schadt, M. and Helfrich, W. Appl. Phys. Letters 1971, 18, 127.

4. Wysocki, J.J., Adams, J. and Haas, W. Phys. Rev. Letters 1968, 20, 1024.

5. Gray, G.W. Plenary Lecture at the Fourth Intern. Liquid Cryst. Conf. at Kent, Ohio, USA, 1972. To be published in Mol. Cryst. Liquid Cryst.

6. van der Veen, J., de Jeu, W.H., Grobben, A.H. and Boven, J. Mol. Cryst. Liquid Cryst. 1972, 17, 291.

7. Kelker, H., Scheurle, B., Hatz, R. and Bartsch, W. Angew. Chem. 1970, 82, 984.

8. Knaak, L.E., Rosenberg, H.M. and Servé, M.P. Mol. Cryst. Liquid Cryst. 1972, 17, 171.

9. Young, W.R., Aviram, A. and Cox, R.J. IBM Res. Rept. 1971, RC 3559, 1 and Angew. Chem. 1971, 83, 399.

10. Malthète, J., Leclercq, M., Gabard, J., Billard, J. and Jacques, J. Compt. Rend. 1971, 273, 265.

11. Van Meter, J.P. and Klanderman, B.H. Papers presented at the Fourth Intern. Liquid Cryst. Conf. at Kent, Ohio, USA, 1972. To be published in Mol. Cryst. Liquid Cryst.

12. Castellano, J.A. US Patent 1928003 (1970).

13. Boller, A., Scherrer, H., Schadt, M. and Wild, P. Proc. IEEE 1972, 60, 1002.

14. E. Merck, Current Information, Licristal, Liquid Crystals 1972, No. 1.

15. Gray, G.W., Hartley, J.B. and Jones, B. J. Chem. Soc. 1955, 1412.

16. Gray, G.W., Hartley, J.B., Ibbotson, A. and Jones, B. J. Chem. Soc. 1955, 4359.

17. Gray, G.W., Harrison, K.J. and Nash, J.A. Electron. Letters 1973, 9(6), 130; Ashford, A., Constant, J., Kirton, J. and Raynes, E.P. Electron. Letters 1973, 9(5), 118.

18. Herbert, A.J. Trans. Faraday Soc. 1967, 63, 555.

19. Gray, G.W. Molecular Structure and the Properties of Liquid Crystals, Academic Press, London, 1962, Chap. 9.

20. Gray, G.W. and Jones, B. J. Chem. Soc. 1953, 4179; Weygand, C. and Gabler, R. Z. Physik. Chem. 1940, B46, 270.

21. Vogel, A.I. Textbook of Practical Organic Chemistry, Longmans, Green and Co., London, New York and Toronto, 1956, p.368.

22. Friedman, L. and Shechter, H. J. Org. Chem. 1961, 26(2), 2522.

23. Le Fèvre, R.J.W. and Turner, E.E. J. Chem. Soc. 1926, 2045.

24. Gray, G.W., Hartley, J.B. and Jones, B. J. Chem. Soc. 1952, 1959.

25. Bach, F.L., Barclay, J.C., Kende, F. and Cohen, E. J. Med. Chem. 1968, 11, 987.

26. Jones, B. and Chapman, F. J. Chem. Soc. 1952, 1829.

27. Jakeman, E. and Raynes, E.P. Paper presented at the Fourth Intern. Liquid Cryst. Conf., Kent, Ohio, USA, 1972.

ELECTRIC FIELD INDUCED DEFORMATION IN NEMATIC PHENYL BENZOATES

G. Baur, A. Stieb, and G. Meier

Institut für Angewandte Festkörperphysik der
Fraunhofer-Gesellschaft, Eckerstrasse 4
D-7800 Freiburg, Germany

INTRODUCTION

Static magnetic and electric field induced deformations in nematic liquid crystals have been studied by several authors (1)-(9). The equilibrium state of the Fréedericksz transition is well understood for both magnetic and electric fields. In the static case the deformation is determined by the magnetic or electric susceptibilities, the elastic constants of the materials used as well as the boundary conditions.

In this paper, we are interested in the dynamics of the Fréedericksz transition where not only electric and elastic but also viscous torques come into play. A calculation of the response times of the magnetic deformation of planar and homeotropic samples was given by F. Brochard et al.(6, 8). Both rise and decay times are dependent on the elastic constants and the viscosities. The rise time additionally depends on the torque induced by the applied field. Long rise times observed in high viscosity materials, e.g. at low temperatures, can therefore be shortened by increasing the applied field, but the decay times are independent of the field. We present measurements of rise and decay times of the electric field induced deformation of parallel oriented layers of phenyl benzoates and describe a method that permits decreasing the decay time in materials where the dielectric anisotropy reverses sign at a frequency of several kHz.

I. EXPERIMENTAL METHOD

A simple method to study a deformation is to observe the absorption or emission of dyes dissolved in the liquid crystal. The guest molecules were chosen to have absorption and emission bands in a wavelength region where the liquid crystal is transparent. The molar concentrations of the solute molecules were less than 10^{-3} mole/liter. The electronic transition moments of the guest molecules were approximately parallel to the long molecular axis. The guest molecules used were dimethylamino-nitrostilbene with a transition moment parallel to the long molecular axis for absorption as well as for emission.

Regarding a parallel oriented sample we introduce a laboratory coordinate system xyz. The z-axis coincides with the optical axis of the liquid crystal and the y-axis lies in the plane of the sample. The incident light propagates along the x-axis (normal to the plane of the sample). μ_z and μ_y are the absorption coefficients for light polarized parallel to the z- and y-directions, respectively.

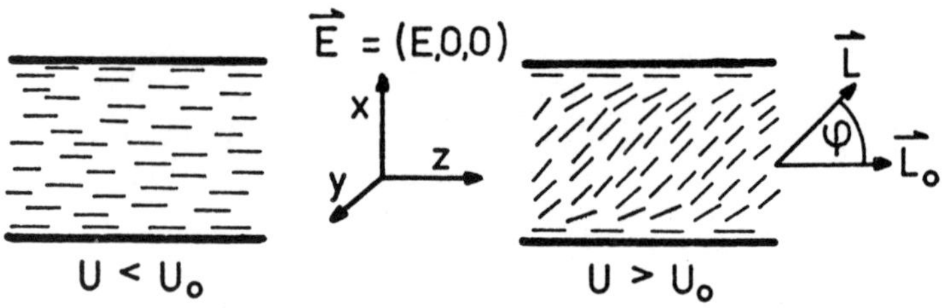

Fig. 1 Geometry of the normal deformation below and above threshold for a parallel oriented sample.

They are order dependent and given by:

$$\mu_{oz} \propto (1 + 2S) / 3$$
$$\mu_{oy} \propto (1 - S) / 3 \qquad (I.1)$$

The analogous expression for the fluorescence intensities are:
i) excitation unpolarized and emission polarized parallel to the z and y axis respectively:

$$I_{ouz} \propto (2S^2 + 5S + 2) /9$$
$$I_{ouy} \propto (-S^2 - S + 2) /9 \qquad (I.2)$$

ii) excitation and emission polarized parallel to the z and y axis respectively:

$$I_{ozz} \propto (4S^2 + 4S + 1) / 9$$
$$I_{oyy} \propto (S^2 - 2S + 1) / 9 \qquad (I.3)$$

S is the order parameter.

Since the order parameter of the liquid crystal is not influenced by the applied magnetic or electric field, the corresponding expressions for a deformed sample are given by the following relations:

Absorption coefficients:

$$\mu_z = \mu_{oz} \langle \cos^2 \varphi \rangle \qquad (I.4)$$
$$\mu_y = \mu_{oy}$$

Fluorescence intensities for polarized excitation:

$$I_{zz} = I_{ozz} \langle \cos^2 \varphi \rangle^2 \qquad (I.5)$$
$$I_{yy} = I_{oyy}$$

φ is the angle between the local optical axes $\vec{L}_o$ and $\vec{L}$ of the undeformed and of the deformed sample respectively. The brackets indicate the spatial averages.

The fluorescence intensities I_{uz} and I_{uy} for unpolarized excitation are not simple functions of φ .

The expressions for the fluorescence intensities given above are valid for small dye concentrations or thin layers.

We define an average angle $\bar{\varphi}$ by

$$\langle \cos^2 \varphi \rangle = \cos^2 \bar{\varphi} \qquad (I.6)$$

Introducing (I. 6) into (I, 4) and (I. 5) gives

$$\bar{\varphi} = \text{arc } \cos \sqrt{\mu_z / \mu_{oz}} \qquad (I.7)$$

and

$$\bar{\varphi} = \text{arc } \cos \sqrt[4]{I_{zz} / I_{ozz}} \qquad (I.8)$$

In our experimental arrangement the area of the specimen penetrated by the light beam had a diameter of 10 millimeters. The absorption measurements were carried out with a Cary 14 spectrophotometer and the fluorescence measurements were made with a Baird-Atomic Fluorispec. Sample thickness was 20 μ.

II. NEMATIC SUBSTANCE

The liquid crystal used was a mixture of some disubstituted benzoyloxybenzoic acid phenyl esters with a mesomorphic range from -10 to 90°C (WI from Merck). This mixture has unusual properties because the dielectric anisotropy $\Delta\varepsilon = \varepsilon_{\parallel} - \varepsilon_{\perp}$ is zero for a frequency f_o (10). $\varepsilon_{\parallel}$ and $\varepsilon_{\perp}$ are the dielectric constants measured parallel and perpendicular to the optical axis respectively. $\Delta\varepsilon$ is positive for frequencies below f_o and negative for frequencies above f_o. This change in sign of the dielectric anisotropy is due to a dispersion of $\varepsilon_{\parallel}$ which occurs in our mixture at 25°C at a frequency of about 10 kHz.

Theoretically the relaxation time τ_R is given by (11)

$$\tau_R = g\tau_o \qquad \text{(II. 1)}$$

with

$$g = (kT/W)\left\{\exp(W/kT) - 1\right\}$$
$$\tau_o = 4\pi\eta a^3/kT \qquad \text{(II. 2)}$$

g is the so-called retardation factor and describes the influence of the nematic order in terms of a nematic potential barrier W. τ_o is the ordinary Debye relaxation time, η is the viscosity and a is a constant of molecular dimensions. Considering the temperature dependence of the viscosity, the relaxation frequency f_R is given by

$$f_R \propto \exp(-A/kT) \qquad \text{(II. 3)}$$

where A is temperature independent.

A logarithmic plot of f_o as a function of $1/T$ is given in fig. 2. f_o was taken as the frequency at which no deformation occurred.

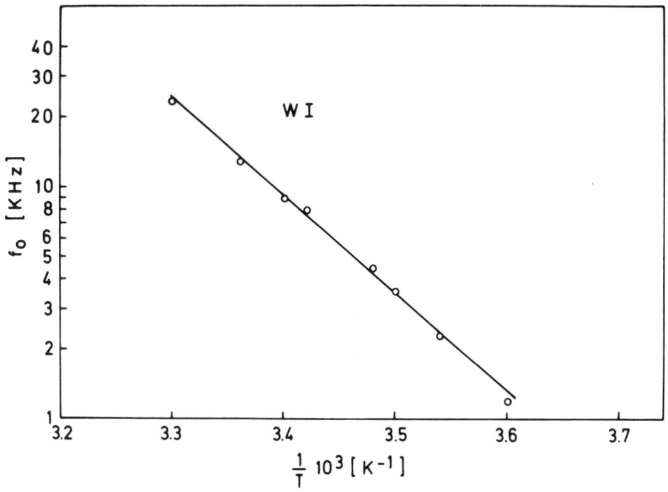

Fig. 2 Temperature dependence of the frequency f_o at which $\Delta\varepsilon$ changes sign in the nematic material WI

III. STATIC DEFORMATION

A theoretical treatment of the static deformation of a liquid crystal in an electric field was given by Deuling (5) and by Gruler, Scheffer and Meier (4). These authors give expressions for the angle φ defined in sect. I which is dependent on the elastic constants K_{11} (splay) and K_{33} (bend), the dielectric constants $\varepsilon_{\parallel}$ and $\varepsilon_{\perp}$, the thickness of the sample d and the applied voltage U. For a parallel oriented sample the deformation begins at a threshold voltage

$$U_o = \pi (K_{11}/\varepsilon_o \Delta\varepsilon)^{1/2} \qquad \text{(III. 1)}$$

In our experiments the deformation was observed using the fluorescence method described in sect. I. It is useful to observe the fluorescence intensity I_{zz} because it is dependent on the fourth power of cos φ and it is therefore more sensitive to small changes in φ than the absorption coefficients. Fig. 3a shows I_{zz} as a function of the applied voltage U for different frequencies. I_{yy} is independent on the applied voltage which is expected from (I. 5). The deviation of I_{zz} from I_{yy} indicates

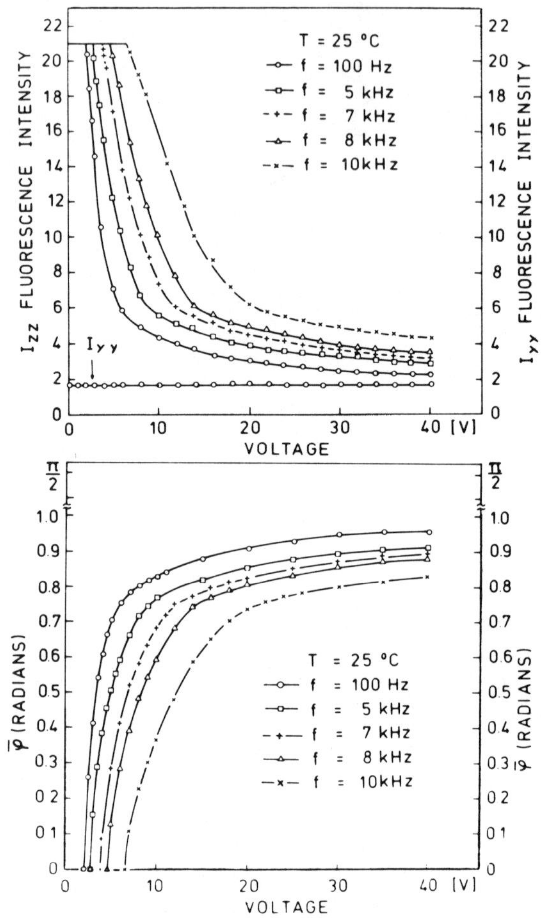

Fig. 3 Upper part: Fluorescence intensities as a function of the applied voltage

Lower part: Mean angle $\bar{\varphi}$ as a function of the applied voltage the deviation of the deformed sample from a homeotropic sample. The angles evaluated from I_{zz} are plotted in fig. 3b. U_o increases with increasing frequency up to f_o. For $f > f_o$ no deformation occurred. In the frequency range near f_o the scattering mode previously reported by de Jeu et al. (10) was observed.

IV. DYNAMICS OF THE DEFORMATION

The dynamics of a deformation can be described in terms of a time dependent local optical axis $\vec{L}$ ($\vec{r}$, t) and a velocity $\vec{V}$ ($\vec{r}$, t) of a fluid flow produced by the gradient of the angular velocity of $\vec{L}$. φ ($\vec{r}$, t) is the time dependent angle between $\vec{L}_o$ and $\vec{L}$. The conservation of angular and linear momentums yields the equations of motion. For a deformation of a parallel oriented sample in an electric field φ ($\vec{r}$, t) is assumed to depend only on the x-direction and the time t: φ (x, t). The only nonzero component of the velocity is assumed to be V_z (x, t). Then the linearised equations of motion are given by

$$\Delta\varepsilon\varepsilon_o E^2\varphi + K_{11}\frac{\partial^2\varphi}{\partial x^2} - (\alpha_3-\alpha_2)\frac{\partial\varphi}{\partial t} - \alpha_3\frac{\partial V}{\partial x} = 0 \qquad \text{(IV. 1)}$$

and

$$\frac{1}{2}(\alpha_3+\alpha_4+\alpha_6)\frac{\partial^2 V}{\partial x^2} + \alpha_3\frac{\partial^2\varphi}{\partial x\partial t} = 0 \qquad \text{(IV. 2)}$$

respectively. Inertial effects are neglected. α_i are the viscosity coefficients defined by Leslie. The solutions

$$\varphi = \varphi_{max}\left(\cos kx - \cos\frac{1}{2}kd\right)e^{st} \qquad \text{(IV. 3)}$$

$$V = V_{max}\left(\sin kx - \frac{2x}{d}\sin\frac{1}{2}kd\right)e^{st}$$

given by Brochard et al. (6) satisfy the boundary conditions

$$\varphi(\pm d/2, t) = V_z(\pm d/2, t) = 0 \qquad \text{for all } t$$

Rise time

The reciprocal time constant s follows from (IV, 3):

$$\frac{1}{\tau_r} = s = \frac{(\frac{\pi}{d})^2 K_{11}}{\beta(\alpha_3-\alpha_2)} \left\{ (\frac{U}{U_o})^2 - 1 \right\} \qquad \text{(IV. 4)}$$

To a good approximation, β is constant for small deformations.

Because of the separation of the variables in (IV. 3) we can set

$$\langle \varphi(t) \rangle = \varphi_{max} \langle \cos kx - \cos\tfrac{1}{2}kd \rangle e^{st} = \varphi_{max} B e^{st} \qquad \text{(IV. 5)}$$

For small angles φ we get a good approximation by assuming

$$\langle \varphi(t) \rangle = \bar{\varphi}(t)$$

Then from (I. 8) we get

$$\arccos \sqrt[4]{I_{zz}/I_{ozz}} = \langle \varphi(t) \rangle = \varphi_{max} B e^{st} \qquad \text{(IV. 6)}$$

or

$$\ln \arccos \sqrt[4]{I_{zz}/I_{ozz}} = c + st \qquad \text{(IV. 7)}$$

where c is a constant.

Equation (IV. 7) offers the opportunity to check the exponential decay law for $\bar{\varphi}$ and to evaluate the time constant $\tau_r = 1/s$ from the fluorescence intensities I_{zz}. A plot of $\ln \bar{\varphi}$ versus t is given in fig. 4.

The voltage dependence of the time constant τ_r is shown in fig. 5. For low voltages $1/\tau_r$ is proportional to $(\frac{U}{U_o})^2 - 1$ which is predicted by the theory (equ. (IV. 4)).

Decay time

Two types of decay times are to be distinguished: the ordinary passive decay time τ_p which follows the turning off of an electric field and the active decay time τ_a which follows a "frequency

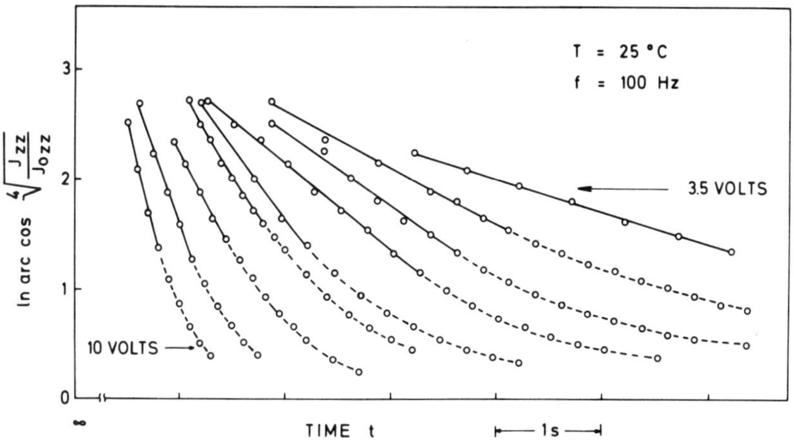

Fig. 4 Mean angle of deformation as a function of time

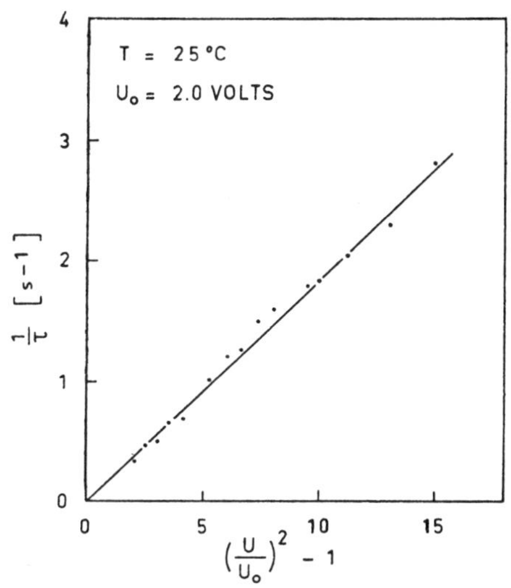

Fig. 5 Voltage dependence of the rise time

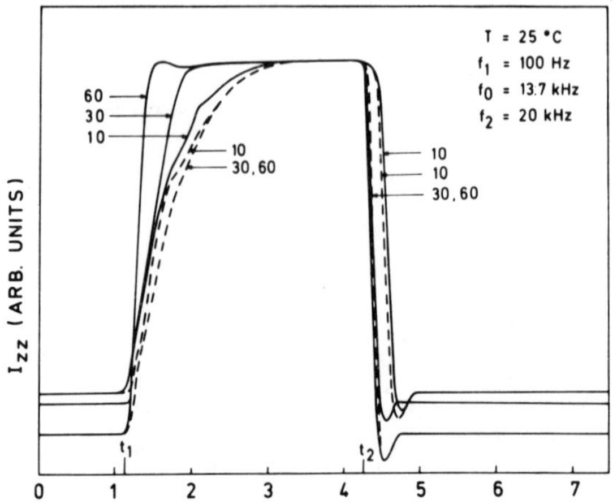

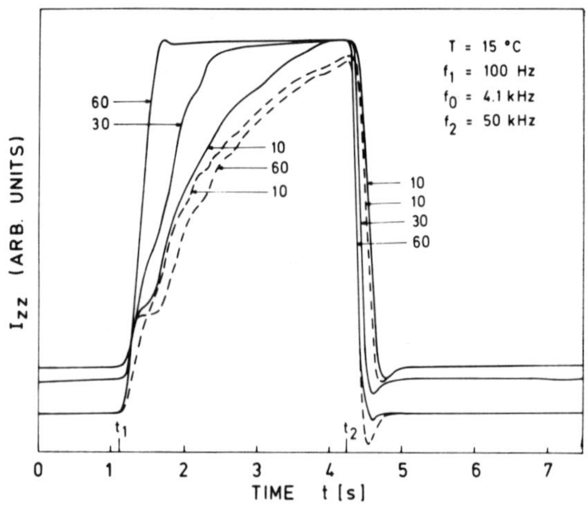

Fig. 6 Dynamics of the deformation measured by the fluorescence intensity I_{zz} at $25^{o}C$ (upper part) and $15^{o}C$ (lower part). Dashed curves indicate the passive decay and rise of the deformation; the electric field is turned off at t_1 and turned on at t_2. Solid curves indicate the active decay and rise ; at t_1 the frequency of the applied field is switched from $f < f_o$ to $f > f_o$ and at t_2 from $f > f_o$ to $f < f_o$. Arrows indicate the voltage at which the field is turned on or off or its frequency switched.

switch" from $f < f_o$ to $f > f_o$ with unchanged voltage.

In the passive case equ. (IV.4) gives for U = 0

$$1/\tau_p = - \frac{(\frac{\pi}{d})^2 K_{11}}{\beta(\alpha_3 - \alpha_2)} \qquad \text{(IV.8)}$$

For the active case we assume that the voltage is unchanged and the frequency is switched from $f < f_o$ to $f > f_o$. The torque induced by the field changes sign and τ_a is given by

$$1/\tau_a = - \frac{(\frac{\pi}{d})^2 K_{11}}{\beta(\alpha_3 - \alpha_2)} \left\{ (\frac{U}{U'_o})^2 + 1 \right\}$$

with $U'_o = \pi(K_{11}/\varepsilon_o \Delta\varepsilon^*)^{1/2}$

where $\Delta\varepsilon^*$ is the dielectric anisotropy at the high frequency used.

$1/\tau_p$ and $1/\tau_a$ are related to the fluorescence intensity I_{zz} in analogy to equation (IV.6). A comparison of the active and passive decay times with the rise time is given in fig. 6. High fluorescence intensities I_{zz} indicate the undeformed parallel oriented sample, low values of I_{zz} indicate the deformed sample.

The "turn off" or the "frequency switch" and the "turn on" of the field are marked by t_1 and t_2 respectively. The curves indicate that even at low temperatures the decay time can be shortened drastically by switching the frequency from $f < f_o$ to $f > f_o$.

Acknowledgement

The technical assistance of F. Windscheid is gratefully acknowledged. We would like to thank Dr. Steinsträsser (E. Merck, Darmstadt) for supplying us with the WI mixture.

REFERENCES

1) H. Zocher, Trans. Faraday Soc. 29, 945 (1933)

2) A. Saupe, Z. Naturforsch. 15a, 815 (1960)

3) M.F. Schiekel and K. Fahrenschon, Appl. Phys. Letters 19, 391 (1971)

4) H. Gruler, T.J. Scheffer, and G. Meier, Z. Naturforsch. 27a, 966 (1972)

5) H. Deuling, Mol. Cryst. and Liq. Cryst. 19, 123 (1972)

6) F. Brochard, P. Pieranski, and E. Guyon, Phys. Rev. Letters 28, 1681 (1972)

7) F. Brochard, P. Pieranski, and E. Guyon, J. Physique 33, 681 (1972)

8) F. Brochard, P. Pieranski, and E. Guyon, J. Physique 34, 35 (1973)

9) R.A. Soref and M.J. Rafuse, J. Appl. Phys. 43, 2029 (1972)

10) W.H. de Jeu, C.J. Gerritsma and W.J.A. Goossens, Phys. Lett. 39A, 355 (1972)

11) G. Meier and A. Saupe, Mol. Cryst. 1, 515 (1966). A more complete theoretical treatment of the dielectric relaxations in nematic liquid crystals was given by A.J. Martin, G. Meier, and A. Saupe, Symposium of the Faraday Society, 1971, No. 5, p. 119

KERR EFFECT IN THE ISOTROPIC PHASE OF p-AZOXYANISOLE

N. V. Madhusudana and S. Chandrasekhar

Raman Research Institute, Bangalore 560006

India

ABSTRACT

The electric birefringence in the isotropic phase of p-azoxyanisole is evaluated in terms of the phenomenological model of de Gennes. It is shown that when the contributions of the polarizability and the permanent dipole moment are taken into account there should occur a reversal of sign of the birefringence a few degrees above the nematic-isotropic transition point T_c. The result is in accordance with experiment.

INTRODUCTION

It has long been known [1] that the magnetic birefringence in the isotropic phase of p-azoxyanisole (PAA) increases rapidly as the temperature approaches the nematic-isotropic transition point T_c. This behavior is due to the effect of short range orientational order which can be described in terms of the phenomenological model of de Gennes [2]. It is natural to expect that the electric birefringence in the isotropic phase should also conform to this simple description, but the measurements of Tsvetkov and Ryumtsev [3] show that this is not the case. The Kerr constant of PAA actually exhibits a reversal of sign at about T_c+5^oK. The suggestion has been made [3] that the sign reversal may be due to an increase of the orientational freedom of the molecule, but as yet there appears to be no quantitative explanation of the phenomenon. The aim of this paper is to show that the observed temperature variation of the electric birefringence can be understood in terms of the phenomenological model when proper allowance is made for the contributions of polarizability and the permanent dipole moment to the free energy.

THEORY

The free energy per mole of the isotropic phase in the presence of an externally applied field (magnetic or electric) is assumed to be of the form [2]

$$F = a(T-T^{*})Q^2 - BQ^3+CQ^4 \ldots + NW(Q) \qquad (1)$$

where $Q = \frac{1}{2}(\overline{3\cos^2\theta-1})$ is the long range order parameter, θ the angle which the molecular long axis makes with the preferred direction, W(Q) the <u>average</u> orientational potential energy of a molecule due to the external field and N the Avogadro number.

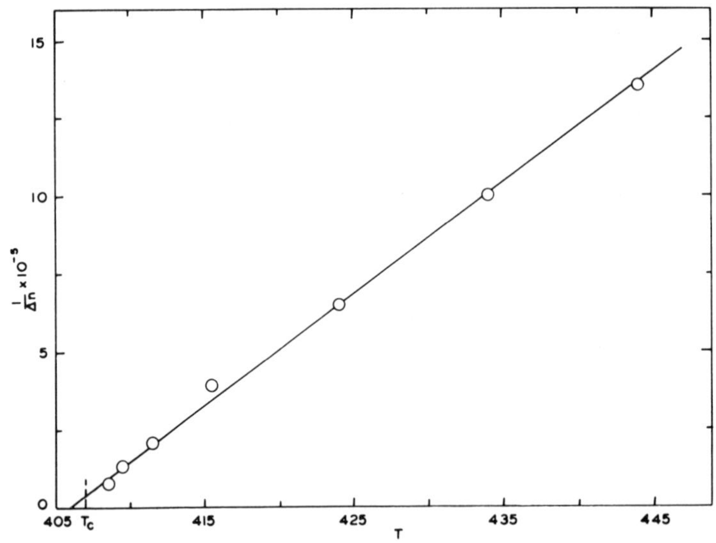

Fig.1: Reciprocal of magnetic birefringence of p-azoxyanisole versus temperature. Circles represent the experimental data of Zadoc-Khan [2] and the full line gives the variation according to Eq.(2) with a = 27.4 J mole^{-1} $^{o}K^{-1}$ and $T_c-T^{*} = 1^{o}K$.

If the external field is magnetic,

$$W(Q) = - \frac{\Delta\eta}{3} H^2 Q ,$$

where $\Delta\eta$ is the anisotropy of the diamagnetic susceptibility of the molecule. Neglecting cubes and higher powers of Q, the magnetic birefringence is given by

$$\Delta n = \frac{\pi N \, \Delta\chi \, \Delta\eta \, H^2 (n^2+2)^2}{27 Vna (T-T^*)} \qquad (2)$$

where $\Delta\chi$ is the anisotropy of optical polarizability of the molecule, n the mean refractive index and V the molar volume. Substituting $\Delta\chi = 25.9 \times 10^{-24} cm^3$ [4], $\Delta\eta = 104.2 \times 10^{-30} cm^3$ [5] and H=33,900 G [1] for PAA, the experimental data of Zadoc-Kahn [1] yield a=27.4 J $mole^{-1}\,{}^oK^{-1}$ and $T_c - T^* = 1^oK$ (Fig.1).

Under the action of an electric field E, the orientational potential energy of the molecule arises from (i) the anisotropy of low frequency polarizability $\Delta\alpha$, and (ii) the net permanent dipole moment μ. Moreover the effect of the cavity field and the reaction field produced in the medium cannot be ignored as in the diamagnetic case. Applying the Onsager theory, Maier and Meier [6] have worked out these contributions to the dielectric properties of the nematic phase. We follow their treatment closely. The average orientational energy due to the induced dipole moment is given by

$$W_1 = - \frac{\Delta\alpha}{3} F h^2 E^2 Q$$

where $h = 3\varepsilon/(2\varepsilon+1)$ is the cavity field factor, ε the average dielectric constant, $F = 1/(1-\alpha f)$ is the reaction field factor, α is the average polarizability and $f = \frac{2\varepsilon-2}{2\varepsilon+1} \frac{4\pi}{3} \frac{N}{V}$. (The effect of the anisotropy of the dielectric constant may be neglected since we are considering the very weakly ordered isotropic phase.)

The average orientational energy due to the permanent dipole moment is

$$W_2 = - \frac{F^2h^2\mu^2E^2}{6kT}(3\cos^2\beta-1)Q$$

where β is the angle which the dipole moment makes with the molecular long axis [see ref.6]. In obtaining this average, the distribution function is supposed to involve only even powers of $\cos\theta$. This is clearly valid in the present case in view of the assumed form of the free energy expression (1). Putting $W(Q) = W_1+W_2$, the free energy in the presence of an electric field is therefore

$$F = a(T-T^{*})Q^2 - \frac{Fh^2E^2}{3}\left[\Delta\alpha - \frac{F\mu^2}{2kT}(1-3\cos^2\beta)\right]Q.$$

The condition $\frac{\partial F}{\partial Q} = 0$ leads to the result

$$Q = \frac{Fh^2E^2\left[\Delta\alpha - \frac{F\mu^2}{2kT}(1-3\cos^2\beta)\right]}{6a(T-T^{*})}$$

and
$$\Delta n = \frac{\pi N\,\Delta\chi(n^2+2)^2Fh^2E^2\left[\Delta\alpha - \frac{F\mu^2}{2kT}(1-3\cos^2\beta)\right]}{27nVa(T-T^{*})}. \quad (3)$$

Evidently, the electric birefringence can be positive or negative depending on the sign of $\left[\Delta\alpha - \frac{F\mu^2}{2kT}(1-3\cos^2\beta)\right]$. For the elongated molecules of nematogenic compounds $\Delta\alpha$ is always positive, but the sign of the permanent dipole contribution depends on the angle β. If β is small, the two terms add to give rise to a strong positive Δn, whereas if β is sufficiently large Δn may be negative. Further, since the second term is proportional to T^{-1} there can occur, in principle, a reversal of sign of Δn with temperature. Qualitatively this accounts for the observation of Tsvetkov and Ryumtsev [3].

We shall now evaluate Δn of PAA from Eq. (3) using the values of a and T_c-T^{*} derived from the magnetic birefringence. The relevant data for PAA reported by Maier and Meier [7] are summarized below:

μ=2.22 D; $\beta \approx 64^{\circ}$ from dielectric anisotropy measurements in the nematic phase and 61° from the Kerr constant in dilute solutions (we shall use the mean value $\beta = 62.5^{\circ}$ in the present calculations), $\Delta\alpha$, the anisotropy of low frequency polarizability extrapolated

from the value in the optical region = 23.0×10^{-24} cm^3, ε=5.65 at T_c+5°K and $E=12 \times 10^3$V cm^{-1}. Inserting these values Δn derived from Eq. (3) has been plotted in Fig.2 as a function of temperature along with the experimental points of Tsvetkov and Ryumtsev. It will be seen that there is in fact a reversal of sign of Δn, though it occurs at $\sim 9^{\circ}$K above T_c. However, it may be emphasized that since there is a competition

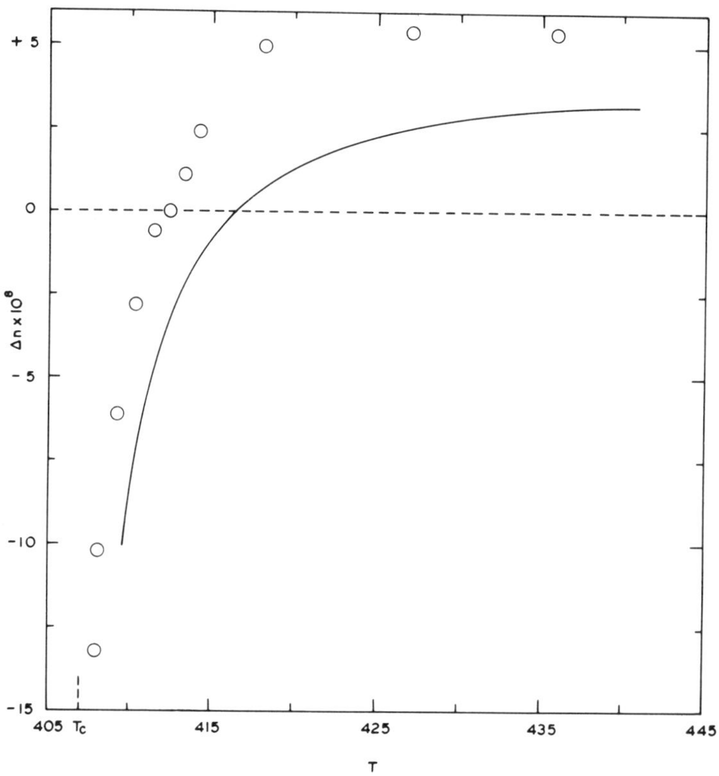

Fig.2: Electric birefringence of p-azoxyanisole versus temperature. Full line represents theoretical variation according to Eq.(3) and circles the experimental data of Tsvetkov and Tyumtsev [3].

between the polarizability and the permanent dipole contributions, even a small error in β will cause an appreciable shift in the temperature at which $\Delta n=0$. For example, if $\beta=62^{\circ}\ 24'$ the temperature of reversal coincides almost exactly with the experimental value. We may conclude therefore that Eq. (3) does broadly explain the observed trend in the temperature variation of the electric birefringence of PAA.

REFERENCES

1. J. Zadoc-Kahn, Ann. Phys. 11, 455 (1936).
2. P. G. de Gennes, Mol. Cryst. and Liq. Cryst. 12, 193 (1971).
3. V. N. Tsvetkov and E. I. Ryumtsev, Soviet Physics Crystallography, 13, 225 (1968).
4. A. Saupe and W. Maier, Z. Naturforsch. 16a, 816 (1961).
5. H. Gasparoux, B. Regaya and J. Prost, Compt. Rend. 272B, 1168 (1971).
6. W. Maier and G. Meier, Z. Naturforsch. 16a, 262 (1961).
7. W. Maier and G. Meier, Z. Naturforsch. 16a, 470 (1961).

CORRELATION TIME OF THE PROTON-ELECTRON INTERACTION IN MBBA WITH TRACES OF NITROXYDE MOLECULES

Jean-Pierre LE PESANT* and Pierre PAPON

Laboratoire de Resonance Magnetique, Universite Paris VI

and ESPCI, 10 rue Vauquelin, 75005 Paris, France

The work I am going to present here, is the continuation of the one which has been presented at the Fourth International Liquid Crystal Conference at Kent, Ohio, last summer.[1]

The double resonance technique can be described briefly as follows. Let us consider a sample containing resonant spins of two different species, in a large static magnetic field and let us irradiate this sample with two electromagnetic waves, the frequencies of which are such that they correspond respectively to the resonance phenomenon for each of the spin species.

If there is an interaction between the two different species, the saturation of the transitions of one of the species gives rise to a change of population on the levels of the other spin system and modifies the transition probabilities of the coupled system.

In the particular case where the resonant spins are respectively protons and electronic spins, the saturation of the electronic transitions can give rise, under some particular experimental conditions to the obtantion of dynamic nuclear polarization effects (D.N.P.). More precisely, this means that the system formed by the hydrogen nuclei and the electronic spins in interaction, has quantum states which are combinations of nuclear states and electronic states. Between these states, the interaction hamiltonian generates transitions, the frequency of which is characteristic of the type of interaction or in other words of the type of dynamic-nuclear polarization.

* Present address : Harvard University, DEAP, Gordon McKay Laboratory, 9 Oxford Street, Cambridge, Mass. 02138

If we measure the nuclear resonance spectrum in media of viscosity ranging from a viscosity characteristic of solids to a viscosity characteristic of liquids, we find that the line amplitude depends on the viscosity and that, for a given viscosity, if we vary the frequency of the electronic radio frequency irradiation, we obtain various nuclear resonance line amplitudes.

The polarization effect can be characterized by the ratio $P = \frac{A-Ao}{Ao}$ where A is the maximum amplitude of the nuclear resonance line in the presence of electronic saturation and Ao the maximum amplitude in the absence of such saturation.

Let us characterize the transition region from liquid effect to solid effect as the region where the displacement of the maximum of P is between 0 and 1 Gauss. In order to simplify the calculations, we remark that in this region, the effective correlation time for the nucleus-electron interaction is given with a good approximation [2] by the relation :

$$P = -\frac{\gamma_S^2 H_{1S}^2}{\frac{1}{T_e^2} + \gamma_S^2 (H_{1S}^2 + \Delta H^2)} \cdot \frac{\frac{2}{\Lambda^2+\Omega^2} - \frac{3}{5}\frac{4\gamma_S \Delta H \omega}{\Lambda^4}}{\frac{7}{5}\frac{2}{\Lambda^2+\Omega^2} + \frac{3}{5}\frac{2}{\Lambda^2}} \left|\frac{\gamma_S}{\gamma_I}\right| \quad (1)$$

where γ_S and γ_I are respectively electronic and nuclear gyromagnetic ratios, H_{1S} is the microwave magnetic field of irradiation, Te electronic relaxation time, $\Delta H = H_0 - \frac{\Omega}{\gamma_S}$,

Λ is defined by the relation :

$$\Lambda = \frac{1}{T_e} + \frac{1}{\tau_c} \quad (2)$$

where τ_c is the correlation time which characterizes the dipole-dipole interaction and consequently the intermolecular movement.

When we plot P as a function of $\Delta H = H_0 - \frac{\Omega}{\gamma_S}$, the point of this curve where $\Delta H = 0$ corresponds to a value given by :

$$P = -\frac{\gamma_S^2 H_{1S}^2}{\frac{1}{T_e^2} + \gamma_S^2 H_{1S}^2} \cdot \frac{5\Lambda^2}{10\Lambda^2 + 3\Omega^2} \left|\frac{\gamma_S}{\gamma_I}\right| \quad (3)$$

From this we can derive Λ :

$$\Lambda^2 = \frac{3}{5} \frac{\Omega^2}{\left|\frac{P_0}{P} \frac{\gamma_S}{\gamma_I}\right| \frac{\gamma_S^2 H_{1S}^2}{\frac{1}{T_e^2} + \gamma_S^2 H_{1S}^2} - 2} \tag{4}$$

Using the experimental values of P measured at $\Delta H = 0$, we can deduce the values of Λ versus temperature and, using the relation (2) we deduce those of τ_c.

Experimental Technique

Our measurements were performed on a double resonance spectrometer which was a combination of a microwave spectrometer and a proton resonance spectrometer.

The static magnetic field was 3300 G which corresponds to a proton resonance frequency of 13,8 MHz and to an electronic resonance frequency of 9870 MHz ; the microwave power was 470 mW, giving an electronic pumping field of 0,64 G.

The data we present here were obtained by dissolving in MBBA the free radical tetramethyl-2-2-6-6 brosylate -4-oxyle 1, abreviated hereafter as T.B.O., at a concentration 1.2 x 10^{-2} mole per liter (i.e. a concentration of 3.2 x 10^{-3} mole TBO per mole MBBA, or a mass concentration of 4.7 x 10^{-3}).

The nuclear resonance spectrum of MBBA, we obtained with our spectrometer was composed of a central line which we attributed mainly to CH_3 and C_4H_9 groups, and of two satellites, symetrically located on each side of the central line. The splitting of the satellites was mainly due to dipolar interaction between the two nearest neighbour protons situated on a same benzenic ring.

Typical D.N.P. curves are shown on fig. 1 and 2.

For the comodity of the experiment the electronic frequency was kept constant and the magnetic field and nuclear frequency varied to keep resonance conditions, which is equivalent to varying the electronic frequency only.

Determination of the Correlation Time τ_c.

If we saturate completely one of the three electronic transitions,

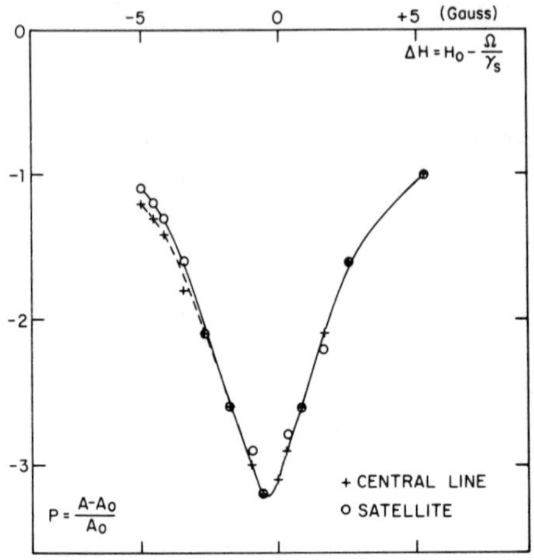

Fig. 1 $T-T_c = -3.5^{o}$

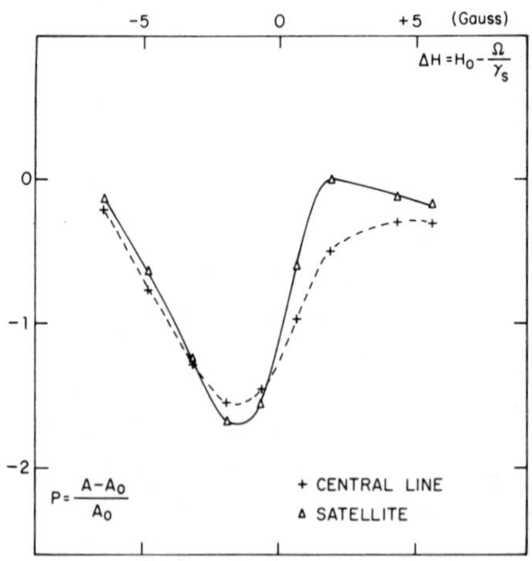

Fig. 2 $T-T_c = -23^{o}$

on one hand we have $\gamma_S^2 H_{1S}^2 \gg \frac{1}{T_e^2}$, and on the other hand the experimental polarization is 1/3 of the one given by the relation (1).

The relation (4) becomes :

$$\Lambda^2 = \frac{3}{5} \frac{\Omega^2}{\left|\frac{1}{3P} \frac{\gamma_S}{\gamma_I}\right| - 2} \quad (5)$$

So that :

$$\frac{1}{\tau_c} = \left[\frac{3}{5} \frac{\Omega^2}{\left|\frac{1}{3P} \frac{\gamma_S}{\gamma_I}\right| - 2} \right]^{\frac{1}{2}} - \frac{1}{T_e} \quad (6)$$

In the experimental cases we encountered $\frac{1}{T_e}$ was always negligible when compared to the first right-hand term in equation (6).

The figure 3 gives the variation of τ_c versus temperature.

The values of polarization P at $\Delta H = 0$ being different for the central line and for the satellites, this gives different values for the correlation time characteristic of the interaction between the electronic spins and the protons belonging to the C_4H_9 and CH_3 on one hand and to the benzenic rings on the other hand. The values of τ_c range from 1.8 to 3.2 x 10^{-10} seconds.

In organic liquids, the variation of τ_c versus temperature is generally represented by the relation :

$$\frac{1}{\tau_c} = A \exp \left(- \frac{W_o}{kT} \right) \quad (7)$$

where Wo is an activation energy.

However, in the nematic phase, this relation does not account for the experimental variation of τ_c and we think that this disagreement is due to the nematic order.

We take into account the order parameter by introducing in (7) a limited expansion of the free energy F containing the order parameter:

$$\frac{1}{\tau_c} = A \exp \left(- \frac{W_o}{kT} - \frac{\Delta F}{kT} \right) \quad (8)$$

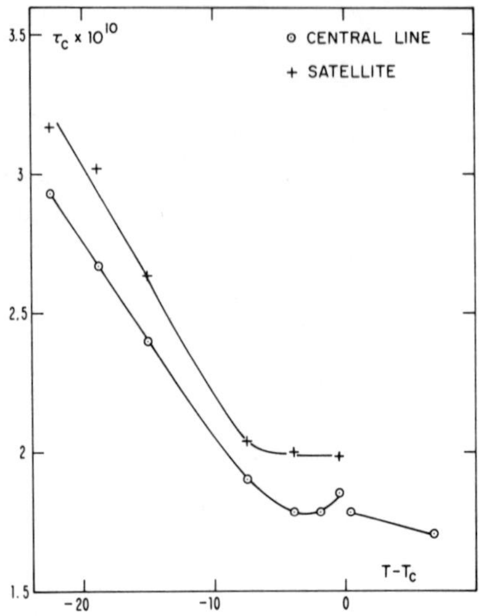

Fig. 3

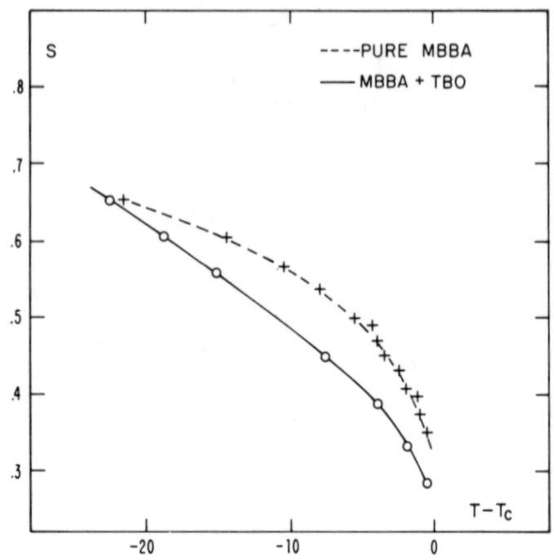

Fig. 4

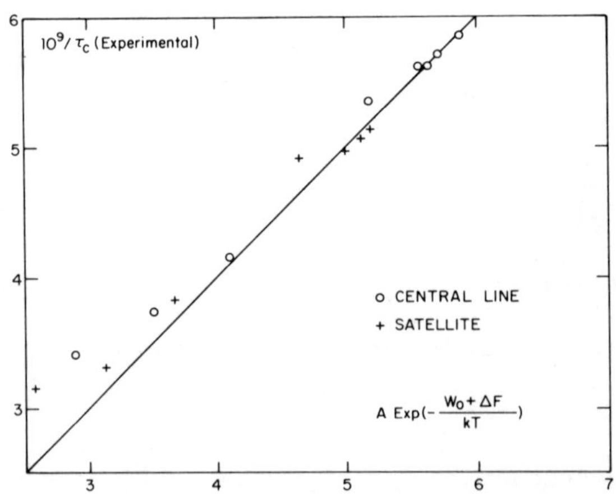

Fig. 5

where $\Delta F = F - F_o = a\,(T-T^*)S^2$

The small quantity of free radicals in solution behave like impurities and so modify the values of the order parameter and the transition temperature. The transition temperature for our samples of pure MBBA was 45°C and it was shifted to 40.5°C in the case of samples containing 1.2×10^{-2} mole/liter of T.B.O. The variation of the order parameter S in these two cases is represented on figure 4.

The value of T* which gives the best agreement with experimental values of τ_c in both cases of central line and satellites is given by :

$$T_c - T^* = 5.4°\ K$$

where T_c is the actual liquid-nematique transition temperature.

Figure 5 shows a fit to experimental values of $1/\tau_c$ derived from equ. (8). For the central line, we find $W_o/k = 637°K$; $A = 42.8 \times 10^9\ s^{-1}$; $a/k = -18.12$; this value of W_o/k describes also the behaviour in the liquid phase where S = 0(the two upper points on fig.5). For the satellites we find $W_o/k = 670°K$ and the same values for A and a/k (this means that only F_o is different).

To conclude, we recall that the values of D.N.P. in liquid crystals is very small and not very different from those found in liquids of similar viscosity but that within this range of values the DNP effect is slightly different for the various type of protons belonging to the same molecule of MBBA. This difference gives rise to a small difference in the values of the correlation time for the two types of proton-electron interaction. The more interesting point is that the correlation time of such an interaction is clearly influenced by the nematic order and that its variation is well described by equation (8).

Aknowledgments : One of the authors, J.P. Le Pesant, acknowledges the National Science Foundation for support through grants GH-33576 and GH-34401 while writing this article.

References :

[1] J.P. Le Pesant and P. Papon. 4th Inter. Liq. Cryst. Conference, Kent. Ohio, Aug. 1972, to be published in Molec. Cryst. and Liq. Cryst.
A complete theory of D.N.P. effects is given in J. Leblond, P. Papon and J. Korringa, Phys. Rev. A4, 1539, (1971)

[2] P. Papon Thèse, Paris 1967 (unpublished). P. Papon, J.L. Motchane, and J. Korringa, Phys. Rev. 175, 641, (1968)

BRILLOUIN SCATTERING IN THE ISOTROPIC PHASE OF MBBA

T. R. Steger, Jr. and J. D. Litster

Department of Physics and Center for Materials Science

and Engineering, Massachusetts Institute of Technology

We report here the results of our study of Brillouin scattering in the isotropic phase of the nematic liquid crystal para-methoxybenzylidene p-n-butylaniline (MBBA). In the nematic phase the anisotropic molecules of this material exhibit long range orientational order with their long axes tending to align parallel, while in the isotropic phase the molecular axes are randomly oriented. Although the isotropic phase shows no net order, the mean squared amplitude of fluctuations in the order becomes large near the nematic-isotropic transition[1]. In this paper we predict the effect of these fluctuations on the spectrum of Brillouin scattered light in the isotropic phase of a nematic liquid crystal. We show the spectrum obtained for Brillouin scattering in the isotropic phase of MBBA, give values of the hypersonic velocity and attenuation measured in the temperature region just above the nematic-isotropic phase transition, and compare these values with ultrasonic data.

The orientational order in the nematic phase of a liquid crystal may be specified by the symmetric traceless tensor[2]

$$Q_{\alpha\beta} = \frac{1}{2} \langle 3\zeta_\alpha \zeta_\beta - \delta_{\alpha\beta} \rangle$$

where ζ_α, ζ_β are the Cartesian components of the symmetry axis of a molecule and the brackets denote an ensemble average over a small volume. The dielectric constant tensor may be expressed in terms of the order parameter as

$$\varepsilon_{\alpha\beta} = \bar{\varepsilon}\, \delta_{\alpha\beta} + \frac{2\Delta\varepsilon}{3} Q_{\alpha\beta}$$

where $\Delta\varepsilon$ is the anisotropy in the dielectric constant for a perfectly ordered liquid crystal.

In a light scattering experiment light with wave vector $\vec{k}$ and polarization $\hat{i}$ is incident on a medium and is scattered at an angle θ with wave vector $\vec{k}'$ and polarization $\hat{f}$. The light is scattered by that spatial Fourier component of the fluctuations in the dielectric constant whose wave vector q conserves momentum between the wave vectors: $\vec{q} = \vec{k}'-\vec{k}$, or $q = 2k \sin(\theta/2)$. The spectral power density of light scattered at angular frequency shift ω is proportional to the spectrum of the dielectric constant fluctuations[3]

$$I(q,\omega) \sim \langle \delta\varepsilon_{if}^2(q,\omega)\rangle = \int_{-\infty}^{\infty} \langle \delta\varepsilon_{if}(q,0)\,\delta\varepsilon_{if}(-q,t)\rangle e^{i\omega t} dt$$

where $\delta\varepsilon_{if} = i_\alpha \delta\varepsilon_{\alpha\beta} f_\beta$. In an ordinary liquid the dielectric constant fluctuations are isotropic, resulting from the independent fluctuations of the temperature and the density[4]. Temperature fluctuations produce the Rayleigh component of the spectrum, centered at the incident light frequency. Density fluctuations give rise to the Brillouin components, symmetrically shifted by the frequency $\omega_B = \pm qC$ where C is the sound velocity.

In the isotropic phase of a nematic liquid crystal, the fluctuations in the dielectric constant tensor may be written

$$\delta\varepsilon_{\alpha\beta}(q,t) = \frac{\partial\bar{\varepsilon}}{\partial\rho}\,\delta\rho(q,t)\,\delta_{\alpha\beta} + \frac{2\Delta\varepsilon}{3}\,Q_{\alpha\beta}(q,t)$$

The fluctuations $\delta\rho$ and $Q_{\alpha\beta}$ are not independent in this case, and a coupling term of the type $\langle\delta\rho(q,0)Q_{zz}(-q,t)\rangle$ must be considered in the spectrum. We use de Gennes' phenomenological model of the nematic-isotropic transition[5] and include the effects of compressibility to predict the spectrum of light scattered by these fluctuations. The method used is similar to that used by Volterra[6] to discuss light scattering due to anisotropic fluctuations in ordinary liquids and the details will be published elsewhere. The spectrum of the scattered light near the Brillouin frequency ω_B should have three main components, as shown in Figure 1. Component I_A is the tail of the Rayleigh line and is due to order parameter fluctuations. Component I_B is a Lorentzian centered at ω_B and is the spectrum of the density fluctuations. Component I_C is a small asymmetric term centered at ω_B and is the contribution of the coupling term.

In our experimental system the light source was a single-mode frequency stabilized He-Ne laser[7] with an output power of 10 mw, short-term stability of 1 MHz, and long-term stability of 3 MHz/hr. The beam was focused in a cylindrical cell 10 cm long containing MBBA, and light scattered at an angle of 6.52° ($q = 1.824 \times 10^4 \text{cm}^{-1}$) was collected using a conical lens. The difficulty involved in

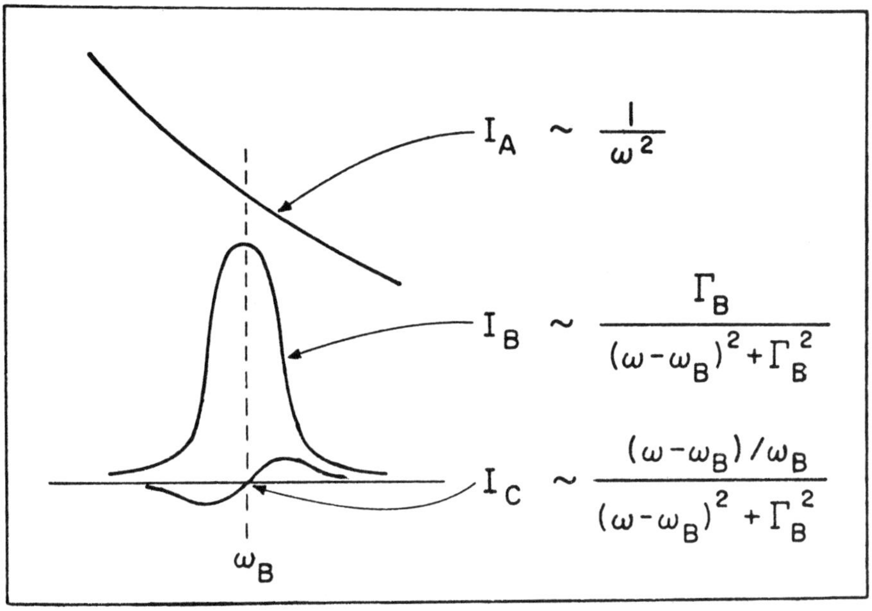

Figure 1. The predicted spectrum of scattered light near the Brillouin frequency ω_B showing the three main components expected.

observing the Brillouin peaks in a liquid crystal is that the Rayleigh scattering intensity is ~10^4 greater than the Brillouin intensity. Since most of the scattered light was from the Rayleigh component, a high contrast frequency filter was necessary to avoid leakage which would swamp the weak Brillouin components. The scattered light was frequency analyzed using a tandem spherical Fabry-Perot interferometer[8] to obtain very high contrast and detected by a photomultiplier and photon-counting equipment. The recorded spectrum was then the convolution of the theoretical spectrum with the tandem Fabry-Perot transmission profile. The temperature of the sample was stabilized to 2 mdeg and was measured with a platinum resistance thermometer.

A representative experimental trace is shown in Fig. 2. This spectrum was first analyzed by removing the Rayleigh tail. The remaining spectrum was then analyzed by numerically convolving the instrumental profile with various Lorentzians until a fit was obtained, as shown in the insert in Fig. 2. Deviation in the tails is probably due to our insufficiently accurate knowledge of the instrumental profile, which falls off extremely rapidly either side of its peak response and is therefore difficult to measure. We did not observe the asymmetric term corresponding to coupling between

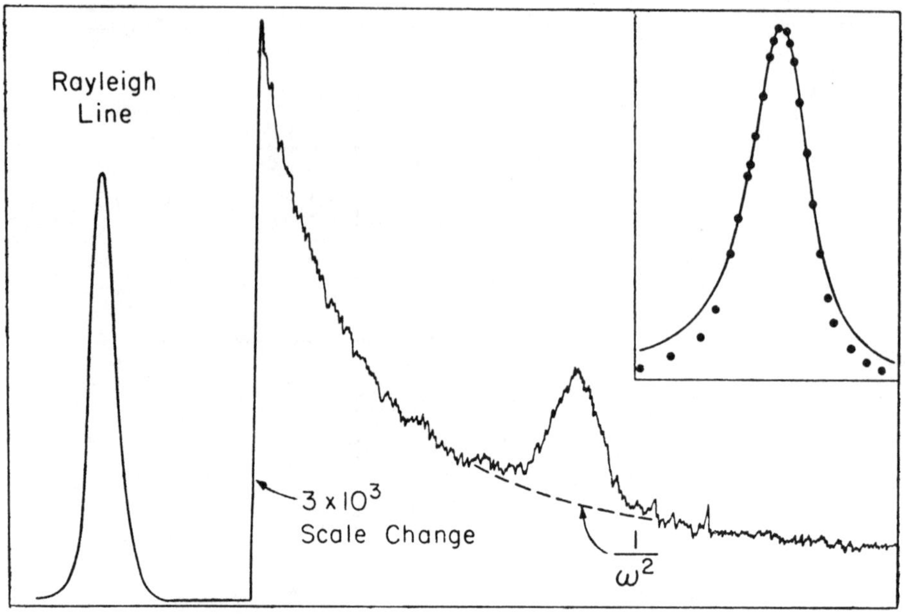

Figure 2. A representative experimental trace showing the observed spectrum. The dashed line indicates the component I_A which was subtracted away. The insert shows the remaining spectrum (points) fit to a Lorentzian (line) corresponding to component I_B. Component I_C was not observed.

the order parameter and sound waves. This term is undoubtedly quite small and our signal levels did not permit much resolution.

We are able to determine the hypersonic frequency ν_s, velocity C, and attenuation Γ_B from our data. These results are summarized in Table 1. The nematic-isotropic transition temperature for our sample was T_K = 44.9°C. We will compare our results to ultrasonic data on MBBA at T = T_K + 4°C. Eden, Garland, and Williamson[9] have measured ultrasonic velocity and absorption between 0.3 and 23 MHz and fit their data to a single relaxation model with ν_r ~ 11 MHz. Their fit gives an infinite-frequency sound velocity of 1485 m/sec, roughly 2% lower than our measured value. From their values of attenuation per wavelength we calculate $\Gamma_B/2\pi$ = 13 MHz at ν = 440 MHz. Martinoty[10,11] has measured ultrasonic absorption in MBBA at frequencies between 15 and 150 MHz and also fits his data to a single relaxation model, with ν_r ~ 17 MHz. From his results we calculate $\Gamma_B/2\pi$ = 14.5 MHz at ν = 440 MHz. At the hypersonic frequency $\nu \gg \nu_r$, we expect our velocity measurement to be nearly C_∞. The 2% discrepancy probably reflects the failure of the single relaxation model close to the nematic-isotropic transition[9]. The

T(°C)	ν_s (MHz)	C(m/sec)	$\Gamma_B/2\pi$ (MHz)
47.10	440 ± 4	1515 ± 15	18.5 ± 2
49.70	446 ± 4	1536 ± 15	15.0 ± 2

Table 1. Hypersonic frequency ν_s, velocity C, and attenuation $\Gamma_B/2\pi$ measured at two temperatures just above the nematic-isotropic transition temperature $T_K = 44.9°C$.

agreement of the attenuation values is fairly good, the differences being probably explained by temperature gradients in our cell. The details of the relaxation model have little effect since nearly 99% of the attenuation at 440 MHz is from the non-relaxing part of the ultrasonic absorption.

In summary, we have studied the spectrum of Brillouin scattered light in the isotropic phase of MBBA. The features of the observed spectrum are in agreement with those of the predicted spectrum except that we do not observe the term due to coupling between the order parameter and sound waves. The velocity and attenuation of hypersound have been measured near the nematic-isotropic phase transition and show fair agreement with ultrasonic values.

It is with pleasure that we acknowledge the valuable assistance of Noel Clark and helpful discussions with John Miller.

References

1. T. W. Stinson and J. D. Litster, Phys. Rev. Letters 25, 503 (1970).
2. P. G. de Gennes, Phys. Letters 30A, 454 (1969).
3. G. B. Benedek, "Thermal Fluctuations and the Scattering of Light" in Brandeis Lectures in Theoretical Physics, 1966, Vol. 2, M. Chretien et al., Eds., Gordon and Breach (1968).
4. R. D. Mountain, CRC Crit. Rev. Solid State Sci. 1, 5 (1970).
5. P. G. de Gennes, Mol. Cryst. and Liq. Cryst. 12, 193 (1971).
6. V. Volterra, Phys. Rev. 180, 156 (1969).
7. N. A. Clark, Ph.D. Thesis, Physics Dept., M.I.T., Cambridge, Mass. (1970).
8. D. S. Cannell, Ph.D. Thesis, Physics Dept., M.I.T., Cambridge, Mass. (1970).
9. D. Eden, C. W. Garland, and R. C. Williamson, J. Chem. Phys. 58, 1861 (1973).
10. P. Martinoty, Thesis, Institute of Physics, Strasbourg, France (1970).
11. P. Martinoty and S. Candau, C.R. Acad. Sci. 271B, 107 (1970).

PERIODIC DISTORTIONS IN CHOLESTERIC LIQUID CRYSTAL

Jacques RAULT*

H.H.Wills Physics Laboratory
University of Bristol, England U.K.

Several kinds of periodic patterns appear in liquid crystal submitted to an external field or to hydrodynamic flow. In cholesteric liquid crystal we show that the static deformation of the planar "Grandjean" texture involves mainly a bend-splay deformation, or a bending of the "cholesteric planes".

The topological properties of these static deformation patterns in cholesteric liquid crystal are explained in cases where the sample is submitted to :

a) steady magnetic field
b) high frequency elastic field
c) mechanical untwisting
d) coupled external field and mechanical untwisting
e) homeotropic anchoring conditions
f) presence of defects

In these different situations, when the distortion patterns appear, two helical axis at right angles could be defined in the sample. In high field or mechanical untwisting, a continuous change from one unperturbated structure, to another at right angles is topologically possible and corresponds to our observations.

* Present address : Physique des Solides, Faculté d'Orsay
France - 91

The periodical pattern observed in all the described cases are not observed in nematic liquid crystal where Frederick transitions usually occurs ; the periodicity is related to the periodicity of the cholesteric, and could be seen as the effect of the competitivity of two cholesteric effects at right angles.

1. STEADY MAGNETIC FIELD

Experimental observations on a cholesteric planar texture submitted to a static field have been made by several authors (1,2,3,4). We restrict ourselves to the experiment of the reference (4). In this case the diamagnetic anisotropy χ_a is positive, the sample is a mixture of M.B.B.A (methoxy-benzilidyne-p-n-butyl aniline) and cholesteric impurities, sandwiched between two parallel rubbed glass slides. Under an applied magnetic field, H parallel to the helical axis, molecules tend to align their long axis parallel to the field. When H is increased slowly enough to avoid lag effects, different kinds of deformation patterns occur according to the value of L/P : L and P being respectively the thickness and the pitch of the sample. The value of H at which this phenomenon appears is called the optical threshold field (H_1).

If H is increased further $H \geq H_2$ a 90° rotation of the helical axis could occur, but in this situation generally the transition cholesteric-nematic occurs because the transition field (7,8) is of the order H_2. The so-called homeotropic geometry is then obtained.

The threshold parameters of the observed distortions are in good agreement with the theoretical model of Helfrich (5) - Hurault (6) but the orientation of the wav - vector of the distortion compared with the boundary conditions are not explained.

1.1. One Dimensional Pattern

Experimentally lines of distortion are seen (figure 1) parallel to the direction of the surface alignment for L/P = 1/2 and perpendicular for L/P = 1, and 3/2. These facts allow us to know which kind of distortion is involved in low field (near H_H). Considering a film of thickness P for example in the "Grandjean" texture of figure 2a, by applying a magnetic field the local optic axis M_o tends to turn along the field direction. Two types of deformations could exist, assuming that the deformations arise in the middle of the sample. Figure 2b represents a splay bend deformation, and figure 2c a twist deformation, the unperturbed part of this model is obtained by rotating model (a) by 90°.

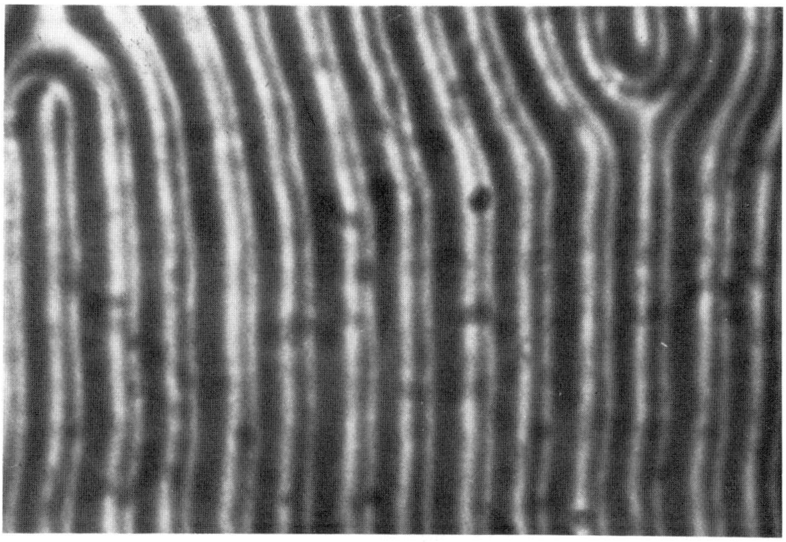

Figure 1

Periodical distorsion of a cholesteric Grandjean texture submitted to a magnetic field $H > H_1$ parallel to the helical axis (perpendicular to the plane of figure) of the unperturbed structure. In high field these lines form pairs. After Rondelez[4] the cholesteric pitch = 50 μm, the sample thickness ≃ 30 μm, the wave-length of the deformation is about 90 μm.

In both cases the periodic variation of the refractive index leads to a foccussing effect which give the optical bright lines. In case (b) for L = P lines are perpendicular to the rubbing direction and in case (c) they are parallel. We conclude therefore that the distortions involved experimentally are splay-bend distortions as sketched in case (b). For L/P = 1 or 3/2, it can be seen that the lines obtained are perpendicular to the rubbing direction as observed experimentally.

When the field is increased further, the lines form pairs which are more closely spaced than before. This phenomenon, shown in figure (1) can be explained by introducing twist deformation, which increases progressively with increasing field. Let us consider in a slice of thickness P/2 in the middle of the sample, two bend deformations in the planes $z = \mp P/4$, and a twist deformation in the plane $z = 0$. In the intermediate planes both

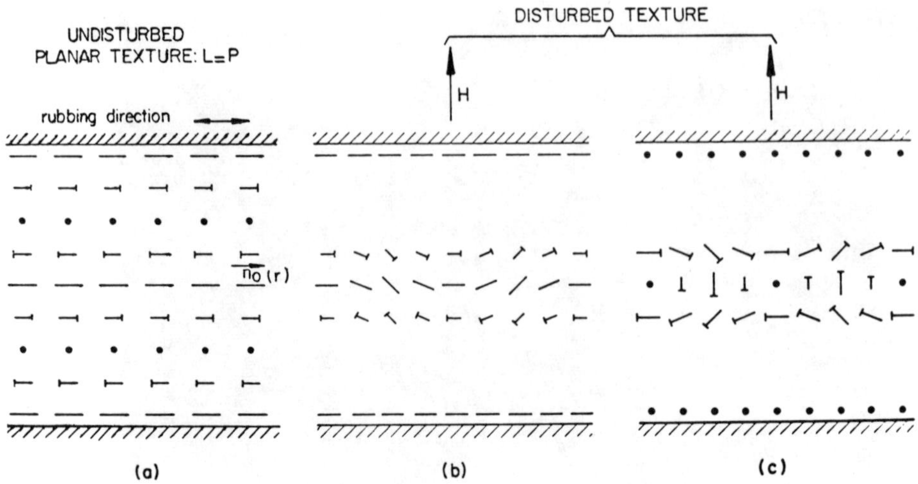

Figure 2
Possible deformations of a cholesteric liquid crystal under a destabilizing field.
a) unperturbed structure
b) periodic splay-bend distorsion
c) periodic twist distorsion.
Dash and dots represent the projections of the directors of the molecules parallel and perpendicular to the plane of figure.

splay-bend and twist are involved. This gives the structure sketched in figure (3). The important fact is that we obtain this distortion model continuously from the previous model of figure 2b,c. The applied field creates coupled dislocation lines. These dislocations with continuous cores are not coupled by a twist distortion (π) of the chilarity of the cholesteric as it appears between $\chi(+1)$ and $\chi(-1)$ dislocations (9), but coupled via the magnetic field by two twist deformation ($\pi/2$) of opposite signs. The double periodicity therefore could be explained essentially in assuming that along the axis ox, there is equipartition of the twist energy. The distance Pg of the region in figure 3 where the twist along ox is left-handed would be greater than the distance Pd where the twist is right handed (opposite to the cholesteric).

The example of pure twist of figure 3 in the middle of the sample correspond to the deformation drawn in

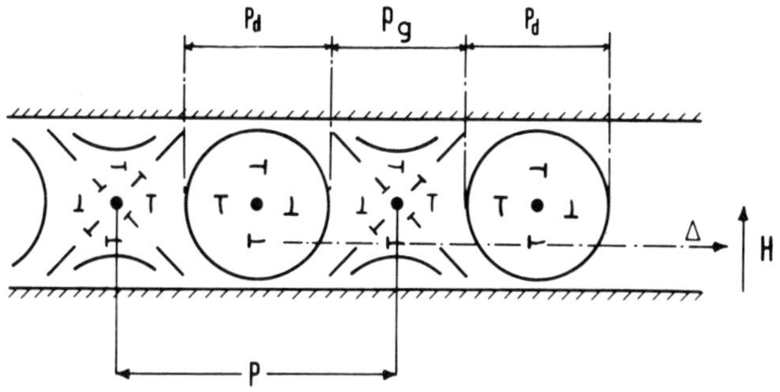

Figure 3
Possible distortion of a cholesteric slice of thickness $\frac{P}{2}$, in a high field. Dislocations of opposite sign ($\mp 2\pi$) are continuously formed. By switching off the field, dislocations disappear and the Grandjean texture with vertical helical axis is obtain.

figure 2b. In fact this kind of distorsion in thick sample must be added to the splay-bend distortion of figure 2b which arrives first in low field.

This kind of pattern of figure 3 but with a twist coupling between opposite dislocations ; complete rotation of π of the directors of the molecules between two lines) could also exist. It could be obtain from the pattern of figure 3 by a rotation of $\pi/2$ of the motif of the negative dislocation (-2π). This model would explain the appearance of dislocations lines observed by Rondelez when passing from the nematic phase to the cholesteric phase by decreasing the magnetic field below the transition field (7). If the field is switched off dislocations lines are still observed. The repulsive coupling (9) between two lines of opposite signs, hinder theses lines to annihilate.

1.2. Two Dimensional Pattern

For large value of P/L a pattern of square grids appear, lines are parallel and perpendicular to the rubbing direction. Figure 4 shows this pattern which is

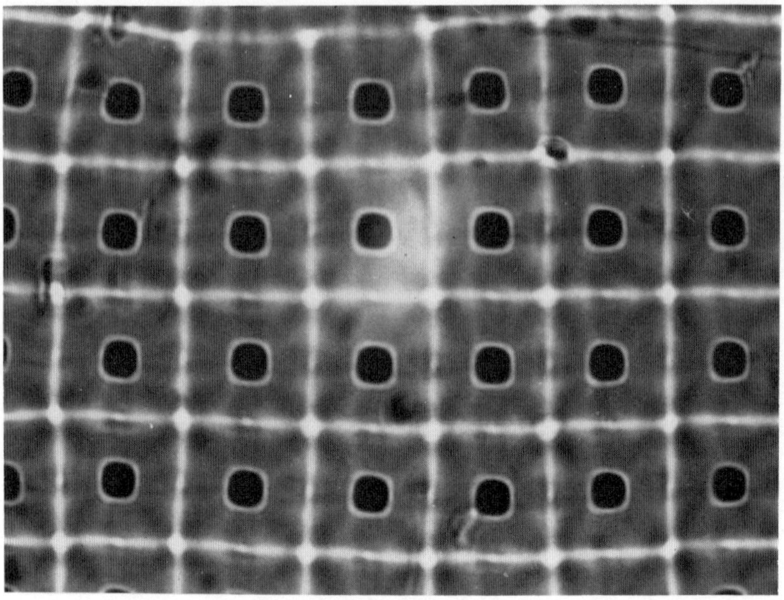

Figure 4
Square grid pattern observed(4) for a thick (L >> P) cholesteric sample in a magnetic field. The distorsion lines are parallel and perpendicular to the rubbing direction. The same pattern is also observed by mechanical untwisting.

topologically explained in the same way by the presence of splay-bend deformations as in figure 2b. A square pattern of twist deformation as drawn in figure 2c is topologically possible, and we have represented it in figure 5. In reality to explain the observed pattern we must replace in this drawing the twist by the splay-bend deformation.

The projection of the directors of the molecules of the cholesteric slab parallel to xoy is given in two vertical planes xoz and yoz. By considering twist distortion in the middle of the sample, two sets of distortion lines parallel and perpendicular to the glass rubbing direction are located at levels 3 = P/4 and z = P/2. As the distortion involved for each set of parallel lines is the same, a perfect square grid pattern is obtained. At the junction of two lines, the focussing effect is maximum and therefore gives optical bright spots.

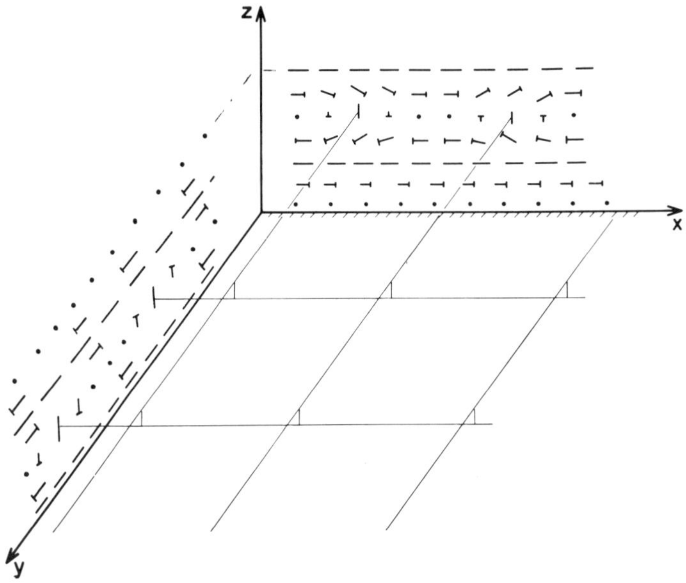

Figure 5
Square grid pattern involving two sets of distortion lines parallel to OX and OY at levels $Z = \frac{P}{4}$ and $Z = \frac{P}{2}$. The projection of directors of molecules are drawn in the planes YOZ and XOZ.

In fact in thick samples, several superposed grid patterns should exist ; the fact that only a single stable grid pattern is observed, means that each lines located at $Z = P/4 + nP/2$, are on the same vertical oz, and therefore coupled. Two bend distortions distant of P/2 in figure 6, are coupled via an intermediate twist deformation. In the case of figure a complete twist of $\pi/2$ along ox in the plane $z = P/2$ is represented. This leads to a wall distortion, in the middle of which directors of molecules are all vertical, and parallel to the magnetic field except near the glass walls. In this ideal model, we can describe this cholesteric structure, as a tilted cholesteric, the tilted angle α between directors and the helical axis of the unperturbated structure being constant along the z axis, but varying continuously from $\pi/2$ in the unperturbed state to o in the middle of the distortion wall. It must be pointed out that because of the change of chirality around ox, both sides of this distortion are not equivalent. Just below

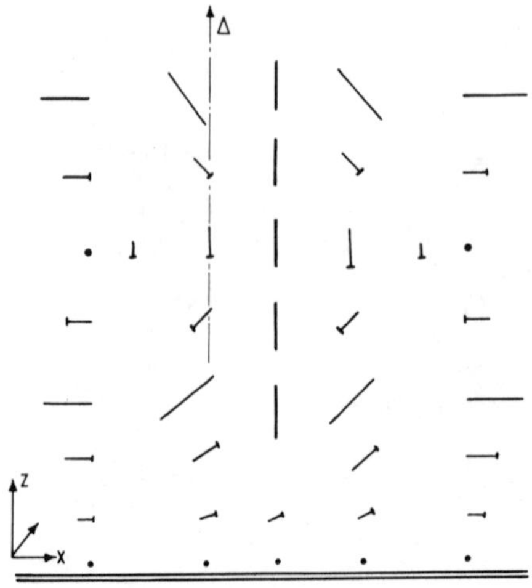

Figure 6
Coupling between two splay-bend deformations at level $z = P/4$, $3P/4$, through a twist deformation, at level $z = P/2$. The angle of tilt along the axis is varying continuously with x and constant with z in the case of figure.

the second threshold field $H < H_2$ the pattern of figure 4 is still observed, but in thick sample it is then possible to observe two differents square patterns by focussing above and below the middle of the sample. The bright spots of the second pattern are in the middle of the grid of the first pattern, and reciprocally.

Above the second threshold field another transition occurs. A rotation of nearly 90° of the helical axis everywhere in the sample is observed. Cholesteric planes are seen edge-on in figure 7, with the same symmetry as before. Similar textures have been described in the literature (10,11). It is to be noticed that the existence of spirals of both sense indicate that the cholesteric planes are tilted in two ways ; toward the top and bottom of the sample. A set of the same sense spirals are focussed near the top, and the other near the bottom. These properties are relevant to the model given in figure 8. In reference (1) this sinuous cholesteric planes have been observed edge-on.

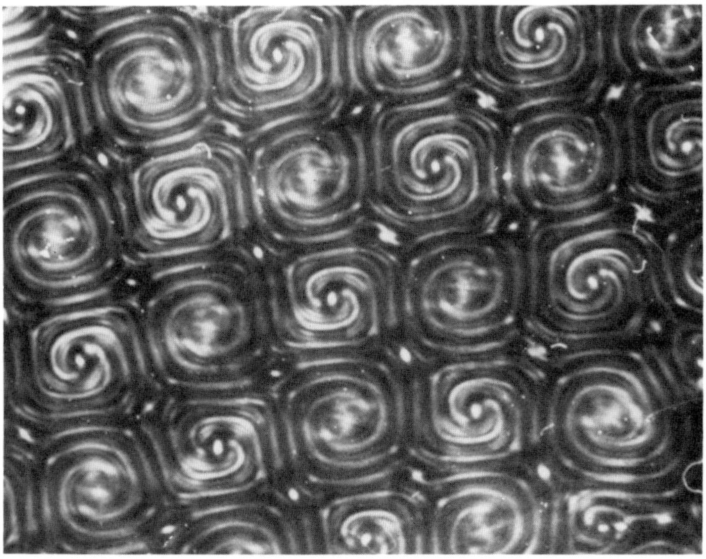

Figure 7
In high magnetic[4] field $H >> H_H$, a 90° rotation of the helical axis is observed, the square grid pattern of figure 4, give rise to this pattern of square spirals. Cholesteric planes are seen edge-on. After Rondelez[4], this structure is also observed in electric field and by mechanical untwisting.

We notice that there is a difference between these two structures shown in figure 4 and 7, above and below the second threshold field ; in the first structure, lines, or squares, the deformation involved is a splay-band deformation. In the second one there is simple deformation of pure bend of the cholesteric planes. In fact secondary twist deformation would occur also.

The notion of "cholesteric plane" is not obvious ; it could be defined in several ways, as pointed out by Frank (18). Postulating that "cholesteric planes" are the places in the space where the directors of molecules make a constant angle with a plane of reference (planes of figures) we see that the intersection of the cholesteric planes with the plane of figure is a straight line in figure 2b and a undulated line in figure 8. The definition of the unique axis of twist is also not obvious. Let us point out only that it could not be defined as the normal to the "cholesteric planes". Distortions in

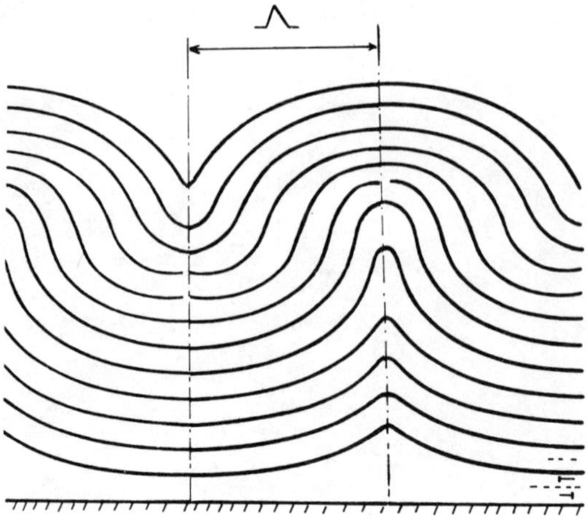

Figure 8
Progressive deformation under an external field of the cholesteric planes in the middle of a sample originally in the "Grandjean structure", continuous lines represents the places in the plane of figure where directors of molecules are parallel to this plane. The magnetic field is apply parallel to the helical axis of the unperturbated structure. The same kind of deformation occur in a plane perpendicular to the plane of figure. The periodicity Λ is the half periodicity of the square grid pattern of figures 4 and 5.

our models in fact could be described locally by two axis of twist at right angle.

The square shape of the distortion seen in figure 7 could be explained by theses arguments. Let us consider a very thin cholesteric slice having the form of the distorbed cholesteric planes of figure 8. In two sections at right angle, which are mirors planes of the distorsion pattern, the deformation due to the magnetic field are different. This implies that there are no symmetry of revolution in the deformation. The section of a "cholesteric plane" in disturbed region by a horizontal plane should not be a circle but a ellipse. The axis of this ellipse rotate helicaly, if the plane of section is displaced along the oz axis. It can be seen that in this model the place in the plane of focussing the microscope where directors of molecules are parallel to this plane, is a square spiral of the shape shown in figure 7.

2. HIGH FREQUENCY ELECTRIC FIELD

Several experiments have been described in the literature using d.c and a.c electric fields. Electro-hydrodynamical effects are present, such effects have been demonstrated in reference (12). To our knowledge the only static deformation of a planar cholesteric texture submitted to an electric field has been recently described by Hervet et al (4). The sample of negative dielectric anisotropy is submitted to a high frequency electric field (f = 10 KH) perpendicular to the helical axis. Above a threshold field, a pattern of stripes perpendicular to the field is observed. We notice that no square grid pattern occurs.

This simple topological property demonstrates that the deformation could only be of the type of figure 2b, involving splay and bend and not the twist deformation of figure 2c. Deformations in the electric field ($\varepsilon_a < 0$) perpendicular to the helical axis, and in the magnetic field ($\chi_a > 0$) parallel to the helical axis appear then for the same value of the equivalent field H as it is observed (4).

The theory of Hurault (4,6) for this case, predicting a threshold field 25 % lower than H_H does not seem to apply. The total distorsion involved in this cas has two components, a component of pure twist, of finite amplitude and a second component of bending of infinitesimal amplitude, these deformations are different from the deformation sketched in figure 2b and 8.

3. MECHANICAL UNTWISTING

Patterns as shown in figures 4 and 7 could be obtained by untwisting a cholesteric planar structure around the helical axis. The untwisting could be achieved by increasing the sample thickness by a vertical displacement of one of the glass slides bounding the upper and the lower surface of the material (10). Because of the translation symmetry in a cholesteric, a rotation of a glass slide around the helical axis of the planar structure normal to the glass slide is equivalent to a displacement parallel to this axis.

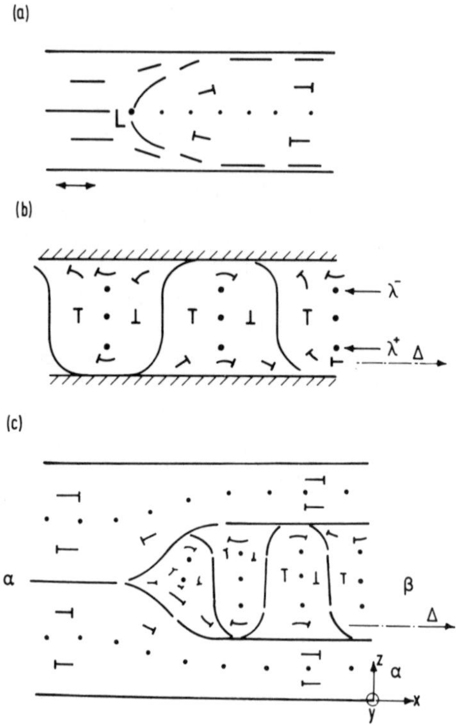

Figure 9

Distortion of a cholesteric planar structure by mechanical untwisting.

a) the thickness of the sample is smaller than the half pitch. An edge dislocation loop is nucleated

b,c) a new cholesteric phase with the Grandjean texture is nucleated continuously, if the thickness is greater than P.

For low values of L/P we do not observe a distortion pattern, but the non equilibrium twisting is relaxed by dislocation lines. Figure 9a shows a thin cholesteric slab ($L \ll P/2$), by increasing L a dislocation loop is nucleated and goes through the whole sample, relaxing the structure, the same process appears if L is increasing to 2P.

Starting now from a sample of thickness $L = nP$, $n \geq 2$ no dislocations are nucleated ; by increasing L to

2nP a periodic square grid pattern similar to that of figure 4 appears. The periodical patterns appear always for the same twist energy $(1/8K_2q_o^2)$ per unit volume.

A 90° tilt of the helical axis is also observed if L is increasing slightly, and gives the same texture as shown in figure 7. We can say that the more L is increased the more the cholesteric planes are tilted, in order to keep the pitch constant.

Recently similar phenomenon have been pointed out in a smectic A liquid crystal (13). A negative pressure tending to separate the layers is equivalent to a destabilizing magnetic field and the samples react by an undulation of its layers. The destabilizing untwisting $q_c = \pi/p_c$ could be compared to the destabilizing threshold field H_1. For weak tilted angle ϕ, of the cholesteric planes, the first order term in ϕ^2 of the twist energy is $1/2K_2(q-q_o)q.\phi^2$. If $q < q_o$ the cholesteric is untwisted, this energy is negative and is analogous to the first order term of the magnetic energy $-1/2(\chi_a H_1^2\phi^2$. The threshold untwist q_c is then related to the Hellfricht-Hurault destabilizing threshold field H_H by the relation $(q_o-q_c)q_c = \chi_a/K_2\ H_1^2$.

Deformation of cholesteric planes by untwisting are of the same nature as the deformation in the presence of an Electric and Magnetic field as described above. However in particular cases we obtain a quite different structure : if the untwisting is very rapid, no square grid pattern is observed, but an array of stripes perpendicular to the rubbing direction appears. These stripes which look like dislocation lines are located near the glass slides. The models proposed in figure 9b and 9c account for these properties. These new structures could be described as a set of disclinations of opposite sign $(\pm\pi)$. We must point out that there is a continuous passage from the unperturbed structure to the structure described above with rapid untwisting. Figure 9c shows the nucleation of the β cholesteric domain in an untwisted cholesteric phase α. The helical axis in the centre of the sample (cases b and c) is horizontal and parallel to the rubbing direction. The case b would exist if no dislocation as in (a) nucleate in the sample.

By going near the walls or near the β domain a helical axis could be defined parallel to the rubbing direction but the structure is no longer the unperturbed one. We are dealing with a tilted cholesteric structure along the Δ lignes, the angle of tilt along this axis decreases continuously from $\pi/2$ in the centre to 0 at a distance $\sim P/2$ from the glass walls (c). This situation is quite equivalent to that of figure 6 obtained in the magnetic field.

Growth Process by Diffusion

The formation of the spiral square pattern by continuous diffusion of cholesteric impurities in a planar structure, could be explained by the untwisting effects. Bouligand (15) has observed the structure shown in figure 7, in a sample of high pitch sandwiched between two glass slides, when a concentrated cholesteric is added. By diffusion cholesteric impurities go in the region of the planar structure and decrease the pitch. By untwisting, the structure square grid patterns are initially formed followed by the spiral square pattern as sketched in figure 8. Another process for generating stripes or square grid pattern ; involving mechanical untwisting could be obtained by varying the temperature. A sample of cholesteric mixture heated or cooled in such a way that the pitch decreases, present these distorsion patterns when sandwitched between two rubbed glass slides (20). It is to be noticed also, that in high external field just below the cholesteric-nematic transition (7,8) the pitch of the cholesteric increase with the field. The untwisting effect (instabilities by uncompression effect) could then occur if the sample sandwitched between two rubbed glass slides is submitted to a magnetic or electric field near the field of the cholesteric-nematic transition. This effect then would enhance the instabilities due to the magnetic field, described by Helfrich and Hurault. We give below in section IV an example where both low external field and untwisting are apply to a cholesteric slab.

4. COUPLED MECHANICAL UNTWISTING AND EXTERNAL FIELD

The deformation described in figure 2b and 8 could occur in a sample submitted both to a low external field (magnetic or electric field parallel to the helical axis) and a mechanical untwisting. This situation arises in one experiment described by Scheffer (3). Our model explains qualitatively the observed properties.

The sample is sandwiched between two glass slides rubbed in the same direction and making an angle for

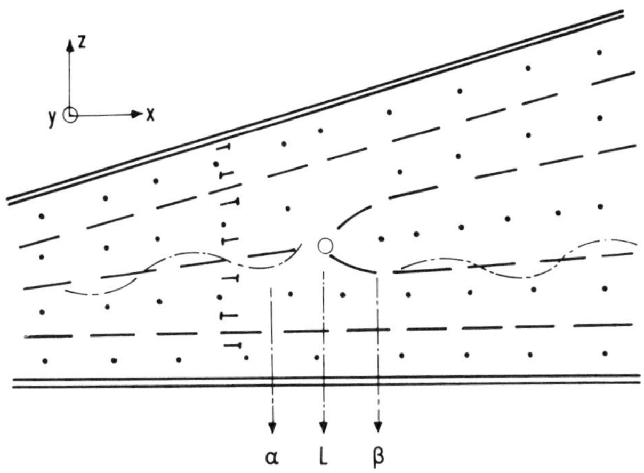

Figure 10
Dislocation line (L) in the "Cano-wedge". The wide region β is compressed, the narrow region α is untwisted. Undulations of cholesteric planes are enhanced in the untwisted region α.

obtaining the so-called Cano wedge. This situation involves the presence of equally-spaced "Grandjean" disclinations. In the middle of the space between two disclination lines, the pitch of the structure is the equilibrium pitch, on the wider side of a disclination line the cholesteric is compressed, on the narrower side it is untwisted (figure 10).

Without an external field no distortion pattern is observed, because the untwisting is too low. By applying an external field the two phenomena which tend to undulate the cholesteric planes, cooperate in the untwisted region (α). The threshold field in this region is lower than the threshold field in the unperturbed region.

On the wider side (β) of the disclination line the compressibility of the structure is obviously acting in the opposite sense, and thus leading to a higher threshold field.

The lines of the square grid pattern are parallel and perpendicular to the rubbing direction. Lines perpendicular to the disclination lines are equally spaced;

this is quite obvious because of the translation symmetry along the line oy parallel to the dislocation line in the wedge. But the direction ox perpendicular to the line is no longer a translation symmetry axis of the geometry ; distortion lines parallel to the disclination lines do not involve the same energy on both sides of the disclination line, because as pointed out above splay-bend is more easy in (α) than in (β) regions, and this leads to a varying periodicity along ox. The width/length ratio of the cell departs from unity in the region of equilibrium pitch to 1.25 on the narrow side and 0.80 on the wider side. The ratio then between the periodicity λ_w and λ_n of the wide and narrow side along ox is then equal to 1.47.

In the equilibrium pitch region, Helfrich's theory gives for the threshold parameter :

$$H^2 = (2\pi)^2\left(\frac{2k_2k_3}{\chi_a}\right)^{1/2}\left(\frac{1}{PL}\right) \qquad \lambda^2 = \left(\frac{2k_3}{k_2}\right)^{1/2}(PL)$$

where k_2 and k_3 are the elastic constants of twist and bend deformations (18), P and L being the pitch and thickness of the cholesteric sample.
In the non equilibrium pitch region, formulae lead to the ratio

$$\frac{H_w}{H_n} = \frac{\lambda_w}{\lambda_n} = \left(\frac{k_{3w}}{k_{3n}}\right)^{1/4}$$

where $k_3(w)$ and $k_3(n)$ are phenological constants equal to k_3 in the equilibrium region, greater and smaller than k_3 in the wide and narrow regions, the experimental value $H_w/H_n = 1.35$ is in good agreement with the value $\lambda_w/\lambda_n = 1.47$.

In conclusion, the observations of Scheffer (3) of the different types of the square grid pattern near the disclinations, in a Cano-wedge submitted to an external field, show that the deformations involved in the two processes described in former paragraphs are of the same kind.

5. HOMEOTROPIC CONDITIONS

The planar structure of a cholesteric could be destroyed also by the anchorage conditions on the glass walls. We report here some topological models of distortion which gives periodic patterns, and that could be explained by the undulations of cholesteric planes.

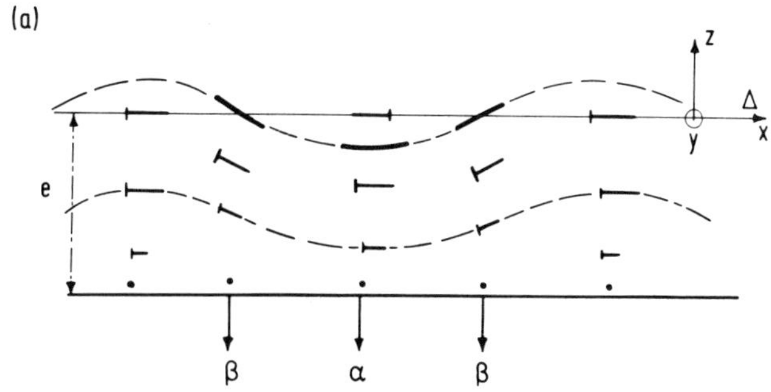

Figure 11
Distortion in a cholesteric film of thickness $\frac{P}{4}$, produced by homeotropic anchorage at the free surface z = 0.

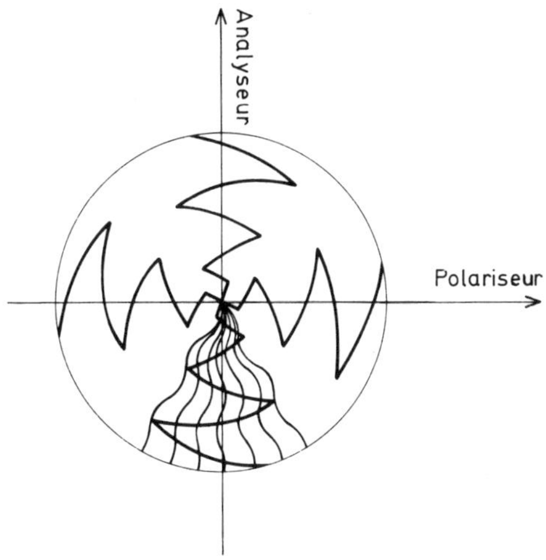

Figure 12
Convergent waved pattern of the projection of the directors of the molecules on a plane z = cste, when the sample of figure 11 present a symmetry of revolution around a vertical axis. The black-cross observed in crossed polarizers has a zig-zag shape.

As before, we must notice that the situation depends on the value of L/P. Several patterns require the presence of dislocations, or surface dislocations. We give an example of a structure involving splay-bend distortion.

An example of weak bending of the cholesteric planes as in the sections 1, 2, 3, sketched in figure 11 explain the periodic comma observed in a thin cholesteric film.

We interpret this phenomenon by assuming a weak anchoring condition on one or two surfaces of the cholesteric sample.

Let us suppose that $L = P/4$, and assume that molecules tend to be tilted on one surface, and that on the other surface there is a perfect anchoring as drawn in figure 11, the tilt could be achieved in two ways, between to β regions of opposite tilt angle there is a region where there is no tilt, as it is necessary to pass continuously from one region β to the other. But now assuming that there is equipartition of elastic energy along ox, we notice that the α regions could not have the equilibrium pitch and is thus periodically compressed and untwisted. Going along ox we pass continuously from splay-bend distortion to twist distortion. Along the surface an axis of helical rotation parallel to ox is defined, and we could say that along this axis we are dealing with a tilted cholesteric. The projection of the directors of the cholesteric at the surface has a waved shape.

Around a dislocation $\chi(+1)$, the pattern obtained has cylindrical symmetry ; between crossed polarizers, the black cross (where the molecules are parallel to the polarizers) has a sinuous form as it is drawn in figure 12 and correspond to the observed distortion shown in figure 13.

Distortion lines in figure 11 are perpendicular to the mean direction of the molecules at the surface. In cases where the free surface is oblique with respect to the cholesteric planes, the direction of molecules on this surface has a bowed shape (11) the distortion line has then the same shape and leads to the periodic comma (1, 12) whose optical properties have been described by Friedel (16).

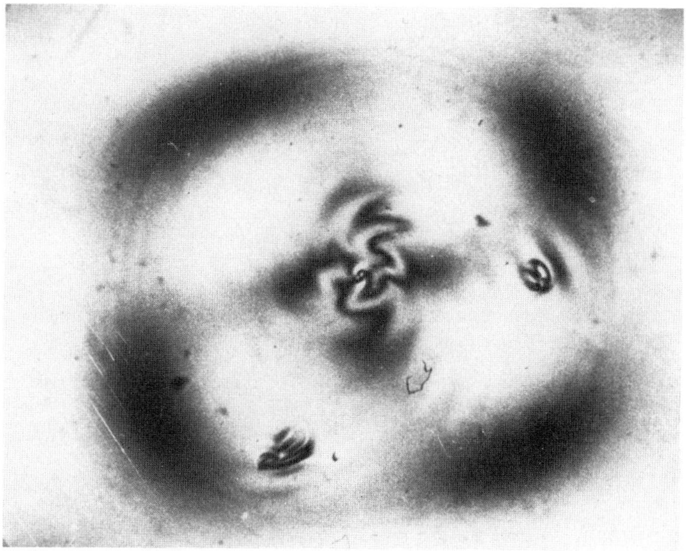

Figure 13
Periodic distortion, around a $\chi(+1)$ dislocation in a thin cholesteric sample-crossed polarizers.

6. PRESENCE OF DEFECTS

As in a magnetic or electric field, the presence of defects could perturbate periodically the planar structure. Two examples of perturbated structures in thick samples are given below, one around an isolated dislocation $\chi(+1)$, the other near a network of dislocations.

6.1. Isolated Dislocation

Figure 14 shows an example of periodic distortion around a dislocation $\chi(+1)$ (disclination 2π). The thickness of the sample is greater than P. Because of the nematic structure of the core of the dislocation (9), a local helical axis near the dislocation line could be defined as being radial and perpendicular to the helical axis of the unperturbated structure far from the line.

Because of the orientation of the molecules along the lines, there are two different cholesteric effects

Figure 14
Periodic distortion rings around a $\chi(+1)$ dislocation in a thick cholesteric sample. ($\frac{P}{2}$ = 20 μ). The distance two rings is of the order of the half pitch.

as sketched in figure 15. Directors of molecules have a tendency to twist helically around the axis Δ and oz, which are respectivelly important near and far from the oz axis. In an intermediate region, both axes of twist could exist, and that is relevant to the model described in figures 3a, 9b, 11a. The distortion involved in this model are mainly splay and bend ; and the twist whatever its direction, respect the chirality of the cholesteric. In transmitted light bright lines are seen at the places L where the directors of molecules are parallel to oz.

By rotating the pattern of figure around oz, a dislocation $\chi(+1)$ is generated. The dislocation along oz is then surrounded by circular distortion lines as it is observed in figure 15. The fact that the distance between two rings is of the order of P/2, and that the rings become more and more faint, far from the line accounts for our model of distortions.

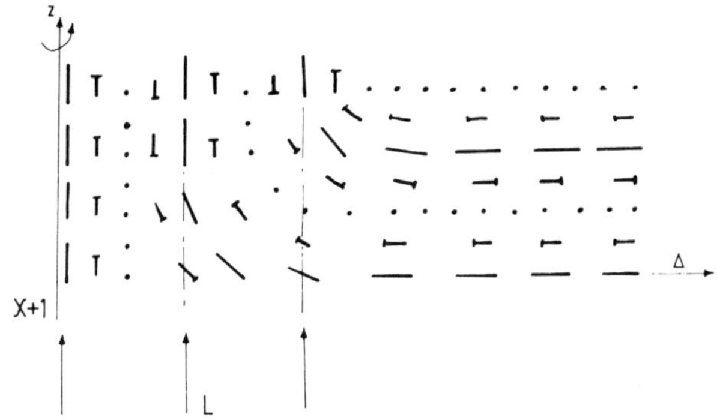

Figure 15
Model of distortion around a dislocation $\chi(+2\pi)$. The cholesteric structure is obtain by rotating this pattern around the OZ axis. In transmitted light focussing effects are observed in position L (see figure 14).

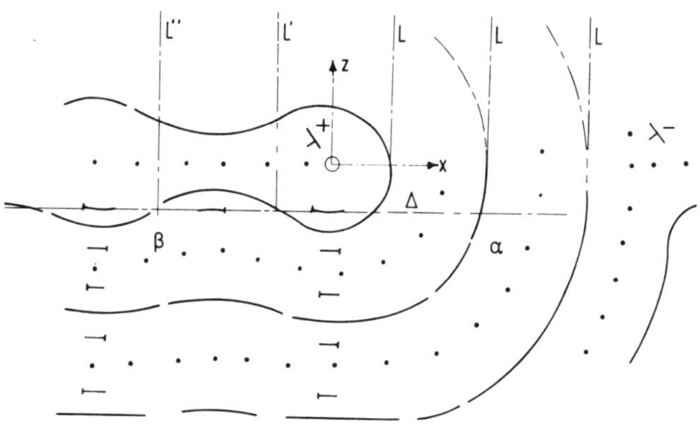

Figure 16
Cross-section of a edge dislocation ; paire $\lambda^+\lambda^-$ of Burgers vector $3\frac{P}{2}$. A decrease of the bend of the "cholesteric plane" at the right of the λ^+ disclination imply a increase of the twist on the left of the disclination. Periodical bending of the cholesteric planes gives periodical lines besides the dislocation line.

6.2. Periodical Distorsion Near Grouped Dislocations

Generally grouped edge dislocations, (pairs) separate two regions of the same pitch, but a different helical axis. Periodical distortion lines are also observed besides these dislocation lines, and could be explained by the "competitivity" between two different twists. The helical axis is vertical in region α and horizontal in region β of figure 16. Places where directors are vertical are seen in transmitted light.

In the particular case here a dislocation line of Burger's vector n P/2 would give n observed bright lines.

Looking at the plane $z = \pm P/4$ we see that along ox, by undulating the c holesteric planes, it is possible to pass continuously from a helical axis perpendicular to oz in (α) region to an axis parallel in β region. This distorsion decreases the splay-bend energy on the right of the λ^+ disclination line and introduces a twist energy on the left, leading then to the equipartition of the elastic energy around the line. Along ox the director of molecules turns around the same chirality as the cholesteric, but the angle of tilt $\alpha(x)$ along this axis decreases from $\pi/2$ in α to 0 in β. Molecules weakly tilted along oz, in position L' and L" give the small focussing effect which decrease with the distance to the lines. Figure 18 shows the rotation of the molecules around the (Δ) axis of figures 12 and 15. The tilt angle α is decreasing from $\pi/2$ to 0. The "helical axis" near oz and far from oz are at right angles.

In conclusion, at the junction of two domains with a planar structure at right angles, there exists an intermediate region where the director of molecules rotate helicoidally around two axes perpendicular. This would be possible by a progressive tilt of the molecules along these axes, yielding in certain cases to cholesteric planes undulations. Typical periodic distorsion besides dislocation lines are shown in figure 17.

Theoretical treatment of the distortion near the dislocation is necessary to establish whether this effect is important.

We think however that this effect is enhanced by the untwisting effect, in the particular case of grouped dislocations. In the cholesteric sample between rubbed glass slides, a network of edge dislocations, parallel

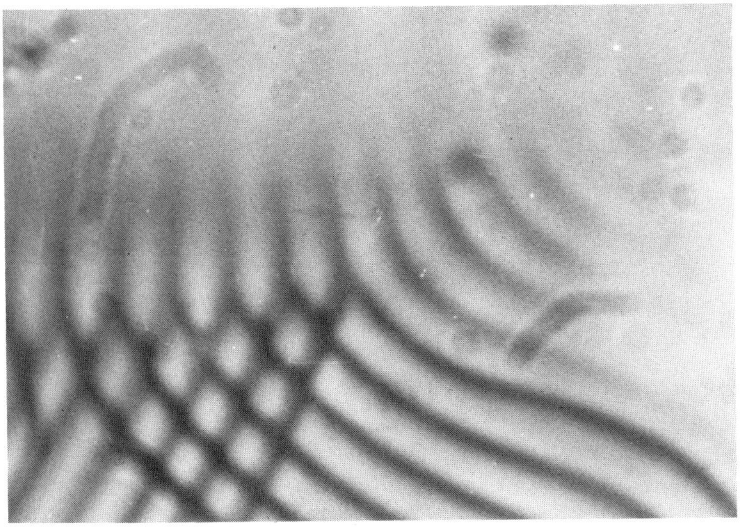

Figure 17
Periodical distortions besides edge dislocation in the right part of the photograph. The distance between two lines (25 μ) is equal to the half pitch the cholesteric.

to the planar texture, change the pitch of the cholesteric. If the cholesteric is untwisted, the effect described in paragraph 3 enhances the undulation of the cholesteric planes, or in other words, the tendency of the directors of the molecules to twist helicoidally around an axis (Δ) perpendicular to the helical axis of the unperturbated structure.

We think that this effect of bending the cholesteric planes by mechanical untwisting must be also involved because this effect on figure 17 does not appear on both sides of the dislocation network. Generally, the network of dislocations is isolated in the cholesteric sample between glass slides, the untwisted side is the side of the λ^- disclination, the supplementary cholesteric planes introduced by an edge dislocation (paire $\lambda^+\lambda^-$) is on the λ^+ side. This untwisted effect then will enhance the periodical distortion in the λ^- side, and reduce on the other side. For convenience, the drawing represented in figure 16 shows the cholesteric planes undulation on side λ^+. The undulations are in opposite phase in this figure, but in the case where the untwisting effect occurs, they must be in phase.

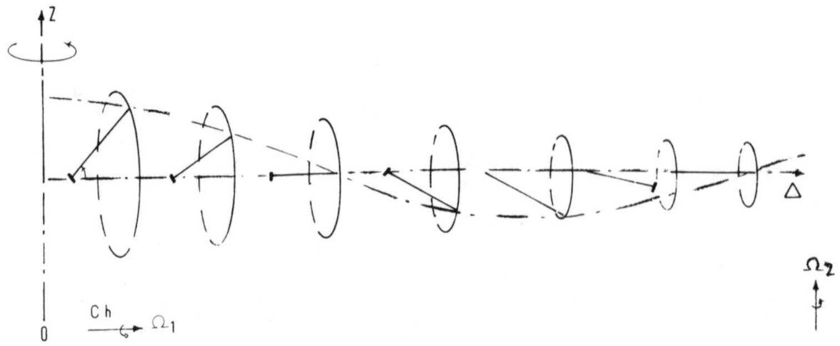

Figure 18
Helical rotation of the directors of the molecules along the axis Δ as shown in figures (6,9,11,14,15), but in this case the angle of tilt between directors and the Δ axis decrease from $\pi/2$ to 0 with the distance from the OZ axis. The projection of directors on the surface Z = C^{ste} has a sinuous form as sketched in figure 12.

6.3. Periodical distortion of grouped dislocations

Often in the Cano-wedge geometry, periodic focus lines besides the dislocations are seen and the lines (figure 19) when numberous are themselves undulating, this undulation along the oz axis perpendicular to the plane of figure could be seen by the difference of focussing on the lines, and also by the focussing effect observed besides the lines. The wave-vector of this distorsion is parallel to the mean direction of the lines instead of being perpendicular as in figure 17. This is due to the fact that now we are dealing with the Cano-wedge geometry, then the untwisting effect is decreasing continuously far from the dislocations.

The undulation of the lines in the direction perpendicular to the line, is related to this phenomenon of untwisting. We show this effect on the special case of one isolated dislocation as shown in figure 10. In the Cano-wedge geometry, in the ideal case where no splay-bend distorsion in the untwisted region (α), the dislocation line L is exactly at the middle of the distance between two unperturbated regions, because the twist energy is the same on the both sides of the dislocation line. Now assuming that the Burger's vector of the dislocation is of the order to the thickness of the sample,

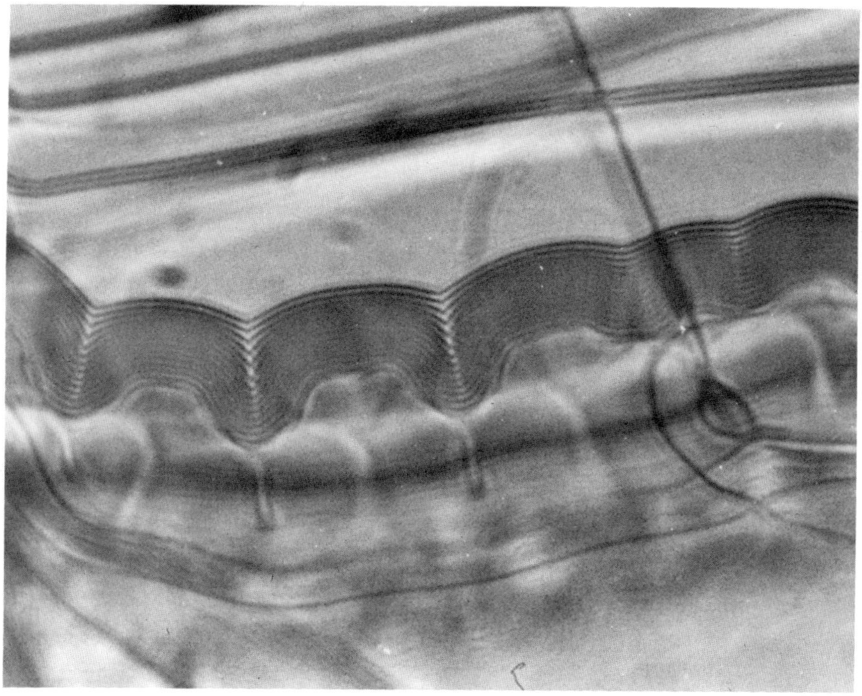

Figure 19

Dislocation of Burgers vector b = 7.P in a cholesteric sample between rubbed glass slides. The thickness of the Cano wedge near these lines is of the orders of 10P. Periodical distor ions in the Grandjean texture seen besides the ondulated lines, are due to the untwisting effect.

then undulations of cholesteric occur on one side of the dislocation if the twist q in this side is lower than the threshold twist q_c.

By untwisting effect in α region of figure 10, splay-bend instabilities when occur, decrease the energy in this side. The energy of the other side does not change, this imply that on the line, exist a force toward the region β. A stable position of the line on the right is reached.

Now let us consider only one-dimensional undulation of wave-vector parallel to line ; periodically along the line in the region α splay-bend distortion occur, and the energy on this side of the dislocation is reduced periodically. This will imply a periodical force on the dislocation in the direction perpendicular to the line (direction ox) and this would give a sinuous form to the dislocation in the plane xoy. This sinuous form is not observed on isolated single dislocations, but is observed one a set of coupled dislocations or a dislocation of high Burgers vector of order the thickness of the sample as shown in figure 19. The undulations of cholesteric planes are clearly seen in this figure, by the observations of focal line near the dislocation lines and by the change of focus on these dislocations lines.

CONCLUSION

Topological properties of the static distortions in magnetic and electric field and by mechanical untwisting are well explained by a model of splay-bend distortions. Broadly speaking this distortion could be described as a bending of the "cholesteric planes".

A continuous passage from the unperturbated structure to another one at right angles observed in high field below the cholesteric-nematic transition involves progressive bending of these cholesteric planes, and in certain cases progressive formation of coupled disclinations of opposite sign and with a continuous core. In weak distortion a unique local helical axis of the structure can not be defined. Two helical axes Ω_1 and Ω_2 at right angles exist, the tilt angles α_1 and α_2 of the directors of the molecules with these axes are varying periodically in the sample, and varying with the strength of the external field and of the untwisting. Two competitive cholesteric effects along Ω_1 and Ω_2 are present, and the relative importance of these two effects depends on the strengh of the external field, or of the untwisting effect.

Two similar situations arrive without external field in cholesteric samples ; when anchoring conditions on the glass walls are in such a way that the direction of the helical axis is varying in the sample, and when dislocation lines introduce in the "Grandjean" texture a local change of $\pi/2$ in the direction of the helical axis.

Acknowledgments

I am indebted to Dr. Rondelez for helpful discussions and for communication of the photographs of Figures 1, 4 and 7. I acknowledge Dr. Frank for critical comments.

REFERENCES

1. J.Rault, P.Cladis, Molecul.Cryst.Liquid.Cryst. 15, 1 (1971)
2. F.Rondelez, J.P.Hulin, Solid State Comm., 10, 1009 (1972)
3. T.J.Scheffer, Phys.Rev.Lett., 28, 593 (1972)
4. H.Hervet, J.P.Hurault, F.Rondelez, to be published F.Rondelez, Thesis, Orsay (1973)
5. W.Helfrich, Appl.Phys.Lett., 17, 531 (1970)
6. J.P.Hurault, to be published in J.Chem.Phys.
7. P.G.de Gennes, Solid State Comm., 6, 163 (1968)
8. R.B.Meyer, Appl.Phys.Lett., 12, 281 (1968)
9. J.Rault, Solid State Comm., 9, 1965 (1971)
10. J.Rault, Thesis, Orsay (1973)
11. Y.Bouligand, J.Physique, 30, C4, 90 (1969)
12. F.Rondelez, H.Arnould, C.J.Gerritsma, Phys.Rev.Letters, 28, 735 (1972)
13. M.Delaye, R.Ribotta, E.Durand, Phys.Letters, 44A, 239 (1973) N.A.Clark, R.B.Meyer, Appl.Phys.Letters, 22, 10 (1973)
14. J.Rault, submitted to Mol. Cryst.
15. Y.Bouligand, J.Phys., 33, 715 (1972)
16. M.Kleman, to be published in Phil.Mag.
17. J.Friedel, Ann.de Phys., 18-273 (1922)
18. F.C.Frank, Disc.Faraday Soc., 25, 1 (1958)
19. F.C.Frank, Private communication
20. C.J.Gerritsma, P.V.Zanten, Phys.Letters, 37, 47 (1971).

ERASURE OF TEXTURES STORED IN NEMATIC-CHOLESTERIC MIXTURES

B. Kerlleñevich* and A. Coche

Centre de Recherches Nucléaires

67037 Strasbourg CEDEX - France

INTRODUCTION

In a previous paper (1) we had already reported results on the erasure by an ac field (1 KHz) of the texture stored in nematic-cholesteric mixtures by application of a dc field. The decay of the scattered light intensity after the dc field removal can be characterized by an erasure time τ_E defined as the time during which this intensity falls from 90 to 10 per cent of its initial value. This time can easily be determined and the variations of τ_E with various parameters have been examined for mixtures of negative dielectric anisotropy consisting of a commercial product (LCI), nematic in the range 5-70°C and of cholesteryl nonanoate (CN). The experimental results showed that the time τ_E decreases when the temperature increases (following a law similar to that of the pure nematic substance)and also when the erasing ac voltage rises, the greater the concentration the faster the decrease. In this paper we have studied in more detail for the same mixtures, the variations of the erasure time τ_E (for 1 KHz) with the dc voltage producing the storage effect.

EXPERIMENTAL CONDITIONS

The experimental set-up is similar to that described in ref. (1). The CN concentration varies between 1 and 15 % (by weight). An homogeneous alignement of the liquid crystal mixture is obtained by rubbing of transparent electrodes. The light transmitted by the cell of 18 μm thickness (and in some cases of 36 μm) impinges on a photomultiplier, the anodic signal of which is registered on an X-Y recorder. The various structures stored have also been observed with a Leitz-Orthoplan polarizing microscope with a photometric device for transmission measurements of a given area. Simultaneously the variations of the layer capacity can be measured with a Wayne-Kerr bridge. In some experiments in order to compare more easily the various structures, we used a technique similar to that of Williams (2): the conducting layer is dissolved on half of the surface of both glasses and two adjacent areas of the liquid crystal, with and without field, can be observed simultaneously with the polarizing microscope.

RESULTS

1) In absence of dc field the samples are optically active even for small (≈1 %) quantities of CN.

2) In a general manner with a dc increasing field, the transmission and the capacity of the cell decrease rapidly above a certain threshold voltage, the greater the CN concentration the higher this threshold. For voltages higher than 50-60 volts (for a thickness of 18 μm) the capacity as well as the transmission vary little. These variations which are shown in Fig. 1 (a) for two

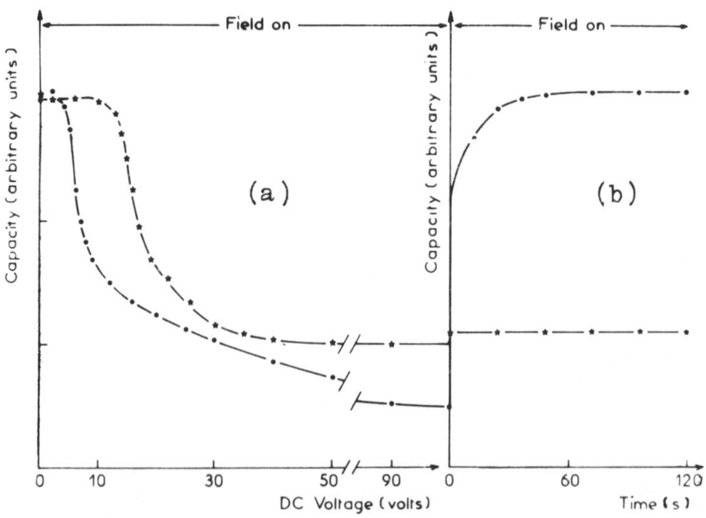

Fig. 1. Variations of capacity as a function of dc applied field (a) and of time after dc field removal (b): • LCI + 1 % CN, * LCI + 15 % CN.

CN concentrations (1 and 15 % by weight) can be compared to those observed by Rondelez (3) and Hulin (4) on mixtures of MBBA and CN submitted to a magnetic field parallel to the helical axis of the cholesteric structure. They indicate a 90° tilt, with respect to their original orientation, of the cholesteric planes which are perpendicular to the electrodes after this tilt.

3) For mixtures containing small quantities of CN (1 %) the variations of capacity and transmission appear at 5-6 volts. Simultaneously for this voltage a "fingerprint" is observed as mentioned by Hulin (4) for mixtures of MBBA and CN under the effect of a magnetic field.

When the dc field is cut-off the capacity and the transmission return to their initial values - without field - in a short time, even for high values of the field. As an example, the variation of capacity after application of a voltage of 100 volts is shown in Fig. 1 (b). Now if an alternative voltage is applied simultaneously to the dc field removal the transmission returns to its initial value with an erasure time τ_E of the order of 0.2 s which is the time characterizing the spontaneous decay of the dynamic scattering for the pure nematic product. For these concentrations τ_E is practically independent of the dc voltage previously applied.

4) For CN concentrations higher than 4 % approximately the behaviour is quite different. The variations of the capacity and of the transmission appear for voltages higher than those of the previous case but differently from it the capacity and the transmission change little after the dc field removal (Fig. 2). Melamed et Rubin (5) have also observed a small variation of the transmission after the dc field cut-off for cholesteric mixtures. In addition it must be mentioned that the variation of capacity that appears between no field and dc field conditions is the same for the various concentrations of CN in the studied range. This fact is probably related to the variation of the dielectric anisotropy $\Delta\varepsilon = \varepsilon_{//} - \varepsilon_{\perp}$ with the CN concentration: we have measured this quantity which has been found constant for CN concentrations higher than 2 %.

For a given erasing voltage the erasure time τ_E decreases rapidly when the dc voltage rises, starting from a certain value of the dc voltage (Vdc). Vdc increases with the CN concentration. For values higher than 50-60 volts, τ_E is approximately constant. This behaviour is illustrated in Fig. 3. Some experiments performed with mixtures of MBBA and CN gave similar results.

By microscopic examination it is possible to observe after the dc field removal and without erasing voltage

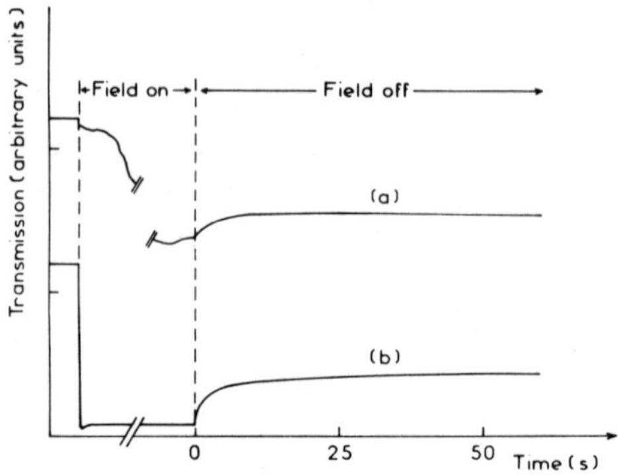

Fig. 2. Variation of cell transmission (LCI + 10 % CN) after dc field removal: (a) with 20 volts dc, (b) with 80 volts dc.

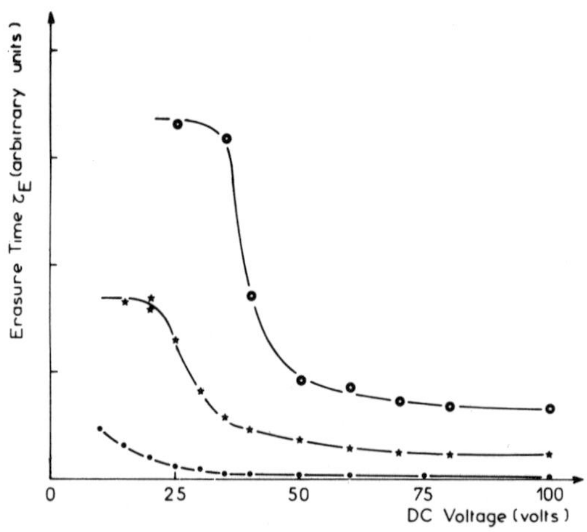

Fig. 3. Variations of erasure time with the dc voltage: ○ LCI + 15 % CN, *LCI + 10 % CN, ● LCI + 5 % CN (the erasing voltage is applied simultaneously with the dc field removal).

a mixed focal-conic-planar texture. In the focal-conic texture small areas which are planar and optically active can be seen. Wysocki (6) has already mentioned such a texture for mixtures of cholesteryl chloride and of cholesteryl nonanoate. The density of these planar domains decreases with an increasing dc field. It rises slowly after the field removal indicating a spontaneous - but partial - transition from focal-conic texture into planar one. Such a transformation has been also observed by Wysocki (6). Our results reported in Fig. 3 show that the relaxation to the initial state of a mixed texture is slower when the density of planar regions is higher.

It is interesting to mention that in the CN concentration range used the transmitted light between crossed polarizers shifts from red to green-blue with an increasing dc field. In these experiments we have also observed that the film which presents a uniform structure before field application, shows after erasure, a number of "disinclinations".

5) In an other run of experiments we have determined the erasure time τ_E when the ac erasing voltage is applied some time T_d after the dc field removal. In Fig. 4 the erasure times τ_E are plotted as a function of T_d for three dc voltages.

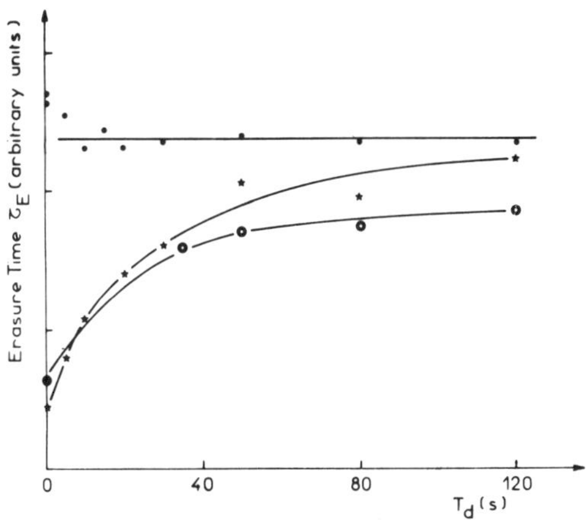

Fig. 4. Variations of erasure time with the time T_d for a CN concentration of 15 % after application of various dc voltages: ● 20 volts, *60 volts, ○ 80 volts.

It appears that

i) for low voltages (20 volts for example) τ_E is practically independent of time T_d ,

ii) for higher voltages, on the contrary, τ_E increases with T_d and tends towards the order of value corresponding to low voltages.

We assume that during the time T_d after the application of a high dc voltage a mixed texture appears progressively with a density of planar regions which tends towards that existing at low voltages. According to this, the relaxation time characterizing the return towards the initial structure must be independent of T_d for textures stored at low voltages. For higher voltages instead, where the initial density of the planar domains is lower, this density rises with the time T_d and consequently the erasure time also rises.

ACKNOWLEDGMENTS

The authors wish to thank A.Stampfler for his valuable help and J.L. Ciffre as well as C. Koehl for their able technical assistance.

REFERENCES

* Universidad Nacional del Sur, Departamento de Física, Bahía Blanca (Argentina).

(1) B. Kerlleñevich and A. Coche, Mol. Liquid Crystals, (in press).

(2) R. Williams, J. Chem. Phys. 39 (1963) 384.

(3) F. Rondelez and J.P. Hulin, Sol. State Commun. 10 (1972) 1009.

(4) J.P. Hulin, Appl. Phys. Lett. 21 (1972) 455.

(5) L. Melamed and D. Rubin, Appl. Phys. Lett. 16 (1970) 149.

(6) J.J. Wysocki, Mol. Liquid Crystals, 14 (1971) 71.

RHEOLOGICAL PROPERTIES OF THERMOTROPIC AND LYOTROPIC MESOPHASES FORMED BY AMMONIUM LAURATE

B. Tamamushi and M. Matsumoto

Nezu Chemical Institute, Musashi University

Nerimaku, Tokyo, Japan

INTRODUCTION

Long chain ammonium carboxylates are a kind of soaps but they are characteristic in their behavior, namely, they not only form anhydrous thermotropic mesophases at relatively low temperatures but they form lyotropic mesophases with water and another amphiphilic substance like normal alcohol at ordinary temperature. They are moreover weak electrolytes which can be considered polar molecules with pretty high dipole moments.

In an early work Lawrence described some characteristic properties of ammonium carboxylates as materials forming both thermotropic and lyotropic mesophases (1). Recently, studies on flow properties and surface tensions of homologous ammonium carboxylates (myristate, palmitate, stearate and oleate) have been reported by one of the present authors (2). The present paper deals with the experimental results on the rheological properties of thermotropic and lyotropic mesophases formed by ammonium laurate, an example of ammonium carboxylates. The lyotropic mesophases studied here concern the ternary system: ammonium laurate + water + n-octanol.

EXPERIMENTAL

Materials

Ammonium laurate was prepared by passing dry ammonia through the ethanol solution of pure lauric acid (m.p. 42.7 - 45.5°C) and the precipitate was recrystallized and repeatedly washed by acetone, whose two transition temperatures were: 75.5°C for the crystal-anisotropic liquid transition and 114.5°C for the anisotropic-isotropic liquid transition. The product was kept in a desiccator filled with ammonia. The samples used were always freshly prepared ones.

n-Octanol was obtained by vaccum distillation of a commercial product of high purity, and its boiling point was 84°C under 9mmHg. Water used was previously treated by ion-exchanger and redistilled.

Rheological Measurement

The rheological measurement was carried out by applying a viscometer of Couette type (L-II type, Iwamoto Co.) which worked in the range of shear rate between 10 - 500sec^{-1} and under shear force up to 5000dyn.cm^{-2} at given constant temperature.

Phase Diagram

The phase diagram of the ternary system under examination was obtained through mixing the three components in various compositions in 10ml messcylinders with glass stoppers, and after strong shaking the mixtures were kept overnight in a thermostat of 30°C. Then the mixtures were visually examined to see whether they are transparent or turbid, homogeneous or heterogeneous, and also to see whether they are anisotropic or isotropic by means of a polarizing micro-scope. Microscopic pictures were taken with samples of various chemical compositions. With some samples of mesomorphic mixtures X-ray diagrams were obtained for determining their crystalline structures. The Leitz-ultramicroscope was also applied to examine if apparently homogeneous samples are truly homogeneous or not.

RESULTS AND DISCUSSION

Flow Properties of Thermotropic Mesophase

Fig. 1, 2 and 3 illustrate the relation between shear rate and shear force at various temperatures, the relation between viscosity and shear force at various temperatures and the relation between viscosity and temperature at various shear rates, respec-tively.

From Fig. 1 and 2 one can notice that the flow of the meso-morphic melt of ammonium laurate is non-Newtonian and plastic so

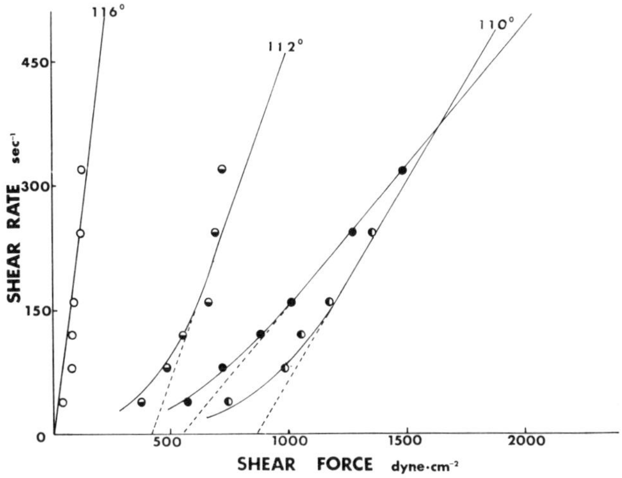

Fig. 1. Shear rate-shear force relation at various temperatures for ammonium laurate

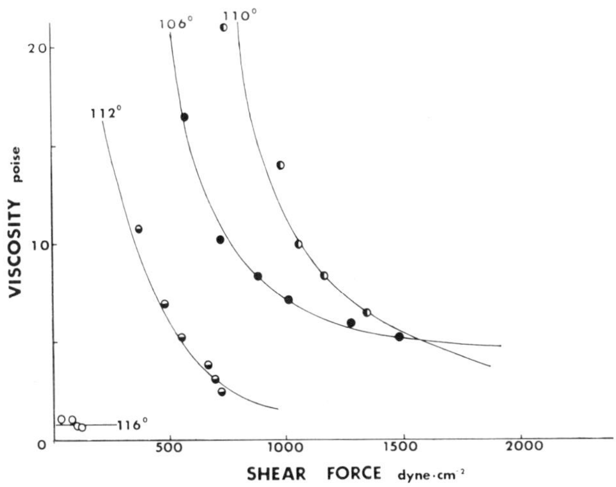

Fig. 2. Viscosity-shear force relation at various temperatures for ammonium laurate

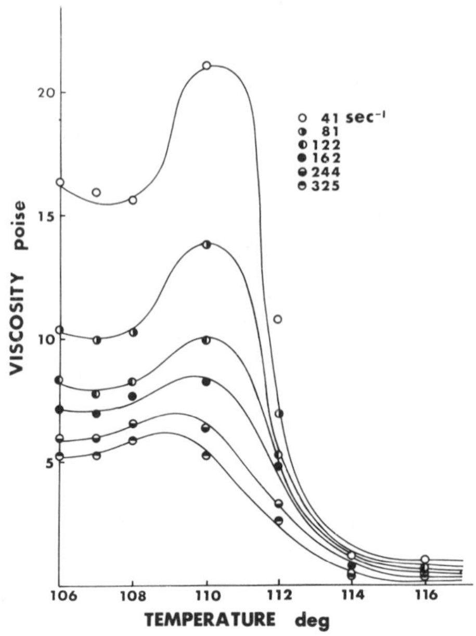

Fig. 3. Viscosity-temperature relation at various shear rates for ammonium laurate

far as the temperature is lower than the anisotropic-isotropic transition point, while the flow of the isotropic melt above the transition point is Newtonian. The plastic behavior represented by Bingham's yield value is fairly remarkable and the value of viscosity is also pretty high, which may be considered characteristic to the smectic mesophase under investigation. In Fig. 3 it is notable that viscosity-temperature curves have maxima at a certain temperature near the transition point and also that the height of the maximum is dependent on shear rate, namely, the higher the shear rate, the lower the maximum. These flow characteristics of the anhydrous melt have also been found for other ammonium carboxylates with greater number of carbon atoms (2).

With thermotropic - cholesteric, nematic or smectic - mesophases many works on their rheological properties have been done by many investigators (3). It was pointed out by Ostwald, in the Faraday Society's General Discussion of 1933, that mesomorphic melts generally do not obey the Hagen-Poiseuille's law connecting viscosity with shear rate, and also that viscosity increases sharply with the rise of temperature and then decreases, giving a maximum at a certain temperature (4). Ostwald's idea on these points was criticized by Lawrence in the same Discussion (5). More recently, Porter and his collaborator discussed the appearance of the viscosity maximum on the basis of their experimental findings on cholesteryl myristate where they demonstrated that the steep viscosity maximum at the cholesteric -isotropic transition temperature is dependent not only on shear rate but also on time. These investigators as well as Lawrence considered that the viscosity maximum near the transition temperature is due to some type of turbulent effect (6).

The result obtained in the present study shown in Fig. 3 is analogous with that of Porter et al. in respect to the dependence of viscosity maximum on shear rate, although the maximum point appears not so sharply in our case. It is here supposed that the increase of viscosity near the smectic-isotropic transition point may be due to the rotational motion of long-shaped molecules in the direction perpendicular to their molecular axes, which will take place suddenly near the transition temperature. This rotational molecular motion will result in strong interactions among neighboring molecules and cause a sharp increase of viscosity. It is also supposed that this increase of viscosity will be lessened by high shear rate which acts to give molecular orientation parallel to their molecular axes.

Phase Diagram of Ternary System

The phase diagram of the ternary system obtained at 30°C is broadly illustrated in Fig. 4.

The region L_1, which means micellar solution, is in this case a small area, because the solubility of ammonium laurate is small at 30°C which is lower than the Kraft point. The region L_2 denotes apparently transparent and homogeneous solution which however shows Tyndall phenomenon and minute particles with Brownian motion under ultra-microscope, so that, this region represents a kind of micro-emulsions (7). The region 2L denotes emulsion and the region 2L + LC means emulsion with liquid crystalline phase. The region L + LC represents two phase region consisting of liquid and liquid crystalline phases, while the region LC represents homogeneous liquid crystalline phase. According to X-ray analysis and microscopic observation, the greater part (lower and right sides) of the LC region corresponds to lamellar (neat) phase, whereas the upper part corresponds to hexagonal (middle) phase. In region L + LC the liquid crystalline phase is either lamellar or hexagonal, or a mixture of lamellar and hexagonal structures.

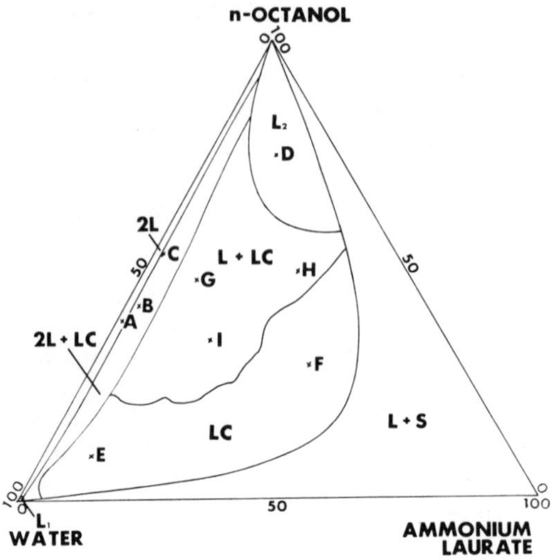

Fig. 4. Phase diagram of ternary system: ammonium laurate + water + n-octanol

Rheological Behavior of Various Regions of Phase Diagram

The rheological measurement was carried out with samples from various regions, whose positions of sampling are denoted by letters A,B,C,D,E,F,G,H and I; the first six positions correspond to apparently homogeneous mixtures, while the last three correspond to apparently two-phase mixtures. The results are listed in Table 1.

In this table the numerical values of composition for points A,B,C,D,E and F represent exact values for those homogeneous mixtures taken for rheological measurement, but the values (in parenthesis) for G,H and I do not represent exact values, because the samples for measurement were taken from the lower parts of the two-phase mixtures after being set overnight in thermostat.

The flow type was classified by the shear rate-shear force curve obtained for each sample, thixotropy being acertained by

Table 1. Properties of samples from various regions of phase diagram of ternary system

Sampling position	Composition H_2O/Am.laurate	Phase-type	Flow-type	Viscosity(P) /shear rate(s^{-1})
A	60.5/1.0	2L (emulsion)	slightly plastic	0.25 /200
B	55.2/2.4	2L + LC (emulsion)	plastic & thixotropic	4 /200
C	44.1/1.9	2L + LC	plastic & thixotropic	2 /200
D	10.3/14.9	L_2 (microemulsion)	slightly plastic	0.2 /200
E	80.8/9.5	LC (lamellar)	plastic & thixotropic	5 /450
F	29.3/41.6	LC (lamellar)	plastic	9 /450
G	(41/12)	L + LC (lamellar)	dilatant	6 /450
H	(20/28)	L + LC (hexagonal)	plastic & thixotropic	15 /450
I	(45/20)	L + LC (lamellar + hexagonal)	plastic	6 /200

the characteristic hysteresis loop of the shear rate-shear force curve. The viscosity was estimated from the viscosity-shear rate or viscosity-shear force curve for each sample. For the sake of comparison, only approximate values of viscosity under comparable shear rates are given in this table.

As noticed in this table, most samples under examination are more or less plastic, the plastic nature expressed by Bingham's yield value being particularly remarkable in regions, E,F,H and I. It is also notable that samples from regions, B,C,E and H are thixotropic, whereas the sample of region G is dilatant. The value of viscosity varies from 0.2 to 15 poises on different regions. The samples from emulsion and microemulsion regions show lower viscosity and samples from liquid crystalline phases have greater viscosity.

The ternary systems of the type: soap + water + amphiphile, have been extensively studied by many investigators (8). Some of these studies concern rheological properties of emulsions or liquid crystalline phases of the systems. Lawrence and coworkers measured viscosity as one of the physical properties related with the phase equilibria of the system: Teepol + water + n-alcohol (9). More recently, Friberg and co-workers made rheological measurement with the ternary system: water + tricaprylin + monocaprylin (10), and also demonstrated the influence of liquid crystalline phases on the rheological behavior of emulsions in the system: water + p-xylene + emulsifier (11).

Comparison of Rheological Behavior of Lyotropic Mesophases with that of Thermotropic Mesophase

It was considered interesting in the present study to compare the flow properties of the thermotropic mesophase with those of the lyotropic mesophases formed by the same substance under examination. Comparing the values of Bingham's yield value and viscosity for the thermotropic mesophase with corresponding values for the lyotropic mesophases, one finds following data shown in Table 2, where numerical values for both yield value and viscosity are approximate ones.

It is remarkable in this table that thermotropic and lyotropic mesophases have comparable viscosity values and also comparable stress yield values. This fact suggests that there may be some similarity in intermolecular structures between thermotropic and lyotropic mesophases. Both are considered to form lamellar molecular arrangement which exhibits high plasticity and great viscosity. The main difference between the two lies in the fact that, in the former

Table 2. Rheological properties of thermotropic and lyotropic mesophases of ammonium laurate

mesophase	temperature (°C)	yield value ($dyn.cm^{-2}$)	viscosity(P) /shear rate(s^{-1})
thermotropic	106 - 112	400 - 850	2 - 20 / 350 - 40
lyotropic	30	300 - 2000	2 - 30 / 450 - 25

case, there is a thermal effect due to high temperature, whereas, in the latter case, there is a solvent effect. Lawrence also pointed out that the action of heat and of solvent are not dissimilar in so far as they both loosen the directive forces holding the molecules in their normal crystal lattice(1). There may be, of course, some difference in the nature of intermolecular forces between the two cases.

The fact that lyotropic mesophases formed by ammonium laurate show thixotropy or dilatancy besides plasticity, suggests that they possess more complex structures than that of thermotropic mesophase of the same substance. Thixotropy may be due to a kind of loose network structures which will be destroyed by shear force and recovered after time. With some samples, such network structures were actually observed in their microscopic pictures. Dilatancy is considered to be due to some heterogeneity - foam mixing in the present case - of the sample under examination.

In regard to a clearer view on the relation between molecular structure and rheological properties of lyotropic mesophases, a further study is being carried out in our laboratory.

Acknowledgement

Our thanks are due to Mr, Y. Sugiura of Department of Polymer Physics, Institute of Physical and Chemical Research, Saitama, Japan, for his advice and assistance in X-ray analysis of the samples.

SUMMARY

With ammonium laurate, rheological properties of its thermotropic mesophase have been measured. The flow type of the anisotropic melt of this compound is evidently plastic in the temperature range of 106 - 112°C, while that of the isotropic liquid above the transition temperature (about 114°C) is Newtonian. Viscosity increases with the rise of temperature near the transition point, showing maximum whose position being dependent on shear rate.

Applying the same compound as one of components in a ternary system containing water and n-octanol as other two components, the phase diagram of this system was obtained. Rheological measurement was made with samples from various regions of the phase diagram where lyotropic mesophases of various types appeared. Some similar properties, namely, high plasticity and great viscosity were found with these lyotropic mesophases as in the case of the thermotropic mesophase of the same substance. However, the rheological behavior of the lyotropic mesophases is generally more complex than that of the thermotropic mesophase, as revealed by such properties as thixotropy and dilatancy.

These experimental results were discussed from the viewpoint of intermolecular forces and structures existing in both these kinds of mesophases.

REFERENCES

1. Lawrence, A.S.C., The Faraday Society's General Discussion (April 1933) 1009.

2. Tamamushi, B., to be published in Biorheology, 10 (1973); Proc. VI. Int. Cong. Surface Active Substances, Carl Hanser Verlag, München (1973).

3. cf. Brown, G.H. and Shaw, W.G., Chem. Rev., 57, 1050 (1957); Porter, R.S. and Johnson, F., Chapter 5 in "Rheology IV" ed. by F. Eirich, NewYork (1968).

4. Ostwald, Wo., The Faraday Society's General Discussion (April 1933) 1002.

5. Lawrence, op. cit., 1080

6. Porter, R.S. and Sakamoto, K., "Liquid Crystals II" Proc. 2nd Int. Liquid Crystal Conf. (1969), 237.

7. Schulman, J.H. and Riley, D.P., J. Colloid Sci., 3 (1948),383.

8. e.g. Dervichian, D.G., The Faraday Society's Discussion, No.18 (1954), 231; Hyde, A.J., Langbridge, D.M. and Lawrence, A.S.C., ibid., 239; Lawrence, A.S.C., The Faraday Society's Discussion, No. 25 (1958), 51; Mandel, L., Fontell, K. and Ekwall, P., "Ordered Fluids and Liquid Crystals", Amer. Chem. Soc., (1967), 89.

9. Hyde, Langbridge and Lawrence, op. cit., 239.

10. Hellström, B., and Friberg, S., Pharmaceutica Suecica, 7(1970), 691

11. Friberg, S. and Solyom, P., Kolloid-Z. u. Z. Polymere, 236 (1970), 173.

LIQUID CRYSTALS II.[1] LIQUID CRYSTALLINE PROPERTIES OF TRANS-CINNAMIC ACID ESTERS

Freeman B. Jones, Jr. and Joseph J. Ratto

Science Center, Rockwell International

Thousand Oaks, California 91360

Research interest in room temperature liquid crystals has grown in recent years due to their increased applications in display devices. A current emphasis in the area has been to obtain nematic materials that are colorless as well as stable to atmospheric contaminants. Several low melting compounds and mixtures from the Schiff-base,[2,3] azo and azoxy benzenes,[4,5] chlorostilbenes,[6] and benzoyloxybenzoate[7,8] series have been investigated for these purposes. Many of these compounds exhibit properties that are undesirable for display applications such as instabilities toward moisture and u.v. radiation in addition to some yellow colorations.

In an attempt to find colorless low-melting liquid crystals stable to the actions of atmospheric contaminants, we have synthesized a number of derivatives of p,p'-disubstituted phenylcinnamates of formula I

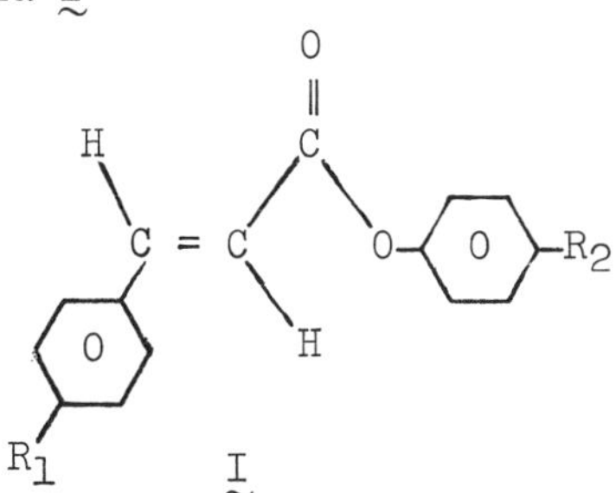

I

in which R_1 and R_2 represent n-alkyl, n-alkoxy, and n-acyloxy groups.

Experimental

All melting points are corrected. Infrared spectra were obtained on a Perkin-Elmer 421 spectrophotometer. Nmr spectra were recorded on a Varian A-60 spectrometer. Phase transitions were analyzed on a calibrated Fisher 300 QDTA. Where the p-alkyl and p-alkoxycinnamic acids were not available commercially, they were prepared by the condensation of malonic acid with the appropriate p-substituted benzaldehyde. The p-acyloxycinnamic acids were prepared by treatment of p-hydroxycinnamic acid with the appropriate acid chloride. The acids were purified and their melting and transition temperatures are shown in Table I. The cinnamoyl chlorides were best prepared by treating the acids with thionyl chloride and reacting *in situ* the resulting acid chlorides with the appropriate p-substituted phenols. Crystalline products were purified by multiple recrystallation from isopropyl alcohol and/or ethanol/water. Samples were studied as thin films between glass cover slips. The structures of the esters were identified by their ir and nmr spectra. The significant bands of their infrared spectra can be exemplified by the spectrum of cinnamate **13**, whose major absorbances are observed at 1720 cm^{-1} (C=O stretching); 1210 cm^{-1} (C-O ester stretching); 976 cm^{-1} (trans CH out-of-plane deformation). The 60 MHz nmr spectrum for compound **11** is shown in Fig. 1; it consists of four major proton absorptions, the integrals of each correspond to the proportional relation of 9:3:2:8. The most significant signals are the chemical shifts for the singlet centered at $\delta 3.8$, attributed to the methoxy protons and the AB quartet centered at $\delta 7.23$. The latter pattern, typical for vinyl trans protons of cinnamic acid esters has ν_{H_A} centered at $\delta 6.59$ and ν_{H_A} centered at $\delta 7.90$ relative to TMS.

Typical synthetic procedures are as follows:

p-n-Butylcinnamic acid. For this preparation the procedure of Gray and Jones[10] was followed. A mixture of p-n-butylbenzaldehyde (16.2 g, 0.1 moles), malonic acid (20.8 g, 0.2 moles), 40 ml of pyridine and 0.5 ml of piperidine was heated for three hours at 100°C. The mixture was allowed to cool to room temperature then poured over 250 g of ice and slowly acidified with concentrated hydrochloric acid. The suspension was filtered and several recrystallizations from ethanol/water afforded 20 g (70%) of a white crystalline product, m.p. 143.8°C.

p-n-Pentanoyloxycinnamic acid. Following a procedure similar to that of Overberger, Luhrs and Chien,[11] pentanoyl chloride (40.4 g, 0.3 moles) was slowly added to a cooled solution (about) 5°C) of p-hydroxycinnamic acid in 175 g of dry pyridine over a period of 30 minutes. The mixture was stirred for four hours while the temperature slowly increased to room temperature. The

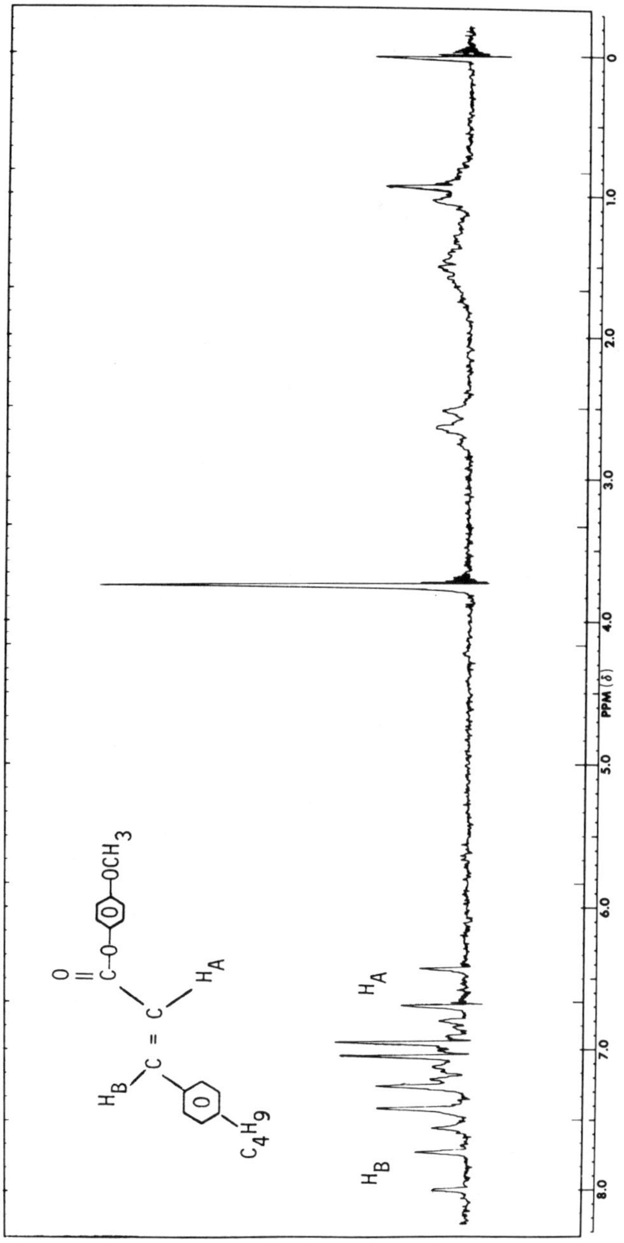

Fig. 1: Nmr spectra of trans-p-n-butylphenyl-p'-methoxycinnamate 13; 15% in toluene-d_8.

Table I. Melting and Transition Points for Trans-Cinnamic Acids

Compound No.	R	M.P. (°C)	Mesomorphic Transition Temperature (°C)
1	C_3H_7[a]	175.5	
2	C_4H_9	143.8	156.1
3	$C_5H_{11}O$[b]	139.5	179.5
4	$C_7H_{15}O$[b]	148	175
5	$C_4H_9CO_2$	145	186
6	$C_5H_{11}CO_2$[c]	150	175.9
7	$C_7H_{15}CO_2$[c]	152.1	180.6

[a]See reference No. 9
[b]Previously synthesized by Gray and Jones[10]
[c]See reference No. 11

resulting solution was poured over ice and slowly acidified with dilute H_2SO_4. The material was allowed to solidify and then filtered. This product was purified by Norite treatment of the first of several recrystallizations from ethanol/water. This procedure afforded 20 g (40%) of a crystalline material, m.p. 149.5°C.

p-Tolyl-p'-n-pentanoyloxycinnamate. Thionyl chloride (1.3 g, 0.11 moles) was slowly added to a well stirred mixture of n-pentanoyloxycinnamic acid (2.6 g, 0.01 mole) and 75 ml of benzene while maintaining the reaction mixture at 50°C. The temperature was slowly raised to reflux and held there until no further evolution of gases occurred, then cooled to 25°C. p-Cresol (1.1 g, 0.01 moles) was added and the reaction mixture was refluxed for two hours. The solution was cooled and the precipitate which formed was collected. Several recrystallizations from isopropyl alcohol gave 2.5 g (74%) of product, m.p. 94.8°C.

Results and Discussion

It has been established by a number of workers[12,13] in the field that intermolecular hydrogen bonding can lead to mesomorphic properties in certain molecules, namely carboxylic acids. Apparently intermolecular hydrogen bonding, which encourages dimer formation, increases the tendency for mesophase formation by greatly increasing the molecular length of the molecule. The results of our investigation which demonstrates the existence of mesomorphism in these p-substituted cinnamic acids (see Table I) is further evidence that hydrogen bonding leads to mesomorphism in liquid crystals.

Esterification of these acids eliminates hydrogen bonding and concomitantly yields colorless and crystalline p-substituted cinnamic acid esters (Table II) with lower mesophase stabilities. A feature evident from Table II is the difference in mesomorphic properties of the alkyl, alkoxy, and acyloxy cinnamates. In cinnamates 8 and 9, where R_1 and R_2 are alkyl groups, low temperature nematic mesophases exist in both compounds beginning at melting points of 59.7°C and 53.6°C, respectively, and continuing until the isotropic state is reached at 69.6°C and 61.3°C. We hypothesize that the lower polarizability of the alkyl groups relative to the remaining members of this cinnamate series is, to a significant degree, responsible for the low melting points of these compounds. In comparison, it is observed that the melting points of alkyl, alkoxy cinnamates 11 - 16 exist at somewhat higher temperatures, but with similar narrow nematic temperature ranges. In the two groups of cinnamates the difference in mesophasic stabilities can be traced to a difference between the polarizabilities and permanent dipoles of the alkoxy and alkyl groups. The polarizability and polar nature of alkyl groups are much less than that of alkoxy groups. For a methyl group the dipole moment is 0.37D as opposed to a moment of 1.28D for the methoxy group.[14] Similarly, increasing the dipolar and polarizable character of the cinnamates, as is the case for acyloxy substituents, result in melting points and nematic-isotropic transitions that are substantially higher than the alkyl derivatives. The comparable melting points of acyloxy and alkoxy cinnamates may be explained by the hypothesis that the preferred conformation of acyloxy compounds reduces their dipolar character until the effective dipole moments of the two cinnamates are nearly equal.

In the homologous series (see Fig. 2) where R_1 was fixed at $C_7H_{15}CO_2$, the first three members of the series show well known nematogenic properties. The remaining members exhibit a monotropic phase transition in addition to the usual smectic and nematic properties. With the exception of the ethyl derivative, all of the members of this series have nematic ranges of about 15°C. Most likely the residual intermolecular dipolar attractions which are important in determining this behavior within the different nematic phases are similar. The lower melting point and broader nematic range for the ethyl compound may be related to end group polarizability which affect the tightness of molecular packing within the crystal structure; the ethyl group adopting a conformation allowing less contiguity of neighboring molecules.

Since smectic properties appear rather early in this series, it became apparent that further increases in chain length would result in increased smectic thermal stability and even shorter nematic ranges. In order to prepare materials with lower melting

Table II.
Transition Temperatures[a] for p-Trans-Cinnamic Acid Esters

No.	R_1	R_2	C-S (°C)	S-N[b] (°C)	N-I (°C)
8	C_3H_7	C_5H_{11}		59.7	69.6
9	C_4H_9	C_4H_9		53.6	61.3
10	C_4H_9	C_5H_{11}		96.4	103.9
11	C_4H_9	CH_3O		81.3	88.8
12	CH_3O	C_3H_7		100.5	103.9
13	CH_3O	C_4H_9		81.8	93.0
14	$C_5H_{11}O$	C_4H_9		96.2	102.0
15	$C_5H_{11}O$	C_5H_{11}		99.1	138.8
16	$C_7H_{15}O$	C_4H_9	72.9	86.1	110.7
17	$C_4H_9CO_2$	C_2H_5O		108.9(88.9)[c]	139.3
18	$C_4H_9CO_2$	C_4H_9O		94.6	140.1
19	$C_5H_{11}CO_2$	CH_3		87.6	120.7
20	$C_5H_{11}CO_2$	C_4H_9		73.2	119.3
21	$C_5H_{11}CO_2$	C_5H_{11}		99.1	106.6
22	$C_5H_{11}CO_2$	CH_3O		69.3	136.6
23	$C_5H_{11}CO_2$	C_2H_5O		95.9	147.3
24	$C_5H_{11}CO_2$	C_4H_9O		101.0	150.5
25	$C_7H_{15}CO_2$	CH_3		94.8	112.1
26	$C_7H_{15}CO_2$	C_2H_5		74.6	108.4
27	$C_7H_{15}CO_2$	C_3H_7		100.2	117.0
28	$C_7H_{15}CO_2$	C_4H_9	70.0(43.0)[c]	96.2	114.0
29	$C_7H_{15}CO_2$	C_5H_{11}	88.7(63.7)[c]	103.0	117.6
30	$C_7H_{15}CO_2$	CH_3O	68.5	75.4	138.8
31	$C_7H_{15}CO_2$	C_2H_5O		94.6	147.0
32	$C_7H_{15}CO_2$	C_4H_9O	83.1	95.1	138.8

[a]C, crystalline; S, smectic; N, nematic; I, isotropic.
[b]If there is no smectic phase, then this is a crystal-nematic transition temperature.
[c]Monotropic transition.

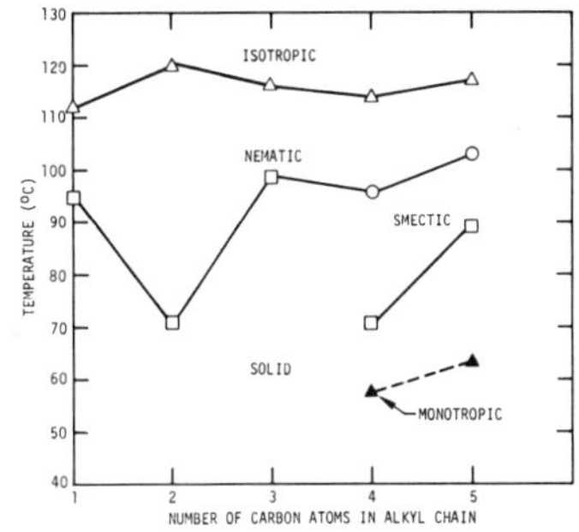

Fig. 2: Phase transition plot for: $C_7H_{15}CO_2C_6H_4CH{:}CHCO_2C_6H_4C_nH_{2n+1}$.

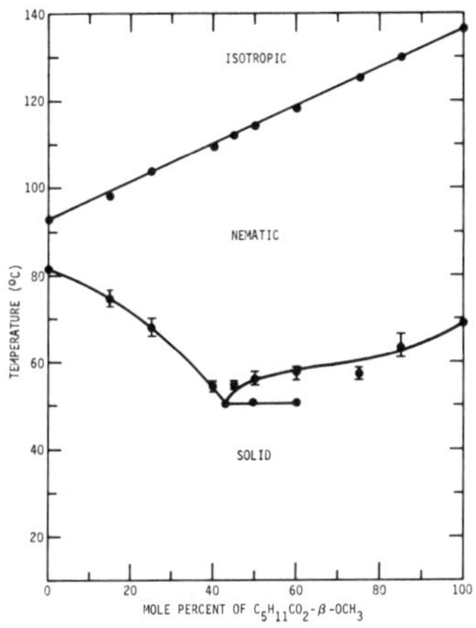

Fig. 3: Phase diagram of cinnamates $C_5H_{11}CO_2$-β-OCH_3 and CH_3O-β-C_4H_9.

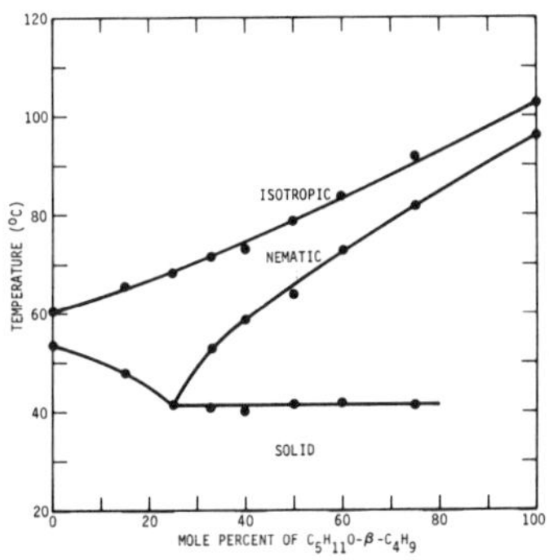

Fig. 4: Phase diagram of cinnamates $C_5H_{11}O$-β-C_4H_9 and C_4H_9-β-C_4H_9.

points and wider nematic ranges, a series of mixtures was investigated. It has been shown in a number of investigations that the nematic-isotropic transition temperatures of two or more mixtures often vary linearly with mixture composition. Normally, compounds similar in molecular structure and shape show less of a deviation from linearity than dissimilar compounds.[12] Conversely, the solid-nematic transition temperatures for binary or other multicomponent mixtures gradually decrease to a eutectic point since the components usually do not pack ideally in the solid phase.

These effects have been explained by Pohl and Steinstrasser[3] through the existence of head-to-tail strands of molecules identified as nematic secondary structures. The phase diagrams of a series of binary mixtures, presented in Figs. 3 and 4, indicate that all compositions of these binary mixtures exhibit a nematic mesophase. In both cases the crystal-nematic transitions are considerably below the melting point of either of the pure components. Furthermore, mixing of these nematic materials did not disrupt the order of the nematic phase since the nematic-isotropic transitions form, in both cases, smooth curve relationships over all binary compositions. The eutectic mixtures of **13** and **22** (57 mole % **13**, 43 mole % **22**) and **9** and **14** (75 mole % **9**, 25 mole % **14**) had nematic temperature ranges from 50.1 - 111°C and 42 - 68°C (Figs. 3 and 4), respectively. Several other mixtures with broad nematic temperature ranges are shown in Table III.

Table III. Nematic Temperature Ranges for Several Cinnamic Acid Esters Multicomponent Mixtures

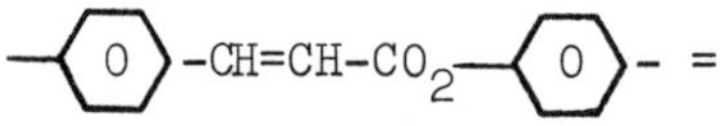
= β

Compound No.	Components	Mole %	Nematic Temperature Range (NTR) °C
A	C_4H_9-β-OCH_3/C_4H_9-β-C_4H_9	47.0/53.0	45 - 70.5
B	C_4H_9-β-OCH_3/$C_5H_{11}O$-β-C_4H_9	69/31	41 - 71.0
C	C_4H_9-β-OCH_3/CH_3O-β-C_4H_9	44.7/55.3	51 - 89.3
D	CH_3O-β-C_4H_9/C_4H_9-β-OCH_3/ C_4H_9-β-C_5H_{11}	43.3/37/19.7	40 - 92
E	$C_5H_{11}CO_2$-β-OCH_3/ $C_7H_{15}CO_2$-β-OCH_3	50/50	62.2 - 138.2

The stability of these materials was compared to that of Schiff-bases by exposing a sample composed of 67 mole % 9 and 33 mole % 20 to laboratory contaminants and a temperature of 80°C for 72 hours. After this time, the color and nematic-isotropic transition temperature of 73.6°C was unchanged. By contrast, the nematic-isotropic transition temperature of the eutectic mixture of p-methoxybenzylidine-p'-n-butylaniline (MBBA) and p-ethoxy-benzylidine-p'-n-butylaniline (EBBA) drops from 61 to 57°C after being exposed for a few minutes to atmospheric contaminants at room temperature.[15]

Additional series of cinnamates are currently in preparation in order to lower even further the mesophasic stabilities of these materials. Other experiments are also underway to determine their electro-optical properties.

References

1. For the first part of this series see: F. Jones and J. Ratto, in press.
2. H. Kelker and B. Scheurle, Angew. Chem. Internat. Edit. 8, 884 (1969).
3. R. Steinstrasser and L. Pohl, Z. Naturforsch. 26b, 87 (1971).
4. J. Van der Veen, W. H. Dejeu, A. H. Grabber, and J. Boren, Mol. Cryst. and Liq. Cryst. 17, 291 (1972).
5. R. Steinstrasser and L. Pohl, Tetrahedron Lett. 1921 (1971).
6. W. R. Young, A. Aviram and R. J. Cox, J. Amer. Chem. Soc. 93, 3976 (1972).
7. Rolf Steinstrasser, Angew. Chem. Internat. Edit. 11, 633 (1972).
8. W. R. Young and D. C. Green, IBM Research Journal, RC4121 and references cited therein.
9. B. P. Smirnov and I. G. Christyakov, Izy. Vyssh. Ucheb. Zaved. Khim. Technol. 13, 217 (1970).
10. G. W. Gray and B. Jones, J. Chem. Soc., 1467 (1954).
11. C. G. Overberger, E. J. Luhrs and P. K. Chien, J. Amer. Chem. Soc. 72, 1200 (1950).
12. G. W. Gray, "Molecular Structure and the Properties of Liquid Crystals", Academic Press, New York, N.Y. (1962).
13. J. E. Goldmacher and M. T. McCaffney, in "Liquid Crystals and Ordered Fluids", Plenum Press, New York, 375 (1970).
14. V. I. Minkin, D. A. Osipov and Y. A. Zhdanov, "Dipol Moments in Organic Chemistry", Plenum Press, New York-London (1970).
15. F. Jones, Jr., R. Chang, and E. P. Parry, U.S. patent filed.

EFFECT OF STRUCTURE ON THE STABILITY OF NEMATIC MESOPHASES

Michael J.S. Dewar, A. Griffin, R.M. Riddle

Department of Chemistry, University of Texas at Austin, Austin, Texas 78712

Studies into the effect of structural modifications on the central ring of liquid crystals of the type

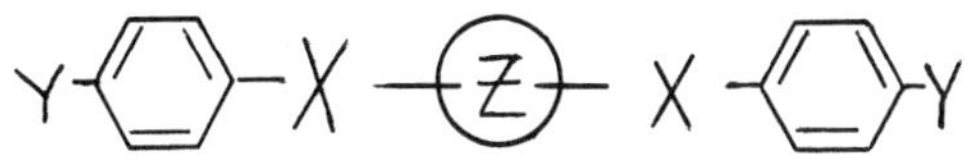

X = -COO, -OOC
Y = -OR

in which the central group is a non-benzenoid ring are rare (1, 2, 3). It was our purpose to extend the range of such ringed compounds and to reinvestigate some previously reported systems. As noted by Young et al (4) molecular modifications must be systematic and must be accompanied by as small a perturbation of the molecular properties as possible if one is to correctly analyze the results only in terms of the specific modification made. Therefore we have employed identical terminal and wing groups and have chosen central rings which insure the same length-to-breadth ratio as found in the molecule chosen as the reference compound bis-(p-methoxyphenyl)-terephthalate.

The recent utility of Differential Thermal Analysis (DTA), Differential Scanning Calorimetry (DSC), and Adiabatic Calorimetry in obtaining thermodynamic parameters for liquid-crystalline phase changes is now well-established (5). From intuitive arguments alone it is readily obvious that these thermodynamic parameters, ΔH (enthalpy of transition) and ΔS (entropy of transition), provide a better insight into the evaluation of intermolecular binding and ordering, respectively, in the mesophase than the value of the transition temperature alone. With this in mind we chose to use thermodynamic values as obtained from DTA tracings as our criteria of mesophase stability.

THE HYDROCARBON CENTRAL GROUPS

From the data in table 1 the most striking fact is that the parent compound Ia has the lowest heat and entropy changes of all the compounds studied when going from the nematic phase to the isotropic liquid. This supports immediately the idea that the mesophase-isotropic transition temperature (the parent has the highest in the series) is not necessarily a good measure of the stability (i.e. binding strength) of the mesophase.

It is noteworthy that the completely saturated systems Ib and Ie have a more strongly bound mesophase than the aromatic ring even though they are lacking a π-election system thought to be responsible for strong attractive forces in the mesophase (1, 2, 6). The cyclohexylene and bicyclooctylene groups should both possess strong local dipoles by virtue of the presence of the ester carbonyls. The inherent polarity of this moiety is not reduced due to interactions (resonance or inductive) with an aromatic ring as is the case in the parent molecule. Thus structures such as that shown in figure 1

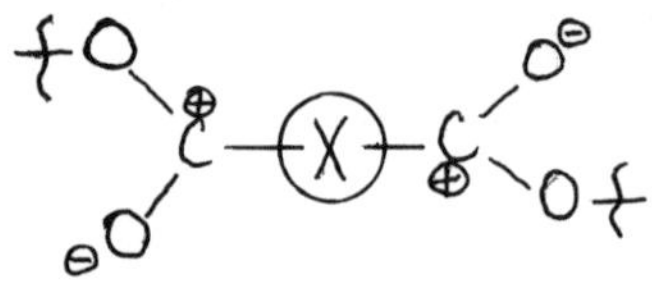

Figure 1. Polarity of Ester Carbonyls.

which should substantially increase intermolecular attractions (7) are more prevalent in the saturated-ring systems. There is another factor which should lead to increased attractions between molecules and that is packing efficiency in the nematic phase. The narrow central portion of the trans-1,4-cyclohexylene unit should allow for a relatively close approach of neighboring molecules permitting increased binding forces. These forces should be especially favored if the cyclohexylene group is not allowed to rotate completely freely about the major axis of the central group. This restricted rotation is not unreasonable to assume in the oriented nematic phase. The bicyclooctylene group should also have its rotation restricted somewhat, even moreso possibly than the less-bulky cyclohexylene. Spatial cavities

Table 1. The Hydrocarbon Central Groups

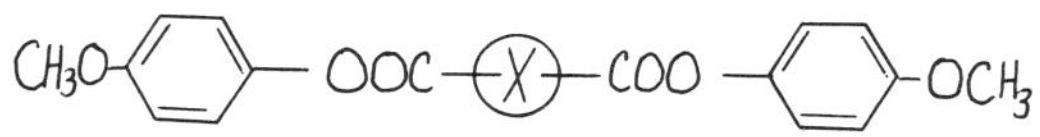

X	$T_{C \to N}$ °C	$T_{N \to I}$ °C	$\Delta H_{N \to I}$ Kcal/mole	$\Delta S_{N \to I}$ e.u.
Ia 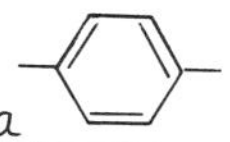	210.6	287.6	0.158	0.282
Ib 	152.0	268.8	0.211	0.389
Ic 	205.6	271.8	0.233	0.428
Id 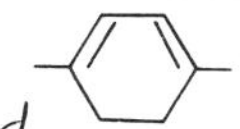	143.7	249.9	0.186	0.356
Ie 	140.2	243.3	0.223	0.432

Table 2. The Heteroaromatic Central Groups

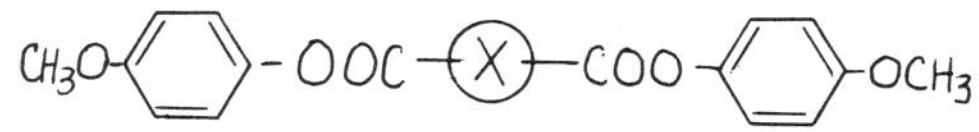

X	$T_{C \to N}$ °C	$T_{N \to I}$ °C	$\Delta H_{N \to I}$ Kcal/mole	$\Delta S_{N \to I}$ e.u.
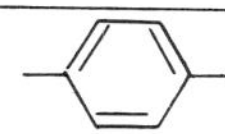	210.6	287.6	0.158	0.282
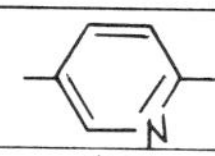	173.2	255.3	0.236	0.447
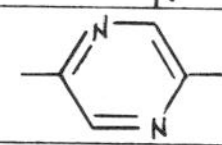	202.6	249.5	0.242	0.463
	—	153 (extrap)	—	—
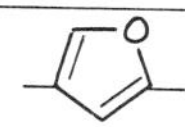	—	140 (mono)	—	—
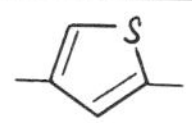	—	86 (extrap)	—	—
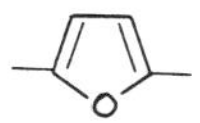	—	62 (extrap)	—	—

created by the absence of a π-electron system allow for closer packing of the bicyclic system as opposed to the aromatic system even though an additional ethylene bridge is present in the former compound. These factors combine to produce greater heats of transition for the saturated compounds. The differences in the dispersion energies for the three structures under consideration is hard to estimate, but one would not expect them to be alone responsible for the relatively large differences noted between the aromatic and saturated groups.

The entropy changes in going from the nematic to isotropic states of the completely saturated and aromatic compounds are, we feel, easily explained. The large entropy change (relative to the parent system Ia) for the cyclohexylene unit is probably due to the increased flexibility of this molecule at the higher temperatures of the isotropic state. It is not afforded such activity in the confines of the nematic arrangement. The freer rotation of the bicyclooctylene group about its major axis in the liquid phase is likewise probably the reason for the increased ΔS in its case. Conjugative constraints on rotational freedom of the terephthalate system persist into the liquid phase and thus the smaller ΔS is to be expected for the parent compound.

In considering comparisons among the parent compound Ia and the cyclohexadienyl central groups, it seems reasonable to assume a "trans" arrangement of the ester carbonyls with respect to each other in all cases. (This is probably necessary for any exhibition of mesomorphism.) The 1,4-cyclohexadienyl group is symmetric and has no net dipole in itself; however, the carbonyl groups have associated with them local dipoles. Although conjugated with a carbon-carbon double bond, these dipoles are stronger than that of the 1,3-cyclohexadienyl unit. The effect of hyperconjugative interaction with the carbonyls by the γ carbon-hydrogen bonds should add to the polar character of the carbonyls. This is not possible in the 1,3-dienyl group. There is a large twist angle between the π-bonds in the 1,3-cyclohexadienyl ring which precludes efficient conjugation between the two. This would result in the structure shown below when considering the

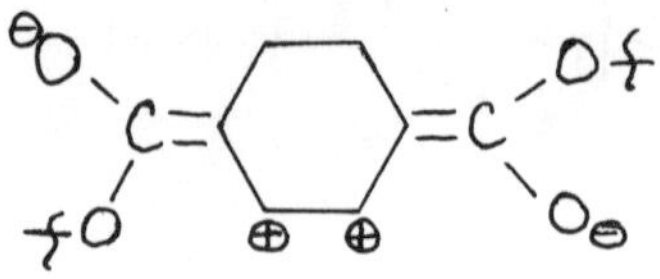

Figure 2. Unfavorable Resonance Structure of 1,3-Diester.

polar structures. This is of course a high-energy resonance form which makes the polar character of the carbonyl group of less importance than in the case of the 1,4-cyclohexadienyl group. Even though the 1,3-cyclohexadiene has a dipole acting laterally across the molecule, the decreased carbonyl polarity causes it to have the more weakly-bound nematic mesophase of the cyclohexadienyl groups (smaller heat of transition). It appears that the effect of an aromatic ring in decreasing carbonyl polarity is strong enough for the parent compound to have the lowest nematic-isotropic heat of transition of all the unsaturated systems.

The entropies of the nematic-isotropic transition for the cyclohexadienes can be explained on the basis of the flexibility of the central ring. The 1,4-cyclohexadiene is the more flexible of the two. More twist freedom is allowed in the isotropic phase than for the 1,3-diester. Since both are under strict orientational constraints in the mesophase which are lifted in the isotropic phase, the change in entropy is expected to be larger for the 1,4-compound. This is indeed observed. As in the case of the saturated groups, both of the cyclohexadienyl diesters go to more "free" (less-ordered) isotropic phases than the parent system which is not allowed such twisting and flexing.

THE HETEROAROMATIC CENTRAL GROUPS

The data in table 2 show some intriguing facts concerning the 6-membered aromatic rings. The parent all-carbon system has again the highest transition temperature and the lowest enthalpy and entropy changes of the group. Using the criterion of transition temperatures as indicators of mesophase stability, the results are quite surprising since lateral dipoles have been added with little perturbation in molecular geometry. This is expected to lead to increased intermolecular forces (dipole-dipole;dipole-induced dipole). The transition heats of the nematic-isotropic phase change show however that there is an increased intermolecular bonding in the aza-aromatics. In the case of the pyridine ring there exists a permanent dipole in the ring as shown in figure 3. This will have an attractive influence which will order molecules above and below it such that their dipoles will orient themselves to maximize ring dipole-dipole attractive forces. These mesophase-strengthening effects lead to a greater heat of transition for this

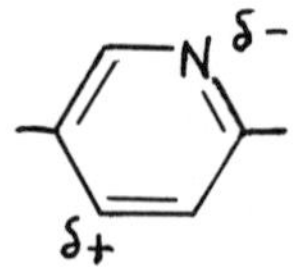

Figure 3. Permanent Dipole in Pyridine Ring.

mono aza-aromatic system than the parent system.

Entropy effects in the pyridine compound are due to the presence of the nitrogen lone pair. The presence of a nitrogen atom in heterocyclic liquid-crystalline Schiff bases is credited with enhancing smectogenic properties by Oh (8). We feel that the lone pair of a nitrogen atom does indeed order the molecules by forcing decreased shearing of molecules past their lateral neighbors due to the unfavorable lone pair-lone pair repulsions during the sliding movement. Adjacent molecules will tend to further orient themselves by having their nitrogens (this applies to the pyridine system only) on opposite sides from each other. The pyrazine ring orients adjacent molecules strongly also. Shearing is reduced since molecules are oriented laterally on both sides of the central ring by a lone pair on each side. Intermolecular bonding is affected in opposite ways by the lone pairs of the di-aza ring. The two lone pairs produce strong attractive coulombic forces, but at the same time the lone pairs repel each other in the more smectic-like arrangement which should increase the intermolecular separations and decrease intermolecular bonding. The net effect of these intermolecular orderings and bindings is that the pyridine and the pyrazine have approximately the same ΔH and ΔS of transition in going from the nematic to isotropic phase.

When the central benzene ring was replaced by furan and thiophene rings, only one of the compounds synthesized showed a mesophase and it was monotropic. Comparison of these results with molecular geometry indicates no general trend except that large deviations from linearity and large deviations of the angles between substituents from 180 degrees hinder the appearance of liquid-crystallinity. Table 3 on the next page shows these compounds. The letter D indicates the angle through which one of the phenyl groups would have to be distorted in order for the axes passing through the 1 and 4 phenyl positions to be parallel (deviation from linearity).

Since these molecules are V-shaped, we feel that in a mesophase the allignment should be as shown below.

	D	Angle Between Substituents
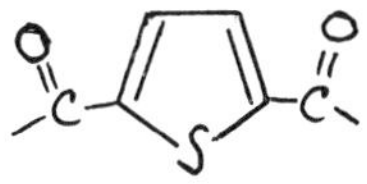	13	32
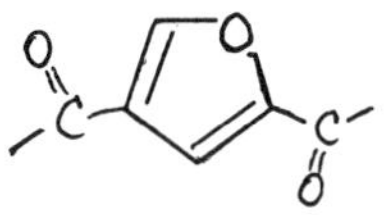	40	20
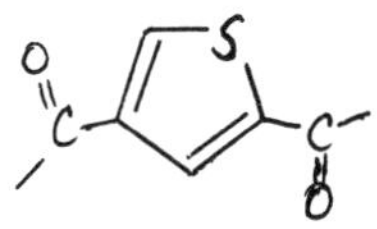	38	40
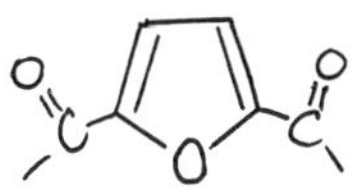	0	53

Table 3. Furan and Thiophene Central Groups.

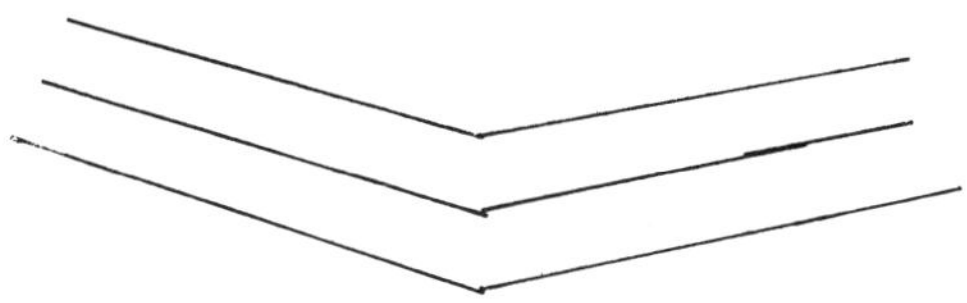

Figure 4. V-Shaped Arrangement.

We predict that this allignment should impose even stricter requirements on rotational freedom and should be reflected in large ΔS values for the mesomorphic-isotropic transitions. We are attempting to prepare such compounds to test this prediction.

CONCLUSIONS

From the data presented above it seems that entropy effects are of extreme importance. In our cases enthalpy and entropy both increase with entropy winning out (determining to a greater extent the temperature of transition). This may be a general trend since even a small degree of additional order (approximately 10%) for these compounds results in much-decreased transition temperatures. In the terephthalate compound the symmetry number is four in both the mesophase and isotropic phase. In the pyridine case for example the symmetry number is less than four, due to the ordering because of the ring dipole in the mesophase, but goes to four in the isotropic phase where dipoles have little influence on molecular ordering. This means that a larger entropy change should occur in the example given than the parent compound. This is what indeed is seen.

Also, from the data presented here and from the observations discussed by Young (9), it is apparent that the mesophase-isotropic transition temperature is not a reliable guide to intermolecular forces and is simply a proportionality variable which forces the equality between the two more fundamental and independent quantities ΔH and ΔS as shown below.

$$\Delta H = T\Delta S$$

Therefore interpretation of structural modifications on a molecular basis by transition temperatures alone should be used only with the utmost care.

REFERENCES

1. M.J.S. Dewar and R.S. Goldberg, J. Amer. Chem. Soc., 92, 1582 (1970).

2. L. Verbit and R.L. Tuggey, Mol. Cryst. Liq. Cryst., 17, 49 (1972).

3. M.J.S. Dewar and R.M. Riddle, unpublished work.

4. W.R. Young, I. Haller, and L. Williams, in Liquid Crystals and Ordered Fluids, J.F. Johnson and R.S. Porter, editors, Plenum Press, N.Y., 1970, p. 383.

5. R.S. Porter, E.M. Barrall, II, and J.F. Johnson, Accounts Chem. Res., 2, 53 (1969).

6. G.W. Gray, Molecular Structure and the Properties of Liquid Crystals, Academic Press, New York, N.Y., Chapter 8, (1962).

7. For a discussion of this point see W. Elser, Mol. Cryst. Liq. Cryst., 8, 230 (1969) and ref. 2.

8. C.S. Oh, Mol. Cryst. Liq. Cryst., 19, 95 (1972).

9. W.R. Young, I. Haller, and D.C. Green, J. Org. Chem., 37, 3707 (1972).

QUANTUM CHEMICAL EVALUATION OF INTERMOLECULAR FORCES IN A COMPOUND PRODUCING LYOTROPIC LIQUID CRYSTAL: PART I

R.K. Mishra and N.K. Roper

Department of Biophysics, All-India Institute of Medical Sciences, New Delhi-110016, INDIA

The structure of lyotropic mesophases is critically influenced by the nature and relative amount of dispersing medium. This fact alone is sufficient to direct one's attention to the nature, disposition and magnitude of intermolecular forces, attractive as well as repulsive, as the probable determinants of stability and phase transitions in these systems. This suspicion is further strengthened by the fact that structural classes in these systems are much fewer than the compounds involved (Mishra, 1972)pointing out again that relatively general forces as opposed to highly specific orbital configurations and characteristics of atoms, might be directly involved in the production of this state. Indeed, the presence of large apolar segments with relatively weak residual atomic charges and polar ends stresses again the need to search in the same direction. It would seem that the dispersing medium brings about the phase changes by affecting the geometric disposition and magnitudes of the intermolecular forces. For a proper calculation of these forces one would require in sequence: analysis of conformation of these molecules, their charge density matrices, evaluation of intermolecular forces vectors as a function of intermolecular distances, many-body interactions, modulation of interactions by the dispersing medium("solvent"), evaluation of energies of various modes of associations. The termination of clusters of molecules to finite sizes in the lyotropic mesophases would then be approached by the summing of permitted deviations in bond lengths and angles and associational strengths leading to the determination of dimensions of ordered clusters possible against random thermal fluctuations. All these calculations constitute a formidable task. In this paper and the next an attempt is made to demonstrate how a beginning in this direction can be made, quantum chemically where currently feasible for at least a partial solution of the problem. It is hoped that when properly refined these methods may be applied to thermotropic liquid crystals and to intramolecular associations of types similar to lyotropic mesophases induced by dispersing medium. In this paper attention is directed towards the determination of conformational hyperspace of dimyristoyl lecithin, while in the Part II, Mishra and Tyagi (1973) deal with the evaluation of intermolecular forces and their modulation.

METHODS

Nomenclature

The numbering of various atoms in the glycerol backbone of dimyristoyl lecithin and substitution points of acylester groups and the polar end are shown in Figure 1. θ_1, θ_2, θ_3, θ_4 are the torsion angles in glycerol while subscripted α, β, γ refer to torsion angles involving the substituents. These are shown in Table 1. As regards handedness of rotation, when looking along bond j, the far bond j+1 rotates clockwise with respect to near bond j-1 in case of the right handed rotation and angles vary from 0° to 360°. The positive angles run from 0° to + 180° and the negative from 0° to -180°. The cis planar configuration of j+1 and j-1 bonds defines the 0° torsion angle. All these designations are those used by Sundaralingam (1972).

Bond lengths, bond angles and torsion angles.

These have been derived for the α-chain (polar end) from the X-ray data on glycerylphosphorylcholine by Abrahamsson (1966) and are set out in Table 2. The starting conformation of β and γ chains is taken to be fully extended and trans to the α-chain.

The standard bond lengths and bond angles have been adopted for the β and γ chains. These provide the internal coordinates of atoms in the whole molecule and then the nuclear Cartesion coordinates are obtained by the procedures of Thompson (1967).

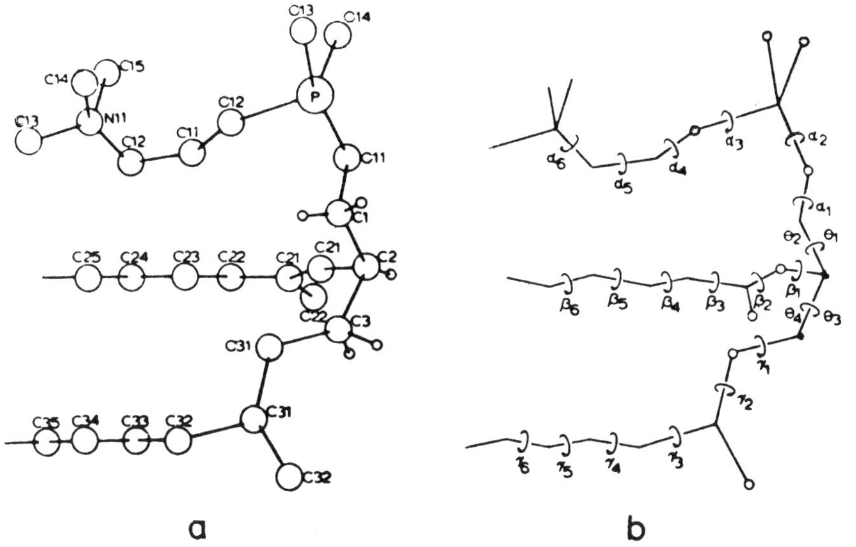

Figure-1:(a) Numbering of various atoms in the glyceryl phosphoryl choline(GPC) backbone of dimyristoyl phosphoryl choline. (b) Notations for relevant torsion angles. (Sundaralingam, 1972).

TABLE 1

NOTATIONS FOR TORSION ANGLES IN PC*

Notation	Torsion Angle	Notation	Torsion Angle
θ_1	O11—C1—C2—C3	β_1	C1—C2—O21—C21
θ_2	O11—C1—C2—O21	β_2	C2—O21—C21—C22
θ_3	C1—C2—C3—O31	β_3	O21—C21—C22—C23
θ_4	O21—C2—C3—O31	β_4	C21—C22—C23—C24
α_1	C2—C1—O11—P	β_{11}	C2(n−3)—C2(N−2)—C2(n−1)—C2n
α_2	C1—O11—P—O12	γ_1	C2—C3—O31—C31
α_3	O11—P—O12—C11	γ_2	C3—O31—C31—C32
α_4	P—O12—C11—C12	γ_3	O31—C31—C32—C33
α_5	O12—C11—C12—N11	γ_4	C31—C32—C33—C34
α_6	C11—C12—N11—C13	γ_{11}	C3(n−3)—C3(n−2)—C3(n−1)—C3n

PC = Phophatidyl choline
*Sundaralingam (1972).

Table 2

Bond lengths, bond angles and torsion angles of various atoms in glycerylphosphorylcholine residue.

S. No.	Atom	Connected to atom	Bond length (Å)	Bond angle (degrees)	Torsion angle (degrees)
1.	C 2	C 2	0.000	000.00	000.00
2.	C 1	C 2	1.554	000.00	000.00
3.	O 11	C 1	1.390	109.13	000.00
4.	P	O 11	1.579	115.80	164.62
5.	O 12	P	1.624	104.02	- 71.38
6.	C 11	O 12	1.418	118.27	- 58.86
7.	C 12	C 11	1.510	112.69	-138.04
8.	N 11	C 12	1.510	115.33	- 72.62
9.	C 13	N 11	1.522	106.34	174.06
10.	C 14	C 13	1.540	109.92	56.81
11.	C 15	C 14	1.522	106.34	174.06
12.	C 3	C 2	1.449	111.65	- 62.88
13.	O 21	O 2	1.433	107.57	- 71.06

Variation of torsion angles

γ-chain is now rotated about the C2 - C3 bond causing variation in the torsion angle θ3 in steps of 10°. The atomic coordinates are computed at each stage and energy of the conformation computed using a potential energy function (see equation (1) below). These are plotted for various relative orientations of β and γ- chains. Setting the torsion angle θ3 corresponding to the lowest energy the process is repeated now for three pairs of successive torsion angles γ_3, γ_4; γ_4, γ_5 and γ_{14}, γ_{15}. Liquori (1968) in calculation of the conformation of acetylcholine considered acylester group to be planar due to partial double bond character of C 31 - O 31 bond. Following this and the work of Sundaralingam (1972) γ_1, γ_2 and β_1, β_2 have been set at 180. β -chain is rotated by varying β_3, β_4. Conformational energy for various orientations of β-chain is obtained with reference to γ-chain.

Atomic charges

Semi-empirical quantum-chemical methods have been used to calculate electron densities at various atoms, viz., Del Re procedure (1958, 1964) for the σ-charges and the variable electronegativity Pariser-Parr-Pople method (Brown and Heffernan, 1958) for the π-charges. Although the β and γ chains in the present case have predominant σ-type electron distributions one does encounter π-distribution at the acylester junction. The net charges represent the sum of σ and π electronic charges.

Conformational energy

Conformational energy can be mathematically expressed as follows:

$$E(\beta,\gamma) = E_\beta(\beta) + E_\gamma(\gamma) + \sum \Big[E_{\ell,jk}(\beta,\gamma) + E_{r,jk}(\beta,\gamma) + E_{c,jk}(\beta,\gamma) + E_{i,jk}(\beta,\gamma) \Big] \quad (1)$$

Where

$E_\beta(\beta)$ and $E_\gamma(\gamma)$ denote the intrinsic torsional barriers about the bonds.

$E_{\ell,jk}(\beta,\gamma)$ denote the London-van der Waals dispersion energy between pairs of atoms j and k.

$E_{r,jk}(\beta,\gamma)$ is the van der Waals repulsive energy.

$E_{c,jk}(\beta,\gamma)$ is the Coulomb interaction energy.

$E_{i,jk}(\beta,\gamma)$ is the energy of polarisation interaction between atom-pair. j,k. The summation is over all pairs of atoms separated by two or more bonds.

The intrinsic torsional potential energy $E_\beta(\beta)$ or $E_\gamma(\gamma)$ has been included to take into account the exchange energy contribution from the bond orbitals associated with the atoms attached to the given bond (Brant and Flory, 1965; Scott and Scheraga, 1965; Epstein and Lipscomb, 1970), including the effects of distortion of these orbitals by rotation (Jorgensen and Allen, 1971).

The $E_{\ell,jk}$ and $E_{r,jk}$ terms can be combined to give the L-J (6-12) potential which constitutes non-bonded energy. This can be written as:

$$E_{L\text{-}J,jk} = -\frac{A_{jk}}{r_{jk}^{6}} + \frac{B_{jk}}{r_{jk}^{12}} \qquad (2)$$

Where r_{jk} is the internuclear distance in A between interacting atoms j and k. The coefficients A are evaluated using the Slater-Kirkwood equation (1931) but by replacing N by an "effective" N as suggested by Pitzer (1959).

$$A_{jk} = \frac{\frac{3}{2}\cdot e\cdot \frac{\hbar}{m^{1/2}}\cdot \alpha_1\cdot\alpha_2}{(\alpha_1/N_1)^{1/2}+(\alpha_2/N_2)^{1/2}} \qquad (3)$$

where α_1 and α_2 are the atomic polarisabilities; e, m are the electronic charges and mass respectively. N_1, N_2, "effective" values of N for atoms j and k respectively. Atomic polarisabilities are from Ketelaar (1958). Values of 'effective' N for various atoms are obtained from the curve given by Scott and Scheraga (1965). The co-efficient B_{jk} which represents repulsive term is obtained by adjusting its value so that the minimum in the potential function $E_{L\text{-}J,jk}$ occurs at a distance R equal to the sum of the van der Waals radii of the interacting atoms j and k. The van der Waals radii of various atoms are obtained from Bondi (1964) and are shown in Table 3.

Table 3

Showing van der Waals radii, polarisabilities and effective values of N for various atomic species

Atoms	van der Waals radii (A)	$\alpha \times 10^{24}$ (Cm^3)	Effective N
H	1.20	0.42	0.9
C	1.70	0.93	5.2
N	1.55	0.87	6.1
O	1.52	0.84	7.0
P	1.90	3.00	12.00

The Coulombic energy of the interacting pair bearing charges q_j and q_k is calculated using the relation:

$$E_{c,jk} = 332 \cdot \frac{q_j \cdot q_k}{\epsilon \cdot r_{jk}} \tag{4}$$

Where ϵ is the local dielectric constant. The numerical factor (332) is used to obtain energies in Kcal/mole. Ramachandran and Sasisekharan (1968) have shown that the effective dielectric constant for the electrostatic interactions between two charges in a molecule lies in the range 2.0 - 4.0 and that it has an upper limit of 5.0. In view of this we have taken the value of ϵ as 3.0 for the computation of Coulomb and polarization energies.

The charge - induced dipole energy (polarisation) for two interacting atoms j and k bearing net atomic charges q_j and q_k is calculated using the relation:

$$E_{i,jk} = -166\left(q_j^2 \cdot \alpha_k + q_k^2 \cdot \alpha_j\right) / \epsilon \cdot r_{jk}^6 \quad (4)$$

The energy so obtained is in Kcal/mole.

Results and discussions

The calculations for the conformation of β and γ chains of dimyristoyl derivative of glyceryl phosphoryl choline (GPC) have been performed by varying torsion angles θ_3 ; γ_3, γ_4 ; γ_4, γ_5 ; γ_{14}, γ_{15} of the γ- chain and torsion angles β_3, β_4 of the β -chain keeping all other torsion angles fixed at 180°. The angular increment for various internal rotation has been taken to be 30°. For θ_3, we have taken the increment to be 10°, since this angle will determine greatly the relative orientation of γ-chain with respect to β-chain.

The results obtained have been illustrated as iso-energy contours scanning the configurational hyperspace of torsion angles γ_3, γ_4 ; γ_4, γ_5 ; γ_{14}, γ_{15} and β_3, β_4 in the Figures 3,4,5 and 6 respectively. Figure 2 shows the variation of conformational energy with reference to torsion angle θ_3 with 10° of angular increment. Minimum energy is found to be at -100°. The trans-configuration is disallowed because there is steric hindrance between carboxyl oxygen atom O 31 of γ-chain and carbonyl oxygen atom O 22 of β-chain. Gauche (g^-) conformation, which corresponds to the experimental torsion angle of carboxyl oxygen of γ- chain in GPC (see Table 2) is not allowed on energy considerations; for it is 10 Kcal/mole above the lowest minimum at -100°.

<u>γ_3 - γ_4 rotation.</u> Figure 3 shows iso-energy contours for the relative energies of various conformations generated by successive variation of γ_3, γ_4. Relative energies are measured with respect to the energy of the lowest minimum taken to be 0. The lowest minimum occurs at (240°, 90°). There are a total of 10 minima. These have been summarised in the Table 4a.

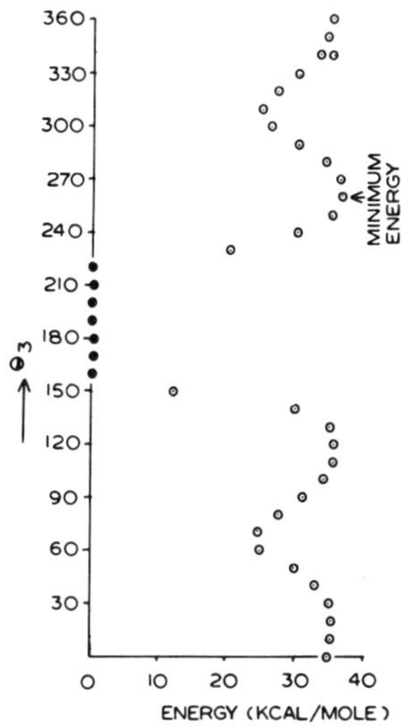

Figure 2: Showing variation of conformational energy with respect to torsion angle θ_3. Minimum energy is indicated by an arrow.

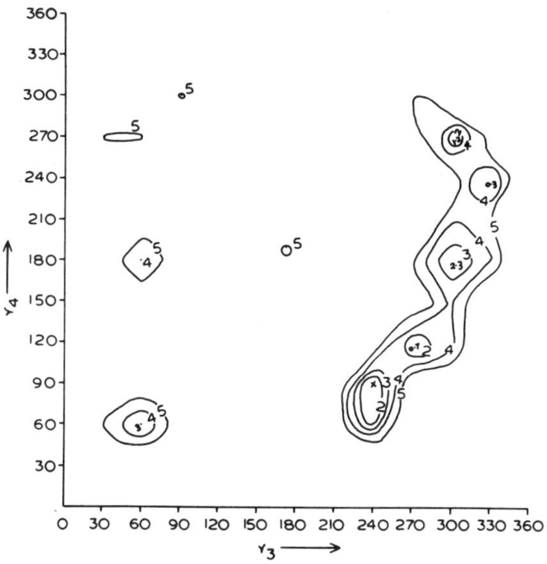

Figure 3: Shows iso-energy contours corresponding to 2,3,4,5,Kcal/mole in the configurational hyperspace of γ_3, γ_4 torsion angles. The lowest observed minima is indicated by a cross (x) mark.

Table 4a

Summary of minima corresponding to $\gamma 3$, $\gamma 4$ rotation.

Conformation	Relative energy (Kcal/mole)	Conformation	Relative energy (Kcal/mole)
(240°, 90°)	0.0	(60°, 180°)	4.0
(330°, 240°)	0.3	(270°, 300°)	4.6
(270°, 120°)	0.7	(30°-60°, 270°)	4.8
(300°, 270°)	1.2	(90°, 300°)	5.0
(60°, 60°)	3.4	(180°, 180°)	5.0

There are 4 local minima at (330°, 240°), (300°, 270°), (270°, 120°) and (270°, 300°) of energies 0.3, 0.7, 1.2, 4.6 Kcal/mole respectively above the lowest observed minimum. These minima are situated amidst regions of low energy ($\leqslant$ 6 Kcal/mole). Thus transitions among the conformations corresponding to these minima are quite probable. Remaining 5 local minima, though of low energies, are almost isolated and surrounded by regions of high energy. Hence these minima are likely to occur only infrequently.

Effect of torsional potential on conformation. Exclusion of torsion potential from conformational energy does not produce any significant effect on the conformational energy minima. The positions and relative energies of various minima are shown in the Table 4b. There are 8 minima in place of 10. The position of the lowest minimum remains at (240°, 90°). The relative energies as well as positions of other two local minima e.g. (330° , 240°) and (270°, 120°) are also unchanged. The local minimum at (300°, 270°) alters its value from 1.2 to 4.73 Kcal/mole. The broad minima at (30°- 60°, - 270°) get split and occur at (30°, 270°) with relative energy of 6.59 Kcal/mole. The local minimum of 4.6 Kcal/mole at (270°, 300°) remains at that configuration but its value increases to 8.08

Table 4b

Showing the effect of exclusion of torsional potential for $\gamma 3 - \gamma 4$ rotations.

Conformation	Relative energy Kcal/mole	Conformation	Relative energy Kcal/mole
(240°, 90°)	0.0	(30°,270°)	6.59
(330°, 240°)	0.3	(270°,300°)	8.08
(270°, 120°)	0.7	(30°, 60°)	8.39
(300°, 270°)	4.73	(90°,300°)	8.53

Kcal/mole. The local minimum at (180°, 180°) disappears and the one at (90°, 300°) gets increased in its relative energy from 5.0 to 8.53 Kcal/mole.

Effect of non-bonded energy alone. It is interesting to note that if non-bonded energy terms alone are considered for determining conformational energy minima, gross features of energy map of Figure 3 remain unaltered. Relative energies of the various minima obtained are shown in the Table 4c.

Table 4c

Showing effect of non-bonded energy alone for $\gamma_3 - \gamma_4$ rotation.

Conformation	Relative energies (Kcal/mole)	Conformation	Relative energies (Kcal/mole)
(240°, 90°)	0.00	(30°, 270°)	6.65
(330°, 240°)	0.31	(270°, 300°)	8.10
(270°, 120°)	0.70	(30°, 60°)	8.45
(300°, 270°)	4.73	(90°, 300°)	8.55

Contribution of Electrostatic energy. Conformational energy has been examined as sum of (i) non-bonded energy, (ii) torsional energy, (iii) electrostatic energy (Coulomb energy + dipole-induced dipole energy). We have observed that energy map of non-bonded interactions alone and those of non-bonded and electrostatic interactions show essentially similar features with regard to position of various minima and their relative energies. Thus the electrostatic energy contributes only slightly. This is so if the molecule is electrically neutral. If it were ionised, electrostatic energy would contribute significantly towards conformational energy.

$\gamma 4 - \gamma 5$ rotation. Figure 4 shows the effect of variation of torsion angles $\gamma 4$, $\gamma 5$ on conformational energy. $\gamma 3$ has been set at 240° corresponding to the lowest minimum of $\gamma 3$, $\gamma 4$ rotations. The lowest minimum occurs at (90°, 180°). This is surrounded by a narrow region of low energy contours. Table 5 indicates the relative energies of various minima for $\gamma 4$ - $\gamma 5$ rotation. There are 3 local minima at (150°, 270°), (90°, 90°) and (180°, 180°) of energies 4.00, 6.52 and 7.52 Kcal/mole respectively above the lowest minima at (90°, 180°). In the Figure only those minima which correspond to relative energies less than 5 Kcal/mole have been displayed.

Table 5

Summary of the minima corresponding to rotation.

Conformation	Relative energy Kcal/mole
(90°, 180°)	0.00
(150°, 270°)	4.00
(90°, 90°)	6.52
(180°, 180°)	7.52

$\gamma 14$ -$\gamma 15$ rotation. Since the lowest minimum of $\gamma 4$ - $\gamma 5$ rotation occurs at (180°, 180°), it is to be expected quite logically that the lowest minimum of

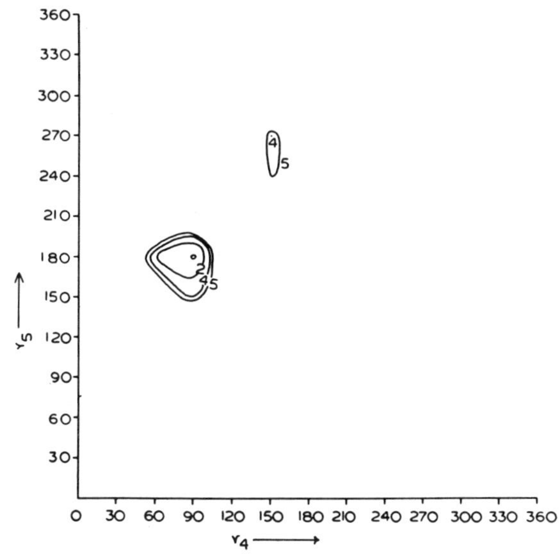

Figure 4: Displays iso-energy contours corresponding to 2,4,5 Kcal/mole. The lowest minimum occurs at (90°, 180°) shown by the mark "o".

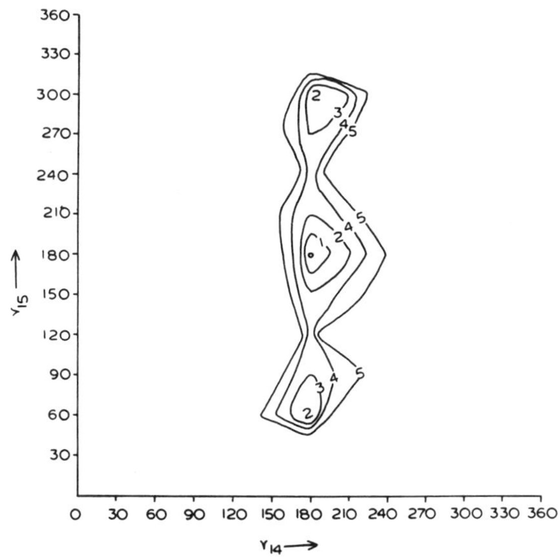

Figure 5: Illustrates iso-energy contours corresponding to 2,3,4,5 Kcal/mole. The lowest minimum is observed at (180°, 180°).

$\gamma 5 - \gamma 6$ rotation shall show (180°, 180°) preferred configuration. In order to test this hypothesis, we have studied the variation of conformational energy with respect to $\gamma 14$ and $\gamma 15$ rotation. This is shown in Figure 5. As expected, the lowest minimum comes up at (180°, 180°) and there are two symmetrically positioned minima at (180°, 60°) (tg^+) and (180°, 300°) (tg^-) of relative energies 2.34 and 2.30 Kcal/mole respectively.

Summary of conformation of γ-chain

$\theta 3$	=	-100°
$\gamma 3$	=	240°
$\gamma 4$	=	90°
$\gamma 5 - \gamma 15$	=	180°
$\gamma 1$	=	180°
$\gamma 2$	=	180°

<u>Rotation of β-chain relative to γ-chain.</u> Having calculated conformation of γ-chain by taking into consideration interaction of non-bonded atoms belonging to γ as well as β-chain, we examine the conformational energy of the molecule by rotating β-chain about appropriate single bonds. Due to partial double bond character of C 21 - O 21 bond, the acylester group is planar (Pauling, 1969), therefore, we set $\beta_1 = \beta_2 =$ 180° and investigate the variation of conformational energy with respect to systematic variation of torsion angles $\beta 3$ and $\beta 4$. Figure 6 illustrates the iso-energy contours obtained in the low energy region ($\leq$ 6Kcal/mole). The Figure shows a solitary minimum at (180°, 180°) surrounded by regions of high energies. Besides this lowest minimum, there are a few local minima of energies greater than 9 Kcal/mole above the lowest minima.

<u>$\beta 4 - \beta 5$ Rotation.</u> Since the lowest minimum of $\beta 3 - \beta 4$ rotation occurs at (180°, 180°), this suggests variation of $\beta 4 - \beta 5$ would show again at (180°, 180°). Thus the entire β-chain prefers to be in extended and trans-configuration.

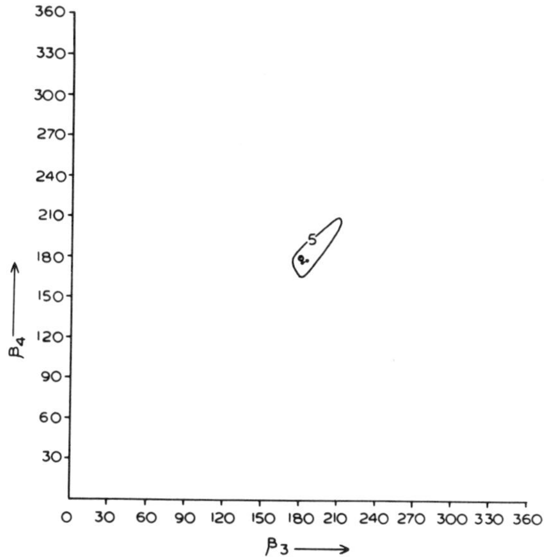

Figure 6: Exhibits iso-energy contour corresponding to 5 Kcal/mole. The lowest minimum occurs at (180°, 180°).

The distances between carbon atoms in equivalent positions of the two chains vary from 4.51 Å to 5.4 Å with an average spacing 4.9 Å. The γ-chain is initially quite apart from the β -chain, then bends and becomes nearly parallel to the latter. In β-tricaprin the two saturated hydrocarbon chains run essentially parallel to each other and the acylester group is more or less in trans-planar configuration (Jensen and Mabis, 1966).

Our calculated conformation of γ-chain, almost parallel to the β-chain, is consistent with that in crystalline lipids the hydrocarbon chains arrange themselves so as to permit maximum van der Waals interaction (Williams and Chapman, 1970).

Laser - Raman spectroscopic studies (Mendelsohn, 1970) indicate that hydrocarbon chains of dipalmitoyl lipid derivative, which contains two saturated hydrocarbon chains of palmitic acid, are predominantly in trans - conformation. This is consistent with our calculated conformation which indicates preference for trans-configuration of the two saturated hydrocarbon chain of myristic acid. Chapman and Fenkett (1966) have also arrived at somewhat similar conclusions from nuclear magnetic resonance studies.

It emerges from our studies that although the two hydrocarbon chains contain several single bonds about which the chains can rotate, there are many but not excessively large probable conformations which the two saturated chains can assume, owing to restrictions imposed by non-bonded interactions. Perhaps the actual final conformation and orientations can be aided by intermolecular forces contingent on packing.

Acknowledgement: This work forms a part of Mishra - Govil project on conformation of phospholipids.

REFERENCES

Abrahamsson, S., Pascher, I., Acta Cryst. 21, 79 (1966).
Bondi, A., J. Phys. Chem. 68, 44 (1964).
Brant, D.A., Flory, P.J., J. Amer. Chem. Soc. 87, 2791 (1965)
Brown, R.D., Heffernan; Trans. Faraday Soc. 54,757(1958)
Chapman, D., Penkett, S.A., Nature (London) 211,1304 (1966)
Del Re, G., J. Chem. Soc. 4031 (1958).
Del Re, G., in "Electronic Aspects of Biochemistry", p 221, B. Pullman, Ed., Academic Press New York, 1964.
Epstein, I.R., Lipscomb, W.N., J. Amer. Chem. Soc. 92, 6094 (1970).
Jensen, L.H.; Mabis, A.J., Acta Cryst. 21, 770(1966).
Jorgensen, W.L., Allen, L.C., J. Amer. Chem. Soc. 93, 567 (1971).
Ketelaar, J., "Chemical Constitution", Elsevier Pub. Co. New York, 1958.
Liquori, A.M., J. Mol. Biol. 33, 445 (1968).
Mendelsohn, R., Biochim. Biophys. Acta 290, 15 (1970).
Mishra, R.K., 4th International Liquid Crystal Conference, Kent, Ohio (1972) (Communicated for publication, 1972).
Mishra, R.K., Tyagi, R.S., This volume (1973).
Chothia, C., Pauling, P., Nature (London). 223, 919 (1969).
Pitzer, K.S. Advan. Chem. Phys. 2, 59 (1959).
Ramachandran, G.N., Sasisekharan, V., Adv. Protein Chem. 23, 283 (1968).
Scott, R.A., Scheraga, H.A., J. Chem. Phys. 42, 2209 (1965).
Slater, J.C., Kirkwood, J.G., Phys. Rev. 37, 682 (1931).

QUANTUM CHEMICAL EVALUATION OF INTERMOLECULAR FORCES IN A COMPOUND PRODUCING LYOTROPIC LIQUID CRYSTAL: PART II, CONSEQUENCES OF INTRODUCTION OF A DISSIMILAR MOLECULE

R.K. Mishra and R.S. Tyagi

Department of Biophysics,
All India Institute of Medical Sciences
New Delhi-110016, INDIA

The salient conformational features and associated energies in dimyristoyl lecithin have been considered elsewhere (Mishra and Roper, 1973). We point out here (a) some avenues for evaluating the intermolecular energies as a function of the type of packing, (b) possible role of dispersing medium in bringing about phase transitions and (c) the influence of a specific molecule, cholesterol, in changing physical properties of the system.

METHODS

The nomenclature of atoms in various chains in dimyristoyl lecithin and the allowed and disallowed regions defining conformational hyperspaces of the substituents on the backbone of the molecule are described in the accompanying communication (Mishra and Roper, 1973).

The intermolecular energies

Energies corresponding to the electrostatic (Keesom), dipole-induced dipole (Debye, Falkenhagen) and the dispersion interactions (London, Eisenschitz; Wang) are calculated by a new procedure by Rein, Claverie and Pollak (1968) and Huron and Claverie(1969) using a dipole-dipole, monopole-dipole and monopole bond polarizability approximation, leading to a consideration of interactions by point charges localised on atoms and dipole sets localised on bonds. In a perturbation treatment without exchange, the long-range interaction energy terms between molecules 1 and 2 are given by

$$E = \langle O_1 O_2 | H' | O_1 O_2 \rangle$$

(electrostatic) (1st order)

$$-\sum_{\psi_1 \neq 0} \frac{|\langle O_1 O_2 | H' | \psi_1 O_2 \rangle|^2}{E\psi_1 - EO_1} \quad -\sum_{\psi_2 \neq 0} \frac{|\langle O_1 O_2 | H' | O_1 \psi_2 \rangle|^2}{E\psi_2 - EO_2}$$

(polarization of 1 by 2) (polarization of 2 by 1)

$$-\sum_{\psi_1 \neq 0} \sum_{\psi_2 \neq 0} \frac{|\langle O_1 O_2 | H' | \psi_1 \psi_2 \rangle|^2}{E\psi_1 + E\psi_2 - (EO_1 + EO_2)}$$

(dispersion) (2nd order)

\+ higher-order terms

where H' is the interaction Hamiltonian (perturbation); $|O_i\rangle, |\psi_i\rangle$ are the eigenfunctions of the molecule i; EO_i, $E\psi_i$ are the corresponding eigenvalues. In the approximation stated, the long-range energy calculation is composed of three terms:

The electrostatic energy. E_{el}.

$$E_{el} = \sum_{\beta_1=1}^{a_1} \sum_{\beta_2=1}^{a_2} \frac{\rho_{\beta_1} \rho_{\beta_2}}{R_{\beta_1 \beta_2}}$$

where a_1 and a_2 are the number of atoms in molecule 1 and 2 respectively; ρ_{β_1} and ρ_{β_2} are the charges on atom β_1, of molecule 1 and β_2 of molecule 2 respectively, separated by the distance $R_{\beta_1 \beta_2}$

The polarization energy. E_{pol}. The expression for polarization energy, E^{pol}_{Rq} which is the energy due to the coupling of static charge on one molecule with the total transition dipole moment in the other, is

$$E_{pol} = \frac{1}{2} \sum_{K}^{b_2} \left[\alpha_K^T (\vec{\xi}_K \cdot \vec{\xi}_K) + \delta_K (\vec{\xi}_K \cdot \alpha_K^{-4})^2 \right]$$

where

$$\vec{\xi}_K = \sum_{\beta=1}^{a_1} \frac{\rho_\beta}{R^3_{\beta K}} \vec{R}_{\beta K}$$

is the electric field created by set of monopoles ρ_β ($\beta = 1, 2, \ldots, a_1$) at the middle of the bond K in the other molecule. ρ_β is charge on atom β.

The dispersion energy. E_{disp}

$$E_{disp} = -\frac{1}{4}\frac{I_1 I_2}{I_1+I_2}\sum_{i=1}^{b_1}\sum_{j=1}^{b_2}\frac{1}{r_{ij}^6} tr\,(\bar{\bar{A}}_{1i}\,\bar{\bar{T}}_{ij}\,\bar{\bar{A}}_{2j}\,\bar{\bar{T}}_{ij})$$

where I_1 and I_2 are the molecular ionization potentials, $\bar{\bar{A}}_{1i}$ and $\bar{\bar{A}}_{2j}$ are the respective bond polarizability tensors, and $\bar{\bar{T}}$ is

$$\bar{\bar{T}} = \left(3\,\frac{\vec{R}}{|R|}\otimes\frac{\vec{R}}{|R|} - 1\right)$$

in which $\otimes$ denotes the dydic product of tensors, $\vec{R}$ is the vector joining the mid-points of the two bonds; b_1 and b_2 are the number of bonds in molecule 1 and 2 respectively.

The trace of the tensors product in the equation above is

$$\begin{aligned} tr(\bar{\bar{A}}_1\bar{\bar{T}}\,\bar{\bar{A}}_2\bar{\bar{T}}) = {} & 6\alpha_1^T\alpha_2^T + \alpha_1^T\,\delta_2\,(3(\vec{\alpha}_2^L.\vec{\gamma})^2+1) \\ & + \alpha_2^T\,\delta_1\,(3(\vec{\alpha}_1^L.\vec{\gamma})^2+1) \\ & + \delta_1\delta_2(3(\vec{\alpha}_1^L.\vec{\gamma})(\vec{\alpha}_2^L.\vec{\gamma}) - (\vec{\alpha}_1^L.\vec{\alpha}_2^L)^2) \end{aligned}$$

$\alpha_1^T, \alpha_1^L, \alpha_2^T, \alpha_2^L$ are the transverse and longitudinal polarizabilities for the bond 1 and 2, $\vec{\gamma}$ is a unit vector in the direction of $\vec{R}$ and δ is

$$\delta = \alpha^L - \alpha^T$$

For evaluation of the expression for the trace of tensor product, we replaced $(\vec{\alpha}_1^L.\vec{\gamma})$ and $(\vec{\alpha}_2^L.\vec{\gamma})$ by their respective average expectation values for that orientation of two bonds in which the dispersion interaction energy is maximum. Short range energy terms are repulsive on the whole and the energy is calculated with the help of a modified Kitaygorodsky (1964) formula following Huron and Claverie (1969).

$$E_{rep.} = 30{,}000\left[\sum_{\beta_1=1}^{a_1}\sum_{\beta_2=1}^{a_2}\exp\left\{-5.5\sqrt{\frac{R_{\beta_1\beta_2}}{W_{\beta_1}W_{\beta_2}}}\right\}\right]$$

where $R_{\beta_1\beta_2}$ is the distance between two atoms β_1 and β_2 with Van der Waal radii values W_{β_1} and W_{β_2} respectively.

Orientations of acyl-chains considered

The conformational analysis of dimyristoyl lecithin suggests that this molecule has the β and γ

chains nearly in a plane and the polar end is oriented perpendicular to this plane, i.e. the molecule when viewed from above the α-carbon, with the plane of two acyl-chains perpendicular to the plane of paper, appears to have an L-shape. The carbon atoms of β and γ chains face mid points of C-C bonds with the plane of their C-C bond sequence twisted with respect to each other at an angle, as shown in Fig.1. The chains are nearly parallel and the minimum distance of approach between them for our analysis is regarded as 5.0 Å for the most part.

When in association, the two molecules of dimyristoyl lecithin orient themselves in such a fashion that the β and γ chains of the two molecules respectively are in association, with the polar ends facing away from each other at 180°. There are two possibilities of orientation of β chain of one molecule with respect to γ chain of the other molecule; one is that the carbon atoms of the two chains are at the same elevation with respect to the polar end (Carbon-Carbon State) and the other in which the carbon atoms of one face mid point of C-C bonds in the other chain (Carbon-bond State), as shown in Fig.2. The interaction energy of the two possibilities are set out in Table I.

TABLE-I

Intermolecular energy of association (Kcal/mole) of two hydrocarbon chains of two different dimyristoyl lecithin molecules.

Geometry	Intermolecular distance	E_{disp}	$E_{rep.}$	$E_{el.}$	E_{TOTAL}
Carbon-Carbon State	4.0 Å	-2.6	0.9	12.0	10.3
Carbon-bond State	4.0 Å	-4.7	0.9	1.6	-2.15

The latter sequence is liable to be preferred. In the former case the bond dipole-dipole dispersion interaction is limited to the bonds facing each other, the neighbouring bonds being too far. Thus the major contribution to the dispersion energy is only from interaction between one bond dipole and the other bond dipole opposite it. The electrostatic energy is also

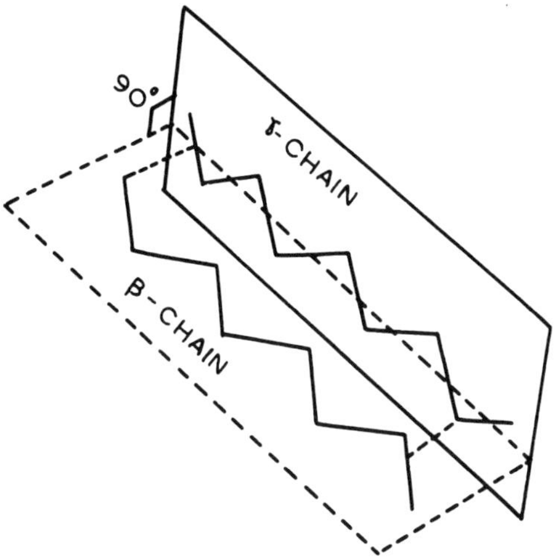

Fig. 1

Relative orientation of the planes of two hydrocarbon chains.

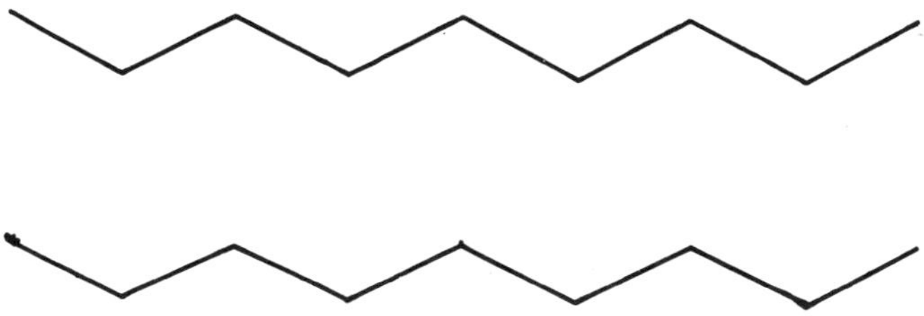

Carbon-Carbon State

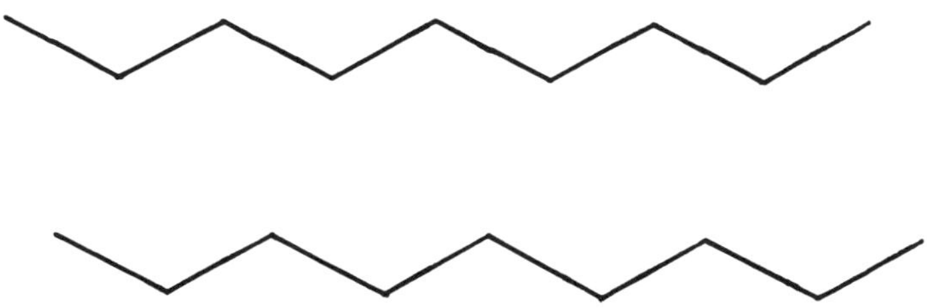

Carbon-bond State

Fig. 2

Possible modes of orientation of two hydrocarbon chains of two different dimyristoyl lecithin molecules in association.

very high in this case; and repulsion predominates even at 4.0Å distance.

In the latter case, where a carbon atom faces a bond, the dispersion energy is contributed by each bond dipole interacting with two bond dipoles. The electrostatic energy is repulsive, but the total energy of association still shows an attraction between two molecules.

Since the polar groups (α chains) of the two molecules have been considered to be oriented at 180° with respect to each other, the repulsion energy at the polar ends is the least. Thus their interaction is markedly reduced by large distances involved and the polarization effects in these are considered negligible. The β and γ chains in the two molecules consist of methylenic units only, which are found to have small residual charges (C, -0.08e; H, +0.04e), and therefore, the polarization effects of these also are not considered here. The order of magnitude of the polarization energy of a β chain due to the presence of a γ chain at a distance of 4 Å is about - 10^{-2} Kcal/mole, one may compare this to the dispersion energy at the same distance ($\sim$ 1 Kcal/mole).

Associations considered

Based on common modes of association in lyotropic liquid crystalline state following were considered:

(a) Lamellar arrangement: The hydrocarbon chains of a row of molecules, as shown in Fig. 3(a), are separated by 4 Å distance from those of the other row. Each chain interacts with the other at a distance of 4 Å . Interactions between chains at distances more than 5 Å are neglected.

(b) Hexagonal arrangement: Arrangement as shown in Fig. 3(b) was considered. The molecules in a row with polar ends on one side are separated by 5 Å distance, whereas the distances between the hydrocarbon chains of the other row of molecules is 4 Å .

(c) Closed-packed hexagonal arrangement: As shown in Fig. 3(c), this arrangement is considered by bringing the opposite sides of the hexagonal arrangement closer so that the distance between the interacting chain becomes 3.5 Å , all the other distances remaining

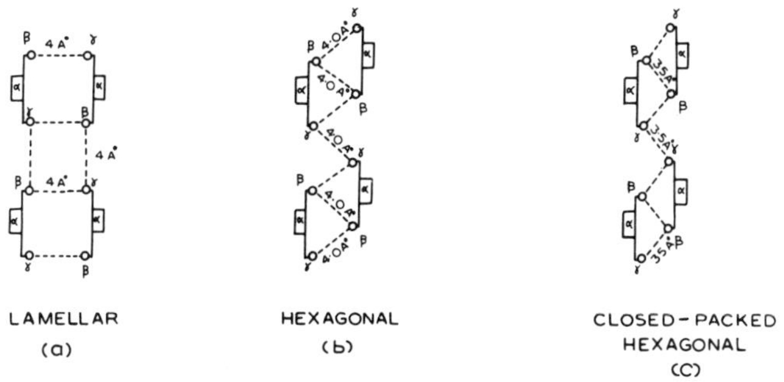

Fig. 3

Geometry considered for association of dimyristoyl lecithin molecules.

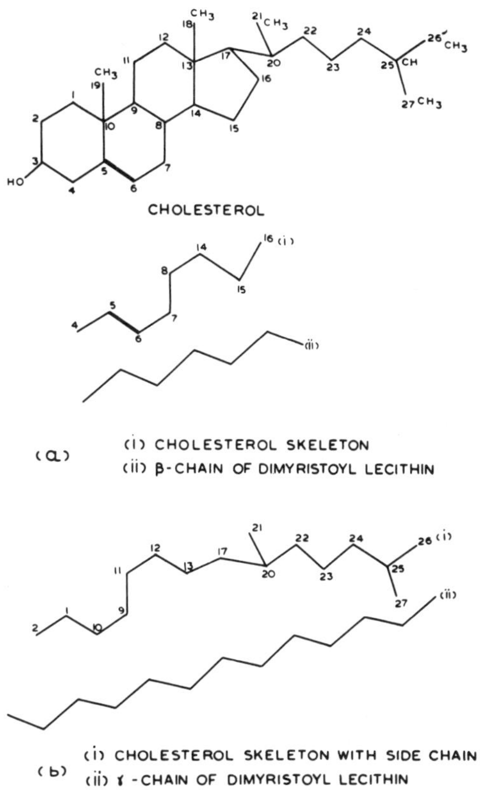

Fig. 4

The orientation of cholesterol with respect to the β and γ chains of dimyristoyl lecithin molecules.

the same. Here the two inter-acting chains have the closest distance of approach.

In figure (3), the dotted lines indicated the interactions considered.

Orientation of cholesterol considered

Because of the possibility of hydrogen bond at hydroxyl group at C3 and the nitrogen of choline residue and further because of the need to maximise interaction with β and γ acyl chains we visualised the following arrangement:

Atoms 4,5,6,7,8,14,15 and 16 of cholesterol with β chain, and atoms 2,1,10,9,11,12,13,17 and 20 to 27 with γ-chain, as shown in Fig. 4.

The end of cholesterol i.e. C_{27} reaches the mid-point of the last bond, i.e. C_{12}-C_{13} of the γ-chain.

Parameters

The Van der Waal radii were those from Pauling, the conformation of polar end was that given by Gupta and Govil (1972). Atomic charges on cholesterol were obtained by Del Re procedure (1958) and are as follows:

C, -0.08e; H, 0.04e; O, -0.50e, where e denotes the electron charge.

The charges on cholesterol were also obtained by Del Re procedure (1958) and are:

C, -0.08e; H, 0.04e; P, -0.8e; N, +1.2e; and O, -0.3e.

The value of ionization potential for both the molecules following Rein, Claverie and Pollak (1968) were assumed to be 10eV, whereas the bond polarizability values were taken from Le Fevre (1965) and are given below in $10^{-23} cm^3$ units:

	α^T	α^L
C-H	0.064	0.064
O-H	0.076	0.061
N-C	0.069	0.057
C-C	0.027	0.099
C=C	0.073	0.280
C-C	0.021	0.224
ar ar		

RESULTS AND DISCUSSIONS

The intermolecular energies of the three modes of packing of acyl chains are shown in Table-II.

TABLE-II

Energies of intermolecular association (Kcal/mole) of a dimyristoyl lecithin molecule in three modes of packing of acyl chains as desired in the text.

Geometry	Intermolecular distance	E_{disp}	E_{rep}	E_{el}	E_{TOTAL}
(a) Lamellar	4.0 Å	-18.8	+3.7	+6.5	-8.6
(b) Hexagonal	4.0 Å	-18.8	+3.7	+6.5	-8.6
(c) Closed packed hexagonal	3.5 Å	-52.0	+18.8	+18.2	-16.0

This shows a clear preference of closed packed hexagonal arrangement over a lamellar one or over another one in which the molecules are translated in the lamella to occupy hexagonal arrangement. The last one has the same energy of association per molecule as in the lamellar case. This is because in both these arrangement each acyl-chain interacts with two more acylchains, each at 4.0Å distance.

It is interesting at this point to consider the three body interactions and see if they could influence association sensitive to the distance between the molecules. The total third order contribution to intermolecular force contributes a term $E_3(abc)$ which may modify pairwise interaction $E_2(ab)$. The third molecule may belong to the dispersing medium. It has been argued by Sinanoğlu et al (1964) that the Van der Waal's dispersion interaction $\xi(ab)$ between two molecules a and b becomes in solution

$$\xi(ab) = E_2(ab) + \xi_3(ab) \cong B' / R^6_{ab}$$

where R is the intermolecular distance between molecules a and b and B' is a modified factor B which takes into account the mean excitation energies and the polarizabilities of the molecules. B' is modified from B by incorporating K, a factor which reduces the coefficient of the expression of London energy between two molecules a and b. The factor K can be considered for two cases (a) K-lattice when the solvent molecules are arranged in a geometrical fashion and (b) K-continuum when they form a continuum. It may be assumed

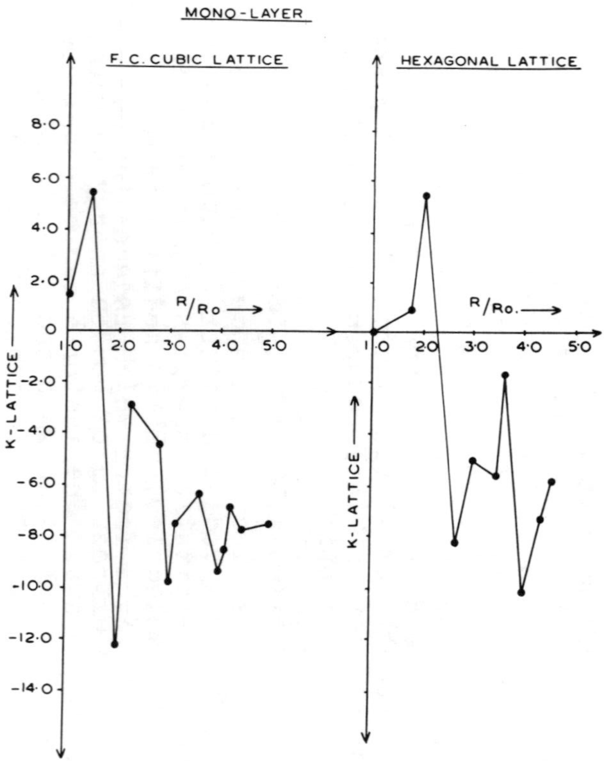

Fig. 5: Variation of the factor K-lattice with distance between two interacting molecules a and b which are arranged in a monolayer.

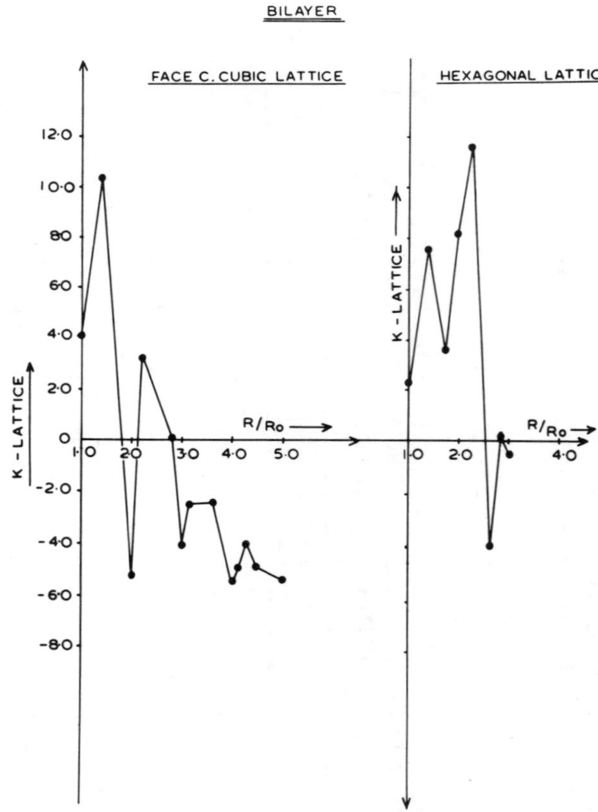

Fig. 6: Variation of the factor K-lattice with distance between two interacting molecules a and b forming the part of a bilayer.

that near the molecules lattic arrangement may prevail and beyond that the medium forms i.e. continuum. The factor K varies from 2 for $1 < (R/R_o) < 5$ and approaches 9.4 when the intermolecular distance R is larger than 5 times R_o the distance between nearest neighbours in the solvent.

For thin layers, however, the factor K exhibits an anamolous behaviour, because it value depends both on the size and the shape of the lattice arrangement of molecules. Calculations for monolayers and bilayer were made, both for the face-centred cubic and the hexagonal arrangement. The figures 5 and 6 indicate how this factor varies with distance in the two cases and indicates how the intermolecular force between two molecules may be modulated by the considerations advanced by Sinanoğlu (1963).

The change in the signature factor of the K-value in thin layers predicts the reversion of the role of the third order interaction contribution even at distances for which $R/R_o \sim 2$, which increase the energy of association between two molecules. This type of behaviour of K vanishes rapidly as the thickness of the layer increases. In case of unrestricted bulk one reproduces the same form of curve as obtained by Sinanoğlu (1963).

It may be recognised that in a stable dispersion the energies of cavity formation and mixture are already dealt with and the interaction and reduction terms above are available for modulation of association.

Association of cholesterol with dimyristoyl lecithin

One of the important features of cholesterol in tissues and membranes is its tendency to associate with phospho-lipids, as there is the general tendency for the strength of association to increase by the addition of cholesterol. The interaction energy of cholesterol as considered in the geometry of association stated above are shown in Table-III. Distinct increase seems to occur by the association of cholesterol.

One may also remark on the possibility of hydrogen bond between the hydroxyl function at carbon 3 of cholesterol and the polar end or the lecithin. Sundaralingam (1972) states that hydrogen bond in phospholipids should be at the acylester and the

TABLE-III

Energy of association of cholesterol (Kcal/mole) with dimyristoyl lecithin.

Geometry	Intermolecular distance	E_{disp}	$E_{el.}$	E_{rep}	E_{TOTAL}
(a)	3.5 Å	-5.9	2.0	3.0	-0.9
(b)	3.5 Å	-17.8	4.5	6.4	-6.9
Cholesterol with Dimyristoyl lecithin	3.5 Å	-23.7	6.5	9.4	7.8

Phosphorus. Because of the positioning of cholesterol close to the β and γ chains in the model of association considered above it is relevant to enquire into the balance of forces between the hydroxyl function at C3 of cholesterol and the terminal methyl groups and N^+ at the polar end, specifically for the availability of space and the feasibility of hydrogen bond formation. The N^+ is surrounded by eleven hydrogen atoms which have each an average residual charge of 0.04 electron of which atmost six form an envelope between the hydrogen of hydroxyl and the nitrogen. The total energy between the OH hydrogen and N^+, -0.3 Kcal/mole, is still weekly attractive at the O(-H)...N distance of 3.2 Å, a distance suggested by Pauling. Since the atomic charge on hydrogen is 0.04e and that on nitrogen 1.2e, the repulsive energy predominates. Because of the attraction of oxygen to the methyl hydrogens totalling -10 Kcal/mole and the strong dispersion term in the chains it would seem that hydroxyl may still remain close to the polar end, but not necessarily make a hydrogen bond to it.

It may be remarked that in face of enormous computational requirements some idealised arrangements have been considered and methods have been used which include several approximations. These results essentially indicate a fruitful general approach that is possible rather than definite conclusions that one can draw at the present time.

REFERENCES

Claverie, P., Rein, R., Int. J. Quantum Chem. 3, 537 (1969).
Del Re, G., J. Chem. Soc., 4031 (1964).
Gupta, S.P., Govil, G., FEBS Letters 27, 68 (1972).

Huron, M.J., Claverie, P., Chem. Phys. Letters 4, 7 (1969).
Kestner, N.R., Sinanoğlu, O., J. Chem. Phys 88,7(1963).
Kitaygorodsky, A.I., Tetrahedron 14, 230 (1964).
Le Fevre, R.J.W., Adv. Phys. Org. Chem. 3, 1 (1965).
Mishra, R.K., Roper, N.K., This volume (1973).
Rein, R., Claverie, P., Pollak, M., Int. J. Quantum Chem. 2, 129 (1968).
Sinanoğlu, O., Abdulnur, S., Kestner, N.R., in 'Electronic Aspects of Biochemistry', p 303, B.Pullman, Ed., Academic Press N.Y. (1964).
Sundaralingam, M., Ann. N.Y. Acad. Sci. 195, 324(1972).

INDEX

The listings in the Index are alphabetical according to author citation. The technical terms are not necessarily in the form given in Chemical Abstracts. The first and subsequently more important uses of the terms are indexed. The individual types of smectic mesophase are listed but other classes, viz. nematics and cholesterics, are cited too numerous to list by type. Lyotropic mesophase systems are listed together. Inorganic compounds are generally not listed unless they have special meaning to the mesophase system. The terms liquid crystal and mesophase are considered equivalent. Compound identification by number and letter, e.g. p or 4, to designate position are cited as in the original. British spelling, where used by the author, has not been changed. Combined or separate words for complex compounds are also indexed as given by the authors.

Roger S. Porter

Julian F. Johnson

INDEX

A

B

C

I

J

K

L

M

T

U

V

X

Z